Microwave Scattering and Emission Models and Their Applications

The Artech House Remote Sensing Library

Fawwaz T. Ulaby, Series Editor

Handbook of Radar Scattering Statistics for Terrain, F. T. Ulaby and M. C. Dobson

Introduction to Electromagnetic Wave Propagation, P. Rohan

Microwave Radiometric Systems: Design and Analysis, N. Skou

Microwave Remote Sensing, Volume I: Fundamentals and Radiometry, F. T. Ulaby, R. K. Moore, and A. K. Fung

Microwave Remote Sensing, Volume II: Radar Remote Sensing and Surface Scattering and Emission Theory, F. T. Ulaby, R. K. Moore, and A. K. Fung

Microwave Remote Sensing, Volume III: From Theory to Applications, F. T. Ulaby, R. K. Moore, and A. K. Fung

Microwave Scattering and Emission Models and Their Applications, Adrian K. Fung

Radargrammetric Image Processing, F. W. Leber

Radar Polarimetry for Geoscience Applications, F. T. Ulaby and C. Elachi, editors

Radar Scattering Statistics for Terrain: Software and User's Manual, F. T. Ulaby and M. C. Dobson

For further information on these and other Artech House titles, contact:

Artech House
685 Canton Street
Norwood, MA 01602
617-769-9750
Fax: 617-762-9230
Telex: 951-659
email: artech@world.std.com

Artech House
Portland House, Stag Place
London SW1E 5XA England
+44 (0) 71-973-8077
Fax: +44 (0) 71-630-0166
Telex: 951-659
email: artech@world.std.com

Microwave Scattering and Emission Models and Their Applications

Adrian K. Fung

Artech House
Boston • London

Library of Congress Cataloging-in-Publication Data

Fung, Adrian K.
Microwave Scattering and Emission Models and Their Applications/Adrian K. Fung
Includes bibliographical references and index.
ISBN 0-89006-523-3
1. Bistatic radar—Mathematical models. 2. Radar cross sections.
I. Title.
TK6592.B57F86 1994 94-41694
621.3848–dc20 CIP

A catalogue record for this book is available from the British Library

685 Canton Street
Norwood, MA 02062

International Standard Book Number: 0-89006-523-3
Library of Congress Catalog Card Number: TK6592.B57F86 1994

10 9 8 7 6 5 4 3 2 1

To my wife Jean, whose support, encouragement, and patience made this work possible

Contents

About the Author

Adrian K. Fung earned his Ph.D. degree from the University of Kansas, Lawrence. Currently director of the Wave Scattering Research Center and professor of Electrical Engineering at the University of Texas at Arlington, Dr. Fung is a Fellow of the Institute of Electrical and Electronic Engineers, member of Sigma Xi and U.S. Commission F of the International Scientific Radio Union. He has been awarded the Halliburton Excellence in Research Award (1987), Distinguished Research Award from the University of Texas at Arlington (1989) and Distinguished Achievement Award from the IEEE Geoscience and Remote Sensing Society (1989). Dr. Fung coauthored Artech House's three volume series on *Microwave Remote Sensing,* and was a contributor to the *Manual of Remote Sensing*. His research interests include wave scattering and emission from irregular surfaces and random media, radar image simulation, numerical simulation of radar scattering and inversion and classification techniques in remote sensing.

Preface

A scattering or emission model for a terrain generally involves both surface and volume scattering. At this point in time the simplest way to integrate these two mechanisms is to use the radiative transfer formulation. This is done in Chapter 1 along with all relevant definitions and properties of scattering and emission. Such a formulation calls for a scattering phase function for volume scatterers and a surface scattering phase function in its boundary conditions. Thus, the development of the subsequent chapters provides a surface scattering phase function and its development, various forms of phase functions for scatterers of specific shapes, simple solutions to the radiative transfer formulation, a generalization of the radiative transfer formulation to a multi-layered medium using matrix doubling and applications of the resulting models.

Over the past forty years many scattering and emission theories for the earth terrain and sea surface have been developed with the hope that practically useful scattering and emission models for users can be derived from them. In general, an exact formulation is too complex to evaluate and an approximate formulation may not have a wide enough range of validity to generate a useful model. Generally, a compromise is needed between how simple a model can get and what type of simplifying assumptions one can tolerate before making the model being developed too restrictive. As a result, only a few models have been developed from various theories and most of them are not simple to use. In organizing this book models are separated into two classes: models in algebraic form or easy to evaluate are referred to as *simple models* and models that are more difficult to evaluate but have a wider range of validity are called *complex models*. The former class of models is given in Chapters 2, 3 and 7, while the latter class is given in Chapters 8, 9 and 11. Readers interested only in model applications may read the application portions of Chapters 2 and 3. The application of a dense medium model is given in Chapter 10. It is anticipated that many users are not interested in the development of a model. Hence, we summarize a bistatic surface scattering model under a single scattering condition in Appendix 2A and the Rayleigh phase matrix for volume scattering in Appendix 2C. In illustrating the use of a simple model we either give or refer to a specific equation representing the model. Only one surface scattering model is reported in this book, and the reasons for selecting this particular surface scattering model are (1) it is able to integrate the standard Kirchhoff and the small perturbation models into one model and (2) it retains an algebraic form under single scattering condition.

Over the last fifteen years, computer simulation methods have been used as an additional means to determine the region of validity of a model. This is a significant step forward. However, a satisfactory expression for the range of validity of a model remains elusive because (1) the simulation method is efficient only for a two dimensional problem, (2) the method has a set of approximations inherent to all numerical methods, for example, number of sampling points per wavelength, number of samples averaged etc. and (3) it is impossible to exhaust all possible cases. Therefore, while it is recognized that there is a range of validity associated with each model, the exact range is rarely known and quite difficult to establish. In practice, only a rule of thumb or some inequalities are provided to serve as a guide to the user. For this reason the derivation of the basic surface scattering model is provided in Chapters 4 and 5 and a computer simulation study of its range of validity is given in Chapter 6. Readers interested in obtaining a better understanding of the validity of the model and its generalization to include multiple scattering may consult these chapters on the fundamentals of the model and all the assumptions used in its derivation.

All the contents of the first eleven chapters relate to the so-called forward problem where the scene condition is assumed known and we want to predict its radiometric response. The final objective of remote sensing, however, is the inverse problem, where given a set of remotely sensed data we want to find some information about the scene. An effective method to deal with the inverse problem is a special form of neural network. A report on the combined use of this method and models developed in this book to retrieve scene parameters or to classify a scene is given in Chapter 12, which was organized by Mr. Dawson who did his thesis on this subject.

All the models given in this book have been developed over the last twenty years and various government agencies have supported such an effort. In particular, the author is grateful to the support of the National Aeronautics and Space Administration, the National Science Foundation, the Office of Naval Research, the Air Force Rome Laboratory and the Army Research Office. Most of the figures and contents of Chapters 8, 9 and 10 were supplied to the author by Dr. Saibun Tjuatja; figures and calculations in Chapters 6 and 7 were supplied by Dr. K. S. Chen, and the polarimetric figures in Chapter 2 and figures in Chapter 11 were provided by Mr. Faouzi Amar. The author is also grateful to Ms. Anita Wofford for reading through the manuscript and polishing its prose and syntax.

Chapter 1
Introduction

1.1 INTRODUCTION

In radar measurements the quantity measured is the radar cross section for an isolated target or the scattering coefficient for an area extensive target. To measure the radar cross section of a target the size of the target must be smaller than the coverage of the radar beam, and the converse is true in measuring the scattering coefficient. Theoretical modeling in radar sensing deals with the modeling of either the radar cross section or the scattering coefficient. Similarly, in passive sensing a radiometer measures the brightness temperature of a target (or its emissivity, if its physical temperature is a constant), and this is the quantity modeled in theoretical studies. In defining the scattering coefficient and the brightness temperature the effects of antenna pattern and range have been removed so that these quantities are influenced only by the target and the exploring electromagnetic wave parameters independent of the particular sensor system used to do the measurement. Some of the reasons for developing theoretical models are the following:

1. To assist data interpretation by providing a relation for the measured quantity as a function of the electromagnetic, geometric and target parameters based on the physics of the problem.
2. To study the sensitivity of the measured quantity to various parameters of interest.
3. To provide a tool for interpolating and extrapolating data.
4. To provide simulated data in simulation studies or training of neural networks for classification or inversion applications.
5. To assist in the design of experiments through model prediction.

In Sections 1.2 and 1.3 we begin with basic definitions in emission and scattering and their interpretations. Then, in Section 1.4 we discuss the general properties associated with the bistatic scattering coefficient and the emissivity based on the reciprocity and duality theorems in electromagnetic theory. A brief development of the radiative transfer formulation in matrix form to account for polarization properties is given in Section 1.5. This formulation integrates naturally both surface and volume scattering in a self-consistent manner. Although the radiative transfer approach cannot account for phase effects in multiple scattering calculations, it does include phase effect in the phase function calculation and it accounts for higher order multiple scattering more effectively than the wave approach. In Section 1.6, a summary account of the meaning of like and cross polarization and polarization states is given. Finally, in Section 1.7 we give expressions for the distribution functions of the scattered power for different scenes and an expression for the distribution function of the phase difference between like polarizations. In polarimetric studies this phase difference provides an additional, independent piece of information to the conventional radar measurement.

1.2 BISTATIC SCATTERING COEFFICIENT

Intuitively, an object can scatter an incident wave into all possible directions with varying strength, and this scattering pattern should vary with the incident direction. To compare between the scattering strengths of objects in a given direction, some common reference is needed. For the radar cross section of an object the common reference is an idealized isotropic scatterer. Thus, the radar cross section of an object observed in a given direction is the cross section of an *equivalent isotropic scatterer* that generates the *same* scattered power density as the object in the observed direction. Mathematically, the *radar cross section* σ of an object observed in a given direction is the ratio of the total power scattered by an equivalent isotropic scatterer to the incident power density on the object. When the radar transmitter and receiver are in the same medium, the intrinsic impedances in the scattered and incident power calculations cancel and the radar cross section of the object is given by

$$\sigma = \frac{4\pi R^2 |E^s|^2}{|E^i|^2} = 4\pi |S|^2 \tag{1.1}$$

where R = range between target and the radar receiver,
E^i = the incident field,
E^s = the scattered field along the direction under consideration,
S = scattering amplitude of the object.

In radar backscattering measurements (transmitter and receiver are co-located) of an isolated target of cross section σ, the returned power is given by [Ulaby et al., 1982]

$$P_r = \frac{P_t}{4\pi R^2} G_t \sigma \frac{A_r}{4\pi R^2} \tag{1.2}$$

where P_t = transmitted power,
G_t = transmitter-antenna power gain,
A_r = receiver-antenna aperture = $\lambda^2 G_r / (4\pi)$,
G_r = receiver-antenna power gain,
λ = radar wavelength.

The above radar equation is not formulated for area-extensive targets and does not include polarization effects. To generalize it to bistatic scattering (transmitter and receiver are located separately) from an area-extensive target such as a sea surface, we shall view the area-extensive target as composed of an infinite collection of statistically identical targets with a statistically averaged cross section of differential size $d\sigma$. Then, we rewrite $d\sigma$ as the product of the averaged *radar cross section per unit area* σ^0 times the differential area ds occupied by each target. Note that we have changed our discussion from a deterministic problem to a statistical problem, and we shall denote the statistical average by the symbol $\langle\ \rangle$. Applying (1.2) to a differential target and recognizing that the ranges from transmitter R_t and receiver R_r to the target may be different, we have

$$dP_r = \frac{P_t G_t G_r \lambda^2 \sigma^0}{(4\pi)^3 R_r^2 R_t^2} ds \tag{1.3}$$

To find the total averaged power we integrate (1.3) over the illuminated area A_0, yielding

$$P_r = \iint_{A_0} \frac{P_t G_t G_r \lambda^2 \sigma^0}{(4\pi)^3 R_r^2 R_t^2} ds \tag{1.4}$$

where σ^0 is also known as the *scattering coefficient*, and it is dimensionless. In terms of the scattered and incident fields, it can be written as

$$\sigma^0 = \frac{\langle\sigma\rangle}{A_0} = \frac{\langle|E^s|^2\rangle}{A_0|E^i|^2/(4\pi R_r^2)} \tag{1.5}$$

which shows σ^0 as the ratio of the statistically averaged, scattered power density to the average incident power density over the surface of the sphere of radius R_r, a definition analogous to the antenna gain pattern. It is not the ratio of the total scattered power to the incident power. Thus, σ^0 may be larger than one in some directions and less than one in others. At a surface boundary, we may also be interested in finding the scattered power into the medium below the boundary. In this case we need to write the more general form of (1.5) by including the intrinsic impedance of the upper medium η_1 and the lower medium η_2. The scattering coefficient becomes a transmission coefficient from medium 1 to medium 2 as

$$\sigma^{21} = \frac{\mathrm{Re}\langle|E^t|^2/\eta_2{}^*\rangle}{[A_0 \mathrm{Re}\,(|E^i|^2/\eta_1{}^*)]/(4\pi R_r^2)} \tag{1.6}$$

where Re is the symbol for the real part operator and * is the complex conjugate operator.

To generalize the above equations for the scattering and transmission coefficients to include polarization effects, note that the quantities dependent on polarization are the electric fields and the corresponding powers. Let p denote the incident polarization and q the scattered polarization. We shall allow p and q to denote either vertical or horizontal polarization. Then, we can add qp as subscripts to the relevant quantities and write the power relation given by (1.4) as

$$P_q = \iint_{A_0} \frac{P_p G_t G_r \lambda^2 \sigma_{qp}^0}{(4\pi)^3 R_r^2 R_t^2} ds \tag{1.7}$$

A method to recover σ_{qp}^0 from the average received power is to take another average power measurement P_q^c under identical conditions on a calibrated target, i.e., a target with a known scattering coefficient σ_{qp}^{0c}. For a radar system with a narrow beamwidth, the scattering coefficient may be taken to be a constant over the width of the beam. If so, it can be removed from under the integral sign and computed in terms of the ratio of the received powers as

$$\sigma_{qp}^0 = \frac{P_q}{P_q^c} \sigma_{qp}^{0c} \tag{1.8}$$

1.3 BRIGHTNESS TEMPERATURE

In accordance with quantum theory all objects at an absolute temperature above zero radiate energy in the form of electromagnetic waves [Kraus, 1966]. Let the self-emitted power density into air from a differential area dA be

$$dp = |\vec{E}|^2 / \eta_0 \tag{1.9}$$

where $\vec{E}$ is the *root mean square* value of the radiated electromagnetic field at a distance R from the area; and η_0 is the intrinsic impedance of air. The *brightness* B (which has the same meaning as the *intensity* to be discussed in Section 1.5) of the differential area observed along the direction (θ, ϕ) at a distance R away is defined as the radiated power density at R per unit solid angle subtended by the differential area as illustrated in Figure 1.1. Mathematically, we have

$$B = \frac{dp}{(dA \cos\theta) / R^2} = \frac{dp}{d\Omega} \tag{1.10}$$

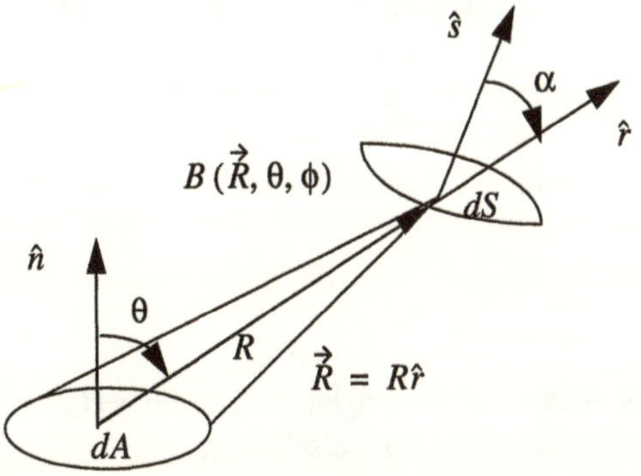

Figure 1.1 Brightness emitted by area dA at a distance R along a given direction. Also illustrated is power flowing through area dS.

Note that if the differential area is part of a random surface, then instead of the power density as given in (1.9) we need to calculate the average power density by performing an ensemble average. This averaged power density should be used in the definition of brightness of the area.

Next, we consider how the brightness of an object is related to its physical temperature. Under thermal equilibrium the physical temperature of an object is a constant. This means that the object must absorb and emit at the same rate to keep its temperature unchanged. Hence, a good absorber is also a good radiator and vice versa. A perfect absorber is an idealized object called a *blackbody* which radiates uniformly in all directions. At microwave frequencies the brightness B_{bb} of a blackbody over a narrow bandwidth Δf follows the Rayleigh-Jeans law [Ulaby et al., 1981] and is proportional to its physical temperature T. That is,

$$B_{bb} = 2k_B (T/\lambda^2)\Delta f \tag{1.11}$$

where k_B = Boltzmann's constant = 1.38×10^{-23} joule K^{-1}; λ is the radiation wavelength in meters in the medium in which the brightness of the body is measured, Δf is in Hz and temperature T is in Kelvin. The Rayleigh-Jeans law is a low frequency approximation to Planck's law. For $T = 300$ K it holds up to a frequency of 117 GHz with less than 1% of deviation from the Planck's law [Ulaby et al., 1981]. Even at 300 GHz the deviation from Planck's law is only about 3%.

A real object is not perfect; and hence it radiates less than the blackbody at the same physical temperature, and its radiation is generally not uniform. To show directional dependence we let $B(\theta, \phi)$ be its brightness, where (θ, ϕ) are angular variables in a spherical coordinate system. Thus, $B(\theta, \phi) < B_{bb}$, and we can define a coefficient called *emissivity* $e(\theta, \phi)$ of an object as the ratio of its brightness to the brightness of the blackbody at the same physical temperature,

$$e(\theta, \phi) = \frac{B(\theta, \phi)}{B_{bb}} \tag{1.12}$$

Under thermal equilibrium the amount of power emitted is the same as that absorbed. We can interpret the emissivity of a body as the fractional power absorbed by the body relative to the total available power. This fractional power is also known as the *absorptivity.*

The *brightness temperature* T_B of an object is the product of its emissivity and physical temperature,

$$T_B = e(\theta, \phi) T \tag{1.13}$$

This definition allows us to write the brightness of an object in terms of its brightness temperature in a form similar to that for a blackbody,

$$B(\theta, \phi) = e(\theta, \phi) B_{bb} = \frac{T_B}{T}[2k(T/\lambda^2)\Delta f] = 2k(T_B/\lambda^2)\Delta f \tag{1.14}$$

Thus, the brightness temperature of a body is always less than its physical temperature and because its use permits brightness of a real body to be expressed by the same formula as that for a blackbody we can interpret it as the blackbody-equivalent radiometric temperature.

1.4 GENERAL PROPERTIES OF T_B AND σ^0

In this section some basic properties of the bistatic scattering coefficients of a rough surface boundary are given. These properties are useful in theoretical model development as well as model applications. Also given is a relation between the scattering coefficient and the emissivity at a rough boundary.

The definition of the bistatic scattering coefficient indicates that it is a function of the magnitude squared of the radiated electromagnetic field. The field must satisfy the reciprocity theorem in electromagnetic theory when the medium and the target are linear and isotropic. Consequently, there is a corresponding relation that the scattering coefficients must satisfy. In a spherical coordinate system as depicted in Figure 1.2 this reciprocal relation for the bistatic scattering coefficient $\sigma^0_{qp}(\theta_s, \phi_s; \theta, \phi)$ is obtained by interchanging the roles of θ, ϕ with θ_s, ϕ_s, and the incident polarization p with the scattered polarization q,

$$\sigma^0_{qp}(\theta_s, \phi_s; \theta, \phi) = \sigma^0_{pq}(\theta, \phi; \theta_s, \phi_s) \tag{1.15}$$

It shows an equality between two quantities indicating that when one quantity is known so is the other. Such a relation is useful for reducing the amount of computation when p is equal to q, i.e., for like polarized scattering coefficients. When p is not equal to q, it gives a relation between two different expressions and allows one expression to be derived from the other or computed in terms of the other. Similarly, there is also a reciprocal relation for the transmission coefficients because the transmitted field must

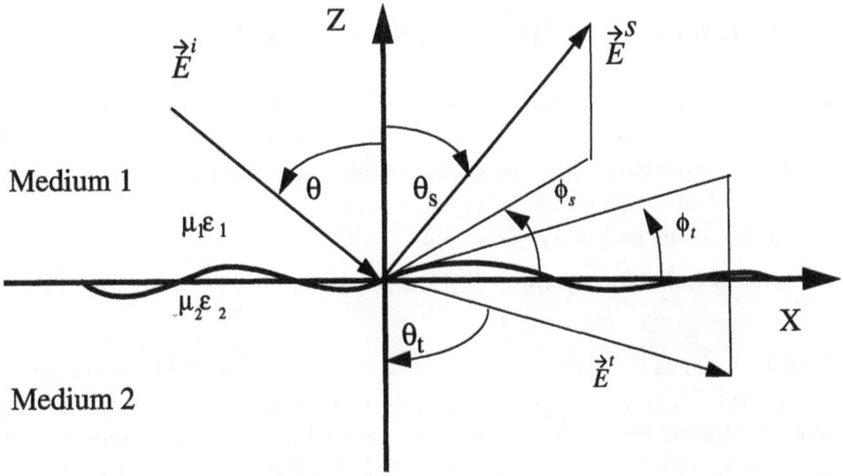

Figure 1.2 Geometry of the scattering problem at a rough surface boundary.

satisfy reciprocity. From (1.6) the bistatic transmission coefficient from medium 1 to medium 2 with incident polarization p along direction (θ, ϕ) and transmitted polarization q along direction (θ_t, ϕ_t) is

$$\sigma_{qp}^{21} = \frac{\mathrm{Re}\langle |E_q^t(\theta_t, \phi_t)|^2 / \eta_2^* \rangle}{[A_0 \mathrm{Re}(|E_p^i(\theta, \phi)|^2 / \eta_1^*)] / (4\pi R_r^2)}$$

and from medium 2 to medium 1 with angles and polarizations interchanged is

$$\sigma_{pq}^{12} = \frac{\mathrm{Re}\langle |E_p^t(\theta, \phi)|^2 / \eta_1^* \rangle}{[A_0 \mathrm{Re}(|E_q^i(\theta_t, \phi_t)|^2 / \eta_2^*)] / (4\pi R_r^2)}$$

Recognizing that, when the incident field amplitudes are the same, the received field amplitudes will also be the same, we see that the difference between the expressions lies in the intrinsic impedances. Hence, we can relate the transmission coefficients as

$$\sigma_{qp}^{21}(\theta_t, \phi_t; \theta, \phi)\,[\mathrm{Re}(1/\eta_1^*)]^2 = \sigma_{pq}^{12}(\theta, \phi; \theta_t, \phi_t)\,[\mathrm{Re}(1/\eta_2^*)]^2 \tag{1.16}$$

This reciprocal relation is not only useful for saving time in computation but also provides an expression for transmission in the opposite direction across a boundary for all polarization combinations. When the media have the same permeability, the above equation simplifies to

$$\sigma_{qp}^{21}(\theta_t, \phi_t; \theta, \phi)\,[\mathrm{Re}(\sqrt{\varepsilon_1^*})]^2 = \sigma_{pq}^{12}(\theta, \phi; \theta_t, \phi_t)\,[\mathrm{Re}(\sqrt{\varepsilon_2^*})]^2 \tag{1.17}$$

Another useful theorem in electromagnetic theory is *duality*. It is well known that the electric and magnetic fields form a dual because they obey the same mathematical equation. More specifically, the dual of an electromagnetic quantity is obtained from it by making the following changes:

1. Change electric field to magnetic field.
2. Change permittivity to permeability and vice versa.
3. Change magnetic field to the negative of the electric field.

It has been shown in [Ulaby et al., 1982] that, after an expression for the horizontally polarized scattering coefficient is developed, the expression for the vertically polarized scattering coefficient can be found from it by duality, i.e., σ_{vv}^0 is the dual of σ_{hh}^0. It is unnecessary to develop the scattering-coefficient expressions for both of these polarizations. Similarly, the dual of σ_{vh}^0 is σ_{hv}^0. Hence, duality lets us derive a desired expression from its dual rather than rework the problem from the beginning. This is true for both scattering and transmission coefficients. In summary, by making the changes as indicated in the previous paragraph the following dual relations can be obtained:

$$\sigma_{vv}^0(\theta_s, \phi_s; \theta, \phi) \leftrightarrow \sigma_{hh}^0(\theta_s, \phi_s; \theta, \phi) \tag{1.18}$$

$$\sigma_{vh}^0(\theta_s, \phi_s; \theta, \phi) \leftrightarrow \sigma_{hv}^0(\theta_s, \phi_s; \theta, \phi) \tag{1.19}$$

$$\sigma_{vv}^{nm}(\theta_t, \phi_t; \theta, \phi) \leftrightarrow \sigma_{hh}^{nm}(\theta_t, \phi_t; \theta, \phi) \tag{1.20}$$

$$\sigma_{vh}^{nm}(\theta_t, \phi_t; \theta, \phi) \leftrightarrow \sigma_{hv}^{nm}(\theta_t, \phi_t; \theta, \phi) \tag{1.21}$$

where n, m may take on 1 or 2. It is important to note that the dual transformation requires the change in both the permeability and the permittivity in an expression. Hence, the common practice of setting the relative permeability in an expression to unity makes the expression invalid for dual transformation.

Although the brightness temperature is directly proportional to the brightness in the same medium, the ratio of brightnesses between two different media is not the same as the ratio of the brightness temperatures in the corresponding media because of the difference in the electrical properties of the media. From (1.14) we have the relation

$$B_1/B_2 = (\varepsilon_1\mu_1 T_{B1})/(\varepsilon_2\mu_2 T_{B2}) \tag{1.22}$$

where ε_1, μ_1 are the permittivity and permeability of medium 1 and a similar definition holds for ε_2, μ_2.

1.4.1 A Relation Between Bistatic Scattering Coefficient and Emissivity

To find a relation between σ^0 and $e(\theta, \phi)$ at a surface boundary we need to find the total scattered power from the boundary in terms of σ^0 and obtain the power absorbed by the medium below as the difference between the incident and scattered power. Then, we apply the thermal equilibrium condition to relate the absorbed power to the emitted power, which relates to $e(\theta, \phi)$.

As shown in Figure 1.2 when a plane wave of polarization p is incident on an irregular surface boundary, part of the incident energy is scattered back and the rest transmitted into the medium below. In general, the rough boundary causes depolarization so that both a p and a q polarized waves appear in scattering and transmission. Thus, the total scattered power density at a distance R from the illuminated area generally consists of two terms resulting from a polarized and a depolarized field as

$$\left[\left|E_{pp}^s\right|^2 + \left|E_{qp}^s\right|^2\right]/\eta_1$$

To find the total scattered power we integrate the power density given above over the upper hemisphere as

$$P_s = \int_0^{2\pi}\int_0^{\pi/2} \{\left[\left|E_{pp}^s\right|^2 + \left|E_{qp}^s\right|^2\right]/\eta_1\}\, R^2 \sin\theta_s d\theta_s d\phi_s \tag{1.23}$$

The fraction of the scattered power to the incident power intercepted by A_0 is the ratio of P_s to the intercepted power,

$$\left(\int_0^{2\pi}\int_0^{\pi/2} \{\left[\left|E_{pp}^s\right|^2 + \left|E_{qp}^s\right|^2\right]/\eta_1\}\, R^2 \sin\theta_s d\theta_s d\phi_s\right) / \left(A_0\left|E_p^i\right|^2 \cos\theta/\eta_1\right)$$

$$= \frac{1}{4\pi}\int_0^{2\pi}\int_0^{\pi/2} \left[\sigma_{pp}^0(\theta_s, \phi_s; \theta, \phi) + \sigma_{qp}^0(\theta_s, \phi_s; \theta, \phi)\right]\frac{\sin\theta_s}{\cos\theta} d\theta_s d\phi_s$$

The fractional power entering the medium below is one minus the scattered power given above and is called the *absorptivity.* Under thermal equilibrium the fractional power entering the medium below when the incident wave is along (θ, ϕ) must be equal to the fractional power emitted along the same direction which by definition is the emissivity for p polarization. That is,

$$e_p(\theta, \phi) = 1 - \frac{1}{4\pi}\int_0^{2\pi}\int_0^{\pi/2} \left[\sigma_{pp}^0(\theta_s, \phi_s; \theta, \phi) + \sigma_{qp}^0(\theta_s, \phi_s; \theta, \phi)\right]\frac{\sin\theta_s}{\cos\theta} d\theta_s d\phi_s \tag{1.24}$$

A special case of the above equation is for a plane boundary that does not depolarize. Hence, the second term under the integral sign is zero and the first term in (1.24) becomes [Ulaby et al., 1986]

$$\sigma_{pp}^{0}(\theta_s, \phi_s; \theta, \phi) = 4\pi\mu R_{pp}(\mu_s, \mu)\,\delta(\mu_s - \mu)\,\delta(\phi_s - \phi) \tag{1.25}$$

where $R_{pp}(\mu_s, \mu)$ is the magnitude squared of the Fresnel reflection coefficient called *reflectivity* for p polarization, $\mu = \cos\theta$ and $\mu_s = \cos\theta_s$. Upon carrying out the integration we obtain

$$e_p(\theta, \phi) = 1 - R_{pp} \tag{1.26}$$

which shows that, at a plane interface between two semi-infinite media, the emissivity is equal to one minus the reflectivity and vice versa.

1.5 THE RADIATIVE TRANSFER FORMULATION

In this section we present the basic development of the radiative transfer theory and its formulation for scattering from an inhomogeneous layer with irregular boundaries. Specific formulations for active and passive sensing are given in subsections.

In the classical formulation of the radiative transfer equation [Chandrasekhar, 1960] the fundamental quantity used is the *specific intensity* I_ν. It is defined in terms of the amount of power dP (watts) flowing along the $\hat{r}$ direction within a solid angle $d\Omega$ through an elementary area dS in a frequency interval $(\nu, \nu + d\nu)$ as follows:

$$dP = I_{\nu}, \cos\alpha dS d\Omega d\nu \tag{1.27}$$

where α is the angle between the outward normal $\hat{s}$ to dS and the unit vector $\hat{r}$. The dimension of I_ν is W m^{-2} sr^{-1} Hz^{-1}. In most remote-sensing applications, the radiation at a single frequency is considered. Thus, it is more convenient to consider the intensity I at frequency ν, which is defined as the integral of I_ν over the frequency interval $(\nu - d\nu/2, \nu + d\nu/2)$. In terms of intensity, the amount of power at a single frequency can be written as

$$dP = I\cos\alpha dS d\Omega \tag{1.28}$$

The dimensions and meaning of I are the same as for the *brightness* discussed earlier in Section 1.3. The *equation of transfer* governs the variation of intensities in a medium that absorbs, emits and scatters radiation. Within the medium, consider a cylindrical volume of unit cross-section and length dl. From a phenomenological viewpoint, energy balance requires that the change in intensity I propagating through the cylindrical volume along the distance dl is due to absorption loss, scattering loss, thermal emission and scattering in the direction of propagation, i.e.,

$$dI = -\kappa_a I dl - \kappa_s I dl + \kappa_a J_a dl + \kappa_s J_s dl \tag{1.29}$$

where κ_a, κ_s are the *volume absorption* and *volume scattering coefficients.* The first two terms on the right-hand side of (1.29) represent, respectively, absorption loss and loss due to scattering away from the direction of propagation, J_a and J_s are the *absorption source function* (or emission source function) and the *scattering source function* representing the intensity scattered into the direction of propagation. Equation (1.29) is the radiative transfer equation in which the definition of J_s is

$$J_s(\theta_s, \phi_s) = \frac{1}{4\pi}\int_0^{2\pi}\int_0^{\pi} P(\theta_s, \phi_s; \theta, \phi)\, I(\theta, \phi)\, \sin\theta d\theta d\phi \tag{1.30}$$

where $P(\theta_s, \phi_s; \theta, \phi)$ is the phase function accounting for scattering within the medium to be defined in the next subsection. It is clear from (1.30) that J_s is not an independent source of the medium but is itself a function of the propagating intensity. On the other hand, J_a is an independent source function proportional to the temperature profile of the medium, i.e., it is the source function in passive remote-sensing problems. As such, it should be dropped in active remote-sensing problems, in which the source is an incident wave from the radar transmitter outside the scattering medium.

For the active sensing problem to be considered in this section, we will treat partially polarized waves by introducing the Stokes parameters. Then, we shall generalize the scalar radiative transfer equation to a matrix equation. In so doing, it is helpful first to establish the relation between the scattered intensity and the incident intensity, and then relate these intensities to the corresponding electric fields.

1.5.1 Stokes Parameters, Phase Matrices, and Radiative Transfer Equations

For an elliptically polarized monochromatic plane wave, $\vec{E} = (E_v\hat{v} + E_h\hat{h})\exp(j\vec{k}\cdot\vec{r})$, propagating through a differential solid angle $d\Omega$ in a medium with intrinsic impedance η, where $\hat{v}$ and $\hat{h}$ are the unit vectors denoting vertical and horizontal polarization, respectively, the *modified Stokes parameters* I_v, I_h, U and V in the dimension of intensity are defined for real η as

$$I_v d\Omega = \langle |E_v|^2 \rangle / \eta \tag{1.31}$$

$$I_h d\Omega = \langle |E_h|^2 \rangle / \eta \tag{1.32}$$

$$U d\Omega = 2\mathrm{Re}\langle E_v E_h^* \rangle / \eta \tag{1.33}$$

$$V d\Omega = 2\mathrm{Im}\langle E_v E_h^* \rangle / \eta \tag{1.34}$$

These four parameters have the same dimensions and hence are more convenient to use than amplitude and phase, which have different dimensions. It has been shown that the amplitude, phase, and polarization state of any elliptically polarized wave can be completely characterized by these parameters [Ishimaru, 1978, pp. 30–33]. If η is complex, we should replace $1/\eta$ in (1.31) through (1.34) by $\mathrm{Re}\,[1/\eta^*]$, where * is the symbol for complex conjugate.

Phase Matrix for Rough Surfaces

To relate the scattered intensity to the incident intensity, consider a plane wave illuminating a rough surface area A. The relation between the vertically and horizontally polarized scattered field components E_v^s, E_h^s and those of the incident field components E_v^i, E_h^i is

$$\begin{bmatrix} E_v^s \\ E_h^s \end{bmatrix} = \frac{e^{jkR}}{R} \begin{bmatrix} S_{vv} & S_{vh} \\ S_{hv} & S_{hh} \end{bmatrix} \begin{bmatrix} E_v^i \\ E_h^i \end{bmatrix} \tag{1.35}$$

where $S_{pq}(p, q = v \text{ or } h)$ is the scattering amplitude in meters, R is the distance from the center of the illuminated area to the point of observation and k is the wave number. Consider $|E_v^s|^2$

$$|E_v^s|^2 = \frac{1}{R^2}\left[|S_{vv}|^2|E_v^i|^2 + |S_{vh}|^2|E_h^i|^2 + 2\mathrm{Re}\,(S_{vv}S_{vh}{}^*E_v^iE_h^{i*})\right]$$

where

$$\begin{aligned} 2\mathrm{Re}\,(S_{vv}S_{vh}{}^*E_v^iE_h^{i*}) &= 2\mathrm{Re}\,\{\,[Re\,(S_{vv}S_{vh}{}^*) + j\mathrm{Im}\,(S_{vv}S_{vh}{}^*)\,] \\ &\qquad [\mathrm{Re}\,(E_v^iE_h^{i*}) + j\mathrm{Im}\,(E_v^iE_h^{i*})]\,\} \\ &= 2\mathrm{Re}\,(S_{vv}S_{vh}{}^*)\,\mathrm{Re}\,(E_v^iE_h^{i*}) - 2\mathrm{Im}\,(S_{vv}S_{vh}{}^*)\,\mathrm{Im}\,(E_v^iE_h^{i*}) \end{aligned}$$

Recognizing the above relation we can obtain the following quantities using (1.35) and (1.31) through (1.34):

$$\begin{aligned} \langle|E_v^s|^2\rangle/\eta &= \langle\,[|S_{vv}|^2 I_v + |S_{vh}|^2 I_h + \mathrm{Re}\,(S_{vv}S_{vh}{}^*)\,U]\rangle d\Omega/R^2 \\ &\quad - \langle\mathrm{Im}\,(S_{vv}S_{vh}{}^*)\,V\rangle d\Omega/R^2 \end{aligned} \tag{1.36}$$

$$\begin{aligned} \langle|E_h^s|^2\rangle/\eta &= \langle\,[|S_{hv}|^2 I_v + |S_{hh}|^2 I_h + \mathrm{Re}\,(S_{hv}S_{hh}{}^*)\,U]\rangle d\Omega/R^2 \\ &\quad - \langle\mathrm{Im}\,(S_{hv}S_{hh}{}^*)\,V\rangle d\Omega/R^2 \end{aligned} \tag{1.37}$$

$$2\mathrm{Re}\langle E_v^s E_h^{s*}\rangle/\eta = \langle [2\mathrm{Re}(S_{vv}S_{hv}^*)I_v + 2\mathrm{Re}(S_{hh}^*S_{vh})I_h]\rangle d\Omega/R^2$$

$$+ \langle [\mathrm{Re}(S_{vv}S_{hh}^* + S_{vh}S_{hv}^*)U - \mathrm{Im}(S_{vv}S_{hh}^* - S_{vh}S_{hv}^*)V]\rangle d\Omega/R^2 \tag{1.38}$$

$$2\mathrm{Im}\langle E_v^s E_h^{s*}\rangle/\eta = \langle [2\mathrm{Im}(S_{vv}S_{hv}^*)I_v + 2\mathrm{Im}(S_{hh}^*S_{vh})I_h]\rangle d\Omega/R^2$$

$$+ \langle [\mathrm{Im}(S_{vv}S_{hh}^* + S_{vh}S_{hv}^*)U - \mathrm{Re}(S_{vv}S_{hh}^* - S_{vh}S_{hv}^*)V]\rangle d\Omega/R^2 \tag{1.39}$$

The left-hand sides of the above equations are in watts per square meter. To convert them to intensity, we use the definition given by (1.10) and divide both sides of the equations by the solid angle subtended by the illuminated area A at the point of observation, $(A\cos\theta_s)/R^2$, where θ_s is the angle between the scattered direction and the direction normal to the area A. Equation (1.36) becomes

$$R^2\langle|E_v^s|^2\rangle/(\eta A\cos\theta_s) = \langle[|S_{vv}|^2 I_v + |S_{vh}|^2 I_h + \mathrm{Re}(S_{vv}S_{vh}^*)U]\rangle d\Omega/(A\cos\theta_s)$$

$$-\langle \mathrm{Im}(S_{hv}S_{hh}^*)V\rangle d\Omega/(A\cos\theta_s) \tag{1.40}$$

The term on the left-hand side of the above equation is the intensity of the scattered field. In view of (1.5) the first and second terms on the right-hand side are recognized as scattering coefficients divided by $4\pi\cos\theta_s$. Similarly, we can convert the left-hand sides of (1.37) through (1.39) into intensities and rewrite all four resulting equations into a matrix equation. This matrix equation relates the scattered intensities $\mathbf{I}^s$ to the incident intensities $\mathbf{I}^i$ through a dimensionless quantity known as the phase matrix $\mathbf{P}$,

$$\mathbf{I}^s = \frac{1}{4\pi}\mathbf{P}\mathbf{I}^i d\Omega \tag{1.41}$$

The components of $\mathbf{I}^i$ are the Stokes parameters as defined by (1.31) to (1.34) for the incident plane wave. The components of the scattered intensity $\mathbf{I}^s$ are also *Stokes parameters* but are defined for spherical waves. They differ from the plane wave definition in the normalizing solid angle $(A\cos\theta_s)/R^2$. For example, the first Stokes parameter in the scattered intensity $\mathbf{I}^s$ is

$$I_v^s = R^2\langle|E_v^s|^2\rangle/(\eta A\cos\theta_s) \tag{1.42}$$

The element of the phase matrix relating I_v^s to I_v^i is $4\pi\ \langle|S_{vv}|^2\rangle/(A\cos\theta_s)$. It can be expressed in terms of the scattering coefficient σ_{vv}^0 defined by (1.5) as $\sigma_{vv}^0/\cos\theta_s$. To sum up all possible incident intensities from all directions contributing to $\mathbf{I}^s$ along a given direction, we integrate over all solid angles, i.e.,

$$\mathbf{I}^s = \frac{1}{4\pi}\int_{4\pi}\mathbf{P}\mathbf{I}^i d\Omega \tag{1.43}$$

Equation (1.43) is the generalized version of (1.30), for partially polarized waves. Here, $\mathbf{I}^s$, $\mathbf{I}^i$ are column vectors whose components are the Stokes parameters. The detailed contents of the phase matrix are summarized below

$$\mathbf{P} = 4\pi\langle\mathbf{M}\rangle / (A\cos\theta_s) \tag{1.44}$$

where the *Stokes matrix* **M** has been provided by Ishimaru [1978],

$$\begin{bmatrix} |S_{vv}|^2 & |S_{vh}|^2 & \mathrm{Re}\,(S_{vv}S_{vh}^*) & -\mathrm{Im}\,(S_{vv}S_{vh}^*) \\ |S_{hv}|^2 & |S_{hh}|^2 & \mathrm{Re}\,(S_{hv}S_{hh}^*) & -\mathrm{Im}\,(S_{hv}S_{hh}^*) \\ 2\mathrm{Re}\,(S_{vv}S_{hv}^*) & 2\mathrm{Re}\,(S_{vh}S_{hh}^*) & \mathrm{Re}\,(S_{vv}S_{hh}^* + S_{vh}S_{hv}^*) & -\mathrm{Im}\,(S_{vv}S_{hh}^* - S_{vh}S_{hv}^*) \\ 2\mathrm{Im}\,(S_{vv}S_{hv}^*) & 2\mathrm{Im}\,(S_{vh}S_{hh}^*) & \mathrm{Im}\,(S_{vv}S_{hh}^* + S_{vh}S_{hv}^*) & \mathrm{Re}\,(S_{vv}S_{hh}^* - S_{vh}S_{hv}^*) \end{bmatrix}$$

Phase Matrix for an Inhomogeneous Medium

Consider a homogeneous medium embedded randomly with particles. Each particle is characterized by a *bistatic radar cross section* σ_p due to a p-polarized ($p = v$ or h) incident intensity, which is defined as 4π times the magnitude squared of the scattering amplitude of the particle [Ishimaru, 1978]. Then, the *scattering cross section of the particle* Q_{sp} is defined as the cross section that would produce the total scattered power surrounding the particle due to a unit incident Poynting vector of polarization p,

$$Q_{sp}(\theta, \phi) = \frac{1}{4\pi}\int_{4\pi} \sigma_p d\Omega_s = \int_{4\pi} \langle |S_{vp}|^2 + |S_{hp}|^2 \rangle d\Omega_s \tag{1.45}$$

where θ, ϕ indicate the incident direction, and integration is over the scattered solid angle. The *volume-scattering coefficient* for the inhomogeneous medium and polarization p is

$$\kappa_{sp} = N_v Q_{sp} \tag{1.46}$$

where N_v is the *number of particles per unit volume* [Ishimaru,Vol.1, 1978]. The scattering coefficient κ_{sp} represents the scattering loss per unit length and has the units of Np m^{-1}. In the case of a continuous, inhomogeneous medium, the scattering amplitudes S_{vp}, S_{hp} are for an effective volume V. The volume scattering coefficient is defined as [Tsang and Kong, 1976]

$$\kappa_{sp} = \frac{1}{V} Q_{sp} \tag{1.47}$$

Another important parameter for characterizing an inhomogeneous medium is its *absorption loss,* represented by the *volume-absorption coefficient* in κ_{ap}. This quantity may be defined in terms of the average relative permittivity ε_{ap} of the medium, where p denotes the incident polarization. Letting k_0 be the free-space wavenumber, we define the absorption coefficient for p polarization as [Tsang and Kong, 1975]

$$\kappa_{ap} = 2k_0\left|\mathrm{Im}\sqrt{\varepsilon_{ap}}\right| \tag{1.48}$$

This equation may be used for either a continuous inhomogeneous medium or a discrete inhomogeneous medium. In the latter case the *absorption cross* section Q_{ap} for one particle and p polarization can be defined as

$$Q_{ap} = \kappa_{ap}/N_v \tag{1.49}$$

From (1.46) and (1.49), the total cross section, also known as the *extinction* cross section, for a particle is

$$Q_{ep} = Q_{ap} + Q_{sp} \tag{1.50}$$

and the *extinction coefficient is* $\kappa_{ep} = N_v Q_{ep}$. In (1.50) Q_{ep} is the effective area that generates the total scattered and absorbed power due to a unit incident Poynting vector of polarization p. Conceptually, either Q_{ep} or Q_{sp} may be used in place of $A\cos\theta_s$ in (1.44) to define the phase matrix of a single particle. However, unlike $A\cos\theta_s$, Q_{ep} and Q_{sp} have polarization dependence in general [Ishimaru and Cheung, 1980] and hence are matrices. Let us denote them as $\mathbf{Q}_e$ and $\mathbf{Q}_S$. The choice of the definition for the phase matrix depends on the assumed form of the scattering source term in (1.29). When the term is written as $\kappa_s\mathbf{J}_s$ the definition is [Van de Hulst, 1957]

$$\mathbf{P}_s = 4\pi\mathbf{Q}_s^{-1}\langle\mathbf{M}\rangle \tag{1.51}$$

If the term is written as $\kappa_e\mathbf{J'}_s$ then the definition should be [Ishimaru, 1978, Vol.1, pp. 157-165]

$$\mathbf{P}_e = 4\pi\mathbf{Q}_e^{-1}\langle\mathbf{M}\rangle \tag{1.52}$$

Both definitions have appeared in the literature. Clearly, the phase matrix is a term created for convenience; the scattering source term is the fundamental quantity. Thus, while (1.51) is not the same as (1.52), the source terms are the same in both cases, as they should be,

$$\kappa_s\mathbf{J}_s = \kappa_e\mathbf{J'}_s = \int_{4\pi} N_v\langle\mathbf{M}\rangle\mathbf{I}\,d\Omega \tag{1.53}$$

In view of (1.53) and (1.29), the radiative transfer equation for *partially* polarized waves in a discrete inhomogeneous medium is

$$\frac{d\mathbf{I}}{dl} = -\kappa_e\mathbf{I} + \frac{\kappa_e}{4\pi}\int_{4\pi}\mathbf{P}_e\mathbf{I}d\Omega + \kappa_a\mathbf{J}_a \tag{1.54}$$

or

$$\frac{d\mathbf{I}}{dl} = -\kappa_e\mathbf{I} + \frac{\kappa_s}{4\pi}\int_{4\pi}\mathbf{P}_s\mathbf{I}d\Omega + \kappa_a\mathbf{J}_a \tag{1.55}$$

In a continuous, inhomogeneous medium, it is more convenient to use the Stokes matrix instead of the phase matrix. Making use of (1.47), we have

$$\frac{d\mathbf{I}}{dl} = -\kappa_e\mathbf{I} + \int_{4\pi}\frac{\langle\mathbf{M}\rangle}{V}\mathbf{I}d\Omega + \kappa_a\mathbf{J}_a \tag{1.56}$$

In (1.56), V is the effective illuminated volume as given in (1.47) and will cancel out upon evaluating <M>.

Most of the studies to date have dealt with the special case of spherical particles. For these particles, or for nonspherical particles with random orientations, κ_e and κ_s reduce to scalars [Ishimaru and Cheung, 1980]. If this is so, it is convenient to convert the independent variable in (1.54) or (1.55) to *optical thickness*, τ. Then (1.54) and (1.55) reduce respectively to

$$\frac{d\mathbf{I}}{d\tau} = -\mathbf{I} + \frac{1}{4\pi}\int_{4\pi}\mathbf{P}_e\mathbf{I}d\Omega + (1-a)\,\mathbf{J}_a \tag{1.57}$$

and

$$\frac{d\mathbf{I}}{d\tau} = -\mathbf{I} + \frac{a}{4\pi}\int_{4\pi}\mathbf{P}_s\mathbf{I}d\Omega + (1-a)\,\mathbf{J}_a \tag{1.58}$$

where $\tau = \int\kappa_e dl$ and a *is* the albedo, which is the ratio of κ_s to κ_e. Because both (1.57) and (1.58) appear in the literature, it may be helpful to note that $\mathbf{P}_e = a\mathbf{P}_s$.

The radiative transfer equation is formulated on the basis of energy balance. Thus, the phase changes of the scattered wave and its cross-correlation terms are ignored in the solution of the transfer equation. Furthermore, the phase function used in practice comes from averaging the magnitude squared of the far-zone scattering amplitudes of an individual scatterer or its equivalent. Hence, for the transfer method to apply, the spacing between scatterers in a discrete medium or adjacent inhomogeneities in a continuous medium must be large. An experimental study by Vasalos [1969] shows that for the transfer equation [as defined by (1.54) through (1.56)] to be applicable, the spacing between scatterers must be larger than $\lambda/3$ and $0.4d$, where λ is the wavelength in the host medium and d is the diameter of the scatterer. Vasalos [1969] considered volume fractions as high as 0.295, d/λ ratios from 0.186 to 1.2, and optical thicknesses in the 0.01–3211 range. Additional studies by Hottel et al. [1971] show that a smaller spacing-to-wavelength ratio ($l/\lambda = 0.117$) can be acceptable if the d/λ ratio is larger than 0.23 at a volume fraction of 0.2195 or less. Theoretical studies on the relationship between the radiative transfer approach and the wave approach using Maxwell's equations have been carried out by Fante [1981]. Related studies have also been reported by Ishimaru and Kuga [1982].

1.5.2 Scattering from an Inhomogeneous Layer with Irregular Boundaries

For bounded media, scattering or reflection may occur at the boundary. Both incident and scattered intensities are needed in the boundary conditions. Therefore, it is necessary to split the intensity matrix into upward $\mathbf{I}^+$ and downward $\mathbf{I}^-$ components and rewrite (1.55) as two equations. For active sensing applications, the thermal source term is not needed. It is also a standard practice to express the slant range in terms of the vertical distance, i.e., let $l = z/\cos\theta$ (see Figure 1.3).

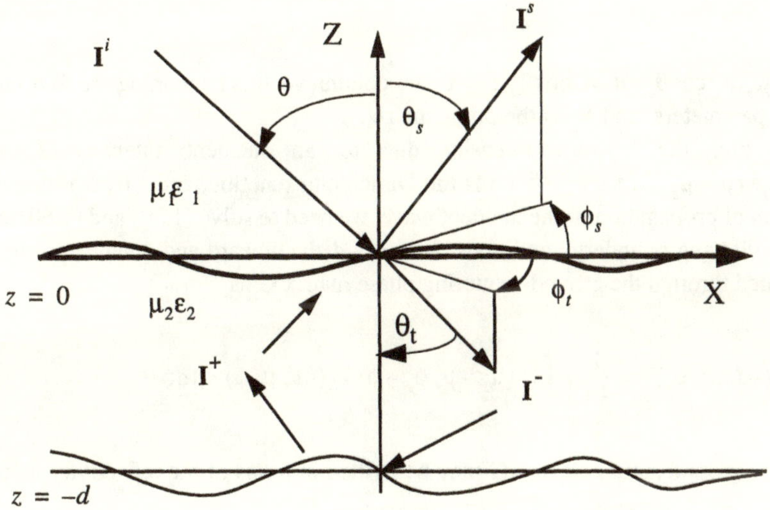

Figure 1.3 Scattering from an inhomogeneous layer above a homogeneous half space.

Consider the problem of a plane wave in air incident upon an inhomogeneous layer above a ground surface. The geometry of the scattering problem is depicted in Figure

1.3. The inhomogeneous layer is assumed to have such characteristics that the *upward intensity* $\mathbf{I}^+$ and the *downward intensity* $\mathbf{I}^-$ satisfy the radiative transfer equation. Upon rewriting (1.55) in terms of these intensities we obtain [Tsang and Kong, 1978]

$$\mu_s \frac{d}{dz}\mathbf{I}^+(z, \mu_s, \phi_s) = -\kappa_e \mathbf{I}^+(z, \mu_s, \phi_s) + \frac{1}{4\pi}\int_0^{2\pi}\int_0^1 \kappa_s \mathbf{P}_s(\mu_s, \mu, \phi_s - \phi)\,\mathbf{I}^+(z, \mu, \phi)\, d\mu d\phi + \frac{1}{4\pi}\int_0^{2\pi}\int_0^1 \kappa_s \mathbf{P}_s(\mu_s, -\mu, \phi_s - \phi)\,\mathbf{I}^-(z, \mu, \phi)\, d\mu d\phi \tag{1.59}$$

$$\mu_s \frac{d}{dz}\mathbf{I}^-(z, \mu_s, \phi_s) = \kappa_e \mathbf{I}^-(z, \mu_s, \phi_s) - \frac{1}{4\pi}\int_0^{2\pi}\int_0^1 \kappa_s \mathbf{P}_s(-\mu_s, \mu, \phi_s - \phi)\,\mathbf{I}^+(z, \mu, \phi)\, d\mu d\phi - \frac{1}{4\pi}\int_0^{2\pi}\int_0^1 \kappa_s \mathbf{P}_s(-\mu_s, -\mu, \phi_s - \phi)\,\mathbf{I}^-(z, \mu, \phi)\, d\mu d\phi \tag{1.60}$$

where $\mu_s = \cos\theta_s$; $\mu = \cos\theta$; $\mathbf{I}^+$, $\mathbf{I}^-$ are column vectors containing the four *modified* Stokes parameters; and $\mathbf{P}_s$ is the phase matrix.

To find the upward intensity due to an incident intensity $\mathbf{I}^i$, where $\mathbf{I}^i = \mathbf{I}_0\delta(\mu - \mu_i)\,\delta(\phi - \phi_i)$, $\delta(\,)$ is the Dirac delta function, and (θ_i, ϕ_i) denotes the direction of propagation of the incident wave, we need to solve (1.59) and (1.60), subject to the following boundary conditions. At $z = -d$ the upward and downward intensities are related through the ground-scattering phase matrix $\mathbf{G}$ as

$$\mathbf{I}^+(-d, \mu_s, \phi_s) = \frac{1}{4\pi}\int_0^{2\pi}\int_0^1 \mathbf{G}(\mu_s, -\mu, \phi_s - \phi)\,\mathbf{I}^-(-d, \mu, \phi)\, d\mu d\phi \tag{1.61}$$

If the ground surface is flat, $\mathbf{G}$ may be written in terms of the reflectivity matrix $\mathbf{R}_g$ as

$$\mathbf{G} = 4\pi\mathbf{R}_g\delta(\mu_s - \mu)\,\delta(\phi_s - \phi) \tag{1.62}$$

At the top boundary $z = 0$, the upward and downward intensities are related through the surface-scattering and transmission phase matrices $\mathbf{S}_R$ and $\mathbf{S}_T$ [Fung and Eom, 1981a,b]

$$\mathbf{I}^-(0, \mu_s, \phi_s) = \frac{1}{4\pi}\int_0^{2\pi}\int_0^1 \mathbf{S}_R(-\mu_s, \mu, \phi_s - \phi)\,\mathbf{I}^+(0, \mu, \phi)\,d\mu d\phi$$

$$+\frac{1}{4\pi}\int_0^{2\pi}\int_0^1 \mathbf{S}_T(-\mu_s, -\mu, \phi_s - \phi)\,\mathbf{I}^i(0, \mu, \phi)\,d\mu d\phi \qquad (1.63)$$

Once $\mathbf{I}^+(0, \mu_s, \phi_s)$ is determined within the inhomogeneous layer, the upward intensity transmitted from the layer into air can be found using the transmission scattering matrix of the surface, $\mathbf{S}_T$ as

$$\mathbf{I}^+(\mu_s, \phi_s) = \frac{1}{4\pi}\int_0^{2\pi}\int_0^1 \mathbf{S}_T(\mu_s, \mu, \phi_s - \phi)\,\mathbf{I}^+(0, \mu, \phi)\,d\mu d\phi \qquad (1.64)$$

The total scattered intensity in air is given by the sum of $\mathbf{I}^+(\mu_s, \phi_s)$ and $\mathbf{I}_s$, where $\mathbf{I}_s$ is the intensity due to random surface scattering by the top layer boundary. The explicit forms of the matrices $\mathbf{R}_g$, $\mathbf{S}_R$ and $\mathbf{S}_T$ are given in Appendix 2A. The expressions for $\mathbf{S}_R$ and $\mathbf{S}_T$ are for an irregular boundary. The $\mathbf{G}$ matrix is assumed to have the same form as $\mathbf{S}_R$. Once the total scattered intensity for a p-polarized component I_p^s of the intensity matrix is found, the scattering coefficient for this component is defined relative to the incident intensity $I_q^i = I_{q0}\delta(\mu - \mu_i)\,\delta(\phi - \phi_i)$ of polarization q along (μ_i, ϕ_i) direction as

$$\sigma_{pq}^0 = (4\pi I_p^s \cos\theta_s)/I_{q0} \qquad (1.65)$$

The transfer equations given by (1.59) and (1.60) can be solved exactly by using numerical techniques [Fung and Chen, 1981]. A solution technique using the matrix doubling approach is given in Part III. To gain some insight into the transfer equation solution, we shall consider first-order solutions in the next two chapters.

1.5.3 Emission from an Inhomogeneous Layer with Irregular Boundaries

Consider the emission problem of an inhomogeneous irregular layer above a homogeneous half space (Figure 1.4). We assume that the incoherent source term in (1.58) is represented by the temperature profile of the layer $\boldsymbol{T}$. The applicable equations are similar to (1.59) and (1.60) except that

1. The incoherent source term in (1.58), $(1-a)\boldsymbol{T} = \kappa_a(\boldsymbol{T}/\kappa_e)$, is included.
2. The intensity matrices are replaced by the temperature matrices, $\mathbf{T}^+$ and $\mathbf{T}^-$, which consist of vertically and horizontally polarized components.
3. We assume that $\boldsymbol{T}$ has no azimuthal dependence.

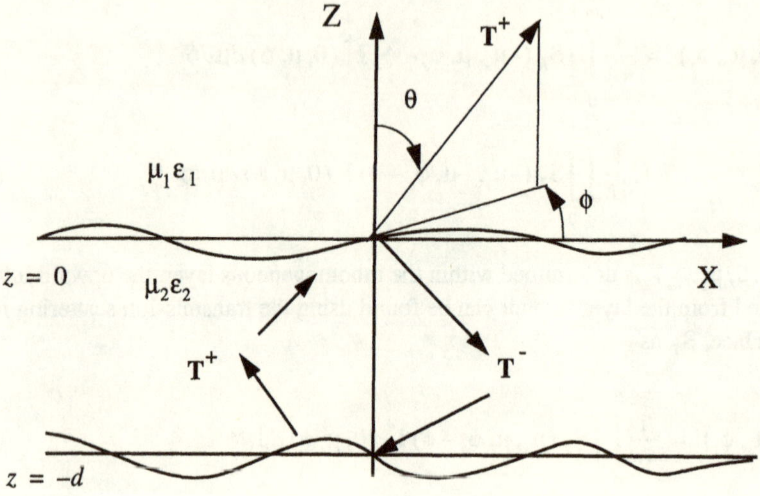

Figure 1.4 Emission from an inhomogeneous layer above a homogeneous half space.

The second point above is the consequence of the incoherent nature of natural emission, which causes the third and the fourth Stokes parameters to vanish. The third point allows us to replace the elements of the phase matrix by their zeroth-order Fourier components in both the transfer equations and the associated boundary conditions. For most problems this assumption is acceptable. If not, a more complete formulation similar to the scattering problem would be needed. Recognizing these changes we can write the governing equations for $\mathbf{T}^+$ and $\mathbf{T}^-$ as follows [Ulaby et al, 1986]:

$$\mu_s \frac{d\mathbf{T}^+}{dz} = -\kappa_e \mathbf{T}^+ + \kappa_a T + \mathbf{F}^+ \tag{1.66}$$

$$\mu_s \frac{d\mathbf{T}^-}{dz} = \kappa_e \mathbf{T}^- - \kappa_a T - \mathbf{F}^- \tag{1.67}$$

where

$$\mathbf{F}^{\pm} = \frac{1}{2}\int_0^1 \kappa_s \mathbf{P}_s(\pm\mu_s, \mu)\, \mathbf{T}^+(z, \mu)\, d\mu$$

$$+ \frac{1}{2}\int_0^1 \kappa_s \mathbf{P}_s(\pm\mu_s, -\mu)\, \mathbf{T}^-(z, \mu)\, d\mu \tag{1.68}$$

and T is the temperature profile of the inhomogeneous layer. In (1.68) $\mathbf{P}_s(\pm\mu_s, \mu)$ is the zeroth-order Fourier component of the phase matrix $\mathbf{P}_s(\mu_s, \mu; \phi_s - \phi)$. The applicable boundary condition at $z = 0$ is

$$\mathbf{T}^-(0, \mu_s) = \frac{1}{4\pi}\int_0^{2\pi}\int_0^1 \mathbf{S}_R(-\mu_s, \mu, \phi_s - \phi)\,\mathbf{T}^+(0, \mu)\,d\mu d\phi$$

$$= \frac{1}{2}\int_0^1 \mathbf{S}_R(-\mu_s, \mu)\,\mathbf{T}^+(0, \mu)\,d\mu \tag{1.69}$$

and at $z = -d$

$$\mathbf{T}^+(-d, \mu_s) = \frac{1}{2}\int_0^1 \mathbf{G}(\mu_s, -\mu)\,\mathbf{T}^-(-d, \mu)\,d\mu + \mathbf{e}_g T_g \tag{1.70}$$

where the last term represents emission from the lower half space into the layer. It is assumed that the lower half space has constant temperature T_g and the emissivity across the boundary is $\mathbf{e}_g$. The first-order solution of the above problem will be given in Chapter 3.

1.6 DEPENDENCE OF THE SCATTERING COEFFICIENT ON POLARIZATION STATES

In previous sections we discussed scattering characteristics of linearly polarized waves. In this section the scattering properties of plane waves at other polarization states are considered.

A standard method to denote the polarization state of a plane wave is to use the polarization ellipse defined in terms of the ellipticity angle τ_t and a pair of orthogonal vectors, $\hat{x}$ and $\hat{y}$. For example, a left-handed elliptic polarization can be represented as

$$\vec{E} = (\hat{x}\cos\tau_t + j\hat{y}\sin\tau_t)\,E_o\exp[j(\omega t - kz)] \tag{1.71}$$

We can generalize the polarization state further by assuming that the major axis of the ellipse in $\hat{x}$ direction is oriented at an angle ψ_t from the $\hat{\phi}_t$ axis of the reference coordinates, $\hat{\phi}_t$ and $\hat{\theta}_t$, of the transmitting antenna defined in Figure 1.5. Thus, the polarization state of the transmitting antenna is defined by its radiating field $\vec{E} = \hat{a}_t E_o\exp[j(\omega t - kz)]$, where the unit polarization vector is

$$\hat{a}_t = \hat{\theta}_t(\cos\tau_t\sin\psi_t - j\sin\tau_t\cos\psi_t) + \hat{\phi}_t(\cos\tau_t\cos\psi_t + j\sin\tau_t\sin\psi_t) \tag{1.72}$$

This unit vector is characterized by the ellipticity angle τ_t and the orientation angle ψ_t. It is clear that the polarization state of the receiving antenna can be characterized in a similar way, and we shall denote it by $\hat{a}_r$. In general, a different set of ellipticity and orientation angles may associate with the receiving antenna. To find the average received power we need to calculate the received voltage using the scattering matrix of the target which is discussed in the next section.

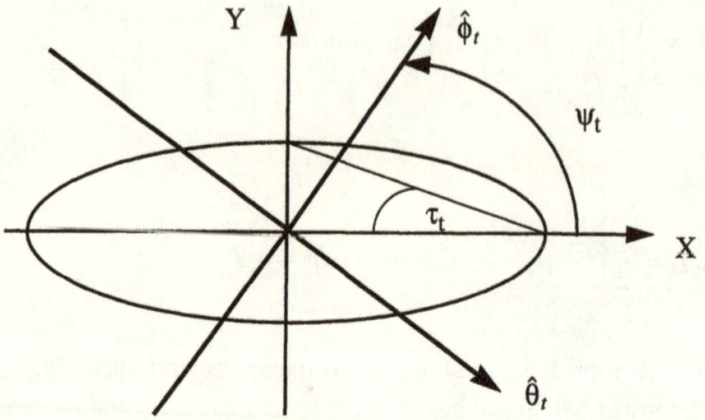

Figure 1.5 Elliptic polarization state.

1.6.1 Target Scattering Matrix and the Averaged Received Power

The polarization unit vector of the transmitting antenna written in matrix form using phasor representation is

$$\begin{bmatrix} a_{t\theta} \\ a_{t\phi} \end{bmatrix} = \begin{bmatrix} \cos\tau_t \sin\psi_t - j\sin\tau_t \cos\psi_t \\ \cos\tau_t \cos\psi_t + j\sin\tau_t \sin\psi_t \end{bmatrix} \tag{1.73}$$

Let us denote the target scattering matrix by $\mathbf{S} = [S_{qp}]$. Then, the scattered field for unit incident amplitude is

$$S = \frac{e^{-jkR}}{R}\left[S_{qp}\right] = \frac{e^{-jkR}}{R}\begin{bmatrix} S_{\theta\theta} & S_{\theta\phi} \\ S_{\phi\theta} & S_{\phi\phi} \end{bmatrix} \tag{1.74}$$

and $S_{qp} = \exp(jkR) RE^s_q(\theta_s, \phi_s)/E^i_p(\theta_i, \phi_i)$ where E^i_p and E^s_q denote the incident and scattered fields with polarization states p and q, respectively. Then, the normalized received voltage for transmitting polarization t and receiving polarization r is

$$V_{rt} = \frac{e^{-jkR}}{R}\begin{bmatrix} a_{r\theta} & a_{r\phi} \end{bmatrix}\begin{bmatrix} S_{\theta\theta} & S_{\theta\phi} \\ S_{\phi\theta} & S_{\phi\phi} \end{bmatrix}\begin{bmatrix} a_{t\theta} \\ a_{t\phi} \end{bmatrix}$$

$$= [a_{r\phi}(S_{\phi\phi}a_{t\phi} + S_{\phi\theta}a_{t\theta}) + a_{r\theta}(S_{\theta\phi}a_{t\phi} + S_{\theta\theta}a_{t\theta})]\,(e^{-jkR}/R) \qquad (1.75)$$

It is clear that both t and r depend on the ellipticity and orientation angles of the transmitting and receiving antennas, respectively. For simplicity we assume that there is no correlation between S_{pq} and S_{pp} or S_{qq}. Then, the expression for the average received power is

$$P_{rt} = \langle |V_{rt}|^2 \rangle$$

$$= \big[|a_{r\phi}|^2\sigma^0_{\phi\phi}|a_{t\phi}|^2 + 2\mathrm{Re}\,(a_{r\phi}a_{r\theta}{}^*\sigma^0_{\phi\phi\theta\theta}a_{t\phi}a_{t\theta}{}^*) + 2\mathrm{Re}\,(a_{r\phi}a_{r\theta}{}^*\sigma^0_{\phi\theta\theta\phi}a_{t\phi}{}^*a_{t\theta})$$

$$+ |a_{r\phi}|^2\sigma^0_{\phi\theta}|a_{t\theta}|^2 + |a_{r\theta}|^2\sigma^0_{\theta\phi}|a_{t\phi}|^2 + |a_{r\theta}|^2\sigma^0_{\theta\theta}|a_{t\theta}|^2\big]A/(4\pi R^2) \qquad (1.76)$$

where * is the symbol for complex conjugate; R is the range from the transmitting antenna to the illuminated area A and $\sigma^0_{\phi\theta}$ and $\sigma^0_{\phi\phi\theta\theta}$ are the scattering coefficients defined as follows:

$$\sigma^0_{\phi\theta\theta\phi} = 4\pi\langle S_{\phi\theta}S_{\theta\phi}{}^*\rangle/A$$

$$\sigma^0_{\phi\phi\theta\theta} = 4\pi\langle S_{\phi\phi}S_{\theta\theta}{}^*\rangle/A$$

Note that once the antenna coordinates, θ and ϕ, are fixed, the target scattering matrix defined in terms of them is fixed irrespective of the choice of the transmitting or receiving antenna polarization state, which is a function of the angles τ and ψ. Hence, it is possible to determine the response to different polarization states with a given set of S_{pq}'s, which is a function of the incident and scattered directions. It is a common practice to calculate the scattering coefficient rather than the average power, which has range dependence. The scattering coefficient for elliptic polarization has the form

$$\sigma^0_{rt} = 4\pi R^2 P_{rt}/A$$

$$= |a_{r\phi}|^2\sigma^0_{\phi\phi}|a_{t\phi}|^2 + 2\mathrm{Re}\,(a_{r\phi}a_{r\theta}{}^*\sigma^0_{\phi\phi\theta\theta}a_{t\phi}a_{t\theta}{}^*) + 2\mathrm{Re}\,(a_{r\phi}a_{r\theta}{}^*\sigma^0_{\phi\theta\theta\phi}a_{t\phi}{}^*a_{t\theta})$$

$$+ |a_{r\phi}|^2\sigma^0_{\phi\theta}|a_{t\theta}|^2 + |a_{r\theta}|^2\sigma^0_{\theta\phi}|a_{t\phi}|^2 + |a_{r\theta}|^2\sigma^0_{\theta\theta}|a_{t\theta}|^2. \qquad (1.77)$$

To do computation with (1.77) we need to know all the scattering coefficients that appear in it.

For backscattering from a randomly rough surface, all the scattering coefficients in (1.77) are given in Appendix 2A. Hence, numerical illustrations of (1.77) will be given in Chapter 2.

Like and Cross Polarizations

Following definitions developed in antenna reception theory [Mott, 1986], we define like polarization to be reception with matched antennas, i.e.,

$$|\hat{a}_r \cdot \hat{a}_t| = 1 \tag{1.78}$$

Note that this definition is for antennas and hence has nothing to do with target properties. The cross or orthogonal polarization is defined to be with zero reception, i.e.,

$$|\hat{a}_r \cdot \hat{a}_t| = 0 \tag{1.79}$$

To illustrate the meaning of (1.78) consider the transmitting antenna with polarization defined by (1.71). Let (1.71) represent the radiated wave from a transmitting antenna. Rewriting (1.71) in time domain with $E_0 = 1$, we have

$$\vec{E}_t = \hat{x}\cos\tau_t \cos(\omega t - kz) - \hat{y}\sin\tau_t \sin(\omega t - kz) \tag{1.80}$$

which is seen to be a left-hand elliptically polarized wave. A sign change in τ_t would make the wave right-handed. If the receiving antenna is chosen to have the same polarization, its radiated field will have the same mathematical form but expressed in coordinates for the receiving antenna. As illustrated in Figure 1.6 transmitting and receiving antenna systems must point in opposite directions. This means that when the radiated field of the receiving antenna is expressed in the coordinate system of the transmitting antenna, its propagation phase must take the form $\omega t + kz$ and the rectangular coordinate system for the receiving antenna may be related to that of the transmitting antenna as follows:

$$\hat{x}' = \hat{x}$$
$$\hat{y}' = -\hat{y} \tag{1.81}$$

Similarly, the polarization base vectors in the receiving antenna coordinate may be selected as

$$\hat{\theta}_r = \hat{\theta}_t$$

and

$$\hat{\phi}_r = -\hat{\phi}_t$$

Thus, the radiating field from the receiving antenna expressed in the coordinates of the transmitting antenna is

$$\vec{E}_r = \hat{x}\cos\tau_r \cos(\omega t + kz) + \hat{y}\sin\tau_r \sin(\omega t + kz) \tag{1.82}$$

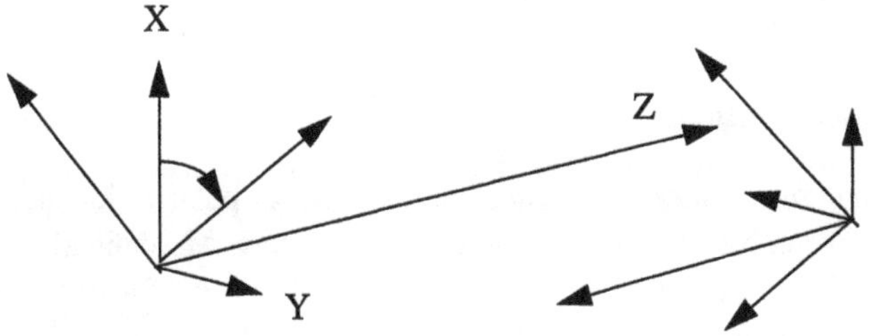

Figure 1.6 Illustration of transmitting and receiving antenna systems.

When we convert to phasor form, it becomes

$$\vec{E}_r = (\hat{x}\cos\tau_r - j\hat{y}\sin\tau_r)\exp[j(\omega t + kz)] \tag{1.83}$$

From (1.71) and (1.83) the polarization unit vectors for transmitting and receiving are, respectively,

$$\hat{a}_t = \hat{x}\cos\tau + j\hat{y}\sin\tau \quad \text{and} \quad \hat{a}_r = \hat{x}\cos\tau - j\hat{y}\sin\tau$$

when we set $\tau_t = \tau_r = \tau$. Clearly, one unit vector is the complex conjugate of the other, $|\hat{a}_r \cdot \hat{a}_t| = 1$ and the polarization states of *both antennas are left-hand elliptic*. Hence, this case is referred to as like polarization.

When we generalize the polarization vector of the transmitting antenna from (1.71) to (1.72), we need to set $\hat{\theta}_r = \hat{\theta}_t$, $\hat{\phi}_r = -\hat{\phi}_t$ and $\psi_r = \psi_t + \pi$ to express the receiving polarization vector in the transmitting antenna coordinates. Thus, the polarization vector for the matched antenna becomes

$$\begin{aligned}\hat{a}_r &= \hat{\theta}_r(\cos\tau_r\sin\psi_r - j\sin\tau_r\cos\psi_r) + \hat{\phi}_r(\cos\tau_r\cos\psi_r + j\sin\tau_r\sin\psi_r)\\ &= \hat{\theta}_t(\cos\tau_t\sin\psi_t + j\sin\tau_t\cos\psi_t) + \hat{\phi}_t(\cos\tau_t\cos\psi_t - j\sin\tau_t\sin\psi_t)\end{aligned} \tag{1.84}$$

To illustrate cross or orthogonal polarization consider the transmitting antenna defined by (1.72) and set $\psi_r = \psi_t - \pi/2$ and $\tau_r = -\tau_t$. Thus, the polarization vector of the receiving antenna becomes

$$\hat{a}_r = \hat{\theta}_t(\cos\tau_t\cos\psi_t + j\sin\tau_t\sin\psi_t) - \hat{\phi}_t(\cos\tau_t\sin\psi_t - j\sin\tau_t\cos\psi 12_t) \tag{1.85}$$

If we take the dot product of (1.72) and (1.85), we find that $\hat{a}_r \cdot \hat{a}_t = 0$. When we check its polarization state, it is right-hand elliptic. Thus, left-hand elliptic and right-hand elliptic polarizations are mutually orthogonal. The special case with $\tau = 0$ and $\psi = 0$ leads to $\hat{a}_r = \hat{\phi}_t$ and $\hat{a}_t = \hat{\theta}_t$, which are obviously orthogonal.

1.6.2 Polarization Synthesis

A way to calculate the scattering coefficients for different polarization states was given by (1.77). An alternative way is to use the Stokes matrix, **M**, defined in Section 1.5. According to Ulaby and Elachi [1990] the scattering coefficient for a general transmitting and receiving antenna polarization may be written as

$$\sigma_{rt}^0(\psi_r, \tau_r; \psi_t, \tau_t) = \frac{4\pi}{A}(\mathbf{I}_r \cdot \mathbf{R} \cdot \mathbf{M} \cdot \mathbf{I}_t) \tag{1.86}$$

where A is the illuminated area, $\mathbf{I}_t$, $\mathbf{I}_r$ are the Stokes vector with unit amplitude and $\mathbf{R}$ is a rotating matrix required to convert the stokes matrix to the modified stokes scattering operator. It is a diagonal matrix given by

$$\mathbf{R} = \begin{bmatrix} 1 & 0 & 0 & 0 \\ 0 & 1 & 0 & 0 \\ 0 & 0 & 0.5 & 0 \\ 0 & 0 & 0 & -0.5 \end{bmatrix} \tag{1.87}$$

The Stokes vectors are expressed in terms of the ellipticity and orientation angles as

$$\mathbf{I}_r = \begin{bmatrix} 0.5\,(1+\cos 2\tau_r \cos 2\psi_r) \\ 0.5\,(1-\cos 2\tau_r \cos 2\psi_r) \\ \cos 2\tau_r \sin 2\psi_r \\ \sin 2\tau_r \end{bmatrix} \quad \text{and} \quad \mathbf{I}_t = \begin{bmatrix} 0.5\,(1+\cos 2\tau_t \cos 2\psi_t) \\ 0.5\,(1-\cos 2\tau_t \cos 2\psi_t) \\ \cos 2\tau_t \sin 2\psi_t \\ \sin 2\tau_t \end{bmatrix}$$

Since **M** is not a function of orientation or ellipticity angle, the scattering coefficient for all incident polarization states can be generated from (1.86) for a given **M**, which depends only on target properties and view angles. For a like polarization set $\tau_r = \tau_t$ and $\psi_r = \psi_t + \pi$, and for a cross polarization set $\psi_r = \psi_t - \pi/2$ and $\tau_r = -\tau_t$.

1.7 DISTRIBUTION FUNCTIONS FOR SCATTERED POWER AND PHASE DIFFERENCE

In previous sections we have discussed the average value of radar or radiometer signals. A more complete description of a statistical signal is given by its distribution function from which both the signal variance and its mean can be computed. In this section we shall consider the distribution functions of the signal amplitude and the phase difference.

The former has been widely used in detection problems, and the latter is needed in computing the average phase difference which is part of the average properties of a signal.

A special case that allows analytical treatment is first presented leading to distribution functions of the signal amplitude and its phase difference. This is the problem when a terrain can be represented by a collection of discrete, isolated point scatterers. The presentation below is a variation of the studies by Barakat [1987], Sarabandi [1992] and Touzi and Lopes [1991]. Models for signal distribution of more general terrains obtained by computer simulation are summarized in Section 1.7.2. Note that the term *signal amplitude* here refers to the amplitude of the scattered power, while *signal phase difference* refers to the phase difference between the like polarized scattered fields. Readers interested in the distribution function of the field amplitude $p(A)$ can recover it from the distribution of the power amplitude $p(\sigma)$ by the following relation:

$$p(A) = 2Ap(\sigma)\Big|_{\sigma = A^2} \tag{1.88}$$

1.7.1 Signal Amplitude and Phase Difference Statistics in Backscattering

Consider a plane wave incident on an area-extensive target consisting of isolated point scatterers. Then, the relation between the incident and scattered field is given by (1.35), which is repeated below for ease of reference,

$$\begin{bmatrix} E_v^s \\ E_h^s \end{bmatrix} = \frac{e^{jkR}}{R}\begin{bmatrix} S_{vv} & S_{vh} \\ S_{hv} & S_{hh} \end{bmatrix}\begin{bmatrix} E_v^i \\ E_h^i \end{bmatrix}$$

The elements of the S matrix above are known as scattering amplitudes representing the scattering strengths of the illuminated area excluding the range effect. They are in general complex quantities. Within an illuminated area there are many point scatterers. Thus, each scattering amplitude contains contributions from all illuminated scatterers. We can write the scattering amplitude in terms of the contributions from the scatterers as

$$S_{qp} = |S_{qp}|\exp(j\phi_{qp}) = \sum_{n=1}^{N} |S_{qp}^n|\exp(j\phi_{qp}^n) \tag{1.89}$$

In (1.89) N is the total number of scatterers randomly located within the illuminated area, each with amplitude S_{qp}^n and phase ϕ_{qp}^n. When N is large enough so that the central limit theorem is applicable, the real and the imaginary parts of S_{qp} are Gaussian distributed random variables with zero mean [Davenport, 1970]. In this case the amplitude of S_{qp} is Rayleigh distributed, and its phase ϕ_{qp} is uniformly distributed [Davenport and Root, 1958] over $(-\pi, \pi)$. In addition, S_{qp} and ϕ_{qp} are independent.

In backscattering the cross polarized scattering amplitudes are equal due to reciprocity. Hence, there are six independent quantities in the scattering matrix resulting from the real and imaginary parts of the scattering amplitudes. Experimental observations indicate that the density function of $\phi_{vh} - \phi_{vv}$ is uniform over $(-\pi, \pi)$ [Ulaby et al., 1991]. If so, S_{hv} and S_{vv}, S_{hh} are independent. The overall joint density function is the product of the joint density functions of the like and cross polarized components. Hence, to find the density functions of the phase sum and difference of like polarized scattering amplitudes, we only need to consider the like polarized components. For large N the real and imaginary parts x_i of the co-polarized elements of the scattering matrix are Gaussian random variables with the joint density function [Davenport, 1970]

$$p(x_1, x_2, x_3, x_4) = \frac{1}{4\pi^2 D^{1/2}} \exp\left[-\frac{1}{2D} \sum_{i,j=1}^{4} a_{ij} x_i x_j\right] \tag{1.90}$$

where a_{ij} are the cofactors of the determinant D of the covariance matrix $\boldsymbol{m}$ with elements

$$\begin{aligned}
m_{11} &= \langle |S_{vv}|^2 \rangle \langle \cos^2\phi_{vv} \rangle = \langle |S_{vv}|^2 \rangle / 2 \\
m_{12} &= \langle |S_{vv}|^2 \rangle \langle \cos\phi_{vv} \sin\phi_{vv} \rangle = m_{21} = 0 \\
m_{13} &= \langle |S_{vv} S_{hh}| \rangle \langle \cos\phi_{vv} \cos\phi_{hh} \rangle = m_{31} \\
m_{14} &= \langle |S_{vv} S_{hh}| \rangle \langle \cos\phi_{vv} \sin\phi_{hh} \rangle = m_{41}
\end{aligned} \tag{1.91}$$

$$\begin{aligned}
m_{22} &= \langle |S_{vv}|^2 \rangle \langle \sin^2\phi_{vv} \rangle = \langle |S_{vv}|^2 \rangle / 2 \\
m_{23} &= \langle |S_{vv} S_{hh}| \rangle \langle \sin\phi_{vv} \cos\phi_{hh} \rangle = m_{32} \\
m_{24} &= \langle |S_{vv} S_{hh}| \rangle \langle \sin\phi_{vv} \sin\phi_{hh} \rangle = m_{42}
\end{aligned} \tag{1.92}$$

$$\begin{aligned}
m_{33} &= \langle |S_{hh}|^2 \rangle \langle \cos^2\phi_{hh} \rangle = \langle |S_{hh}|^2 \rangle / 2 \\
m_{34} &= \langle |S_{hh}|^2 \rangle \langle \sin\phi_{hh} \cos\phi_{hh} \rangle = m_{43} = 0
\end{aligned} \tag{1.93}$$

$$m_{44} = \langle |S_{hh}|^2 \rangle \langle \sin^2\phi_{hh} \rangle = \langle |S_{hh}|^2 \rangle / 2 \tag{1.94}$$

The elements of the covariance matrix indicate that in general there is correlation between the real and imaginary parts of the co-polarized scattering elements but there is no correlation between the real and imaginary parts of the same element. The joint density function is known, if all the m_{ij}'s are known. Next, we note that the above elements also appear in the Stokes matrix elements given in Section 1.5.1. In particular, by denoting the elements in the Stokes matrix as M_{ij} we have the following relations:

$$M_{43} - M_{34} = 2(m_{23} - m_{14}) = 2\langle |S_{vv}S_{hh}| \rangle \langle \sin(\phi_{vv} - \phi_{hh}) \rangle$$

$$M_{33} + M_{44} = 2(m_{13} + m_{24}) = 2\langle |S_{vv}S_{hh}| \rangle \langle \cos(\phi_{vv} - \phi_{hh}) \rangle$$

$$M_{11} = 2m_{11} = \langle |S_{vv}|^2 \rangle = 2m_{22}$$

$$M_{22} = 2m_{33} = \langle |S_{vv}|^2 \rangle = 2m_{44} \quad (1.95)$$

When a target is given we assume that its Stokes matrix is known. Hence, our task here is to express m_{ij} in terms of M_{ij}. There are four unknowns in the first two equations in (1.95). Unless we can find additional conditions we shall not be able to determine all the m_{ij}. Upon examining the meaning of m_{13} and m_{24} we see that the former represents the correlation between the real parts of the scattering amplitudes S_{vv} and S_{hh}, while the latter represents the correlation between their imaginary parts. For two complex quantities, their correlation is defined as the average of one quantity times the complex conjugate of the other. Hence, their correlation must depend only upon their phase difference. Therefore, we have the following averages,

$$\langle \cos\phi_{vv} \cos\phi_{hh} \rangle = \frac{\langle \cos(\phi_{vv} + \phi_{hh}) \rangle + \langle \cos(\phi_{vv} - \phi_{hh}) \rangle}{2} = \frac{\langle \cos(\phi_{vv} - \phi_{hh}) \rangle}{2}$$

$$\langle \sin\phi_{vv} \sin\phi_{hh} \rangle = \frac{\langle \cos(\phi_{vv} - \phi_{hh}) \rangle - \langle \cos(\phi_{vv} + \phi_{hh}) \rangle}{2} = \frac{\langle \cos(\phi_{vv} - \phi_{hh}) \rangle}{2}$$

$$\langle \cos\phi_{vv} \sin\phi_{hh} \rangle = \frac{\langle \sin(\phi_{vv} + \phi_{hh}) \rangle - \langle \sin(\phi_{vv} - \phi_{hh}) \rangle}{2} = -\frac{\langle \sin(\phi_{vv} - \phi_{hh}) \rangle}{2}$$

$$\langle \sin\phi_{vv} \cos\phi_{hh} \rangle = \frac{\langle \sin(\phi_{vv} + \phi_{hh}) \rangle + \langle \sin(\phi_{vv} - \phi_{hh}) \rangle}{2} = \frac{\langle \sin(\phi_{vv} - \phi_{hh}) \rangle}{2}$$

The equality of $\langle \cos\phi_{vv} \sin\phi_{hh} \rangle = -\langle \sin\phi_{vv} \cos\phi_{hh} \rangle$ shows that

$$m_{14} = -m_{23} = (M_{43} - M_{34})/4 \quad (1.96)$$

and the equality of $\langle \cos\phi_{vv} \cos\phi_{hh} \rangle = \langle \sin\phi_{vv} \sin\phi_{hh} \rangle$ indicates that

$$m_{13} = m_{24} = (M_{33} + M_{44})/4 \quad (1.97)$$

With the addition of the above two conditions all the m_{ij}'s are known. We can now find all the cofactors and the determinant of the covariance matrix,

$$a_{11} = m_{33} D^{-1/2}$$

$$a_{12} = a_{21} = 0$$

$$a_{13} = a_{31} = -m_{13} D^{-1/2}$$

$$a_{14} = a_{41} = -m_{14}D^{-1/2} \tag{1.98}$$

$$a_{22} = m_{33}D^{-1/2}$$

$$a_{23} = a_{32} = m_{14}D^{-1/2}$$

$$a_{24} = a_{42} = -m_{13}D^{-1/2} \tag{1.99}$$

$$a_{33} = m_{11}D^{-1/2}$$

$$a_{34} = a_{43} = 0 \tag{1.100}$$

$$a_{44} = m_{11}D^{-1/2} \tag{1.101}$$

where D is the determinant given by

$$D = (m_{11}m_{33} - m_{13}^2 - m_{14}^2)^2 \tag{1.102}$$

Substituting the cofactors and the determinant into (1.90) we obtain the joint density function written in terms of the scattering amplitudes and phase difference as

$$p(S_{vv}, S_{hh}, \phi_{vv} - \phi_{hh}) = \frac{S_{vv}S_{hh}}{4\pi^2 D^{1/2}} \exp\left[-\frac{1}{2D^{1/2}}(m_{33}S_{vv}^2 + m_{11}S_{hh}^2 - 2mS_{vv}S_{hh})\right] \tag{1.103}$$

where

$$m = m_{13}\cos(\phi_{vv} - \phi_{hh}) - m_{14}\sin(\phi_{vv} - \phi_{hh})$$

From (1.103) we can derive the marginal density functions for either the phase difference $\phi_{vv} - \phi_{hh}$ or the power amplitude $S_{vv,\,hh}^2$.

Phase Difference Distribution

To find the joint density function for ϕ_{vv} and ϕ_{hh} we need to integrate with respect to S_{vv} and S_{hh}. The differential for integration is $S_{vv}S_{hh}dS_{vv}dS_{hh}$. It is convenient to make the following change of variables:

$$\xi^2 = (m_{33}S_{vv}^2)/(2D^{1/2})$$

$$\eta^2 = (m_{11}S_{hh}^2)/(2D^{1/2})$$

$$2\xi d\xi = (m_{33}S_{vv}dS_{vv})/D^{1/2}$$

$$2\eta d\eta = (m_{11}S_{hh}dS_{hh})/D^{1/2}$$

which leads to the joint density for ϕ_{vv} and ϕ_{hh} as

$$p(\phi_{vv}, \phi_{hh}) = \frac{D^{1/2}}{\pi^2 m_{11}m_{33}}\int_0^\infty\int_0^\infty \xi\eta\exp\left[-\xi^2-\eta^2+\frac{2m\xi\eta}{\sqrt{m_{11}m_{33}}}\right]d\xi d\eta$$

A further change of variables to polar coordinates, $\xi = r\cos\theta, \eta = r\sin\theta$ leads to a simple integral in r and θ as

$$\begin{aligned}
p(\phi_{vv}, \phi_{hh}) &= \frac{D^{1/2}}{2\pi^2 m_{11}m_{33}}\int_0^{\pi/2}\int_0^\infty r^3\sin2\theta\exp\left[-r^2+\frac{mr^2\sin2\theta}{\sqrt{m_{11}m_{33}}}\right]drd\theta \\
&= \frac{D^{1/2}}{4\pi^2 m_{11}m_{33}}\int_0^\pi 0.5\sin\alpha\left(1-\frac{m\sin\alpha}{\sqrt{m_{11}m_{33}}}\right)^{-1/2}d\alpha \\
&= \frac{D^{1/2}}{4\pi^2 m_{11}m_{33}}\left[\frac{1}{1-b^2}+\frac{2b}{(1-b^2)^{1.5}}\tan^{-1}\left(\frac{1+b}{\sqrt{1-b^2}}\right)\right] \\
&= \frac{1-\overline{m}_{13}^2-\overline{m}_{14}^2}{4\pi^2}\left[\frac{1}{1-b^2}+\frac{2b}{(1-b^2)^{1.5}}\tan^{-1}\left(\frac{1+b}{\sqrt{1-b^2}}\right)\right] \qquad (1.104)
\end{aligned}$$

where

$$b = \overline{m}_{13}\cos(\phi_{vv}-\phi_{hh}) - \overline{m}_{14}\sin(\phi_{vv}-\phi_{hh})$$

$$\overline{m}_{13} = m_{13}/\sqrt{m_{11}m_{33}} = [2\langle|S_{vv}S_{hh}|\rangle\langle\cos\phi_{vv}\cos\phi_{hh}\rangle]/[\langle|S_{vv}|^2\rangle\langle|S_{hh}|^2\rangle]^{1/2}$$

$$\overline{m}_{14} = m_{14}/\sqrt{m_{11}m_{33}} = [2\langle|S_{vv}S_{hh}|\rangle\langle\cos\phi_{vv}\sin\phi_{hh}\rangle]/[\langle|S_{vv}|^2\rangle\langle|S_{hh}|^2\rangle]^{1/2}$$

An examination of the above relations indicates that $\overline{m}_{13}$ represents the correlation coefficient between the real parts of the vertically and horizontally polarized scattering amplitudes, and $\overline{m}_{14}$ represents the cross correlation coefficient between the real and imaginary parts of those scattering amplitudes. As such, $\overline{m}_{13}$ is generally less than or equal to one, and $\overline{m}_{14}$ is less than $\overline{m}_{13}$. It follows from the above equation that as the correlation coefficients become smaller, i.e., as $\overline{m}_{14}$ and $\overline{m}_{13}$ approach zero, the joint density function becomes uniformly distributed. Hence, both the phase-difference and the phase-sum density functions approach uniform distribution.

The density functions of the phase sum $\phi^+ = \phi_{vv}+\phi_{hh}$ and the phase difference $\phi^- = \phi_{vv}-\phi_{hh}$ are given, respectively, by [Papoulis, 1965]

$$p(\phi^+) = \int_{-\pi}^{\pi} p(\phi^+ - \phi_{hh}, \phi_{hh})\, d\phi_{hh}$$

$$= 1/(2\pi) \tag{1.105}$$

and

$$p(\phi^-) = \int_{-\pi}^{\pi} p(\phi^- + \phi_{hh}, \phi_{hh})\, d\phi_{hh}$$

$$= \frac{1 - \bar{m}_{13}^2 - \bar{m}_{14}^2}{2\pi}\left[\frac{1}{1-b^2} + \frac{2b}{(1-b^2)^{1.5}}\tan^{-1}\left(\frac{1+b}{\sqrt{1-b^2}}\right)\right] \tag{1.106}$$

Thus, the phase-sum density function is uniformly distributed over $(-\pi, \pi)$. It is independent of target parameters and hence does not carry information about the target. On the other hand, there is an expression for the phase-difference density function that is dependent on target parameters. In particular, we can compute the mean phase difference as

$$\langle\phi^-\rangle = \int_{-\pi}^{\pi} \phi^- p(\phi^-)\, d\phi^- \tag{1.107}$$

The phase-difference density function depends only on $\bar{m}_{13}$ and $\bar{m}_{14}$. It is plotted in Figure 1.7 and Figure 1.8 for various values of $\bar{m}_{13}$ and $\bar{m}_{14}$, respectively. In general, this density function has a Gaussian appearance shifted to the right hand side of the origin by an amount proportional to the value of the cross correlation coefficient between the real and imaginary parts of the vertically and horizontally polarized scattering amplitudes, $\bar{m}_{14}$. The width of the lobe of the Gaussian-like function is controlled mainly by $\bar{m}_{13}$, which represents the cross correlation coefficient between the real parts of the scattering amplitudes. As $\bar{m}_{13}$ decreases, the width of the lobe increases and its peak value decreases (Figure 1.7). This behavior indicates that, when the correlation between the scattering amplitudes becomes smaller, the phase difference tends to spread out over a wider range of angles. The other parameter $\bar{m}_{14}$ does not have much influence on the width of the lobe but is the cause of the loss in symmetry of the phase difference density function. As $\bar{m}_{14}$ increases, the center of the lobe is shifted gradually to the right hand side of the origin. Meanwhile, there is also a small decline in the peak value and a broadening of the lobe. However, since this parameter is not expected to be large, the changes are generally small. The influences of the $\bar{m}_{14}$ parameter are illustrated in Figure 1.8.

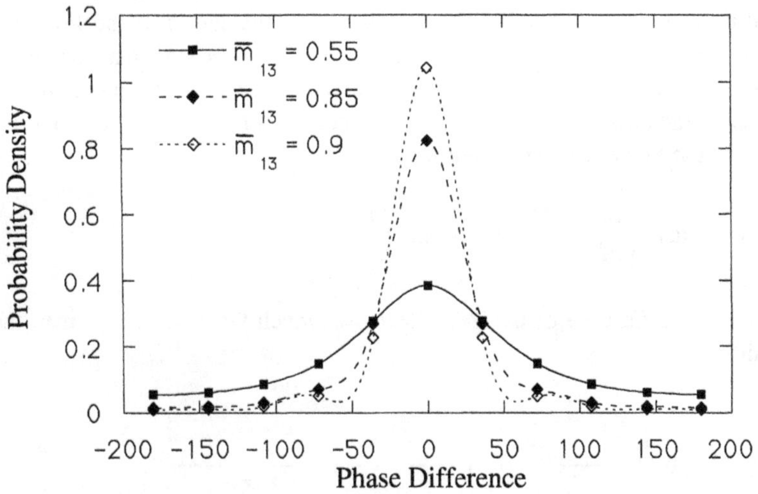

Figure 1.7 Dependence of phase-difference density function on $\overline{m}_{13}$ when $\overline{m}_{14} = 0$.

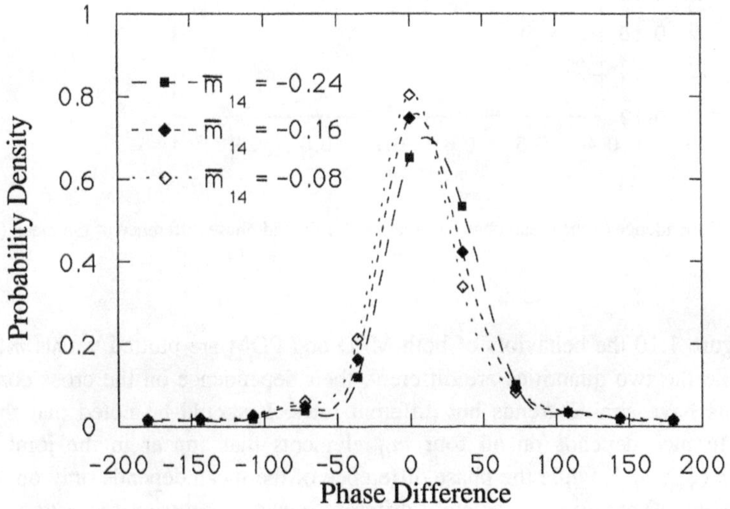

Figure 1.8 Dependence of phase-difference density function on $\overline{m}_{14}$ when $\overline{m}_{13} = 0.85$.

Other behavior of interest is the dependence of the mean phase difference and the phase difference of the mean on the correlation coefficients $\overline{m}_{13}$ and $\overline{m}_{14}$. These are shown in Figures 1.9 through 1.11. From Figure 1.9 it is seen that the mean phase difference (in radians) decreases with an increase in $\overline{m}_{13}$. In the literature the phase difference of the mean (PDM) has also been reported. It has been defined in terms of the elements of the Stokes matrix as follows

$$PDM = \tan^{-1}\left(\frac{M_{43} - M_{34}}{M_{33} + M_{44}}\right) = -\tan^{-1}\left(\frac{m_{14}}{m_{13}}\right) \tag{1.108}$$

This phase difference of the mean decreases much faster with m_{13} than the mean phase difference.

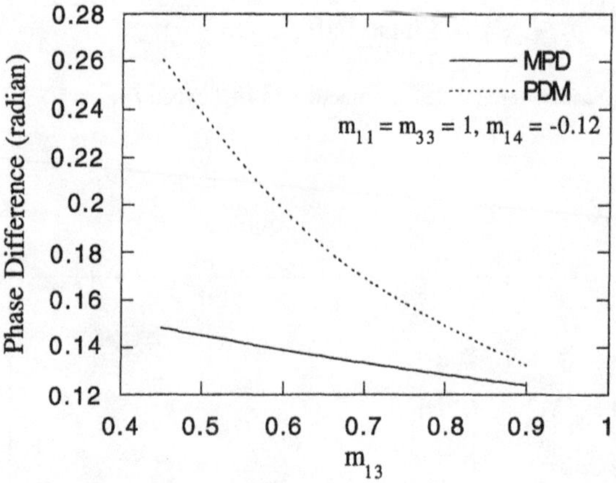

Figure 1.9 Dependence of the mean phase difference (MPD) and phase difference of the mean (PDM) on $\overline{m}_{13}$.

In Figure 1.10 the behaviors of both MPD and PDM are plotted versus m_{14}. Here again while the two quantities are different, their dependence on the cross correlation coefficients have similar trends but different rates. It should be noted that the mean phase difference depends on all four m_{ij} elements that appear in the joint density function $p(\phi_{vv}, \phi_{hh})$, while the phase difference of the mean depends only on the ratio of m_{14} to m_{13}. Therefore, completely different behavior between these two types of phase difference will show up when we consider the effect of m_{11}. To illustrate the last point we plot in Figure 1.11, the mean phase difference (MPD) and the phase difference of the mean (PDM) as a function of m_{11}. As expected PDM does not depend on m_{11} and appears as a constant while MPD decreases with m_{11}.

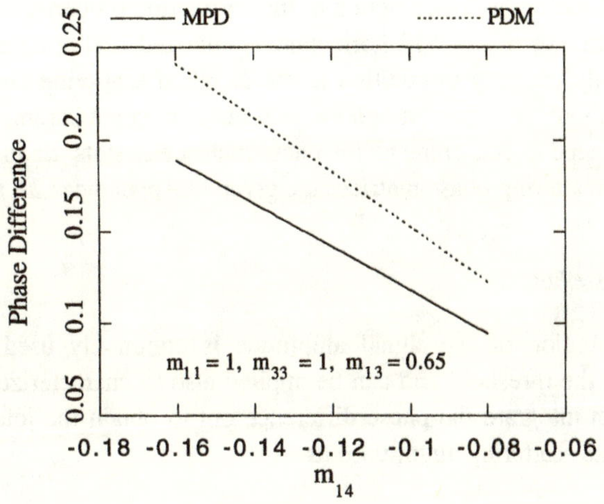

Figure 1.10 Dependence of the mean phase difference (in radians) on $\overline{m}_{14}$.

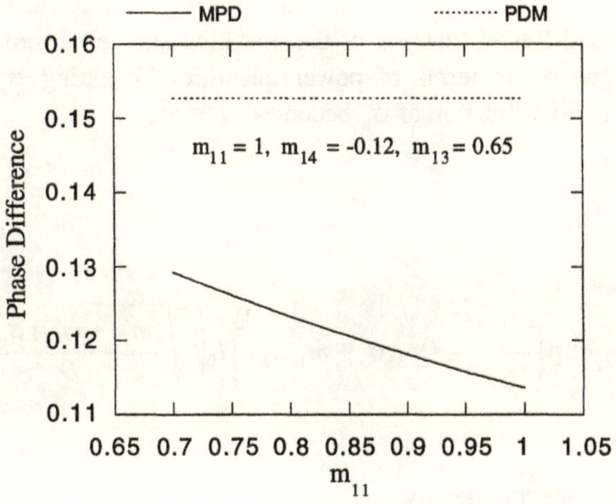

Figure 1.11 A comparison between mean phase difference (MPD) and phase difference of the mean (PDM).

In the above we considered the dependence of the mean phase difference of the like polarized scattering amplitudes as a function of the correlation properties of the real and imaginary parts of the vertically and horizontally polarized scattering amplitudes. It appears to be the only quantity in addition to the standard scattering coefficients that carries information about the target. In order to find out how target parameters affect the phase difference we need to determine all the phase matrix elements. Examples of rough surface and volume scattering phase matrices are given in Appendices 2A and 2C.

Signal Amplitude Distribution

The statistical distribution of the signal amplitude is commonly used in detection problems for setting the threshold and can be applied also to characterize terrain type. From (1.103) we can integrate the phase difference out to obtain the joint probability density function of the scattering amplitudes as

$$p(S_{vv}, S_{hh}) = 2\pi \int_{-\pi}^{\pi} p(S_{vv}, S_{hh}, \phi_{vv} - \phi_{hh})\, d(\phi_{vv} - \phi_{hh})$$

$$= \frac{S_{vv}S_{hh}}{D^{1/2}} \exp\left[-\frac{1}{2D^{1/2}}(m_{33}S_{vv}^2 + m_{11}S_{hh}^2)\right] I_0\left(\frac{\sqrt{(m_{13}^2 + m_{14}^2)}}{D^{1/2}} S_{vv}S_{hh}\right) \tag{1.109}$$

where I_0 is the modified Bessel function of the first kind and zeroth order. Next, we rewrite the above equation in terms of power quantities by letting $\sigma_v = S_{vv}^2$ and $\sigma_h = S_{hh}^2$. Then, the density function of σ_v becomes

$$p(\sigma_v) = \int_0^{\infty} p(\sigma_v, \sigma_h)\, d\sigma_h$$

$$= \int_0^{\infty} \frac{1}{4D^{1/2}} \exp\left[-\frac{1}{2D^{1/2}}(m_{33}\sigma_v + m_{11}\sigma_h)\right] I_0\left(\sqrt{\frac{(m_{13}^2 + m_{14}^2)\,\sigma_v\sigma_h}{D}}\right) d\sigma_h \tag{1.110}$$

Applying the following integral identity,

$$\int_0^{\infty} \exp[-\alpha x]\, I_0(2\sqrt{\beta x})\, dx = \frac{\exp(\beta/\alpha)}{\alpha} \tag{1.111}$$

we have

$$p(\sigma_v) = \frac{1}{2m_{11}}\exp\left[-\frac{\sigma_v}{2m_{11}}\right] \tag{1.112}$$

Since σ_v and σ_h occur symmetrically in (1.110), the above equation shows that the polarized scattered power has an exponential distribution. Clearly, such a conclusion is true only for the assumed type of terrain, i.e., a terrain that scatters like a collection of independent, isotropic scatterers. For other types of terrain there is no simple analytic expression for the distribution of the scattered power. In the next subsection we shall list the results of the signal distribution functions for a variety of terrain types based on computer simulation.

1.7.2 Distribution Functions for Terrains

For more complicated terrains than the one described in the previous subsection, it is more efficient to determine an empirical signal distribution function by comparing known distribution functions or their transformed versions with computer simulated data and select the one that comes closest to the data in the least squares sense. This section summarizes signal distribution models (functions) for radar returns from a bare ground, a bare ground with a hard target, a forested area, a forested area above a rough ground surface and a forest-covered ground with a hard target. The inputs to the various signal distribution models are the model parameters, which can be computed in terms of the mean and variance of the scattered signal from the terrain of interest. If a hard target is present, an independent estimate is needed, depending upon the target radar cross section and the illuminated area under consideration. The proper signal distribution model for each scene will be given in a later subsection. The backscattering data, including mean and standard deviation of the normalized scattered power for ground and trees at several frequencies and incident angles may be found in Ulaby and Dobson [1989].

The Signal Distribution Models

For ease of reference the common signal distribution models for describing signals scattered from a terrain are summarized in this subsection. Density functions listed for the signal amplitude, x, are the standard Rayleigh, Weibull, gamma and lognormal functions. Assuming that the density function for x is $p_x(x)$, additional density functions for the variable $y = x^2$ can be derived from probability theory with the relation

$$p_y(y) = p_x(\sqrt{y})/(2\sqrt{y}) \tag{1.113}$$

We shall refer to the derived density functions as transformed density functions. For example, the one derived from Rayleigh is referred to as the *transformed Rayleigh distribution*. Note that all model parameters given below are estimated from the mean μ and variance σ^2 of the random variable x.

(a). The Rayleigh and transformed Rayleigh distributions are

$$P_R(x) = \frac{x}{\alpha^2}\exp\{-\frac{1}{2}\left(\frac{x}{\alpha}\right)^2\} \tag{1.114}$$

and

$$P_{TR}(y) = \frac{1}{2\alpha^2}\exp\{-\frac{1}{2}\left(\frac{y}{\alpha^2}\right)\},\ y > 0 \tag{1.115}$$

where the scale parameter α is given by

$$\alpha = \mu\sqrt{\frac{2}{\pi}}$$

(b). The lognormal and transformed lognormal distributions are

$$P_{ln}(x) = \frac{1}{x\alpha\sqrt{2\pi}}\exp\{-\frac{1}{2}\left(\frac{\ln x - \lambda}{\alpha}\right)^2\} \tag{1.116}$$

and

$$P_{Tln}(y) = \frac{1}{2y\alpha\sqrt{2\pi}}\exp\{-\frac{1}{2}\left(\frac{\ln\sqrt{y} - \lambda}{\alpha}\right)^2\},\ y > 0 \tag{1.117}$$

The parameters λ and α can be related to the mean μ and variance σ^2 of x as

$$\lambda = \frac{1}{2}\ln\left(\frac{\mu^4}{\sigma^2 + \mu^2}\right)$$

$$\alpha = \sqrt{\ln\left(\frac{\sigma^2}{\mu^2}\right)}$$

(c). The gamma and transformed gamma distributions are

$$P_G(x) = \frac{1}{\alpha\Gamma(\lambda)}\left(\frac{x}{\alpha}\right)^{\lambda - 1}\exp\left(-\frac{x}{\alpha}\right) \tag{1.118}$$

and

$$P_{TG}(y) = \frac{1}{2\sqrt{y}\alpha\Gamma(\lambda)}\left(\frac{\sqrt{y}}{\alpha}\right)^{\lambda - 1}\exp\left(-\frac{\sqrt{y}}{\alpha}\right) \tag{1.119}$$

where $\Gamma(x)$ is the gamma function, α is a scale parameter and λ is a shape parameter given by

$$\lambda = \frac{\mu^2}{\sigma^2}$$

$$\alpha = \frac{\sigma^2}{\mu}$$

(d). The Weibull and transformed Weibull distributions are

$$P_W(x) = \frac{\lambda}{\alpha}\left(\frac{x}{\alpha}\right)^{\lambda-1} \exp\{-\left(\frac{x}{\alpha}\right)^{\lambda}\} \qquad (1.120)$$

and

$$P_{TW}(y) = \frac{\lambda}{2\alpha\sqrt{y}}\left(\frac{\sqrt{y}}{\alpha}\right)^{\lambda-1} \exp\{-\left(\frac{\sqrt{y}}{\alpha}\right)^{\lambda}\} \qquad (1.121)$$

where $p_W(x) = 0$, if $x < 0$ or x is complex; α and λ are scale and shape parameters, respectively, related to the mean and variance of x as

$$\mu = \alpha\Gamma\left(1 + \frac{1}{\lambda}\right)$$

$$\sigma^2 = \alpha^2\left[\Gamma\left(1 + \frac{2}{\lambda}\right) - \Gamma^2\left(1 + \frac{1}{\lambda}\right)\right]$$

Note that to obtain the necessary model parameters from mean and variance one may apply standard algorithms [Gerald and Wheatly, 1984] to solve the above nonlinear equations. From the above it is seen that the Rayleigh distribution is a special case of the Weibull distribution when $\lambda = 2$ and α in (1.120) is replaced by $\sqrt{2}\alpha$.

It is important to note that as distribution functions, either *p(x)* or *p(y)* may be applied to describe either the signal amplitude or its power. This being the case care needs to be exercised in reading the literature.

Scenes and Associated Models

The distributions of the backscattered power from a forested area have been investigated by Karam et al. [1992] using numerical simulation. Note that in computer simulation only scattering from a *resolution cell* is simulated. Their studies include backscattering from a half-space of (1) disc-shaped and needle-shaped leaves, (2) branches, and (3) a combination of leaves and branches. Let a be the radius of a cylindrically shaped object and h be half of its height. Two leaf sizes and one branch size are considered in the simulation results given below. The parameters used for each case are as follows:

1. Disk-shaped leaves ($ka = 0.52, 1.3, kh = 0.02$ cm, $\varepsilon_r = 15.3 - j1.7$)

2. Needle-shaped leaves ($ka = 0.01$ cm, $kh = 0.52, 1.3$ cm, $\varepsilon_r = 15.3 - j1.7$)

3. Cylindrical branches ($ka = 0.5, kh = 6.5, \varepsilon_r = 27.3 - j8.4$)

The statistical model for the amplitude of the voltage backscattered from a rough ground has been shown to be Weibull [Schleher, 1976; Fung and Chen, 1992]. Hence, the transformed Weibull distribution is the correct distribution for power. In addition, when a hard target is present on the ground, Rice's method can be used to derive the resulting density function [Nilsson and Glisson, 1980]. Comparisons of simulated results with statistical density functions will be given in Section 1.7.3

The results of the above studies for the backscattered power are summarized below:

1. Vegetation

(a). A half-space of disc-shaped leaves: transformed gamma, transformed Rayleigh distribution.

(b). A half-space of needle-shaped leaves: transformed Rayleigh distribution.

(c). A half-space of branches: transformed gamma distribution.

The transformed gamma distribution is given above for most cases, because it is the best for large and medium-size scatterers. For small needle- or disc-shaped leaves (small compared with the wavelength), the transformed Rayleigh distribution (or exponential distribution) gives a better agreement. When both leaves and branches cause significant scattering we anticipate the transformed gamma distribution to be applicable.

2. Ground

(a). A bare ground surface: transformed Weibull model. That is,

$$p_{TW}(y) = \frac{\lambda}{2\alpha\sqrt{y}}\left(\frac{\sqrt{y}}{\alpha}\right)^{\lambda-1}\exp\left[-\left(\frac{\sqrt{y}}{\alpha}\right)^{\lambda}\right] \qquad (1.122)$$

(b). Ground surface with a hard target: the density function of this type of clutter can be derived by Rice's method yielding [Nilsson and Glisson, 1980]

$$p_{RW}(y) = \frac{1}{2\pi}\int_{-\pi}^{\pi}\left[p_W\left(\sqrt{y - A\sin^2\phi} - \sqrt{A}\cos\phi\right)\right.$$

$$\left. + p_W\left(-\sqrt{y - A\sin^2\phi} - \sqrt{A}\cos\phi\right)\right]\sqrt{y/(y - A\sin^2\phi)}\,d\phi \qquad (1.123)$$

where y is the scattering coefficient and A is the ratio of the radar cross section of the hard target divided by the illuminated area. Note that the operation described in the previous sentence is needed to put y and A on the same reference. In (1.123) $p_W(\sqrt{y})$ is the Weibull probability density function for the scattered field amplitude from the ground surface.

3. Forest-covered ground

The total contribution to scattering consists of two major parts: (i) surface scattering attenuated by the vegetation layer, and (ii) volume scattering from the vegetation layer. Thus, the signal distribution model $\sigma_{qp}{}^0$ for the scattered power from a forest covered ground can be constructed approximately in terms of the signal distribution model of the ground $\sigma_{qp}{}^0(s)$ and that of the vegetation $\sigma_{qp}(v)$ in the following form,

$$\sigma_{qp}{}^0 = \sigma_{qp}(s)\exp(-2\tau_d/\cos\theta) + \sigma_{qp}(v) \tag{1.124}$$

where

$$\tau_d = k_e d$$

k_e is the extinction coefficient and d is the physical depth of the forest layer. The vegetation and ground models are as defined in 1 and 2 above and $k_e = 0.5 f^{0.75}$ dB/m, where f is frequency in GHz [Ulaby et al, 1986].

1.7.3 Computer Simulation of Signal Distribution

In this section we show samples of computer simulated results leading to the signal statistical distributions recommended in the previous section.We shall show results first in random surface scattering and then volume scattering.

Signal Distribution in Surface Scattering

In this section we show comparisons between computer simulated backward scattered signal in the plane of incidence and the signal statistical models. The incident wave is elliptically polarized with 30 degrees ellipticity and 45 degrees orientation. Receiving polarizations have the same ellipticity and are oriented 45 degrees and 135 degrees corresponding to like and cross polarizations respectively. The incident angle is 75 degrees and the observation angle is 80 degrees. The surface correlation length along the look direction is 10 cm and the rms height is 0.86 cm [Fung and Chen, 1992]. The radar wavelength used in computation is 20 cm. In Figure 1.12 we show the comparison between the like polarized power distribution and the transformed density functions. Both transformed Weibull and gamma density functions give close agreement to the simulated distribution. However, the peak of the transformed Weibull appears to line up a bit better with the simulation. Note that, in application to threshold detection, small signal levels are not of interest. In the large signal region all distribution curves appear exponential. This may be the reason why the exponential function (or transformed Rayleigh function) has been commonly used to describe power distribution, although the overall signal power distribution is not exponential.

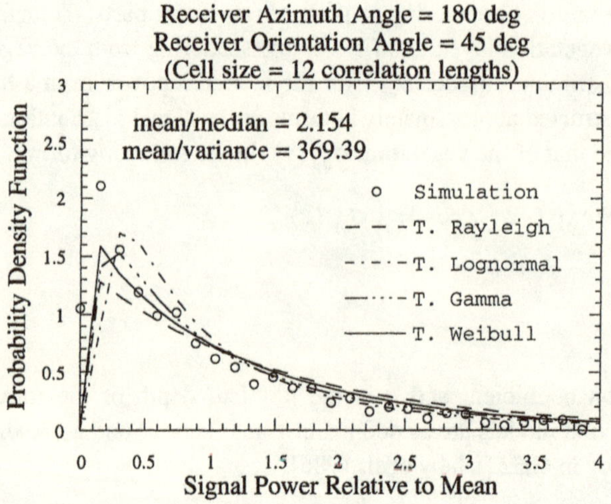

Figure 1.12 Comparisons between computer simulated distribution and statistical density functions. Like polarization in backward surface scattering.

Next, we show a similar comparison for cross polarized scattering in Figure 1.12. Here, again both the transformed Weibull and transformed gamma density functions are close to the simulation results. However, if we compute the rms errors, the transformed Weibull has a somewhat smaller error. In forward scattering the transformed gamma distribution is significantly better than the transformed Weibull and other density functions [Fung and Chen, 1992].

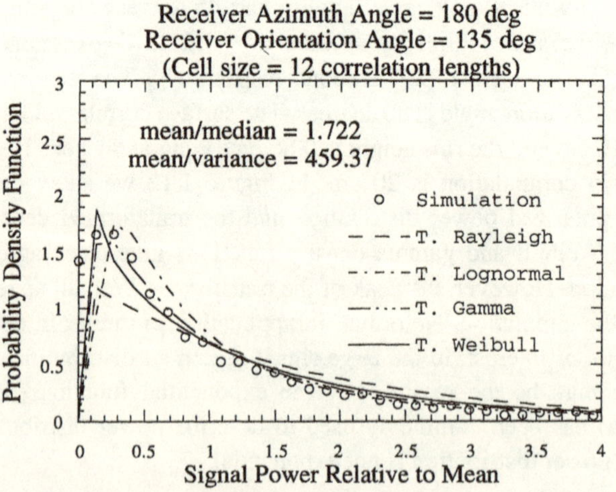

Figure 1.13 Comparisons between computer simulated distribution and statistical density functions. Cross polarization in backward surface scattering.

In the figures shown the simulated distributions are formed from 1100 samples. When the number of samples is limited, it will be very difficult to determine the distribution function. For the cases shown it is clearly difficult to differentiate between the distribution functions through measured data. This is because the number of independent samples must be very large and resolution has to be sufficiently fine.

Signal Distribution in Volume Scattering

Here, we consider scattering from a half-space of vegetation-like scatterers by simulating vegetation components as cylindrical objects with symmetry defined by having a diameter of $2a$ and a length of $2h$. We shall simulate disc-shaped leaves, needle-shaped leaves and cylindrical branches, each having a finite dielectric constant. The vegetation components are assumed to be randomly distributed and oriented within a sufficiently large volume (at least a five-wavelength cube) to generate enough phase change due to the spacing between scatterers. Each signal sample is a sum of the scattered fields from all the scatterers within this volume. Over 2000 samples are calculated and used to form a signal distribution function.

From theoretical studies in Section 1.7.1 we expect scattering from a collection of randomly distributed and oriented small scatterers to be Rayleigh distributed in amplitude and exponentially distributed in power. This is so irrespective of the specific shape of the scatterer, if it is sufficiently small. This is, indeed, the case as shown in Figure 1.12 for backscattering from needle-shaped leaves and in Figure 1.15 for disc-shaped leaves. Note that the disc-shaped leaf has a much larger volume than the needle-shaped leaf if the radius of the disc is the same as the length of the needle. Hence, we have to use a smaller *ka* in Figure 1.15 to realize the Rayleigh distribution.

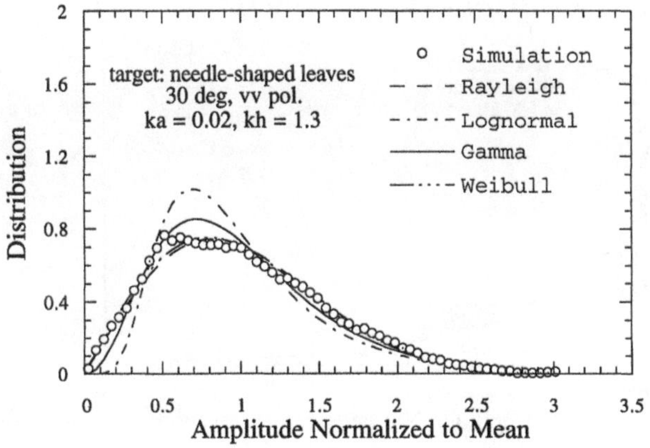

Figure 1.14 Comparisons between computer simulated voltage distribution and statistical density functions for backscattering from needle-shaped leaves.

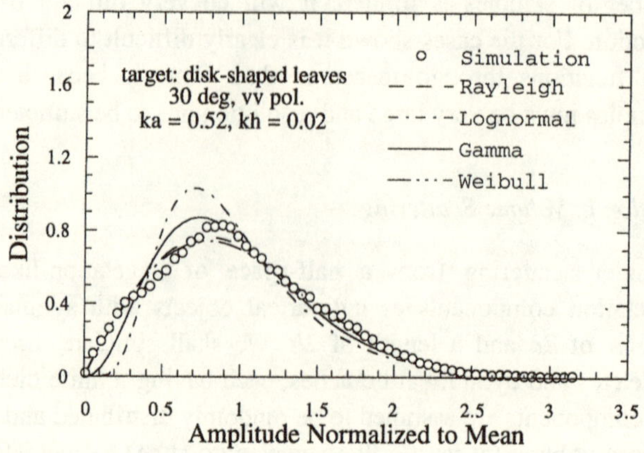

Figure 1.15 Comparisons between computer simulated voltage distribution and statistical density functions for backscattering from disc-shaped leaves.

When we increase *ka* for the disc-shaped leaves to 1.3, the distribution becomes gamma as shown in Figure 1.16. This indicates that the size of the volume of the scatterer is important. We expect that when we use a large cylinder instead of a needle, the distribution also changes into gamma.

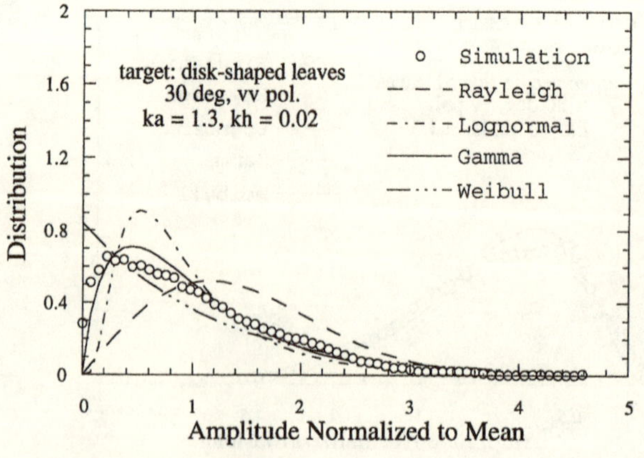

Figure 1.16 Comparisons between computer simulated voltage distribution and statistical density functions for backscattering from disc-shaped leaves.

As shown in Figure 1.17 the distribution of cylindrical branches is close to gamma. In all the cases considered above, scatterers have been assumed to be randomly located and oriented. It is expected that signal distribution may depend on orientation distribution that is not considered in the above discussion.

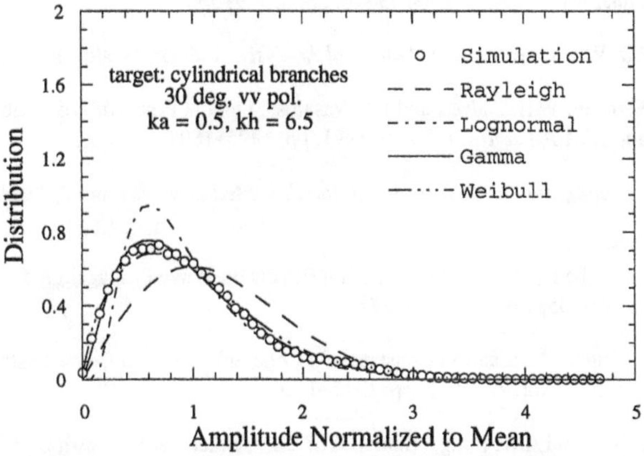

Figure 1.17 Comparisons between computer simulated voltage distribution and statistical density functions for backscattering from cylindrical branches.

In summary, the signal distribution for the instantaneous power scattered backward from a randomly rough surface has been found to be a transformed Weibull function. In the forward direction, it is expected to be the transformed gamma function. In backscattering from a forested area, the shape of the distribution function is dependent on the size of the scatterer relative to the exploring wavelength. When the scatterers are small, the distribution of the received voltage is Rayleigh and power is exponential. However, if the size of the scatterer is large, the received voltage is gamma distributed.

REFERENCES

Barakat, R., "Statistics of the Stokes Parameters," *J. Opt. Soc. Am.* A, Vol. 4, no. 7, 1987, pp. 1256–1263.

Chandrasekhar, S., *Radiative Transfer*, Dover Publications, New York, 1960.

Davenport, W.B., *Probability and Random Processes*, McGraw-Hill, New York, 1970.

Davenport, W.B., and W.L. Root, *Random Signals and Noise*, McGraw-Hill, New York, 1958, Chapter 8.

Fante, R.,L., "Relationship Between Radiative Transport Theory and Maxwell's Equations in Dielectric Media," *J. Opt. Soc. Am.*, Vol. 71, 1981, pp. 460–468.

Fung, A.K., and M.F. Chen, "Scattering from a Rayleigh Layer with an Irregular Interface," *Radio Science*, Vol. 16, 1981, pp. 1337–1347.

Fung, A.K., and K.S. Chen, "Bistatic Signal Statistics of Randomly Rough Surfaces," *IGARSS' 92*, Vol. 1,

Fung, A.K., and K.S. Chen, "Bistatic Signal Statistics of Randomly Rough Surfaces," *IGARSS' 92*, Vol. 1, 1992, pp. 183–185.

Fung, A.K, and H.J. Eom, "Multiple Scattering and Depolarization by a Randomly Rough Kirchhoff Surface," *IEEE Trans. Ant. Prop.*, Vol. 29, 1981a, pp. 463–471.

Fung, A.K., and H.J. Eom, "A Theory of Wave Scattering from an Inhomogeneous Layer with an Irregular Interface," *IEEE Trans. Ant. Prop.*, Vol. 29, 1981b, pp. 899–910.

Gerald, C.F., and P.O. Wheatley, *Applied Numerical Analysis*, Addison-Wesley, Reading, MA, 1984.

Hottel, H.C., A.F. Sarofim, W.H. Dalzell and I.A. Vasalos, "Optical Properties of Coatings, Effect of Pigment Concentration," *AIAA Journal*, Vol. 9, 1971, pp. 1895–1898.

Ishimaru, A., *Wave Propagation and Scattering in Random Media*, Vols. 1 and 2, Academic Press, New York, 1978.

Ishimaru, A., and R.L. Cheung, "Multiple Scattering Effects on Wave Propagation due to Rain," *Ann. Telecommunications*, Vol. 35, 1980, pp. 373–378.

Ishimaru, A., and Y. Kuga, "Attenuation Constant of a Coherent Field in a Dense Distribution of Particles," *J. Opt. Soc. Am.*, Vol. 72, 1982, pp. 1317–1320.

Karam, M.A., K.S. Chen and A.K. Fung, "Statistics of Backscatter Radar Return from Vegetation," *IGARSS'92*, Vol. 1, 1992, pp. 242–244.

Karam, M.A., A.K. Fung, R.H. Lang and N.S. Chauhan, "A Microwave Scattering for Layered Vegetation," *IEEE Trans. Geoscience and Remote Sensing*, Vol. 30, no. 4, July, 1992, pp. 767–784.

Kraus, J.D., *Radio Astronomy*, McGraw-Hill, New York, 1966, Chapter 3.

Mott, H., *Polarization in Antennas and Radar*, John Wiley & Sons, New York, 1986, Chapter 3.

Nilsson, A.A., and T.H. Glisson, "On the Derivation and Numerical Evaluation of the Weibull-Rician Distribution," *IEEE Trans. Aerospace and Electronic Systems*, Vol. AES-12, no. 6, 1980, pp. 864–867.

Papoulis, A., *Probability, Random Variables and Stochastic Processes*, McGraw-Hill, New York, 1965.

Sarabandi, K., "Derivation of Phase Statistics from the Mueller Matrix," *Radio Science*, Vol. 27, no. 5, 1992, pp. 553–562.

Schleher, D.C., "Radar Detection in Weibull Clutter," *IEEE Trans. Aerospace and Electronic Systems*, Vol. AES-12, no. 6, 1976, pp. 736–743.

Touzi, R., and A. Lopes, "Distribution of the Stokes Parameters and the Phase Difference in Polarimetric SAR," *IGARSS'91*, Vol. 2, 1991, pp. 103–106.

Tsang, L., and J.A. Kong, "Radiative Transfer Theory for Active Remote Sensing of Half Space Random Media," *Radio Science*, Vol. 13, 1978, pp. 763–773.

Tsang, L., and J.A. Kong, "The Brightness Temperature of a Half Space Random Medium with Nonuniform Temperature Profile," *Radio Science*, Vol. 10, 1975, pp. 1025–1033.

Tsang, L., and J.A. Kong, "Thermal Microwave Emission from Half Space Random Media," *Radio Science*, Vol. 11, 1976, pp. 599–609.

Ulaby, F.T., and M.C. Dobson, *Handbook of Radar Scattering Statistics for Terrain,* Artech House,

Dedham, MA, 1989.

Ulaby, F.T., and C. Elachi, *Radar Polarimetry for Geoscience Applications*, Artech House, Dedham, MA, 1990.

Ulaby, F.T., R.K. Moore and A.K. Fung, *Microwave Remote Sensing: Active and Passive*, Vol. 1, Artech House, Dedham, MA, 1981, Chapter 4.

Ulaby, F.T., R.K. Moore and A.K. Fung, *Microwave Remote Sensing: Active and Passive*, Vol. 2, Artech House, Dedham, MA, 1982, Chapter 12.

Ulaby, F.T., R.K. Moore and A.K. Fung, *Microwave Remote Sensing: Active and Passive*, Vol. 3, Artech House, Dedham, MA, 1986, Chapter 13, p. 1197.

Ulaby, F.T., J. Stiles, K. Sarabandi and Y. Oh, "Model and Measurements of the Phase Statistics of the Scattering Matrix of Distributed Surfaces," *IGARSS'91*, Vol. 2, 1991, p. 401.

Van de Hulst, H.C., *Light Scattering by Small Particles*, John Wiley and Sons, New York, 1957.

Vasalos, I.A. "Effects of Separation Distance on the Optical Properties of Dense Dielectric Particles Suspensions," Ph D. Dissertation, Chemical Engineering Department, Massachusetts Institute of Technology, 1969.

Chapter 2

First-Order Radiative Transfer Solution—Active Sensing

2.1 INTRODUCTION

The problem of wave scattering from an inhomogeneous layer with irregular boundaries can be solved by noting that the intensity propagation and scattering within the layer satisfy the radiative transfer equation. A direct solution of the radiative transfer equation subject to appropriate boundary conditions has been done by either reformulating it as a pair of matrix equations or applying numerical methods [Ulaby et al., 1986]. In either case it is difficult to obtain a physical meaning out of the solutions. To get more insight into the solutions regarding possible scattering mechanisms it is instructive to carry out a first-order solution analytically in the volume scattering coefficient, κ_s. In what follows we shall begin with the formulation of the scattering problem from an inhomogeneous layer with irregular boundaries as given in Chapter 1. Next, we convert the differential radiative transfer equations into integral equations. Then, we incorporate the boundary conditions into the integral equations from which we derive the first-order solution for an irregular layer. This solution consists of coherent and incoherent components including surface scattering, volume scattering and surface-volume interaction for different polarizations.

Simplified scattering models and their illustrations for two important special cases of the general first-order solution are carried out in Section 2.5. The first case is polarized backscattering from a randomly rough surface for which multiple scattering effects can be ignored. An explicit expression for this special model is given. Properties of surface backscattering, its polarization properties and its applications to laboratory and field measurements are discussed and illustrated. The impact of medium inhomogeneities is also shown and applied to data analysis of soil surfaces. The second case is backscattering from an irregular, inhomogeneous layer where surface roughness consideration is included only in surface scattering terms and ignored in volume scattering calculations. The backscattering properties of such a layer model are illustrated. Then, the model is applied to backscattering from a sea ice layer with and without snow cover. More general surface scattering models are given in Chapters 5 and 7. More accurate models for scattering from irregular, single and multilayer media are treated in Chapter 8.

2.2 RADIATIVE TRANSFER EQUATIONS

We now consider the same scattering problem as formulated in Section 1.5.2. We can write the pair of transfer equations for the upward and downward intensity matrices denoted by $\mathbf{I}^+$ and $\mathbf{I}^-$ as follows:

$$\frac{d}{dz}\mathbf{I}^+(z) = -\kappa_{es}\mathbf{I}^+(z) + \mathbf{F}^+(z) \tag{2.1}$$

$$\frac{d}{dz}\mathbf{I}^-(z) = \kappa_{es}\mathbf{I}^-(z) - \mathbf{F}^-(z) \tag{2.2}$$

where $\kappa_{es} = \kappa_e / \cos\theta_s = \kappa_e / \mu_s$

$\mu_s = \cos\theta_s,\ \mu = \cos\theta$

$\mathbf{I}^+(z) = \mathbf{I}(z, \mu_s, \phi_s)$

$\mathbf{I}^-(z) = \mathbf{I}(z, -\mu_s, \phi_s)$

$$\mathbf{F}^{\pm}(z) = \frac{\kappa_{ss}}{4\pi}\int_0^{2\pi}\int_0^1 \mathbf{P}(\pm\mu_s, \mu, \phi_s - \phi)\,\mathbf{I}^+ d\mu d\phi$$
$$+ \frac{\kappa_{ss}}{4\pi}\int_0^{2\pi}\int_0^1 \mathbf{P}(\pm\mu_s, -\mu, \phi_s - \phi)\,\mathbf{I}^- d\mu d\phi \tag{2.3}$$

In (2.3) κ_s, κ_e are the volume scattering and extinction coefficient matrices taken to be diagonal [Ulaby et al., 1986, Chapter 13], $\kappa_{ss} = \kappa_s / \cos\theta_s$ and $\mathbf{P}(\mu_s, \mu, \phi_s - \phi)$ is the scattering phase matrix. The above differential equations given by (2.1) and (2.2) are of the first-order and have standard solutions of the form

$$\mathbf{I}^+(z) = \mathbf{I}^+(-d)\,e^{-\kappa_{es}(z+d)} + \int_{-d}^{z} \mathbf{F}^+(z')\,e^{-\kappa_{es}(z-z')}\,dz' \tag{2.4}$$

$$\mathbf{I}^-(z) = \mathbf{I}^-(0)\,e^{\kappa_{es}z} + \int_{z}^{0} \mathbf{F}^-(z')\,e^{\kappa_{es}(z-z')}\,dz' \tag{2.5}$$

These solutions are not real solutions because $\mathbf{F}^{\pm}$ are functions of the upward and downward propagating intensities. Hence, (2.4) and (2.5) are really integral equations for these intensities. To cast them into a form suitable for iterative solution we need to incorporate the applicable boundary conditions. Recall that at the upper boundary $z = 0$ the condition is

$$\mathbf{I}^-(0) = \frac{1}{4\pi}\int_0^{2\pi}\int_0^1 \mathbf{S}_r(-\mu_s, \mu, \phi_s - \phi)\,\mathbf{I}^+ d\mu d\phi$$

$$+\frac{1}{4\pi}\int_0^{2\pi}\int_0^1 \mathbf{S}_t(-\mu_s, -\mu, \phi_s - \phi)\,\mathbf{I}^i d\mu d\phi \tag{2.6}$$

where $\mathbf{S}_r(-\mu_s, \mu, \phi_s - \phi)$, $\mathbf{S}_t(-\mu_s, -\mu, \phi_s - \phi)$ are the surface scattering and transmission phase matrices, and $\mathbf{I}^i = \mathbf{I}_0\delta(\mu - \mu_i)\,\delta(\phi - \phi_i)$ is the intensity of the incident plane wave. At the lower boundary $z = -d$, we have

$$\mathbf{I}^+(-d) = \frac{1}{4\pi}\int_0^{2\pi}\int_0^1 \mathbf{G}(\mu_s, -\mu, \phi_s - \phi)\,\mathbf{I}^-(-d)\,d\mu d\phi \tag{2.7}$$

where $\mathbf{G}$ is the scattering phase matrix for the lower surface boundary. Substituting the above boundary conditions into (2.4) and (2.5) we have

$$\mathbf{I}^+(z) = \left(\frac{1}{4\pi}\int_0^{2\pi}\int_0^1 \mathbf{G}(\mu_s, -\mu, \phi_s - \phi)\,\mathbf{I}^-(-d)\,d\mu d\phi\right) e^{-\kappa_{es}(z+d)}$$

$$+\int_{-d}^{z} \mathbf{F}^+(z')\,e^{-\kappa_{es}(z-z')}\,dz' \tag{2.8}$$

$$\mathbf{I}^-(z) = e^{\kappa_{es}z}\left[\frac{1}{4\pi}\int_0^{2\pi}\int_0^1 \mathbf{S}_r(-\mu_s, \mu, \phi_s - \phi)\,\mathbf{I}^+ d\mu d\phi\right]$$

$$+\,e^{\kappa_{es}z}\left[\frac{1}{4\pi}\mathbf{S}_t(-\mu_s, -\mu_i, \phi_s - \phi_i)\,\mathbf{I}_0\right]$$

$$+\int_{z}^{0} \mathbf{F}^-(z')\,e^{\kappa_{es}(z-z')}\,dz' \tag{2.9}$$

The above two equations are the starting integral equations suitable for deriving an iterative solution. The objective of the problem is to solve for the upward propagating intensity at $z = 0$ and then transmit it into the upper medium above the inhomogeneous layer. The approach is to assume that the albedo is small and volume scattering is treated as first order, while surface scattering at the layer boundaries is treated as zero order. Iteration begins with the incident intensity entering the layer medium from above.

2.3 UPWARD SCATTERING INTENSITY WITHIN THE LAYER

To seek a first-order solution in the volume scattering coefficient κ_s, let us consider the significant terms in (2.9). Of the three terms in (2.9) the first term represents a downward intensity due to scattering of an upward intensity by the top boundary. This upward intensity should come mainly from the scattering of the incoming intensity by volume inhomogeneities or by the lower boundary. Scattering and absorption losses naturally reduce this term to a level much smaller than the original incoming intensity generated by the incident wave, which is represented by the second term. The major portion of the third term in (2.9) represents the incoming intensity scattered downward by volume inhomogeneities. Again due to scattering and absorption losses this term is also smaller than the incoming intensity but it is larger than the first term. Let us obtain a more explicit form of the stated portion of the third term in (2.9):

$$\begin{aligned}
\int_z^0 \mathbf{F}^-(z')\, e^{\kappa_{es}(z-z')}\, dz' &\approx \int_z^0 \left(\frac{\kappa_{ss}}{4\pi} \int_0^{2\pi}\int_0^1 \mathbf{P}(-\mu_s, -\mu, \phi_s - \phi)\, \mathbf{I}^-\, d\mu\, d\phi \right) e^{\kappa_{es}(z-z')}\, dz' \\
&\approx \int_z^0 \left(\frac{\kappa_{ss}}{4^2\pi^2} \int_0^{2\pi}\int_0^1 \mathbf{P}(-\mu_s, -\mu, \phi_s - \phi)\, e^{\kappa_e z'/\mu}\, \mathbf{S}_t(-\mu, -\mu_i, \phi - \phi_i)\, \mathbf{I}_0\, d\mu\, d\phi \right) e^{\kappa_{es}(z-z')}\, dz' \\
&= \frac{\kappa_s}{4^2\pi^2} \int_0^{2\pi}\int_0^1 \frac{\mu\left(e^{\kappa_{es} z} - e^{\kappa_e z/\mu}\right)}{\kappa_e(\mu_s - \mu)} \mathbf{P}(-\mu_s, -\mu, \phi_s - \phi)\, \mathbf{S}_t(-\mu, -\mu_i, \phi - \phi_i)\, \mathbf{I}_0\, d\mu\, d\phi
\end{aligned} \tag{2.10}$$

The last expression in (2.10) shows that it is a first-order term in scattering. Thus, it may be needed for second-order or surface-volume interaction calculations. For a layer with a low dielectric constant, this source term will be responsible for generating surface-volume interaction from the lower boundary of the layer. In summary, the major downward propagating terms are represented by the second term in (2.9) and a smaller term given by (2.10).

Now, we consider the upward propagating terms. In view of (2.8) there are two contributing terms to the upward propagating intensity $\mathbf{I}^+(0)$ at the top boundary. To the first order in scattering, these two terms are

1. The upward propagating intensity due to scattering from the lower boundary responding to the incoming intensity arriving at $z = -d$. From (2.9) the incoming intensity originated from the incident wave arriving at the lower boundary is

$$\mathbf{I}^{-}(-d) \approx e^{-\kappa_{es}d}\left[\frac{1}{4\pi}\mathbf{S}_t(-\mu_s, -\mu_i, \phi_s - \phi_i)\,\mathbf{I}_0\right] \tag{2.11}$$

2. Upward scattering within the layer due to the propagation of the incoming intensity within the layer. From (2.9) the incoming propagating intensity within the layer is

$$\mathbf{I}^{-}(z) \approx e^{\kappa_{es}z}\left[\frac{1}{4\pi}\mathbf{S}_t(-\mu_s, -\mu_i, \phi_s - \phi_i)\,\mathbf{I}_0\right] \tag{2.12}$$

In the following subsections we shall first use (2.11) and (2.12) to calculate direct contributions from volume and surface and then calculate the interaction term with (2.10).

2.3.1 Direct Scattering from Volume and Lower Boundary

For the development to follow we shall consider volume scattering from the layer and direct scattering from the lower boundary. After substituting (2.11) and (2.12) into (2.8), we obtain the two contributing terms to the first-order upward intensity at $z = 0$ before transmitting out of the layer as

$$\begin{aligned}
\mathbf{I}^{+}(0, \mu_s, \phi_s) &\approx \frac{e^{-\kappa_{es}d}}{4^2\pi^2}\int_0^{2\pi}\int_0^{1}\mathbf{G}(\mu_s, -\mu, \phi_s - \phi)\, e^{-\kappa_e d/\mu}\,\mathbf{S}_t(-\mu, -\mu_i, \phi - \phi_i)\,\mathbf{I}_0\, d\mu\, d\phi \\
&+ \int_{-d}^{0}\left(\frac{\kappa_{ss}}{4^2\pi^2}\int_0^{2\pi}\int_0^{1}\mathbf{P}(\mu_s, -\mu, \phi_s - \phi)\, e^{\kappa_e z'/\mu}\,\mathbf{S}_t(-\mu, -\mu_i, \phi - \phi_i)\,\mathbf{I}_0\, d\mu\, d\phi\right) e^{\kappa_{es} z'}\, dz' \\
&= \frac{e^{-\kappa_{es}d}}{4^2\pi^2}\int_0^{2\pi}\int_0^{1}\mathbf{G}(\mu_s, -\mu, \phi_s - \phi)\, e^{-\kappa_e d/\mu}\,\mathbf{S}_t(-\mu, -\mu_i, \phi - \phi_i)\,\mathbf{I}_0\, d\mu\, d\phi \\
&+ \frac{\kappa_s}{16\kappa_e\pi^2}\int_0^{2\pi}\int_0^{1}\{1 - \exp[-\kappa_{es}d\,(\mu_s + \mu)/\mu]\} \\
&\qquad \frac{\mathbf{P}(\mu_s, -\mu, \phi_s - \phi)}{(\mu_s + \mu)}\,\mu\,\mathbf{S}_t(-\mu, -\mu_i, \phi - \phi_i)\,\mathbf{I}_0\, d\mu\, d\phi \\
&\equiv \mathbf{I}_g + \mathbf{I}_v
\end{aligned} \tag{2.13}$$

The first term $\mathbf{I}_g$ in (2.13) represents scattering from the lower boundary and is controlled by lower boundary scattering and the loss in the layer. This term is important when the scattering by the lower boundary is high and the loss within the layer is low.

The second term $\mathbf{I}_v$ is the scattering from the layer volume. It is important when the scattering albedo of the layer is high and the optical depth of the layer is large.

If the top boundary can be treated as a plane, then the rough surface transmission phase function $\mathbf{S}_t(-\mu, -\mu_i, \phi - \phi_i)$ can be replaced by the plane boundary transmission function, $4\pi \mathbf{T}_{t1}(-\mu, -\mu_i)\,\delta(\mu - \mu_t)\,\delta(\phi - \phi_i)$, and the incoming intensity simplifies to

$$e^{\kappa_e z/\mu}\,[\mathbf{T}_{t1}(-\mu, -\mu_i)\,\delta(\mu - \mu_t)\,\delta(\phi - \phi_i)\,\mathbf{I}_0] \qquad (2.14)$$

where $\mathbf{T}_{t1}(-\mu, -\mu_i)$ is the transmissivity from the top boundary into the layer and μ_t is the cosine of the angle of transmission. In this case the upward intensity at the top boundary inside the layer is a simplified version of (2.13),

$$\mathbf{I}^+(0, \mu_s, \phi_s) \approx \frac{e^{-\kappa_{es} d(1 + \mu_s/\mu_t)}}{4\pi}\mathbf{G}(\mu_s, -\mu_t, \phi_s - \phi_i)\,\mathbf{T}_{t1}(-\mu_t, -\mu_i)\,\mathbf{I}_0$$

$$+ \frac{\kappa_s}{4\pi\kappa_e}\{1 - \exp[-\kappa_{es} d(\mu_s + \mu_t)/\mu_t]\}\frac{\mathbf{P}(\mu_s, -\mu_t, \phi_s - \phi_i)}{(\mu_s + \mu_t)}\mu_t \mathbf{T}_{t1}(-\mu_t, -\mu_i)\,\mathbf{I}_0$$

$$(2.15)$$

In the above we have discussed either a plane or a rough boundary generating coherent and noncoherent scattering, respectively. A more general problem is one where the boundary scattering is partly coherent and partly noncoherent. That is, $\mathbf{S}_t = \mathbf{S}_t^c + \mathbf{S}_t^n$ and $\mathbf{G} = \mathbf{G}^c + \mathbf{G}^n$, where the superscript c represents coherent and n represents noncoherent scattering. In the literature the coherent scattering portion of the scattering phase matrix has been approximated as

$$\mathbf{S}_t^c(\mu, \mu_i, \phi - \phi_i) \approx 4\pi L_{t1}(-\mu, -\mu_i)\,\mathbf{T}_{t1}(-\mu, -\mu_i)\,\delta(\mu - \mu_t)\,\delta(\phi - \phi_i) \qquad (2.16)$$

where $L_{t1}(-\mu, -\mu_i) = \exp\left[-\sigma_1^2(k_{rt}\mu - k_r\mu_i)^2\right]$, σ_1 is the rms height of the top boundary; k_r and k_{rt} are the real parts of the wave numbers in the incident medium and the layer, respectively. Let σ_2 and $\mathbf{R}$ be the rms height and the reflectivity of the lower boundary, respectively, and $L_r(\mu_s, -\mu) = \exp[-\sigma_2^2 k_{rt}^2(\mu_s^2 + \mu^2)]$. Then, the coherent part of $\mathbf{G}$ is approximated as

$$\mathbf{G}^c(\mu_s, \mu, \phi_s - \phi) \approx 4\pi L_r(\mu_s, -\mu)\,\mathbf{R}(\mu_s, -\mu)\,\delta(\mu_s - \mu)\,\delta(\phi_s - \phi) \qquad (2.17)$$

An illustration of the geometry of this general problem is given in Figure 2.1 Note that, except for the attenuation factor, (2.16) and (2.17) have the same mathematical form as for a plane boundary. Thus, we may view the coherent scattering as an equivalent plane reflection and transmission problem with a modification in the reflection and transmission coefficients by the loss factors $L_r(\mu_s, -\mu)$, $L_{t1}(-\mu, -\mu_i)$ representing boundary roughness effects. For this more general condition, the incoming intensity within the layer consists of two terms after crossing the top boundary. Its representation can be found by modifying (2.12) using (2.16) to

$$\bar{\mathbf{I}}(z) \approx e^{\kappa_e z/\mu} \{ \frac{1}{4\pi} \mathbf{S}_t^n (-\mu, -\mu_i, \phi - \phi_i)$$

$$+ \mathbf{T}_{t1} (-\mu, -\mu_i)\, \delta(\mu - \mu_t)\, \delta(\phi - \phi_i)\, L_{t1} (-\mu, -\mu_i) \} \mathbf{I}_0 \quad (2.18)$$

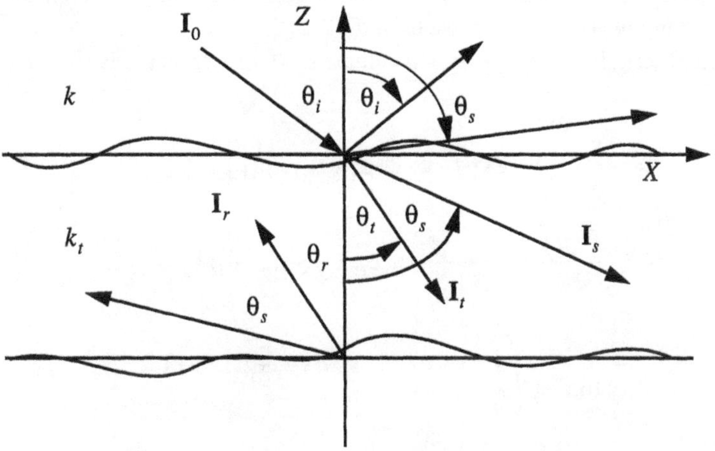

Figure 2.1 An illustration of coherent and incoherent scattering at layer boundaries.

Now we can apply (2.18) to give a more explicit form of (2.13) including coherent and noncoherent contributions. Consider the first term in (2.13), which represents scattering of the incoming intensity by the lower boundary. It expands into four terms as follows

$$\mathbf{I}_g(\mu_s, \phi_s) = \frac{e^{-\kappa_{es} d}}{4^2 \pi^2} \int_0^{2\pi} \int_0^1 (\mathbf{G}^c + \mathbf{G}^n)\, e^{-\kappa_e (d/\mu)} \left(\mathbf{S}_t^c + \mathbf{S}_t^n \right) \mathbf{I}_0 \, d\mu \, d\phi$$

$$= \frac{e^{-\kappa_{es} d}}{4^2 \pi^2} \int_0^{2\pi} \int_0^1 \mathbf{G}^n (\mu_s, -\mu, \phi_s - \phi)\, e^{-\kappa_e (d/\mu)}\, \mathbf{S}_t^n (-\mu, -\mu_i, \phi - \phi_i)\, \mathbf{I}_0 \, d\mu \, d\phi$$

$$+ \frac{e^{-\kappa_{es} d (1 + \mu_s/\mu_t)}}{4\pi} \mathbf{G}^n (\mu_s, -\mu_t, \phi_s - \phi_i)\, L_{t1} (-\mu_t, -\mu_i)\, \mathbf{T}_{t1} (-\mu_t, -\mu_i)\, \mathbf{I}_0$$

$$+\frac{e^{-2\kappa_{es}d}}{4\pi}\mathbf{R}(\mu_s,-\mu_s)\,L_r(\mu_s,-\mu_s)\,\mathbf{S}_t^n(-\mu_s,-\mu_i,\phi_s-\phi_i)\,\mathbf{I}_0$$

$$+e^{-\kappa_{es}d(1+\mu_s/\mu_t)}L_rL_{t1}\mathbf{R}(\mu_s,-\mu_t)\,\mathbf{T}_{t1}(-\mu_t,-\mu_i)\,\delta(\mu_s-\mu_t)\,\delta(\phi_s-\phi_i)\,\mathbf{I}_0 \qquad (2.19)$$

Of the four terms in (2.19) the first term is dominating if surface boundaries are very rough, and the last term is dominating if the boundaries are very smooth and the dielectric contrast between the incident medium and the layer is small. In general, we have to deal with all four terms. Recognizing that the arguments of L_r, L_{t1} must match those of $\mathbf{R}(\mu_s,-\mu_t)$ and $\mathbf{T}_{t1}(-\mu_t,-\mu_i)$, respectively; we shall omit their arguments to simplify writing as we did in the last term of (2.19).

The other term in (2.13) represents volume scattering from the layer. It expands into two terms as

$$\mathbf{I}_v(\mu_s,\phi_s) = \frac{\kappa_s}{4\pi\kappa_e}\{1-\exp[-\kappa_{es}d(\mu_s+\mu_t)/\mu_t]\}$$

$$\frac{\mathbf{P}(\mu_s,-\mu_t,\phi_s-\phi_i)}{(\mu_s+\mu_t)}\mu_tL_{t1}\mathbf{T}_{t1}(-\mu_t,-\mu_i)\,\mathbf{I}_0$$

$$+\frac{\kappa_s}{16\pi^2\kappa_e}\int_0^{2\pi}\int_0^1\{1-\exp[-\kappa_{es}d(\mu_s+\mu)/\mu]\}$$

$$\frac{\mathbf{P}(\mu_s,-\mu,\phi_s-\phi)}{(\mu_s+\mu)}\mu\mathbf{S}_t^n(-\mu,-\mu_i,\phi-\phi_i)\,\mathbf{I}_0d\mu d\phi \qquad (2.20)$$

For a smooth top boundary and a low dielectric contrast at the boundary, only the first term in (2.20) is important. If the boundary is very rough, then only the last term in (2.20) is needed. While such simplifications are generally not possible, good estimates can be obtained in a vegetated medium and a dry snow layer because these media have a small dielectric constant relative to air. For a sea ice medium such an estimate is not as satisfactory but it can still provide a good understanding of the relative contribution between surface and volume scattering.

2.3.2 Surface-Volume Interaction

When an inhomogeneous layer has a low dielectric constant such as a snow layer, most of the energy is transmitted coherently into the layer. For dry snow the transmitted energy will reach the snow-ground or snow-ice interface because the extinction in the snow layer is small. As a result it is of interest to consider surface-volume interaction. This can be done by first calculating the downward intensity using (2.10) simplified by incorporating (2.16) into it, yielding

$$= \frac{\kappa_s}{4\pi}\int_0^{2\pi}\int_0^1 \frac{\mu\,(e^{\kappa_{es}z} - e^{\kappa_e z/\mu})}{\kappa_e\,(\mu_s - \mu)}\mathbf{P}\,(-\mu_s, -\mu, \phi_s - \phi)\,L_{t1}T_{t1}\,(-\mu, -\mu_i)$$

$$\delta\,(\mu - \mu_t)\,\delta\,(\phi - \phi_i)\,\mathbf{I}_0 d\mu d\phi$$

$$= \frac{\kappa_s}{4\pi}\frac{\mu_t\,(e^{\kappa_{es}z} - e^{\kappa_e z/\mu_t})}{\kappa_e\,(\mu_s - \mu_t)}\mathbf{P}\,(-\mu_s, -\mu_t, \phi_s - \phi_i)\,L_{t1}T_{t1}\,(-\mu_t, -\mu_i)\,\mathbf{I}_0 \tag{2.21}$$

The interaction of the above volume scattering term with the lower boundary to generate an upward intensity at the top surface boundary due to coherent reflection is easily determined. Let us define the direction of incidence on the upper boundary inside the layer as θ_0 and ϕ_0 with $\mu_0 = \cos\theta_0$ and $\kappa_{e0} = \kappa_e/\cos\theta_0$. Then, the upward intensity at $z = 0$ is

$$\mathbf{I}^+_{vg}\,(0, \mu_0) = e^{-\kappa_{e0}d} L_r R\,(\mu_0, -\mu_0)\,\frac{\kappa_s}{4\pi}\frac{\mu_t\,(e^{-\kappa_{e0}d} - e^{-\kappa_e d/\mu_t})}{\kappa_e\,(\mu_0 - \mu_t)}$$

$$\mathbf{P}\,(-\mu_0, -\mu_t, \phi_0 - \phi_i)\,L_{t1}T_{t1}\cdot(-\mu_t, -\mu_i)\,\mathbf{I}_0 \tag{2.22}$$

In part (a) of Figure 2.2 we illustrated the scattering processes in (2.22). Clearly, there is a term similar to it as illustrated in part (b) of Figure 2.2.

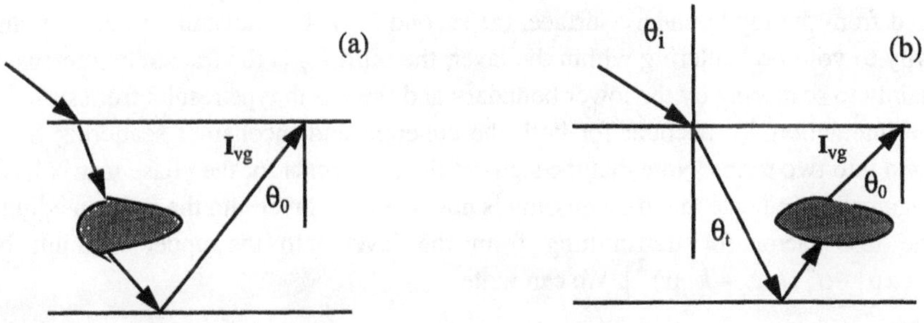

Figure 2.2 An illustration of the interactions between volume and surface scattering within a layer.

Thus, the total interaction terms are

$$\mathbf{I}^+_{vg}\,(0, \mu_0) = e^{-\kappa_{e0}d} L_r\,(\mu_0, -\mu_0)\,R\,(\mu_0, -\mu_0)\,\frac{\kappa_s}{4\pi}\frac{\mu_t\,(e^{-\kappa_{e0}d} - e^{-\kappa_e d/\mu_t})}{\kappa_e\,(\mu_0 - \mu_t)}$$

$$\mathbf{P}\,(-\mu_0, -\mu_t, \phi_0 - \phi_i)\,L_{t1}\,(-\mu_t, -\mu_i)\,T_{t1}\,(-\mu_t, -\mu_i)\,\mathbf{I}_0$$

$$+ e^{-\kappa_{et} d} L_r R(\mu_t, -\mu_t) \frac{\kappa_s \mu_t}{4\pi} \frac{(e^{-\kappa_{e0} d} - e^{-\kappa_e d/\mu_t})}{\kappa_e (\mu_0 - \mu_t)} \mathbf{P}(\mu_0, \mu_t, \phi_0 - \phi_t) L_{t1} T_{t1}(-\mu_t, -\mu_i) \mathbf{I}_0$$

(2.23)

In backscattering we expect the above two terms to sum up as twice what is given in (2.22).

2.4 TOTAL SCATTERED INTENSITY FROM AN INHOMOGENEOUS LAYER

The total scattered intensity from an inhomogeneous layer with irregular boundaries consists of two contributions: one comes from scattering by the top rough interface and the other from the inhomogeneous layer, i.e., (2.19), (2.20) and (2.23), transmitted into the upper medium. That is,

$$\begin{aligned} \mathbf{I}(\mu_s, \phi_s) &= \frac{1}{4\pi} \mathbf{S}_s(\mu_s, \mu_i, \phi_s - \phi_i) \mathbf{I}_0 \\ &\quad + \frac{1}{4\pi} \int_0^{2\pi} \int_0^1 \mathbf{S}_t(\mu_s, \mu, \phi_s - \phi) [\mathbf{I}_v(\mu, \phi) + \mathbf{I}_g(\mu, \phi) + \mathbf{I}_{vg}(\mu, \phi)] \, d\mu d\phi \\ &\equiv \mathbf{I}_s + \mathbf{I}_{vt} + \mathbf{I}_{gt} + \mathbf{I}_{vgt} \end{aligned} \tag{2.24}$$

In (2.24) we identify four types of intensity terms: the first one $\mathbf{I}_s$ represents the intensity scattered from the top boundary surface, the second $\mathbf{I}_{vt}$ is the transmitted intensity due primarily to volume scattering within the layer, the third $\mathbf{I}_{gt}$ is the transmitted intensity due mainly to scattering by the lower boundary and the fourth type results from surface-volume interaction. To account for both the coherent and incoherent scattering $\mathbf{S}_t$ is separated into two terms. Note that the signs of the arguments for the phase matrix have been changed to indicate that transmission is now from the layer into the upper medium. Let the loss factor for transmitting from the layer into the upper medium be $L_{1t} = \exp[-\sigma_1^2 (k_r \mu_s - k_{rt} \mu)^2]$. We can write

$$\mathbf{S}_t = \mathbf{S}_t^n(\mu_s, \mu, \phi_s - \phi) + 4\pi \mathbf{T}_{1t}(\mu_s, \mu) \delta(\mu - \mu_0) \delta(\phi - \phi_s) L_{1t} \tag{2.25}$$

where the angles associated with $\mathbf{T}_{1t}(\mu_s, \mu)$ appearing in $\mu_0 = \cos\theta_0$ and $\mu_s = \cos\theta_s$ are related through Snell's law, i.e., $k \sin\theta_s = k_t \sin\theta_0$.

2.4.1 Direct Surface and Volume Scattering Terms Above the Layer

Substituting (2.25) into the middle two terms in (2.24) we have

$$\mathbf{I}_{vt}(\mu_s, \phi_s) + \mathbf{I}_{gt}(\mu_s, \phi_s) = \mathbf{T}_t(\mu_s, \mu_0) L_{1t} [\mathbf{I}_v(\mu_0, \phi_s) + \mathbf{I}_g(\mu_0, \phi_s)]$$

$$+\frac{1}{4\pi}\int_0^{2\pi}\int_0^1 \mathbf{S}_t^n(\mu_s,\mu,\phi_s-\phi)\,[\mathbf{I}_v(\mu,\phi)+\mathbf{I}_g(\mu,\phi)]\,d\mu d\phi \tag{2.26}$$

These two terms account for direct scattering contribution from the inhomogeneous layer to the total scattering in the upper medium. Recall that (2.19) and (2.20) contain two terms that are double integrals. Thus, when (2.24) is expanded, it will have two fourfold integrals, six twofold integrals and four algebraic terms.

For the purpose of treating special cases it is helpful to show all the terms in $\mathbf{I}_{vt}(\mu_s,\phi_s)$ and $\mathbf{I}_{gt}(\mu_s,\phi_s)$. We consider each one of them below.

The Volume Scattering Term

$$\mathbf{I}_{vt}(\mu_s,\phi_s) = \mathbf{T}_{1t}(\mu_s,\mu_0)\,L_{1t}\{\frac{\kappa_s}{4\pi\kappa_e}\{1-\exp\left[-\frac{\kappa_e d(\mu_0+\mu_t)}{\mu_t\mu_0}\right]\}$$

$$\frac{\mathbf{P}(\mu_0,-\mu_t,\phi_s-\phi_i)}{(\mu_0+\mu_t)}\mu_t L_{t1}\mathbf{T}_{t1}(-\mu_t,-\mu_i)\,\mathbf{I}_0$$

$$+\frac{\kappa_s}{16\pi^2\kappa_e}\int_0^{2\pi}\int_0^1\{1-\exp\left[-\frac{\kappa_e d(\mu_0+\mu)}{\mu\mu_0}\right]\}$$

$$\frac{\mathbf{P}(\mu_0,-\mu,\phi_s-\phi)}{(\mu_0+\mu)}\mu\mathbf{S}_t^n(-\mu,-\mu_i,\phi-\phi_i)\,\mathbf{I}_0 d\mu d\phi\,\}$$

$$+\frac{1}{4\pi}\int_0^{2\pi}\int_0^1\mathbf{S}_t^n(\mu_s,\bar{\mu},\phi_s-\bar{\phi})\,\{\frac{\kappa_s}{4\pi\kappa_e}\{1-\exp\left[-\frac{\kappa_e d(\bar{\mu}+\mu_t)}{\bar{\mu}\mu_t}\right]\}$$

$$\frac{\mathbf{P}(\bar{\mu},-\mu_t,\bar{\phi}-\phi_i)}{(\bar{\mu}+\mu_t)}\mu_t L_{t1}\mathbf{T}_{t1}(-\mu_t,-\mu_i)\,\mathbf{I}_0$$

$$+\frac{\kappa_s}{16\pi^2\kappa_e}\int_0^{2\pi}\int_0^1\{1-\exp\left[-\frac{\kappa_e d(\bar{\mu}+\mu)}{\mu\bar{\mu}}\right]\}$$

$$\frac{\mathbf{P}(\bar{\mu},-\mu,\bar{\phi}-\phi)}{\bar{\mu}+\mu}\mu\mathbf{S}_t^n(-\mu,-\mu_i,\phi-\phi_i)\,\mathbf{I}_0 d\mu d\phi\,\}\,d\bar{\mu}d\bar{\phi} \tag{2.27}$$

In (2.27) every term is proportional to the albedo κ_s/κ_e. It represents volume scattering from the layer. When the scattering albedo is small $\mathbf{I}_{vt}(\mu_s,\phi_s)$ becomes unimportant. The scattering processes in the first two terms in (2.27) are illustrated in Figure 2.3, where the graph on the left is for the first term. Note that the transmitted intensity into the lower medium is coherent in the first term and noncoherent in the second term. After scattering noncoherently by the volume inhomogeneities along with an appropriate amount of attenuation as the intensity propagates in the layer, the transmission back into the upper medium is coherent for both terms. In the graphs we use a plane boundary to indicate coherent process and an irregular boundary to denote noncoherent process. The last two terms in (2.27) may be similarly illustrated. They differ from the first two terms only in transmission back into the upper medium. Unlike the first two terms this transmission is noncoherent.

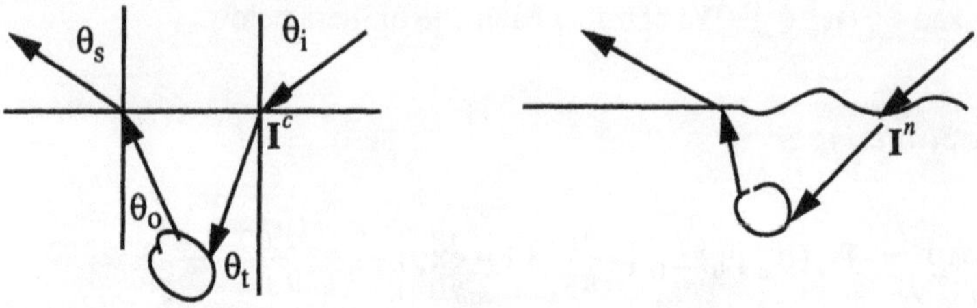

Figure 2.3 Scattering processes of two volume scattering terms. $\mathbf{I}^c$ is the coherently transmitted intensity and $\mathbf{I}^n$ is the noncoherently transmitted intensity.

The terms contained in $\mathbf{I}_{gt}(\mu_s, \phi_s)$ may also be separated into terms with coherent and noncoherent transmissions back into the upper medium. There are four terms in each case as shown in the following equation.

The Ground Scattering Term

$$\mathbf{I}_{gt}(\mu_s, \phi_s) = \mathbf{T}_{1t}(\mu_s, \mu_0) L_{1t} \left\{ \frac{e^{-\kappa_{e0}d}}{4^2\pi^2} \int_0^{2\pi}\int_0^1 \mathbf{G}^n(\mu_0, -\mu, \phi_s - \phi) \right.$$

$$e^{-\kappa_e d/\mu} \mathbf{S}_t^n(-\mu, -\mu_i, \phi - \phi_i) \mathbf{I}_0 d\mu d\phi$$

$$+ \frac{e^{-\kappa_{e0}d(1+\mu_0/\mu_t)}}{4\pi} \mathbf{G}^n(\mu_0, -\mu_t, \phi_s - \phi_i) L_{t1} \mathbf{T}_{t1}(-\mu_t, -\mu_i) \mathbf{I}_0$$

$$\left. + \frac{e^{-2\kappa_{e0}d}}{4\pi} \mathbf{R}(\mu_0, -\mu_0) L_r \mathbf{S}_t^n(-\mu_0, -\mu_i, \phi_s - \phi_i) \mathbf{I}_0 \right\}$$

$$+ \mathbf{T}_{1t}(\mu_i, \mu_t) L_{1t} e^{-2\kappa_{et}d} L_r L_{t1} \mathbf{R}(\mu_t, -\mu_t) \mathbf{T}_{t1}(-\mu_t, -\mu_i) \delta(\phi_s - \phi_i) \mathbf{I}_0$$

$$+ \frac{1}{4\pi} \int_0^{2\pi}\int_0^1 \mathbf{S}_t^n(\mu_s, \bar{\mu}, \phi_s - \bar{\phi}) \left\{ \frac{e^{-\kappa_e(d/\bar{\mu})}}{4^2\pi^2} \int_0^{2\pi}\int_0^1 \mathbf{G}^n(\bar{\mu}, -\mu, \bar{\phi} - \phi) \right.$$

$$e^{-\kappa_e d/\mu} \mathbf{S}_t^n (-\mu, -\mu_i, \phi - \phi_i) \mathbf{I}_0 d\mu d\phi$$

$$+ \frac{e^{-\kappa_{et} d (1 + \mu_t/\bar{\mu})}}{4\pi} \mathbf{G}^n (\bar{\mu}, -\mu_t, \bar{\phi} - \phi_i) L_{t1} \mathbf{T}_{t1} (-\mu_t, -\mu_i) \mathbf{I}_0$$

$$\left. + \frac{e^{-2\kappa_e (d/\bar{\mu})}}{4\pi} \mathbf{R} (\bar{\mu}, -\bar{\mu}) L_r \mathbf{S}_t^n (-\bar{\mu}, -\mu_i, \bar{\phi} - \phi_i) \mathbf{I}_0 \right\} d\bar{\mu} d\bar{\phi}$$

$$+ \frac{1}{4\pi} \mathbf{S}_t^n (\mu_s, \mu_t, \phi_s - \phi_i) e^{-2\kappa_{et} d} L_r L_{t1} \mathbf{R} (\mu_t, -\mu_t) \mathbf{T}_{t1} (-\mu_t, -\mu_i) \mathbf{I}_0 \qquad (2.28)$$

The scattering processes for the first four terms in (2.28) with coherent transmission back into the upper medium are illustrated in Figure 2.4. The upper two graphs from left to right are for the first two terms and the lower two graphs are for the third and fourth terms, respectively. The fourth term is a purely coherent term and hence can contribute only in the specular direction. The other three terms involve noncoherent scattering process and hence can contribute to scattering in any direction. The first term includes two noncoherent scattering processes. It should be the dominant term if the coherent intensity component is small. Conversely, it is an unimportant term if the boundary is relatively smooth or the dielectric constant of the layer is small and hence the coherently scattered intensity is large.

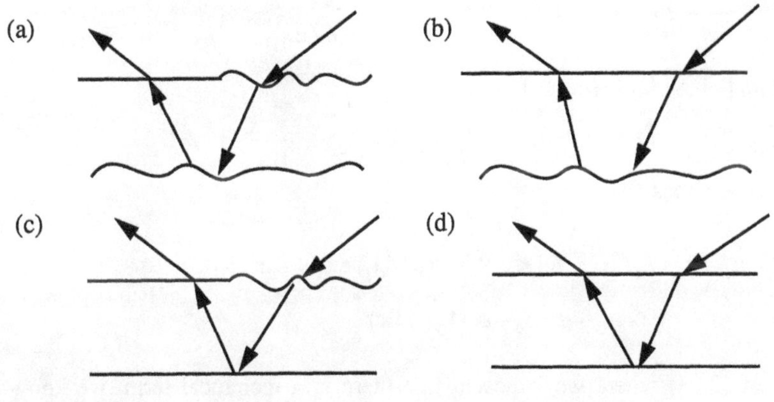

Figure 2.4 Scattering processes of the first four terms in (2.28). Irregular boundaries are used for indicating noncoherent processes and plane boundaries for coherent processes.

The last four terms in (2.28) go through similar scattering processes as those illustrated in Figure 2.4 except in transmission back into the upper medium. For these terms the process is noncoherent transmission. All terms in (2.28) depend on the attenuation inside the layer and the strength of scattering from the lower boundary. They become unimportant if loss in the layer is high or the lower boundary discontinuity is small.

For many problems some of the terms in the above two equations will become negligible due to a change in condition. For example, a vegetation layer and a dry snow layer may have a negligible top boundary and a cloud layer does not have either boundary. It is, therefore, of interest to consider some special cases in the above two equations. These are given in the next section.

2.4.2 Surface-Volume Interaction Term Above the Layer

From (2.22) the intensity coherently transmitted across the top boundary into air is

$$\mathbf{I}_{vg}^{+}(0, \mu_s) = L_{1t}(\mu_s, \mu_0)\, T_{1t}(\mu_s, \mu_0)\, L_r(\mu_0, -\mu_0)\, R(\mu_0, -\mu_0)\, \frac{\kappa_s \mu_t \exp(-2\kappa_{e0} d)}{4\pi\kappa_e(\mu_0 - \mu_t)}$$

$$\left[1 - e^{-\kappa_e d(\mu_0 - \mu_t)/(\mu_t \mu_0)}\right] \mathbf{P}(-\mu_0, -\mu_t, \phi_0 - \phi_i)\, L_{t1}(-\mu_t, -\mu_i)\, T_{t1}(-\mu_t, -\mu_i)\, \mathbf{I}_0 \quad (2.29)$$

Note that we calculated only coherent reflection and transmission in the above interaction term because we assume the layer dielectric constant is low. If this is not so, the above term may not be the dominant term and a we would have to deal with both coherent and noncoherent calculations.

A case of special interest is when we consider backscattering. In this case $\mu_s = \mu_i$ and $\mu_0 = \mu_t$. The following reductions occur in (2.29)

$$\mu_t \{[1 - e^{-\kappa_e d(\mu_0 - \mu_t)/(\mu_t \mu_0)}]/(\mu_0 - \mu_t)\} \approx \kappa_e d/\mu_0$$

$$L_{1t}(\mu_i, \mu_t) = L_{t1}(-\mu_t, -\mu_i) = L_t$$

$$T_{1t}(\mu_i, \mu_t) = T_{t1}(-\mu_t, -\mu_i) = T_t$$

and (2.29) becomes

$$\mathbf{I}_{vg}^{+}(0, \mu_i) = (L_t T_t)^2 L_r R(\kappa_s/\kappa_e)(\kappa_e d/\mu_t) \exp(-2\kappa_{et} d)$$

$$\mathbf{P}(-\mu_0, -\mu_t, \phi_0 - \phi_i)\, \mathbf{I}_0/(4\pi) \quad (2.30)$$

In view of (2.23) where we indicated that there is a reciprocal term, we know that another term similar to (2.30) exists. Hence, the total backscattering intensity due to surface-volume interaction is twice the value of (2.30).

2.5 APPLICATIONS TO SPECIAL CASES

Two special simplified models are considered in the following subsections. One is for backscattering from randomly rough surfaces accounting only for single scattering; the other is for backscattering from an irregular inhomogeneous Rayleigh layer. The regions of applicability of these approximate models are not well defined. Qualitative guidelines and rules of thumb are given to provide readers with some references. These are given in the beginning part of the subsection that introduces each model. The discussion of each model is followed by model characteristics and then its application to measurements.

2.5.1 Backscattering from Randomly Rough Surfaces

For a soil or water surface we can usually treat it as a half space with an irregular boundary. For this problem only the first term in (2.24) is needed. Let the surface root mean square (rms) height σ and correlation length L be normalized to the incident wave number k, i.e., we use $k\sigma$ and kL as roughness parameters. Let the incident angle be θ.

Model for Small and Medium Roughness

From Appendix 2A the like polarized backscattering coefficients useful for surfaces with small or medium size roughness are given by

$$\sigma_{pp}^{0} = 4\pi\cos\theta_s [I_p(\mu_s, \phi_s)/I^i_p]\Big|_{\theta_s=\theta}$$

$$= \frac{k^2}{2}\exp(-2k_z^2\sigma^2)\sum_{n=1}^{\infty}\left|I_{pp}^{n}\right|^2\frac{W^{(n)}(-2k_x, 0)}{n!} \tag{2.31}$$

where $k_z = k\cos\theta$, $k_x = k\sin\theta$, and $pp = vv$ or hh,

$$I_{pp}^{n} = (2k_z\sigma)^n f_{pp}\exp(-k_z^2\sigma^2) + \frac{(k_z\sigma)^n[F_{pp}(-k_x, 0) + F_{pp}(k_x, 0)]}{2} \tag{2.32}$$

and the symbol $W^{(n)}(-2k_x, 0)$ is the Fourier transform of the nth power of the surface correlation coefficient. Examples of this function are given in Appendix 2B.

For a specific polarization we can find f_{pp}, $F_{pp}(k_x, 0)$ and $F_{pp}(-k_x, 0)$ from Appendix 2A. For ease of reference we shall list the special forms of these coefficients that are valid under the additional assumption that *we can replace the local angle of incidence in the Fresnel reflection coefficients by the incident angle*. The range of validity of this assumption varies with the type of surface correlation function used and the value of the dielectric constant. At this point in time a complete study on the range of validity has not been carried out for dielectric surfaces. Studies given in Chapter 6 indicate that the range of validity is most restrictive for the Gaussian correlation and progressively more relaxed as the correlation function approaches exponential. For a surface with permittivity ε_r, we require

$$(k\sigma)(kL) < 1.2\sqrt{\varepsilon_r} \tag{2.33}$$

For the 1.5-power correlation function that has a functional form between Gaussian and exponential, the condition is more relaxed. It becomes (see Chapter 6)

$$(k\sigma)\,(kL) < 1.6\sqrt{\varepsilon_r} \tag{2.34}$$

Under the above conditions, we can now write the field coefficients for the surface scattering model as follows:

$$f_{vv} = 2R_{\parallel}/\cos\theta;\; f_{hh} = -2R_{\perp}/\cos\theta \tag{2.35}$$

$$F_{vv}(-k_x, 0) + F_{vv}(k_x, 0)$$
$$= \frac{2\sin^2\theta\,(1+R_{\parallel})^2}{\cos\theta}\left[\left(1-\frac{1}{\varepsilon_r}\right)+\frac{\mu_r\varepsilon_r-\sin^2\theta-\varepsilon_r\cos^2\theta}{\varepsilon_r^2\cos^2\theta}\right] \tag{2.36}$$

$$F_{hh}(-k_x, 0) + F_{hh}(k_x, 0)$$
$$= -\frac{2\sin^2\theta\,(1+R_{\perp})^2}{\cos\theta}\left[\left(1-\frac{1}{\mu_r}\right)+\frac{\mu_r\varepsilon_r-\sin^2\theta-\mu_r\cos^2\theta}{\mu_r^2\cos^2\theta}\right] \tag{2.37}$$

where $R_{\perp}, R_{\parallel}$ are the Fresnel reflection coefficients for horizontal and vertical polarizations. *These coefficients in (2.35) should be evaluated at normal incidence for surfaces satisfying the tangent plane approximation where kL is large* [Ulaby et al., 1982, Chapter 12]. At this time a satisfactory transition of the local angle of incidence in the Fresnel reflection coefficients from the incident angle to normal incidence is not known. It is expected that, for a Gaussian correlated surface, we should evaluate the Fresnel reflection coefficients at normal incidence when kL is larger than 5. This leads to a different expression for the surface model. However, for exponentially correlated surface, a much larger kL may be used as indicated by (2.34). Such a transition is not critical if the dielectric constant of the surface is large. Otherwise, the range of validity of the model should be restricted as indicated by (2.33) and (2.34).

For perfectly conducting surfaces, the restrictions indicated above do not apply. We show in Chapter 6 that the model is generally very accurate for a variety of surfaces with different scales of roughness. Due to the lack of an accurate shadowing function we expect the model to be increasingly inaccurate as the incident angle increases towards grazing incidence.

Model for Large Roughness or High Frequency

In the high frequency region where $k\sigma$ is large, it is not convenient to use (2.31). Instead its limiting form in the high frequency region is more appropriate. It will be shown in Chapter 5 that in the high frequency region the terms in (2.32) involving F_{pp} may be neglected, and only one term is left in the limit. For this particular term, which is the standard Kirchhoff model [Ulaby et al., 1982, Chapter 12], two classes of surface may be considered: Gaussian height distributed surfaces and modified exponentially distributed surfaces [Fung, 1984]. The definition of the modified exponential density function is

$$p(z) = \frac{|z|^{\mu} c^{\mu+1} K_{\mu}(c|z|)}{2^{\mu}\sqrt{\pi}\Gamma(\mu+0.5)} \tag{2.38}$$

where $c = \sqrt{2\nu}/\sigma$, $\mu = \nu - 0.5$, $0.75 < \nu < 1$, Γ and K_{μ} are the gama and modified Bessel functions and σ is the surface rms height. This density function and Gaussian density function have the following characteristic functions

$$\langle \exp[jk_z z] \rangle = \begin{bmatrix} \exp(-g/2), & \text{Gaussian} \\ 1/[1+g/(2\upsilon)]^{\nu}, & \text{med} \end{bmatrix} \tag{2.39}$$

$$\langle \exp[jk_z(z-z')] \rangle = \begin{bmatrix} \exp\{-g[1-\rho(\xi)]\}, & \text{Gaussian} \\ \{1+g[1-\rho(\xi)]/(2\upsilon)\}^{-2\upsilon}, & \text{med} \end{bmatrix} \tag{2.40}$$

where $g = (k_z\sigma)^2$ and $\rho(\xi)$ is the normalized surface correlation function. In the high frequency region or for large g, we can expand the surface correlation about its origin and approximate it by its first two nonzero terms. This process leads to geometric optics solutions that have analytic forms for the Gaussian distribution [Ulaby et al., 1982] and the modified exponential distribution (when $\upsilon = 0.75, 1.0$) as follows:

$$\sigma^0_{pp}(\theta) = \begin{bmatrix} [|R(0)|^2/(2\sigma_s^2\cos^4\theta)]\exp[-\tan^2\theta/(2\sigma_s^2)], & \text{Gaussian} \\ [3|R(0)|^2/(2\sigma_s^2\cos^4\theta)]\exp(-\sqrt{3}\tan\theta/\sigma_s), & \upsilon, 0.75 \\ [2|R(0)|^2\sin\theta/(\sigma_s^3\cos^5\theta)]K_{-1}(-2\tan\theta/\sigma_s), & \upsilon, 1.0 \end{bmatrix} \tag{2.41}$$

where $\sigma_s^2 = \sigma^2|\rho''(0)|$ is the variance of the surface slope; $|\rho''(0)|$ is the second derivative of the normalized surface correlation at the origin and $K_n(\ )$ is the modified Bessel function of the second kind and of order n. Equation (2.41) is independent of the special form of the surface correlation function. It is useful for surfaces with large scale roughnesses that are larger than the exploring wavelength and do not have smaller scale roughnesses, so that the geometric optics condition can be realized. Note that we did not develop the corresponding modified exponential density cases at lower frequencies or smaller roughness because, under those conditions, the surface height distribution is not the governing factor in noncoherent scattering. Instead, it is the surface correlation function that is important. A number of surface correlation functions are summarized in Appendix 2B. We shall first plot the expressions in (2.41) to show the differences in the backscattering behaviors of these scattering coefficients. In Figure 2.5 we see that there is not much difference between $\upsilon = 0.75$ and $\upsilon = 1.0$, expressions at large angles of incidence. At small angles, $\upsilon = 1.0$ expression gives the highest values and the Gaussian expression the lowest values. For a very large rms slope (Figure 2.5 (c)), the Gaussian expression actually peaks at an angle away from normal incidence, a situation that does not occur with modified exponential distribution.

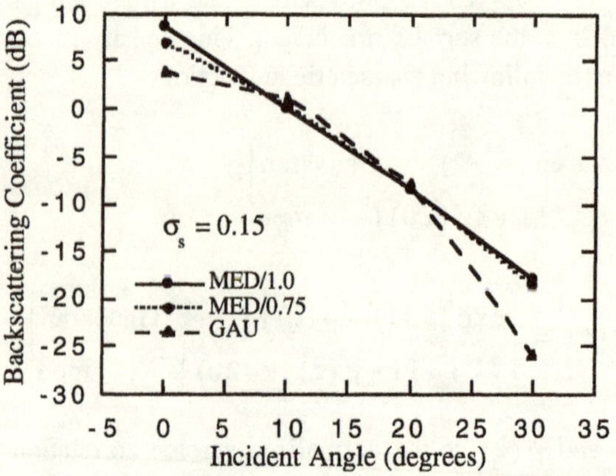

Figure 2.5 (a) Comparisons of the backscattering coefficients in (2.36), for slope values of 0.15.

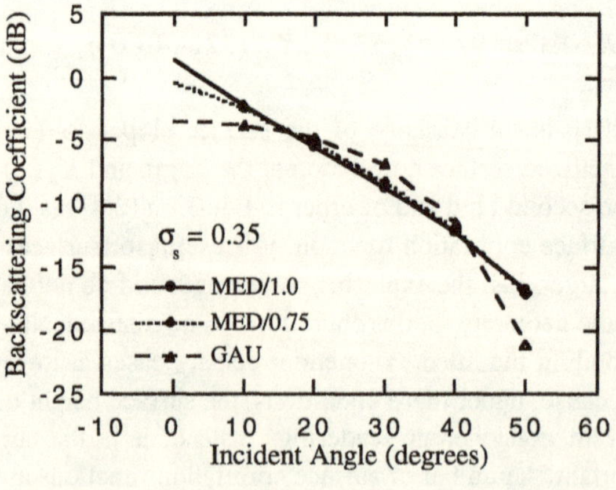

Figure 2.5 (b) Comparisons of the backscattering coefficients in (2.36) when the rms slope is 0.35.

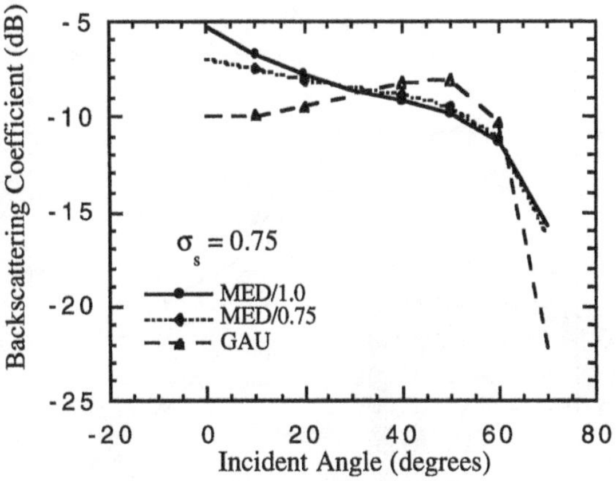

Figure 2.5 (c) Comparisons of the backscattering coefficients in (2.41) when the rms slope is 0.75.

Next, we shall illustrate the dependence of the model given by (2.31) on the type of correlation function, roughness scales and the dielectric constant.

To show the effect of the surface correlation function on surface scattering we show in Figure 2.6 the backscattering coefficients corresponding to exponential, 1.5-power, and Gaussian correlation functions (Appendix 2B). At large angles of incidence the exponential function gives the highest level and the Gaussian gives the lowest level among the three functions. The exponential correlation function is commonly used in theoretical models to compare with measurements. In most cases it gives a better agreement than the Gaussian correlation especially for small incident angles or for surface roughness values that fall into the low or intermediate frequency regions. However, it is not differentiable at the origin and hence is not an acceptable correlation function at high frequencies and may give poor estimates at large angles of incidence. It follows that a surface with an exponential correlation function does not have a slope distribution. Thus, it is incompatible with theoretical model development. For the purpose of theoretical analysis it should be viewed as an approximation to a more complicated correlation function, which is differentiable at the origin (see Appendix 2B).

The curvatures of the three sets of backscattering curves corresponding to three different correlation functions in Figure 2.6 are different. More details can be seen after we replot them over a smaller angular region (0 to 30 degrees) as shown in Figure 2.7. Here, we see that the Gaussian correlation leads to a bell shaped curve as expected, while the 1.5-power function generates a fairly straight line and the exponential correlation function produces an exponential shaped angular curve. The latter two curves coincide almost exactly at zero and 30 degrees and the 1.5-power curve is clearly higher at in between angles. In applications, the 1.5-power function may be a good alternative to the popular exponential function for some cases.

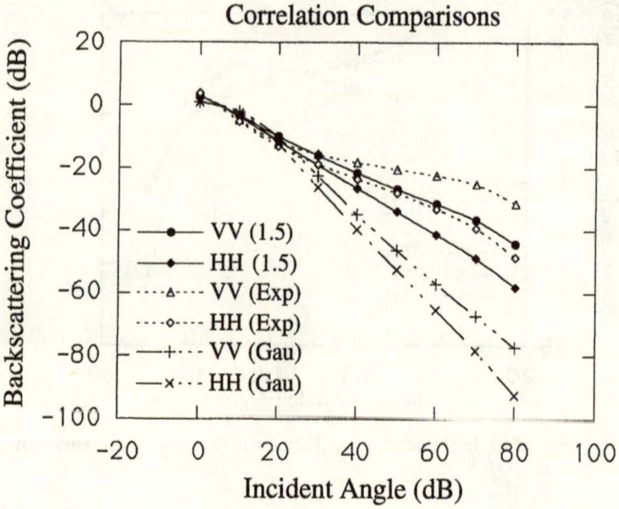

Figure 2.6 Effects of surface correlation on backscattering, $k\sigma = 0.2$, $kL = 5$, dielectric constant = 15.

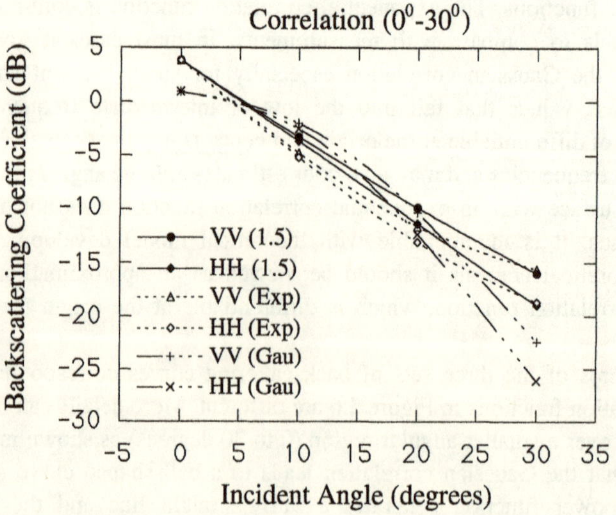

Figure 2.7 An enlarged comparison of the effects of surface correlation function in the small angular region. $k\sigma = 0.2$, $kL = 5$, $\varepsilon_r = 15$.

Another set of correlation functions may be defined as $\exp[-(\xi/L)^n]$, where n is an even number between 1.2 and 2. It is clear that this correlation function varies between the regular exponential and Gaussian correlation functions. When n is selected to be an even number, the function is always differentiable. Hence, this is another function that can provide an exponential-type scattering curve in addition to some other correlation functions shown in Appendix 2B. Like many others, however, it does not have a closed form Fourier transform, and scattering calculations with it have to be done numerically. An illustration using this set of correlation functions is shown in Figure 2.8, where the $n = 1.2$ curves appear to be exponential, while the $n = 1.8$ curves begin to assume a Gaussian appearance in the small angular region. Of course, the $n = 1.8$ curves drop off slower over large incident angles than the Gaussian curves, which are defined by the $n = 2.0$ curves. Similarly, the $n = 1.2$ curves do not rise as fast as the simple exponential curves defined by $n = 1$ at large incident angles. This transition from exponential to Gaussian form will depend not only on the value of n but also on kL. For small kL the transition region for n is wide as shown in Figure 2.8, where the change from exponential to Gaussian is gradual from $n = 1.2$ to 1.8. If kL is large the transition region is expected to be compressed towards the $n = 2.0$ region. This means that the backscattering curves will continue to look exponential until n is much closer to 2. For example, if the surface parameters in Figure 2.8 are multiplied by a factor of 6, we expect the backscattering curves to appear exponential over the small angular region up to $n = 1.8$.

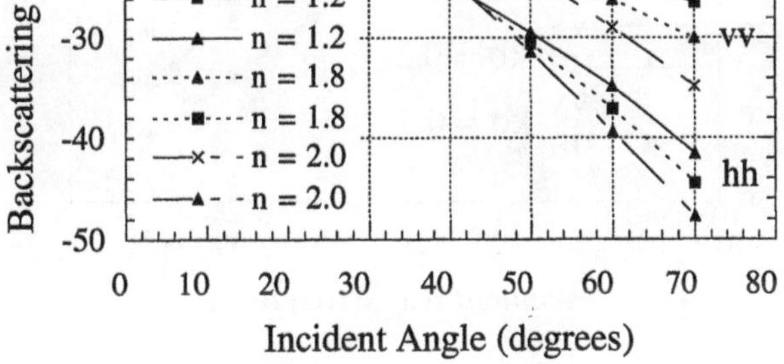

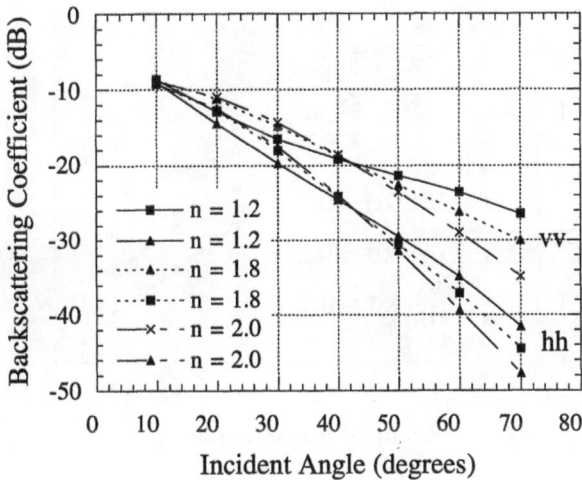

Figure 2.8 Effects of correlation function on backscattering: $k\sigma = 0.126$, $kL = 2.64$, correlation function is equal to $\exp\left[-(\xi/L)^n\right]$ and $\varepsilon_r = 15.5 - j3.71$.

Next, we show the effect of the surface parameter $k\sigma$. This is illustrated in Figure 2.9 and Figure 2.10. Here, the 1.5-power correlation function is used. The parameter kL is chosen to be 4. In Figure 2.9 the backscattering coefficient is seen to rise in level as $k\sigma$ increases from 0.1 to 0.5. The angular trends remain almost unchanged except for a gradual narrowing of the spacing between vertical and horizontal polarizations. This means that for $k\sigma$ smaller than 0.5, its influence is primarily on the level of the backscattering curve and to a much lesser extent slows down the angular trends for both vertical and horizontal polarization. However, as $k\sigma$ increases further from 0.5 to 1.2, Figure 2.10 shows that the angular trend of the backscattering coefficient begins to level off significantly. The level of the backscattering curve at small angles of incidence begins to drop after $k\sigma$ reaches 0.8, while that at large angles continues to rise. The angular region where the backscattering curves with small $k\sigma$ values cross over those with large $k\sigma$ values is between 15 and 20 degrees. In summary, the effects of $k\sigma$ on backscattering vary depending on its value. When $k\sigma$ is increasing up to around 0.5, it causes a rise in the backscattering curve over all angles of incidence as shown in Figure 2.9. Then, as $k\sigma$ increases further to 0.8, it causes a gradual leveling off of the angular trend by actually lowering the backscattering at small angles of incidence. This leveling off becomes significant as $k\sigma$ increases beyond 0.8. In the meantime the spacing between vertical and horizontal polarizations decreases with an increase in $k\sigma$.

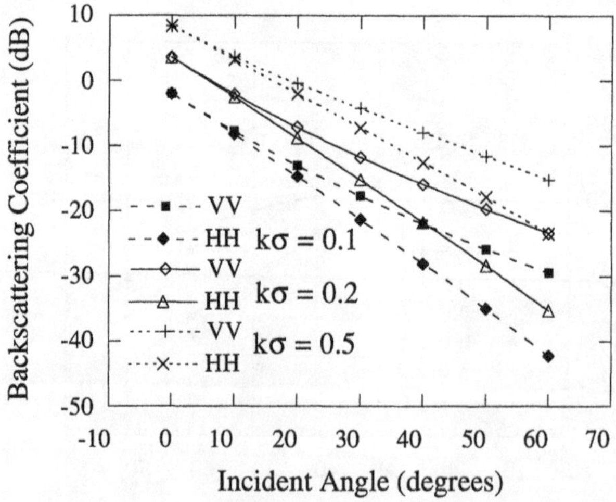

Figure 2.9 Effects of $k\sigma$ variation on backscattering.

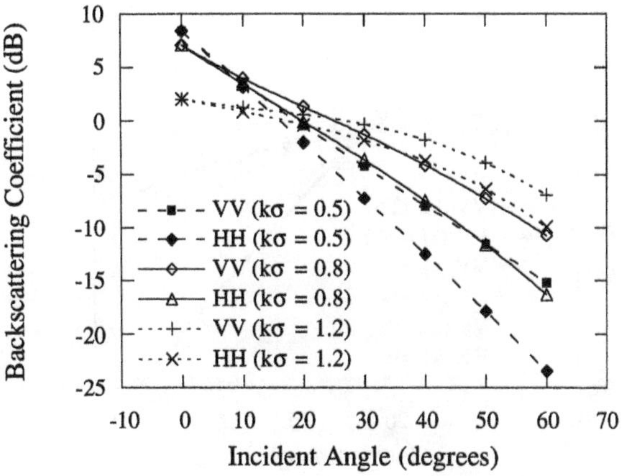

Figure 2.10 Effects of larger rms surface height variation on backscattering.

Next, we consider the effects of kL variation on the backscattering coefficient, when $k\sigma$ is fixed at 0.5 in Figure 2.11. An increase in kL is seen to cause a narrowing of spacing between vertical and horizontal polarizations. In general, the backscattering coefficient drops off faster as kL increases. With the choice of 1.5-power correlation function the angular trends are mostly linear when $k\sigma$ is 0.5. When we increase the $k\sigma$ value to unity, the shape of the angular curve has a larger drop off at large angles of incidence as shown in Figure 2.12. In addition, the spacing between VV and HH polarizations is even smaller than the corresponding case in Figure 2.11. This means that the current model approaches the Kirchhoff model faster when both kL and $k\sigma$ get large.

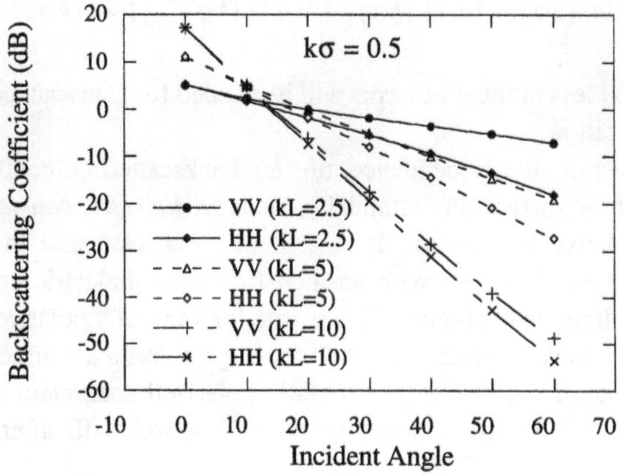

Figure 2.11 Effect of variation in kL on the backscattering coefficient. $k\sigma = 0.5$, $\varepsilon_r = 64$, $\rho = (1 + x^2)^{-1.5}$.

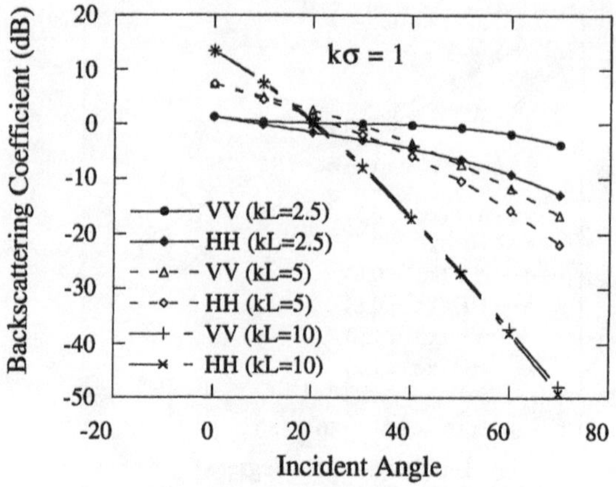

Figure 2.12 Effects of kL variation on backscattering coefficient when $k\sigma = 1$, $\varepsilon_r = 64$, $\rho = (1 + x^2)^{-1.5}$.

The surface scattering model given by (2.31) is in the form of a series. In applications it is necessary to know how many terms are needed to achieve convergence. To provide a reference, computations are carried out for different number of terms and a given value of $k\sigma$ until the change in the backscattering coefficients is less than 0.2 of a dB. The number of terms needed is plotted versus the $k\sigma$ value in Figure 2.13. These estimates are based on the use of the 1.5-power correlation function with kL equal to 10 and a dielectric constant of 81. It is anticipated that for a smaller kL, the number of terms needed may be some what smaller. An equation that describes the curve in Figure 2.13 is

$$n = -1.47 + 16.96\,(k\sigma) - 15.71\,(k\sigma)^2 + 9.488\,(k\sigma)^3 - 1.47\,(k\sigma)^4$$

It is also expected that less number of terms will be needed for Gaussian correlation than for exponential correlation.

Finally, we illustrate the dependence of the backscattering coefficient on the dielectric constant of the surface. For simplicity, the term *dielectric constant* will always refer to its value relative to vacuum. It is generally expected that the level of the backscattering coefficient increases with an increase in the dielectric constant and its effect on the angular trend is negligible. This is true for vertically polarized wave. For a horizontally polarized wave, both the level and the angular trend are affected. As shown in Figure 2.14, the rate of increase in horizontally polarized coefficient decreases with the incident angle. At 60 degrees the change is only around 5 dB after the dielectric constant is changed by a factor of 12.

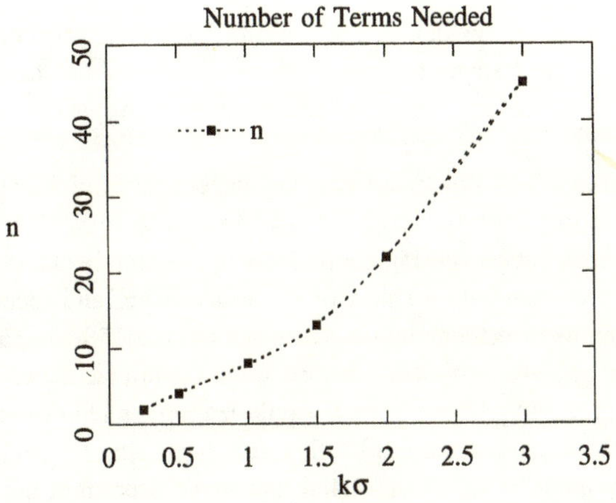

Figure 2.13 Terms needed in computing the scattering coefficient.

Due to this difference in polarization behavior, the spacing between the VV and HH polarizations is seen to widen as the surface dielectric constant increases. This means that the horizontally polarized coefficient drops off faster with the incident angle as surface dielectric constant becomes larger.

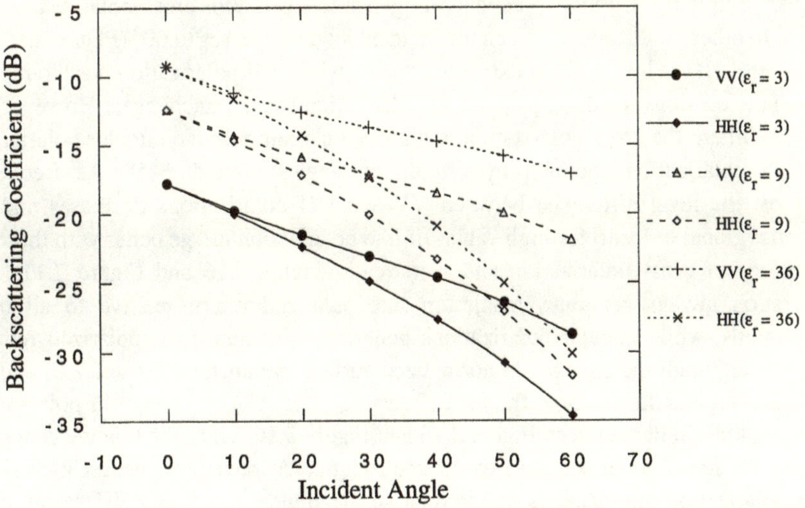

Figure 2.14 Effect of change in the surface dielectric constant on backscattering. Calculation uses $k\sigma = 0.125$, $kL = 1.4$ and 1.5-power correlation function.

2.5.2 Polarization Dependence in Backscattering

Theoretical calculations of the dependence of the backscattering coefficient on the polarization states of the transmitting and receiving antenna are carried out in this section using (1.77) and the scattering coefficients given in Appendix 2A for randomly rough surfaces with a Gaussian correlation.

Figures 2.15 through 2.17 illustrate frequency dependence at incident angles of 30° and 60° with $k\sigma$ changing from 0.25 to 2.0 and kL from 0.69 to 5.58, a factor of 8. The like and cross polarized backscattering coefficients in real numbers are plotted in three dimensions versus the polarization states of the transmitting and receiving antennas, taken to be the same or in orthogonal. In view of Figure 2.15(a) for a slightly rough surface the like polarized coefficient reaches a maximum at 0° ellipticity and 0° orientation angle (i.e., $\tau = 0^{\circ}$ and $\psi = 0^{\circ}$ or VV polarization) and has minima at locations around $\pm 90^{\circ}$ orientation angles and $\pm 30^{\circ}$ ellipticity. Scattering by circularly polarized waves (at ellipticity angle = $45^{\circ}$for left-hand and = -45° for right-hand) is generally close to the minimum value, while horizontally polarized wave HH (at $\tau = 0^{\circ}$ and $\psi = \pm 90^{\circ}$) is at a local maximum along the ellipticity axis. In a general sense the cross polarized scattering has its maximum and minimum values occurring around the locations where the like polarization has its minimum and maximum values respectively; i.e, in the region where the like polarization is high, the cross polarization is low and vice versa. For cross polarization it generally has high values when it is circularly polarized and it dips at the locations of VV and HH in the polarization space. In other words, the cross polarized reception at 30° incidence takes on its maximum values around $\pm 45^{\circ}$ ellipticity and minimum values at 0° and $\pm 90^{\circ}$ orientation angles along 0° ellipticity. Circular polarizations generate maximum cross polarized returns relative to other polarizations. When the incident angle increases to 60° (Figure 2.15(b)), the value of HH is low in the polarization space, while VV polarization has the highest value. The locations for the dips in cross polarization remain unchanged. However, the locations where the cross polarization is high include not only the circular polarization but also polarizations defined by orientation angles around $\pm 45^{\circ}$. As frequency increases, the level difference between VV and HH polarizations decreases and the locations (global or local) for high values of like polarization merge better with those for low values of cross polarization and vice versa (Figure 2.16 and Figure 2.17). VV polarization always generates maximum like polarized return relative to all other polarizations, while circular polarizations generate maximum cross polarized returns. Note that although the changes in normalized surface parameters, $k\sigma$ and kL, and the backscattering coefficient over frequency are large, the relative change in polarization states remains similar between Figure 2.15 and Figure 2.16. Thus, the relative change in polarization states is not sensitive to surface roughness. Indeed, for backscattering the entire polarization plot depends on the relative magnitude of VV and HH, because the cross polarized scattering is very small for the cases considered. Major changes in the shape of the polarization graph occur when $k\sigma$ reaches 2.0, i.e., when we are in the high frequency region (Figure 2.17). Here, the difference between VV and HH is small at large angles of incidence and becomes negligible at small angles of incidence. Instead of having a peak at 0° ellipticity and 0° orientation angles for like polarization (Figure 2.15 and Figure 2.16), we now have a simple V-shaped graph (Figure 2.17(a)).

For circular polarizations, the cross polarized scattering can be higher than the like polarized scattering in surface backscattering (Figure 2.15(a)). Thus, in the polarization space there are regions where the cross polarized return is higher than the like and vice versa. Furthermore, there are regions in the polarization space where both like and cross polarized return are low (Figure 2.15(b)). These polarization properties may be used to suppress background effects when there is a target of interest sitting above the irregular surface [Swartz et al, 1988; Dubois and van Zyl, 1989]. For example, in Figure 2.15(b) both the like and the cross polarized return at $\tau = 0^o$, $\Psi = \pm 90^o$ are significantly lower than at other polarization states. A complex target that sits above such a surface is not likely to have a similar polarization property. Thus, we can enhance the signal from this target relative to background clutter by operating at $\tau = 0^o$, $\Psi = \pm 90^o$ and at large angles of incidence. If the surface is very rough or the frequency is high, we should use the cross polarized mode and linear polarization for suppressing the surface effects.

2.5.3 Comparisons with Backscattering Measurements from Surfaces

In this section we show comparisons of the backscattering model described in the previous section with measurements from constructed and soil surfaces. The man-made surfaces are perfectly conducting surfaces so that pure surface scattering is a certainty. Soil surfaces on the other hand may have some volume scattering occurring at the same time due to inhomogeneities present in the soil medium that may not be negligible compared to the incident wavelength. If so, a surface scattering model alone may not be a suitable model to use except at low enough frequencies, where volume scattering albedo is so small that it can be neglected. When it is not possible to ignore the volume scattering effect, we should consider scattering from an inhomogeneous half space, which is another special case of scattering from a layered medium. This calls for adding a volume scattering term to the surface scattering term given below in this section. Another possible effect in dealing with natural surfaces is that it may be a multiscale surface, i.e., it has more than one scale of roughness and can be characterized only by a multi parameter correlation function as opposed to a single parameter correlation (which is what we have been discussing so far). We shall show that when a good match between model and data can be achieved by adding a volume term, it can also be achieved to some extent with a multiparameter correlation. Thus, it may remain uncertain from modeling study alone whether the surface has more than one scale of roughness or there is volume scattering. The impact of a multiscale surface has one additional important implication; i.e., at large angles of incidence the smaller scale roughness is the dominant contributor. If this is true, then at a given frequency we can sense only some and not necessarily all the roughness scales of a multiscale surface. As a result we should not expect to use the true surface parameters in all cases when dealing with a multi-parameter surface. This idea is intuitively plausible but lacks a rigorous justification.

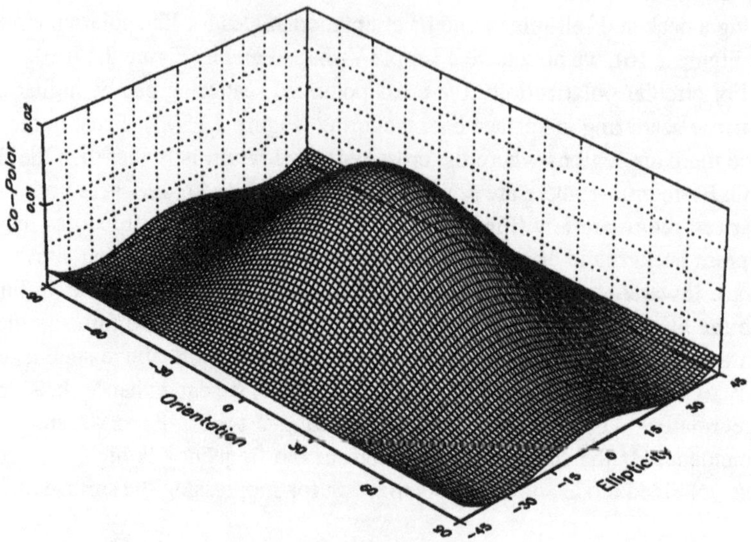

Like Polarization

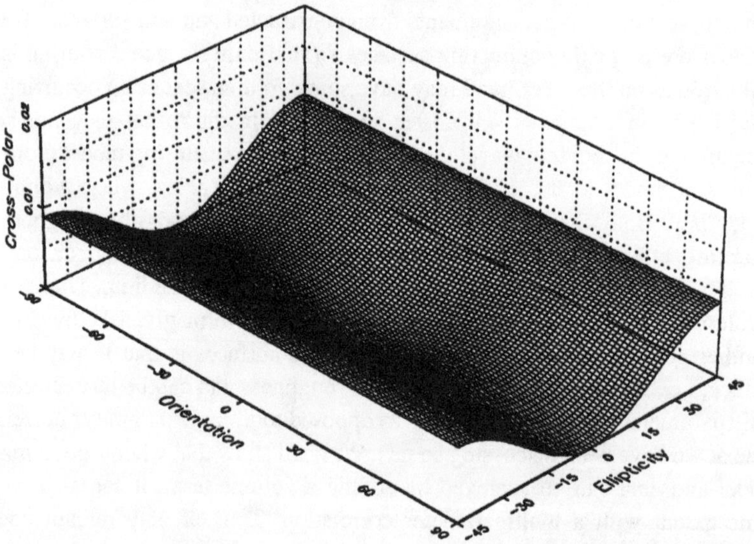

Cross Polarization

(a)

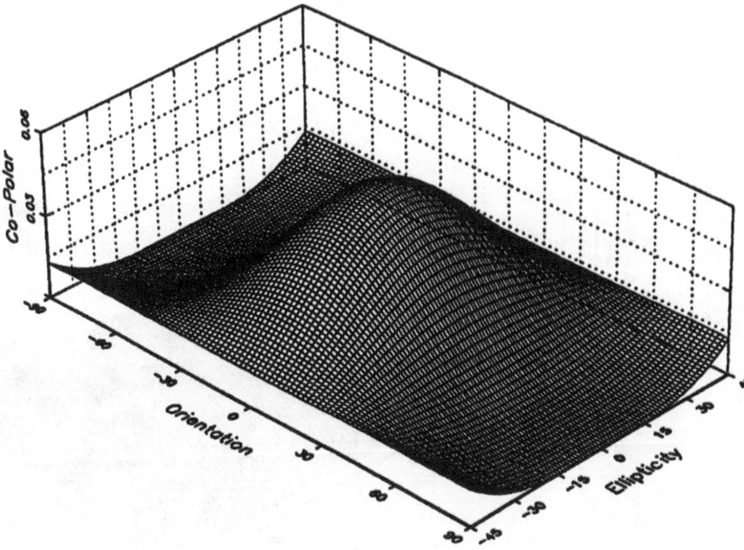

Like Polarization

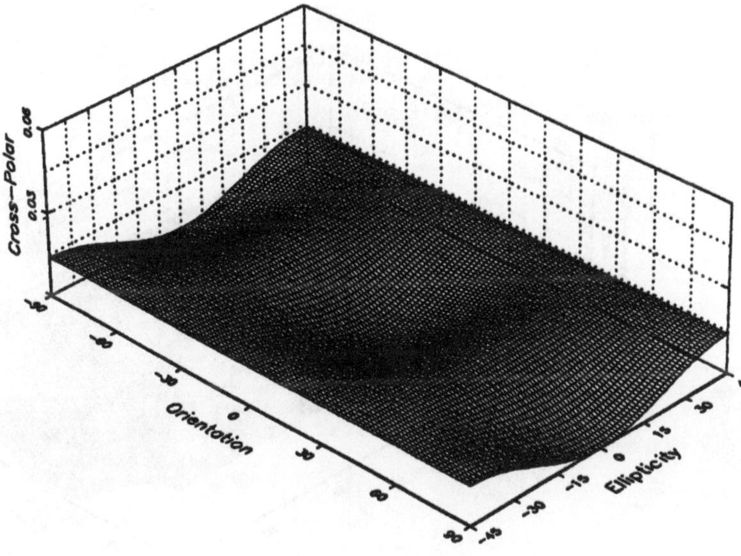

Cross Polarization

(b)

Figure 2.15 Backscattering coefficients as a function of polarization states at (a) 30° and (b) 60° incident angle, when $k\sigma = 0.25$, $kL = 0.69$

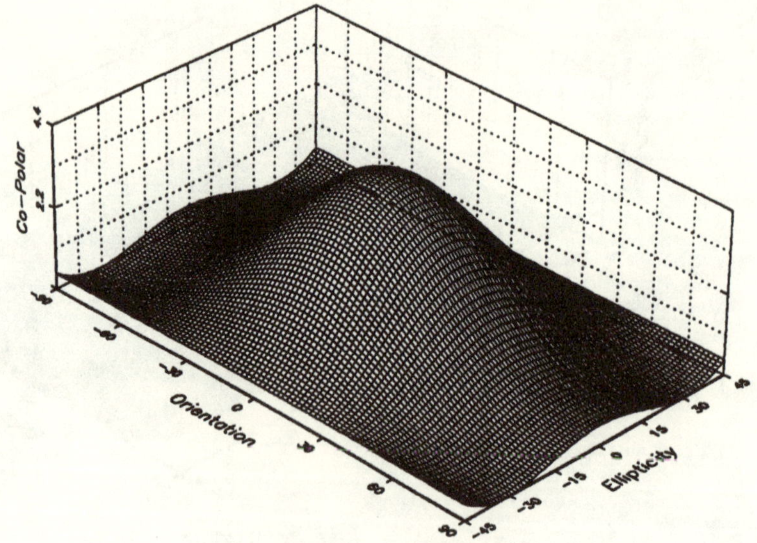

Like Polarization

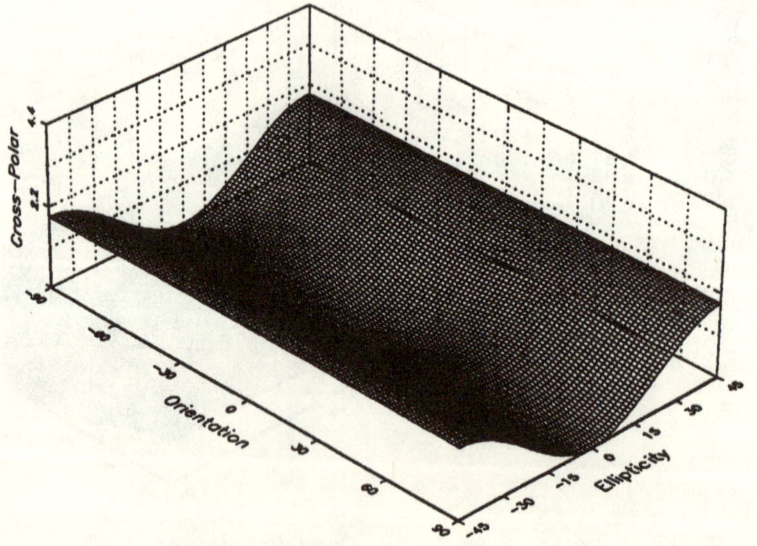

Cross Polarization

(a)

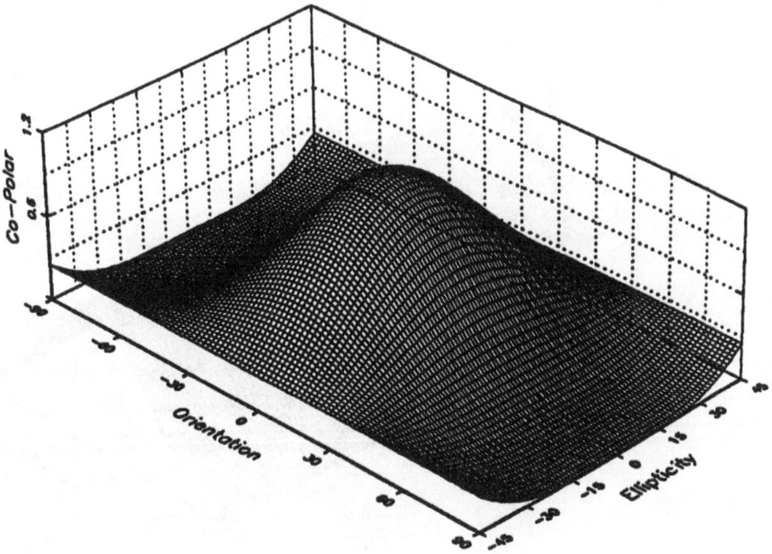

Like Polarization

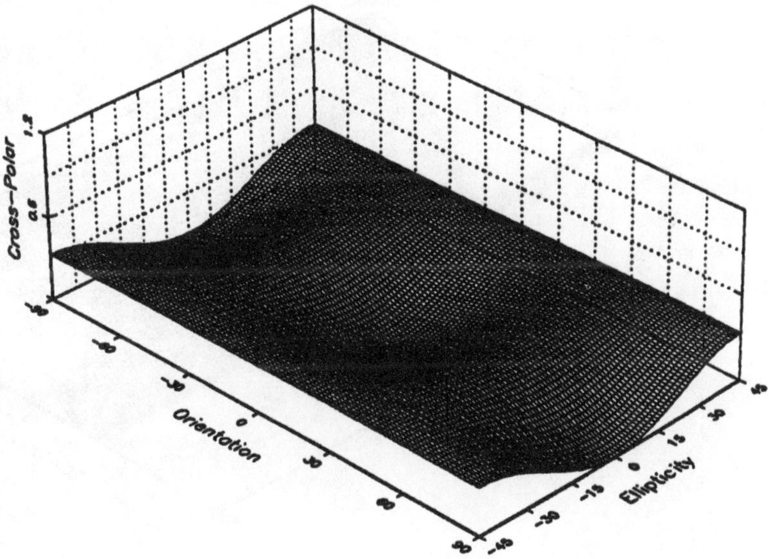

Cross Polarization

(b)

Figure 2.16 Backscattering coefficients as a function of polarization states at (a) 30° and (b) 60° incident angle, when $k\sigma = 0.75$, $kL = 2.07$

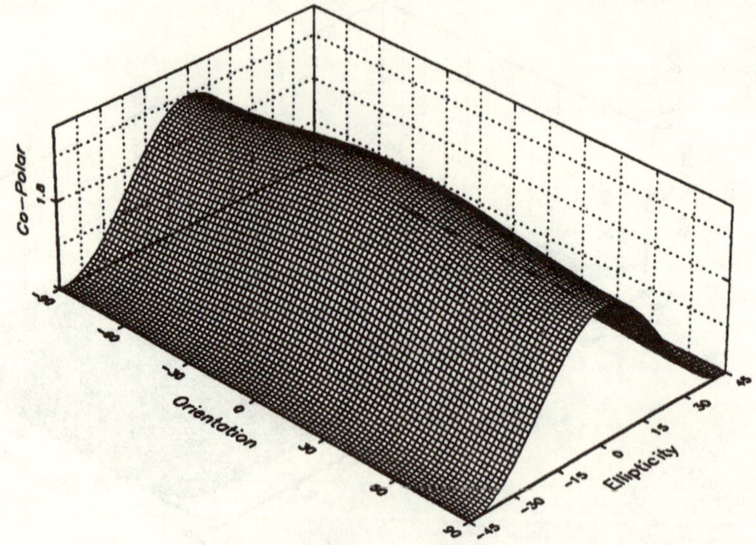

Like Polarization

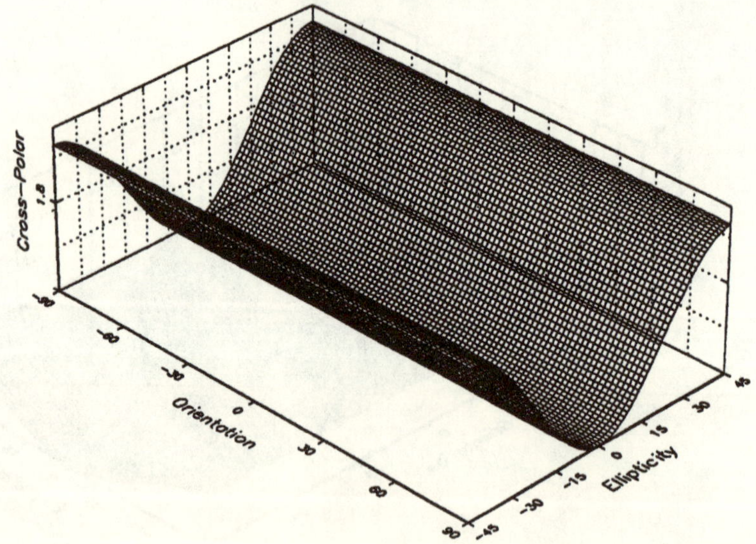

Cross Polarization

(a)

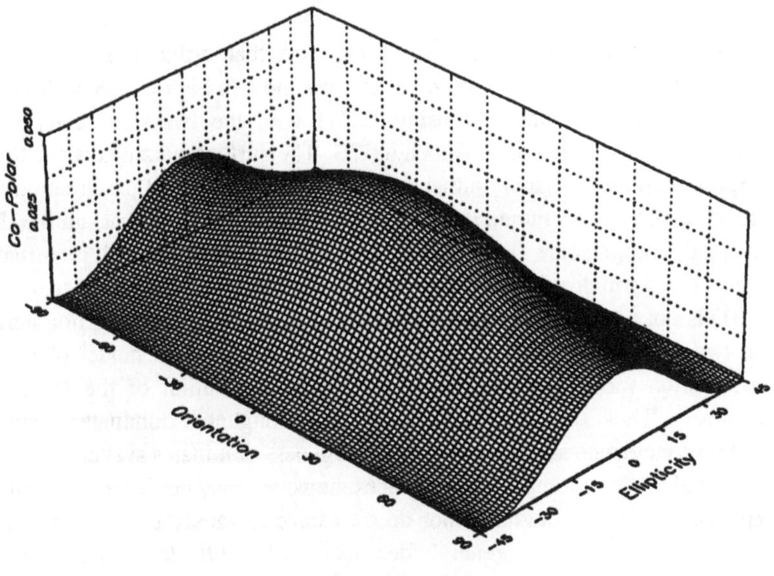

Like Polarization

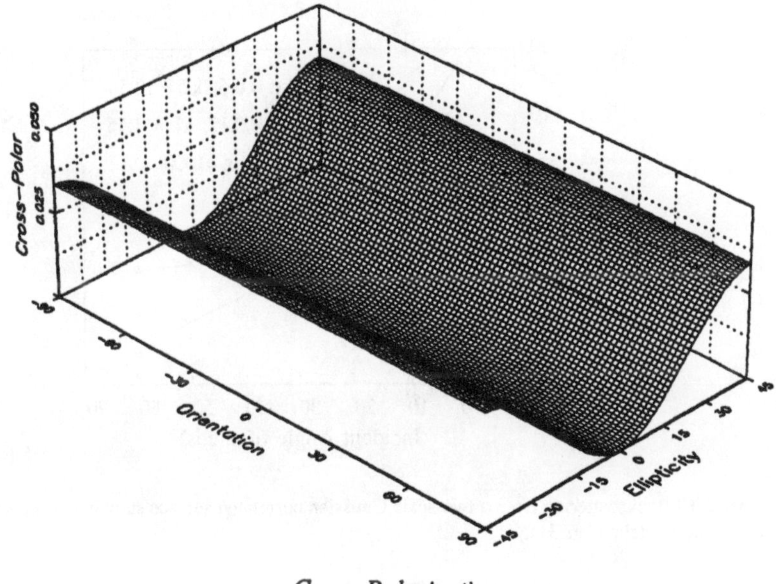

Cross Polarization

(b)

Figure 2.17 Backscattering coefficients as a function of polarization states at (a) 30° and (b) 60° incident angle, when $k\sigma = 2$, $kL = 5.58$

As an illustration in support of the idea that different scale roughnesses may dominant at different angles of incidence, we show in Figure 2.18 backscattering from a two-scale surface calculated using (2.31) with the two-scale Gaussian correlation function given by (2B.5) in Appendix 2B. The surface parameters for this surface are purposely chosen to satisfy either the Kirchhoff or the small perturbation models in the literature [Chen and Fung, 1988; Ulaby et al., 1982]. Note that, unlike the conditions obtained through some theoretical studies, for the first-order small perturbation model to apply, the normalized correlation length, kL, should be smaller than 3.0. Although (2.31) is not a two-scale model, as shown in Figure 2.18 its prediction agrees well with the two-scale model defined as the sum of the Kirchhoff model plus the first order perturbation model averaged over the slope distribution of the large scale surface roughness. Thus, the idea that the large scale roughness dominates scattering at small angles of incidence and the small scale roughness dominates scattering at large angles is supported by this example. Real surfaces, however, may not have well-defined and well-separated scales of roughness nor do they have to satisfy the assumptions of the two-scale model. While the general idea seems plausible, the problem with multiscale surfaces is yet to be investigated.

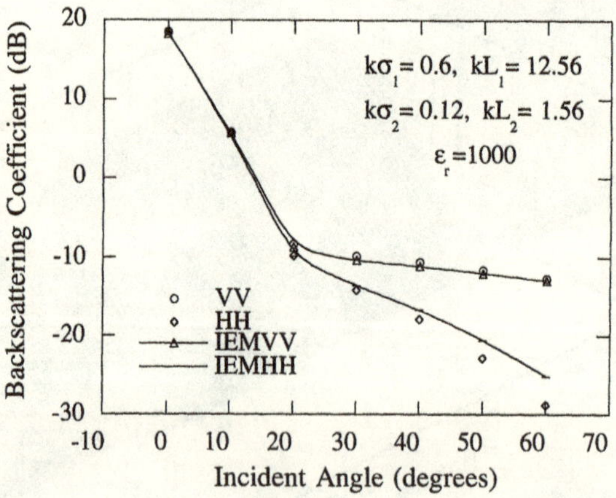

Figure 2.18 Backscattering from a two-scale Gaussian correlated surface showing comparisons between a two-scale model and (2.31).

Constructed Surfaces

First we consider backscattering from constructed surfaces with known surface statistics. Here, no selection of surface parameters is needed. The first surface considered has Gaussian height distribution, its correlation and rms height properties are shown in Figure 2.19. Its correlation function happens to agree very well with the 1.5-

power function given in Appendix 2B. It has a kL of 3.25 and a $k\sigma$ of 0.44 at 25 GHz. Thus, the surface parameters are somewhat larger than those required by the first-order perturbation assumptions but they are not large enough for the Kirchhoff model to apply. The backscatter measurements taken at 25 GHz from this surface [Ulaby et al., 1982] are shown in Figure 2.20, in which we have also plotted the predictions based on the first-order perturbation model, the Kirchhoff model and the IEM model. It is seen that the Kirchhoff model does not agree with measurements as expected. The first-order perturbation model shows fairly good agreement with the vertically polarized measurements but has a progressively worse agreement with the horizontally polarized measurements as the incident angle increases. The angular drop-off of the first-order perturbation model is too fast for horizontal polarization. On the other hand, (2.31) is in good agreement with the data in level and angular trends for both vertical and horizontal polarizations. This indicates that one test for the applicability of a scattering model is its ability to predict the correct spacing between the like polarizations.

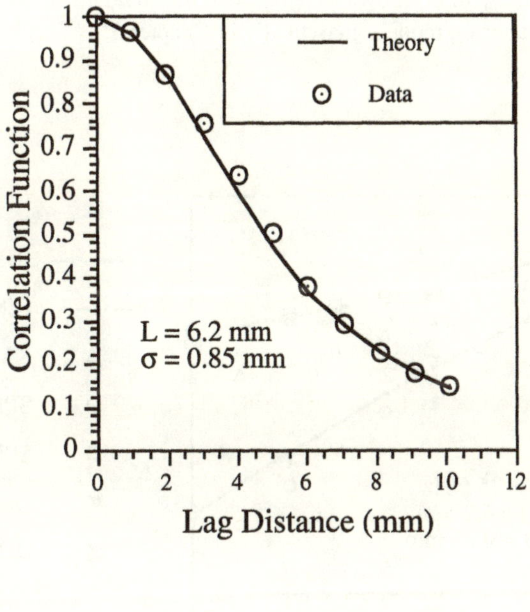

Figure 2.19 Correlation function of a constructed random surface. It fits the function, $\rho = (1 + \xi^2)^{-1.5}$.

The next surface is perfectly conducting, isotropic, Gaussian distributed and Gaussian correlated. It has a rms height of 0.25 cm and a correlation length of 2 cm. Backscattering measurements were taken over a wide range of frequencies, 5 GHz to 10 GHz, so that $k\sigma$ and kL values that fall into the low and the intermediate frequency ranges are included [Nance et al., 1990; Nance, 1992]. Three incident angles were used, ranging from 15 to 55 degrees. As shown in Figure 2.21 excellent agreement in level and frequency trends over the frequency range is obtained at 15 degree incidence for both vertical and horizontal polarizations. The predictions are within the 95% confidence interval of the data. At 35 degree incidence the agreement between the IEM model and data remains excellent for HH polarization, but for VV polarization (Figure 2.22) the theoretical predictions at 9.5 and 10 GHz lie outside the confidence interval by about 0.5 of a dB. All frequency trends are in agreement with the data. Similar agreement is obtained at 55 degrees for VV polarization (Figure 2.23), but for HH polarization the agreement is not as good because the data are contaminated by the edge effects of the target and show some oscillations. However, only one frequency point at 55 degrees lies outside the confidence interval by more than 1 dB. These comparisons with controlled laboratory experiments indicate that the scattering model given by (2.31) works well in the low and the intermediate frequency regions. Note that in the low and high frequency regions the IEM model will be shown to agree with the standard small perturbation and Kirchhoff surface scattering models analytically in Chapter 5.

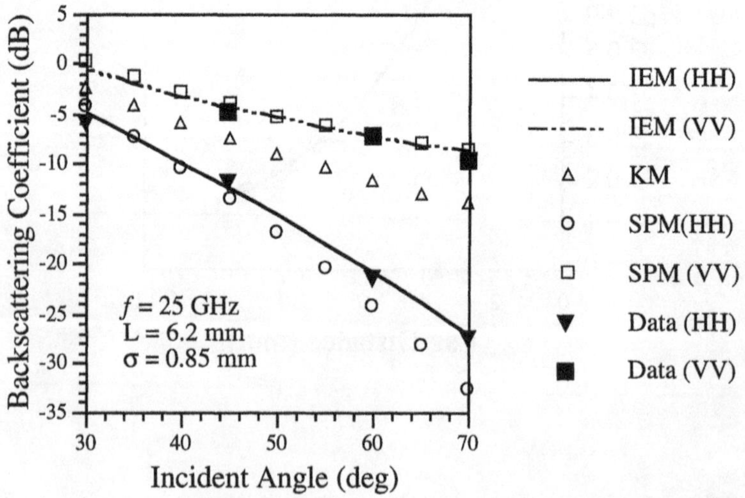

Figure 2.20 Comparisons of IEM, SPM and KM with measurements from a randomly rough surface.

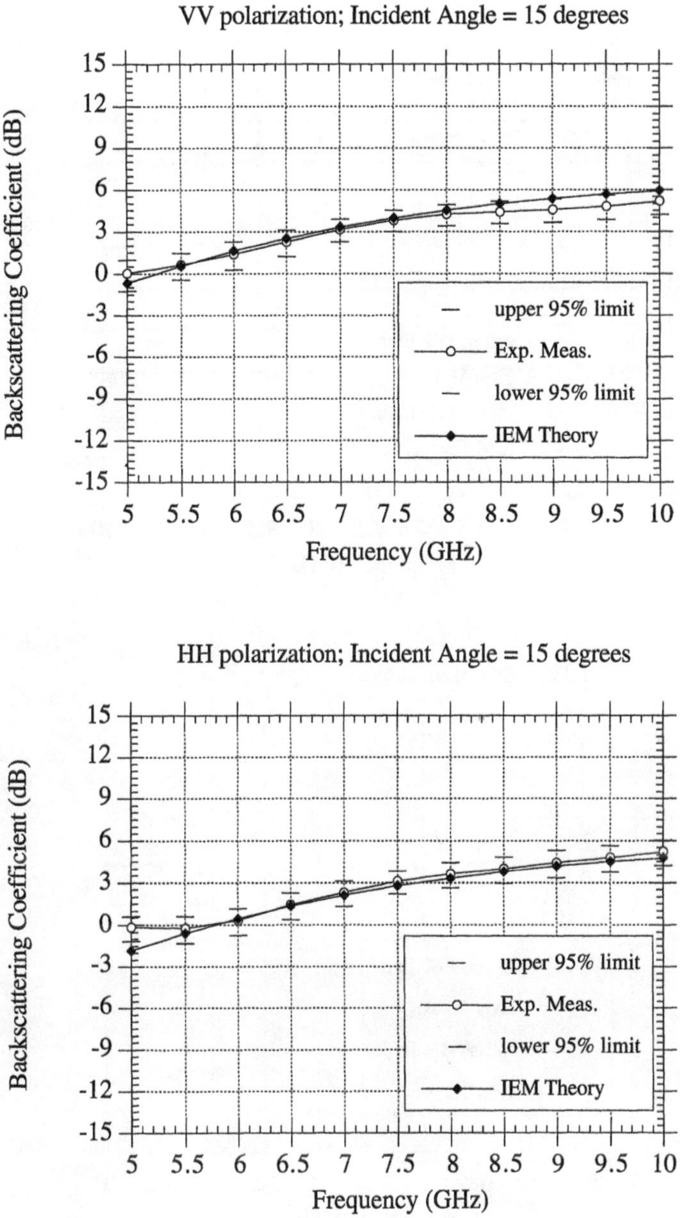

Figure 2.21 Comparisons between the IEM model and laboratory controlled measurements from a Gaussian correlated surface with a rms height of 0.25 cm and a correlation length of 2 cm at 15 degrees incidence.

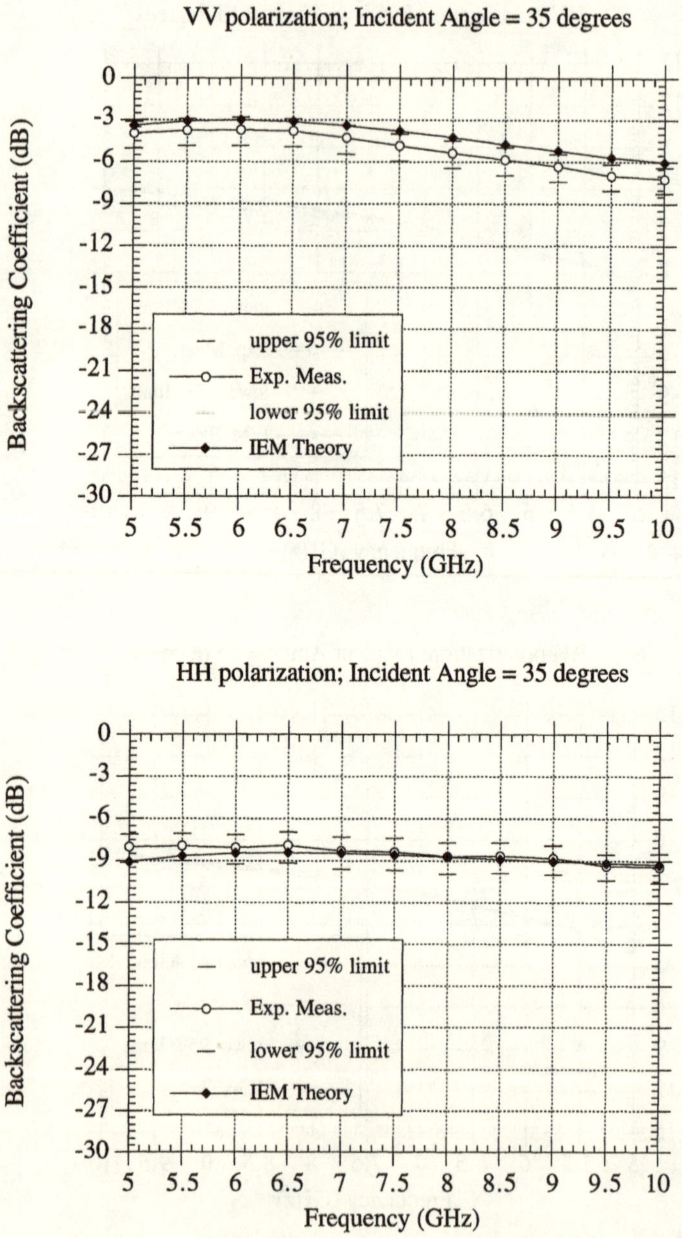

Figure 2.22 Comparisons between the IEM model and laboratory controlled measurements from a Gaussian correlated surface with a rms height of 0.25 cm and a correlation length of 2 cm at 35 degree incidence.

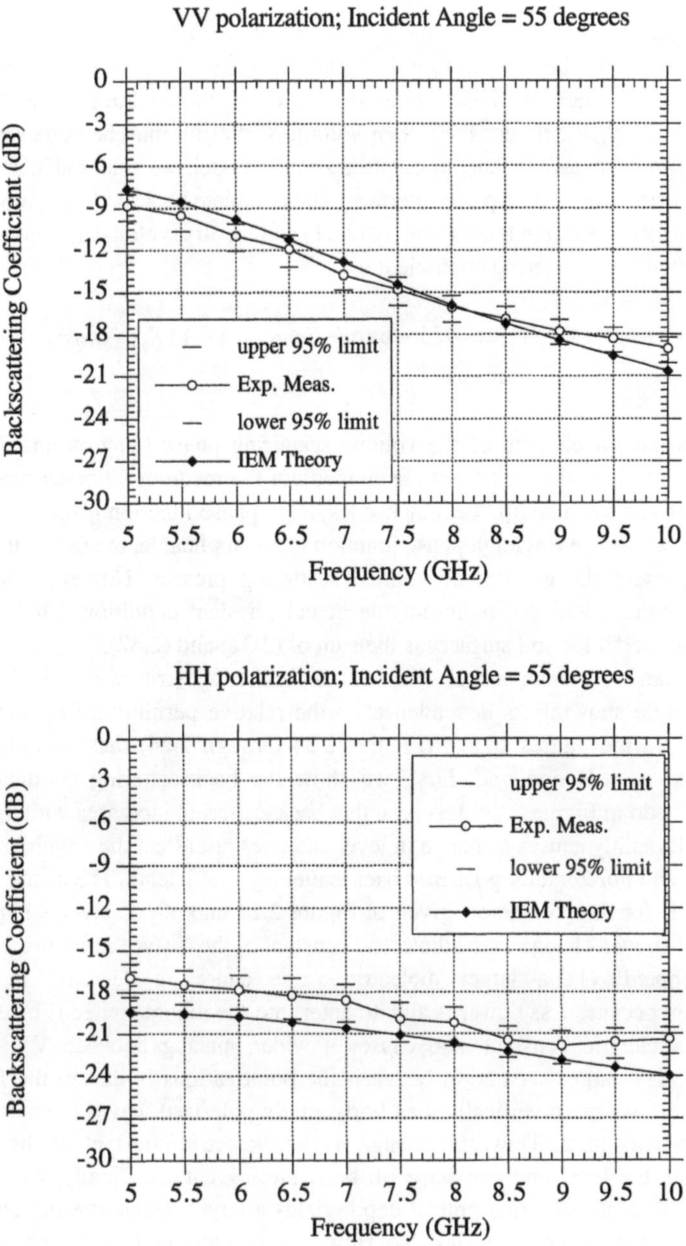

Figure 2.23 Comparisons between the IEM model and laboratory controlled measurements from a Gaussian correlated surface with a rms height of 0.25 cm and a correlation length of 2 cm at 55 degree incidence.

Soil Surfaces

For soil surfaces we should allow the possibility of having some contribution from volume scattering. If the soil surface is highly reflecting, the volume term will be unimportant because not much energy can penetrate into the medium below. If the soil surface has a low dielectric constant, then volume scattering may be generated by the coherent transmission term. Thus, to obtain a simple model, we shall calculate volume scattering contribution using a plane interface. The complete backscattering model is the sum of the surface scattering term given by (2.31) and the first volume scattering term in (2.27) converted into scattering coefficient as

$$\sigma^0_{vpp} = 0.5\,(\kappa_s/\kappa_e)\,T_{1t}T_{t1}\cos\theta\,[1-\exp(-2\kappa_e d/\cos\theta_t)\,]\,P_{pp}\,(\cos\theta_t, -\cos\theta_t;\,\pi) \tag{2.42}$$

where P_{pp} is the *pp* element of the volume scattering phase function and T_{ij} is the Fresnel power transmission coefficient from medium *j* to medium *i*. For scatterers small compared with the wavelength, we may use Rayleigh phase function given in Appendix 2C. For cases where the Rayleigh phase function is not applicable, the use of it in (2.42) may still represent the amount of volume scattering present. However, the model parameters obtained will not represent the actual physical condition. The suggested backscattering model for soil surfaces is the sum of (2.31) and (2.42).

To understand the behavior of this volume scattering term we plot (2.42) as a function of angle showing its dependence on the relative permittivity ε_r, albedo and optical depth. In the figures shown (Figures 2.24 through 2.26) these parameters are abbreviated as ε_r, *alb* and *opd*. First, we show the backscattering coefficient as a function of albedo in Figure 2.24. It is seen that backscattering increases with the albedo as expected. It mainly causes a change in level and does not affect the angular shapes of the vertically and horizontally polarized backscattering coefficients. The change in level is about 10 dB for the parameters given in Figure 2.24 and VV is always higher than HH. Next, we show a change in the dielectric constant of the surface in Figure 2.25. Two effects are noted: (1) a larger dielectric value causes a decrease in volume backscattering, because less power is able to enter into the inhomogeneous medium and (2) a larger dielectric constant also causes a wider spacing between VV and HH polarizations. The larger separation between the polarizations is due to the Brewster angle, which allows more vertically than horizontally polarized waves to enter into the inhomogeneous medium. Thus, the change in the dielectric constant of the medium influences both the level and the shape of the scattering curves. Finally, we show the effect of optical depth. A larger optical depth yields a larger backscattering coefficient because more scatterers become available. However, saturation effects are obvious after the optical depth exceeds 1.0 because less amount of energy will be able to penetrate to deeper levels and scatterers that are too deep will not be seen by the incident wave. Similar to the effect of an increase in albedo, an increase in optical depth does not affect the shape of the angular curves.

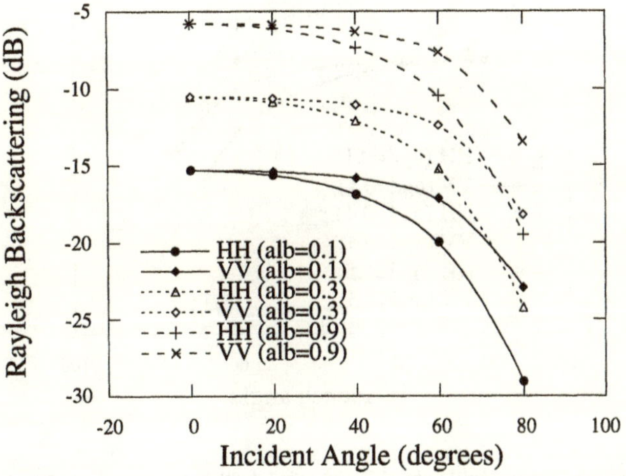

Figure 2.24 Dependence of volume backscattering on albedo variation. $\varepsilon_r = 3.4 - j0.2$, *opd* = 0.32.

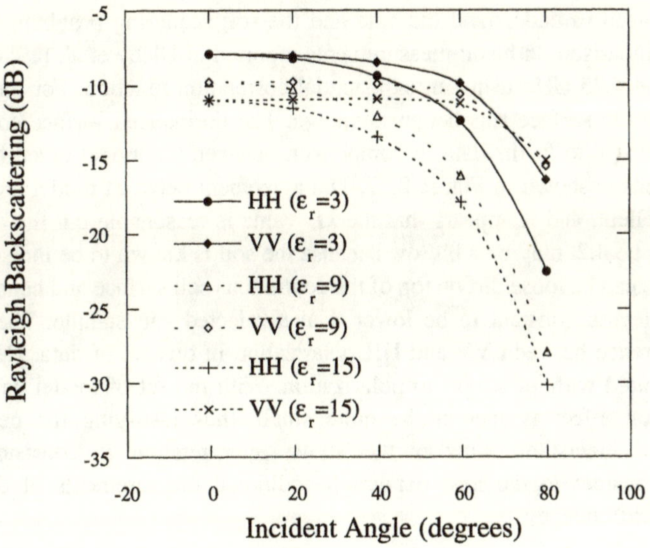

Figure 2.25 Dependence of volume backscattering on dielectric constant with *opd* = 0.32 and *alb* = 0.53.

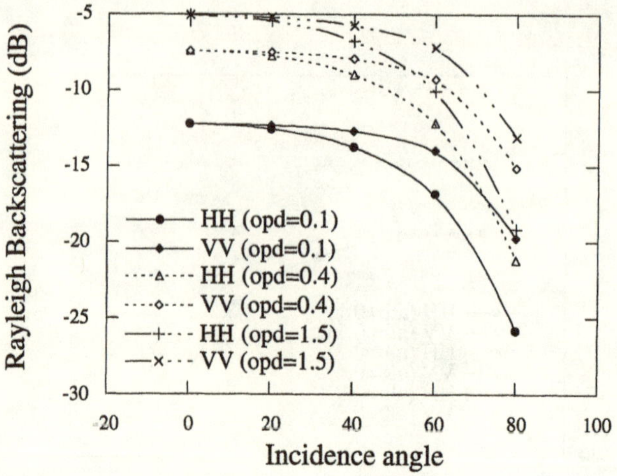

Figure 2.26 Dependence of volume backscattering on optical depth with $\varepsilon_r = 3.4 - j0.2$ and $alb = 0.53$.

In the rest of this section we shall show comparisons between scattering models given by (2.31) and (2.42) and backscattering measurements from soil surfaces. The purpose is not to justify how well these models work but rather to show how scattering models can be used to understand the data and the soil scattering problem. We shall begin with a comparison with soil measurements reported in Ulaby et al. [1986, Figure 21.20] at 1.1 and 4.25 GHz using an exponential correlation function. For the surface considered, only the surface rms height is known. The normalized surface correlation length, kL, is chosen to fit the data. A comparison between the model given by (2.31) and measurements is shown in Figure 2.27. The agreement between model predictions and data is excellent, and it appears that the kL value is reasonable but the dielectric value, chosen to be 4.2, may be a bit low because the soil is known to be moist. On the other hand, there can be loose dirt on top of the continuous soil surface and hence causes the average dielectric constant to be lower than a selected soil sample. There is no significant difference between VV and HH polarization in this set of data. Hence, the data were presented without regard to polarization. With the set of model parameters used, polarization effect is seen to be quite small, thus justifying the neglect of polarization. To understand whether the above interpretation is consistent over frequency we consider in the next paragraph additional measurements of the same surface at another frequency.

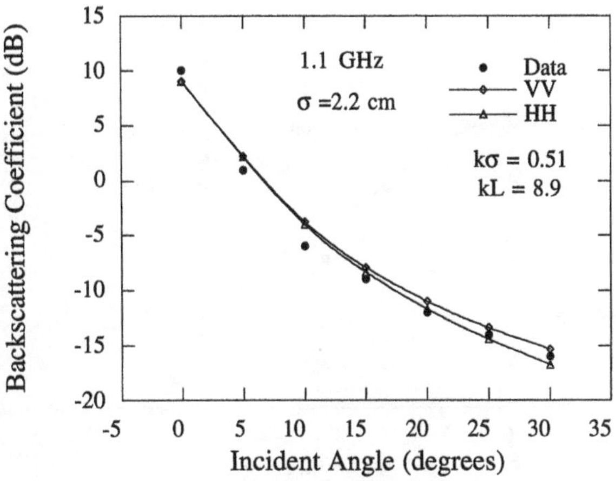

Figure 2.27 Comparison between the IEM model and soil measurements at 1.1 GHz using a relative dielectric constant of 4.2. The surface rms height is 2.2 cm.

The next set of measurements taken on the same soil surface is at 4.25 GHz. Upon examining the data we see that they are at a higher level relative to those at 1.1 GHz at large angles of incidence after frequency dependence is accounted for. In view of the low dielectric value obtained through data comparison at 1.1 GHz, it appears that the data level at 4.25 GHz is more consistent with moist soil. Hence, we use a higher dielectric value for the 4.25 data set. If the surface correlation is a single parameter exponential function, then we should use $k\sigma = 1.96$ and $kL = 34.3$. Another point to note before carrying out data comparison is that soil surfaces are generally porous near the top. At 4.25 GHz the albedo may become appreciable. Hence, we should include the volume scattering term in model comparison, i.e., we should use both (2.31) and (2.42). By choosing a dielectric constant of 9.1 we obtain the fit in Figure 2.28, which shows very good agreement between model predictions and data. The albedo chosen is 0.4 using the Rayleigh phase function. In Figure 2.28 the solid lines represent total scattering; scattering due to surface alone is denoted in dashed lines as VVS and HHS for the vertical and horizontal polarizations. Clearly, the levels of VVS and HHS are low at beyond 15 degrees. Scattering due to the volume term alone is denoted as VOL in the figure. It is seen that, while volume scattering is low, it can cause a 2 dB change at large angles to realize a good fit to the data. The surface parameters used in Figure 2.28 are in violation of the approximate validity condition given earlier. The surface model given by (2.31) is only marginally valid. Indeed, the VVS values are as much as 1 dB lower than those of HHS, which is not characteristic of surface scattering.

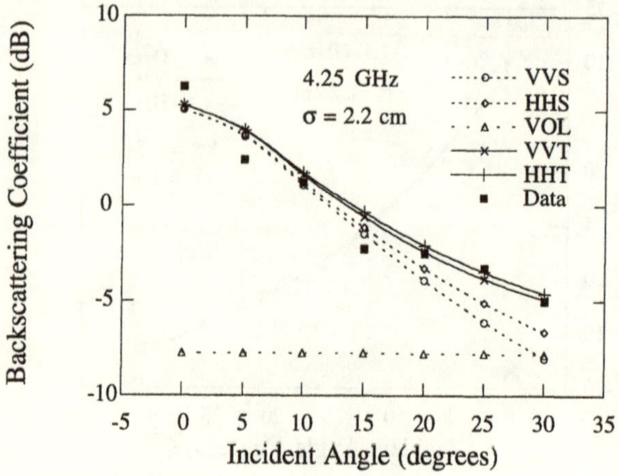

Figure 2.28 Comparison between the (2.31) and soil measurements at 4.25 GHz using $\varepsilon_r = 9.1$, $k\sigma = 1.96$, $kL = 34.3$ and albedo = 0.4.

As stated at the beginning of this section, an alternative interpretation of the data in Figure 2.28 is possible using a multiparameter correlation function. That is, we assume the surface has two scales of roughness and that it does not generate any volume scattering. As shown in Appendix 2B, we can find an approximate surface spectrum for a two-scale surface where the large scale is exponentially correlated and the small scale is Gaussian correlated. This is probably the most useful surface spectrum since it provides an additional control over the large angular region. As shown in Figure 2.29 good agreement between model predictions and measurements similar to those in Figure 2.28 is obtained. Note that in order to keep approximately the same surface height standard deviation, we use a $k\sigma_1 = 1.90$ for the large scale, which is slightly smaller than the value in Figure 2.28, and add another scale to the surface. The additional scale roughness has a $k\sigma_2 = 0.08$ and a $kL_2 = 1.56$. Although this is a very small scale, it is able to raise the level of scattering in the 20 to 30 degree region to produce a good fit. The dielectric constant used is the same as in Figure 2.28. Here again, due to the use of large values for the surface parameters in (2.31) the surface scattering model is only marginally valid. We observe that horizontally polarized scattering is again higher than that of the vertically polarized scattering for incident angles 15 degrees or larger. In one respect this two-scale surface interpretation is worse than the volume scattering interpretation given in Figure 2.28, since it has a 1 dB separation between the vertical and horizontal polarizations, whereas the data do not show a dependence on polarization state.

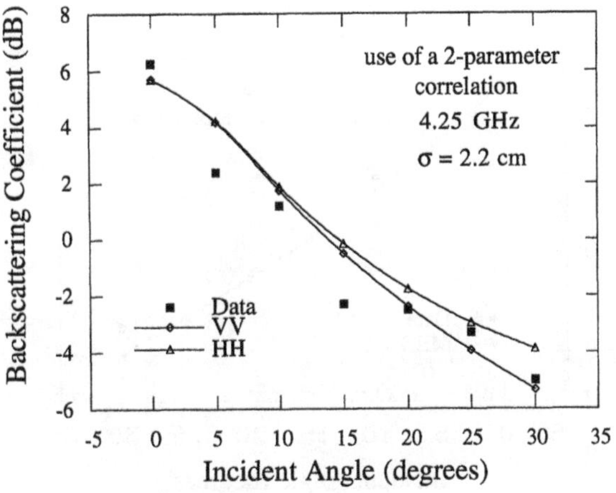

Figure 2.29 Comparison of (2.31) with the same measurement as in Figure 2.28 using a two-parameter correlation function: $k\sigma_1 = 1.90$, $kL_1 = 34.3$, $k\sigma_2 = 0.08$, $kL_2 = 1.56$.

Measurements at 7.25 GHz were also acquired for the surface considered in Figure 2.29. It is clear that the scattering models we used in the previous two figures are no longer appropriate. At this frequency, scattering should be taking place in the high frequency region. Hence, we shall use the model given by (2.41). The likely candidate is the modified exponential model with $\upsilon = 1.0$. We shall denote the model as MED/1.0 in Figure 2.30 where a very good agreement between model and measurements is obtained. The selected surface parameters are $\varepsilon_r = 4.2$ and a rms slope of 0.55. It is interesting to note that this dielectric value is the same as the one used at 1.1 GHz. As stated before it is possible that the average dielectric constant is lower than the one in a soil sample because near the top of the surface there may be some loose dirt. The rms slope used in Figure 2.30 is much larger than the one that can be calculated from the surface parameters in Figure 2.29. A possible explanation is that natural soil surfaces generally contain more than one scale of roughness and may not be continuous surfaces. At low frequencies these effects are smoothed by the incident wavelength. Indeed, the small scale roughnesses may not be able to cause an appreciable phase change to have a significant impact on scattering. At higher frequencies system resolution is finer and the smaller scales of roughness may now produce enough phase change to affect total scattering. As stated earlier the scattering problem of multiscale surfaces and their frequency dependence is not well understood. While these ideas may be intuitively plausible, a rigorous theoretical explanation is lacking.

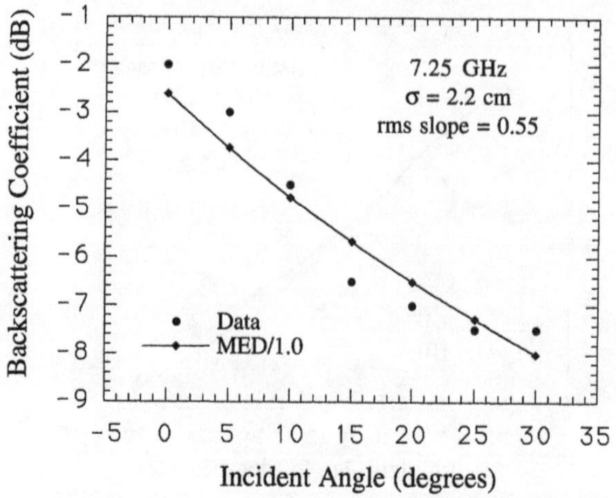

Figure 2.30 Comparison between (2.41) and soil measurements at 7.25 GHz using $\varepsilon_r = 4.2$ and a rms slope of 0.55.

In studying the above surfaces not much is known about them. Although it is possible to obtain a good fit to the data, we find many unanswered questions associated with the interpretations of these surfaces. The study may shed more light on data interpretation, if we have data taken on the same surface over a much wider angular range. Next, we consider a set of backscattering measurements at three different frequencies from asphalt surface [Ulaby et al., 1986, Chapter 21, Figure 21.8]. Here, the surface is known to be fairly flat and dry but there is no quantitative information on surface dielectric constant. What we can demonstrate is whether there is consistency in the data level or trend. As it turns out, the surface roughness parameters seem to be self-consistent as frequency changes but not the data level. As shown in Figure 2.31 we obtain a very good fit using (2.31) and the 1.5-power correlation function at 8.6 GHz. The surface parameters look very reasonable but the relative dielectric constant is 5, which may be somewhat high for dry asphalt. Keeping approximately the same surface parameters at 17 GHz, we obtain another fit that is good but with an even higher dielectric constant, which is 9 in Figure 2.32. Thus, we believe either the data level is too high or the asphalt is really not very dry. When we increase the incident frequency to 35.6 GHz, the surface scattering model becomes marginally valid. However, the surface parameters remain self-consistent and another good fit is achieved with the dielectric constant remaining at 9.0 as shown in Figure 2.33. Thus, we conclude from the analysis that either the data level or the dielectric property of the surface is questionable. The surface parameters appear to be well defined.

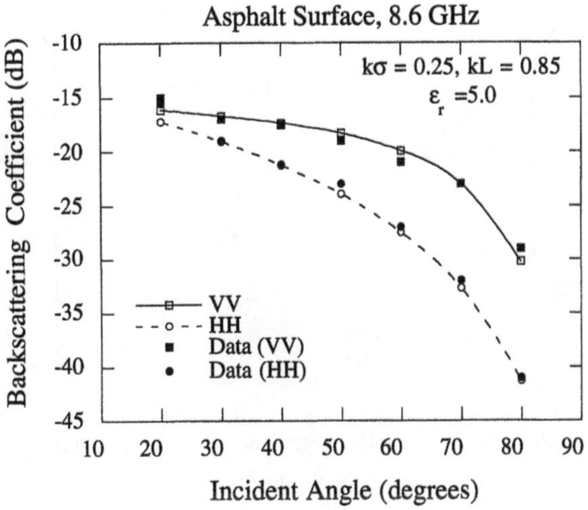

Figure 2.31 A comparison between (2.31) and backscattering data of an asphalt surface at 8.6 GHz.

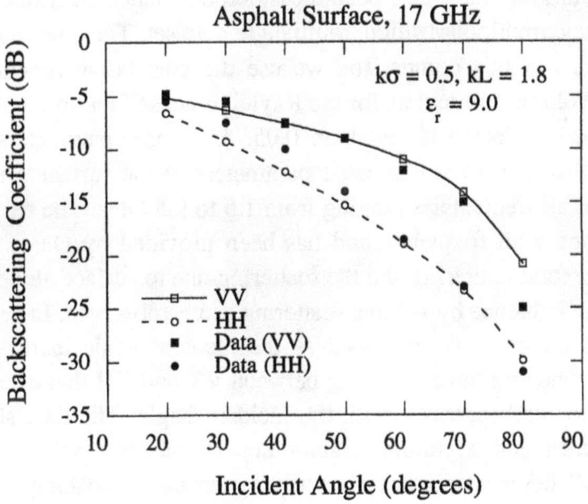

Figure 2.32 A comparison between (2.31) and backscattering data of an asphalt surface at 17 GHz.

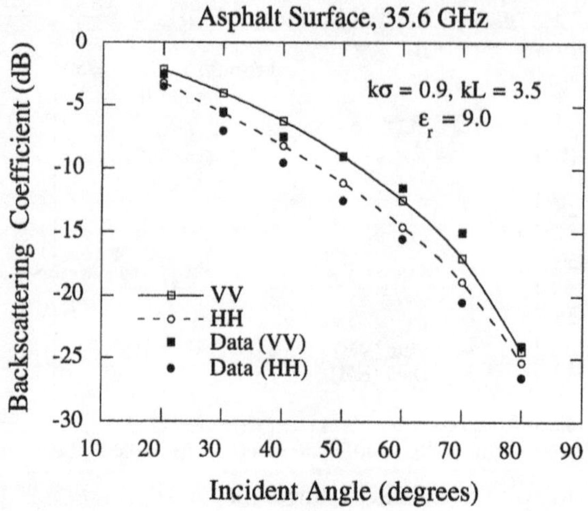

Figure 2.33 A comparison between (2.31) and backscattering data of an asphalt surface at 35.6 GHz.

In what follows we give another example of comparisons between scattering models and soil data reported by Oh et al. [1992]. This is a very special set of soil measurement because surface parameters have also been measured and hence are known. In addition, it is a multifrequency, multipolarization, multiangle data set. Thus, no model parameter selection is needed. For this comparison we use the correlation function defined as $\exp[-(\xi/L)^{1.2}]$. Volume parameters for the Rayleigh phase function, however, are not known but the albedo selected is less than 0.05. All model parameters are listed in Figure 2.34. Note that, since the geometric parameters of the surface are known, they remain the same for all frequencies ranging from 1.5 to 9.5 GHz. The surface dielectric constant does change with frequency and has been provided by Oh et al. [1992]. In Figure 2.34 both the total scattering and the scattering due to surface alone are shown so that the extent of the influence by volume scattering can be observed. In general, volume scattering becomes increasingly important as the incident angle increases. Note that surface scattering alone will have a spacing between VV and HH that decreases with an increase in frequency and increases with the incident angle. The data show a similar frequency dependence and a similar angular dependence but only up to about 60 degrees. Beyond 60 degrees there is no further increase in spacing. This seems to indicate that a different scattering mechanism has come in. With volume scattering included excellent agreement is obtained between the model predictions and the data, indicating that the model does provide the correct level and the general trend in frequency, polarization and incident angle.

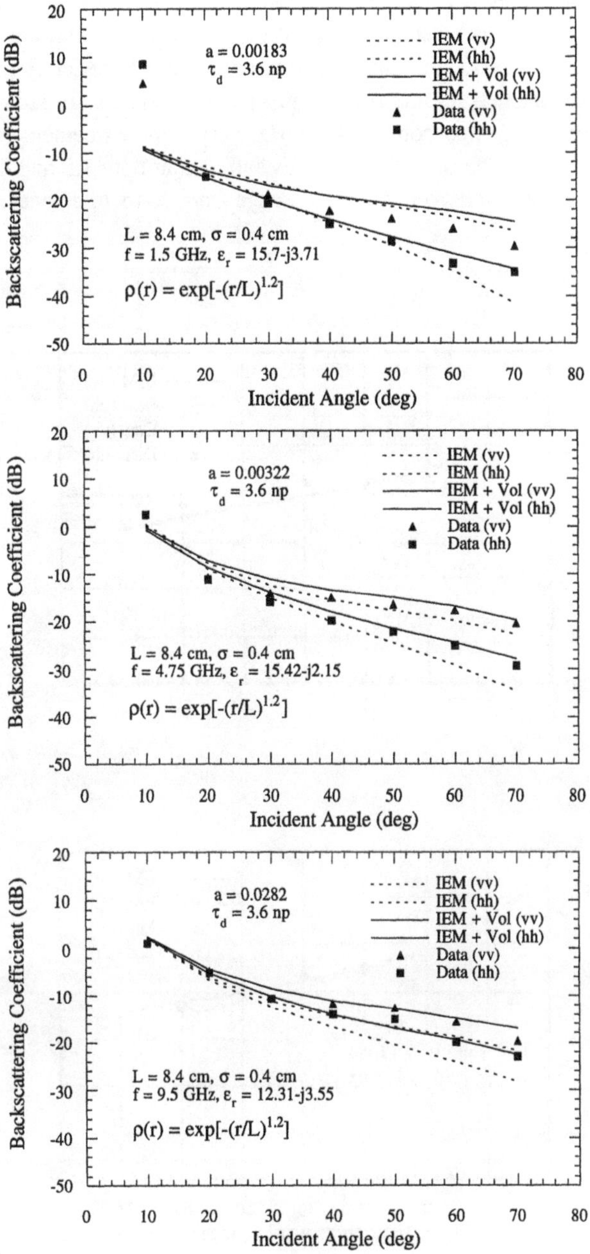

Figure 2.34 Comparison between the sum of (2.31) and (2.42) and backscattering measurements from a known soil surface.

Finally, we show a comparison between the surface model given by (2.31) and a rougher surface reported by Oh et al. [1992]. This surface also has known parameters so that selection of model parameters is again unnecessary. In addition, its surface scattering strength is high enough relative to volume scattering at each frequency that it is not necessary to consider volume scattering. The comparisons are shown in Figure 2.35. Except for the 50 degree point at 1.5 GHz, very good agreement is realized in levels and trends between the model predictions and data at both 1.5 and 4.75 GHz, in VV and HH polarizations, and over an angular range from 20 to 70 degrees.

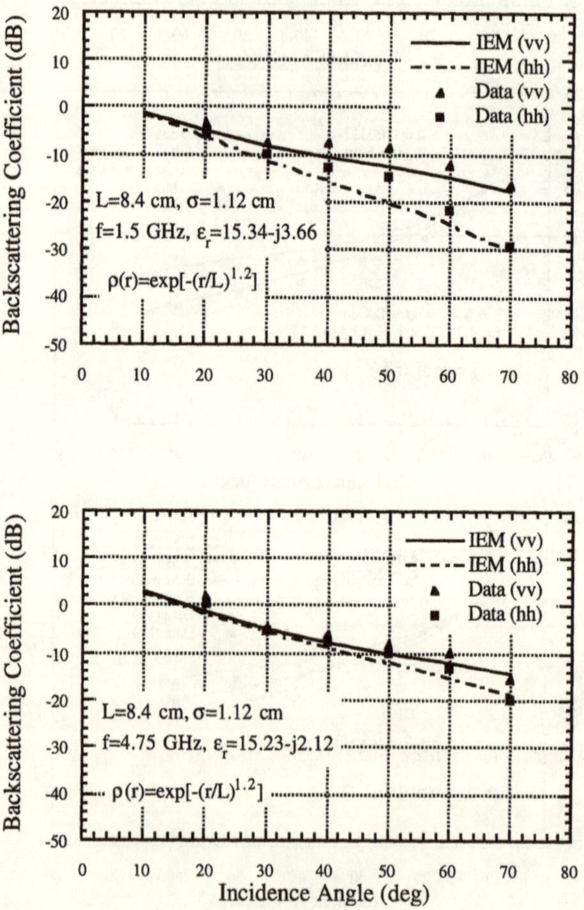

Figure 2.35 Comparison between the surface model (2.31) and backscattering measurements from a known soil surface.

2.5.4 A Layer Scattering Model and Its Applications

For backscattering from a layered medium, we shall select the significant terms in (2.24) to form a simple model. The first term in (2.24) is the surface scattering term accounting for scattering from the top boundary. We shall use (2.31) for it. The second term in (2.24) is the volume scattering term. We approximate it by (2.42). The third term is dominated by noncoherent scattering from the bottom boundary surface attenuated by the layer. We approximate it by using (2.31) and accounting for crossing of top surface boundary and attenuation due to propagation loss through the layer

$$\sigma^0_{gpp} = \cos\theta T_{1t}(\theta, \theta_t) T_{t1}(\theta_t, \theta) \exp(-2\kappa_e d/\cos\theta_t) \sigma^0_{pp}(\theta_t)/\cos\theta_t \qquad (2.43)$$

where $\sigma^0_{pp}(\theta_t)$ is the surface scattering coefficient for the lower boundary evaluated using parameters of the layer medium; $\kappa_e d$ is the optical depth of the layer; θ and θ_t are the angles of incidence and transmission from the medium above into the layer. The fourth term in (2.24) is the interaction term between volume inhomogeneities and the lower boundary of the layer. From (2.30) its backscattering coefficient is given by

$$\sigma^0_{vgpp} = (4\pi\cos\theta I^s)/I_0 = \cos\theta (L_{1t} T_{t1})^2 L_r R (\kappa_s/\kappa_e)(\kappa_e d/\cos\theta_t)$$
$$[P_{pp}(-\mu_t, -\mu_t, \phi_t - \phi_i) + P_{pp}(\mu_t, \mu_t, \phi_t - \phi_i)] \exp(-2\kappa_e d/\mu_t) \qquad (2.44)$$

where R and T_{1t} are the reflectivity and transmissivity at the bottom and top layer boundaries, respectively; κ_s/κ_e is the albedo; $\mu_t = \cos\theta_t$, $\mu_i = \cos\theta$ and L_{t1}, L_r are the loss factors due to boundary roughness given by

$$L_{t1}(-\mu_t, -\mu_i) = \exp[-\sigma_1^2 (k_{rt}\mu_t - k_r\mu_i)^2]$$
$$L_r(\mu_t, -\mu_t) = \exp[-\sigma_2^2 k_{rt}^2 (\mu_t^2 + \mu_t^2)]$$

In the above relations σ_1, σ_2 are the rms heights of the top and bottom boundaries; k_r denotes the real part of k and k_{rt} is the real part of the wave number inside the layer.

The complete backscattering model for the layer is the sum of (2.31), (2.42), (2.43) and (2.44). This model makes three major assumptions:

1. Only single scattering is important.
2. Transmission across the top boundary can be accounted for by using the Fresnel power transmission, T_{t1}.
3. Reflection at the lower boundary for the surface-volume interaction term can be calculated using the Fresnel power reflection coefficient, R.

The first assumption is usually acceptable in computing like polarized backscattering coefficients. The second assumption is expected to work when the coherently transmitted power across the top boundary is approximately the same as the total transmitted power, and this happens for low dielectric layers. The third assumption is an estimate of the total energy incident on the volume inhomogeneities from the lower boundary. In real situations both coherent and noncoherent energy are being scattered or transmitted at an irregular boundary. Thus, the theoretical estimates for the coherent energy alone using $L_r R$ or $L_t T_t$ is smaller than the actual energy. Since the loss factors are themselves estimates, in practice fairly good results are obtained by replacing $L_r R$ by R and $L_t T_t$ by T_t.

In dealing with an irregular scattering layer there are four major contributing terms. They originate from the top surface, the volume of the layer, the bottom layer boundary and the combined effect of surface and volume. Generally, the fourth term is smaller than the first three. In what follows we shall examine the conditions under which one or more of these terms will either dominate or be negligible.

Assuming that the first three terms, given by (2.31), (2.42), (2.43), make a significant contribution to the total backscattering, we expect scattering from the top and bottom surface boundaries to dominate over the small incident angular range, and the layer volume scattering to dominate over the large incident angles. An example of such a situation is shown in Figure 2.36, where the Rayleigh phase function given in Appendix 2C is used for volume scattering and an exponential correlation is used in (2.31) and (2.43) for surface scattering calculations. Both the total and each of the individual contributing terms are shown to illustrate the exact region where each term is dominating or negligible. In Figure 2.36 scattering from the top boundary surface appears to be higher than that from the lower boundary over the first 5 degrees. Then the bottom boundary contribution is the highest over about 5 degrees to 25 degrees. Beyond 25 degrees volume scattering begins to dominate. This figure illustrates the situation where all three mechanisms of scattering have their region of dominance. Clearly, this is the result of the particular choice of model parameters and medium properties. The use of other parameters may completely change the situation. In general, surface scattering has a more rapid drop off with the incident angle than volume scattering. This is the reason why volume scattering is likely to dominate only at large angles of incidence.

In Figure 2.36 if volume scattering is at a higher level, say, 5 dB more due to higher albedo value, then the bottom surface contribution will not be significant and volume scattering will dominate over the range from 10 to 70 degrees. Top surface scattering remains important between 0 and 10 degrees. In this situation only the top boundary and the layer volume scattering are important. A similar condition can occur not by having a strong volume scattering but a weak lower boundary contribution. One possible cause for this to happen is when the layer has a high attenuation, and another possible cause is when the lower boundary does not have a high discontinuity. The latter situation can be realized by lowering the permittivity of the half space from 36 to 5. This condition is illustrated in Figure 2.37.

In Figure 2.37 the scattering from the lower boundary is significantly lower than that from the top interface for all angles of incidence. The level of volume scattering by the layer exceeds that of surface scattering after 20 degrees. Hence, this is the case where we can ignore the bottom boundary altogether. Furthermore, if we are interested only in the large angular region ($\theta > 40$ degrees), we do not need to consider surface scattering. As stated earlier the same situation can be realized by having a thick lossy layer.

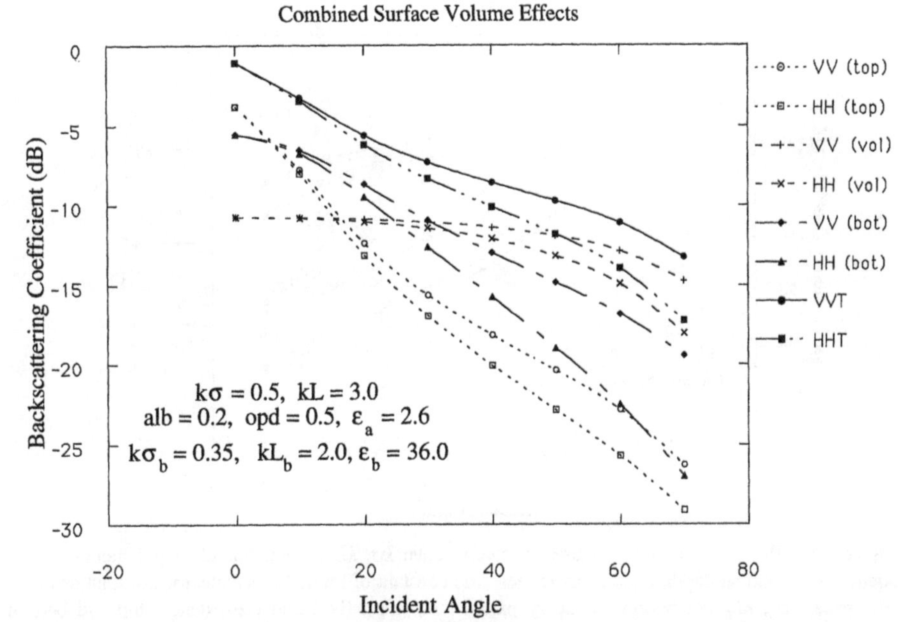

Figure 2.36 Illustration of layer scattering components: $k\sigma$, kL are top boundary parameters; *alb* = albedo, *opd* = optical depth, ε_a = average dielectric constant of layer, ε_b = dielectric constant of lower half space; $k\sigma_b$, kL_b are bottom boundary parameters; (top), (bot) and (vol) denote top and bottom boundaries and layer volume, respectively.

Examples that may parallel the situation depicted in Figure 2.36, are a thick saline ice layer and a moist ground surface. These are cases where scattering from the bottom layer boundary has been attenuated by the loss in the layer.

Another interesting case is when the bottom rather than the top layer boundary dominates scattering at small angles of incidence. This happens when the layer has a small average dielectric constant such as a snow layer above an irregular ground surface. The same situation may also occur for an inhomogeneous ice layer even though its dielectric constant may be twice that of snow. The requirement here is for the ice layer to be thin and the bottom layer interface to have a large dielectric discontinuity. This is the case when there is a thin ice layer floating above water or a thick lake ice layer that has a low loss. These situations are not uncommon in nature. An illustration of such a case is shown in Figure 2.38, where the top boundary scattering is seen to be at a lower level than that from the lower boundary over all angles and is considerably lower than the volume scattering term after the incident angle exceeds 20 degrees. The change made here relative to Figure 2.36 is in raising the dielectric constant of the lower half space from 5 to 25.

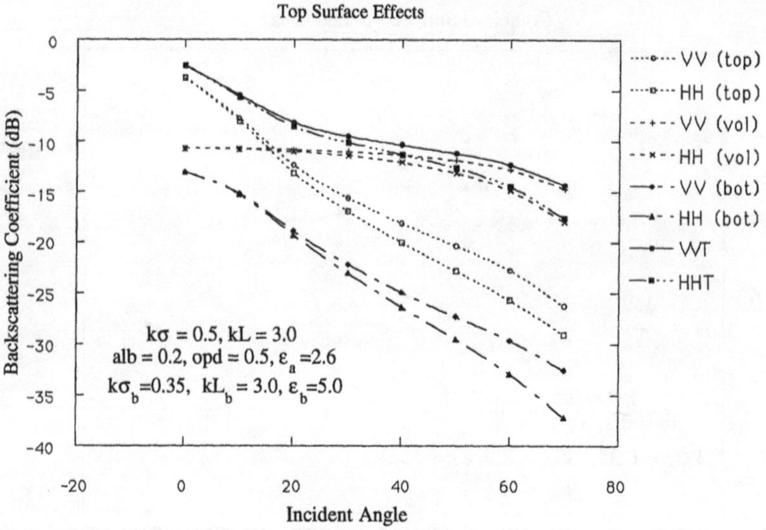

Figure 2.37 Illustration of layer scattering components: $k\sigma$, kL are top boundary parameters; *alb* = albedo, *opd* = optical depth, ε_a = average dielectric constant of layer, ε_b = dielectric constant of lower half space; $k\sigma_b$, kL_b are bottom boundary parameters; (top), (bot) and (vol) denote top and bottom boundaries and layer volume, respectively.

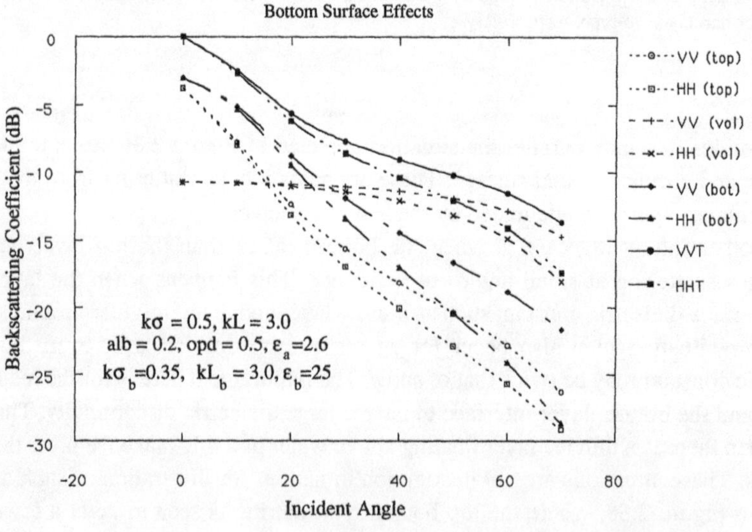

Figure 2.38 Illustration of layer scattering components: $k\sigma$, kL are top boundary parameters; *alb* = albedo, *opd* = optical depth, ε_a = average dielectric constant of layer, ε_b = dielectric constant of lower half space; $k\sigma_b$, kL_b are bottom boundary parameters; (top), (bot) and (vol) denote top and bottom boundaries and layer volume, respectively.

We have seen cases where the top surface scattering alone dominates. This happens often in scattering from natural soil surfaces considered in the early part of this section. It is also possible to have a case dominated by volume backscattering. A thick snow layer is an example. Another example is the problem of scattering from low density sea ice, where the layer dielectric is smaller than the regular ice and is highly inhomogeneous. Thus, surface scattering is not particularly strong. Furthermore, many ice surfaces are fairly smooth around 5 GHz region. Thus, a strong surface reflection occurs in the specular direction and not much energy is scattered by the surface boundary in the backscattering direction. Examples where the bottom surface boundary is important are when we have dry snow covered ice, snow covered ground and vegetation covered ground. As stated before surface scattering generally dominates only over small angles of incidence when there is a significant amount of volume scattering.

Comparisons with Measurements

The simplified layer scattering model defined as the sum of (2.31), (2.42), (2.43) and (2.44) is now applied to a set of backscattering measurements from a saline ice layer above saltwater. These are laboratory measurements taken at 5 and 10 GHz with horizontal and vertical polarizations over the angular range 10 to 60 degrees. Most of the physical parameters of this ice layer have been reported and are summarized in Table 2.1. Other quantities that depend on salinity, temperature and frequency such as the layer dielectric constant and saline water permittivity can be estimated from the formulae given in Appendix E of Ulaby et al. [1986]. Several of these parameters are known to lie in certain ranges. For example, the ice layer is not uniform and hence its thickness lies in a range depending on spatial location. In this situation an average value is selected. Similar statements can be made about the surface rms height and correlation length. The specific values of these parameters used in modeling are given in the figures showing the comparisons between model predictions and data.

Table 2.1
Ice Layer Parameters

Ice Parameters	*Measured/Observed Values*
Thickness	6.5–8 cm
Air-ice interface rms height, σ	0.02–0.048 cm
Air-ice interface correlation length, L	0.669–1.77 cm
Air-ice interface temperature	-16°C
Ice-water interface temperature	-1°C
Ice salinity	7–11.5 ‰
Water salinity	24 ‰

The surface correlation function selected here is again the exponential function. The comparison with measurements at 5 GHz is shown in Figure 2.39. In addition, the individual components contributing to the total backscattering are also shown. Note that the volume scattering term given in the figure includes surface-volume interaction term. The results in the figure indicate that scattering is dominated by the lower boundary of the layer at small angles of incidence and both layer volume scattering and scattering from the top boundary make a contribution at large angles of incidence. The reason why this happens is because the scattering from both the top boundary and the layer volume are varying much more slowly with the incident angle than that from the bottom boundary. The agreement between horizontally polarized prediction and data is excellent. In vertical polarization the prediction is lower but is within 1 dB of the data.

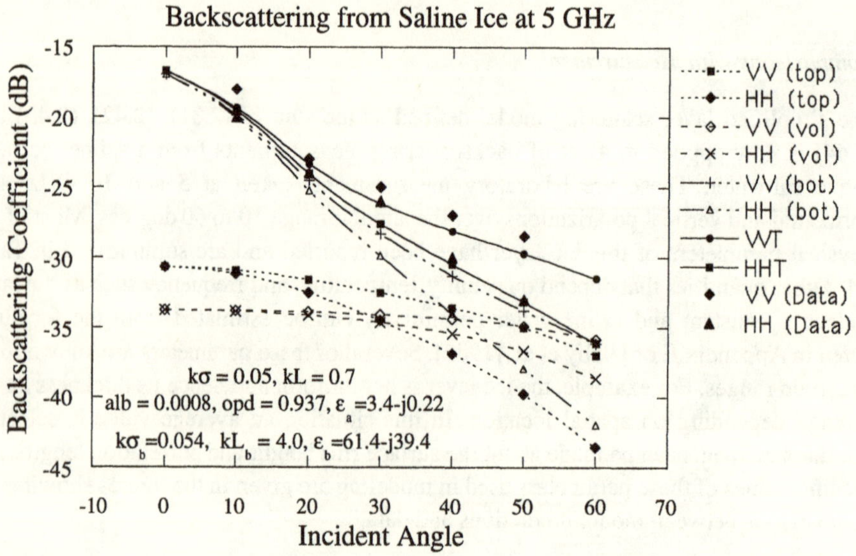

Figure 2.39 Comparisons between the layer model and saline ice measurements at 5 GHz.

Next, we consider backscattering at 10 GHz. Here, all parameter change follows their frequency variation (Figure 2.40). Except at 10 degrees incidence, model prediction for horizontal polarization is again in excellent agreement with data. The vertically polarized prediction is lower than data at some angles. Results indicate that scattering is dominated by the top boundary for most angles, except at 10 degrees where lower boundary is also a factor, and 60 degrees, where volume scattering also contributes.

Another set of comparisons involve a 12 cm saline ice layer above water with and without some snow cover. Measurements have been obtained for bare ice, ice with 5 cm and 12 cm snow cover [Gogineni, private communication]. The interesting question about such a problem is what happens to the ice when there is snow cover. It is believed that brine wicks up into the snow and the snow-ice interface. This action could cause both a rise in the dielectric discontinuity of the snow-ice interface and a roughening effect. At this point in time it is difficult to verify which effect is dominant or if both should be acting. To illustrate this issue comparisons between model predictions and data are shown in Figures 2.41 through 2.45 with the assumption that the change in dielectric constant rather than roughness is important, and then in Figures 2.46 and 2.47 we assume that only the roughening effect is important. We shall see that either assumption leads to a reasonable overall fit to the data but they do carry different implications which have to be resolved by well controlled experimental investigations.

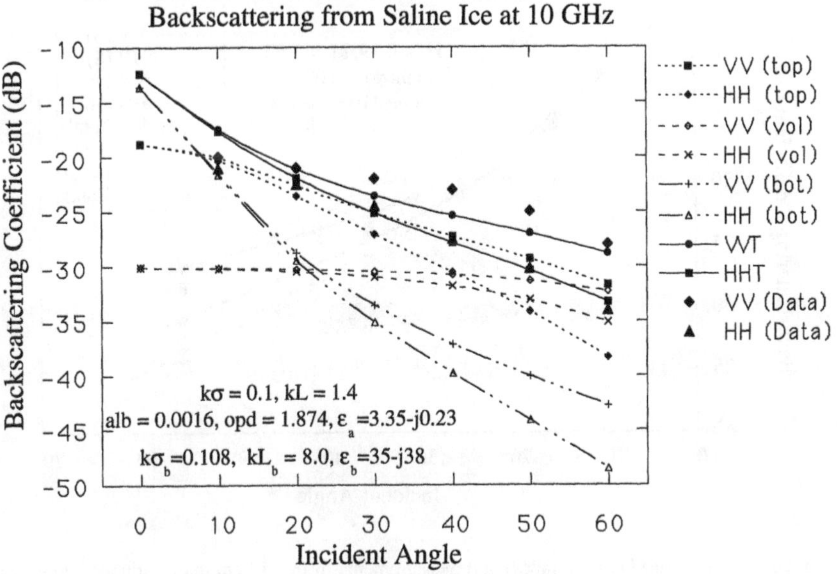

Figure 2.40 Comparisons between the layer model and saline ice measurements at 10 GHz.

The case under consideration is a 12 cm thick saline ice layer with a normalized rms height of 0.19 and a correlation length of 3.92 at 13.5 GHz. The surface correlation is known to be closely exponential based on calculations from several of its profiles. The ice relative permittivity is 3.38 - *j* 0.23. When we choose an albedo of 0.0015 we obtain an excellent fit to the data, as shown in Figure 2.41. The optical depth of the ice is so thick that the scattering acts as if it is from a half space. In Figure 2.41 we show surface contribution from air-ice interface and volume scattering in addition to the total scattering. It is seen that over small angles of incidence surface scattering is dominating leading to VV higher than HH in backscattering. At large angles (30 degrees or more) volume scattering begins to make a contribution resulting in narrowing the spacing between VV and HH polarizations. Note that data collected at 10 degrees and 15 degrees should not have a significant difference between VV and HH polarizations. The reversal in VV and HH at these small angles indicate instability in data that may result from a small number of samples, antenna pattern effects, background effect, etc. What is indicative, however, is that at large angles of incidence the data do not show a large separation between VV and HH polarization. If this is so, then it implies that the scattering is not just by surface. In addition, we note that beyond 30 degrees VV is always higher than HH. This indicates that surface scattering is dominating but some volume scattering effect is present because the separation between these polarizations is smaller and the level of scattering is higher than what should happen with a pure surface scattering phenomenon.

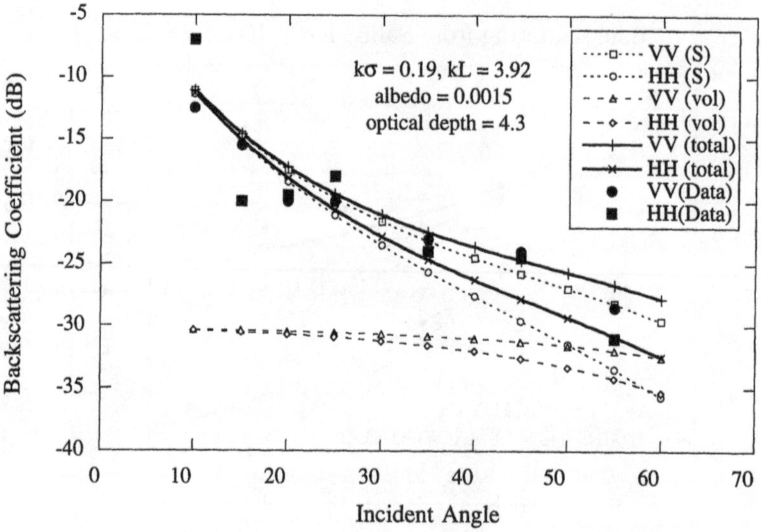

Figure 2.41 Comparisons between model and measurements from a 12 cm thick saline ice layer with a rough top boundary.

In Figure 2.42 we show model comparisons with data from the same ice with a 12 cm thick snow cover. In this case all ice surface parameters remain unchanged except for the ice surface permittivity, which has been increased from 3.38 - j0.23 to 4.38 - j4.88 to account for the wicking effect of the brine. In Figure 2.42 we show the same $k\sigma$ and kL for the snow-ice boundary as those in Figure 2.41 which are for the air-ice boundary. In actual computation the wave number in air must be changed to that in snow. Note that beyond 15 degrees the measured HH is always higher than VV, a reversal from the polarization behavior shown in Figure 2.41 over large angles of incidence. Theoretically, this comes about because of surface-volume interaction resulting from scattering by the snow layer and the snow-ice interface. Such an interaction raises HH but not VV because of the Brewster angle effect. This can happen only if the snow-ice interface is fairly smooth, generating significant coherent scattering. Because snow has a relative permittivity of 1.67- j 0.00026 the air snow interface does not contribute much to scattering. To see how good the fit is we replot the same result in Figure 2.43 without the air-snow scattering term. It is clear from Figure 2.43 that the agreement between model and data is excellent. To demonstrate what happens if we reject the idea that there is wicking of the brine by keeping the ice permittivity the same as without snow cover, then we obtain the fit as shown in Figure 2.44. In this case the model prediction is clearly significantly lower than the data especially at small angles of incidence where surface scattering should dominate the return. At large angles there is the support from volume scattering and therefore the general agreement is not affected very much. Thus, if the wicking of the brine does not show up as causing a higher permittivity, it must have increased surface roughness in some way. Finally, we further justify the consistency in our modeling interpretation by considering the same ice with a 5 cm thick snow cover. In this case we only change the optical depth of snow by the same proportion as the change in physical depth, i.e., from 0.05 to 0.021. All other model parameters remain the same as those shown in Figure 2.42. The resulting fit is given in Figure 2.45. Again excellent agreement is obtained.

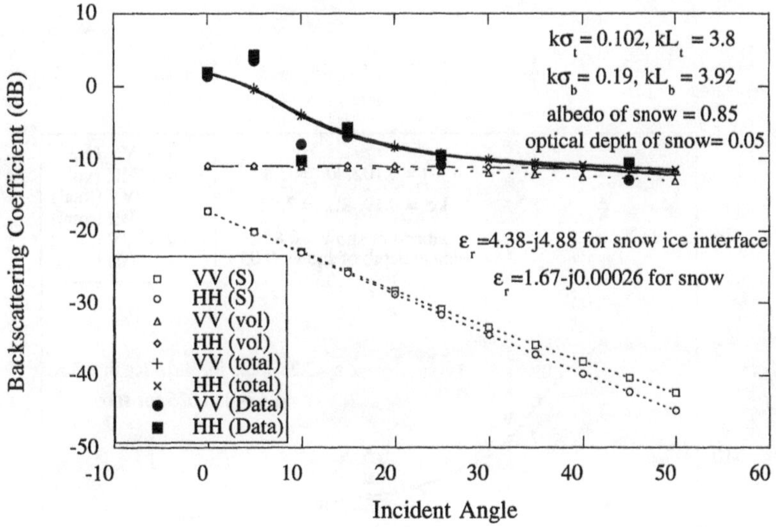

Figure 2.42 Comparisons between model and measurements from a 12 cm thick saline ice layer with a 12 cm snow cover using a larger dielectric constant.

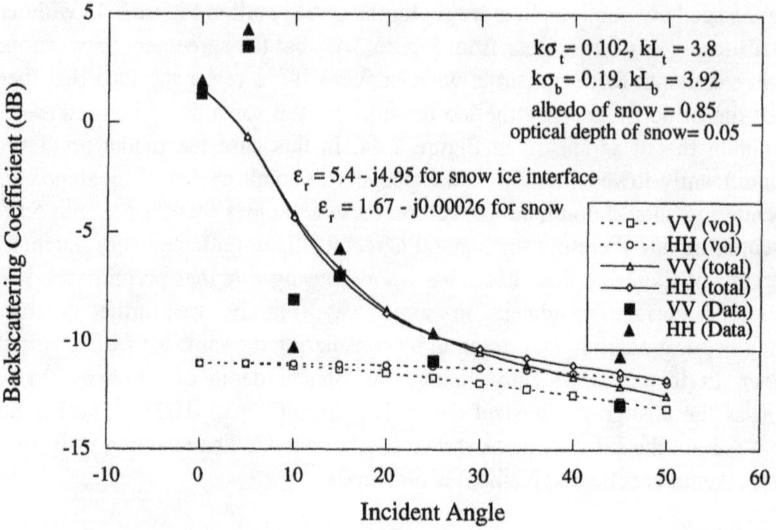

Figure 2.43 Comparisons between model and measurements from a 12 cm thick saline ice layer with a 12 cm snow cover without showing the top surface contributions.

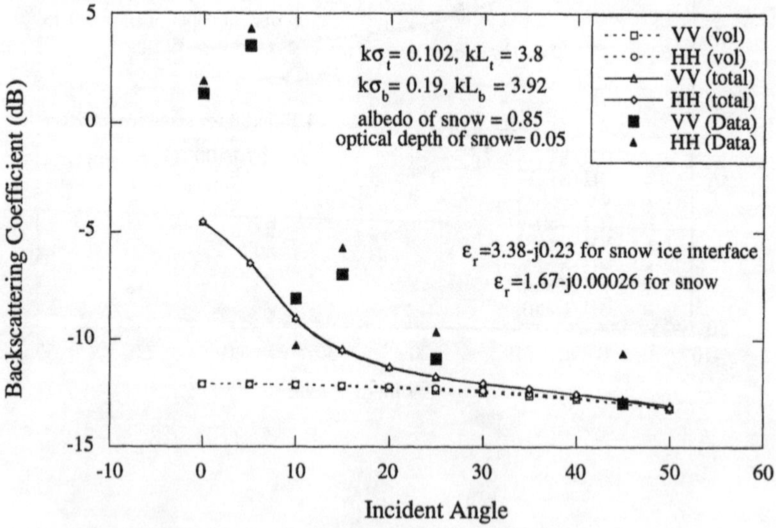

Figure 2.44 Comparisons between model and measurements from a 12 cm thick saline ice layer with a 12 cm snow cover using the bare ice dielectric constant not including wicking effect.

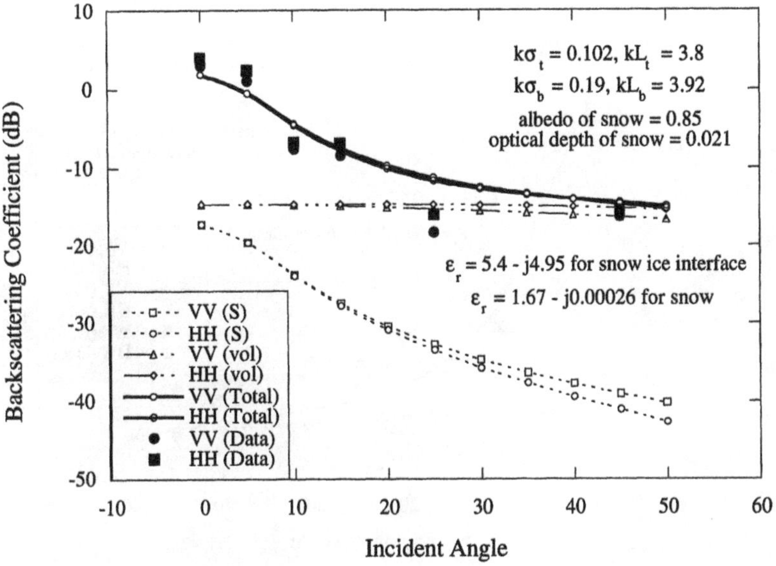

Figure 2.45 Comparisons between model and measurements from a 12 cm thick saline ice layer with a 5 cm snow cover using a larger dielectric constant.

Next, we assume that the wicking effect leads to a rougher snow-ice interface rather than a higher dielectric constant. This assumption comes about because measurements of coherent return at normal incidence indicate a decrease of about 7 dB when there is snow cover relative to the bare ice. Note that the addition of snow to an ice surface introduces two effects: (a) it reduces the dielectric discontinuity at the snow-ice interface and (b) it magnifies roughness by a factor equal to the square root of the snow permittivity. However, these two effects cause a decrease of only 4.8 dB when we keep all surface parameters unchanged. It turns out that if we increase $k\sigma$ from 0.19 to 0.33 and kL by the same proportion, we will have a decrease of 7 dB in coherent scattering with snow cover, and the resulting model predictions are in good agreement with data as shown in Figure 2.46 for the 12 cm snow cover. In part this is because there is a significant fluctuation in the data over a small range of angles thus leaving the angular trend with more latitude. At any rate, there is a very good overall agreement between model predictions and measurements. The volume scattering effect indicated in the figure is similar to that in Figure 2.43. It shows that volume scattering is very important after 30 degrees. To further test the consistency of the model we decrease the physical depth of the snow layer from 12 to 5 cm (this corresponds to a decrease of optical depth from 0.05 to 0.021) and keep all other parameters unchanged. We realize a fit with data given in Figure 2.47, which also looks very good.

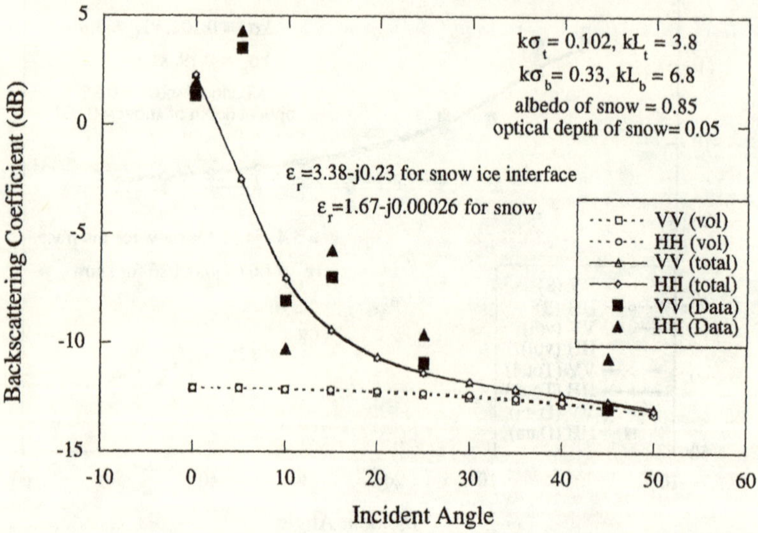

Figure 2.46 Comparisons between model and measurements from a 12 cm thick saline ice layer with a 12 cm snow cover using larger roughness parameters.

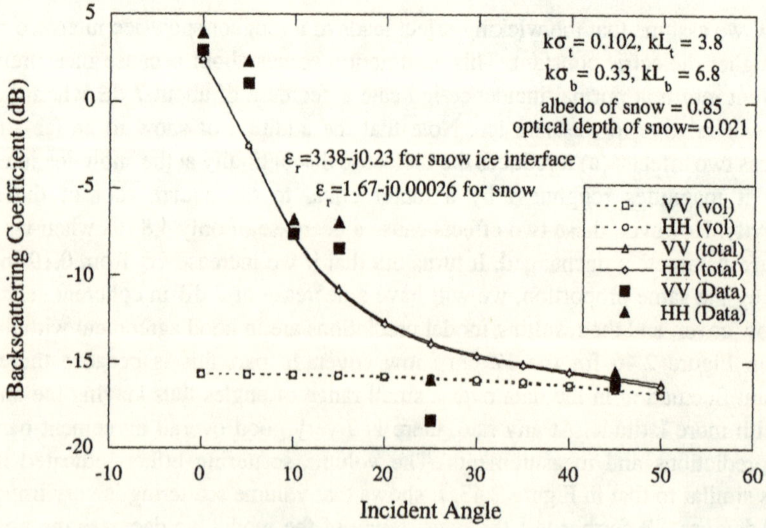

Figure 2.47 Comparisons between model and measurements from a 12 cm thick saline ice layer with a 5 cm snow cover using larger roughness parameters.

When we compare Figure 2.46 with Figure 2.43 we see that there are significant differences in the meaning of the predictions. First of all, there is a very small separation between VV and HH when the wicking effect is to generate more surface roughness. Secondly, the VV polarization remains higher than HH polarization. Thirdly, with rougher surface conditions, the angular drop off is much faster, suggesting that at large angles of incidence volume scattering should be more important.

In summary, we believe that surface scattering is the dominant mechanism for saline ice at small angles of incidence. At large incident angles volume scattering can become significant. For snow covered sea ice there is wicking of the brine, causing an increase in either permittivity discontinuity or rougher condition at the snow-ice interface. This uncertainty can be resolved by conducting controlled experiments to determine the polarization characteristics of the data and to obtain better defined angular trends at small angles of incidence. If there is a definite flip over of polarization in going from bare to snow covered ice, then there must be an increase in the boundary discontinuity coupled with a smaller angular drop-off between 10 and 20 degrees. On the other hand, if there is negligible difference between polarizations and there is a large angular drop-off over small angles of incidence, then surface roughness must have increased.

REFERENCES

Chen, M.F., and A.K. Fung, "A Numerical Study of the Regions of Validity of the Kirchhoff and Small Perturbation Rough Surface Models," *Radio Science,* Vol. 23, no. 2, 1988, pp. 163–170.

Dubois, P. C., and J.J. van Zyl, "Polarization Filtering of SAR Data," *IGARSS'89 Digest*, Vol. 3, 1989, pp. 1816–1819.

Fung, A.K., "Estimation of Random Surface Parameters," *Electromagnetics*, Vol. 4, 1984, pp. 229–243.

Nance, C. E., A.K. Fung, J.W. Bredow, "Comparison of Integral Equation Predictions and Experimental Backscattering Measurements from Random Conducting Rough Surfaces," *IGARSS'90 Digest*, Vol. 1, 1990, pp. 477–480.

Nance, C. E., "Scattering and Image Analysis of Conducting Rough Surfaces," Ph.D. Dissertation, E.E. Department, University of Texas at Arlington, 1992.

Oh, Y., K. Sarabandi and F.T. Ulaby, "An Empirical Model and an Inversion Technique for Radar Scattering from Bare Soil Surfaces," *IEEE Trans. Geoscience and Remote Sensing*, Vol. 30, no. 2, 1992, pp. 370–381.

Swartz, A.A., H.A. Yueh, J.K. Kong, L.M. Novak and R.T. Shin, "Optimal Polarizations for Achieving Maximum Contrast in Radar Images," *J. Geophysical Research*, Vol. 93, no. B12, 1988, pp. 15252–15260.

Ulaby, F.T., R.K. Moore and A.K. Fung, *Microwave Remote Sensing*, Vol. 2, Artech House, Dedham, MA, 1982.

Ulaby, F.T., R.K. Moore and A.K. Fung, *Microwave Remote Sensing*, Vol. 3, Artech House, Dedham, MA, 1986.

Appendix 2A

Summary of a Bistatic Surface Model

This appendix provides a summary of the surface scattering model based on the integral equation method developed in Chapter 4. Only the portion of the model accounting for single scattering is given below. Shadowing effects are not included.

2A.1 ROUGH SURFACE SCATTERING MATRICES

The bistatic single-scatter scattering coefficient matrix in the upper medium is given by

$$\sigma^{\circ} = \frac{4\pi}{A}\langle \mathbf{S} \rangle \tag{2A.1}$$

where A is the illuminated area and the elements of $\langle \mathbf{S} \rangle$ are equal to

$$\begin{bmatrix} \langle |S_{vv}|^2 \rangle & \langle |S_{vh}|^2 \rangle & \mathrm{Re}\langle S_{vv} S_{vh}{*} \rangle & -\mathrm{Im}\langle S_{vv} S_{vh}{*} \rangle \\ \langle |S_{hv}|^2 \rangle & \langle |S_{hh}|^2 \rangle & \mathrm{Re}\langle S_{hv} S_{hh}{*} \rangle & -\mathrm{Im}\langle S_{hv} S_{hh}{*} \rangle \\ 2\mathrm{Re}\langle S_{vv} S_{hv}{*} \rangle & 2\mathrm{Re}\langle S_{vh} S_{hh}{*} \rangle & \mathrm{Re}\langle S_{vv} S_{hh}{*} + S_{vh} S_{hv}{*} \rangle & \mathrm{Im}\langle S_{vh} S_{hv}{*} - S_{vv} S_{hh}{*} \rangle \\ 2\mathrm{Im}\langle S_{vv} S_{hv}{*} \rangle & 2\mathrm{Im}\langle S_{vh} S_{hh}{*} \rangle & \mathrm{Im}\langle S_{vv} S_{hh}{*} + S_{vh} S_{hv}{*} \rangle & \mathrm{Re}\langle S_{vv} S_{hh}{*} - S_{vh} S_{hv}{*} \rangle \end{bmatrix}$$

Each of the above matrix elements has the form

$$\langle S_{qp} S_{rs}{*} \rangle = \frac{k^2 A}{8\pi} \exp\left[-\sigma^2 (k_z^2 + k_{sz}^2)\right] \sum_{n=1}^{\infty} \sigma^{2n} (I_{qp}^n I_{rs}^{n\,*}) \frac{W^{(n)}(k_{sx} - k_x, k_{sy} - k_y)}{n!} \tag{2A.2}$$

where

$$I_{\alpha\beta}^n = (k_{sz} + k_z)^n f_{\alpha\beta} \exp(-\sigma^2 k_z k_{sz}) + \frac{(k_{sz})^n F_{\alpha\beta}(-k_x, -k_y) + (k_z)^n F_{\alpha\beta}(-k_{sx}, -k_{sy})}{2} \tag{2A.3}$$

In (2A.3) the Kirchhoff field coefficients, $f_{\alpha\beta}$, are given by (2A.4) through (2A.7),

$$f_{vv} = \frac{2R_{\parallel}}{\cos\theta + \cos\theta_s} [\sin\theta\sin\theta_s - (1+\cos\theta\cos\theta_s)\cos(\phi_s - \phi)] \tag{2A.4}$$

$$f_{hh} = -\frac{2R_{\perp}}{\cos\theta + \cos\theta_s} [\sin\theta\sin\theta_s - (1+\cos\theta\cos\theta_s)\cos(\phi_s - \phi)] \tag{2A.5}$$

$$f_{hv} = 2R\sin(\phi_s - \phi) \tag{2A.6}$$

$$f_{vh} = 2R\sin(\phi - \phi_s) \tag{2A.7}$$

where $R = (R_{\parallel} - R_{\perp})/2$. For simplicity of writing we shall use the following notations in writing the complementary field coefficients, F_{qp}:

$$\begin{aligned}
s &= \sin\theta, \quad ss = \sin\theta_s \\
cs &= \cos\theta, \quad css = \cos\theta_s \\
sf &= \sin(\phi_s - \phi), \quad csf = \cos(\phi_s - \phi) \\
sq &= (\mu_r\varepsilon_r - \sin^2\theta)^{1/2}, \quad sqs = (\mu_r\varepsilon_r - \sin^2\theta_s)^{1/2} \\
r &= (sq\ css)/(sqs\ cs) \\
\mu_r &= \mu_t/\mu, \quad \varepsilon_r = \varepsilon_t/\varepsilon
\end{aligned}$$

$$\begin{aligned}
c_1 &= (csf - s\ ss)/(sq\ css) \\
c_{1s} &= (csf - s\ ss)/(sqs\ cs) \\
c_2 &= s(ss - s\ csf)/css \\
c_{2s} &= ss(s - ss\ csf)/cs \\
T_v &= 1 + R_{\parallel}, \quad T_{vm} = 1 - R_{\parallel} \\
T_h &= 1 + R_{\perp}, \quad T_{hm} = 1 - R_{\perp} \\
T_p &= 1 + R, \quad T_m = 1 - R
\end{aligned}$$

Using the above notations we can write the complementary field coefficients as

$$F_{vv}(-k_x, -k_y) = -(csT_{vm} - sqT_v/\varepsilon_r)(T_v csf + T_{vm}\varepsilon_r c_1) + (T_{vm}^2 - csT_vT_{vm}/sq)\,c_{2s} \tag{2A.8}$$

$$F_{hh}(-k_x, -k_y) = (csT_{hm} - sqT_h/\mu_r)(T_h csf + T_{hm}\mu_r c_1) - (T_{hm}^2 - csT_hT_{hm}/sq)\,c_{2s} \tag{2A.9}$$

$$F_{hv}(-k_x, -k_y) = (csT_m - sqT_p/\varepsilon_r)(T_p/css + T_m\varepsilon_r/sq)\,sf + (T_m^2 - csT_pT_m/sq)\,s^2 sf \tag{2A.10}$$

$$F_{vh}(-k_x, -k_y) = (csT_p - sqT_m/\mu_r)(T_m/css + T_p\mu_r/sq)\,sf + (T_p^2 - csT_pT_m/sq)\,s^2 sf \tag{2A.11}$$

$$F_{vv}(-k_{xs}, -k_{sy}) = -(cssT_{vm} - sqsT_v/\varepsilon_r)(T_v csf + T_{vm}\varepsilon_r c_{1s}) + (T_v^2 - cssT_vT_{vm}/sqs)\,c_{2s} \tag{2A.12}$$

$$F_{hh}(-k_{xs}, -k_{sy}) = (cssT_{hm} - sqsT_h/\mu_r)(T_h csf + T_{hm}\mu_r c_{1s}) - (T_h^2 - cssT_hT_{hm}/sqs)\,c_{2s} \tag{2A.13}$$

$$F_{hv}(-k_{sx}, -k_{sy}) = -(cssT_p - sqsT_m/\mu_r)(T_m/cs + T_p\mu_r/sqs)\,sf - (T_m^2 - cssT_pT_m/sqs)\,ss^2 sf \tag{2A.14}$$

$$F_{vh}(-k_{sx}, -k_{sy}) = -(cssT_m - sqsT_p/\varepsilon_r)(T_p/cs + T_m\varepsilon_r/sqs)\,sf - (T_p^2 - cssT_pT_m/sqs)\,ss^2 sf \tag{2A.15}$$

$$R_{\parallel} = \frac{\varepsilon_r cs - sq}{\varepsilon_r cs + sq} \tag{2A.16}$$

$$R_{\perp} = \frac{\mu_r cs - sq}{\mu_r cs + sq} \tag{2A.17}$$

$$R_r = (1 - R)/(1 + R) \tag{2A.18}$$

$$k_x = k\sin\theta\cos\phi;\; k_y = k\sin\theta\sin\phi;\; k_z = k\cos\theta \tag{2A.19}$$

$$k_{sx} = k\sin\theta_s\cos\phi_s;\; k_{sy} = k\sin\theta_s\sin\phi_s;\; k_{sz} = k\cos\theta_s \tag{2A.20}$$

2A.2 TRANSMISSION PHASE MATRICES

For transmission a similar set of expressions exist for the elements of the scattering amplitude matrix. The main expression corresponding to (2A.2) is

$$\langle S_{qp}S_{rs}{}^*\rangle = \frac{k_t^2 A}{8\pi\eta_r}\exp[-\sigma^2(k_z^2 + k_{tz}^2)]\sum_{n=1}^{\infty}\sigma^{2n}(I_{qp}^n I_{rs}^{n*})\frac{W^{(n)}(k_{tx} - k_x, k_{ty} - k_y)}{n!} \tag{2A.21}$$

where

$$I_{\alpha\beta}^{n} = (k_z - k_{tz})^n f_{t\alpha\beta} \exp(\sigma^2 k_z k_{tz})$$

$$+ \frac{(-k_{tz})^n F_{t\alpha\beta}(-k_x, -k_y) + (k_z)^n F_{t\alpha\beta}(-k_{tx}, -k_{ty})}{2} \quad \text{(2A.22)}$$

and $\eta_r = \eta_t/\eta$ is the ratio of the intrinsic impedances of the lower to the upper medium; $k_{tz} = k_t \cos\theta_t$ and θ_t is measured from the negative z-axis in the lower medium. Let us introduce the following notations for some quantities in the lower medium similar to what we did in the previous section:

$$st = \sin\theta_t, \quad cst = \cos\theta_t$$

$$sf = \sin\phi_t, \quad csf = \cos\phi_t$$

$$Z_x = \frac{k_t st\ csf - ks}{k_t cst - kcs}$$

$$Z_y = \frac{k_t st\ sf}{k_t cst - kcs}$$

$$a = (cs + sZ_x)\, csf + sf\ s\ Z_y$$

$$b = cst\ csf + Z_x st$$

$$c = cst\ sf\,(cs + sZ_x) + Z_y\,(st\ cs - s\ cst\ csf)$$

$$\text{rem} = (\mu_r \varepsilon_r)^{1/2}$$

$$qt = \text{rem}\ cst, \quad sqt = (1 - \mu_r \varepsilon_r \sin^2\theta_t)^{1/2}$$

In Appendix 4D the Kirchhoff field coefficients are given as

$$f_{tvv} = aT_{vm} + bT_v \eta_r \quad \text{(2A.23)}$$

$$f_{thh} = -bT_h - aT_{hm}\eta_r \quad \text{(2A.24)}$$

$$f_{thv} = cT_m + sf\ T_p \eta_r \quad \text{(2A.25)}$$

$$f_{tvh} = sf\ T_m + cT_p \eta_r \quad \text{(2A.26)}$$

$$a_1 = (st\ s - \text{rem}\ csf) / (sq\ cst)$$

$$a_2 = s\,(st - s\ csf/\text{rem}) / cst$$

$$a_{1t} = (\text{rem}\ st\ s - csf) / (sqt\ cs)$$

$$a_{2t} = st\,(s - st\ csf\ \text{rem}) / cs$$

Using the above notations we can write the complementary field coefficients as

$$F_{tvv}(-k_x, -k_y) = -(csT_{vm} - sqT_v/\varepsilon_r)(T_v csf + T_{vm}a_1) - (T_{vm}^2 - csT_vT_{vm}/sq)\,a_2 \tag{2A.27}$$

$$F_{thh}(-k_x, -k_y) = [(csT_{hm} - sqT_h/\mu_r)(T_h csf + T_{hm}a_1)](\text{rem}/\varepsilon_r) + (T_{hm}^2 - csT_hT_{hm}/sq)\,a_2(\text{rem}/\varepsilon_r) \tag{2A.28}$$

$$F_{thv}(-k_x, -k_y) = (csT_m - sqT_p/\varepsilon_r)(T_m\text{rem}/sq - T_p/cst)\,sf - (T_m^2 - csT_pT_m/sq)\,s^2 sf/\text{rem} \tag{2A.29}$$

$$F_{tvh}(-k_x, -k_y) = [(csT_p - sqT_m/\mu_r)(T_p\text{rem}/sq - T_m/cst)]\,sf\ \text{rem}/\varepsilon_r - (T_p^2 - csT_pT_m/sq)\,s^2 sf/\varepsilon_r \tag{2A.30}$$

$$F_{tvv}(-k_{xs}, -k_{sy}) = -(sqtT_{vm} - qtT_v/\varepsilon_r)(T_v csf + T_{vm}a_{1t}) - (cstT_v^2/sqt - T_vT_{vm}/\text{rem})\,\mu_r a_{2t} \tag{2A.31}$$

$$F_{thh}(-k_{xs}, -k_{sy}) = (sqtT_{hm} - qtT_h/\mu_r)(T_h csf + T_{hm}a_{1t})\,\text{rem}/\varepsilon_r + (cstT_h^2/sqt - T_hT_{hm}/\text{rem})\,\text{rem}\ a_{2t} \tag{2A.32}$$

$$F_{thv}(-k_{sx}, -k_{sy}) = (sqtT_p - qtT_m/\mu_r)(T_p/sqt - T_m/cs)\,\mu_r sf/\text{rem} + (qtT_m^2/sqt - T_mT_p)\,\text{rem}\ st^2 sf \tag{2A.33}$$

$$F_{tvh}(-k_{sx}, -k_{sy}) = (sqtT_m - qtT_p/\varepsilon_r)(T_m/sqt - T_p/cs)\,sf + (qtT_p^2/sqt - T_mT_p)\,\mu_r st^2 sf \tag{2A.34}$$

$$k_x = k\sin\theta\cos\phi;\ k_y = k\sin\theta\sin\phi;\ k_z = k\cos\theta \tag{2A.35}$$

$$k_{tx} = k_t\sin\theta_t\cos\phi_t;\ k_{ty} = k_t\sin\theta_t\sin\phi;\ k_{tz} = k_t\cos\theta_t \tag{2A.36}$$

Appendix 2B

Surface Correlation Functions and Roughness Spectra

Three commonly used surface correlation functions, some of their sums, and their associated roughness spectra are given below.

2B.1 GAUSSIAN CORRELATION FUNCTION

For an isotropically rough surface a single-parameter Gaussian correlation function normalized to the surface height variance is of the form

$$\rho(\xi) = \exp\left[-\frac{\xi^2}{L^2}\right] \tag{2B.1}$$

If a random surface is generated by a Gaussian process, its derivative (i.e., its slope) is also Gaussian and its slope variance σ_s^2 is related to the second derivative of its correlation at the origin:

$$\sigma_s^2 = -\sigma^2 \frac{d^2}{d\xi^2}\rho(\xi)\bigg|_{\xi=0} = \frac{2\sigma^2}{L^2} \tag{2B.2}$$

The existence of a surface slope distribution, therefore, requires the existence of the second derivative of the correlation function at the origin. Many studies have shown that in the high frequency limit the backscattering coefficient is proportional to the surface slope distribution. When a surface correlation is not differentiable, the surface must have discontinuities, and it is not possible to apply such a correlation function to do high frequency analysis.

The roughness spectrum corresponding to a Gaussian correlation is

$$W(K) = \int_0^\infty \rho(\xi) J_0(K\xi)\xi d\xi = \frac{L^2}{2}\exp\left[-\left(\frac{KL}{2}\right)^2\right] \tag{2B.3}$$

and the Fourier transform of the nth power of the correlation function is

$$W^{(n)}(K) = \int_0^\infty \rho^n(\xi) J_0(K\xi)\xi d\xi = \frac{L^2}{2n}\exp\left[-\frac{(KL)^2}{4n}\right] \tag{2B.4}$$

For a two-scale rough surface, it may be characterized by a two-parameter correlation function. If the correlation function is Gaussian and the roughness scales are independent, then we have a correlation function of the form,

$$\rho(\xi) = a\exp\left[-\frac{\xi^2}{L_1^2}\right] + b\exp\left[-\frac{\xi^2}{L_2^2}\right] \tag{2B.5}$$

Its nth power may be written as

$$\begin{aligned} \rho^n(\xi) &= \left\{a\exp\left[-\frac{\xi^2}{L_1^2}\right] + b\exp\left[-\frac{\xi^2}{L_2^2}\right]\right\}^n \\ &= \sum_{m=0}^{n} \frac{n!a^{(n-m)}b^m}{(n-m)!m!}\exp\left[-\frac{(n-m)L_2^2 + mL_1^2}{L_1^2L_2^2}\xi^2\right] \end{aligned} \tag{2B.6}$$

where $a = \sigma_1^2/\sigma^2$, $b = \sigma_2^2/\sigma^2$ and σ_1^2, σ_2^2 are the variances of the two scales while σ^2 is the total variance of the surface. The associated roughness spectra are

$$W(K) = \int_0^\infty \rho(\xi) J_0(K\xi)\xi d\xi = \frac{aL_1^2}{2}\exp\left[-\left(\frac{KL_1}{2}\right)^2\right] + \frac{bL_2^2}{2}\exp\left[-\left(\frac{KL_2}{2}\right)^2\right] \tag{2B.7}$$

and

$$W^{(n)}(K) = \int_0^\infty \rho^n(\xi) J_0(K\xi)\xi d\xi = \sum_{m=0}^{n} \frac{n!a^{(n-m)}b^m L_e^2}{(n-m)!m!\,2}\exp\left[-\frac{(KL_e)^2}{4}\right] \tag{2B.8}$$

where the square of the equivalent correlation length is

$$L_e^2 = \frac{L_1^2L_2^2}{(n-m)L_2^2 + mL_1^2} \tag{2B.9}$$

In practice, the bell-shaped curve generated by the Gaussian correlation function does not occur very often. Instead, many angular scattering coefficient curves appear to follow an exponential shape generated by the exponential correlation function described below.

2B.2 EXPONENTIAL CORRELATION FUNCTION

This is a correlation function used very often in practice. Theoretically, this correlation function has the undesirable property of not being differentiable at the origin because it has to be made into an even function to qualify as a correlation. Not being differentiable means that the surface it characterizes does not have a rms slope and hence the surface does not allow theoretical analysis. In practical applications we may view the exponential correlation function as an approximation to a more complex function which is differentiable at the origin. Two examples of such functions are

$$f_1 = \exp\left[-\frac{|\xi|}{L} + \frac{|\xi|}{L}\exp\left(-\frac{|\xi|}{d}\right)\right] \tag{2B.10}$$

Near the origin this function behaves like

$$f_1 \approx \exp\left(-\frac{\xi^2}{dL}\right)$$

Another function is

$$f_2 = \exp\left[-\xi^2/\sqrt{(d^2+\xi^2 L^2)}\right] \tag{2B.11}$$

which behaves like $f_2 \approx \exp(-\xi^2/d^2)$ near the origin. A third example is

$$f_3 = \exp[-(\xi/L)^n] \tag{2B.12}$$

where n ranges from 1.2 to 1.8. None of these functions, however, has a simple analytic form for its spectrum. Thus, it is simpler to use the exponential in practice. It is understood that in using an exponential correlation function, the region near $\xi = 0$ must not be important for the problem under consideration. This requirement is usually satisfied under low frequency conditions or over small incident angles in the backscattering problem.

For a single-parameter correlation function of the form, $\exp(-|\xi|/L)$ the corresponding roughness spectrum is

$$W(K) = L^2[1+(KL)^2]^{-1.5} \tag{2B.13}$$

and the spectrum for the correlation function to the nth power is

$$W^{(n)}(K) = \int_0^\infty \rho^n(\xi) J_0(K\xi)\xi d\xi = \left(\frac{L}{n}\right)^2\left[1+\left(\frac{KL}{n}\right)^2\right]^{-1.5} \tag{2B.14}$$

For a two-scale isotropically rough surface with a correlation of the form

$$\rho(\xi) = a\exp\left[-\frac{|\xi|}{L_1}\right] + b\exp\left[-\frac{|\xi|}{L_2}\right] \tag{2B.15}$$

its nth power can be represented as

$$\rho^n(\xi) = \sum_{m=0}^{n} \frac{n!a^{(n-m)}b^m}{(n-m)!m!}\exp\left[-\frac{(n-m)L_2 + mL_1}{L_1L_2}\xi\right] \tag{2B.16}$$

The corresponding roughness spectra are given respectively as follows:

$$W(K) = aL_1^2[1 + (KL_1)^2]^{-1.5} + bL_2^2[1 + (KL_2)^2]^{-1.5} \tag{2B.17}$$

$$W^{(n)}(K) = \sum_{m=0}^{n} \frac{n!a^{(n-m)}b^m}{(n-m)!m!}L_t^2[1 + (KL_t)^2]^{-1.5} \tag{2B.18}$$

where $L_t = (L_1L_2)/[(n-m)L_2 + mL_1]$.

Upon comparing the backscattering angular curves using the Gaussian and the exponential correlation functions it is seen that the former leads to bell-shaped curves that drop off very fast at large angles of incidence, while the latter leads to exponential curves that tend to level at large angles. A correlation function that can generate angular scattering curves that lie between the Gaussian and the exponential types is a 1.5-power function to be discussed in Section 2B.4 or the f_3 function given by (2B.12).

2B.3 ANOTHER TWO-PARAMETER CORRELATION FUNCTION

Consider a correlation function that is a combination of an exponential and a Gaussian function, i.e.,

$$\rho(\xi) = a\exp\left[-\frac{|\xi|}{L_1}\right] + b\exp\left[-\frac{\xi^2}{L_2}\right] \tag{2B.19}$$

This is perhaps the more useful function in practical applications because it naturally can bridge the gap between an exponential and a Gaussian function. The associated roughness spectrum is

$$W(K) = aL_1^2[1 + (KL_1)^2]^{-1.5} + \frac{bL_2^2}{2}\exp\left[-\left(\frac{KL_2}{2}\right)^2\right] \tag{2B.20}$$

but the Bessel transform of its nth power does not have a simple closed form. One possible approximate form which is useful when $L_1 \gg L_2$ is

$$W^{(n)}(K) = a^n\left(\frac{L_1}{n}\right)^2\left[1+\left(\frac{KL_1}{n}\right)^2\right]^{-1.5} + b^n\left(\frac{L_2^2}{2n}\right)\exp\left[-\frac{(KL_2)^2}{4n}\right]$$

$$+\sum_{m=1}^{n-1}\frac{n!a^{(n-m)}b^m L_2^2}{(n-m)!m!2m}\exp\left[-\frac{(KL_2)^2}{4m}\right] \tag{2B.21}$$

2B.4 A 1.5-POWER CORRELATION FUNCTION

A correlation function that forms a Bessel transform pair with the exponential function is

$$\rho(\xi) = (1+\xi^2/L^2)^{-1.5} \tag{2B.22}$$

For an isotropically rough surface the corresponding power spectrum is found by taking the Bessel transform of this function yielding

$$W(K) = \int_0^\infty \rho(\xi)\,J_0(K\xi)\,\xi d\xi = L^2 e^{-KL} \tag{2B.23}$$

The power spectrum of its *n*th power is

$$W^{(n)}(K) = \int_0^\infty \rho^n(\xi)\,J_0(K\xi)\,\xi d\xi = \frac{L^2 K^{1.5n-1} K_{-(1.5n-1)}(K)}{2^{1.5n-1}\Gamma(1.5n)} \tag{2B.24}$$

where $\Gamma(\ )$ is the gamma function and $K_{-v}(\ )$ is the Bessel function of the second kind and of order v with the imaginary argument. Unlike the exponential correlation, this function is clearly differentiable and the slope variance of such a surface is proportional to the second derivative of $\rho(\xi)$ at the origin given by

$$\sigma_s^2 = -\sigma^2\frac{d^2}{d\xi^2}\rho(\xi)\bigg|_{\xi=0} = \frac{3\sigma^2}{L^2} \tag{2B.25}$$

Note that for surfaces with the same rms heights and correlation lengths, surfaces with a 1.5-power correlation function have larger rms slopes than those with Gaussian correlations.

Appendix 2C
The Rayleigh Phase Matrix

2C.1 RAYLEIGH PHASE MATRIX

The Rayleigh phase matrix is the scattering phase matrix for a sphere small compared with the incident wavelength. A Rayleigh scattering layer may be characterized by two parameters, the optical depth τ and the albedo a. These parameters are related to the sphere radius r_s, its dielectric constant $\varepsilon_s = \varepsilon_s' - j\varepsilon_s''$, the host medium dielectric constant $\varepsilon_b = \varepsilon_b' - j\varepsilon_b''$, the layer depth d and the volume fraction f of scatterers through the volume scattering and absorption coefficients κ_s, κ_a as follows:

$$a = \kappa_s / (\kappa_a + \kappa_s) \tag{2C.1}$$

$$\tau = (\kappa_a + \kappa_s) d \tag{2C.2}$$

where

$$\kappa_a = \kappa_{ab} + \kappa_{as}$$

$$\kappa_{ab} = 2\kappa_b''(1-f), \kappa_{as} = f k_b' \frac{\varepsilon_s''}{\varepsilon_b'} \left| \frac{3\varepsilon_b}{\varepsilon_s + 2\varepsilon_b} \right|^2$$

$$k_b = k\sqrt{\varepsilon_b} = k_b' - jk_b''$$

$$\kappa_s = \frac{8}{3}\pi N (k_b')^4 r_s^6 \left| \frac{\varepsilon_s - \varepsilon_b}{\varepsilon_s + 2\varepsilon_b} \right|^2 \tag{2C.3}$$

where N is the number density related to the volume fraction and the volume of the sphere as

$$N = f / (4\pi r_s^3 / 3) \tag{2C.4}$$

The Rayleigh phase matrix defined in terms of the scattering cross section in accordance with (1.51) is given in this appendix. It is written in terms of its Fourier components which terminate at two. This form is particularly convenient for incorporation in the *matrix doubling method* to be described in Chapter 8. Its arguments are $\mu = \cos\theta$, $\mu_s = \cos\theta_s$ and $\Delta\phi = \phi_s - \phi$. Equations (2C.5) through (2C.7) give the detailed contents of the Fourier components of the Rayleigh phase matrix:

$$\mathbf{P}_0(\mu_s, \mu; \Delta\phi) = \frac{3}{4}\begin{bmatrix} 2\sin^2\theta_s\sin^2\theta + \cos^2\theta\cos^2\theta_s & \cos^2\theta_s & 0 & 0 \\ \cos^2\theta & 1 & 0 & 0 \\ 0 & 0 & 0 & 0 \\ 0 & 0 & 0 & 2\cos\theta\cos\theta_s \end{bmatrix} \tag{2C.5}$$

$$\mathbf{P}_1(\mu_s, \mu; \Delta\phi) = \frac{3\sin\theta\sin\theta_s}{4}\begin{bmatrix} 4\cos\theta\cos\theta_s\cos\phi_s & 0 & 2\cos\theta_s\sin\phi_s & 0 \\ 0 & 0 & 0 & 0 \\ -4\cos\theta\sin\phi_s & 0 & 2\cos\phi_s & 0 \\ 0 & 0 & 0 & 2\cos\phi_s \end{bmatrix} \tag{2C.6}$$

$$\mathbf{P}_2(\mu_s, \mu; \Delta\phi) = \frac{3}{4}\begin{bmatrix} \cos^2\theta\cos^2\theta_s\cos 2\phi_s & -\cos^2\theta_s\cos 2\phi_s & \cos\theta\cos^2\theta_s\sin 2\phi_s & 0 \\ -\cos^2\theta\cos 2\phi_s & \cos 2\phi_s & -\cos\theta\sin 2\phi_s & 0 \\ -2\cos^2\theta\cos\theta_s\sin 2\phi_s & 2\cos\theta_s\sin 2\phi_s & 2\cos\theta\cos\theta_s\cos 2\phi_s & 0 \\ 0 & 0 & 0 & 0 \end{bmatrix} \tag{2C.7}$$

where the incident azimuthal angle has been taken to be zero. If this is not true ϕ_s should be replaced by $\phi_s - \phi$ in the above expressions.

For some applications it is more convenient to use the polarization components of the Rayleigh phase matrix. These can be found by summing up the corresponding Fourier components. Such results are summarized in the next section.

2C.2 POLARIZATION COMPONENTS

The polarization components of the Rayleigh phase matrix are listed below:

$$VV = 0.75\{2\sin^2\theta_s\sin^2\theta + \cos^2\theta\cos^2\theta_s[1 + \cos 2(\phi_s - \phi)] + 4\sin\theta\sin\theta_s\cos\theta\cos\theta_s\cos(\phi_s - \phi)\} \tag{2C.8}$$

$$HH = 0.75[1 + \cos 2(\phi_s - \phi)] \tag{2C.9}$$

$$VH = 0.75\cos^2\theta_s\,[1 - \cos 2\,(\phi_s - \phi)] \tag{2C.10}$$

$$HV = 0.75\cos^2\theta\,[1 - \cos 2\,(\phi_s - \phi)] \tag{2C.11}$$

Chapter 3
First-Order Radiative Transfer Solution—Passive Sensing

3.1 INTRODUCTION

A first-order solution of wave emission from an inhomogeneous layer with irregular boundaries can be solved by using an iterative technique as discussed in Chapter 2 and the formulation given in Section 1.5.3 of Chapter 1. The objective here is to obtain some insight into the solutions regarding possible emission mechanisms using a first-order solution. As is done in Chapter 2, the term first-order solution refers to a first-order scattering in κ_s. In what follows we shall start with the formulation of the emission problem from an inhomogeneous layer with irregular boundaries as given in Chapter 1. Next, we convert the differential radiative transfer equations into integral equations. Then, we incorporate the boundary conditions into the integral equations from which we derive the zeroth- and first-order solutions. In Section 3.5 theoretical emission characteristics are illustrated and Section 3.6 shows comparisons of a simplified emission model with measurements. A comparison between backscattering and emission characteristics and their dependence on surface parameters is given in Section 3.7, where it is shown that emission has a strong dependence on coherent energy. This dependence is the major cause of the difference between backscattering and emission in their sensitivity to surface parameters.

3.2 RADIATIVE TRANSFER EQUATIONS

Recall from Section 1.5.3 that a major difference between the emission and scattering problems is that for emission the temperature vector in most problems can be approximated by using only the first two Stokes parameters. In addition, the source of emission is the temperature profile, which is a scalar quantity generally independent of the azimuthal angle. For simplicity, we assume that the upward and downward propagating temperatures are also independent of the azimuthal angle in the analysis to follow. For a more general problem, the assumption of independence in azimuth may not apply [Tsang, 1991; Nghiem et al., 1991]. If so, a Fourier expansion method carried out in the scattering (active) problem in Chapter 8 would have to be followed and the approach described here is good for every Fourier component.

In the formulation below matrix notation is used. It should be understood that the two matrix elements in the temperature profile matrix are identical and hence the sum of the like and cross phase matrix elements is involved from the very beginning of the emission problem. This is different from the scattering problem where the incident wave is assumed to be either horizontally or vertically polarized and hence the incident source matrix has only one element.

We now consider the same emission problem as formulated in Section 1.5.3. We can write the pair of transfer equations for the upward and downward temperature matrices denoted by $\mathbf{T}^+$ and $\mathbf{T}^-$ as follows:

$$\frac{d}{dz}\mathbf{T}^+(z) = -\kappa_{es}\mathbf{T}^+(z) + \mathbf{F}^+(z) + \kappa_{as}T \tag{3.1}$$

$$\frac{d}{dz}\mathbf{T}^-(z) = \kappa_{es}\mathbf{T}^-(z) - \mathbf{F}^-(z) - \kappa_{as}T \tag{3.2}$$

where $\kappa_{es} = \kappa_e / \cos\theta_s$; $\kappa_{as} = \kappa_a / \cos\theta_s$

$\mu_s = \cos\theta_s$

$\mathbf{T}^+(z) = \mathbf{T}(z, \mu_s, \phi_s)$

$\mathbf{T}^-(z) = \mathbf{T}(z, -\mu_s, \phi_s)$

T is the temperature profile in the layer assumed to be independent of the azimuthal angle.

$$\mathbf{F}^{\pm}(z) = \frac{\kappa_{ss}}{4\pi}\int_0^{2\pi}\int_0^1 \mathbf{P}(\pm\mu_s, \mu, \phi_s - \phi)\,\mathbf{T}^+ d\mu d\phi$$

$$+ \frac{\kappa_{ss}}{4\pi}\int_0^{2\pi}\int_0^1 \mathbf{P}(\pm\mu_s, -\mu, \phi_s - \phi)\,\mathbf{T}^- d\mu d\phi$$

$$= \frac{\kappa_{ss}}{2}\int_0^1 \mathbf{P}_0(\pm\mu_s, \mu)\,\mathbf{T}^+ d\mu + \frac{\kappa_{ss}}{2}\int_0^1 \mathbf{P}_0(\pm\mu_s, -\mu)\,\mathbf{T}^- d\mu \tag{3.3}$$

In (3.1) and (3.3) κ_s, κ_e are the volume scattering and extinction coefficient matrices taken to be diagonal [England, 1975; Tsang and Kong, 1975], $\kappa_{ss} = \kappa_s / \cos\theta_s$ and $\mathbf{P}_0(\mu_s, \mu)$ is the zeroth-order Fourier component of the scattering phase matrix. The above differential equations are of the first-order and have standard solutions of the form

$$\mathbf{T}^+(z) = \mathbf{T}^+(-d)\,e^{-\kappa_{es}(z+d)} + \int_{-d}^{z}\left[\mathbf{F}^+(z') + \kappa_{as}T\right]e^{-\kappa_{es}(z-z')}dz' \tag{3.4}$$

$$\mathbf{T}^-(z) = \mathbf{T}^-(0)\,e^{\kappa_{es} z} + \int_z^0 \left[\mathbf{F}^-(z') + \kappa_{as} T\right] e^{\kappa_{es}(z-z')}\,dz' \tag{3.5}$$

These solutions are not real solutions because $\mathbf{F}^{\pm}$ are functions of the upward and downward propagating temperatures. They are really integral equations for these temperatures. To cast them into a form suitable for iterative solution we need to incorporate the applicable boundary conditions. Recall that at the upper boundary $z = 0$ the condition is

$$\mathbf{T}^-(0) = \frac{1}{4\pi}\int_0^{2\pi}\int_0^1 \mathbf{S}(-\mu_s, \mu, \phi_s - \phi)\,\mathbf{T}^+(0)\,d\mu\,d\phi$$

$$= \frac{1}{2}\int_0^1 \mathbf{S}_0(-\mu_s, \mu)\,\mathbf{T}^+(0)\,d\mu \tag{3.6}$$

where $\mathbf{S}_0(-\mu_s, \mu)$ is the zeroth-order Fourier component of the surface scattering phase matrix. At the lower boundary $z = -d$, we have

$$\mathbf{T}^+(-d) = \frac{1}{2}\int_0^1 \mathbf{G}_0(\mu_s, -\mu)\,\mathbf{T}^-(-d)\,d\mu + \mathbf{e}T_g \tag{3.7}$$

where $\mathbf{G}_0(\mu_s, -\mu)$ is the zeroth-order Fourier component of the scattering phase matrix for the lower surface boundary, T_g is the temperature of the half space below the layer and $\mathbf{e}$ is the emissivity of the lower boundary. Substituting the above boundary conditions into (3.4) and (3.5) we have

$$\mathbf{T}^+(z) = \left[\frac{1}{2}\int_0^1 \mathbf{G}_0(\mu_s, -\mu)\,\mathbf{T}^-(-d)\,d\mu + \mathbf{e}T_g\right] e^{-\kappa_{es}(z+d)}$$

$$+ \int_{-d}^{z} \left[\mathbf{F}^+(z') + \kappa_{as} T\right] e^{-\kappa_{es}(z-z')}\,dz' \tag{3.8}$$

$$\mathbf{T}^-(z) = e^{\kappa_{es} z}\left[\frac{1}{2}\int_0^1 \mathbf{S}_0(-\mu_s, \mu)\,\mathbf{T}^+(0)\,d\mu\right]$$

$$+ \int_z^0 \left[\mathbf{F}^-(z') + \kappa_{as} T\right] e^{\kappa_{es}(z-z')}\,dz' \tag{3.9}$$

The above two equations are the starting equations for iteration. The objective of the problem is to solve for the upward propagating temperature at $z = 0$ and then transmit it into the medium above the inhomogeneous layer.

3.3 UPWARD EMITTED TEMPERATURE WITHIN THE LAYER

To seek a first-order iterative solution with respect to the contributing sources, let us consider the source terms in (3.8) and (3.9). First, we note that the sources of emission are the temperature profile in the layer and the emission from the lower half space $\mathbf{e}T_g$. Let us evaluate the source terms in upward and downward temperatures by assuming a temperature profile of the form $T = t_0 + t_1 z$. The source terms in (3.8) are the upward temperature originating from the layer temperature profile,

$$\int_{-d}^{z} \kappa_{as} T e^{-\kappa_{es}(z-z')} dz' = \frac{\kappa_a}{\kappa_e} t_0 \left[1 - e^{-\kappa_{es}(z+d)}\right]$$

$$+ (\kappa_{as} t_1 / \kappa_{es}^2) \left[(1 + \kappa_{es} d) e^{-\kappa_{es}(z+d)} - 1 + \kappa_{es} z \right] \qquad (3.10)$$

plus the contribution from the lower half space $\mathbf{e}T_g e^{-\kappa_{es}(z+d)}$. The source term in (3.9) is the downward temperature,

$$\int_{z}^{0} \kappa_{as} T e^{\kappa_{es}(z-z')} dz' = \frac{\kappa_a t_0}{\kappa_e} (1 - e^{\kappa_{es} z}) + (\kappa_{as} t_1 / \kappa_{es}^2) \left[(1 + \kappa_{es} z) - e^{\kappa_{es} z} \right] \qquad (3.11)$$

All other terms in (3.8) that depend on the upward and downward temperatures can be evaluated approximately using these three source terms. Our major interest is in the total upward propagating temperature evaluated at $z = 0$. From (3.8) we see that there are two more terms to be evaluated: one term is due to volume scattering by the inhomogeneities in the layer and the other is due to upward scattering of the downward temperature by the lower layer boundary. Let us introduce the following notations for the ratios of the volume scattering and absorption coefficients to the extinction coefficient:

$$a = \kappa_s / \kappa_e$$

and

$$\bar{a} = \kappa_a / \kappa_e$$

The term accounting for volume scattering effect in (3.8) can be evaluated approximately using the three major source terms for the upward and downward temperatures. Using (3.3), (3.10) and (3.11), we can carry out the z-integration in this term as follows:

$$\mathbf{V}(0,\mu_s) = \int_{-d}^{0} \mathbf{F}^+(z')\, e^{\kappa_{es}z'}\, dz'$$

$$\approx \frac{\bar{a}a}{2}\int_0^1 \mathbf{P}_0(\mu_s,\mu)\left[C_0 \boldsymbol{t}_0 + \left(\frac{\mathbf{e}\boldsymbol{T}_g}{\bar{a}} - \boldsymbol{t}_0\right) C_1 e^{-\kappa_e d/\mu}\right] d\mu$$

$$+ \frac{\bar{a}a}{2}\int_0^1 \mathbf{P}_0(\mu_s,\mu)\, \boldsymbol{t}_1 \left\{\left[(\mu + k_e d)\, C_1 e^{-\kappa_e d/\mu} - \mu C_0 + k_e C\right]/\kappa_e\right\} d\mu$$

$$+ \frac{\bar{a}a}{2}\int_0^1 \mathbf{P}_0(\mu_s,-\mu)\, \boldsymbol{t}_0 (C_0 - C_2)\, d\mu$$

$$+ \frac{\bar{a}a}{2}\int_0^1 \mathbf{P}_0(\mu_s,-\mu)\, \boldsymbol{t}_1 (1/\kappa_e)\, [\mu C_0 + k_e C - \mu C_2]\, d\mu \tag{3.12}$$

where

$$C_0 = 1 - \exp(-k_{es} d)$$

$$C_1 = \{1 - \exp[-k_{es} d (1 - \mu_s/\mu)]\} / (1 - \mu_s/\mu)$$

$$C_2 = \{1 - \exp[-k_{es} d (1 + \mu_s/\mu)]\} / (1 + \mu_s/\mu)$$

$$C = \left[(1 + k_{es} d)\, e^{-k_{es} d} - 1\right] / k_{es}$$

Note that because $\boldsymbol{t}_0$ is a constant, the matrix product $\mathbf{P}_0(\mu_s,\mu)\,\boldsymbol{t}_0$ results in summing the zeroth-order Fourier components of P_{vv} and P_{vh} for vertical polarization and P_{hv} and P_{hh} for horizontal polarization. For Rayleigh phase matrix given in Appendix 2C we have

$$P_{0vv} + P_{0vh} = 0.75\left[2(1-\mu_s^2)(1-\mu^2) + (1+\mu^2)\mu_s^2\right] \text{ and}$$

$$P_{0hv} + P_{0hh} = 0.75(1+\mu^2)$$

Another term in (3.8) accounting for upward scattering of the downward temperature by the lower boundary can also be evaluated approximately using the source term given by (3.11) as

$$\mathbf{SG}(0, \mu_s) = \left[\frac{1}{2}\int_0^1 \mathbf{G}_0(\mu_s, -\mu)\,\mathbf{T}^-(-d)\,d\mu\right] e^{-\kappa_{es} d}$$

$$\approx \left[\frac{\bar{a}}{2}\int_0^1 \mathbf{G}_0(\mu_s, -\mu)\,\{t_0(1 - e^{-\kappa_e d/\mu}) + t_1[(\mu/\kappa_e)(1 - e^{-\kappa_e d/\mu}) - d]\}\,d\mu\right] e^{-\kappa_{es} d}$$

(3.13)

The complete expression for the upward temperature at $z = 0$ within the layer is given by

$$\mathbf{T}^+(0, \mu_s) \approx \bar{a}\,\{t_0\left[1 - e^{-\kappa_{es} d}\right] + (t_1/\kappa_{es})\left[(1 + \kappa_{es} d)\,e^{-\kappa_{es} d} - 1\right]\}$$

$$+ \mathbf{e}T_g e^{-\kappa_{es} d} + \mathbf{V}(0, \mu_s) + \mathbf{SG}(0, \mu_s) \qquad (3.14)$$

The terms on the first line of (3.14) represent direct contributions from the layer temperature profile. The first term on the second line is the contribution from the lower half space attenuated by the layer. The last two terms are defined by (3.12) and (3.13) as the volume scattering term and the lower boundary scattering term. Since iteration once allows only single scattering, the volume scattering term does not account for cross polarization. Higher order terms are needed to give a more accurate estimate of the volume scattering term when albedo is larger than approximately 0.2 and optical depth larger than 0.5 [Ulaby et al., 1986, Chapter 13]. However, a significant error in the volume scattering term may not be significant to total emission.

3.4 EMISSION INTO THE UPPER MEDIUM

Here, we give an expression for emission into the upper half space through an irregular surface boundary. As in the previous chapter we denote the surface scattering phase matrix by $S_{1t}^c + S_{1t}^n$, where the first symbol is for the transmission of the coherent component and the second for the noncoherent component.

$$\mathbf{T}^+(\mu_s) = \frac{1}{4\pi}\int_0^{2\pi}\int_0^1 [\mathbf{S}_{1t}^c(\mu_s, \mu, \phi_s - \phi) + \mathbf{S}_{1t}^n(\mu_s, \mu, \phi_s - \phi)]\,\mathbf{T}^+(0, \mu)\,d\mu d\phi$$

$$= \frac{1}{4\pi}\int_0^{2\pi}\int_0^1 [4\pi\mathbf{T}_{1t}(\mu_s, \mu)\,\delta(\mu - \mu_0)\,\delta(\phi - \phi_s)\,L_{1t}(\mu)\,]\,\mathbf{T}^+(0, \mu)\,d\mu d\phi$$

$$+\frac{1}{4\pi}\int_0^{2\pi}\int_0^{1}\mathbf{S}^n_{1t}(\mu_s,\mu,\phi_s-\phi)\,\mathbf{T}^+(0,\mu)\,d\mu d\phi$$

$$= \mathbf{T}_{1t}(\mu_s,\mu_0)\,L_{1t}(\mu_0)\,\mathbf{T}^+(0,\mu_0)+\frac{1}{2}\int_0^1\mathbf{S}^n_{01t}(\mu_s,\mu)\,\mathbf{T}^+(0,\mu)\,d\mu \tag{3.15}$$

where the angles in $\mu_0 = \cos\theta_0$ and $\mu_s = \cos\theta_s$ are related by Snell's law; $\mathbf{T}_{1t}$ is the transmissivity for plane boundary; $\mathbf{S}^n_{01t}(\mu_s,\mu)$ is the zeroth-order Fourier component of transmission phase matrix transmitting from the layer into the upper medium and $L_{1t}(\mu) = \exp\{-(k\sigma)^2[\mu_s-(\varepsilon_{tr})^{0.5}\mu]^2\}$ is the factor representing loss of power in coherent transmission due to surface roughness (see Chapter 2, Section 2.4). Here, k is the wave number in the upper half space; ε_{tr} is the real part of the relative dielectric constant of the layer and σ is the rms height of the boundary surface. It is clear from (3.15) that if the boundary is a plane only the first term is needed while for a very rough boundary where $k\sigma$ is large only the second term is important. At this time an accurate separation between coherent and noncoherent power due to the roughness effect is not known. The above expression represents a first-order approximation that is generally unsatisfactory from the standpoint of energy conservation. Only for large scale roughness, where the coherent component is negligible, can energy conservation across an irregular boundary be demonstrated satisfactorily [Fung and Eom, 1981].

3.5 THEORETICAL EMISSION CHARACTERISTICS

In the application of (3.15) there are two complications. First, only for a plane or an equivalent plane boundary, can double integration be avoided in emission calculations. Second, there are difficulties in preserving energy conservation across a rough boundary because simplifying assumptions are generally used in analytic evaluations of coherent and noncoherent power. In general, even for the first-order, emission calculations involve double integrations and an additional error due to inaccuracy in conserving energy. In the special case when the roughness scale is large relative to the electromagnetic wavelength so that coherent scattering is negligible, surface multiple scattering can be included and energy conservation can be demonstrated [Fung and Eom, 1981]. For a slightly or moderately rough surface both coherent and noncoherent scattering components are present, and each has single and multiple scattering terms. Accurate representations of these components in both backward scattering and forward transmission are needed to realize energy conservation. This is the reason why it is difficult to preserve energy conservation in dealing with a rough surface boundary.

Unlike backscattering coefficient calculations where the coherent scattering component is removed, the coherently transmitted component is needed in emission calculations and can be of dominating importance. For example, at interfaces between air and vegetation, dry or frozen soil, snow layer or ice layer, we expect to see a permittivity discontinuity around 4 or smaller. If so, the energy in the noncoherent terms will be around 10% or smaller, and emission is dominated by coherent transmission, i.e., the first term in (3.15). For these cases, double integrations can be avoided without causing significant error. In addition, it is worth noting that for most soil surfaces there is a thin layer of loose dirt over the more solid or claylike background. The average permittivity discontinuity between air and this thin layer may again be smaller than 4, even though the permittivity of the soil at a deeper level may be larger than 4.

In this section we examine emission behavior for slightly rough surfaces using only the first term in (3.15) and consider cases where the air-medium interface has a permittivity discontinuity less than 4. That is, the emission model is restricted to coherent transmission across layer boundaries, but one-dimensional integration needed to evaluate noncoherent volume scattering within an inhomogeneous layer given in (3.14) is included. For soil surfaces we model the loose dirt on top by a plane parallel homogeneous transition layer with a thickness smaller than 0.4 of the operating wavelength and a Rayleigh layer immediately below it. This means that we interpret the $\mathbf{T}_{1t}(\mu_s, \mu_0)$ in (3.15) as the transmission coefficient for a thin plane layer and volume scattering by the loose dirt or air pockets in the soil medium is accounted for by a Rayleigh scattering layer. A more complete study of surface roughness effect on emission including noncoherent transmission is given in Section 3.7.

3.5.1 Emission from an Inhomogeneous Layer with a Transition Top Boundary

In this section we show theoretical emission behavior of an inhomogeneous layer with a thin transition boundary whose relative permittivity is less than 4. This type of model is expected to be applicable to soil surfaces covered with loose dirt and fresh snow surfaces where the snow density is smaller near the air-snow interface than at deeper levels. We shall let ε_r, ε_L and ε_h represent the permittivities of the transition layer, the inhomogeneous layer and the lower half space, respectively. The normalized depth of the transition layer is kd, where k is the wave number. The optical depth and albedo of the inhomogeneous (Rayleigh) layer are denoted by τ and a, respectively. We shall illustrate the effects of the transition layer on emission by varying ε_r and kd for two different values of the layer and half-space permittivities.

Effects of the Transition Layer

In Figure 3.1 the geometry of the emission problem is illustrated along with a transition layer. Note that although irregular interfaces are shown, due to low values of layer permittivity coherent transmission is assumed to dominate and only the rms height of the interface has an effect on transmission. This means that an equivalent plane transition layer is used to compute the coherently transmitted energy and roughness effect is accounted for only by the rms height of the irregular interface. The transmission of the noncoherent energy is neglected in the studies in this section and Section 3.6. More complete studies including coherent and noncoherent calculations are formulated in Chapter 8 and illustrated in Chapter 9.

When we show the effect of the transition layer, we choose the permittivity of the lower half space to be the same as the average permittivity of the inhomogeneous layer. In application to soil emission this choice means that below the transition layer is homogeneous soil with some inhomogeneities due to the presence of air pockets. These inhomogenieties are represented by a Rayleigh phase function. In the microwave region, volume scattering within the soil is weak but usually noticeable at large angles of incidence. In illustrations a reference set is denoted by setting the permittivity of the transition layer to unity, i.e., $\varepsilon_r = 1$. This is the emission in the absence of the transition layer. In the studies from Figures 3.2 through 3.7, surface boundary roughness effects are kept very small by letting the normalized rms surface height, $k\sigma$, equal to 0.01.

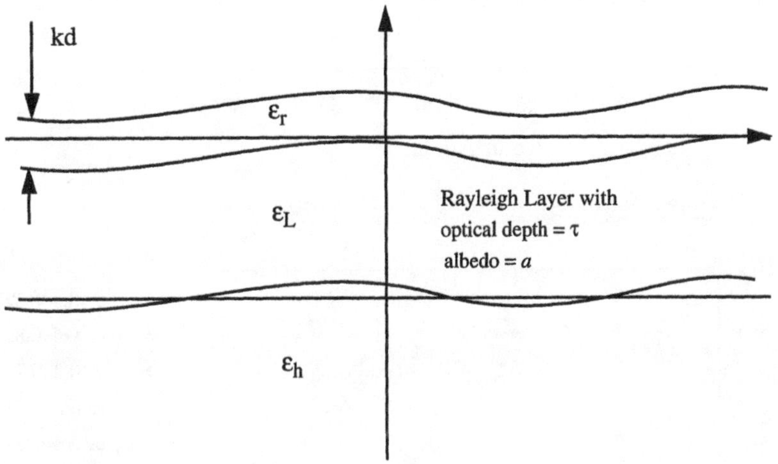

Figure 3.1 Geometry of the layer emission problem including a transition layer with permittivity ε_r.

In Figure 3.2 we show the effect of the permittivity and the depth of the transition layer on emission, when the average permittivity of the inhomogeneous layer is 25 - *j*5.5. The emission from the same medium without the transition layer is denoted in the figure as emission with $\varepsilon_r = 1$. Thus, in each figure the change in emission due to the transition layer is obvious. In Figure 3.2 (a) we see that the emission without the transition layer is strongly influenced by the Brewster angle and increases monotonically with the nadir angle in vertical polarization. The general shape of the emission curves does not change much when we add a transition layer with a small permittivity, $\varepsilon_r = 1.2 - j0.05$. As the permittivity ε_r increases, the vertically polarized emission curve begins to flatten and eventually decrease at large nadir angles. In the mean time the spacing between the two polarization curves becomes narrower. As a result, substantially different angular curves are obtained.

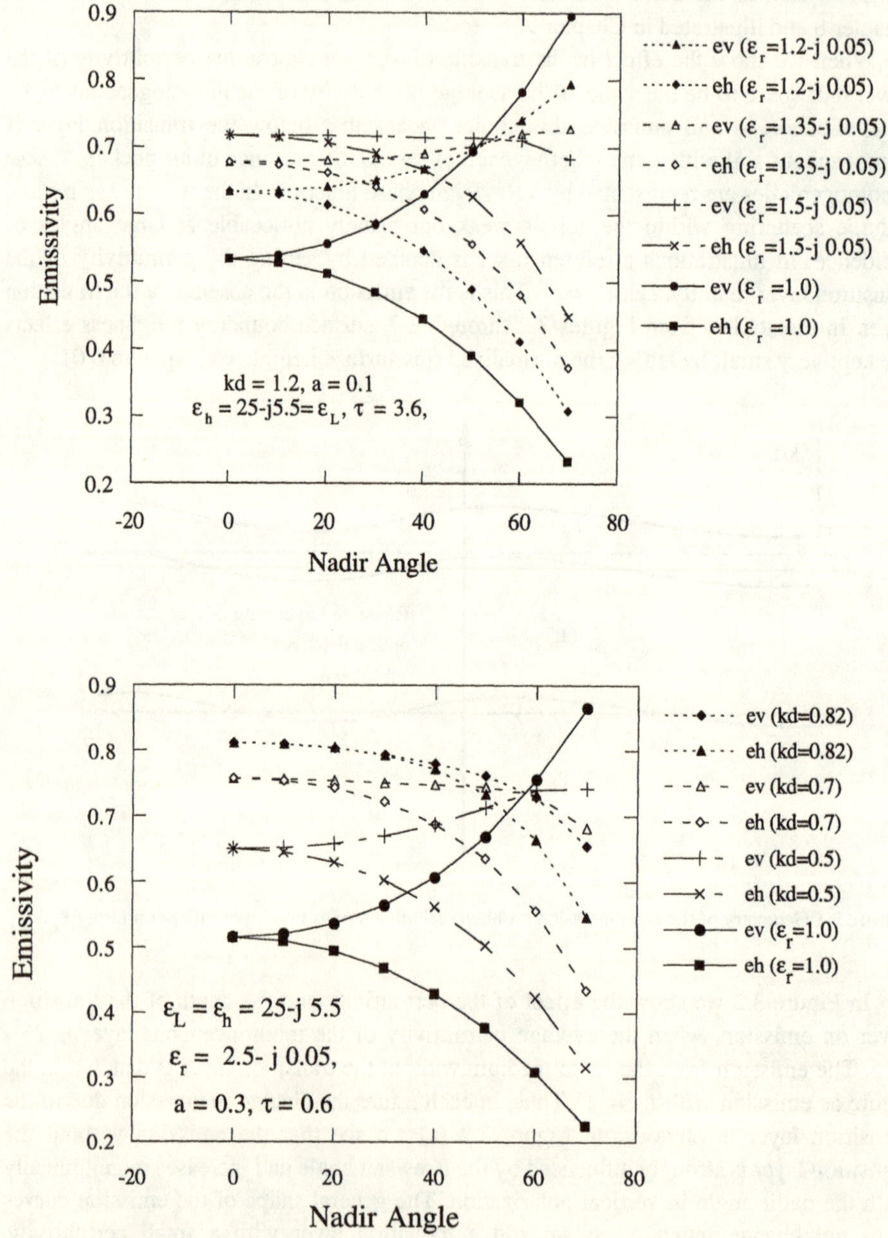

Figure 3.2 Effects of (a) permittivity and (b) thickness of a transition layer on emission from an inhomogeneous Rayleigh layer with albedo *a* and optical depth τ.

Next, we change the thickness of the transition layer in part (b). When the layer is thin ($kd = 0.5$), angular emission characteristics are more similar to those of the $\varepsilon_r = 1$ case except for a difference in level and a significantly slower rise in the vertically polarized emission. As the layer thickness increases, the rise in emission of the vertical polarization with the nadir angle begins to level off and the spacing between the two polarizations decreases. Finally, the vertically polarized emission curve drops with the nadir angle. The general effects of a larger permittivity and layer thickness on emission are similar for the cases examined. In summary, the overall effects of a larger permittivity and layer thickness on emission are to raise the level of the angular curves and narrow the spacing between the polarization components. In all situations, the case without the transition layer is the lowest.

In Figure 3.3 we reexamine the effects of the permittivity of the transition layer on emission, when the permittivity of the inhomogeneous layer is $4 - j0.2$. As expected, this smaller permittivity value raises all the emission curves relative to those in Figure 3.2. Unlike the trend in Figure 3.2, an increase in transition layer permittivity may cause emission to decrease instead of increase. Furthermore, from Figure 3.3 it is the larger rather than the smaller permittivity case that gives an emission characteristic more similar to the case without the transition layer. Apparently, the reason for this behavior is because the large value of the transition layer permittivity is too close to that of the inhomogeneous layer itself, while the smaller values permit the transition layer to act as it is intended. Relative to Figure 3.2 similar reversal also occurs in the layer thickness effect given in Figure 3.4. Here, it is the thinner rather than the thicker layer that causes a greater increase in emission and a narrower spacing between polarization components than the emissions without the transition layer. Clearly, this reversal has its dependence on the choice of transition layer permittivity. In summary, similar changes in the shapes of the emission curves are seen in Figures 3.2, 3.3 and 3.4. These changes appear to be proportional to an increase in permittivity or layer thickness in Figure 3.2 and to a decrease in the same parameters in Figures 3.3 and 3.4. This apparent contradictory result may be explained by the fact that the real cause of larger or smaller emission is not the permittivity or thickness of the transition layer but the impedance matching between the transition and the inhomogeneous layer.

Effects of the Inhomogeneous Layer

In Figure 3.5, we illustrate the effects of optical depth and albedo of the inhomogeneous layer on emission. For an inhomogeneous layer with permittivity, $\varepsilon_L = 25 - j5.5$ above a homogeneous half space with the same permittivity, an increase in either the optical depth or the albedo results in lower emission. This is because an increase in albedo causes a decrease in absorption, while an increase in optical depth lowers the emission from the lower half space. Although the emission by the inhomogeneous layer may increase with an increase in optical depth when the albedo of the layer is small, it can not compensate for the loss in emission from the lower half space, for the case shown in Figure 3.5.

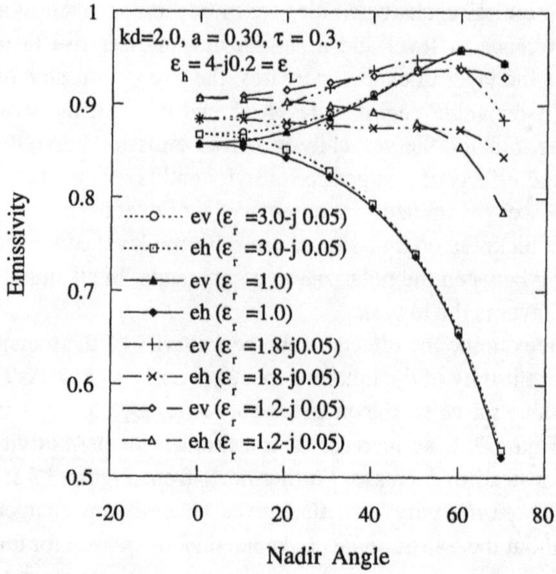

Figure 3.3 Effects of permittivity of a transition layer on emission from an inhomogeneous Rayleigh layer with albedo a and optical depth τ, when the transition layer permittivity is 4 - j 0.2.

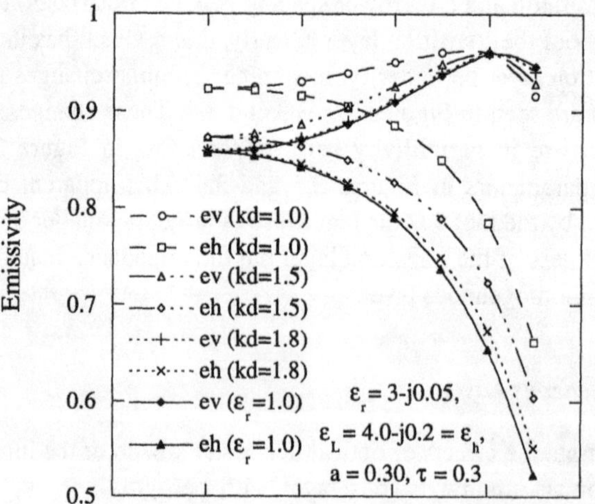

Figure 3.4 Effects of thickness of a transition layer on emission from an inhomogeneous Rayleigh layer with albedo a and optical depth τ, when the transition layer permittivity is 4 - j 0.2.

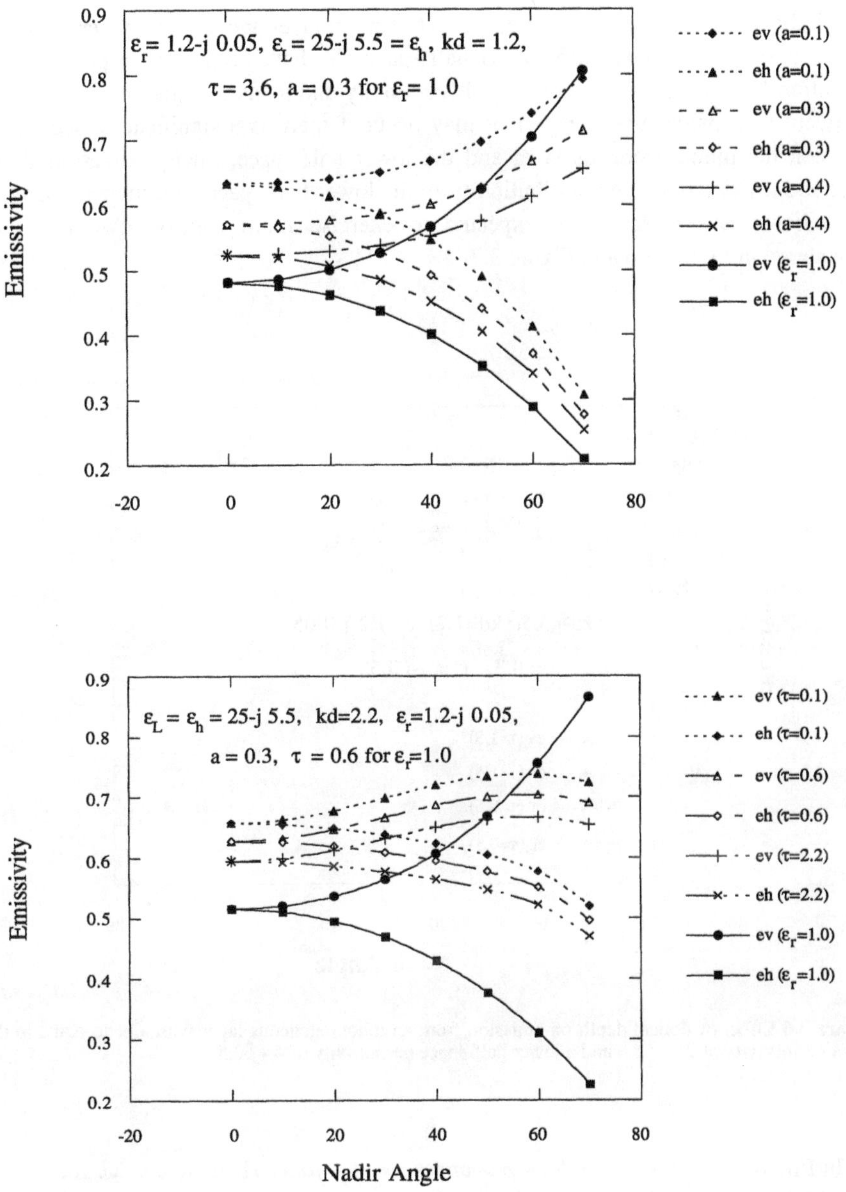

Figure 3.5 Effects of (a) albedo and (b) optical depth of an inhomogeneous Rayleigh layer on emission from a layered medium shown in Figure 3.1.

In the above we considered the effects of a transition layer and an inhomogeneous layer with a permittivity same as the lower half space. We have seen that the presence of a transition layer reduces the boundary discontinuity and leads to higher emission. Now, we want to consider the change that may occur if there is a significant discontinuity between the inhomogeneous layer and the lower half space. Such a situation should substantially decrease the contribution from the lower half space, and the parameters in the inhomogeneous layer are expected to exercise more control. Two cases are considered in Figure 3.6 and Figure 3.7.

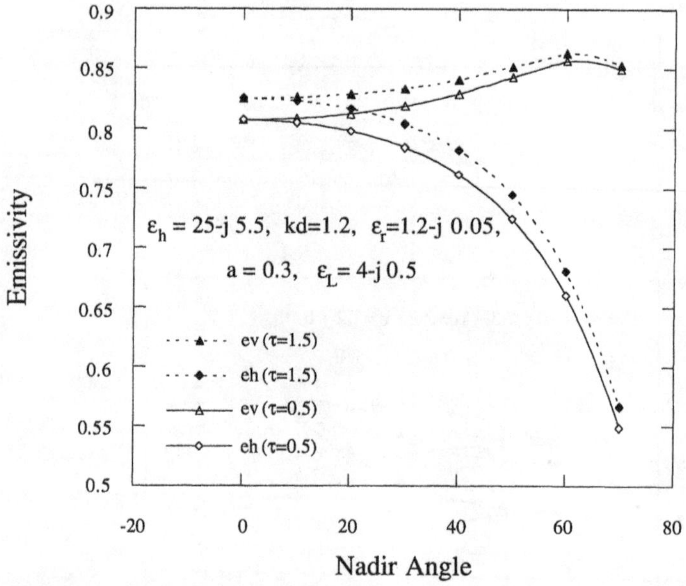

Figure 3.6 Effect of optical depth on emission from an inhomogeneous layer with albedo equal to 0.3 and a permittivity of 25 - j 5.5 and a lower half-space permittivity of 4 - j 0.5.

In Figure 3.6 we let $\varepsilon_L = 25 - 5.5$ and $\varepsilon_h = 4 - j0.5$. Then we change the optical depth τ from 0.5 to 1.5 while the albedo is kept at 0.3. We note that unlike the case when $\varepsilon_L = \varepsilon_h$, as the optical depth increases, emission now increases also. Apparently, this is because the contribution from the inhomogeneous layer is now dominating over that from the half space. If this is not so we expect the behavior to reverse again. To illustrate this point we now repeat the same calculation but with albedo set to 0.7 to decrease the emission from the inhomogeneous layer. As seen in Figure 3.7, as the optical depth increases, emission decreases. Thus, the variation of emission with optical depth is a function of permittivity discontinuity between the lower half space and the inhomogeneous layer and the albedo of the layer.

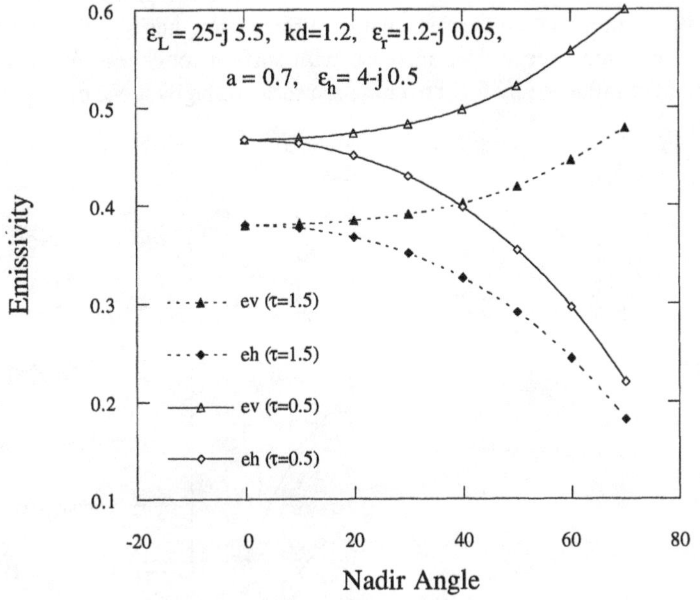

Figure 3.7 Effect of optical depth on emission from an inhomogeneous layer with albedo equal to 0.7 and a permittivity of 25 - j 5.5 and a lower half-space permittivity of 4 - j 0.5.

Effects of Boundary Roughness

In previous illustrations in this section, boundary roughness has been chosen to be small. We shall see that, for emission, surface roughness is important when there is a significant permittivity discontinuity at the boundary. If the discontinuity is small, the roughness effect is also small. In Figure 3.8 we use normalized rms height $k\sigma$ equal to 0.01 for both boundaries of the inhomogeneous layer as a reference. Then, we increase $k\sigma$ to 0.2 at the top boundary of the inhomogeneous layer as one case and repeat the same calculations for $k\sigma$ equal to 0.2 at the lower boundary as another case. It is seen that when roughness is increased at the top boundary there is a small drop in emission along with a small narrowing of spacing between the polarization components, because the boundary discontinuity is not very large due to the presence of a transition layer. When the same increase in roughness is made at the lower instead of the upper boundary, a much larger decrease in emission and narrowing of spacing between polarization components occur along with a slower rise in vertical polarization and a slower drop in horizontal polarization. Note that the emission levels in these calculations are *not correct* because we did not take into account the transmission of the noncoherent energy. However, the impact of roughness on angular trends and the dependence on boundary discontinuity is expected to be real. Experimental results from both land and sea indicate that emission increases with surface roughness. Theoretically, noncoherently transmitted energy does increase with surface roughness. What is not known is the extent of influence by foam over the sea surface and by loose dirt over bare soil on emission.

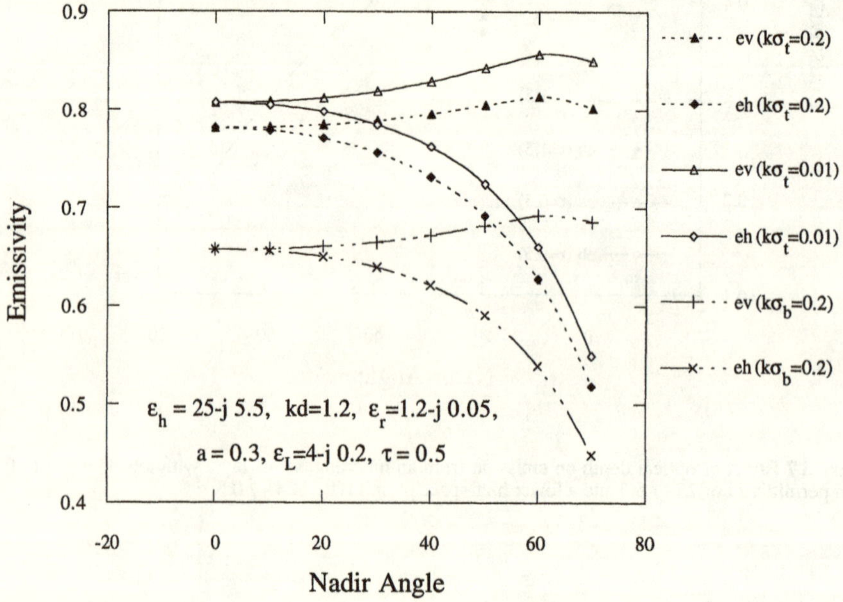

Figure 3.8 Effects of the top and lower boundary roughness of the inhomogeneous layer on angular trends of emissivity.

3.5.2 Parameter Effects on Emission Components

In the previous section we considered parameter effects on emission from an inhomogeneous layer with or without a transition layer and effects of the inhomogeneous layer parameters. We showed the total emissivity in each case without indicating the relative contributions from different contributors as given in (3.14). In this section we shall show the total emissivity and contributions from

1. direct upward emission by the inhomogeneous layer;
2. emission through volume scattering;
3. emission by the lower half space;
4. downward emission within the layer reflected upward by the layer-ground interface.

In the plots of Figures 3.9 through 3.12 we shall consider only a Rayleigh layer above a homogeneous half space. Figure 3.9 is used as a reference with respect to which we alter albedo, optical depth and interface discontinuity and observe the changes in the emission components. In these figures the downward emission denoted as number 4 above is not shown directly in the figures. It is the difference between the total and the other three components in the figure. In Figure 3.9 the contribution from the lower half space is the largest and is denoted as *ground* in the figure. It is followed by direct upward emission denoted as *layer* and by emission through *volume* scattering.

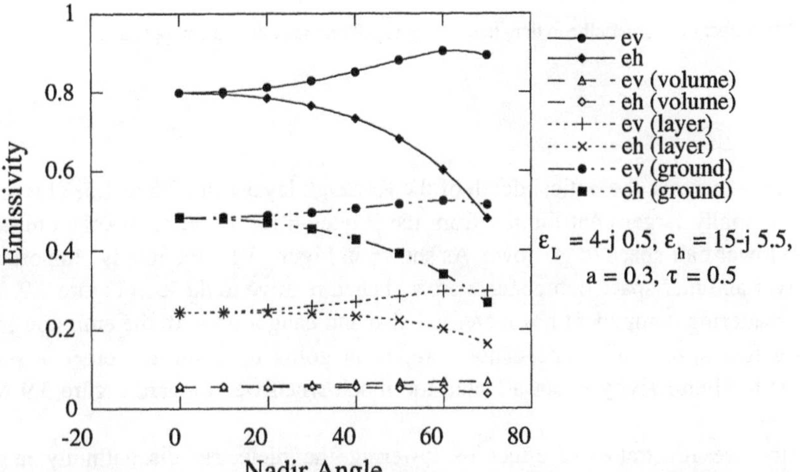

Figure 3.9 Emission components of a Rayleigh layer above a homogeneous half space with plane boundaries.

In Figure 3.10 we increase the albedo to 0.7 while keeping all other parameters unchanged. In this case, the layer emission is lowered significantly while the volume scattering component is increased. These two components are now close in level to each other. The contribution from the lower half space is unaffected because we keep the same optical depth for the inhomogeneous layer. As expected, the overall effect is a lower emissivity.

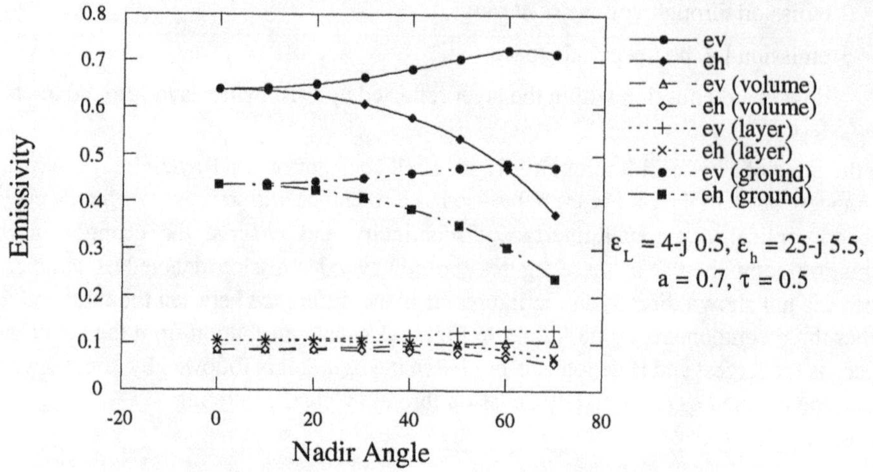

Figure 3.10 Effect of a larger albedo than that in Figure 3.9 on the emission components.

Next, we increase the optical depth of the Rayleigh layer from 0.5 to 1.5. This leads to a substantially larger contribution from the Rayleigh layer and causes the emission from the lower half space to go down. As shown in Figure 3.11, the relative importance of the layer and half space components are switched relative to those in Figure 3.9. The volume scattering component has increased also and caught up with the emission from the lower half space. With one major component going up while the other is going down, the total emissivity is not affected too much when we compare Figure 3.9 with Figure 3.11.

Finally, we illustrate the effect of lowering the dielectric discontinuity at the interface between the Rayleigh layer and the lower half space in Figure 3.12. As expected, the contribution from the lower half space increases and becomes the dominant component. The direct contribution from the Rayleigh layer is not affected. Some increase in the volume scattering component is compensated by the decrease in the reflected component. The total emissivity is higher than a comparable case with a larger discontinuity.

3.6 COMPARISONS WITH EMISSION MEASUREMENTS

In this section comparisons of the simple model discussed in the previous section is applied to data interpretation. Emissions from frozen and wet soil surfaces reported by Wegmuller [1990] and from bare soil surfaces of different roughnesses by Mo et al.

[1987] are considered first. This is followed by comparisons with emissions from snow and ice surfaces acquired in the experiments conducted at the U.S. Army Cold Region Research and Engineering Laboratory in 1988 [Carsey, 1992] and the measurements by Matzler et al. [1980].

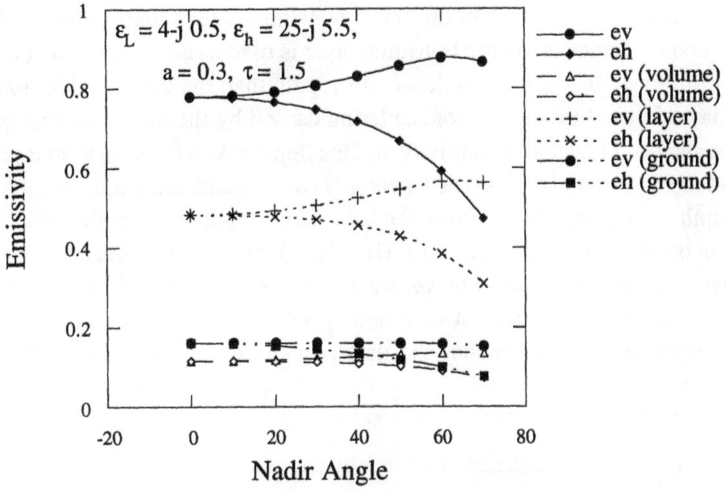

Figure 3.11 Effect of a larger optical depth than that in Figure 3.9 on the emission components.

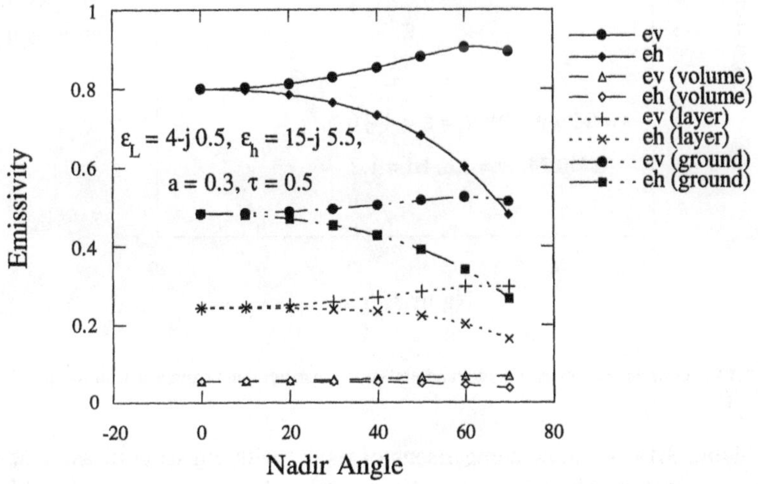

Figure 3.12 Effect of a smaller discontinuity at the interface between layer and half space than that in Figure 3.9 on the emission components.

3.6.1 Comparisons with Emissions from Soil

The measurements taken by Wegmuller [1990] were over a bare wheat field in 1988. Based upon the information of moisture condition and soil properties, estimates of soil permittivity and roughness were made and provided. In the comparisons given in Figures 3.13 through 3.15 the dielectric values used are the same as given in Wegmuller [1990]. The loose dirt on top of the much denser soil surface and the roughness of the soil surface are modeled by a thin dielectric transition layer situated above a Rayleigh layer. The homogeneous dielectric transition layer is used to account for the fact that the average permittivity is lower in the loose dirt region than the dense soil region and the Rayleigh layer is used to account for scattering caused by the loose dirt and associated surface roughness. Thus, the permittivity and the thickness of the transition layer are not known and are chosen to fit data. The Rayleigh layer is characterized by an albedo a and optical depth τ. Figure 3.13 shows the comparison between model and emission measurements from a frozen soil at 3.1 GHz. The normalized thickness and permittivity of the transition layer are denoted by kd, wave number times physical thickness, and ε_r. The curves with $\varepsilon_r = 1$ represent emission without the transition layer. Clearly, the excellent agreement between model and data is realized only with the addition of the transition layer.

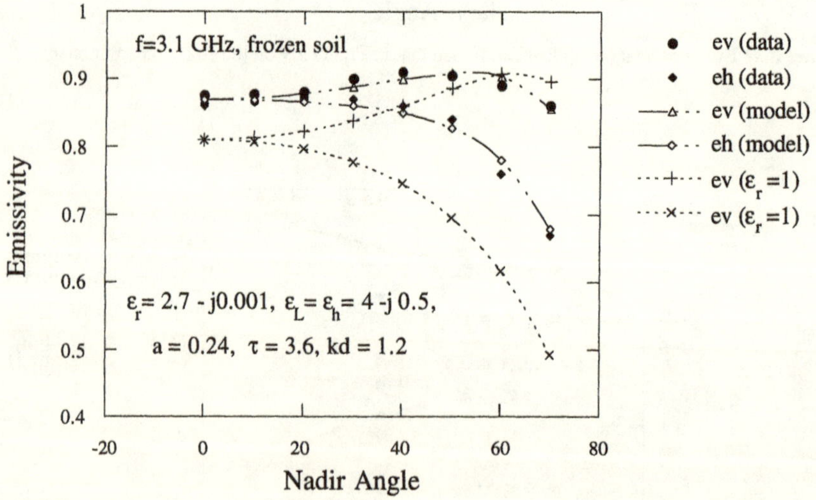

Figure 3.13 Comparisons between model and emission measurements from a frozen soil at 3.1 GHz.

In Figure 3.14 we show comparisons of model with emission measurements from the same soil surface when it was wet. Again, an excellent agreement is possible when a transition layer is included. The major change for wet soil is that the transition layer should have a larger permittivity than it did when the soil was frozen. Due to the large permittivity of the soil and the Brewster angle effect, the use of a plane boundary without a transition layer leads to a much lower level in emission and a very fast rise in vertical polarization and a fast drop in horizontal polarization. Thus, there is a lack of agreement in level and angular trends for both vertical and horizontal polarizations. While the level of emission without the transition layer may be raised somewhat by setting the albedo to zero, the emission level remains substantially lower than the measurement.

Now we consider emission from the same wet soil surface as in Figure 3.14 but at 11 GHz. This change in frequency leads to a substantial change in the angular emission curves as shown in Figure 3.15. The overall level of emission is higher and the spacing between vertical and horizontal polarization is smaller. The penetration depth at higher frequency should be smaller, implying that the upper portion of the medium should be dominating in contribution. This also implies that the transition layer is playing a more important role. The permittivity in the transition layer is smaller because its physical thickness is smaller. As illustrated in the previous section, there is some similarity between the effect of the permittivity ε_r and that of the normalized thickness *kd* of the transition layer (see Section 3.5). Thus, it is possible to determine a combination of *kd* and ε_r by a data fit but not each parameter uniquely by itself. In Figure 3.15 the agreement between model and measurements is very good except for one data point. The need for a transition layer is apparent from the large difference between the models with and without the transition layer.

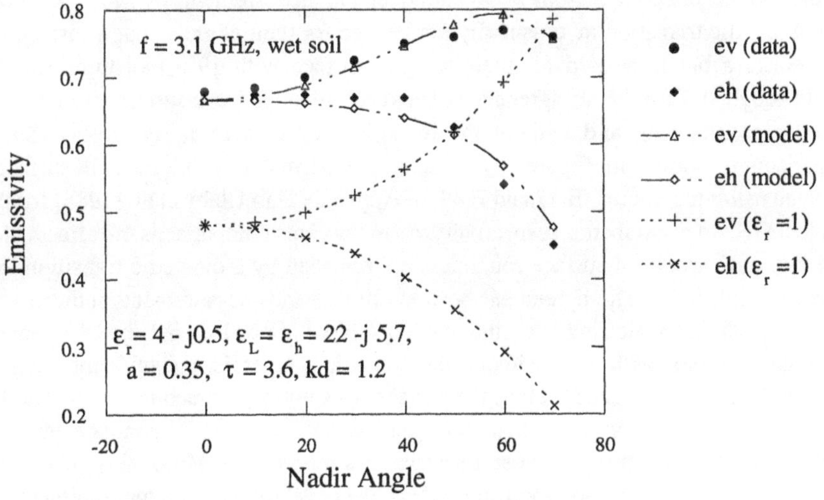

Figure 3.14 Comparisons between model and emission measurements from a wet soil surface at 3.1 GHz.

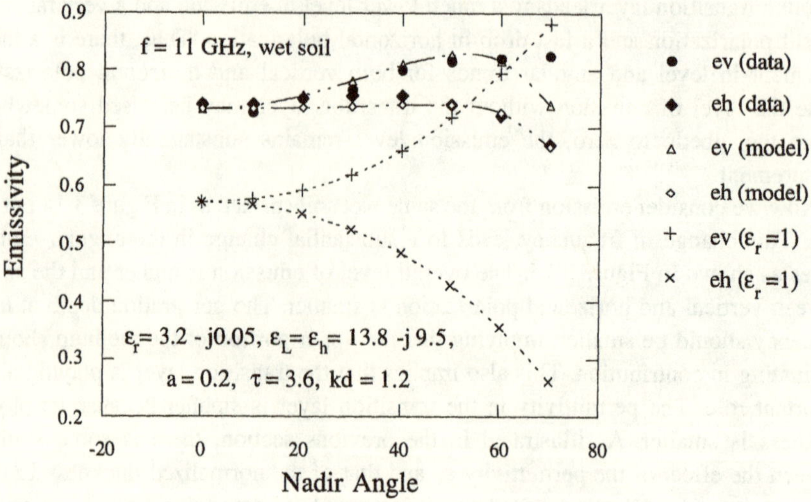

Figure 3.15 Comparison between model and emission measurements from a wet soil at 11 GHz.

In the comparisons shown above the difference between with and without a transition layer is significant. This appears to indicate that the soil type is such that there was a substantial amount of loose dirt in addition to a much denser soil medium. When we consider similar comparisons between model and measurements by Mo et al. [1987], we find that the transition layer is really not needed for their slightly rough surface with 11% moisture but is needed for their rougher surface with 19% moisture. For their slightly rough surface, better agreements between model and measurements occur only when the results with and without the transition layer agree fairly closely. Such a comparison is shown in Figure 3.16. Here, the permittivity of the soil surface is estimated using the figures, E.47 and E.48, of Appendix E in Ulaby et al. [1986] for their 1.4 and 5 GHz measurements, respectively. As in previous comparisons the effects of the loose dirt and associated surface roughness are replaced by a dielectric transition layer and a Rayleigh layer. The agreement between model and measurements indicates that there is significant scattering (i.e., the Rayleigh layer is needed) but the difference in model calculations with and without the transition layer is insignificant (i.e., the transition layer is not needed). Thus, the surface does not have much loose dirt but does have air pockets. An attempt to use a homogeneous half space with boundary roughness to model emission from the same soil surface was reported by Mo et al. [1987]. Their results indicate that there is good agreement between model and measurements for horizontal polarization but for vertical polarization the calculated values are always higher than the measurement and the angular trends are different for all the surfaces they considered at both 1.4 and 5 GHz.

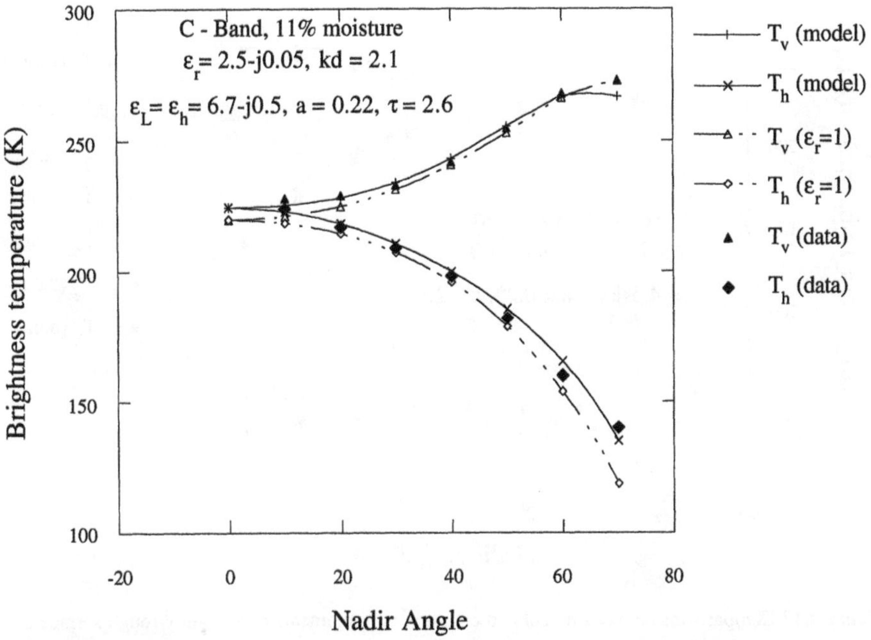

Figure 3.16 Comparisons between model and emission measurements of a slightly rough surface at 5 GHz.

In Figure 3.17 we show the comparison between model and emission measurements of the slightly rough surface at 1.4 GHz. Again, there is no significant difference between calculations with and without a transition layer for this surface. However, a significant amount of scattering is needed to obtain a good match in vertical polarization. As illustrated in Figure 3.17 the level of volume scattering for vertical polarization is over 40 degrees and is very flat, thus causing the overall angular trend to slow down. The information derived from the comparisons shown in Figure 3.16 and Figure 3.17 is the same as it should be since we are dealing with the same surface.

Next, we consider a rougher surface with 19% of moisture at 1.4 and 5 GHz. For this surface we expect the need for a transition layer because it is a much rougher field. Mo et al. [1987] have compared their homogeneous soil model with measurements and observed large disagreement (over 40 degrees) between model and measurements especially at 1.4 GHz in vertical polarization. At 5 GHz the disagreement in vertical polarization can exceed 20 degrees. However, their model is in very good agreement with measurements in horizontal polarization. By allowing for volume scattering and a dielectric transition layer we show in Figures 3.18 and 3.19 that both polarizations can

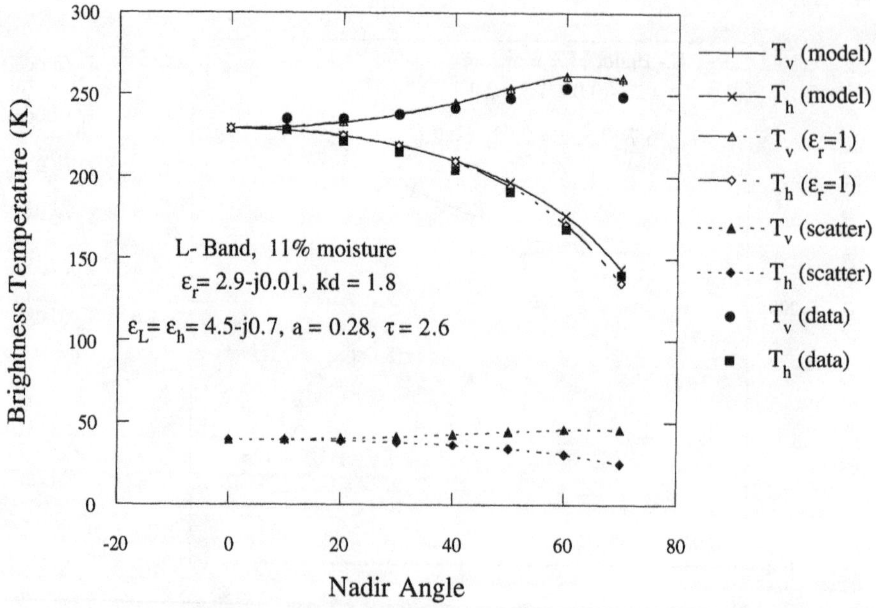

Figure 3.17 Comparisons between model and emission measurements of a slightly rough surface at 1.4 GHz.

be matched very well. As seen in Figure 3.18, the difference in level and angular trends between with and without a transition layer is very large and hence a transition layer is needed. Note that the albedo at 1.4 GHz is larger than that at 5 GHz because the albedo and optical depth here are effective values and the penetration depth is larger at 1.4 GHz than at 5 GHz. The volume scattering component at L-band (1.4 GHz) is also illustrated in Figure 3.19. The level of its vertically polarized component exceeds 50 degrees.

3.6.2 Comparisons with Snow and Ice Emission

In this section we shall consider comparisons of this simplified model with emissions from snow [Shiue et al., 1978; Matzler et al., 1980], sea ice covered with a snow layer [Grenfell, 1992] and saline ice [Carsey, 1992]. Snow and sea ice are inhomogeneous media with fairly smooth boundaries in general. For snow media even if the air-snow boundary is not smooth, roughness is still not very important at microwave frequencies because the permittivity of snow is small. Thus, emission into air is dominated by coherent transmission. This is also true of ice media considered in this section. On the other hand, the snow-ice, ice-water and the snow-ground boundaries are generally not smooth and the permittivity discontinuities at these boundaries are larger than those at the corresponding upper boundaries. We expect noncoherent scattering to be significant at the lower boundary but the inhomogeneities within the layer and the transmission

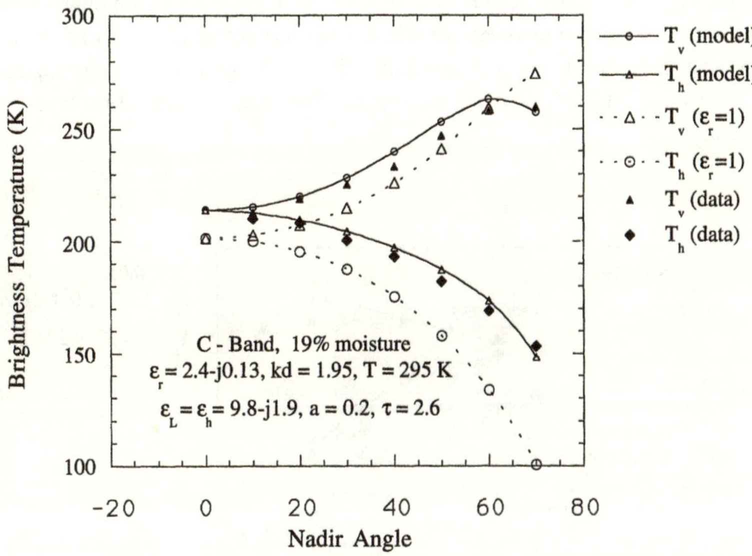

Figure 3.18 Comparisons between model and emission measurements of a rough surface at 5 GHz.

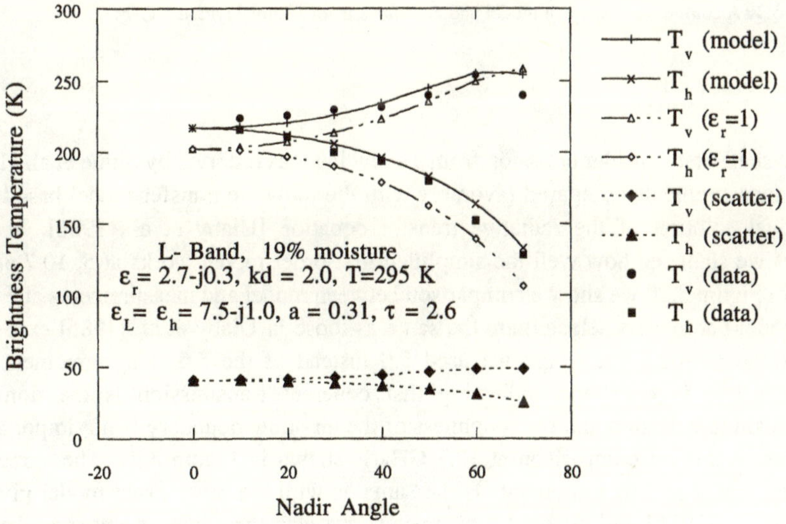

Figure 3.19 Comparisons between model and emission measurements of a rough surface at 1.4 GHz.

property of the upper boundary appear to dominate the angular emission characteristics to the point that no significant error occurs when we approximate the lower boundary by a plane. For all the comparisons in this section the model used is an inhomogeneous (Rayleigh) layer above a homogeneous half space. The idea of a transition layer is not applicable here.

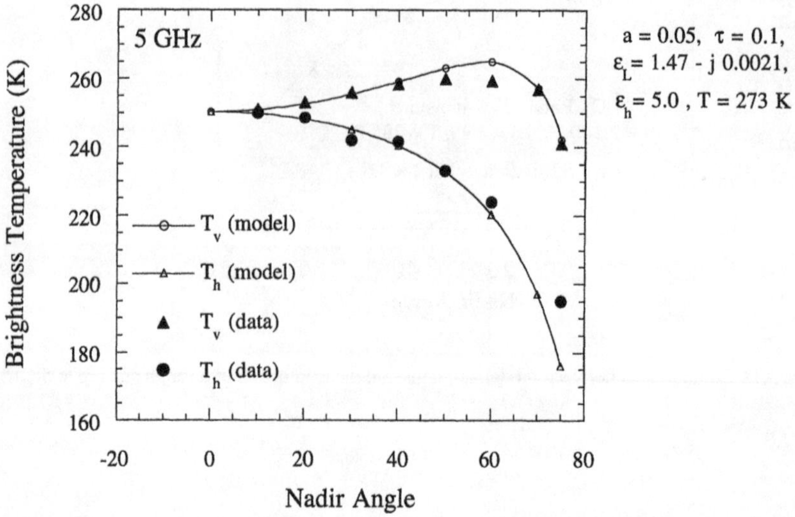

Figure 3.20 A comparison between model and emissions from a snow layer at 5 GHz.

We shall first consider emission from a snow layer as reported by Shiue et al. [1978]. These measurements compared favorably with the radiative transfer model based upon numerical solution of the radiative transfer equation [Ulaby et al., 1986]. In what follows we shall see how well the simplified first-order model works at 5, 10.7 and 18 GHz. In Figure 3.20 we show a comparison between model and measurements at 5 GHz. The model parameters selected are the same as those in Ulaby et al. [1986] except for the ground permittivity, where we used 5.0 instead of the 5.5. The agreement with measurements is excellent, indicating that coherent transmission is the dominant mechanism in emission and the roughness of the air-snow boundary is not important as expected. A similar comparison at 10.7 GHz is shown in Figure 3.21. The agreement between model and measurements is the same as with the more exact model given in Ulaby et al. [1986] and all model parameters are also the same except the albedo is selected to be 0.2 instead of 0.3. Again, the agreement seems to justify the use of a plane boundary.

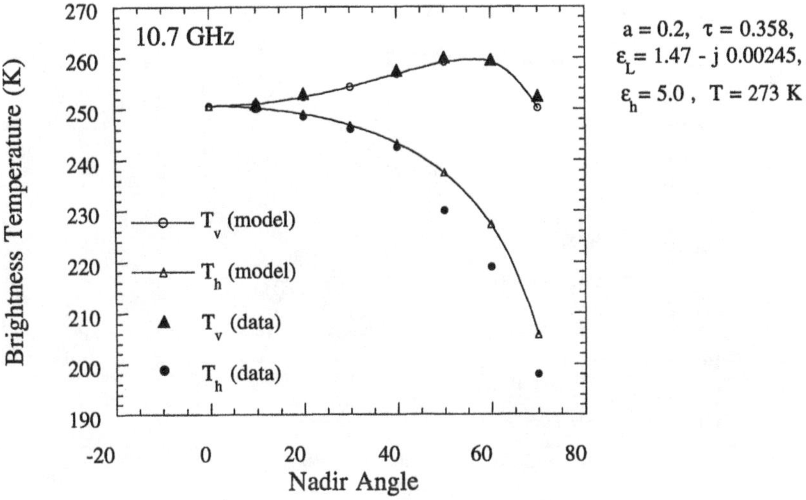

Figure 3.21 A comparison between model and emission measurements at 10.7 GHz.

When we consider emission at 18 GHz, the effect of volume scattering becomes much more important. It should no longer be possible to ignore multiple scattering within the inhomogeneous layer. Since our simple model accounts for only single scattering, signs of failure of this model should appear. In Figure 3.22 we see that the comparison is not satisfactory in at least two aspects. First, the spacing between vertical and horizontal polarization calculated from the model is too wide, and second, the level of the horizontally polarized emission is too low at large angles of incidence. Although the measurements at large nadir angles may be unreasonably high, the calculated values are lower than the expected trend projected from the data based on measurements at small nadir angles. The contribution from multiple scattering is expected to slow down the angular trend of horizontal emission but not vertical polarization because the latter is very flat. It is clear that the simple model is not suitable for cases where multiple scattering is significant.

Now we consider model comparisons with emission measurements from snow taken by Matzler et al. [1980] at 10.4 GHz. The grain size of this snow is such that the albedo of the medium is around 0.35, which is about the maximum value that a single scattering model can tolerate. The comparison is shown in Figure 3.23. The agreement is excellent for vertical polarization. In horizontal polarization there is a significant amount of fluctuation in the data. Overall, the agreement between model and measurement looks very good indicating that the assumption of plane layer interfaces is acceptable.

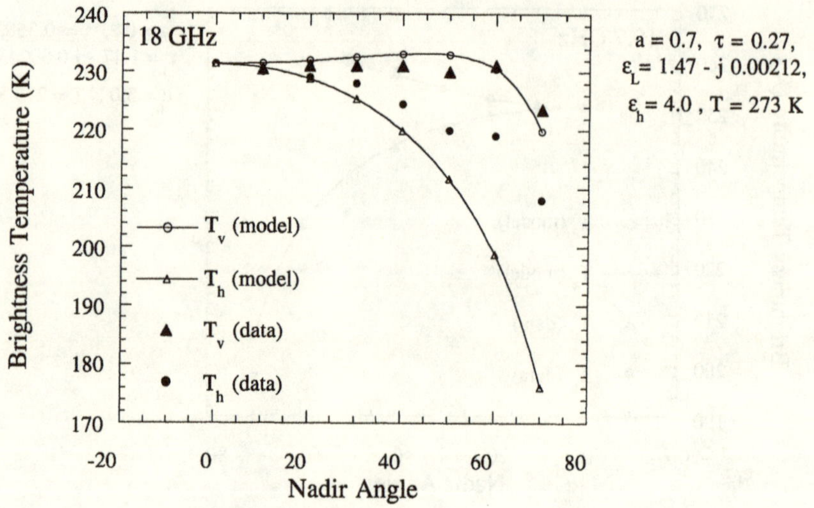

Figure 3.22 A comparison between model and emission measurements at 18 GHz.

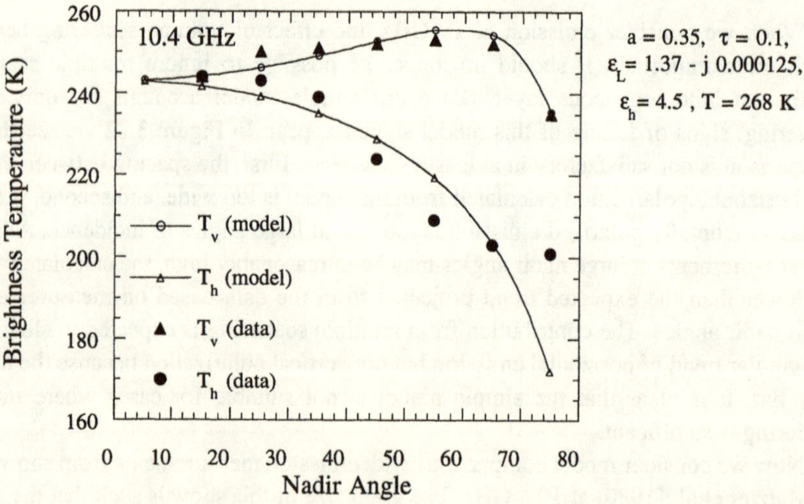

Figure 3.23 A comparison between model and emission from a snow field acquired by Matzler et al. [1980].

In summary, we are able to fit most of the snow emission using a plane Rayleigh layer above a homogeneous half space. This simple model incorporates only a single scattering mechanism. It is found to give good agreements whenever the albedo is small and/or the layer is thin so that multiple scattering does not significantly influence snow emission.

A case of emission from snow covered sea ice has been reported by Grenfell [1992]. The properties of this snow covered sea ice are detailed in Chapter 10, where comparisons are made using the more exact emission model based on matrix doubling. Although the snow is dry and has a large albedo, its optical depth is very thin and does not support an appreciable amount of multiple scattering. For this reason the simple model works well for this data set. The ice below the snow is a meltpond and hence quite homogeneous. The comparison is shown in Figure 3.24 where very good agreement is obtained between model and measurements. We shall see in Chapter 10 that better agreement can be realized using the more exact model, because at 18.7 GHz there is some multiple scattering effect.

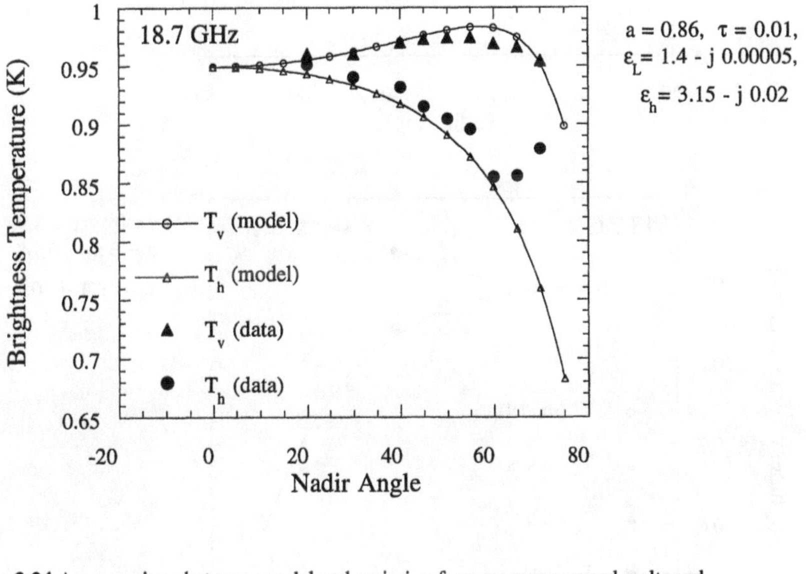

Figure 3.24 A comparison between model and emission from a snow covered meltpond.

Finally, we consider emission from a saline ice layer taken at the U.S. Army Cold Region Research Laboratory reported in Chapter 8 of Carsey [1992]. Comparisons between model and measurements are shown in Figure 3.25. This is a highly lossy ice layer but it is thin and below it is saline water. We do not expect much emission from water because of the large ice-water discontinuity and the large loss through the saline ice layer. Dominant emission contributions should be directly from the ice layer, which has very smooth boundaries. As seen from Figure 3.25 very good agreement is obtained at both 6.7 and 18.7 GHz using the simple layer model. There is, however, some inconsistency in the relative value of the layer permittivity in that it is not slightly larger at 6.7 GHz than 18.7 GHz.

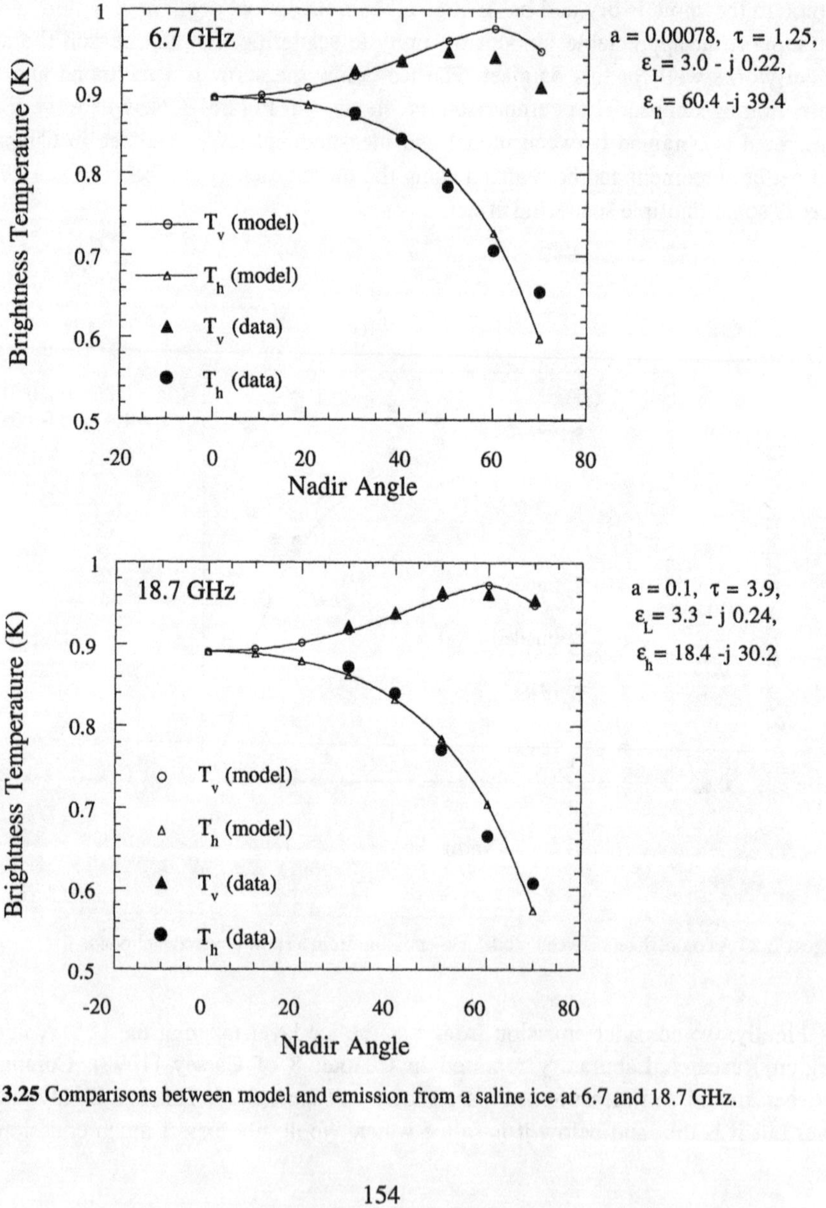

Figure 3.25 Comparisons between model and emission from a saline ice at 6.7 and 18.7 GHz.

3.7 A COMPARISON BETWEEN BACKSCATTERING AND EMISSION

In the previous section we avoided the consideration of the noncoherent transmission across a rough surface boundary. Furthermore, we assumed that horizontal scale roughness is not important in emission. In this section we examine the effect of noncoherent transmission and the effect of the horizontal scale roughness through reciprocity and investigate the effect of surface parameters on both backscattering and emission. The approach is to compute backscattering using the formula given in Appendix 2A of Chapter 2 and the emission across a rough surface is taken to be one minus the reflectivity of the surface boundary (see Chapter 1, Section 1.4). The reflectivity consists of two terms: a coherent term and a noncoherent term that is obtained by integrating the bistatic scattering coefficient defined in Appendix 2A of Chapter 2. In the next two subsections we shall calculate both emission and backscattering characteristics using the same set of surface parameters so that we can see how the change in a given parameter affects backscattering and emission.

3.7.1 Emission Characteristics

The emission from a rough surface represents the total transmitted power from the medium below the surface. From reciprocity this is the same as one minus the reflectivity from the same surface in the region above the surface. In Figure 3.26 we show the dependence of surface emission on the normalized vertical scale roughness, $k\sigma$, wave number times the rms surface height, for a random surface with a correlation function of the form (Appendix 2B),

$$\rho(\xi) = (1+\xi^2/L^2)^{-1.5} \tag{3.16}$$

As $k\sigma$ increases surface emission is seen to increase over small values of the nadir angle. At the same time the rate of increase in vertically polarized emission with nadir angle decreases. There is some decrease in the spacing between vertical and horizontal polarizations with increasing $k\sigma$. The normalized correlation length is denoted by kL and surface permittivity is denoted by ε_r in Figure 3.26.

Next, we show the effect of kL in Figure 3.27. In addition, we also show the total emission and emission calculated using the coherent reflectivity alone. The point to note is that for the chosen value of the surface permittivity an increase in kL makes the surface appear smoother and the angular behavior of the emission curve is more like that of a plane surface. When we calculate using the coherent reflectivity alone the level of the corresponding emission curve is higher because we did not subtract from unity the noncoherent power. However, the angular curves look similar whether we subtract or do not subtract the noncoherent term. If the permittivity were smaller, the noncoherent term would be smaller and the surface effect would be less. As a result the difference in emission between including and excluding noncoherent reflectivity will be small when permittivity is small.

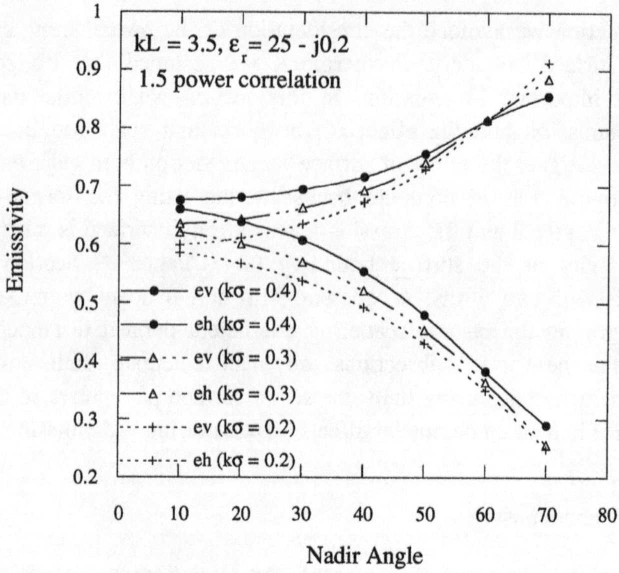

Figure 3.26 Dependence of surface emission on $k\sigma$ variation.

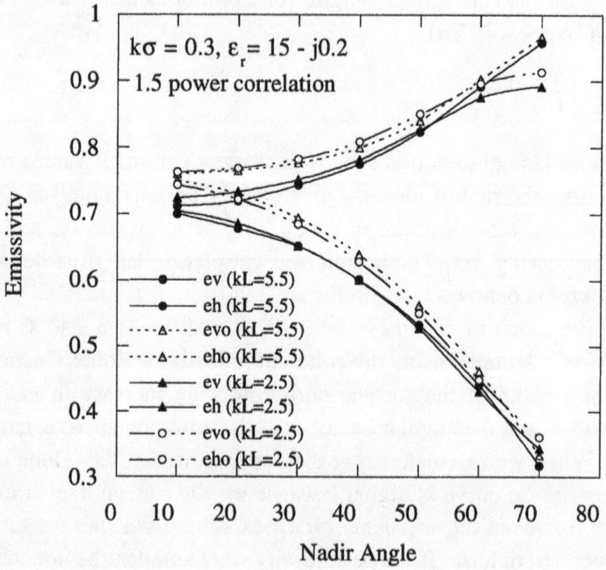

Figure 3.27 Dependence of surface emission on kL variation. The notations *evo* and *eho* represent emission calculations using only the coherent reflectivity.

Now we examine the effect of changing surface permittivity. This is shown in Figure 3.28. Again, calculations using only the coherent reflectivity are also shown to indicate that as permittivity decreases the difference between the calculations including and excluding the noncoherent reflectivity decreases. Furthermore, there is a change in the angular shape of the emission curves as permittivity decreases for surfaces with identical surface roughnesses.

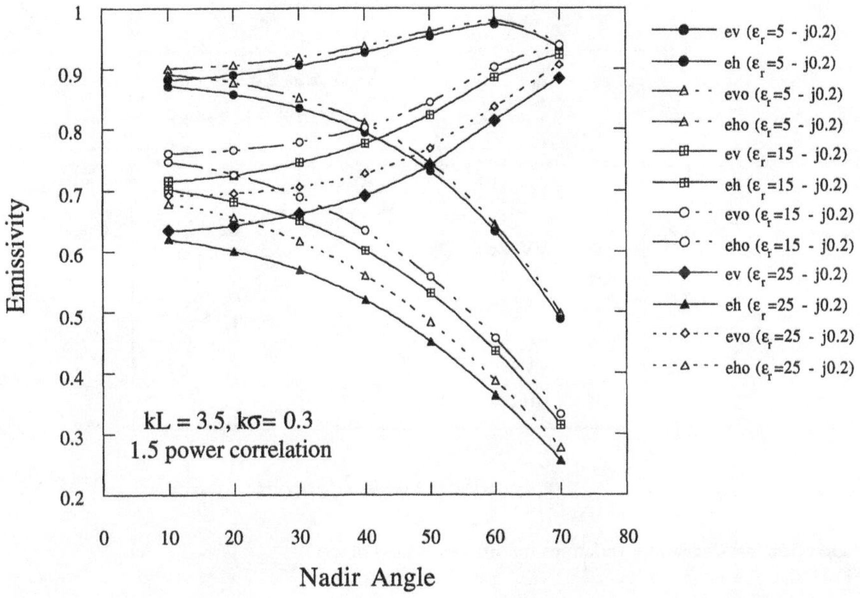

Figure 3.28 Dependence of surface emission on ε_r variation.

In conclusion, the general effect of surface roughness on emission is to narrow the spacing between vertical and horizontal polarizations and decrease the rise in vertical polarization with the nadir angle. These changes rely on the amount of noncoherently transmitted power, which has a significantly different angular behavior compared with that of the coherently transmitted power. When the surface permittivity is small, the surface roughness effect is less important because there is not enough discontinuity to generate a significant amount of noncoherently transmitted power.

3.7.2 Backscattering Characteristics

In this subsection we illustrate backscattering due to the same change in surface parameters as in emission given in the previous subsection. In Figure 3.29 we show backscattering variations of vertical and horizontal polarizations with $k\sigma$. It is seen that unlike emission it causes a change only in level but not in the shape of the angular curves. This in part is due to the use of a small $k\sigma$. It is interesting to note that *both emission and backscattering increase* with vertical roughness over the small nadir and incident angular range! Hence, the idea that backscattering and emission must go in opposite directions in all cases is false.

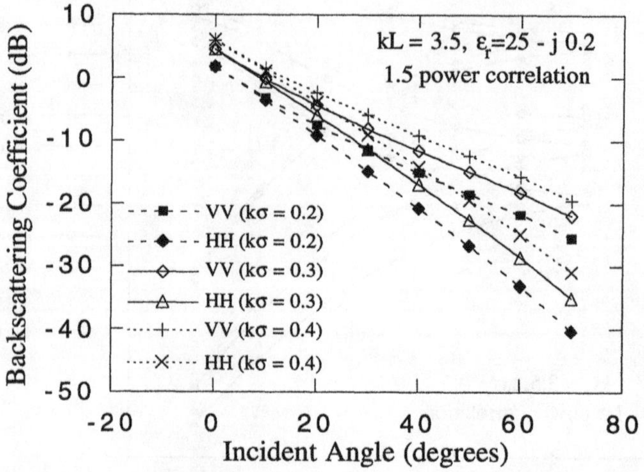

Figure 3.29 Backscattering variations for different values of $k\sigma$.

In Figure 3.30 we consider the variation in backscattering due to *kL*. Here, a change in *kL* causes significant variations in both the shape and level of the angular curves for both vertical and horizontal polarizations. A larger value of *kL* causes a significant drop off in the angular curves for both polarizations. In contrast, when we examine Figure 3.27, we see very little change in emission resulting from the same change in *kL*. This difference in sensitivity is due to the fact that emission depends strongly on coherent transmission while backscattering is sensitive only to noncoherent scattering.

Finally, we show the effect of a change in permittivity on backscattering in Figure 3.31. An increase in permittivity causes an increase in backscattering and some change in the angular curve. In emission (Figure 3.28) the same increase in permittivity generates a large decrease in emission. Thus, with respect to this parameter the change in backscattering and emission is in opposite direction. Furthermore, the changes in both backscattering and emission with the surface permittivity are significant. Thus, surface permittivity provides a useful test for both active and passive sensors.

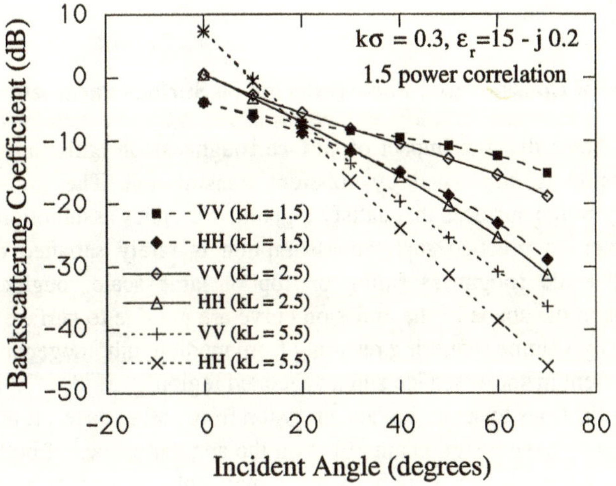

Figure 3.30 Backscattering variations for different values of *kL*.

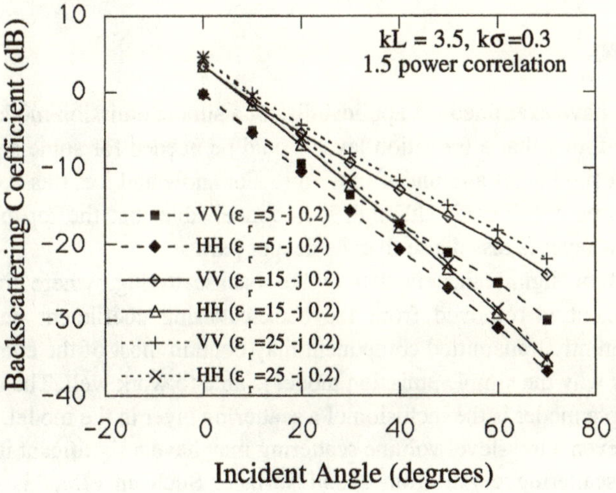

Figure 3.31 Backscattering variations for different values of surface permittivity.

In conclusion, the changes in backscattering and emission are not always in opposite directions. Such changes are dependent on the specific surface parameter. With respect to $k\sigma$, they are changing in the same direction; over small nadir and incident angles emission is not very sensitive while backscattering is very sensitive to changes in *kL*. It is with respect to surface permittivity that backscattering and emission are changing in opposite directions.

3.7.3 Dependence of Emission and Backscattering on Surface Parameters

The above study shows that the impact of surface roughness on emission is not very much when emission is dominated by coherent transmission. The impact can be significant for very rough surfaces that satisfy a geometric optics assumption [Ulaby et al., 1986]. However, in practice such an assumption is rarely satisfied due to the presence of small scale roughness riding on top of large scale roughness. If so, significant changes in the shape of the emission curve are not due to surface roughness but may be caused by volume scattering resulting from medium inhomogeneity. Such an effect is clearly present in snow, sea ice and a vegetated region.

In practical applications to scattering and emission from soil surfaces, it is found that volume scattering may have a significant effect on the angular shapes of both emission and backscattering curves, even though its level may only be one-fifth of the total emission or surface backscattering at large angles of incidence. Two possible reasons for the presence of volume scattering in a soil medium may be stated: (1) it can be caused by the loose dirt on top of soil surfaces and (2) the volume fraction of soil usually ranges from 50 to 70% and hence some air pockets are present.

3.8 DISCUSSIONS

In this chapter we have examined the applicability of a simple emission model as shown in Figure 3.1. We found that a transition layer would be needed for some soil surfaces, where there was a significant amount of loose dirt. For snow and ice, this layer was not needed for the cases considered, although it is conceivable that the top portion of a freshly fallen snow layer is less dense than its deeper part.

Another point of significance is that unlike backscattering, where the coherent scattering component is removed from the backscattering coefficient definition, in emission the coherently transmitted component may contain most of the energy. This is one of the reasons why the simple emission model is able to work well. The other factor that helps the simple model is the inclusion of a scattering layer in the model. As pointed out in Chapter 2, even a low-level volume scattering may have a significant influence on the angular backscattering curve from a soil surface. Such an effect is even more important in emission.

Finally, we showed in Section 3.7 that surface scattering and emission are influenced in different ways by different surface parameters. It is not always true that whenever emission increases backscattering must decrease or vice versa. Generally, the influence of surface roughness parameters is strong in backscattering but much less in emission because backscattering is due to noncoherent scattering while emission may be dominated by coherent transmission.

REFERENCES

Carsey, F.D., *Microwave Remote Sensing of Sea Ice*, AGU Monograph 68, 1992.

England, A.W., "Thermal Microwave Emission from a Scattering Layer," *J. Geophysical Res.*, Vol. 80, 1975, pp. 4484–4496.

Fung, A.K., and H.J. Eom, "Emission from a Rayleigh Layer with Irregular Boundaries," *J. Quart. Spectrosc. Rad. Transfer*, Vol. 26, 1981, pp. 397–409.

Fung, A.K., and H.J. Eom, "Multiple Scattering and Depolarization by a Randomly Rough Kirchhoff Surface," *IEEE Trans. Ant. and Prop.*, Vol. AP-29, no. 3, May 1981, pp. 463–471.

Grenfell, T.C., "Surface-Based Passive Microwave Study of Multiyear Sea Ice," *J. Geophysical Res.*, Vol. 86, no. C9, 1992, pp. 3485–3501.

Matzler, C., E. Schanda, R. Hofer and W. Good, "Microwave Signatures of the Natural Snow Cover at Weissfluhjoch," *NASA Conference Publication 2153*, 1980, pp. 203–223.

Mo, T., T.J. Schmugge and J.R. Wang, "Calculations of the Microwave Brightness Temperature of Rough Soil Surfaces: Bare Soil," *IEEE Trans. Geoscience and Remote Sensing*, Vol. 25, no. 1, 1987, pp. 47–54.

Nghiem, S.V., M.E. Veysoglu, J.A. Kong, R.T. Shin, P. O'Neill and A.W. Lohanick, "Polarimetric Passive Remote Sensing of a Periodic Spoil Surface: Microwave Measurements and Analysis," *J. Electromagnetic Waves and Applications*, Vol. 5, no. 9, 1991, pp. 997–1005.

Shiue, J.C., A.T. Chang, H. Boyne and D. Ellerbruch, "Remote Sensing of Snowpack with Microwave Radiometers for Hydrologic Applications," *Proc. of the Twelfth International Symposium on Remote Sensing of the Environment*, Vol. 2, 1978, pp. 877–886.

Tsang, L., "Polarimetric Passive Microwave Remote Sensing of Random Discrete Scatterers and Rough Surfaces," *J. Electromagnetic Waves and Applications*, Vol. 5, no. 1, 1991, pp. 41–57.

Tsang, L. and J.A. Kong, "The Brightness Temperature of a Half-Space Random Medium with Nonuniform Temperature Profile," *Radio Science*, 10, 1975, pp. 1025–1033.

Ulaby, F.T., R.K. Moore and A.K. Fung, *Microwave Remote Sensing: Active and Passive, From Theory to Applications*, Vol. 3, Artech House, Dedham, MA, 1986.

Wegmuller, U., "The Effect of Freezing and Thawing on the Microwave Signatures of Bare Soil," *Remote Sensing of Environment*, Vol. 33, 1990, pp. 123–135.

Chapter 4

Formulation of the Surface Scattering Problem

4.1 INTRODUCTION

The problem of wave scattering from a randomly rough surface has been studied using both low- and high-frequency approximations. Useful results were obtained in the 1960's [Kovalev and Pozknyak, 1961; Hagfors, 1966; Kodis, 1966; Fung, 1967 a, b, 1968; Valenzuela, 1967; Barrick, 1968] and were applied to the interpretation of measurements from land and sea surfaces. At the time it was recognized that radar backscattering at small angles of incidence seemed to follow the high-frequency solution based on the Kirchhoff formulation [Beckmann and Spizzichino, 1963; Sancer, 1969]. At large angles of incidence the first-order solution of the small perturbation method, a low-frequency solution, appeared to explain the measurements better, especially with regard to their polarization characteristics [Valenzuela, 1967; Fung 1967 a, b]. Towards the end of the 1960's and throughout the 1970's, attempts were made to unite the two approximate solutions with the hope that better agreement with measurements would result. This led to the development of two-scale models [Semyonov, 1966; Wright, 1968; Valenzuela, 1968; Leader, 1978; Brown, 1978], which did result in better agreement with measurements [Fuks, 1966; Bass and Fuks, 1979; Wu and Fung, 1972] and remain as the practical model throughout the 1980's [Ulaby et al., 1982].

There is however an obvious problem with the two-scale model. Real random surfaces may not have a two-scale distribution. Instead, they may have a continuous distribution of surface roughness. In addition, there is no simple method to separate a random surface into two scales of roughness. Beginning in the 1970's and throughout the 1980's many attempts have been made to obtain models with a wider range of validity than the two-scale model [DeSanto, 1974; Gray, et al., 1978; Garcia, et al., 1979; Brown, 1985; Bahar, 1985; Eftimiu, 1986; Fung and Pan, 1986, 1987; Winebrenner and Ishimaru, 1985]. The study by Fung and Pan [1986, 1987] gives expressions for the scattering coefficients directly in terms of the surface parameters, and these scattering coefficients reduce correctly in both the high- and low-frequency limits. Hence, their surface scattering model could be a possible candidate for a practical model. However, their scattering model is restricted to perfectly conducting surfaces with rms slopes less than 0.4.

In the late 1980's the surface backscatter enhancement phenomenon has been reported both experimentally and by computer simulation [Mendez and O'Donnell, 1987; Dainty et al., 1988; O'Donnell and Mendez, 1987; Maradudin et al., 1989; Gu, et al., 1989; Soto-Crespo and Nieto-Vesperinas, 1989]. This phenomenon occurs most clearly for surfaces with rms slopes close to unity and large rms heights. Hence, another model is needed that accounts for dielectric surfaces with large rms heights and rms slopes around unity or more. To do so a formulation suitable for estimating the tangential surface fields on a dielectric random surface is needed. Then, the scattered fields can be computed in terms of these surface fields. Subsequently, the average scattered power and the scattering coefficients can be obtained. The purpose of this chapter is to formulate such a model and discuss the simplifying assumptions made to realize a sufficiently simple form so that it is practical to use.

4.2 THE SURFACE FIELD INTEGRAL EQUATIONS FOR DIELECTRIC SURFACES

In this section we want to reformulate integral equations for tangential surface fields on a dielectric interface. The purpose of this reformulation is to obtain estimates of the tangential surface fields that are more general than the existing Kirchhoff or perturbation surface fields and reduce to known results under special conditions such as a perfectly conducting surface [Li and Fung, 1990]. Consider an incident plane wave defined by the following electric and magnetic fields,

$$\vec{E}^i = \hat{p}E_0 \exp[-j(\vec{k}_1 \cdot \vec{r})] = \hat{p}E^i \tag{4.1}$$

$$\vec{H}^i = \hat{k}_i \times (\hat{p}E^i)/\eta \tag{4.2}$$

where $\vec{k} = \hat{k}_i k$; $k = \omega\sqrt{\varepsilon_1\mu_1}$; $\hat{k}_i$ is the unit vector in the incident direction; $\hat{p}$ is the unit polarization vector; E_0 is the amplitude of the electric field, and η is the intrinsic impedance of region 1. In the field expressions a time factor, $\exp(j\omega t)$, is understood and the geometry of the problem is shown in Figure 4.1.

The governing equations for the tangential surface fields in medium 1 on a dielectric surface have been given by Poggio and Miller [1973] as

$$\hat{n} \times \vec{E} = 2\hat{n} \times \vec{E}^i - \frac{2}{4\pi}\hat{n} \times \int \vec{\mathcal{E}} ds' \tag{4.3}$$

and

$$\hat{n} \times \vec{H} = 2\hat{n} \times \vec{H}^i + \frac{2}{4\pi}\hat{n} \times \int \vec{\mathcal{H}} ds' \tag{4.4}$$

and in medium 2 we have

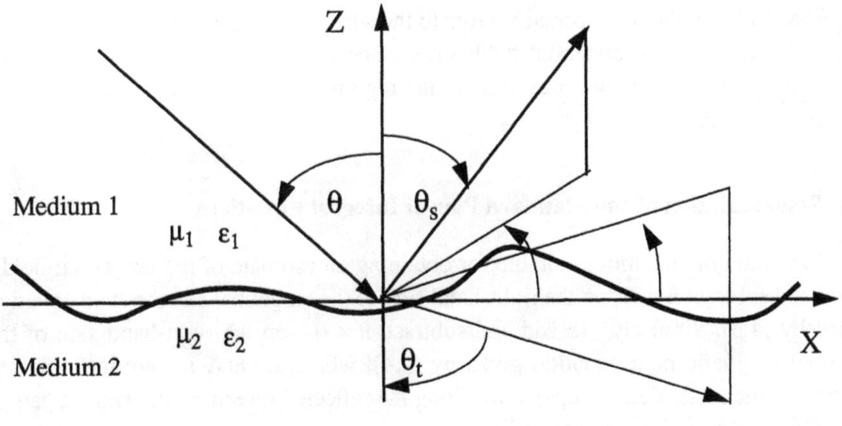

Figure 4.1 Geometry of the surface scattering problem.

$$\hat{n}_t \times \vec{E}_t = -\frac{2}{4\pi}\hat{n}_t \times \int \vec{\mathcal{E}}_t ds' \tag{4.5}$$

$$\hat{n}_t \times \vec{H}_t = \frac{2}{4\pi}\hat{n}_t \times \int \vec{\mathcal{H}}_t ds' \tag{4.6}$$

where

$$\vec{\mathcal{E}} = jk\eta\,(\hat{n}' \times \vec{H}')\,G - (\hat{n}' \times \vec{E}') \times \nabla' G - (\hat{n}' \cdot \vec{E}')\,\nabla' G \tag{4.7}$$

$$\vec{\mathcal{H}} = \frac{jk}{\eta}(\hat{n}' \times \vec{E}')\,G + (\hat{n}' \times \vec{H}') \times \nabla' G + (\hat{n}' \cdot \vec{H}')\,\nabla' G \tag{4.8}$$

and the fields in the lower medium can be written in terms of the fields in the upper medium by applying the boundary conditions on the continuity of the tangential fields as follows

$$\begin{aligned} \vec{\mathcal{E}}_t &= jk_t\eta_t\,(\hat{n}' \times \vec{H}'_t)\,G_t - (\hat{n}'_t \times \vec{E}'_t) \times \nabla' G_t - (\hat{n}'_t \cdot \vec{E}'_t)\,\nabla' G_t: \\ &= -\left[jk_t\eta_t\,(\hat{n}' \times \vec{H}')\,G_t - (\hat{n}' \times \vec{E}') \times \nabla' G_t - (\hat{n}' \cdot \vec{E}')\,\nabla' G_t\,(1/\varepsilon_r)\right] \end{aligned} \tag{4.9}$$

$$\begin{aligned} \vec{\mathcal{H}}_t &= \frac{jk_t}{\eta_t}(\hat{n}_t' \times \vec{E}'_t)\,G_t + (\hat{n}_t' \times \vec{H}'_t) \times \nabla' G_t + (\hat{n}_t' \cdot \vec{H}'_t)\,\nabla' G_t \\ &= -\left[\frac{jk_t}{\eta_t}(\hat{n}' \times \vec{E}')\,G_t + (\hat{n}' \times \vec{H}') \times \nabla' G_t + (\hat{n}' \cdot \vec{H}')\,\nabla' G_t\,(1/\mu_r)\right] \end{aligned} \tag{4.10}$$

$\hat{n}, \hat{n}', \hat{n}_t$ and $\hat{n}'_t$ are the unit normal vectors to the surface and $\hat{n}_t = -\hat{n}$, $\hat{n}'_t = -\hat{n}'$, $\hat{n} \times \vec{E}$ and $\hat{n} \times \vec{H}$ are the total tangential fields on surface in region 1. Similar interpretation applies to quantities with subscript t in region 2; $\varepsilon_r = \varepsilon_2/\varepsilon_1$, $\mu_r = \mu_2/\mu_1$, $\eta_t = \sqrt{\mu_2/\varepsilon_2}$ and $k_t = \omega\sqrt{\varepsilon_2\mu_2}$.

4.2.1 Reformulation of the Standard Pair of Integral Equations

To find the integral equations suitable for obtaining an estimate of the tangential fields, we add and subtract $\hat{n} \times \vec{E}^r$ on the right-hand side of the tangential electric field equation defined by (4.3). Similarly, we add and subtract $\hat{n} \times \vec{H}^r$ on the right-hand side of the tangential magnetic field equation given by (4.4), where $\vec{E}^r$ and $\vec{H}^r$ are the reflected electric and magnetic fields propagating along the reflected direction, $\hat{k}_r$. Thus, a pair of surface integral equations is obtained as follows

$$\hat{n} \times \vec{E} = (\hat{n} \times \vec{E})_k + (\hat{n} \times \vec{E})_c \tag{4.11}$$

and

$$\hat{n} \times \vec{H} = (\hat{n} \times \vec{H})_k + (\hat{n} \times \vec{H})_c \tag{4.12}$$

where $(\hat{n} \times \vec{E})_k$ and $(\hat{n} \times \vec{H})_k$ are the tangential fields under the Kirchhoff approximation, i.e.,

$$(\hat{n} \times \vec{E})_k = \hat{n} \times (\vec{E}^i + \vec{E}^r) \tag{4.13}$$

and

$$(\hat{n} \times \vec{H})_k = \hat{n} \times (\vec{H}^i + \vec{H}^r) \tag{4.14}$$

and $(\hat{n} \times \vec{E})_c$ and $(\hat{n} \times \vec{H})_c$ are defined as

$$(\hat{n} \times \vec{E})_c = \hat{n} \times (\vec{E}^i - \vec{E}^r) - \frac{2}{4\pi}\hat{n} \times \int \vec{\mathcal{E}} ds' \tag{4.15}$$

$$(\hat{n} \times \vec{H})_c = \hat{n} \times (\vec{H}^i - \vec{H}^r) + \frac{2}{4\pi}\hat{n} \times \int \vec{\mathcal{H}} ds' \tag{4.16}$$

The first term on the right-hand side of (4.11) is identical to the tangential Kirchhoff field. It requires a complementary term, which is the second term on the right-hand side of (4.11) to form the total tangential surface field. Thus, it is clear that the Kirchhoff tangential field alone cannot provide a good estimate to the total tangential surface field. Similar remarks are applicable to (4.12). To use (4.11) and (4.12) for estimating the total tangential field we need to express both the Kirchhoff field and the complementary field in terms of the incident field components and the surface reflectivity properties. Then seek an approximation to the total tangential field by removing the local angle dependence in the Fresnel reflection coefficients that appear in both the Kirchhoff term and the complementary term.

4.2.2 Representation in Terms of Local Coordinate Vectors and Fresnel Reflection Coefficients

To rewrite (4.11) and (4.12) in a form convenient for application we need to express the reflected fields in terms of the incident fields and the Fresnel reflection coefficients via a local coordinate system [Ulaby et al., 1982] defined below

$$\hat{t} = (\hat{k}_i \times \hat{n}) / |\hat{k}_i \times \hat{n}| \tag{4.17}$$

$$\hat{d} = \hat{k}_i \times \hat{t} \tag{4.18}$$

$$\hat{k}_i = \hat{t} \times \hat{d} \tag{4.19}$$

This set of coordinates is illustrated in Figure 4.2. In terms of it the incident field can be decomposed into locally horizontal and vertical components so that the corresponding reflected field components can be related to the incident field components through the Fresnel reflection coefficients. Note that the Fresnel reflection coefficient for horizontal polarization is defined for the electric field while that for vertical polarization is defined for the magnetic field. This point will become evident in the derivation given below.

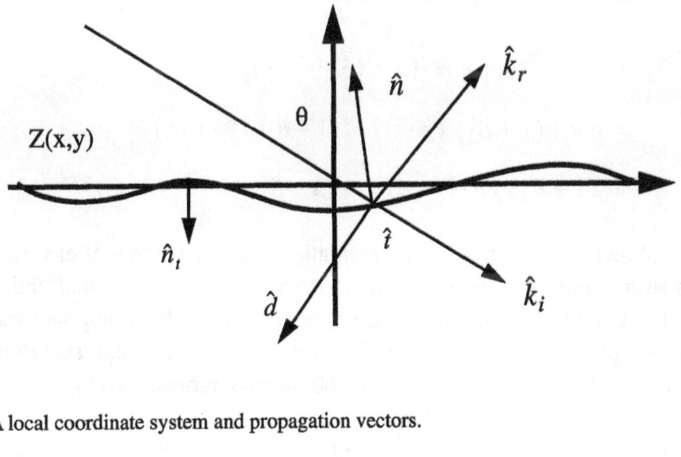

Figure 4.2 A local coordinate system and propagation vectors.

Let the incident electric field be decomposed into locally horizontally and vertically polarized components respectively as

$$\vec{E}^i = \vec{E}^i_\perp + \vec{E}^i_\parallel = (\hat{p}\cdot\hat{t})\hat{t}E^i + (\hat{p}\cdot\hat{d})\hat{d}E^i \tag{4.20}$$

and similarly the incident magnetic field is decomposed as

$$\vec{H}^i = \vec{H}^i_\perp + \vec{H}^i_\parallel = [(\hat{p}\cdot\hat{t})\hat{d}E^i - (\hat{p}\cdot\hat{d})\hat{t}E^i]/\eta \tag{4.21}$$

Kirchhoff Surface Fields

To rewrite the first term in (4.11) we shall consider the two polarization components separately as follows:

$$(\hat{n}\times\vec{E}_\perp)_k = \hat{n}\times\vec{E}^i_\perp(1+R_\perp) = \hat{n}\times\hat{t}(\hat{p}\cdot\hat{t})E^i(1+R_\perp) \tag{4.22}$$

and

$$\begin{aligned}(\hat{n}\times\vec{E}_\parallel)_k &= \hat{n}\times(\vec{E}^i_\parallel + \vec{E}^r_\parallel) = \hat{n}\times(-\hat{k}_i\times\vec{H}^i_\parallel\eta - \hat{k}_r\times\vec{H}^r_\parallel\eta)\\ &= \left[(\hat{n}\cdot\hat{k}_i)\vec{H}^i_\parallel + (\hat{n}\cdot\hat{k}_r)\vec{H}^i_\parallel R_\parallel\right]\eta = (1-R_\parallel)(\hat{n}\cdot\hat{k}_i)\eta\vec{H}^i_\parallel\\ &= -(1-R_\parallel)(\hat{n}\cdot\hat{k}_i)(\hat{p}\cdot\hat{d})\hat{t}E^i = (1-R_\parallel)(\hat{p}\cdot\hat{d})(\hat{n}\times\hat{d})E^i\end{aligned} \tag{4.23}$$

Finally, upon combining the two polarization results we obtain the Kirchhoff term or the first term in (4.11) rewritten in terms of the Fresnel reflection coefficients, local coordinate vectors and the incident fields as

$$\begin{aligned}(\hat{n}\times\vec{E})_k &= \hat{n}\times(\vec{E}^i + \vec{E}^r) = [\hat{n}\times(\vec{E}_\perp + \vec{E}_\parallel)]_k\\ &= \hat{n}\times[(1+R_\perp)(\hat{p}\cdot\hat{t})\hat{t} + (1-R_\parallel)(\hat{p}\cdot\hat{d})\hat{d}]E^i\\ &= (1+R_\perp)(\hat{p}\cdot\hat{t})\hat{n}\times\hat{t}E^i - (1-R_\parallel)(\hat{p}\cdot\hat{d})(\hat{n}\cdot\hat{k}_i)\hat{t}E^i\end{aligned} \tag{4.24}$$

We are showing two different representations for the tangential electric field in the above equation: one representation is in terms of the orthogonal unit vectors, $\hat{t}$ and $\hat{n}\times\hat{t}$, lying in the tangent plane; the other in terms of the orthogonal vectors, $\hat{t}$ and $\hat{d}$, lying in the plane of the incident field vector. Thus, the first representation is suitable for representing tangential fields; while the second representation is convenient for considering the incident field vectors.

Next, we want to express the tangential magnetic field in terms of the Fresnel reflection coefficients. Following the same approach as applied to the electric field we can write the first term in (4.12) multiplied by η as

$$\eta\,(\hat{n}\times\vec{H})_k = -[\,(1-R_\perp)\,(\hat{p}\cdot\hat{t})\,(\hat{n}\cdot\hat{k}_i)\,\hat{t} + (1+R_\parallel)\,(\hat{p}\cdot\hat{d})\,(\hat{n}\times\hat{t})\,]\,E^i$$

$$= \hat{n}\times[\,(1-R_\perp)\,(\hat{p}\cdot\hat{t})\,\hat{d} - (1+R_\parallel)\,(\hat{p}\cdot\hat{d})\,\hat{t}]\,E^i \tag{4.25}$$

The above results indicate that any tangential field can be defined in terms of two orthogonal vectors in the tangent plane. Here, the two preferred orthogonal unit vectors are $\hat{t}$ and $\hat{n}\times\hat{t}$. Note that $\hat{n}\times\hat{d} = -(\hat{n}\cdot\hat{k}_i)\,\hat{t}$ is not a unit vector. With this rewrite every quantity in these expressions is known.

Complementary Surface Field

To express the complementary or the second terms in (4.11) and (4.12) in a similar way, note that the contents of these terms are defined in (4.15) and (4.16). We would like to use an integral representation for the first terms in (4.15) and (4.16) so that the two terms on the right-hand side of (4.15) or (4.16) can be combined. To do so we need an integral representation of the tangential incident fields, which can be found by adding (4.3) to (4.5) and (4.4) to (4.6). The terms on the left-hand sides of the equations sum to zero because of the boundary conditions yielding

$$\hat{n}\times\vec{E}^i = \frac{1}{4\pi}\hat{n}\times\int(\vec{\mathcal{E}}-\vec{\mathcal{E}}_t)\,ds'$$

and (4.26)

$$\hat{n}\times\vec{H}^i = -\frac{1}{4\pi}\hat{n}\times\int(\vec{\mathcal{H}}-\vec{\mathcal{H}}_t)\,ds' \tag{4.27}$$

Recall that the first terms in (4.11) and (4.12) can be written in terms of two unit vectors in the tangent plane. Thus, we should expand (4.26) and (4.27) in terms of these unit vectors. The expansion of the term on the left-hand side of (4.26) yields

$$\hat{n}\times\vec{E}^i = (\hat{p}\cdot\hat{t})\,\hat{n}\times\hat{t}E^i - (\hat{n}\cdot\hat{k}_i)\,(\hat{p}\cdot\hat{d})\,\hat{t}E^i \tag{4.28}$$

Now we expand the right-hand side of (4.26) as

$$\hat{n}\times\int(\vec{\mathcal{E}}-\vec{\mathcal{E}}_t)\,ds' = (\hat{n}\times\hat{t})\,(\hat{n}\times\hat{t})\cdot\hat{n}\times\int(\vec{\mathcal{E}}-\vec{\mathcal{E}}_t)\,ds' + \hat{t}\hat{t}\cdot\hat{n}\times\int(\vec{\mathcal{E}}-\vec{\mathcal{E}}_t)\,ds' \tag{4.29}$$

Substituting (4.28) and (4.29) into (4.26) and equating vector components we obtain the following relations:

$$(\hat{p}\cdot\hat{t})\,E^i = \frac{1}{4\pi}\Big[(\hat{n}\times\hat{t})\cdot\hat{n}\times\int(\vec{\mathcal{E}}-\vec{\mathcal{E}}_t)\,ds'\Big] \tag{4.30}$$

and

$$(\hat{n}_1\cdot\hat{k}_i)\,(\hat{p}\cdot\hat{d})\,E^i = -\frac{1}{4\pi}\Big[\hat{t}\cdot\hat{n}\times\int(\vec{\mathcal{E}}-\vec{\mathcal{E}}_t)\,ds'\Big] \tag{4.31}$$

Thus, analogous to (4.24) we can write

$$\begin{aligned}\hat{n}\times(\vec{E}^i-\vec{E}^r) &= (1-R_\perp)\,(\hat{p}\cdot\hat{t})\,\hat{n}\times\hat{t}E^i-(1+R_\parallel)\,(\hat{n}\cdot\hat{k}_i)\,(\hat{p}\cdot\hat{d})\,\hat{t}E^i\\ &= (1-R_\perp)\,\hat{n}\times\hat{t}\,\{\frac{1}{4\pi}\Big[(\hat{n}\times\hat{t})\cdot\hat{n}\times\int(\vec{\mathcal{E}}-\vec{\mathcal{E}}_t)\,ds'\Big]\}\\ &\quad+(1+R_\parallel)\,\hat{t}\,\{\frac{1}{4\pi}\Big[\hat{t}\cdot\hat{n}\times\int(\vec{\mathcal{E}}-\vec{\mathcal{E}}_t)\,ds'\Big]\}\end{aligned} \tag{4.32}$$

Similarly, we can show that

$$\begin{aligned}\hat{n}\times(\vec{H}^i-\vec{H}^r) &= -(1+R_\parallel)\,\frac{1}{4\pi}\hat{t}\Big[\hat{t}\cdot\hat{n}\times\int(\vec{\mathcal{H}}-\vec{\mathcal{H}}_t)\,ds'\Big]\\ &\quad-(1-R_\parallel)\,\frac{1}{4\pi}\,(\hat{n}\times\hat{t})\,\Big[\hat{n}\times\hat{t}\cdot\hat{n}\times\int(\vec{\mathcal{H}}-\vec{\mathcal{H}}_t)\,ds'\Big]\end{aligned} \tag{4.33}$$

Now that the first terms in (4.15) and (4.16) are represented in integral form we can combine them with the second terms in these equations to yield

$$\begin{aligned}(\hat{n}\times\vec{E})_c &= -\frac{1}{4\pi}\,(\hat{n}\times\hat{t})\,\{\hat{n}\times\hat{t}\cdot\hat{n}\times\int\big[(1+R_\perp)\,\vec{\mathcal{E}}+(1-R_\perp)\,\vec{\mathcal{E}}_t\big]ds'\}\\ &\quad-\frac{1}{4\pi}\hat{t}\,\{\hat{t}\cdot\hat{n}\times\int\big[(1-R_\parallel)\,\vec{\mathcal{E}}+(1+R_\parallel)\,\vec{\mathcal{E}}_t\big]ds'\}\end{aligned} \tag{4.34}$$

$$\begin{aligned}(\hat{n}\times\vec{H})_c &= \frac{1}{4\pi}\,(\hat{n}\times\hat{t})\,\{\hat{n}\times\hat{t}\cdot\hat{n}\times\int\big[(1+R_\parallel)\,\vec{\mathcal{H}}+(1-R_\parallel)\,\vec{\mathcal{H}}_t\big]ds'\}\\ &\quad+\frac{1}{4\pi}\hat{t}\,\{\hat{t}\cdot\hat{n}\times\int\big[(1-R_\perp)\,\vec{\mathcal{H}}+(1+R_\perp)\,\vec{\mathcal{H}}_t\big]ds'\}\end{aligned} \tag{4.35}$$

Equations (4.34) and (4.35) represent the second or the complementary terms on the right-hand sides of (4.11) and (4.12) with the reflected fields replaced by the local coordinate vectors and the Fresnel reflection coefficients. Unlike the Kirchhoff terms there are unknown tangential surface fields under the integral sign. Thus, the complementary terms are yet to be determined. We shall consider approximate representations for these terms in the next section.

In summary, we now have expressions for the tangential surface field equations written in terms of the Fresnel reflection coefficients, local coordinate vectors and the incident fields as follows:

$$\hat{n}\times\vec{E} = \hat{n}\times[(1+R_\perp)(\hat{p}\cdot\hat{t})\hat{t}+(1-R_\parallel)(\hat{p}\cdot\hat{d})\hat{d}]E^i$$
$$-\frac{1}{4\pi}(\hat{n}\times\hat{t})\{\hat{n}\times\hat{t}\cdot\hat{n}\times\int[(1+R_\perp)\vec{\mathcal{E}}+(1-R_\perp)\vec{\mathcal{E}}_t]ds'\}$$
$$-\frac{1}{4\pi}\hat{t}\{\hat{t}\cdot\hat{n}\times\int[(1-R_\parallel)\vec{\mathcal{E}}+(1+R_\parallel)\vec{\mathcal{E}}_t]ds'\} \tag{4.36}$$

$$\hat{n}\times\vec{H} = \hat{n}\times[(1-R_\perp)(\hat{p}\cdot\hat{t})\hat{d}-(1+R_\parallel)(\hat{p}\cdot\hat{d})\hat{t}](E^i/\eta)$$
$$+\frac{1}{4\pi}(\hat{n}\times\hat{t})\{\hat{n}\times\hat{t}\cdot\hat{n}_1\times\int[(1+R_\parallel)\vec{\mathcal{H}}+(1-R_\parallel)\vec{\mathcal{H}}_t]ds'\}$$
$$+\frac{1}{4\pi}\hat{t}\{\hat{t}\cdot\hat{n}\times\int[(1-R_\perp)\vec{\mathcal{H}}+(1+R_\perp)\vec{\mathcal{H}}_t]ds'\} \tag{4.37}$$

The above equations are in the desired form for estimating the tangential surface fields. In these equations the fields, $\vec{\mathcal{H}}$, $\vec{\mathcal{H}}_t$, and $\vec{\mathcal{E}}$, $\vec{\mathcal{E}}_t$ are given by (4.7) through (4.10). From Poggio and Miller [1973] the normal components of the fields relate to their tangential components through the surface divergence operator as follows:

$$\hat{n}\cdot\vec{E} = \frac{j\eta}{k}\nabla_s\cdot(\hat{n}\times\vec{H}) \tag{4.38}$$

where the subscript s for the del operator denotes surface divergence and

$$\hat{n}\cdot\vec{H} = -\frac{j}{k\eta}\nabla_s\cdot(\hat{n}\times\vec{E}) \tag{4.39}$$

Hence, these fields are themselves functions of the tangential fields and the above equations are integral equations.

It is interesting to note that if the lower medium is a perfect conductor, then the tangential electric field vanishes because $\vec{\mathcal{E}}_t = 0$, $R_\perp = -1$ and $R_\parallel = 1$. In the magnetic field equation terms involving $\vec{\mathcal{H}}_2$ are zero because of the reflection coefficients. Using (4.21) we can show that the term without the integral sign reduces to $2\hat{n}\times\vec{H}^i$. From (4.8) $\vec{\mathcal{H}} = (\hat{n}'\times\vec{H}')\times\nabla'G$. Hence, the final form of the magnetic field integral equation for a perfectly conducting surface is

$$\hat{n}\times\vec{H} = 2\hat{n}\times\vec{H}^i+\frac{1}{2\pi}\hat{n}\times\int(\hat{n}'\times\vec{H}')\times\nabla'G\,ds' \tag{4.40}$$

4.3 ESTIMATION OF TANGENTIAL SURFACE FIELDS

Although (4.36) and (4.37) can be used to estimate the tangential surface fields, it is possible to obtain much simpler approximate expressions if we are willing to use different expressions for different polarizations. In this section we shall derive these simpler expressions. We begin by noting that (4.24) can be written in two other forms,

$$
\begin{aligned}
(\hat{n}\times\vec{E})_k &= \hat{n}\times[(1+R_\perp)(\hat{p}\cdot\hat{t})\hat{t}+(1-R_\parallel)(\hat{p}\cdot\hat{d})\hat{d}]E^i \\
&= \hat{n}\times[(1+R_\perp)\hat{p}-(R_\perp+R_\parallel)(\hat{p}\cdot\hat{d})\hat{d}]E^i \\
&= \hat{n}\times[(1-R_\parallel)\hat{p}+(R_\parallel+R_\perp)(\hat{p}\cdot\hat{t})\hat{t}]E^i
\end{aligned}
\tag{4.41}
$$

Similarly, (4.25) can also be written in other forms,

$$
\begin{aligned}
\eta(\hat{n}\times\vec{H})_k &= \hat{n}\times[(1-R_\perp)(\hat{p}\cdot\hat{t})\hat{d}-(1+R_\parallel)(\hat{p}\cdot\hat{d})\hat{t}]E^i \\
&= \hat{n}\times\{\hat{k}_i\times[(1-R_\perp)(\hat{p}\cdot\hat{t})\hat{t}+(1+R_\parallel)(\hat{p}\cdot\hat{d})\hat{d}]\} \\
&= \hat{n}\times\{\hat{k}_i\times[(1-R_\perp)\hat{p}+(R_\perp+R_\parallel)(\hat{p}\cdot\hat{d})\hat{d}]\} \\
&= \hat{n}\times\{\hat{k}_i\times[(1+R_\parallel)\hat{p}+(R_\parallel+R_\perp)(\hat{p}\cdot\hat{t})\hat{t}]\}
\end{aligned}
\tag{4.42}
$$

Simplified Kirchhoff Fields

For vertically polarized incident and scattering calculations, $\hat{p} = \hat{v}$ and $(\hat{v}\cdot\hat{t})$ is small compared with unity and $(R_\parallel + R_\perp)$ is generally much smaller than unity except near grazing. Hence, we ignore the last term in (4.41) and let

$$(\hat{n}\times\vec{E})_{kv} \approx (1-R_\parallel)\,\hat{n}\times\hat{v}E^i \tag{4.43}$$

This approximation leads to a significantly simpler expression because it no longer depends on local coordinate vectors (i.e., $\hat{d}$ and $\hat{t}$). If we consider the vertically polarized scattered field, the above approximation is fully justified. However, if we want to receive horizontally polarized scattering, then the second term in (4.42) cannot be ignored and a different treatment is needed. For the tangential magnetic field, similar consideration from (4.42) leads to an approximate tangential magnetic field as follows:

$$\eta(\hat{n}\times\vec{H})_{kv} \approx (1+R_\parallel)\,\hat{n}\times(\hat{k}_i\times\hat{v})E^i \tag{4.44}$$

Next, we consider horizontal polarization. Here, $\hat{p} = \hat{h}$ and $(\hat{h}\cdot\hat{d})$ is small compared with unity. Hence, this time we approximate (4.41) by

$$(\hat{n}\times\vec{E})_{kh} \approx (1+R_\perp)\,\hat{n}\times\hat{h}E^i \tag{4.45}$$

The above result for horizontal polarization is similar to the corresponding one for vertical polarization. Indeed, the same concept is used in the above simplification. It follows that we can approximate the magnetic tangential surface field in an analogous manner yielding

$$\eta\,(\hat{n}\times\vec{H})_{kh} \approx (1-R_{\perp})\,\hat{n}\times(\hat{k}_i\times\hat{h})\,E^i \tag{4.46}$$

For cross polarized scattering the above approximations do not apply because we cannot make one term small compared with the other in the scattered field calculation. For this case we should return to (4.24) and (4.25) or include the $(R_{\parallel}+R_{\perp})$ terms in the above tangential field expressions. However, when the local unit vector is included, results are much more complicated. To obtain a compromise in estimating the tangential fields for cross polarization we shall use the average of the two possible forms in (4.41) and drop the $(R_{\parallel}+R_{\perp})$ term, yielding

$$(\hat{n}\times\vec{E})_{kc} \approx (1-R)\,(\hat{n}\times\hat{p})\,E^i \tag{4.47}$$

where $R = (R_{\parallel}-R_{\perp})/2$. The error in dropping the $(R_{\parallel}+R_{\perp})$ term is larger than similar cases in like polarization. However, it is the same for both cross polarized components. Similarly, for the tangential magnetic field in cross polarized calculation we use

$$\eta\,(\hat{n}\times\vec{H})_{kc} \approx (1+R)\,\hat{n}\times(\hat{k}_i\times\hat{p})\,E^i \tag{4.48}$$

The above relations provide estimates of the Kirchhoff tangential fields in (4.11) and (4.12). It is interesting to note that for perfectly conducting surfaces, these approximate expressions become exact with $(\hat{n}\times\vec{H})_k = 2\hat{n}\times\vec{H}^i$ because the $(R_{\parallel}+R_{\perp})$ terms are zero.

Simplified Complementary Fields

In order to obtain estimates of the complementary terms we shall use the above estimates for the unknown tangential fields under the integral signs in (4.34) and (4.35). Thus, all unknown quantities on the right-hand sides of (4.34) and (4.35) are removed. Next, we want to further simplify (4.34) and (4.35) by developing polarization dependent expressions for like and cross polarized scattering as we did for the Kirchhoff expressions.

Let us begin with vertically polarized case. On the right-hand side of (4.34) we add and subtract two terms of the following forms:

$$(1-R_{\parallel})\,\frac{1}{4\pi}\,(\hat{n}\times\hat{t})\,(\hat{n}\times\hat{t}\cdot\hat{n}\times\int\vec{\mathcal{E}}_v ds') \tag{4.49}$$

and

$$(1+R_{\|})\frac{1}{4\pi}(\hat{n}\times\hat{t})\ (\hat{n}\times\hat{t}\cdot\hat{n}\times\int\vec{\mathcal{E}}_v ds') \tag{4.50}$$

As a result (4.34) becomes

$$\begin{aligned}(n\times\vec{E}_v)_c &= -(1-R_{\|})\frac{1}{4\pi}\hat{n}\times\int\vec{\mathcal{E}}_v ds' - (1+R_{\|})\frac{1}{4\pi}\hat{n}\times\int\vec{\mathcal{E}}_{vt}\,ds' \\ &\quad -(R_{\|}+R_{\perp})\frac{1}{4\pi}\hat{n}\times\hat{t}\left[\hat{n}\times\hat{t}\cdot\hat{n}\times\int(\vec{\mathcal{E}}_v-\vec{\mathcal{E}}_{vt})\,ds'\right] \\ &\approx -\frac{1}{4\pi}\hat{n}\times\int\left[(1-R_{\|})\,\vec{\mathcal{E}}_v+(1+R_{\|})\,\vec{\mathcal{E}}_{vt}\right]ds'\end{aligned} \tag{4.51}$$

We ignore the $(R_{\|}+R_{\perp})$ term in the above equation because when we replace the integral factor using (4.26) we obtain the dot product

$$(\hat{n}\times\hat{t})\cdot(\hat{n}\times\vec{E}^i) = (\hat{n}\times\hat{t})\cdot(\hat{n}\times\hat{v})\,E^i \tag{4.52}$$

which is a small quantity. As mentioned earlier the sum of the Fresnel reflection coefficients is also small over most of the angular regions except near grazing incidence. Thus, we expect the approximation to be very good. Similarly, if we add and subtract the following two terms,

$$(1+R_{\|})\frac{1}{4\pi}\hat{t}\,(\hat{t}\cdot\hat{n}\times\int\vec{\mathcal{H}}_v\,ds') \tag{4.53}$$

$$(1-R_{\|})\frac{1}{4\pi}\hat{t}\,(\hat{t}\cdot\hat{n}\times\int\vec{\mathcal{H}}_{vt}\,ds')\ 12 \tag{4.54}$$

on the right-hand side of (4.35), we obtain

$$\begin{aligned}(\hat{n}\times\vec{H}_v)_c &= (1+R_{\|})\frac{1}{4\pi}(\hat{n}\times\int\vec{\mathcal{H}}_v\,ds') + (1-R_{\|})\frac{1}{4\pi}(\hat{n}\times\int\vec{\mathcal{H}}_{vt}\,ds') \\ &\quad -(R_{\perp}+R_{\|})\frac{1}{4\pi}\hat{t}\left[\hat{t}\cdot\hat{n}\times\int(\vec{\mathcal{H}}_v-\vec{\mathcal{H}}_{vt})\,ds'\right] \\ &\approx \frac{1}{4\pi}\hat{n}\times\int\left[(1+R_{\|})\,\vec{\mathcal{H}}_v+(1-R_{\|})\,\vec{\mathcal{H}}_{vt}\right]ds'\end{aligned} \tag{4.55}$$

The above relations provide the tangential fields for the complementary terms under vertical polarization. By analogy for horizontal polarization we shall have the following estimates

$$(n\times\vec{E}_h)_c = -\frac{1}{4\pi}\hat{n}\times\int\left[(1+R_\perp)\vec{\mathcal{E}}_h + (1-R_\perp)\vec{\mathcal{E}}_{ht}\right]ds'$$

$$+(R_\perp + R_\parallel)\,\frac{1}{4\pi}\hat{t}\left[\hat{t}\cdot\hat{n}\times\int(\vec{\mathcal{E}}_h - \vec{\mathcal{E}}_{ht})\,ds'\right]$$

$$\approx -\frac{1}{4\pi}\hat{n}\times\int\left[(1+R_\perp)\vec{\mathcal{E}}_h + (1-R_\perp)\vec{\mathcal{E}}_{ht}\right]ds' \tag{4.56}$$

$$(\hat{n}\times\vec{H}_h)_c = \frac{1}{4\pi}\hat{n}\times\int\left[(1-R_\perp)\vec{\mathcal{H}}_h + (1+R_\perp)\vec{\mathcal{H}}_{ht}\right]ds'$$

$$+\,(R_\parallel + R_\perp)\frac{1}{4\pi}\hat{n}\times\hat{t}\left[\hat{n}\times\hat{t}\cdot\hat{n}\times\int(\vec{\mathcal{H}}_h - \vec{\mathcal{H}}_{ht})\,ds'\right]$$

$$\approx \frac{1}{4\pi}\hat{n}\times\int\left[(1-R_\perp)\vec{\mathcal{H}}_h + (1+R_\perp)\vec{\mathcal{H}}_{ht}\right]ds' \tag{4.57}$$

For cross polarization we should use (4.34) and (4.35) or include the $(R_\parallel + R_\perp)$ terms in the above relations. Again, the results are very complicated. Similar estimates for the tangential electric and magnetic fields can also be found for cross polarization but they are less accurate. The reasoning is based on the fact that one can write $(\hat{n}\times\vec{E}_p)_c$ in two different forms and we shall use the average of the two forms as approximate estimates. That is,

$$(\hat{n}\times\vec{E}_p)_c \approx -\frac{1}{4\pi}\hat{n}\times\int\left[(1-R)\vec{\mathcal{E}}_p + (1+R)\vec{\mathcal{E}}_{pt}\right]ds' \tag{4.58}$$

and

$$(\hat{n}\times\vec{H}_p)_c \approx \frac{1}{4\pi}\hat{n}\times\int\left[(1+R)\vec{\mathcal{H}}_p + (1-R)\vec{\mathcal{H}}_{pt}\right]ds' \tag{4.59}$$

In this section much simpler but approximate expressions for the tangential fields are obtained for like and cross polarized calculations. These expressions are polarization dependent so that instead of two expressions, one for the electric field and one for the magnetic field, we have six expressions: two for vertical polarization, two for horizontal polarization and two for cross polarization. This is true for the Kirchhoff term as well as its complementary term.

4.4 FAR-ZONE SCATTERED FIELDS

Now that estimates for the surface fields are available, we can calculate the far-zone scattered field in terms of them. In accordance with the Stratton-Chu integral [Ulaby et al., 1982, Chapter 12] we can write the far-zone scattered field in the medium above the rough surface as

$$E_{qp}^{s} = C\int \{\hat{q}\cdot[\hat{k}_s\times(\hat{n}\times\vec{E}_p)+\eta(\hat{n}\times\vec{H}_p)]\}\exp[j(\vec{k}_s\cdot\vec{r})]\,ds$$

$$= C\int[\hat{q}\times\hat{k}_s\cdot(\hat{n}\times\vec{E}_p)+\eta\hat{q}\cdot(\hat{n}\times\vec{H}_p)]\exp[j(\vec{k}_s\cdot\vec{r})]\,ds \tag{4.60}$$

where $C = -\frac{jk}{4\pi R}\exp(-jkR)$

$$\vec{k}_s = k\hat{k}_s = k(\hat{x}\sin\theta_s\cos\phi_s+\hat{y}\sin\theta_s\sin\phi_s+\hat{z}\cos\theta_s)$$

$$= \hat{x}k_{sx}+\hat{y}k_{sy}+\hat{z}k_{sz} \tag{4.61}$$

In (4.60) the incident polarization is denoted by p and the receiving polarization by q. The polarization vector $\hat{p}$ may be equal to $\hat{h}$ or $\hat{v}$ defined as follows (Figure 4.1):

$$\hat{h} = \hat{\phi} = -\hat{x}\sin\phi+\hat{y}\cos\phi \tag{4.62}$$

$$\hat{v} = -\hat{\theta} = \hat{x}\cos\theta\cos\phi+\hat{y}\cos\theta\sin\phi+\hat{z}\sin\theta \tag{4.63}$$

The polarization vector $\hat{q}$ may be equal to $\hat{h}_s$ or $\hat{v}_s$ defined below:

$$\hat{h}_s = -\hat{\phi}_s = \hat{x}\sin\phi_s-\hat{y}\cos\phi_s \tag{4.64}$$

$$\hat{v}_s = -\hat{\theta}_s = -\hat{x}\cos\theta_s\cos\phi_s-\hat{y}\cos\theta_s\sin\phi_s+\hat{z}\sin\theta_s \tag{4.65}$$

The above choice of the polarization vectors is such that in backscattering, $\theta_s = \theta$ and $\phi_s = \phi\mp\pi$, $\hat{v}_s = \hat{v}$ and $\hat{h}_s = \hat{h}$. Since estimates of the tangential surface fields have been obtained in the previous section, we can now calculate the scattered field. Recall that we have two components for the tangential surface fields, the Kirchhoff component and its complementary component. Hence, we define two corresponding components for the scattered fields as follows:

$$E_{qp}^{s} = E_{qp}^{k}+E_{qp}^{c} \tag{4.66}$$

where

$$E_{qp}^{k} = CE_0\int f_{qp}\exp\{j[(\vec{k}_s-\vec{k}_i)\cdot\vec{r}]\}\,dxdy \tag{4.67}$$

and

$$E_{qp}^{c} = \frac{CE_0}{8\pi^2}\int F_{qp}\exp\{ju(x-x')+jv(y-y')+j\vec{k}_s\cdot\vec{r}'-j\vec{k}_i\cdot\vec{r}'\}\,dxdydx'dy'dudv \tag{4.68}$$

In the above scattered field expressions the definitions of f_{qp} and F_{qp} are as follows:

$$f_{qp} = [\hat{q}\times\hat{k}_s\cdot(\hat{n}\times\vec{E}_p)_k + \eta\hat{q}\cdot(\hat{n}\times\vec{H}_p)_k]\,(D_1/E^i) \tag{4.69}$$

with $D_1 = (Z_y^2+Z_x^2+1)^{1/2}$ and the tangential electric and magnetic fields replaced by the corresponding tangential Kirchhoff fields and

$$F_{qp}{}' = 8\pi^2[\hat{q}\times\hat{k}_s\cdot(\hat{n}\times\vec{E}_p)_c + \eta\hat{q}\cdot(\hat{n}\times\vec{H}_p)_c]\,(D_1/E^i) \tag{4.70}$$

where the tangential fields are the complementary tangential fields given by (4.51) and (4.55) through (4.59). The complementary field coefficient F_{qp} is equal to $F_{qp}{}'$ after the Green's function and its gradient are replaced by spectral representations and after the phase factor $\exp[ju(x-x')+jv(y-y')]$ in the Green's function and u, v, x', y' integrations are factored out. In general, both f_{qp} and F_{qp} are dimensionless, complicated expressions and depend on spatial variables. We shall seek approximate representations for them in the next two subsections so that they become independent of spatial variables. Their final representations are summarized in Appendix 4B. The approaches and major steps leading to these approximate expressions are given in the next two subsections.

4.4.1 Approximate Expressions for f_{qp}

In this section we consider the major steps in the derivation of approximate expressions for f_{qp}. Our intention is to obtain a scattered field and subsequently a scattering coefficient that is applicable to randomly rough surfaces with roughness scales that may be small, comparable or large compared with the incident wavelength. Thus, we cannot use approximations based on the size of roughness or the size of the incident wavelength. In the rough surface scattering literature, the Fresnel reflection coefficients have been approximated as a function of the incident angle at low frequencies and as a function of the specular angle at high frequencies. It is important to note that the Fresnel reflection coefficients become independent of the spatial variables in both approximations. Thus, we shall treat them as independent of the spatial variables and set the local angle of incidence in them as equal to either the specular angle or the incidence angle depending on the ranges of validity of these approximations. Details on the ranges of validity of these approximations are given in Chapter 6.

To obtain a more explicit expression for f_{qp} let us start from (4.69) and substitute the tangential Kirchhoff fields from (4.43) through (4.46) into polarized expressions and (4.41), (4.42) into cross polarized expressions yielding

$$f_{vv} \approx -[(1-R_{\parallel})\,\hat{h}_s\cdot(\hat{n}\times\hat{v}) + (1+R_{\parallel})\,\hat{v}_s\cdot(\hat{n}\times\hat{h})]\,D_1 \tag{4.71}$$

$$f_{hh} \approx [(1-R_{\perp})\,\hat{h}_s\cdot(\hat{n}\times\hat{v}) + (1+R_{\perp})\,\hat{v}_s\cdot(\hat{n}\times\hat{h})]\,D_1 \tag{4.72}$$

$$
\begin{aligned}
f_{hv} &= [\hat{h}_s \times \hat{k}_s \cdot (\hat{n} \times \vec{E}_v) + \eta \hat{h}_s \cdot (\hat{n} \times \vec{H}_v)]\, D_1 / E^i \\
&= \hat{v}_s \cdot \{\hat{n} \times [(1 - R_{\|})\, \hat{v} + (R_{\|} + R_{\perp})\, (\hat{v} \cdot \hat{t})\, \hat{t}]\, D_1\} \\
&\quad -\hat{h}_s \cdot \{\hat{n} \times [(1 + R_{\|})\, \hat{h} + (R_{\perp} + R_{\|})\, (\hat{v} \cdot \hat{t})\, \hat{d}]\, D_1\} \\
&= [(1 - R_{\|})\, \hat{v}_s \cdot (\hat{n} \times \hat{v}) - (1 + R_{\|})\, \hat{h}_s \cdot (\hat{n} \times \hat{h})]\, D_1 \\
&\quad + (R_{\perp} + R_{\|})\, (\hat{v} \cdot \hat{t})\, [\hat{v}_s \cdot (\hat{n} \times \hat{t}) - \hat{h}_s \cdot (\hat{n} \times \hat{d})]\, D_1
\end{aligned}
\tag{4.73}
$$

$$
\begin{aligned}
f_{vh} &= \left[\hat{v}_s \times \hat{k}_s \cdot (\hat{n} \times \vec{E}_h) + \eta \hat{v}_s \cdot (\hat{n} \times \vec{H}_h)\right] D_1 / E^i \\
&= -\hat{h}_s \cdot \{\hat{n} \times [(1 + R_{\perp})\, \hat{h} - (R_{\|} + R_{\perp})\, (\hat{h} \cdot \hat{d})\, \hat{d}]\, D_1\} \\
&\quad + \hat{v}_s \cdot \{\hat{n} \times [(1 - R_{\perp})\, \hat{v} - (R_{\perp} + R_{\|})\, (\hat{h} \cdot \hat{d})\, \hat{t}]\, D_1\} \\
&= [(1 - R_{\perp})\, \hat{v}_s \cdot (\hat{n} \times \hat{v}) - (1 + R_{\perp})\, \hat{h}_s \cdot (\hat{n} \times \hat{h})]\, D_1 \\
&\quad - (R_{\perp} + R_{\|})\, (\hat{h} \cdot \hat{d})\, [\hat{v}_s \cdot (\hat{n} \times \hat{t}) - \hat{h}_s \cdot (\hat{n} \times \hat{d})]\, D_1
\end{aligned}
\tag{4.74}
$$

In the above we did not use approximate tangential fields for the cross polarization. It is clear from the above relations that in general f_{vh} is the dual of f_{hv} and f_{hh} is the negative dual of f_{vv}. For backscattering significant simplification occurs for the f_{qp}'s. Since this is a case of practical importance, we list below the f_{qp}'s under this condition:

$$
f_{vv} \approx [(1 - R_{\|})\, (\hat{n} \cdot \hat{k}_i) - (1 + R_{\|})\, (\hat{n} \cdot \hat{k}_i)]\, D_1 = -2R_{\|}\, (\hat{n} \cdot \hat{k}_i)\, D_1 \tag{4.75}
$$

$$
f_{hh} \approx [(1 + R_{\perp})\, (\hat{n} \cdot \hat{k}_i) - (1 - R_{\perp})\, (\hat{n} \cdot \hat{k}_i)]\, D_1 = 2R_{\perp}\, (\hat{n} \cdot \hat{k}_i)\, D_1 \tag{4.76}
$$

$$
\begin{aligned}
f_{vh} = f_{hv} &= -2\, (R_{\perp} + R_{\|})\, (\hat{h} \cdot \hat{d})\, (\hat{n} \cdot \hat{k}_i)\, (\hat{h} \cdot \hat{t})\, D_1 \\
&= 2\, (R_{\perp} + R_{\|})\, (\hat{v} \cdot \hat{t})\, (\hat{n} \cdot \hat{k}_i)\, (\hat{h} \cdot \hat{t})\, D_1 \\
&= 2\, (R_{\perp} + R_{\|})\, (\hat{v} \cdot \hat{t})\, (\hat{n} \cdot \hat{k}_i)\, (\hat{v} \cdot \hat{d})\, D_1 \\
&= -2\, (R_{\perp} + R_{\|})\, (\hat{h} \cdot \hat{d})\, (\hat{n} \cdot \hat{k}_i)\, (\hat{v} \cdot \hat{d})\, D_1
\end{aligned}
\tag{4.77}
$$

Note that for the cross polarized field coefficients the choice of backscattering causes the major term in the tangential field expression to vanish and the much smaller $(R_{\perp} + R_{\|})$ term to stay on. The equality $f_{vh} = f_{hv}$ indicates that in backscattering the reciprocity theorem holds. The value of the cross polarized field coefficient f_{vh} in backscattering is much smaller than that of the like polarized field coefficients. In fact, at normal incidence it is zero and it is also zero for perfectly conducting surface. Thus, this cross polarized field coefficient is usually ignored in backscattering calculations. Since the Kirchhoff field accounts for single scattering only, the above discussion implies that cross polarized backscattering is not the result of single scattering. To account for cross polarized backscattering we must consider multiple surface scattering that is contained in the complementary term. The approximations given by (4.47) and (4.48) for cross polarized scattering will force $f_{vh} = f_{hv} = 0$ in backscattering but will provide nonzero values in the complementary tangential field. Hence, they may still be valid in practical applications.

As an illustration of the Kirchhoff field coefficients for a specific incidence, let $\hat{k}_i = \hat{x}\sin\theta - \hat{z}\cos\theta = \hat{x}k_x - \hat{z}k_z$, $\hat{h} = \hat{y}$, and $\hat{v} = \hat{x}\cos\theta + \hat{z}\sin\theta$. Then, we can express all the dot and cross vector products as follows:

$$(\hat{h}\cdot\hat{d}) = -(\hat{v}\cdot\hat{t}) = Z_y/D_2 \tag{4.78}$$

$$(\hat{h}\cdot\hat{t}) = (\hat{v}\cdot\hat{d}) = -(\sin\theta - Z_x\cos\theta)/D_2 \tag{4.79}$$

$$\hat{t} = -[\hat{x}Z_y\cos\theta + \hat{y}(\sin\theta - Z_x\cos\theta) + \hat{z}Z_y\sin\theta]/D_2 \tag{4.80}$$

$$\hat{d} = \hat{k}_i\times\hat{t} = [\hat{y}Z_y - \hat{z}\sin\theta(\sin\theta - Z_x\cos\theta) - \hat{x}\cos\theta(\sin\theta - Z_x\cos\theta)]/D_2 \tag{4.81}$$

$$\hat{n}\cdot\hat{k}_i = -(\cos\theta + Z_x\sin\theta)/D_1 \tag{4.82}$$

$$\hat{n}\cdot\hat{d} = \left[Z_x\cos\theta(\sin\theta - Z_x\cos\theta) - Z_y^2 - \sin\theta(\sin\theta - Z_x\cos\theta)\right]/(D_1D_2) \tag{4.83}$$

$$\hat{n}\times\hat{t} = \hat{n}\times(\hat{d}\times\hat{k}_i) = (\hat{n}\cdot\hat{k}_i)\hat{d} - (\hat{n}\cdot\hat{d})\hat{k}_i \tag{4.84}$$

$$\hat{n}\times\hat{d} = \hat{n}\times(\hat{k}_i\times\hat{t}) = (\hat{n}\cdot\hat{t})\hat{k}_i - (\hat{n}\cdot\hat{k}_i)\hat{t} = -(\hat{n}\cdot\hat{k}_i)\hat{t} \tag{4.85}$$

$$\hat{n}\times\hat{v} = (-\hat{x}Z_y\sin\theta + \hat{y}[\cos\theta + Z_x\sin\theta] + \hat{z}(Z_y\cos\theta))/D_1 \tag{4.86}$$

$$\hat{n}\times\hat{h} = -(\hat{x} + \hat{z}Z_x)/D_1 \tag{4.87}$$

where $D_2 = \left[Z_y^2 + (\sin\theta - Z_x\cos\theta)^2\right]^{1/2}$. The approximate expressions for the Kirchhoff field coefficients are

$$\begin{aligned} f_{vv} \approx\ & (1 - R_{\|})[Z_y\sin\theta\sin\phi_s + (\cos\theta + Z_x\sin\theta)\cos\phi_s] \\ & -(1 + R_{\|})(\cos\theta_s\cos\phi_s - Z_x\sin\theta_s) \end{aligned} \tag{4.88}$$

$$f_{hh} \approx [(1 + R_{\perp})(\cos\theta_s\cos\phi_s - Z_x\sin\theta_s)]$$

$$-(1-R_{\perp})\ [Z_y \sin\theta \sin\phi_s + (\cos\theta + Z_x \sin\theta)\cos\phi_s] \tag{4.89}$$

For cross polarization using the approximation given by (4.47) and (4.48), we have

$$f_{hv} \approx (1-R)\ [Z_y \sin\theta \cos\theta_s \cos\phi_s - (\cos\theta + Z_x \sin\theta)\cos\theta_s \sin\phi_s]$$
$$+ (1-R)\, Z_y \cos\theta \sin\theta_s + (1+R) \sin\phi_s \tag{4.90}$$

$$f_{vh} \approx (1-R) \sin\phi_s + (1+R)\, Z_y \cos\theta \sin\theta_s$$
$$+ (1+R)\ [Z_y \sin\theta \cos\theta_s \cos\phi_s - (\cos\theta + Z_x \sin\theta)\cos\theta_s \sin\phi_s] \tag{4.91}$$

Although we have approximated the Fresnel reflection coefficients to be independent of the spatial variables, we see that there is still spatial dependence through the slope terms in the above four equations. However, this dependence can be removed by performing integration by parts and then ignoring the edge term.

As an example, consider the integral given by (4.67) involving one slope term,

$$E^k_{qp} = CE_0 \int Z_x \exp\{j[(k_{sx}-k_x)x + k_{sy}y + (k_{sz}-k_z)z]\}\, dxdy \tag{4.92}$$

where

$$k_{sx} - k_x = k(\sin\theta_s \cos\phi_s - \sin\theta)$$
$$k_{sy} = k \sin\theta_s \sin\phi_s$$
$$k_{sz} - k_z = k(\cos\theta_s + \cos\theta) \tag{4.93}$$

The limits of the above integral are from one end of the illuminated area to the other. Upon applying integration by parts we have

$$E^k_{qp} = -(jCE_0 \exp\{j[(k_{sx}-k_x)x + k_{sy}y + (k_{sz}-k_z)z]\})/(k_{sz}-k_z)$$
$$-CE_0 \int \frac{k_{sx}-k_x}{k_{sz}-k_z} \exp\{j[(k_{sx}-k_x)x + k_{sy}y + (k_{sz}-k_z)z]\}\, dxdy \tag{4.94}$$

The integrated term in the above equation is evaluated at the edge of the illuminated area and hence is negligible. The integral term has the same appearance as (4.92) with the surface slope along x-axis replaced by the negative of the ratio of the wave numbers along the x-axis to those along the z-axis, i.e.,

$$Z_x = -\frac{k_{sx}-k_x}{k_{sz}-k_z} \tag{4.95}$$

Similar arguments will indicate that the surface slope along y-axis can be replaced by setting

$$Z_y = -\frac{k_{sy}}{k_{sz} - k_z} \tag{4.96}$$

This means that the field coefficients f_{qp} become constants of the spatial integration variables after we replace the partial derivatives of the surface in accordance with (4.95) and (4.96). The final forms of f_{qp}'s are given in Appendix 4B.

4.4.2 Approximate Expressions for F_{qp}

In this section we discuss the reduction of the complementary field coefficients. They have many special forms under different conditions. Only the major steps taken are described. Various special forms are summarized in Appendix 4B.

Before we deal with F_{qp} we need to simplify (4.7) through (4.10). To do so we shall use a spectral representation for the Green's function and its gradient, i.e.,

$$G = \left(-\frac{1}{2\pi}\right)\int \frac{j}{q} \exp\left[ju(x-x') + jv(y-y') - jq|z-z'|\right] du dv \tag{4.97}$$

where $q = (k^2 - u^2 - v^2)^{1/2}$ and

$$\nabla' G = \left(-\frac{1}{2\pi}\right)\int \frac{\vec{g}}{q} \exp\left[ju(x-x') + jv(y-y') - jq|z-z'|\right] du dv \tag{4.98}$$

In (4.98) $\vec{g} = \hat{x}u + \hat{y}v \pm \hat{z}q$. In the transmitted medium we shall replace k by k_t in the above two equations and add a subscript, t, to the corresponding quantities. For the purpose of average power calculations from the scattered field we shall show in Chapter 6 that we can drop the phase term with the absolute sign in both the Green's function and its gradient and also the last term in $\vec{g}$ with the $\pm$ sign. The reasons why we can do all these without significantly altering the average power or the scattering coefficient calculations are (1) if the surface points are close together then the difference $z - z'$ will be small; if the two surface points are far apart then they are not correlated and hence will not contribute to the average power; and (2) the term with $\pm$ tends to cancel itself out when we perform ensemble averaging. As we shall see later, our final analytic expressions for the scattering coefficients reduce to known analytic results where such approximations are not made, thus further justifying the use of these simplifying assumptions. Now we rewrite the simplified expressions of (4.7) through (4.10) as follows:

$$\vec{\mathcal{E}}_p = \int \frac{P}{2\pi q}\left[k\eta\,(\hat{n}' \times \vec{\mathcal{H}}_p) + (\hat{n}' \times \vec{\mathcal{E}}_p) \times \vec{g} + (\hat{n}' \cdot \vec{\mathcal{E}}_p)\,\vec{g}\right] du dv \tag{4.99}$$

$$\vec{\mathcal{E}}_{pt} = -\int \frac{P}{2\pi q_t}[k_t\eta_t(\hat{n}' \times \vec{H}'_p) + (\hat{n}' \times \vec{\mathcal{E}}_p) \times \vec{g} + (\hat{n}' \cdot \vec{\mathcal{E}}_p)(\vec{g}/\varepsilon_r)]dudv \quad (4.100)$$

$$\vec{\mathcal{H}}_p = \int \frac{P}{2\pi q}\left[\left(\frac{k}{\eta}\right)(\hat{n}' \times \vec{\mathcal{E}}_p) - (\hat{n}' \times \vec{H}'_p) \times \vec{g} - (\hat{n}' \cdot \vec{H}'_p)\vec{g}\right]dudv \quad (4.101)$$

$$\vec{\mathcal{H}}_{pt} = -\int \frac{P}{2\pi q_t}\left[\left(\frac{k_t}{\eta_t}\right)(\hat{n}' \times \vec{\mathcal{E}}'_p) - (\hat{n}' \times \vec{H}'_p) \times \vec{g} - (\hat{n}' \cdot \vec{H}'_p)\left(\frac{\vec{g}}{\mu_r}\right)\right]dudv \quad (4.102)$$

where $\vec{g} = \hat{x}u + \hat{y}v$ and $P = \exp[ju(x-x') + jv(y-y')]$.

The expressions for the F_{qp}'s are rather lengthy. We shall begin by developing intermediate expressions for them in terms of $\vec{\mathcal{E}}_p, \vec{\mathcal{E}}_{pt}$ and $\vec{\mathcal{H}}_p, \vec{\mathcal{H}}_{pt}$. From (4.70), (4.51) and (4.55) we define F_{vv}' as F_{vv} before we factor out the phase factor P and remove the integrations with respect to x', y' and u, v. That is,

$$\begin{aligned}
F_{vv}' &= 8\pi^2[-\hat{h}_s \cdot (\hat{n} \times \vec{\mathcal{E}}_v)_c + \eta\hat{v}_s \cdot (\hat{n} \times \vec{H}_v)_c]D_1/E^i \\
&= \frac{2\pi D_1}{E^i}\{\hat{h}_s \cdot \hat{n} \times \int[(1-R_\parallel)\vec{\mathcal{E}}_v + (1+R_\parallel)\vec{\mathcal{E}}_{vt}]ds'\} \\
&\quad + \frac{2\pi\eta D_1}{E^i}\{\hat{v}_s \cdot \hat{n} \times \int[(1+R_\parallel)\vec{\mathcal{H}}_v + (1-R_\parallel)\vec{\mathcal{H}}_{vt}]ds'\} \\
&= \frac{D_1}{E^i}\hat{h}_s \cdot \hat{n} \times \iint\{\left[(1-R_\parallel)\frac{k}{q} - (1+R_\parallel)\frac{k_t\eta_t}{q_t\eta}\right]\eta(\hat{n}' \times \vec{H}_v) \\
&\quad + [(1-R_\parallel)/q - (1+R_\parallel)/q_t](\hat{n}' \times \vec{\mathcal{E}}_v) \times \vec{g} \\
&\quad + [(1-R_\parallel)/q - ((1+R_\parallel)/(\varepsilon_r q_t))](\hat{n}' \cdot \vec{\mathcal{E}}_v)\vec{g}\}PD'_1dx'dy'dudv \\
&\quad + \frac{D_1}{E^i}\hat{v}_s \cdot \hat{n} \times \iint\{\left[(1+R_\parallel)\frac{k}{q} - (1-R_\parallel)\frac{k_t\eta}{q_t\eta_t}\right](\hat{n}' \times \vec{\mathcal{E}}_v) \\
&\quad - [(1+R_\parallel)/q - (1-R_\parallel)/q_t]\eta(\hat{n}' \times \vec{H}_v) \times \vec{g} \\
&\quad - [(1+R_\parallel)/q - (1-R_\parallel)/(\mu_r q_t)]\eta(\hat{n}' \cdot \vec{H}_v)\vec{g}\}PD'_1dx'dy'dudv \quad (4.103)
\end{aligned}$$

After we remove the phase factor and the four-fold integral above, we obtain F_{vv} as

$$\begin{aligned}
F_{vv} &= \frac{D'_1D_1}{E^i}\hat{h}_s \cdot \hat{n} \times \{\left[(1-R_\parallel)\frac{k}{q} - (1+R_\parallel)\frac{k_t\eta_t}{q_t\eta}\right]\eta(\hat{n}' \times \vec{H}_v) \\
&\quad + [(1-R_\parallel)/q - (1+R_\parallel)/q_t](\hat{n}' \times \vec{E}_v) \times \vec{g}
\end{aligned}$$

$$+ [(1-R_{\|})/q - (1+R_{\|})/(\varepsilon_r q_t)] (\hat{n}' \cdot \vec{E}_v) \vec{g} \}$$

$$+ \frac{D'_1 D_1}{E^i} \hat{v}_s \cdot \hat{n} \times \{ \left[(1+R_{\|}) \frac{k}{q} - (1-R_{\|}) \frac{k_t \eta}{q_t \eta_t} \right] (\hat{n}' \times \vec{E}_v)$$

$$-[(1+R_{\|})/q - (1-R_{\|})/q_t] \eta (\hat{n}' \times \vec{H}_v) \times \vec{g}$$

$$-[(1+R_{\|})/q - (1-R_{\|})/(\mu_r q_t)] \eta (\hat{n}' \cdot \vec{H}_v) \vec{g} \} \qquad (4.104)$$

The vector products in the above equation with $\hat{h} = \hat{y}$, $\hat{v} = \hat{x}\cos\theta + \hat{z}\sin\theta$ and $\hat{k}_i = \hat{x}\sin\theta - \hat{z}\cos\theta$ are given in Appendix 4A. The tangential and normal field components in F_{vv} can be approximated by the tangential Kirchhoff fields as shown in Appendix 4A yielding

$$F_{vv} \approx \hat{h}_s \cdot \vec{n} \times \{ - \left[(1-R_{\|}) \frac{k}{q} - (1+R_{\|}) \frac{k_t \eta_t}{q_t \eta} \right] (1+R_{\|}) (\hat{n}' \times \hat{h})$$

$$+ [(1-R_{\|})/q - (1+R_{\|})/q_t] (1-R_{\|}) (\hat{n}' \times \hat{v}) \times \vec{g}$$

$$+ [(1-R_{\|})/q - ((1+R_{\|})/(\varepsilon_r q_t))] (1+R_{\|}) (\hat{n}' \cdot \hat{v}) \vec{g} \}$$

$$+ \hat{v}_s \cdot \vec{n} \times \{ \left[(1+R_{\|}) \frac{k}{q} - (1-R_{\|}) \frac{k_t \eta}{q_t \eta_t} \right] (1-R_{\|}) (\hat{n}' \times \hat{v})$$

$$+ [(1+R_{\|})/q - (1-R_{\|})/q_t] (1+R_{\|}) (\hat{n}' \times \hat{h}) \times \vec{g}$$

$$+ [(1+R_{\|})/q - ((1-R_{\|})/(\mu_r q_t))] (1-R_{\|}) (\hat{n}' \cdot \hat{h}) \vec{g} \} \qquad (4.105)$$

where $\vec{n} = \hat{n} D_1$ and $\vec{n}' = \hat{n}' D'_1$. In (4.105) there are surface slope terms that can be removed through integration by parts as we did in the previous subsection. In this case, we have the following substitutions for the partial derivatives of the random surface

$$Z_x = -\frac{k_{sx} + u}{k_{sz}} \qquad (4.106)$$

$$Z_y = -\frac{k_{sy} + v}{k_{sz}} \qquad (4.107)$$

$$Z'_x = \frac{k_x + u}{k_z} \qquad (4.108)$$

$$Z'_y = \frac{k_y + v}{k_z} \qquad (4.109)$$

Thus, F_{vv} becomes independent of the spatial variables also. Unlike f_{vv} it still depends on the u, v variables.

The cross polarized field coefficient F_{hv} can be found in a similar way using the approximate tangential and normal field estimates as

$$\begin{aligned}F_{hv} \approx \hat{v}_s \cdot \vec{n} \times \{ & \left[(1-R)\frac{k}{q} - (1+R)\frac{k_t \eta_t}{q_t \eta} \right] (1+R) (\vec{n}' \times \hat{h}) \\ & - [(1-R)/q - (1+R)/q_t] (1-R) (\vec{n}' \times \hat{v}) \times \vec{g} \\ & - [(1-R)/q - (1+R)/(\varepsilon_r q_t)] (1+R) (\vec{n}' \cdot \hat{v}) \vec{g} \} \\ & + \hat{h}_s \cdot \vec{n} \times \{ \left[(1+R)\frac{k}{q} - (1-R)\frac{k_t \eta}{q_t \eta_t} \right] (1-R) (\vec{n}' \times \hat{v}) \\ & + [(1+R)/q - (1-R)/q_t] (1+R) (\vec{n}' \times \hat{h}) \times \vec{g} \\ & + [(1+R)/q - (1-R)/(\mu_r q_t)] (1-R) (\vec{n}' \cdot \hat{h}) \vec{g} \} \end{aligned} \tag{4.110}$$

and

$$\begin{aligned}F_{vh} \approx \hat{h}_s \cdot \vec{n} \times \{ & \left[(1-R)\frac{k}{q} - (1+R)\frac{k_t \eta_t}{q_t \eta} \right] (1+R) (\vec{n}' \times \hat{v}) \\ & + [(1-R)/q - (1+R)/q_t] (1-R) (\vec{n}' \times \hat{h}) \times \vec{g} \\ & + [(1-R)/q - (1+R)/(\varepsilon_r q_t)] (1+R) (\vec{n}' \cdot \hat{h}) \vec{g} \} \\ & + \hat{v}_s \cdot \vec{n} \times \{ \left[(1+R)\frac{k}{q} - (1-R)\frac{k_t \eta}{q_t \eta_t} \right] (1-R) (\vec{n}' \times \hat{h}) \\ & - [(1+R)/q - (1-R)/q_t] (1+R) (\vec{n}' \times \hat{v}) \times \vec{g} \\ & - [(1+R)/q - (1+R)/(\mu_r q_t)] (1-R) (\vec{n}' \cdot \hat{v}) \vec{g} \} \end{aligned} \tag{4.111}$$

Finally, the complementary field coefficient for horizontally polarized scattering is

$$\begin{aligned}F_{hh} \approx \hat{v}_s \cdot \vec{n} \times \{ & -\left[(1+R_\perp)\frac{k}{q} - (1-R_\perp)\frac{k_t \eta_t}{q_t \eta} \right] (1-R_\perp) (\vec{n}' \times \hat{v}) \\ & - [(1+R_\perp)/q - (1-R_\perp)/q_t] (1+R_\perp) (\vec{n}' \times \hat{h}) \times \vec{g} \\ & - [(1+R_\perp)/q - (1-R_\perp)/(\varepsilon_r q_t)] (1-R_\perp) (\vec{n}' \cdot \hat{h}) \vec{g} \} \\ & + \hat{h}_s \cdot \vec{n} \times \{ \left[(1-R_\perp)\frac{k}{q} - (1+R_\perp)\frac{k_t \eta}{q_t \eta_t} \right] (1+R_\perp) (\vec{n}' \times \hat{h}) \\ & - [(1-R_\perp)/q - (1+R_\perp)/q_t] (1-R_\perp) (\vec{n}' \times \hat{v}) \times \vec{g} \end{aligned}$$

$$-[(1-R_\perp)/q-(1+R_\perp)/(\mu_r q_t)]\,(1+R_\perp)\,(\vec{n}'\cdot\hat{v})\,\vec{g}\,\} \tag{4.112}$$

All the triple cross products in the complementary field coefficients can be rewritten as dot products or triple scalar products using the vector identities,

$$\hat{q}\cdot\vec{n}\times(\vec{n}'\times\hat{p}) = (\vec{n}'\cdot\hat{q})\,(\vec{n}\cdot\hat{p}) - (\vec{n}'\cdot\vec{n})\,(\hat{q}\cdot\hat{p}) \tag{4.113}$$

$$\hat{q}\cdot\vec{n}\times(\vec{n}'\times\hat{p})\times\vec{g} = (\vec{n}\cdot\vec{g})\,(\vec{n}'\cdot\hat{p}\times\hat{q}) - (\hat{q}\cdot\vec{g})\,(\hat{p}\cdot\vec{n}\times\vec{n}') \tag{4.114}$$

The above results complete the formulation of the rough surface scattering problem by providing an expression for the scattered field. Since the field coefficients are now independent of spatial variables, it is possible to provide a general expression for the average scattered power. All special cases such as backscattering, perfectly conducting surface, etc. are represented by special field coefficients.

The expressions of F_{qp} with vector products carried out and slope terms replaced by (4.106) through (4.109) are given in Appendix 4B.

4.5 THE FAR-ZONE TRANSMITTED FIELDS

The far-zone transmitted fields into the medium below the rough interface are generated by the same tangential surface currents as those that cause the backward scattered fields. Thus, the Stratton-Chu formula is also applicable to calculate the transmitted scattered field. The major difference is that we are now interested in the radiated field in the medium below the surface boundary (see Figure 4.3).

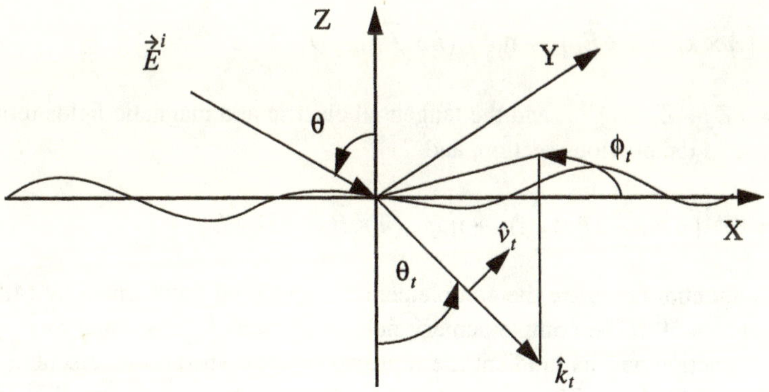

Figure 4.3 Geometry for the transmission problem.

Let the propagation unit vector $\hat{k}_t$, and the polarization unit vectors $\hat{v}_t$, $\hat{h}_t$ be defined for the transmitted field as (Figure 4.3)

$$\hat{v}_t = \hat{x}\cos\theta_t\cos\phi_t + \hat{y}\cos\theta_t\sin\phi_t + \hat{z}\sin\theta_t \tag{4.115}$$

$$\hat{h}_t = \hat{x}\sin\phi_t - \hat{y}\cos\phi_t \tag{4.116}$$

$$\hat{k}_t = \hat{x}\sin\theta_t\cos\phi_t + \hat{y}\sin\theta_t\sin\phi_t - \hat{z}\cos\theta_t \tag{4.117}$$

Then, transmitted scattered field expression is mathematically the same as that given by (4.60) with k_s replaced by k_t and η replaced by η_t

$$E^s_{qp} = C_t \int [\hat{q}\times\hat{k}_t\cdot(\hat{n}\times\vec{E}_p) + \eta_t\hat{q}\cdot(\hat{n}\times\vec{H}_p)]\exp(j\vec{k}_t\cdot\vec{r})\,ds \tag{4.118}$$

where $C_t = -\dfrac{jk_t}{4\pi R}\exp(-jk_t R)$

The above result suggests that we can again rewrite it as the sum of a Kirchhoff and a complementary field similar to (4.67) and (4.68),

$$E^k_{qpt} = C_t E_0 \int f_{qpt}\exp\{j[(\vec{k}_t - \vec{k}_i)\cdot\vec{r}]\}\,dxdy \tag{4.119}$$

$$E^c_{qpt} = \frac{C_t E_0}{8\pi^2}\int F_{qpt}\exp\{ju(x-x') + jv(y-y') + j\vec{k}_t\cdot\vec{r} - j\vec{k}_i\cdot\vec{r}'\}\,dxdydx'dy'dudv \tag{4.120}$$

where

$$f_{qpt} = [\hat{q}\times\hat{k}_t\cdot(\hat{n}\times\vec{E}_p)_k + \eta_t\hat{q}\cdot(\hat{n}\times\vec{H}_p)_k]D_1/E^i \tag{4.121}$$

with $D_1 = (Z_y^2 + Z_x^2 + 1)^{1/2}$ and the tangential electric and magnetic fields remain the same as those in the previous section, and

$$F_{qpt}' = 8\pi^2[\hat{q}\times\hat{k}_t\cdot(\hat{n}\times\vec{E}_p)_c + \eta_t\hat{q}\cdot(\hat{n}\times\vec{H}_p)_c]D_1/E^i \tag{4.122}$$

where the tangential fields are the complementary tangential fields given by (4.51) and (4.55) through (4.59). The complementary field coefficient F_{qpt} is equal to F_{qpt}' after the Green's function and its gradient are replaced by their spectral representations and after the phase factor $\exp[ju(x-x') + jv(y-y')]$ in the Green's function and u, v, x', y' integrations are factored out. More explicit representations of these field coefficients are given in Appendix 4D.

4.6 THE AVERAGE POWER AND SCATTERING COEFFICIENT

As shown in the previous two sections the forms of the scattered fields in both the upper and lower medium are the same. Thus, it is sufficient to consider only the calculation of the average power and the associated scattering coefficients for the backward scattered field into the upper medium. Results for the corresponding transmitted scattered field can be found from analogy.

With the field expression given by (4.66) the average scattered power is given by

$$\langle E^s_{qp} E^{s\,*}_{qp} \rangle = \langle E^k_{qp} E^{k\,*}_{qp} \rangle + 2\mathrm{Re}\langle E^c_{qp} E^{k\,*}_{qp} \rangle + \langle E^c_{qp} E^{c\,*}_{qp} \rangle \tag{4.123}$$

where Re is the real part operator and * is the symbol for complex conjugate. To obtain the incoherent power, we have to subtract the mean-squared power from the total power. That is,

$$\begin{aligned}\langle E^s_{qp} E^{s\,*}_{qp} \rangle - \langle E^s_{qp} \rangle \langle E^s_{qp} \rangle^* &= \langle E^k_{qp} E^{k\,*}_{qp} \rangle - \langle E^k_{qp} \rangle \langle E^k_{qp} \rangle^* \\ &+ 2\mathrm{Re}\left[\langle E^c_{qp} E^{k\,*}_{qp} \rangle - \langle E^c_{qp} \rangle \langle E^k_{qp} \rangle^* \right] + \langle E^c_{qp} E^{c\,*}_{qp} \rangle - \langle E^c_{qp} \rangle \langle E^c_{qp} \rangle^*\end{aligned} \tag{4.124}$$

To carry out the average operation we must make an assumption about the type of surface height distribution. For the purpose of illustration we shall assume Gaussian height distribution here. The averages corresponding to a more general, non-Gaussian distribution is considered in Appendix 4C. Recall we have defined $\vec{k}_s = \hat{x}k_{sx} + \hat{y}k_{sy} + \hat{z}k_{sz}$ and $\vec{k}_i = \hat{x}k_x + \hat{y}k_y - \hat{z}k_z$. In what follows we shall treat each term in (4.124) separately. Consider first the Kirchhoff term:

$$\begin{aligned}P^k_{qp} &= \langle E^k_{qp} E^{k\,*}_{qp} \rangle - \langle E^k_{qp} \rangle \langle E^k_{qp} \rangle^* \\ &= |CE_0 f_{qp}|^2 \{ \langle \iint \exp[j(\vec{k}_s - \vec{k}_i) \cdot (\vec{r} - \vec{r}')]\, dx'dy'dxdy \rangle \\ &- \left| \langle \int \exp\{ j[(\vec{k}_s - \vec{k}_i) \cdot \vec{r}] \}\, dxdy \rangle \right|^2 \}\end{aligned} \tag{4.125}$$

where $|C|^2 = k^2/(4\pi R)^2$ and from Ulaby et al. [1982]

$$\langle \exp\{ j[(k_{sz} + k_z)z] \} \rangle = \exp\left[-(k_{sz} + k_z)^2 (\sigma^2/2) \right] \tag{4.126}$$

$$\langle \exp[j(k_{sz} + k_z)(z - z')] \rangle = \exp\{ -(k_{sz} + k_z)^2 \sigma^2 [1 - \rho(x - x', y - y')] \} \tag{4.127}$$

In (4.126) and (4.127) σ^2 is the variance of the surface and $\rho(x - x', y - y')$ is the normalized surface autocorrelation function. For simplicity we shall refer to ρ as the surface correlation. In addition, we shall assume that the surface is generated by a stationary random process. Hence, we let $\xi = x - x', \varsigma = y - y'$ and A_0 be the illuminated area. After integration with respect to x' and y', we have

$$P_{qp}^{k} = |CE_0 f_{qp}|^2 A_0 \exp\left[-\sigma^2 (k_{sz} + k_z)^2\right]$$

$$\int \{\exp\left[\sigma^2 (k_{sz} + k_z)^2 \rho(\xi, \varsigma)\right] - 1\} \exp\left[j (k_{sx} - k_x) \xi + j (k_{sy} - k_y) \varsigma\right] d\xi d\varsigma \tag{4.128}$$

Next, we consider the cross term,

$$P_{qp}^{kc} = 2\mathrm{Re}\left[\langle E_{qp}^{c} E_{qp}^{k*}\rangle - \langle E_{qp}^{c}\rangle \langle E_{qp}^{k}\rangle^*\right] = |(CE_0)/(2\pi)|^2 \mathrm{Re}\{\int (F_{qp} f_{qp}^*)$$

$$\iint \langle \exp\left[j\vec{k}_s \cdot (\vec{r} - \vec{r}'') + j\vec{k}_i \cdot (\vec{r}'' - \vec{r}') + ju(x - x') + jv(y - y')\right]\rangle$$

$$-\langle \iint \exp\left[j(\vec{k}_s \cdot \vec{r}) - (j\vec{k}_i \cdot \vec{r}') + ju(x - x') + jv(y - y')\right]\rangle$$

$$\langle \int \exp\{j[(\vec{k}_i - \vec{k}_s) \cdot \vec{r}'']\} dx dy dx' dy' dx'' dy'' du dv\rangle \} \tag{4.129}$$

The averages of the following quantities have been shown in Appendix 4 C to be

$$\langle \exp\left[jk_{sz}(z - z'') - jk_z(z'' - z')\right]\rangle = \exp\{\sigma^2 [k_{sz}(k_{sz} + k_z) \rho(\xi, \varsigma)$$

$$+ k_z(k_{sz} + k_z) \rho(\xi', \varsigma')]\} \exp\{-\left[k_{sz}^2 + k_z^2 + k_z k_{sz}(1 + \rho(\xi - \xi', \varsigma - \varsigma'))\right]\sigma^2\} \tag{4.130}$$

$$\langle \exp\left[jk_{sz} z + jk_z z'\right]\rangle = \exp\{-\left[k_{sz}^2 + k_z^2 + 2k_z k_{sz} \rho(\xi - \xi', \varsigma - \varsigma')\right](\sigma^2/2)\} \tag{4.131}$$

$$\langle \exp\left[j(k_{sz} + k_z) z''\right]\rangle = \exp\left[-(k_{sz} + k_z)^2 (\sigma^2/2)\right] \tag{4.132}$$

where $\xi = x - x'', \varsigma = y - y'', \xi' = x' - x'', \varsigma' = y' - y''$. Thus, after integrating x'' and y'', the incoherently scattered power for the cross term is

$$P_{qp}^{kc} = |(CE_0)/(2\pi)|^2 A_0 \mathrm{Re}\{\int (F_{qp} f_{qp}^*) \exp\left[-\sigma^2 (k_{sz}^2 + k_z^2 + k_z k_{sz})\right]$$

$$\iint \exp\left[-\sigma^2 k_z k_{sz} \rho(\xi - \xi', \varsigma - \varsigma')\right] \{\exp\left[\sigma^2 k_{sz}(k_{sz} + k_z) \rho(\xi, \varsigma)\right.$$

$$+ \sigma^2 k_z (k_{sz} + k_z) \rho (\xi', \varsigma') \ ] - 1 \} \exp [jk_{sx}\xi + jk_{sy}\varsigma + ju (\xi - \xi') + jv (\varsigma - \varsigma')]$$

$$\exp [- jk_x\xi' - jk_y\varsigma'] \, d\xi d\varsigma d\xi' d\varsigma' du dv \} \tag{4.133}$$

Finally, the incoherent power for the complementary term is

$$P_{qp}^c = \langle E_{qp}^c E_{qp}^{c\,*} \rangle - \langle E_{qp}^c \rangle \langle E_{qp}^c \rangle^* = \left| (CE_0) / (8\pi^2) \right|^2 \{ \iint F_{qp} F_{qp}^* $$

$$\iiiint \langle \exp \left[j\vec{k}_s \cdot (\vec{r} - \vec{r}) + j\vec{k}_i \cdot (\vec{r}''' - \vec{r}') + ju (x - x') - ju' (x'' - x''') \right] \rangle$$

$$\exp [jv (y - y') - jv' (y'' - y''')] \, dx dx' dx'' dx''' dy dy' dy'' dy''' du dv du' dv'$$

$$- \left| \langle \iiint F_{qp} \exp \left[j\vec{k}_s \cdot \vec{r} - j\left(\vec{k}_i \cdot \vec{r}\right) + ju (x - x') + jv (y - y') \right] dx dx' dy dy' du dv \rangle \right| \tag{4.134}$$

The averages of the following quantities are

$$\langle \exp (jk_{sz}z + jk_z z') \rangle = \exp \{ -\left[k_{sz}^2 + k_z^2 + 2k_z k_{sz}\rho (\xi - \xi', \varsigma - \varsigma') \right] (\sigma^2/2) \} \tag{4.135}$$

$$\langle \exp [jk_{sz} (z - z'') + jk_z (z' - z''')] \rangle = \exp \left[-\sigma^2 (k_{sz}^2 + k_z^2) - \sigma^2 k_z k_{sz}\rho (\tau, \kappa) \right]$$

$$\exp \left[-\sigma^2 k_z k_{sz}\rho (\tau + \xi - \xi', k + \varsigma - \varsigma') \right] \exp \left[\sigma^2 k_{sz}^2 \rho (\xi, \varsigma) + \sigma^2 k_z^2 \rho (\xi', \varsigma') \right]$$

$$\exp \{ \sigma^2 k_z k_{sz} [\rho (\xi + \tau, \kappa + \varsigma) + \rho (\xi' - \tau, \varsigma' - \kappa)] \} \tag{4.136}$$

where $\xi = x - x''$, $\varsigma = y - y''$, $\xi' = x' - x'''$, $\varsigma' = y' - y'''$, $\tau = x'' - x'''$, $\kappa = y'' - y'''$
Using the above relations for the averages and integrating x''' and y''', we have

$$P_{qp}^c = \left| (CE_0) / (8\pi^2) \right|^2 A_0 \{ \iint F_{qp} F_{qp}^* \exp \left[-\sigma^2 (k_{sz}^2 + k_z^2) \right]$$

$$\exp \left[-\sigma^2 k_z k_{sz}\rho (\tau + \xi - \xi', \kappa + \varsigma - \varsigma') - \sigma^2 k_z k_{sz}\rho (\tau, \kappa) \right] \{ \exp \left[\sigma^2 k_{sz}^2 \rho (\xi, \varsigma) \right]$$

$$\exp \{ \sigma^2 k_z^2 \rho (\xi', \varsigma') + \sigma^2 k_z k_{sz} [\rho (\xi + \tau, \kappa + \varsigma) + \rho (\xi' - \tau, \varsigma' - \kappa)] \} - 1 \}$$

$$\exp \{ j [(k_{sx} + u) \xi + (k_{sy} + v) \varsigma - (k_x + u) \xi' - (k_y + v) \varsigma'$$

$$+ (u - u') \tau + (v - v') \kappa] \} \, d\xi d\varsigma d\xi' d\varsigma' d\tau d\kappa du dv du' dv' \tag{4.137}$$

The bistatic scattering coefficient σ_{qp}^0 is related to the average power expression as

$$\sigma_{qp}^{0} = (4\pi R^2 P_{qp}) / (E_0^2 A_0) \tag{4.138}$$

Therefore, we can summarize the bistatic scattering coefficient as

$$\sigma_{qp}^{0} = \sigma_{qp}^{k} + \sigma_{qp}^{kc} + \sigma_{qp}^{c} \tag{4.139}$$

where

$$\sigma_{qp}^{k} = \frac{k^2}{4\pi}|f_{qp}|^2 \exp[-\sigma^2 (k_{sz} + k_z)^2]$$

$$\int \{\exp[\sigma^2 (k_{sz} + k_z)^2 \rho(\xi, \varsigma)] - 1\} \exp[j(k_{sx} - k_x)\xi + j(k_{sy} - k_y)\varsigma]\, d\xi d\varsigma \tag{4.140}$$

$$\sigma_{qp}^{kc} = \frac{k^2}{16\pi^3} \mathrm{Re}\{\int (F_{qp} f_{qp}{}^*) \exp[-\sigma^2 (k_{sz}^2 + k_z^2 + k_z k_{sz})]$$

$$\iint \exp[-\sigma^2 k_z k_{sz} \rho(\xi - \xi', \varsigma - \varsigma')] \{\exp[\sigma^2 k_{sz} (k_{sz} + k_z) \rho(\xi, \varsigma)$$

$$+ \sigma^2 k_z (k_{sz} + k_z) \rho(\xi', \varsigma')] - 1\} \exp[jk_{sx}\xi + jk_{sy}\varsigma + jv(\varsigma - \varsigma') + ju(\xi - \xi')]$$

$$\exp[-jk_x\xi' - jk_y\varsigma']\, d\xi d\varsigma d\xi' d\varsigma' du dv\} \tag{4.141}$$

$$\sigma_{qp}^{c} = |k/(16\pi^{2.5})|^2 \{\iint F_{qp} F_{qp}{}^* \exp[-\sigma^2 (k_{sz}^2 + k_z^2)]$$

$$\exp[-\sigma^2 k_z k_{sz} \rho(\tau + \xi - \xi', \kappa + \varsigma - \varsigma') - \sigma^2 k_z k_{sz} \rho(\tau, \kappa)] (\exp[\sigma^2 k_{sz}^2 \rho(\xi, \varsigma)]$$

$$\exp\{\sigma^2 k_z^2 \rho(\xi', \varsigma') + \sigma^2 k_z k_{sz} [\rho(\xi + \tau, \kappa + \varsigma) + \rho(\xi' - \tau, \varsigma' - \kappa)]\} - 1)$$

$$\exp\{j[(k_{sx} + u)\xi + (k_{sy} + v)\varsigma - (k_x + u)\xi' - (k_y + v)\varsigma']\}$$

$$\exp\{j[(u - u')\tau + (v - v')\kappa]\}\, d\xi d\varsigma d\xi' d\varsigma' d\tau d\kappa du dv du' dv' \tag{4.142}$$

The above equations give the complete representation of the bistatic scattering coefficient for a randomly rough surface. In deriving these results we have used approximate expressions for the tangential fields based on the Kirchhoff approximation. Thus, the effect of shadowing of the incident intensity by the roughness of the surface should be included. The method of correction to the backscattering coefficient under single-scatter condition is simple. We need only to multiply the incident intensity by the shadowing function, $S(\theta, \sigma_s)$, which depends on the incident angle and the surface rms slope, σ_s [Beckmann, 1965; Sancer, 1969; Brown, 1984]. However, it is a more complex problem to include shadowing in multiple scattering [Fung and Eom, 1981]. In addition, existing shadowing functions are accurate only under the geometric optics condition. Hence, a general shadowing function and a completely satisfactory way to include shadowing in multiple scattering are still subjects for further research. A possible way to approach this problem will be discussed in the next chapter after we separate singly scattered terms from multiply scattered terms. For ease of reference we give the two shadowing functions derived by Smith [1967] below. The first one represents the probability that a point on the surface will not be shadowed:

$$R_1(\theta, \sigma_s) = \left[1 - \frac{1}{2} erfc\left(\frac{\cot\theta}{\sigma_s\sqrt{2}}\right)\right] [1 + f(\theta, \sigma_s)]^{-1} \tag{4.143}$$

$$f(\theta, \sigma_s) = \frac{1}{2}\left\{\sqrt{\frac{2}{\pi}}\frac{\sigma_s}{\cot\theta}\exp\left(-\frac{\cot\theta^2}{2\sigma_s^2}\right) - erfc\left(\frac{\cot\theta}{\sigma_s\sqrt{2}}\right)\right\} \tag{4.144}$$

where σ_s is the rms slope of the surface; θ is the incident angle and *erfc* is the error-function complement related to the error function *erf* by

$$\text{erfc}(z) = 1 - \text{erf}(z) = 1 - \frac{2}{\sqrt{\pi}}\int_0^z \exp(-t^2)\, dt \tag{4.145}$$

The other shadowing function appropriate for the Kirchhoff term is $R_2(\theta, \sigma_s)$, defined as the conditional probability that a point on the surface will not lie in the shadow given that its local slope is perpendicular to the incident beam, is

$$R_2(\theta, \sigma_s) = [1 + f(\theta, \sigma_s)]^{-1} \tag{4.146}$$

The bistatic scattering coefficient given by (4.139) with shadowing included is expected to be applicable to a general dielectric rough surface over a wide range of roughness scales or frequencies. The detailed range of validity of this model [Chen et al., 1989] is discussed in Chapter 6. Approximate evaluations of the integrals in (4.140), (4.141) and (4.142) are given in the next chapter where various special cases of this model are also discussed.

4.7 TRANSMITTED SCATTERING COEFFICIENT

By analogy the transmitted scattering coefficients can also be obtained from (4.140), (4.141) and (4.142) by making appropriate changes in corresponding quantities. The advantage of doing this, however, is limited. It is more useful to find the transmitted scattering coefficient after we evaluate the integrals. This will be done in the next chapter.

REFERENCES

Bahar, E., "Scattering by Anisotropic Model of Composite Rough Surface-Full Wave Solution," *IEEE Transaction on Antennas and Propagation*, Vol. 33, 1985, pp. 106–112.

Barrick, D. E., "Rough Surface Scattering Based on the Specular Point Theory," *IEEE Transaction on Antennas and Propagation*, Vol. 16, 1968, pp. 449–454.

Bass, F.G., and I. M. Fuks, *Wave Scattering from Statistically Rough Surfaces*, Pergamon Press, New York, 1979, Chapters 3, 7 and 10.

Beckmann, P., "Shadowing of Random Rough Surface," *IEEE Transaction on Antennas and Propagation*, Vol. 13, 1965, pp. 384–388.

Beckmann, P., and A. Spizzichino, *The Scattering of Electromagnetic Waves from Rough Surfaces*, Macmillan, New York, 1963.

Brown, G. S., "Backscattering from a Gaussian Distributed Perfectly Conducting Rough Surface," *IEEE Transaction on Antennas and Propagation*, Vol. 26, 1978, pp. 472–482.

Brown, G. S., "The Validity of Shadowing Correction in Rough Surface Scattering," *Radio Science*, Vol. 19, 1984, pp. 1461–1468.

Brown, G. S., "Simplification in the Stochastic Fourier Transform Approach to Random Surface Scattering," *IEEE Transaction on Antennas and Propagation*, Vol. 33, 1985, pp. 48–55.

Chen, M. F., K. S. Chen and A.K. Fung, "A Study of the Validity of the Integral Equation Model by Moment Method Simulation – Cylindrical Case," *Remote Sensing of Environment*, Vol. 29, 1989, pp. 217–288.

Dainty, J.C., M.J. Kim and A.J., Sant, "Measurements of Enhanced Back Scattering of Light from One and Two Dimensional Random Rough Surfaces: Notes for Workshop Recent Progress in Surface and Volume Scattering," Madrid, September 14–16, and for the International Working Group Meeting on Wave Propagation in Random Media, Tallinn, September 19–23, 1988.

DeSanto, J. A., "Scattering from a Random Rough Surface Diagram Method for Elastic Media," *Journal of Mathematical Physics*, Vol. 15, 1974, pp. 283–288.

Eftimiu, C., "Scattering by Rough Surfaces: A Simple Model," *IEEE Transaction on Antennas and Propagation*, Vol. 34, 1986, pp. 626–630.

Fuks, I.M., "Theory of Radio-Wave Scattering at a Rough Surface," *Soviet Radiophysics*, Vol. 9, 1966, pp. 513–519.

Fung, A.K., "Theory of Cross-Polarized Power Returned from a Random Surface," *Applied Science Research*, Vol. 18, 1967a, pp. 50–60.

Fung, A.K., "Character of Wave Depolarization by a Perfectly Conducting Rough Surface and its Application to Earth and Moon Experiments," *Planetary and Space Science*, Vol. 15, 1967b, pp. 1337–1347.

Fung, A.K., "Mechanism of Polarized and Depolarized Scattering from Rough Dielectric Surface," *Journal of the Franklin Institute*, Vol. 285, 1968, pp. 125–133.

Fung, A.K., and J.H. Eom, "Multiple Scattering and Depolarization by a Randomly Rough Kirchhoff Surface," *IEEE Trans. Ant. Prop.*, Vol. 29, no. 3, 1981, pp. 463–471.

Fung, A. K., and G.W. Pan, "An Integral Equation Method for Rough Surface Scattering," *Proceedings of the International Symposium on Multiple Scattering of Wave in Random Media and Random Surface*, held at the Pennsylvania State University from 1 July to 1 August 1985 (Pennsylvania State University Press), 1986, pp. 701–714.

Fung, A.K., and G.W. Pan, "A Scattering Model for Perfectly Conducting Random Surface: I. Model Development," *International Journal of Remote Sensing*, Vol. 8, no. 11, 1987, pp. 1579–1593.

Garcia, N., V. Celli and M. Nieto-Vesperinas, "Exact Multiple Scattering of Waves from Random Rough Surfaces, " *Optics Communications*, Vol. 30, 1979, pp. 279–281.

Gray, E.P., R.W. Hart and R.A. Farrall, "An Application of a Variational Principle for Scattering by Random Rough Surfaces, " *Radio Science*, Vol. 13, 1978, pp. 333–343.

Gu, Zu-Han, R.S. Dummer, A.A. Maradudin and A.R. McGurn, "Experimental Study of the Opposition Effect in the Scattering of Light from a Randomly Rough Metal Surface, " *Appl. Optics*, Vol. 28, 1989, pp. 537–543.

Hagfors, T., "Relationship of Geometric Optics and Autocorrelation Approaches to the Analysis of Lunar and Planetary Radar, " *Journal of Geophysical Research*, Vol. 71, 1966, pp. 379–383.

Ishimaru, A., and J.S. Chen, "Scattering from Very Rough Metallic and Dielectric Surfaces: A Theory Based on the Modified Kirchhoff Approximation," *Waves in Random Media*, Vol. 1, no. 1, 1991, pp. 21–34.

Kodis, R.O., "A Note on the Theory of Scattering from an Irregular Surface," *IEEE Transactions on Antennas and Propagation*, Vol. 14, 1966, pp. 77–82.

Kovalev, A. A., and S. I. Pozknyak, "Electromagnetic Wave Scattering from a Statistically Rough Surface with Finite Conductivity," *Radiotekhnika*, Vol. 16, 1961, pp. 31–36.

Leader, J.C., "Incoherent Backscatter from Rough Surfaces: the Two-Scale Model Reexamined, " *Radio Science*, Vol. 13, 1978, pp. 441–457.

Li, Z., and A.K. Fung, "A Reformulation of the Surface Field Integral Equation," *Journal of Electromagnetic Waves and Application*, Vol. 5, 1991, pp. 195–203.

Maradudin, A.A., E. R. Mendez and T. Michel, "Backscattering Effects in the Elastic Scattering of P-Polarized Light from a Large-Amplitude Random Metallic Grating," *Optical Letters*, Vol. 14, no. 1, 1989, p. 151.

Mendez, E.R. and K.O. O'Donnell, "Observation of Depolarization and Backscattering Enhancement in Light Scattering from Gaussian Random Surfaces, " *Optics Communications*, Vol. 61, 1987, p. 91.

O'Donnell, K.O., and E.R.Mendez, "Experimental Study of Scattering from Characterized Random Surface, " *J. Opt. Soc Am.*, Vol. A4, 1987, p. 1194.

Pan, G.W., and A.K. Fung, "A Scattering Model for Perfectly Conducting Random Surface: II. Range of Validity," *International Journal of Remote Sensing*, Vol. 8, no. 11, 1987, pp. 1595–1605.

Poggio, A.J., and E.K. Miller, "Integral Equation Solution of Three Dimensional Scattering Problems," *Computer Techniques for Electromagnetics*, Pergamon, New York, 1973, Chapter 4.

Rice, S.O., "Reflection of Electromagnetic Waves from Slightly Rough Surface," *Communications in Pure and Applied Mathematics*, Vol. 4, 1951, pp. 361–378.

Sancer, M.I., "Shadow-Corrected Electromagnetic Scattering from a Randomly Rough Surface," *IEEE Transactions on Antenna and Propagation*, Vol. 17, 1969, pp. 577–589.

Semyonov, G., "Approximate Computation of Scattering of Electromagnetic Waves by Rough Surface Contours," *Radio Engineering and Electronic Physics*, Vol. 11, 1966, pp. 1179–1187.

Smith, B.G., "Geometrical Shadowing of a Random Rough Surface," *IEEE Transactions on Antenna and Propagation*, Vol. AP-15, 1967, pp. 668–671.

Soto-Crespo, J.M., and M. Nieto-Vesperinas, "Electromagnetic Scattering from Very Rough Random Surface and Deep Reflection Gratings," *J. Opt. Soc Am.*, Vol. 6, no. 3, 1989, pp. 376–384.

Thorsos, E.I. and D. Winebrenner, "An Examination of the Full-Wave Method for Rough Surface Scattering in the Case of Small Roughness," *Journal of Geophysical Research*, Vol. 19, no. C9, 1991, pp. 17107–17121.

Ulaby, F.T., R.K. Moore and A.K. Fung, *Microwave Remote Sensing*, Vol. 2, Artech House, Norwood, MA, 1982, Chapter 12.

Valenzuela, G.R., "Depolarization of EM Waves by Slightly Rough Surfaces," *IEEE Transactions on Antenna and Propagation*, Vol. 15, 1967, pp. 552–557.

Valenzuela, G.R., "Scattering of Electromagnetic Waves from a Tilted Slightly Rough Surface," *Radio Science*, Vol. 3, 1968, pp. 1057–1066.

Wagner, R.J., "Shadowing of Randomly Rough Surface," *J. Acoust. Soc. Am.*, Vol. 41, 1967, pp. 138–147.

Winebrenner, D., and A. Ishimaru, "Investigation of a Surface Field Phase Perturbation Technique for Scattering from Rough Surfaces," *Radio Science*, Vol. 20, no. 2, 1985, pp. 161–170.

Wright, J.W., "A New Model for Sea Clutter," *IEEE Transactions on Antenna and Propagation*, Vol. 16, 1968, pp. 217–223.

Wu, S.T., and A.K. Fung, "A Noncoherent Model for Microwave Emission and Backscattering from the Sea Surface," *Journal of Geophysical Research*, Vol. 77, 1972, pp. 5917–5929.

Appendix 4A

Estimates of Surface Fields

The vector products in F_{qp} and the final estimates of the tangential surface fields are given in this appendix.

4A.1 VECTOR IDENTITIES FOR F_{qp}

For $\hat{h} = \hat{y}$, $\hat{v} = \hat{x}\cos\theta + \hat{z}\sin\theta$, $\hat{n} = (-\hat{x}Z_x - \hat{y}Z_y + \hat{z})/D_1$, $\vec{g} = \hat{x}u + \hat{y}v$ and $\hat{k}_i = \hat{x}\sin\theta - \hat{z}\cos\theta$, we have the following vector identities:

$$\hat{n}' \times \hat{h} = -(\hat{x} + \hat{z}Z_x')/(D_1') \tag{4A.1}$$

$$\hat{n}' \times \hat{v} = -[\hat{x}Z_y'\sin\theta - \hat{y}(\cos\theta + Z_x'\sin\theta) - \hat{z}Z_y'\cos\theta]/(D_1') \tag{4A.2}$$

$$(\hat{n}' \times \hat{h}) \times \vec{g} = [\hat{x}vZ_x' - \hat{y}uZ_x' - \hat{z}v]/(D_1') \tag{4A.3}$$

$$(\hat{n}' \times \hat{v}) \times \vec{g} = -\{\hat{x}vZ_y'\cos\theta - \hat{y}uZ_y'\cos\theta + \hat{z}[vZ_y'\sin\theta + u(\cos\theta + Z_x'\sin\theta)]\}/(D_1') \tag{4A.4}$$

$$\hat{n} \times \vec{g} = -[\hat{x}v - \hat{y}u + \hat{z}(vZ_x - uZ_y)]/D_1 \tag{4A.5}$$

$$\hat{n} \times (\hat{n}' \times \hat{h}) = [\hat{x}Z_yZ_x' - \hat{y}(1 + Z_x'Z_x) + \hat{z}Z_y]/(D_1D_1') \tag{4A.6}$$

$$\hat{n} \times (\hat{n}' \times \hat{v}) = -\{\hat{x}[Z_yZ_y'\cos\theta + (\cos\theta + Z_x'\sin\theta)] + \hat{y}(Z_y'\sin\theta - Z_xZ_y\cos\theta) + \hat{z}[Z_x(\cos\theta + Z_x'\sin\theta) + Z_yZ_y'\sin\theta]\}/(D_1D_1') \tag{4A.7}$$

$$\hat{n} \times [(\hat{n}' \times \hat{h}) \times \vec{g}] = [\hat{x}(vZ_y + uZ_x') + \hat{y}v(Z_x' - Z_x) + \hat{z}(uZ_xZ_x' + vZ_yZ_x')]/(D_1D_1') \tag{4A.8}$$

$$\hat{n} \times [(\hat{n}' \times \hat{v}) \times \vec{g}] = \{\hat{x}[Z_y(vZ_y'\sin\theta + u(\cos\theta + Z_x'\sin\theta)) - uZ_y'\cos\theta] - \hat{y}[vZ_y'\cos\theta + Z_x(vZ_y'\sin\theta + u(\cos\theta + Z_x'\sin\theta))] - \hat{z}(uZ_xZ_y' + vZ_yZ_y')\cos\theta\}/(D_1D_1') \tag{4A.9}$$

$$\hat{n}' \cdot \hat{v} = (\sin\theta - Z_x'\cos\theta) / (D_1') \tag{4A.10}$$

$$\hat{n}' \cdot \hat{h} = -(Z_y') / (D_1') \tag{4A.11}$$

The above identities are applicable to all cases. However, the relations below are polarization dependent. In Chapter 4 we have indicated that expressions for the tangential fields are much simplified if we let them be polarization dependent. In what follows we shall summarize the estimates for both the tangential and normal field components.

4A.2 SURFACE FIELD ESTIMATES FOR CROSS POLARIZATION

In Section 4.3 the tangential fields for cross polarization under the Kirchhoff approximation have been further approximated to yield

$$\hat{n} \times \vec{E}_{v1} \approx (1 - R)\ (\hat{n} \times \hat{v})\, E^i \tag{4A.12}$$

where R = $(R_{\parallel} - R_{\perp})/2$. Similar considerations lead to the following approximations:

$$\eta \hat{n} \times \vec{H}_{v1} \approx (1 + R)\ [\hat{n} \times (-\hat{h})]\, E^i \tag{4A.13}$$

$$\hat{n} \times \vec{E}_{h1} \approx (1 - R)\ (\hat{n} \times \hat{h})\, E^i \tag{4A.14}$$

$$\eta \hat{n} \times \vec{H}_{h1} \approx (1 + R)\ (\hat{n} \times \hat{v})\, E^i \tag{4A.15}$$

for estimating the tangential fields in $\vec{\mathcal{E}}_{1,2}$ and $\vec{\mathcal{H}}_{1,2}$. These tangential field estimates leads to the following estimates for the normal field components:

$$\begin{aligned} \hat{n}' \cdot \vec{E}_{v1} &= j\frac{\eta_1}{k_1}\nabla_s' \cdot (\hat{n}' \times \vec{H}_{v1}) \\ &\approx j\frac{(1+R)}{k_1}\nabla_s' \cdot (\hat{n}' \times \eta_1 \vec{H}^i) \\ &= (1+R)\ (\hat{n}' \cdot \hat{v})\, \vec{E} \end{aligned} \tag{4A.16}$$

$$\eta \hat{n}' \cdot \vec{H}_{v1} = -j\left[\frac{1}{k_1}\nabla_s' \cdot (\hat{n}' \times \vec{E}_{v1})\right]$$

$$\approx -j(1-R)\left[\frac{1}{k_1}\nabla_s' \cdot (\hat{n}' \times \vec{E}^i)\right]$$

$$= (1-R)\,(\hat{n}' \cdot \hat{h})\,E^i \tag{4A.17}$$

Similar considerations yield the other normal field components:

$$\hat{n}' \cdot \vec{E}_{h1} \approx (1+R)\,(\hat{n}' \cdot \hat{h})\,E^i \tag{4A.18}$$

$$\eta \hat{n}' \cdot \vec{H}_{h1} \approx (1-R)\,(\hat{n}' \cdot \hat{v})\,E^i \tag{4A.19}$$

4A.3 SURFACE FIELD ESTIMATES FOR LIKE POLARIZATION

The corresponding set of tangential and normal field estimates for polarized scattering under the Kirchhoff approximation are summarized below. First, the tangential field estimates are obtained from Section 4.3 as

$$\hat{n} \times \vec{E}_{v1} \approx (1-R_{\parallel})\,(\hat{n} \times \hat{v})\,E^i \tag{4A.20}$$

$$\hat{n} \times \vec{H}_{v1} \approx (1+R_{\parallel})\,[\hat{n} \times (-\hat{h})]\,(E^i/\eta) \tag{4A.21}$$

$$\hat{n} \times \vec{E}_{h1} \approx (1+R_{\perp})\,(\hat{n} \times \hat{h})\,E^i \tag{4A.22}$$

$$\hat{n} \times \vec{H}_{h1} \approx (1-R_{\perp})\,(\hat{n} \times \hat{v})\,(E^i/\eta) \tag{4A.23}$$

The corresponding normal field components for polarized scattering are

$$\hat{n}' \cdot \vec{E}_{v1} = (1+R_{\parallel})\,(\hat{n}' \cdot \hat{v}E^i) \tag{4A.24}$$

$$\hat{n}' \cdot \vec{H}_{v1} = (1-R_{\parallel})\,(\hat{n}' \cdot \hat{h}E^i)/\eta \tag{4A.25}$$

$$\hat{n}' \cdot \vec{E}_{h1} = (1-R_{\perp})\,(\hat{n}' \cdot \hat{h}E^i) \tag{4A.26}$$

$$\hat{n}' \cdot \vec{H}_{h1} = (1+R_{\perp})\,(\hat{n}' \cdot \hat{v}E^i)/\eta \tag{4A.27}$$

Appendix 4B
The Field Coefficients

4B.1 THE KIRCHHOFF AND COMPLEMENTARY FIELD COEFFICIENTS

The purpose of this appendix is to provide explicit representions for the field coefficients in the backward scattered field expression. The complementary field expressions are lengthy and they have to be integrated in multiple scattering calculations and evaluated along different directions in single-scatter calculations. Hence, several forms of these coefficients are needed when we calculate the scattering coefficients.

The Kirchhoff field coefficients in (4.72) to (4.75) can be explicitly written by using (4.96), (4.97) as

$$f_{vv} = \frac{2R_{\parallel}}{\cos\theta + \cos\theta_s}\left[\sin\theta\sin\theta_s - (1 + \cos\theta\cos\theta_s)\cos(\phi_s - \phi)\right] \tag{4B.1}$$

$$f_{hh} = -\frac{2R_{\perp}}{\cos\theta + \cos\theta_s}\left[\sin\theta\sin\theta_s - (1 + \cos\theta\cos\theta_s)\cos(\phi_s - \phi)\right] \tag{4B.2}$$

$$f_{hv} = 2R\sin(\phi_s - \phi) \tag{4B.3}$$

$$f_{vh} = 2R\sin(\phi - \phi_s) \tag{4B.4}$$

By substituting (4.107)–(4.110) into (4.106) and (4.113) and carrying out vector products, the complementary field coefficients for like polarized scattering are found to be

$$\begin{aligned} F_{vv}(u, v) = &-\left[(1 - R_{\parallel})\frac{1}{q} - (1 + R_{\parallel})\frac{\mu_r}{q_t}\right](1 + R_{\parallel})C_1 \\ &+ \left[(1 - R_{\parallel})/q - (1 + R_{\parallel})/q_t\right](1 - R_{\parallel})C_2 \\ &+ \left[(1 - R_{\parallel})/q - (1 + R_{\parallel})/(\varepsilon_r q_t)\right](1 + R_{\parallel})C_3 \\ &+ \left[(1 + R_{\parallel})\frac{1}{q} - (1 - R_{\parallel})\frac{\varepsilon_r}{q_t}\right](1 - R_{\parallel})C_4 \\ &+ \left[(1 + R_{\parallel})/q - (1 - R_{\parallel})/q_t\right](1 + R_{\parallel})C_5 \\ &+ \left[(1 + R_{\parallel})/q - (1 - R_{\parallel})/(\mu_r q_t)\right](1 - R_{\parallel})C_6 \end{aligned} \tag{4B.5}$$

$$F_{hh}(u,v) = -\left[(1+R_\perp)\frac{1}{q} - (1-R_\perp)\frac{\mu_r}{q_t}\right](1-R_\perp)C_4$$

$$-[(1+R_\perp)/q - (1-R_\perp)/q_t](1+R_\perp)C_5$$

$$-[(1+R_\perp)/q - (1-R_\perp)/(\varepsilon_r q_t)](1-R_\perp)C_6$$

$$+\left[(1-R_\perp)\frac{1}{q} - (1+R_\perp)\frac{\varepsilon_r}{q_t}\right](1+R_\perp)C_1$$

$$-[(1-R_\perp)/q - (1+R_\perp)/q_t](1-R_\perp)C_2$$

$$-[(1-R_\perp)/q - (1+R_\perp)/(\mu_r q_t)](1+R_\perp)C_3 \qquad (4B.6)$$

In the above $q = (k^2 - u^2 - v^2)^{1/2}$ and $q_t = (k_t^2 - u^2 - v^2)^{1/2}$. The C coefficients are the results of the triple cross products and dot products defined below. For each C coefficient we shall provide its general form and its form for backscattering.

$$C_1(u,v,k_x,k_y,k_{sx},k_{sy}) = k\hat{h}_s \cdot \vec{n} \times \vec{n}' \times \hat{h}$$

$$= k\cos\phi_s\left\{\cos\phi - \frac{(k_{sx}+u)}{k_{sz}k_z}[(k_x+u)\cos\phi + (k_y+v)\sin\phi]\right\}$$

$$+ k\sin\phi_s\left\{\sin\phi - \frac{(k_{sy}+v)}{k_{sz}k_z}[(k_x+u)\cos\phi + (k_y+v)\sin\phi]\right\} \qquad (4B.7)$$

$$C_1(u,v,k_x,0,-k_x,0)$$

$$= -k\left[1 - \frac{u-k_x}{k_z^2}(k_x+u)\right] \qquad (4B.8)$$

$$C_2(u,v,k_x,k_y,k_{sx},k_{sy}) = \hat{h}_s \cdot \vec{n} \times (\vec{n}' \times \hat{v}) \times \vec{g}$$

$$= -\cos\phi_s\left\{\frac{\cos\phi\cos\theta}{k_{sz}}(k_{sx}+u)u + \frac{\sin\phi\cos\theta}{k_{sz}}(k_{sx}+u)v\right.$$

$$+ \frac{\sin\phi\cos\theta}{k_z}(k_x+u)v - \frac{\cos\phi\cos\theta}{k_z}(k_y+v)v$$

$$\left. + \frac{\sin\theta}{k_{sz}k_z}[(k_{sx}+u)(k_x+u)u + (k_{sx}+u)(k_y+v)v]\right\}$$

$$-\sin\phi_s\left\{\frac{\cos\phi\cos\theta}{k_{sz}}(k_{sy}+v)u + \frac{\sin\phi\cos\theta}{k_{sz}}(k_{sy}+v)v\right.$$

$$+\frac{\cos\phi\cos\theta}{k_z}(k_y+v)\,u-\frac{\sin\phi\cos\theta}{k_z}(k_x+u)\,u$$

$$-\frac{\sin\theta}{k_{sz}k_z}[\,(k_{sy}+v)\,(k_y+v)\,v+(k_x+u)\,(k_{sy}+v)\,u]\,\}\tag{4B.9}$$

$$C_2(u,v,k_x,0,-k_x,0)\;=\;\frac{1}{k}\left[\,(u-k_x)\,u-v^2\right]+\frac{\sin\theta}{k_z^2}(u-k_x)\left[u\,(u+k_x)+v^2\right]\tag{4B.10}$$

$$C_3(u,v,k_x,k_y,k_{sx},k_{sy})\;=\;\hat{h}_s\cdot\vec{n}\times(\vec{n}'\cdot\hat{v})\,\vec{g}$$

$$=-(u\cos\phi_s+v\sin\phi_s)$$

$$[\sin\theta-(\,(k_x+u)\cos\phi\cos\theta)/k_z-(\,(k_y+v)\sin\phi\cos\theta)/k_z]\tag{4B.11}$$

$$C_3(u,v,k_x,0,-k_x,0)\;=\;-u^2/k\tag{4B.12}$$

$$C_4(u,v,k_x,k_y,k_{sx},k_{sy})\;=\;k\hat{v}_s\cdot\vec{n}\times\vec{n}'\times\hat{v}$$

$$=k\,[\cos\phi_s\cos\theta_s\,\{\cos\phi\cos\theta\,[1-(k_{sy}+v)\,(k_y+v)/(k_{sz}k_z)\,]$$

$$+\sin\phi\cos\theta\,[\,(k_x+u)\,(k_{sy}+v)/(k_{sz}k_z)\,]+\sin\theta\,(k_x+u)/k_z\,\}$$

$$+\sin\phi_s\cos\theta_s\,\{\cos\phi\cos\theta\,[\,(k_{sx}+u)\,(k_y+v)/(k_{sz}k_z)\,]$$

$$+\sin\phi\cos\theta\,[1-(k_{sx}+u)\,(k_x+u)/(k_{sz}k_z)\,]+\sin\theta\,(k_y+v)/k_z\,\}$$

$$+\frac{\sin\theta_s}{k_{sz}k_z}\{\cos\phi\cos\theta\,(k_{sx}+u)\,k_z+\sin\phi\cos\theta\,(k_{sy}+v)\,k_z$$

$$+\sin\theta\,[\,(k_{sx}+u)\,(k_x+u)+(k_{sy}+v)\,(k_y+v)\,]\,\}\,]\tag{4B.13}$$

$$C_4(u,v,k_x,0,-k_x,0)\;=\;-k\left[\left(\cos^2\theta-\frac{v^2}{k^2}\right)+\sin\theta\frac{u-k_x}{k}\right]$$

$$+k\frac{\sin^2\theta}{k_z^2}\left[\,(u^2-k_x^2)+v^2\right]\tag{4B.14}$$

$$C_5(u,v,k_x,k_y,k_{sx},k_{sy})\;=\;\hat{v}_s\cdot\vec{n}\times(\vec{n}'\times\hat{h})\times\vec{g}$$

$$=\cos\phi_s\cos\theta_s\left[\frac{\cos\phi}{k_{sz}}(k_{sy}+v)\,v-\frac{\cos\phi}{k_z}(k_x+u)\,u\right.$$

$$-\frac{\sin\phi}{k_z}(k_y+v)\,u-\frac{\sin\phi}{k_{sz}}(k_{sy}+v)\,u\Big]$$

$$-\sin\phi_s\cos\theta_s\Big[\frac{\cos\phi}{k_z}(k_x+u)\,v+\frac{\cos\phi}{k_{sz}}(k_{sx}+u)\,v$$

$$+\frac{\sin\phi}{k_z}(k_y+v)\,v-\frac{\sin\phi}{k_{sz}}(k_{sx}+u)\,u\Big]$$

$$-\frac{\sin\theta_s}{k_{sz}k_z}\{\cos\phi\,[\,(k_{sx}+u)\,(k_x+u)\,u+(k_x+u)\,(k_{sy}+v)\,v]$$

$$+\sin\phi\,[u\,(k_{sx}+u)\,(k_y+v)+v\,(k_{sy}+v)\,(k_y+v)\,]\,\} \tag{4B.15}$$

$$C_5(u,v,k_x,0,-k_x,0)\;=\;\frac{1}{k}\big[\,(k_x+u)\,u-v^2\big]-\frac{\sin\theta}{k_z^2}\big[\,(u-k_x)\,u+v^2\big]\,(k_x+u) \tag{4B.16}$$

$$C_6(u,v,k_x,k_y,k_{sx},k_{sy})\;=\;\hat{v}_s\cdot\vec{n}\times(\vec{n}'\cdot\hat{h})\,\vec{g}\;=\;\frac{-1}{k_z}[\,(k_x+u)\sin\phi-(k_y+v)\cos\phi]$$

$$[v\cos\phi_s\cos\theta_s-u\sin\phi_s\cos\theta_s+\sin\theta_s\,(k_{sx}v-k_{sy}u)\,/k_{sz}] \tag{4B.17}$$

$$C_6(u,v,k_x,0,-k_x,0)\;=\;\frac{-1}{k_z}v^2\,[\cos\theta_s-\sin\theta_s\,(k_{sx}/k_{sz})\,]\;=\;-\frac{v^2}{k_z} \tag{4B.18}$$

The cross polarized field coefficients given by (4.111) and (4.112) can also be evaluated in a similar way as

$$F_{hv}(u,v)\;=\;\Big[\,(1-R)\frac{1}{q}-(1+R)\frac{\mu_r}{q_t}\Big]\,(1+R)\,B_1$$

$$-[\,(1-R)/q-(1+R)/q_t]\,(1-R)\,B_2$$

$$-[\,(1-R)/q-(1+R)/(\varepsilon_r q_t)\,]\,(1+R)\,B_3$$

$$+\Big[\,(1+R)\frac{1}{q}-(1-R)\frac{\varepsilon_r}{q_t}\Big]\,(1-R)\,B_4$$

$$+[\,(1+R)/q-(1-R)/q_t]\,(1+R)\,B_5$$

$$+[\,(1+R)/q-(1-R)/(\mu_r q_t)\,]\,(1-R)\,B_6 \tag{4B.19}$$

$$F_{vh}(u,v) = \left[(1-R)\frac{1}{q} - (1+R)\frac{\mu_r}{q_t}\right](1+R)B_4$$
$$+ [(1-R)/q - (1+R)/q_t](1-R)B_5$$
$$+ [(1-R)/q - (1+R)/(\varepsilon_r q_t)](1+R)B_6$$
$$+ \left[(1+R)\frac{1}{q} - (1-R)\frac{\varepsilon_r}{q_t}\right](1-R)B_1$$
$$-[(1+R)/q - (1-R)/q_t](1+R)B_2$$
$$-[(1+R)/q - (1-R)/(\mu_r q_t)](1-R)B_3 \qquad (4B.20)$$

where the B coefficients are

$$B_1(u, v, k_x, k_y, k_{sx}, k_{sy}) = k\hat{v}_s \cdot \vec{n} \times (\vec{n}' \times \hat{h})$$
$$= k\{\cos\phi_s \cos\theta_s \left[\frac{\cos\phi}{k_{sz}k_z}(k_x+u)(k_{sy}+v) - \sin\phi + \frac{\sin\phi}{k_{sz}k_z}(k_{sy}+v)(k_y+v)\right]$$
$$-\sin\phi_s \cos\theta_s \left[\frac{\cos\phi}{k_{sz}k_z}(k_{sx}+u)(k_x+u) - \cos\phi + \frac{\sin\phi}{k_{sz}k_z}(k_{sx}+u)(k_y+v)\right]$$
$$+\frac{\sin\theta_s}{k_{sz}}[(k_{sy}+v)\cos\phi - (k_{sx}+u)\sin\phi]\} \qquad (4B.21)$$

$$B_1(u, v, k_x, 0, -k_x, 0) = -(uv)/(k\cos\theta) \qquad (4B.22)$$

$$B_2(u, v, k_x, k_y, k_{sx}, k_{sy}) = \hat{v}_s \cdot \vec{n} \times (\vec{n}' \times \hat{v}) \times \vec{g}$$
$$= \cos\phi_s \cos\theta_s \{\frac{\cos\phi\cos\theta}{k_{sz}}(k_{sy}+v)u - \frac{\sin\phi\cos\theta}{k_{sz}}(k_{sy}+v)v$$
$$+\frac{\cos\phi\cos\theta}{k_z}(k_y+v)u - \frac{\sin\phi\cos\theta}{k_z}(k_x+u)u$$
$$+\frac{\sin\theta}{k_{sz}k_z}[(k_x+u)(k_{sy}+v)u + (k_{sy}+v)(k_y+v)v]\}$$
$$-\sin\phi_s\cos\theta_s\{\frac{\cos\phi\cos\theta}{k_{sz}}(k_{sx}+u)u + \frac{\sin\phi\cos\theta}{k_{sz}}(k_{sx}+u)v$$
$$+\frac{\sin\phi\cos\theta}{k_z}(k_x+u)v - \frac{\cos\phi\cos\theta}{k_z}(k_y+v)v$$
$$+\frac{\sin\theta}{k_{sz}k_z}[(k_{sx}+u)(k_x+u)u + (k_{sx}+u)(k_y+v)v]\}$$

$$+\frac{\sin\theta_s}{k_{sz}k_z}\{\cos\phi\cos\theta\,[\,(k_{sx}+u)\ (k_y+v)\,u+(k_{sy}+v)\ (k_y+v)\,v]$$

$$-\sin\phi\cos\theta\,[\,(k_{sx}+u)\ (k_x+u)\,u+(k_x+u)\ (k_{sy}+v)\,v]\ \} \tag{4B.23}$$

$$B_2\,(u,v,k_x,0,-k_x,0) = -(2uv)/(k\cos\theta) \tag{4B.24}$$

$$B_3\,(u,v,k_x,k_y,k_{sx},k_{sy}) = \hat{v}_s\cdot\vec{\hat{n}}\times(\vec{\hat{n}}'\cdot\hat{v})\,\vec{\hat{g}}$$

$$= \left[\sin\theta-\frac{\cos\phi\cos\theta}{k_z}(k_x+u)-\frac{\sin\phi\cos\theta}{k_z}(k_y+v)\right]$$

$$\left[\frac{\sin\theta_s}{k_{sz}}(k_{sx}v-k_{sy}u)+\cos\theta_s\,(v\cos\phi_s-u\sin\phi_s)\right] \tag{4B.25}$$

$$B_3\,(u,v,k_x,0,-k_x,0) = (uv)/(k\cos\theta) \tag{4B.26}$$

$$B_4\,(u,v,k_x,k_y,k_{sx},k_{sy}) = k\hat{h}_s\cdot\vec{\hat{n}}\times(\vec{\hat{n}}'\times\hat{v})$$

$$= k\cos\phi_s\left[\frac{\cos\phi\cos\theta}{k_{sz}k_z}(k_{sx}+u)\ (k_y+v)+\sin\phi\cos\theta\right.$$

$$\left.-\frac{\sin\phi\cos\theta}{k_{sz}k_z}(k_{sx}+u)\ (k_x+u)+\frac{\sin\theta}{k_z}(k_y+v)\right]$$

$$-k\sin\phi_s\left[\cos\phi\cos\theta-\frac{\cos\phi\cos\theta}{k_{sz}k_z}(k_{sy}+v)\ (k_y+v)\right.$$

$$\left.+\frac{\sin\phi\cos\theta}{k_{sz}k_z}(k_x+u)\ (k_{sy}+v)+\frac{\sin\theta}{k_z}(k_x+u)\right] \tag{4B.27}$$

$$B_4\,(u,v,k_x,0,-k_x,0) = -(uv)/(k\cos\theta) \tag{4B.28}$$

$$B_5\,(u,v,k_x,k_y,k_{sx},k_{sy}) = \hat{h}_s\cdot\vec{\hat{n}}\times(\vec{\hat{n}}'\times\hat{h})\times\vec{\hat{g}}$$

$$= -\cos\phi_s\left[\frac{\cos\phi}{k_{sz}}(k_{sx}+u)\,v-\frac{\sin\phi}{k_{sz}}(k_{sx}+u)\,u\right.$$

$$\left.+\frac{\cos\phi}{k_z}(k_x+u)\,v+\frac{\sin\phi}{k_z}(k_y+v)\,v\right]$$

$$-\sin\phi_s\left[\frac{\cos\phi}{k_{sz}}(k_{sy}+v)\,v-\frac{\sin\phi}{k_{sz}}(k_{sy}+v)\,u\right.$$

$$\left.-\frac{\cos\phi}{k_z}(k_x+u)\,u-\frac{\sin\phi}{k_z}(k_y+v)\,u\right] \tag{4B.29}$$

$$B_5(u, v, k_x, 0, -k_x, 0) = (2uv)/(k\cos\theta) \tag{4B.30}$$

$$B_6(u, v, k_x, k_y, k_{sx}, k_{sy}) = \hat{h}_s \cdot \vec{n} \times (\vec{n}' \cdot \hat{h})\vec{g}$$
$$= [(k_y + v)\cos\phi - (k_x + u)\sin\phi]\,[(u\cos\phi_s + v\sin\phi_s)/k_z] \tag{4B.31}$$

$$B_6(u, v, k_x, 0, -k_x, 0) = -(uv)/(k\cos\theta) \tag{4B.32}$$

4B.2 SPECIAL CASES

The C coefficients in (4B.7) through (4B.18) evaluated at $u = -k_x = -k\sin\theta\cos\phi$ and $v = -k_y = -k\sin\theta\sin\phi$ are

$$C_1(-k_x, -k_y, k_x, k_y, k_{sx}, k_{sy}) = k\cos(\phi_s - \phi) \tag{4B.33}$$

$$C_2(-k_x, -k_y, k_x, k_y, k_{sx}, k_{sy}) = k\sin\theta\cos\theta\,[\sin\theta_s - \sin\theta\cos(\phi_s - \phi)]/\cos\theta_s \tag{4B.34}$$

$$C_3(-k_x, -k_y, k_x, k_y, k_{sx}, k_{sy}) = k\sin^2\theta\cos(\phi_s - \phi) \tag{4B.35}$$

$$C_4(-k_x, -k_y, k_x, k_y, k_{sx}, k_{sy}) = k\cos\theta\,[\cos(\phi_s - \phi) - \sin\theta\sin\theta_s]/\cos\theta_s \tag{4B.36}$$

$$C_5(-k_x, -k_y, k_x, k_y, k_{sx}, k_{sy}) = C_6(-k_x, -k_y, k_x, k_y, k_{sx}, k_{sy}) = 0 \tag{4B.37}$$

Another set of C coefficients needed is evaluated at $u = -k_{sx} = -k\sin\theta_s\cos\phi_s$ and $v = -k_{sy} = -k\sin\theta_s\sin\phi_s$.

$$C_1(-k_{sx}, -k_{sy}, k_x, k_y, k_{sx}, k_{sy}) = k\cos(\phi_s - \phi) \tag{4B.38}$$

$$C_2(-k_{sx}, -k_{sy}, k_x, k_y, k_{sx}, k_{sy}) = 0 \tag{4B.39}$$

$$C_3(-k_{sx}, -k_{sy}, k_x, k_y, k_{sx}, k_{sy}) = k\sin^2\theta_s\cos(\phi_s - \phi) \tag{4B.40}$$

$$C_4(-k_{sx}, -k_{sy}, k_x, k_y, k_{sx}, k_{sy}) = \frac{k\cos\theta_s\,[\cos(\phi_s - \phi) - \sin\theta\sin\theta_s]}{\cos\theta} \tag{4B.41}$$

$$C_5(-k_{sx}, -k_{sy}, k_x, k_y, k_{sx}, k_{sy}) = \frac{k\sin\theta_s\cos\theta_s\,[\sin\theta - \sin\theta_s\cos(\phi_s - \phi)]}{\cos\theta} \tag{4B.42}$$

$$C_6(-k_{sx}, -k_{sy}, k_x, k_y, k_{sx}, k_{sy}) = 0 \tag{4B.43}$$

Upon comparing between the C coefficients evaluated at $-k_x, -k_y$ and those at $-k_{sx}, -k_{sy}$, it is seen that each differs from its corresponding one by an interchange between θ and θ_s. Apparently, we do not need to perform an interchange between the azimuthal angles because cosine is an even function.

For cross polarization we need to evaluate the B coefficients. For $u = -k_x = -k\sin\theta\cos\phi$ and $v = -k_y = -k\sin\theta\sin\phi$ they are

$$B_1(-k_x, -k_y, k_x, k_y, k_{sx}, k_{sy}) = k\sin(\phi_s - \phi)/\cos\theta_s \tag{4B.44}$$

$$B_2(-k_x, -k_y, k_x, k_y, k_{sx}, k_{sy}) = -k\sin^2\theta\cos\theta\sin(\phi_s - \phi) \tag{4B.45}$$

$$B_3(-k_x, -k_y, k_x, k_y, k_{sx}, k_{sy}) = k\sin^2\theta\sin(\phi_s - \phi)/\cos\theta_s \tag{4B.46}$$

$$B_4(-k_x, -k_y, k_x, k_y, k_{sx}, k_{sy}) = -k\cos\theta\sin(\phi_s - \phi) \tag{4B.47}$$

$$B_5(-k_x, -k_y, k_x, k_y, k_{sx}, k_{sy}) = B_6(-k_x, -k_y, k_x, k_y, k_{sx}, k_{sy}) = 0 \tag{4B.48}$$

The vanishing of the last two coefficients is similar to the C coefficients. For $u = -k_{sx} = -k\sin\theta_s\cos\phi_s$ and $v = -k_{sy} = -k\sin\theta_s\sin\phi_s$, the other set of the B coefficients are

$$B_1(-k_{sx}, -k_{sy}, k_x, k_y, k_{sx}, k_{sy}) = k\cos\theta_s\sin(\phi_s - \phi) \tag{4B.49}$$

$$B_2(-k_{sx}, -k_{sy}, k_x, k_y, k_{sx}, k_{sy}) = k\sin^2\theta_s\cos\theta_s\sin(\phi_s - \phi) \tag{4B.50}$$

$$B_3(-k_{sx}, -k_{sy}, k_x, k_y, k_{sx}, k_{sy}) = 0 \tag{4B.51}$$

$$B_4(-k_{sx}, -k_{sy}, k_x, k_y, k_{sx}, k_{sy}) = -k\sin(\phi_s - \phi)/\cos\theta \tag{4B.52}$$

$$B_5(-k_{sx}, -k_{sy}, k_x, k_y, k_{sx}, k_{sy}) = 0 \tag{4B.53}$$

$$B_6(-k_{sx}, -k_{sy}, k_x, k_y, k_{sx}, k_{sy}) = k\sin^2\theta_s\sin(\phi_s - \phi)/\cos\theta \tag{4B.54}$$

Here again the B coefficients evaluated at $-k_x, -k_y$ are related to those evaluated at $-k_{sx}, -k_{sy}$ by an interchange of θ and θ_s and ϕ and ϕ_s in the corresponding coefficients. The above B and C coefficients are those needed in evaluating single scattering calculations discussed in (5.40) through (5.43). The more general B and C coefficients are required only for multiple scattering terms where integrations over u and v variables have to be carried out.

4B.2.1 Special $F_{qp}(-k_x, -k_y)$ Coefficients

With special C and B coefficients known we can find the corresponding special F_{qp} coefficients, $F_{vv}(-k_x, -k_y)$, $F_{hh}(-k_x, -k_y)$, $F_{hv}(-k_x, -k_y)$, $F_{vh}(-k_x, -k_y)$. For simplicity of writing we shall use the following notations in writing the complementary field coefficients.

$$s = \sin\theta, \ ss = \sin\theta_s$$

$$cs = \cos\theta, \ css = \cos\theta_s$$

$$sf = \sin(\phi_s - \phi), \ csf = \cos(\phi_s - \phi)$$

$$sq = (\mu_r \varepsilon_r - \sin^2\theta)^{1/2}$$

$$\mu_r = \mu_t/\mu, \varepsilon_r = \varepsilon_t/\varepsilon$$

$$c_1 = (csf - s\ ss)/(sq\ css)$$

$$c_2 = s(ss - s\ csf)/css$$

$$T_v = 1 + R_{\parallel}, T_{vm} = 1 - R_{\parallel}$$

$$T_h = 1 + R_{\perp}, T_{hm} = 1 - R_{\perp}$$

$$T_p = 1 + R, T_m = 1 - R$$

$$R = (R_{\parallel} - R_{\perp})/2$$

With the above notations, we can write the $C_i(-k_x, -k_y)$ coefficients as

$$C_1/k = csf$$

$$C_2/k = s\ cs(ss - s\ csf)/css$$

$$C_3/k = s^2 csf$$

$$C_4/k = cs(csf - s\ ss)/css$$

$$C_5 = C_6 = 0$$

Then, we have

$$F_{vv}(-k_x, -k_y) = -\left[\frac{T_{vm}}{cs} - \frac{T_v \mu_r}{sq}\right] T_v C_1(-k_x, -k_y)/k$$

$$+ \left[\frac{T_{vm}}{cs} - \frac{T_v}{\varepsilon_r sq}\right] T_v C_3(-k_x, -k_y)/k$$

$$+ \left[\frac{T_{vm}}{cs} - \frac{T_v}{sq}\right] T_{vm} C_2(-k_x, -k_y)/k$$

$$+\left[\frac{T_v}{cs}-\frac{T_{vm}\varepsilon_r}{sq}\right]T_{vm}C_4(-k_x,-k_y)/k$$

$$= -(csT_{vm}-sqT_v/\varepsilon_r)\,T_v csf - (csT_{vm}-sqT_v/\varepsilon_r)\,T_{vm}\varepsilon_r(csf - s\,ss)/(sq\ css)$$

$$+ (T_{vm}^2 - T_v T_{vm} cs/sq)\,s\,(ss - s\ csf)/css$$

$$= -(csT_{vm}-sqT_v/\varepsilon_r)\,(T_v csf + T_{vm}\varepsilon_r c_1) + (T_{vm}^2 - T_v T_{vm} cs/sq)\,c_2 \qquad (4B.55)$$

In the above, we have first combined C_1 and C_3 terms and then the resultant with the C_4 term to obtain the first term in (4B.55). The coefficient $F_{hh}(-k_x,-k_y)$ can be calculated in a similar way. However, since it is the negative dual of $F_{vv}(-k_x,-k_y)$, we can write it down immediately using (4B.55), i.e.,

$$F_{hh}(-k_x,-k_y) = (csT_{hm}-sqT_h/\mu_r)\,(T_h csf + T_{hm}\mu_r c_1)$$
$$-(T_{hm}^2 - T_h T_{hm} cs/sq)\,c_2 \qquad (4B.56)$$

To find the cross polarized coefficients we list the $B_i(-k_x,-k_y)$ coefficients in simplified notations as follows

$$B_1/k = sf/css$$
$$B_2/k = -s^2 sf\ cs$$
$$B_3/k = s^2 sf/css$$
$$B_4/k = -sf\ cs$$
$$B_5 = B_6 = 0$$

Then, we have

$$F_{hv}(-k_x,-k_y) = \left[\frac{T_m}{cs}-\frac{T_p\mu_r}{sq}\right]T_p B_1(-k_x,-k_y)/k$$

$$-\left[\frac{T_m}{cs}-\frac{T_p}{sq}\right]T_m B_2(-k_x,-k_y)/k$$

$$-\left[\frac{T_m}{cs}-\frac{T_p}{\varepsilon_r sq}\right]T_p B_3(-k_x,-k_y)/k$$

$$+\left[\frac{T_p}{cs}-\frac{T_m\varepsilon_r}{sq}\right]T_m B_4(-k_x,-k_y)/k$$

$$= (csT_m - sqT_p/\varepsilon_r)\,T_p sf/css + (csT_m - sqT_p/\varepsilon_r)\,T_m\varepsilon_r sf/sq$$

$$+ (T_m^2 - T_p T_m cs/sq)\,s^2 sf$$

$$= (csT_m - sqT_p/\varepsilon_r)\,(T_p/css + T_m\varepsilon_r/sq)\,sf + (T_m^2 - T_pT_m cs/sq)\,s^2 sf \tag{4B.57}$$

In the above, we have first combined B_1 and B_3 terms and then the resultant with the B_4 term to obtain the first term in (4B.57). The coefficient $F_{vh}(-k_x, -k_y)$ can be calculated in a similar way. However, since it is the dual of $F_{hv}(-k_x, -k_y)$, we can write it down immediately using (4B.57), i.e.,

$$\begin{aligned} F_{vh}(-k_x, -k_y) &= (csT_p - sqT_m/\mu_r)\,(T_m/css + T_p\mu_r/sq)\,sf \\ &\quad + (T_p^2 - T_pT_m cs/sq)\,s^2 sf \end{aligned} \tag{4B.58}$$

Specialized $F_{qp}(-k_x, -k_y)$ Coefficients

Note that in principle the argument of the Fresnel reflection coefficients is a function of the local angle of incidence. However, under small roughness conditions or for high dielectric surfaces, we can approximate the local incident angle by the incident angle, i.e.,

$$R_{\parallel} = \frac{\varepsilon_r\cos\theta - \sqrt{\mu_r\varepsilon_r - \sin^2\theta}}{\varepsilon_r\cos\theta + \sqrt{\mu_r\varepsilon_r - \sin^2\theta}}; \quad R_{\perp} = \frac{\mu_r\cos\theta - \sqrt{\mu_r\varepsilon_r - \sin^2\theta}}{\mu_r\cos\theta + \sqrt{\mu_r\varepsilon_r - \sin^2\theta}}$$

Under this assumption further specialization of the complementary field coefficients is possible, we have

$$F_{vv}(-k_x, -k_y) = \frac{4\,(\mu_r\varepsilon_r - \sin^2\theta - \varepsilon_r\cos^2\theta)}{[\varepsilon_r\cos\theta + \sqrt{\mu_r\varepsilon_r - \sin^2\theta}]^2}\,\frac{\sin\theta\,[\sin\theta_s - \sin\theta\cos(\phi_s - \phi)]}{\cos\theta_s}$$

$$= \frac{(\mu_r\varepsilon_r - \sin^2\theta - \varepsilon_r\cos^2\theta)\,(1 + R_{\parallel})^2}{(\varepsilon_r\cos\theta)^2\cos\theta_s}\sin\theta\,[\sin\theta_s - \sin\theta\cos(\phi_s - \phi)] \tag{4B.59}$$

$$F_{hh}(-k_x, -k_y) = -\left[\frac{4\,(\mu_r\varepsilon_r - \sin^2\theta - \mu_r\cos^2\theta)}{[\mu_r\cos\theta + \sqrt{\mu_r\varepsilon_r - \sin^2\theta}]^2}\right]\frac{\sin\theta\,[\sin\theta_s - \sin\theta\cos(\phi_s - \phi)]}{\cos\theta_s}$$

$$= -\frac{(\mu_r\varepsilon_r - \sin^2\theta - \mu_r\cos^2\theta)}{(\mu_r\cos\theta)^2\cos\theta_s}(1 + R_{\perp})^2\sin\theta\,[\sin\theta_s - \sin\theta\cos(\phi_s - \phi)] \tag{4B.60}$$

For the cross polarized coefficients, we need another relation for reduction in addition to the use of the incident angle in the Fresnel coefficients, $1 - R = T(1 + R)$, where

$$T = \frac{f(\theta)\,[\mu_r\cos\theta + f(\theta)] + \mu_r\cos\theta\,[\varepsilon_r\cos\theta + f(\theta)]}{\varepsilon_r\cos\theta\,[\mu_r\cos\theta + f(\theta)] + f(\theta)\,[\mu_r\cos\theta + f(\theta)]} \tag{4B.61}$$

$$f(\theta) = (\mu_r\varepsilon_r - \sin^2\theta)^{1/2} \tag{4B.62}$$

With the above definitions, we have

$$F_{hv}(-k_x, -k_y) = \left\{ \frac{(\varepsilon_r T)\cos\theta\cos\theta_s(\varepsilon_r T - \sin^2\theta) - (\mu_r\varepsilon_r - \sin^2\theta)}{T\varepsilon_r\cos\theta_s(\mu_r\varepsilon_r - \sin^2\theta)^{1/2}} \right.$$

$$\left. + \frac{\cos\theta}{\cos\theta_s} - 1 + T\sin^2\theta \right\} (1 - R^2)\sin(\phi_s - \phi) \tag{4B.63}$$

$$F_{vh}(-k_x, -k_y) = \left\{ \frac{\mu_r\cos\theta\cos\theta_s(\mu_r - T\sin^2\theta) - T^2(\mu_r\varepsilon_r - \sin^2\theta)}{T\mu_r(\mu_r\varepsilon_r - \sin^2\theta)^{1/2}\cos\theta_s} \right.$$

$$\left. + \frac{\cos\theta}{\cos\theta_s} - 1 + \frac{\sin^2\theta}{T} \right\} (1 - R^2)\sin(\phi_s - \phi) \tag{4B.64}$$

$F_{vh}(-k_x, -k_y)$ is the dual of $F_{hv}(-k_x, -k_y)$.

Complementary Field Coefficients in Backscattering

The complementary field coefficients in backscattering are of special interest in applications. The simple forms given below are again restricted to small roughness parameters or high dielectric values.

$$F_{vv}(k_x, 0) = \frac{2(\varepsilon_r - 1)\sin^2\theta\,(1 - R_{\parallel}^2)}{\sqrt{\mu_r\varepsilon_r - \sin^2\theta}} = \frac{2(\varepsilon_r - 1)\sin^2\theta}{\varepsilon_r\cos\theta}(1 + R_{\parallel})^2 \tag{4B.65}$$

$$F_{vv}(-k_x, 0) = \frac{2\sin^2\theta}{\cos\theta}(1 - R_{\parallel})^2\left(1 - \frac{\varepsilon_r\cos^2\theta}{\varepsilon_r - \sin^2\theta}\right) \tag{4B.66}$$

$$F_{hh}(k_x, 0) = -\frac{2(\mu_r - 1)\sin^2\theta}{\mu_r\cos\theta}(1 + R_{\perp})^2 \tag{4B.67}$$

$$F_{hh}(-k_x, 0) = \frac{8\sin^2\theta}{\cos\theta}R_{\perp} \tag{4B.68}$$

When the low frequency limit is taken, the combination of Kirchhoff field coefficients and complementary field coefficients can be reduced to well-known small perturbation results. The reduction is given below.

$$\alpha_{vv} = f_{vv} + \frac{1}{4}[F_{vv}(k_x, 0) + F_{vv}(-k_x, 0)]$$

$$= R_{\parallel}\cos^2\theta + \frac{T_{\parallel}^2(\varepsilon_r - 1)}{2\varepsilon_r}\sin^2\theta \tag{4B.69}$$

$$\alpha_{hh} = f_{hh} + \frac{1}{4}[F_{hh}(k_x, 0) + F_{hh}(-k_x, 0)] = R_{\perp}\cos^2\theta \tag{4B.70}$$

Equations (4B.69), (4B.70) are exactly the field coefficients in the expressions of the small perturbation model [Ulaby et al., 1982, Chapter 12).

The Kirchhoff field coefficients for the cross polarized component is zero in the backscattering direction, i.e., $f_{vh} = f_{hv} = 0$. Hence, we should consider the complementary field coefficients which reduce to

$$F_{hv} = F_{vh} = \frac{uv}{k\cos\theta}\{2[(1-R)/q - (1+R)/q_t](1-R)$$
$$+2[(1+R)/q - (1-R)/q_t](1+R)$$
$$-[(1-R)/q - (1+R)\mu_r/q_t](1+R)$$
$$-[(1+R)/q - (1-R)\varepsilon_r/q_t](1-R)$$
$$-[(1-R)/q - (1+R)/(q_t\varepsilon_r)](1+R)$$
$$-[(1+R)/q - (1-R)/(q_t\mu_r)](1-R)\} \tag{4B.71}$$

For a perfectly conducting surface F_{vh} and F_{hv} from (4B.71) reduce to

$$F_{hv} = F_{vh} = 8uv/(qk\cos\theta) \tag{4B.72}$$

The above results show that the cross polarized coefficients satisfy the reciprocity theorem in electromagnetic theory discussed in Chapter 1 and (4B.72) is in agreement with the second-order perturbation result under the small roughness condition.

4B.2.2 The $F_{qp}(-k_{sx}, -k_{sy})$ Coefficients

When the arguments of the field coefficients change, a different set of results is expected. This is particularly obvious in view of the changes in the C and B coefficients. Thus, we need to recompute these field coefficients going through a routine similar to the previous section. For this section $k_{sx} = k\sin\theta_s\cos\phi_s$, $k_{sy} = k\sin\theta_s\sin\phi_s$, $q = k\cos\theta_s$ and $q_t = k(\mu_r\varepsilon_r - \sin^2\theta_s)^{1/2}$.

Some additional notations analogous to those in the previous section are as follows:

$$sqs = (\mu_r\varepsilon_r - \sin^2\theta_s)^{1/2}$$

$$c_{1s} = (csf - s\ ss)/(sqs\ cs)$$

$$c_{2s} = ss\,(s - ss\ csf)/cs$$

The set of $C_i(-k_{sx}, -k_{sy})$ coefficients becomes

$$C_{s1}/k = csf$$

$$C_{s3}/k = ss^2 csf$$

$$C_{s4}/k = css\,(csf - s\ ss)/cs$$

$$C_{s5}/k = ss\ css\,(s - ss\ csf)/cs$$

$$C_{s2} = C_{s6} = 0$$

Using the above coefficients, we have

$$\begin{aligned} F_{vv}(-k_{sx}, -k_{sy}) &= -\left[\frac{T_{vm}}{css} - \frac{T_v\mu_r}{sqs}\right]T_v C_{s1}(-k_{sx}, -k_{sy})/k \\ &+ \left[\frac{T_{vm}}{css} - \frac{T_v}{\varepsilon_r sqs}\right]T_v C_{s3}(-k_{sx}, -k_{sy})/k \\ &+ \left[\frac{T_v}{css} - \frac{T_{vm}\varepsilon_r}{sqs}\right]T_{vm} C_{s4}(-k_{sx}, -k_{sy})/k \\ &+ \left[\frac{T_v}{css} - \frac{T_{vm}}{sqs}\right]T_v C_{s5}(-k_{sx}, -k_{sy})/k \\ &= -\left(cssT_{vm} - \frac{sqsT_v}{\varepsilon_r}\right)T_v csf - \left(cssT_{vm} - \frac{sqsT_v}{\varepsilon_r}\right)T_{vm}\varepsilon_r\frac{(csf - s\ ss)}{sqs\ cs} \\ &+ (T_v^2 - T_vT_{vm}css/sqs)\,ss\,(s - ss\ csf)/cs \\ &= -(cssT_{vm} - sqsT_v/\varepsilon_r)(T_v csf + T_{vm}\varepsilon_r c_{1s}) + (T_v^2 - T_vT_{vm}css/sqs)\,c_{2s} \end{aligned} \tag{4B.73}$$

In the above, we have first combined C_{s1} and C_{s3} terms and then the resultant with the C_{s4} term to obtain the first term in (4B.55). The coefficient $F_{hh}(-k_{sx}, -k_{sy})$ can be calculated in a similar way. However, since it is the negative dual of $F_{vv}(-k_{sx}, -k_{sy})$, we can write it down immediately using (4B.55), i.e.,

$$F_{hh}(-k_{sx}, -k_{sy}) = (cssT_{hm} - sqsT_h/\mu_r)\,(T_h csf + T_{hm}\mu_r c_{1s})$$
$$- (T_h^2 - T_h T_{hm} css/sqs)\, c_{2s} \tag{4B.74}$$

To find the cross polarized coefficients we list the $B_i(-k_{sx}, -k_{sy})$ coefficients in simplified notations as follows

$$B_{s1}/k = sf\ css$$
$$B_{s2}/k = s^2 sf\ css$$
$$B_{s4}/k = -sf/cs$$
$$B_{s6}/k = ss^2 sf/cs$$
$$B_{s5} = B_{s3} = 0$$

Then, we have

$$F_{hv}(-k_{sx}, -k_{sy}) = \left[\frac{T_m}{css} - \frac{T_p\mu_r}{sqs}\right] T_p B_{s1}(-k_{sx}, -k_{sy})/k$$
$$-\left[\frac{T_m}{css} - \frac{T_p}{sqs}\right] T_m B_{s2}(-k_{sx}, -k_{sy})/k$$
$$+\left[\frac{T_p}{css} - \frac{T_m}{\mu_r sqs}\right] T_m B_{s6}(-k_{sx}, -k_{sy})/k$$
$$+\left[\frac{T_p}{css} - \frac{T_m\varepsilon_r}{sqs}\right] T_m B_{s4}(-k_{sx}, -k_{sy})/k$$
$$= -(cssT_p - sqsT_m/\mu_r)\,T_m sf/cs - (cssT_p - sqsT_m/\mu_r)\,T_p\mu_r sf/sqs$$
$$- (T_m^2 - T_p T_m css/sqs)\, ss^2 sf$$
$$= -(cssT_p - sqsT_m/\mu_r)\,(T_m/cs + T_p\mu_r/sqs)\, sf - (T_m^2 - T_p T_m css/sqs)\, ss^2 sf \tag{4B.75}$$

In the above, we have first combined B_{s4} and B_{s6} terms and then the resultant with the B_{s1} term to obtain the first term in (4B.55). The coefficient $F_{vh}(-k_{sx}, -k_{sy})$ can be calculated in a similar way. However, since it is the dual of $F_{hv}(-k_{sx}, -k_{sy})$, we can write it down immediately using (4B.55), i.e.,

$$F_{vh}(-k_{sx}, -k_{sy}) = -(cssT_m - sqsT_p/\varepsilon_r)\,(T_p/cs + T_m\varepsilon_r/sqs)\, sf$$
$$-(T_p^2 - T_p T_m css/sqs)\, ss^2 sf \tag{4B.76}$$

Appendix 4C

Evaluation of Ensemble Averages

This appendix provides ensemble averages needed in the scattering coefficient calculations for a rough surface with a Gram-Charlier height distribution. This distribution reduces to the Gaussian height distribution when we set higher order statistics (beyond the second order) to zero. The Gram-Charlier height distribution is appropriate for the sea surface.

4C.1 BASIC DEFINITIONS

Given a set of n random variables $\{z_1, z_2, \ldots, z_n\}$ and their joint probability density function $p(z_1, z_2, \ldots, z_n)$, their joint characteristic function is defined as

$$\phi(t_1, t_2, \ldots, t_n) \equiv \langle e^{j(z_1 t_1 + z_2 t_2 + \ldots + z_n t_n)} \rangle \tag{4C.1}$$

That is

$$\phi(t_1, t_2, \ldots, t_n) = \int \exp[j(z_1 t_1 + z_2 t_2 + \ldots + z_n t_n)]\, p(z_1, z_2, \ldots, z_n)\, dz_1 dz_2 \ldots dz_n \tag{4C.2}$$

where <> denotes the ensemble average operator.

The joint moments of order $m = q_1 + q_2 + \ldots + q_n$ are given as

$$\mu_{q_1 q_2 \ldots q_n} \equiv \langle z_1^{q_1} z_2^{q_2} \ldots z_n^{q_n} \rangle = (-j)^m \frac{\partial^m}{\partial t_1^{q_1} \ldots \partial t_n^{q_n}} \phi(t_1, t_2, \ldots, t_n) \Bigg|_{t_1 = t_2 \ldots = t_n = 0} \tag{4C.3}$$

The cumulant generating function is defined as

$$K(t_1, t_2, \ldots, t_n) \equiv \ln[\phi(t_1, t_2, \ldots, t_n)] \tag{4C.4}$$

Then, the joint cumulants of order m can be found from the relation

$$\kappa_{q_1 q_2 \ldots q_n} = (-j)^m \frac{\partial^m}{\partial t_1^{q_1} \ldots \partial t_n^{q_n}} K(t_1, t_2, \ldots, t_n) \Bigg|_{t_1 = t_2 = \ldots = t_n = 0} \tag{4C.5}$$

Therefore, the joint characteristic function can be expressed in terms of joint cumulants as

$$\phi(t_1, t_2, \ldots, t_n) = \sum_{(q_1 \ldots q_n) \neq 0} (jt_1)^{q_1} \ldots (jt_n)^{q_n} \frac{\kappa_{q_1} \ldots \kappa_{q_n}}{q_1! \ldots q_n!} \quad (4C.6)$$

where the summation extends over all nonnegative q's.

The joint cumulants are related to the joint moments of n random variables via

$$\ln\left[1 + \frac{\mu_{q_1 00 \ldots 0}}{1!0! \ldots 0!} t_1^{q_1} + \ldots + \frac{\mu_{q_1 q_2 \ldots q_m}}{q_1! q_2! \ldots q_n!} t_1^{q_1} t_2^{q_2} \ldots t_n^{q_n}\right]$$

$$= \frac{\kappa_{q_1 00 \ldots 0}}{1!0! \ldots 0!} t_1^{q_1} + \ldots + \frac{\kappa_{q_1 q_2 \ldots q_m}}{q_1! q_2! \ldots q_n!} t_1^{q_1} t_2^{q_2} \ldots t_n^{q_n} \quad (4C.7)$$

The cumulants can be given in terms of a series of moments by expanding the natural logarithm on the left side of (4C.7) and equating them to the corresponding coefficient on the right. For univariate case,

$$\kappa_1 = \mu_1$$

$$\kappa_2 = \mu_2 - \mu^2{}_1$$

$$\kappa_3 = \mu_3 - 3\mu_2\mu_1 + \mu_1^3$$

$$\kappa_4 = \mu_4 - 4\mu_3\mu_1 - 3\mu_2^2 + 12\mu_2\mu_1^2 - 6\mu_1^4$$

and for bivariates case,

$$\kappa_{10} = \mu_{10}$$

$$\kappa_{20} = \mu_{20} - \mu_{10}^2$$

$$\kappa_{11} = \mu_{11} - \mu_{10}\mu_{01}$$

$$\kappa_{30} = \mu_{30} - 3\mu_{20}\mu_{10} + 2\mu_{10}^3$$

$$\kappa_{21} = \mu_{21} - \mu_{20}\mu_{01} - 2\mu_{11}\mu_{10} + 2\mu_{10}^2\mu_{11}$$

Proceeding similarily for multivariate, a formula for cumulants in terms of a series of moments can be derived with laborious algebra manipulations. However, if $\langle z_1 z_2 \ldots z_n \rangle = 0$, i.e., the mean is zero, then the cumulants and moments are identical for $m \leq 3$.

(i) Second-order cumulants $(m = 2)$:

$$\kappa_{0 \ldots q_i \ldots 0} = \sigma^2, \; i = 1, 2, \ldots, n \quad (4C.8)$$

$$\kappa_{0\ldots q_i \ldots q_j \ldots 0} = \sigma^2 \rho_{ij}, \quad i, j = 1, 2, \ldots, n, \quad i \neq j \tag{4C.9}$$

(ii) Third-order cumulants $(m = 3)$:

$$\kappa_{0\ldots q_i \ldots 0} = \sigma^3, \quad i = 1, 2, \ldots, n \tag{4C.10}$$

$$\kappa_{0\ldots q_i \ldots q_j \ldots 0} = \sigma^3 s_{ij}, \quad q_i < q_j, \quad i, j = 1, 2, \ldots, n; \quad i \neq j \tag{4C.11}$$

$$\kappa_{0\ldots q_i \ldots q_j \ldots q_k \ldots 0} = \sigma^3 s_{ijk}, \quad i, j, k = 1, 2, \ldots, n; \quad i \neq j \neq k \tag{4C.12}$$

In the above equations, ρ, s, denote the correlation function and skewness function, respectively, and σ is the standard deviation of the random variable. The subscripts of ρ, s denote the sequence of random variables. From (4C.11) and (4C.12), it is seen that the former is a special case of the latter. Note that while the correlation is a centrosymmetric function, the skewness is an odd function along any given direction, i.e., $\rho_{ij} = \rho_{ji}$; $s_{ij} = -s_{ji}$.

The power spectrum W and bispectrum B of order n are defined as the Fourier transforms of the correlation function and skewness function of order n, respectively. As an example, let $z(x, y)$ be the random surface generated by a spatially stationary process. Then, we have

$$\langle z(x, y) z(x + \tau_x, y + \tau_y) \rangle = \mu_{12}(\tau_x, \tau_y) = \sigma^2 \rho(\tau_x, \tau_y) \tag{4C.13}$$

Note that for spatially stationary process, the surface correlation depends only on the spacing between two points instead of two specific points. Hence, we can drop the subscripts associated with it.

$$\begin{aligned} &\langle z(x, y) z(x + \tau_x, y + \tau_y) z(x + \zeta_x, y + \zeta_y) \rangle \\ &\quad = \mu_{123}(\tau_x, \tau_y; \zeta_x, \zeta_y) = \sigma^3 s_{123}(\tau_x, \tau_y; \zeta_x, \zeta_y) \end{aligned} \tag{4C.14}$$

$$W(k_x, k_y) = \frac{1}{(2\pi)} \int \sigma^2 \rho(\tau_x, \tau_y) e^{-j(k_x \tau_x + k_y \tau_y)} d\tau_x d\tau_y \tag{4C.15}$$

$$\begin{aligned} B_{ijk}(k_x, k_y; \bar{k}_x, \bar{k}_y) &= \frac{1}{(2\pi)^2} \int \sigma^3 s_{ijk}(\tau_x, \tau_y; \zeta_x, \zeta_y) \\ &\quad \exp[-j(k_x \tau_x + k_y \tau_y + \bar{k}_x \zeta_x + \bar{k}_y \zeta_y)] \, d\tau_x d\tau_y d\zeta_x d\zeta_y \end{aligned} \tag{4C.16}$$

A special case of the skewness function, as mentioned above, occurs when $\zeta_x = \tau_x$ and $\zeta_y = \tau_y$. Then, we get

$$\langle z(x, y) z^2 (x+\tau_x, y+\tau_y)\rangle = \mu_{12}(\tau_x, \tau_y) = \sigma^3 s_{12}(\tau_x, \tau_y) \tag{4C.17}$$

and the bispectrum becomes

$$B_{12}(k_x, k_y) = \frac{1}{(2\pi)}\int \sigma^2 s_{12}(\tau_x, \tau_y) e^{-j(k_x\tau_x + k_y\tau_y)} d\tau_x d\tau_y \tag{4C.18}$$

4C.2 ENSEMBLE AVERAGES

By the method of characteristic function [Longuet-Higgins, 1963] as given in (4C.6), we can find the following ensemble averages up to the third power in t as

$$\langle e^{jz_1t_1}\rangle = \exp\left[\frac{1}{2!}\kappa_2 (jt_1)^2 + \frac{1}{3!}\kappa_3 (jt_1)^3\right] \tag{4C.19}$$

$$\langle e^{j(z_1t_1+z_2t_2)}\rangle = \exp\{\frac{1}{2!}[\kappa_{20}(jt_1)^2 + 2\kappa_{11}(jt_1)(jt_2) + \kappa_{02}(jt_2)^2]$$

$$-\frac{j}{3!}[\kappa_{30}t_1^3 + 3\kappa_{21}t_1^2t_2 + 3\kappa_{12}t_1t_2^2 + \kappa_{03}t_2^3]\} \tag{4C.20}$$

$$\langle e^{j(z_1t_1+z_2t_2+z_3t_3)}\rangle = \exp\{-[\kappa_{110}t_1t_2 + \kappa_{101}t_1t_3 + \kappa_{011}t_2t_3]$$

$$+\frac{1}{2!}[\kappa_{200}(jt_1)^2 + \kappa_{020}(jt_2)^2 + \kappa_{002}(jt_3)^2 + \kappa_{210}(jt_1)^2(jt_2) + \kappa_{201}(jt_1)^2(jt_3)$$

$$+\kappa_{120}(jt_1)(jt_2)^2 + \kappa_{021}(jt_2)^2(jt_3) + \kappa_{102}(jt_1)(jt_3)^2 + \kappa_{012}(jt_2)(jt_3)^2]$$

$$+\frac{1}{3!}[\kappa_{300}(jt_1)^3 + \kappa_{030}(jt_2)^3 + \kappa_{003}(jt_3)^3] + \kappa_{111}(jt_1)(jt_2)(jt_3)\} \tag{4C.21}$$

$$\langle e^{j(z_1t_1+z_2t_2+z_3t_3+z_4t_4)}\rangle = \exp\{-[\kappa_{1100}t_1t_2 + \kappa_{1010}t_1t_3 + \kappa_{1001}t_1t_4 + \kappa_{0110}t_2t_3$$

$$+\kappa_{0101}t_2t_4 + \kappa_{0011}t_3t_4] - \frac{1}{2}(\kappa_{2000}t_1^2 + \kappa_{0200}t_2^2 + \kappa_{0020}t_3^2 + \kappa_{0002}t_4^2)$$

$$-j(\kappa_{1110}t_1t_2t_3 + \kappa_{1101}t_1t_2t_4 + \kappa_{1011}t_1t_3t_4 + \kappa_{0111}t_2t_3t_4)$$

$$-\frac{j}{2!}(\kappa_{2100}t_1^2t_2 + \kappa_{2010}t_1^2t_3 + \kappa_{2001}t_1^2t_4 + \kappa_{1200}t_1t_2^2 + \kappa_{0210}t_2^2t_3 + \kappa_{0201}t_2^2t_4$$

$$+\kappa_{1020}t_1t_3^2 + \kappa_{0120}t_2t_3^2 + \kappa_{0021}t_3^2t_4 + \kappa_{1002}t_1t_4^2 + \kappa_{0102}t_2t_4^2 + \kappa_{0012}t_3t_4^2)$$

$$-\frac{j}{3!}(\kappa_{3000}t_1^3 + \kappa_{0300}t_2^3 + \kappa_{0030}t_3^3 + \kappa_{0003}t_4^2)\} \tag{4C.22}$$

By recasting equations (4C.8) through (4C.12) in the above equations into the forms of interest we have the following:

$$\langle e^{-j2k_z z}\rangle = exp\left[-2k_z^2\sigma^2 + j\frac{4}{3}k_z^3\sigma^3\right] \tag{4C.23}$$

$$\langle e^{-j2k_z(z-\tilde{z})}\rangle = \exp\left[4k_z^2\sigma^2(\rho_{12}-1) + j8k_z^3\sigma^3 s_{12}\right] \tag{4C.24}$$

$$\langle e^{-jk_z(z+z')}\rangle = \exp\left[-k_z^2\sigma^2(\rho_{23}+1) + \frac{j}{3}k_z^3\sigma^3\right] \tag{4C.25}$$

$$\langle e^{jk_z(-z-z'+2\tilde{z})}\rangle = \exp\left[k_z^2\sigma^2(2\rho_{13}+2\rho_{23}-\rho_{12}-3) + jk_z^3\sigma^3(3s_{13}+3s_{23}-2s_{123}-1)\right] \tag{4C.26}$$

$$\langle e^{jk_z(-z-z'+\tilde{z}+\tilde{z}')}\rangle = \exp\left[-k_z^2\sigma^2(\rho_{12}+\rho_{34}-\rho_{13}-\rho_{14}-\rho_{23}-\rho_{24}+2) + jk_z^3\sigma^3(s_{13}+s_{14}+s_{23}+s_{24}-s_{123}-s_{124}+s_{134}+s_{234})\right] \tag{4C.27}$$

Appendix 4D

The Transmitted Field Coefficients

This appendix provides the field coefficients for the transmitted scattered field. It is analogous to Appendix 4B except it does not have some special cases given in Appendix 4B for backscattering.

4D.1 THE KIRCHHOFF TRANSMITTED FIELD COEFFICIENTS f_{tqp}

The Kirchhoff field coefficients in (4.121) can be explicitly written as

$$\begin{aligned} f_{tvv} = {} & (1-R_{\parallel})\,[\,(\cos\theta + Z_x\sin\theta)\cos(\phi_t-\phi) + Z_y\sin\theta\sin(\phi_t-\phi)\,] \\ & + (1+R_{\parallel})\,(\cos\theta_t\cos(\phi_t-\phi) + Z_x\sin\theta_t)\,\eta_r \end{aligned} \tag{4D.1}$$

$$\begin{aligned} f_{thh} = {} & -(1+R_{\perp})\,(\cos\theta_t\cos(\phi_t-\phi) + Z_x\sin\theta_t) \\ & -(1-R_{\perp})\,[\,(\cos\theta + Z_x\sin\theta)\cos(\phi_t-\phi) + Z_y\sin\theta\sin(\phi_t-\phi)\,]\,\eta_r \end{aligned} \tag{4D.2}$$

$$\begin{aligned} f_{thv} = {} & (1-R)\,\{\,(\cos\theta + Z_x\sin\theta)\cos\theta_t\sin(\phi_t-\phi) \\ & + Z_y\,[\sin\theta_t\cos\theta - \sin\theta\cos\theta_t\cos(\phi_t-\phi)]\,\} + (1+R)\,\eta_r\sin(\phi_t-\phi) \\ f_{tvh} = {} & (1-R)\sin(\phi_t-\phi) + (1+R)\,\eta_r\{\,(\cos\theta + Z_x\sin\theta)\cos\theta_t\sin(\phi_t-\phi) \\ & + Z_y\,[\sin\theta_t\cos\theta - \sin\theta\cos\theta_t\cos(\phi_t-\phi)]\,\} \end{aligned} \tag{4D.3}$$

where $\eta_r = \eta_t/\eta = \sqrt{\mu_r/\varepsilon_r}$ and the surface slopes may be replaced by

$$Z_x = \frac{\sqrt{\varepsilon_r\mu_r}\sin\theta_t\cos(\phi_t-\phi) - \sin\theta}{\sqrt{\varepsilon_r\mu_r}\cos\theta_t - \cos\theta}$$

$$Z_y = \frac{\sqrt{\varepsilon_r\mu_r}\sin\theta_t\sin(\phi_t-\phi)}{\sqrt{\varepsilon_r\mu_r}\cos\theta_t - \cos\theta}$$

4D.2 COMPLEMENTARY TRANSMITTED FIELD COEFFICIENTS

Since the mathematical forms of the scattered and transmitted scattered fields are the same, it is possible to recover all the complementary transmitted field coefficients from the corresponding ones in Appendix 4B by analogy. We can account for all necessary sign changes by comparing $\hat{v}_s, \hat{h}_s, \hat{k}_s$ with $\hat{v}_t, \hat{h}_t, \hat{k}_t$. Results can be checked versus direct evaluation based on (4.123) by carrying out vector products. The complementary field coefficients for like polarized scattering are found to be

$$F_{tvv}(u,v) = -\left[(1-R_{\|})\frac{1}{q} - (1+R_{\|})\frac{\mu_r}{q_t}\right](1+R_{\|})C_{t1}$$
$$+ [(1-R_{\|})/q - (1+R_{\|})/q_t](1-R_{\|})C_{t2}$$
$$+ [(1-R_{\|})/q - (1+R_{\|})/(\varepsilon_r q_t)](1+R_{\|})C_{t3}$$
$$+\left[(1+R_{\|})\frac{1}{q} - (1-R_{\|})\frac{\varepsilon_r}{q_t}\right](1-R_{\|})\eta_r C_{t4}$$
$$+ [(1+R_{\|})/q - (1-R_{\|})/q_t](1+R_{\|})\eta_r C_{t5}$$
$$+ [(1+R_{\|})/q - (1-R_{\|})/(\mu_r q_t)](1-R_{\|})\eta_r C_{t6} \tag{4D.4}$$

$$F_{thh}(u,v) = -\left[(1+R_{\perp})\frac{1}{q} - (1-R_{\perp})\frac{\mu_r}{q_t}\right](1-R_{\perp})C_{t4}$$
$$-[(1+R_{\perp})/q - (1-R_{\perp})/q_t](1+R_{\perp})C_{t5}$$
$$-[(1+R_{\perp})/q - (1-R_{\perp})/(\varepsilon_r q_t)](1-R_{\perp})C_{t6}$$
$$+\left[(1-R_{\perp})\frac{1}{q} - (1+R_{\perp})\frac{\varepsilon_r}{q_t}\right](1+R_{\perp})\eta_r C_{t1}$$
$$-[(1-R_{\perp})/q - (1+R_{\perp})/q_t](1-R_{\perp})\eta_r C_{t2}$$
$$-[(1-R_{\perp})/q - (1+R_{\perp})/(\mu_r q_t)](1+R_{\perp})\eta_r C_{t3} \tag{4D.5}$$

In the above $q = (k^2 - u^2 - v^2)^{1/2}$ and $q_t = (k_t^2 - u^2 - v^2)^{1/2}$. The C_t coefficients are the results of the triple cross products and dot products defined below.

$$C_{t1}(u,v,k_x,k_y,k_{tx},k_{ty}) = k\hat{h}_t \cdot \vec{n} \times \vec{n}' \times \hat{h}$$
$$= k\cos\phi_t\left\{\cos\phi + \frac{(k_{tx}+u)}{k_{tz}k_z}[(k_x+u)\cos\phi + (k_y+v)\sin\phi]\right\}$$
$$+ k\sin\phi_t\left\{\sin\phi + \frac{(k_{ty}+v)}{k_{tz}k_z}[(k_x+u)\cos\phi + (k_y+v)\sin\phi]\right\} \tag{4D.6}$$

$$C_{t2}(u,v,k_x,k_y,k_{tx},k_{ty}) = \hat{h}_t \cdot \vec{n} \times (\vec{n}' \times \hat{v}) \times \vec{g}$$

$$= \cos\phi_t \{ \frac{\cos\phi\cos\theta}{k_{tz}} (k_{tx} + u)\, u + \frac{\sin\phi\cos\theta}{k_{tz}} (k_{tx} + u)\, v$$

$$+ \frac{\cos\phi\cos\theta}{k_z} (k_y + v)\, v - \frac{\sin\phi\cos\theta}{k_z} (k_x + u)\, v$$

$$+ \frac{\sin\theta}{k_{tz}k_z} [\, (k_{tx} + u)\ (k_x + u)\, u + (k_{tx} + u)\ (k_y + v)\, v]\ \}$$

$$+ \sin\phi_t \{ \frac{\cos\phi\cos\theta}{k_{tz}} (k_{ty} + v)\, u + \frac{\sin\phi\cos\theta}{k_{tz}} (k_{ty} + v)\, v$$

$$+ \frac{\sin\phi\cos\theta}{k_z} (k_x + u)\, u - \frac{\cos\phi\cos\theta}{k_z} (k_y + v)\, u$$

$$- \frac{\sin\theta}{k_{tz}k_z} [\, (k_{ty} + v)\ (k_y + v)\, v + (k_x + u)\ (k_{ty} + v)\, u]\ \} \tag{4D.7}$$

$$C_{t3} (u, v, k_x, k_y, k_{sx}, k_{sy}) = \hat{h}_t \cdot \vec{n} \times (\vec{n}' \cdot \hat{v}) \times \vec{g}$$

$$= -(u\cos\phi_t + v\sin\phi_t)$$

$$[\sin\theta - (k_x + u) \cos\phi\cos\theta / k_z - (k_y + v) \sin\phi\cos\theta / k_z] \tag{4D.8}$$

$$C_{t4} (u, v, k_x, k_y, k_{tx}, k_{ty}) = k\hat{v}_t \cdot \vec{n} \times \vec{n}' \times \hat{v}$$

$$= k [\cos\phi_t \cos\theta_t \{ \cos\phi\cos\theta [-1 - (k_{ty} + v)\ (k_y + v) / (k_{tz}k_z)]$$

$$+ \sin\phi\cos\theta [\, (k_x + u)\ (k_{ty} + v) / (k_{tz}k_z)] - \sin\theta\, (k_x + u) / k_z \}$$

$$+ \sin\phi_t \cos\theta_t \{ \cos\phi\cos\theta [\, (k_{tx} + u)\ (k_y + v) / (k_{tz}k_z)]$$

$$+ \sin\phi\cos\theta [-1 - (k_{tx} + u)\ (k_x + u) / (k_{tz}k_z)] - \sin\theta\, (k_y + v) / k_z \}$$

$$- \frac{\sin\theta_t}{k_{tz}k_z} \{ \cos\phi\cos\theta\, (k_{tx} + u)\, k_z + \sin\phi\cos\theta\, (k_{ty} + v)\, k_z$$

$$+ \sin\theta [\, (k_{tx} + u)\ (k_x + u) + (k_{ty} + v)\ (k_y + v)]\ \}\ ] \tag{4D.9}$$

$$C_{t5} (u, v, k_x, k_y, k_{tx}, k_{ty}) = \hat{v}_t \cdot \vec{n} \times (\vec{n}' \times \hat{h}) \times \vec{g}$$

$$= \cos\phi_t \cos\theta_t \Big[\frac{\cos\phi}{k_{tz}} (k_{ty} + v)\, v + \frac{\cos\phi}{k_z} (k_x + u)\, u$$

$$+ \frac{\sin\phi}{k_z} (k_y + v)\, u - \frac{\sin\phi}{k_{tz}} (k_{ty} + v)\, u \Big]$$

$$-\sin\phi_t\cos\theta_t\left[\frac{\cos\phi}{k_{tz}}(k_{tx}+u)\,v-\frac{\cos\phi}{k_z}(k_x+u)\,v\right.$$

$$\left.-\frac{\sin\phi}{k_{tz}}(k_{tx}+u)\,u-\frac{\sin\phi}{k_z}(k_y+v)\,v\right]$$

$$-\frac{\sin\theta_t}{k_{tz}k_z}\{\cos\phi\,[\,(k_{tx}+u)\,(k_x+u)\,u+(k_x+u)\,(k_{ty}+v)\,v]$$

$$+\sin\phi\,[u\,(k_{tx}+u)\,(k_y+v)+v\,(k_{ty}+v)\,(k_y+v)\,]\,\}$$

(4D.10)

$$C_{t6}(u,v,k_x,k_y,k_{tx},k_{ty}) = \hat{v}_t\cdot\hat{n}\times(\hat{n}'\cdot\hat{h})\,\hat{g} = -\frac{1}{k_z}[\,(k_x+u)\sin\phi-(k_y+v)\cos\phi]$$

$$[v\cos\phi_t\cos\theta_t-u\sin\phi_t\cos\theta_t+\sin\theta_t(\,(k_{tx}v-k_{ty}u)/k_{tz})\,] \tag{4D.11}$$

The cross polarized field coefficients can also be evaluated in a similar way as

$$F_{thv}(u,v) = \left[(1-R)\frac{1}{q}-(1+R)\frac{\mu_r}{q_t}\right](1+R)\,B_{t1}$$

$$-[\,(1-R)/q-(1+R)/q_t]\,(1-R)\,B_{t2}$$

$$-[\,(1-R)/q-(1+R)/(\varepsilon_r q_t)\,]\,(1+R)\,B_{t3}$$

$$+\left[(1+R)\frac{1}{q}-(1-R)\frac{\varepsilon_r}{q_t}\right](1-R)\,\eta_r B_{t4}$$

$$+[\,(1+R)/q-(1-R)/q_t]\,(1+R)\,\eta_r B_{t5}$$

$$+[\,(1+R)/q-(1-R)/(\mu_r q_t)\,]\,(1-R)\,\eta_r B_{t6} \tag{4D.12}$$

$$F_{tvh}(u,v) = \left[(1-R)\frac{1}{q}-(1+R)\frac{\mu_r}{q_t}\right](1+R)\,B_{t4}$$

$$+[\,(1-R)/q-(1+R)/q_t]\,(1-R)\,B_{t5}$$

$$+[\,(1-R)/q-(1+R)/(\varepsilon_r q_t)\,]\,(1+R)\,B_{t6}$$

$$+\left[(1+R)\frac{1}{q}-(1-R)\frac{\varepsilon_r}{q_t}\right](1-R)\,\eta_r B_{t1}$$

$$-[\,(1+R)/q-(1-R)/q_t]\,(1+R)\,\eta_r B_{t2}$$

$$-[\,(1+R)/q-(1-R)/(\mu_r q_t)\,]\,(1-R)\,\eta_r B_{t3} \tag{4D.13}$$

where the B_t coefficients are

$$B_{t1}(u, v, k_x, k_y, k_{tx}, k_{ty}) = k\hat{v}_t \cdot \vec{\hat{n}} \times (\vec{\hat{n}}' \times \hat{h})$$

$$= k \left\{ \cos\phi_t \cos\theta_t \left[\frac{\cos\phi}{k_{tz}k_z} (k_x + u)(k_{ty} + v) + \sin\phi + \frac{\sin\phi}{k_{tz}k_z} (k_{ty} + v)(k_y + v) \right] \right.$$

$$- \sin\phi_t \cos\theta_t \left[\frac{\cos\phi}{k_{tz}k_z} (k_{tx} + u)(k_x + u) + \cos\phi + \frac{\sin\phi}{k_{tz}k_z} (k_{tx} + u)(k_y + v) \right]$$

$$\left. - \frac{\sin\theta_t}{k_{tz}} \left[(k_{ty} + v)\cos\phi - (k_{tx} + u)\sin\phi \right] \right\} \tag{4D.14}$$

$$B_{t2}(u, v, k_x, k_y, k_{tx}, k_{ty}) = \hat{v}_t \cdot \vec{\hat{n}} \times (\vec{\hat{n}}' \times \hat{v}) \times \vec{\hat{g}}$$

$$= \cos\phi_t \cos\theta_t \left\{ \frac{\cos\phi\cos\theta}{k_{tz}} (k_{ty} + v)u + \frac{\sin\phi\cos\theta}{k_{tz}} (k_{ty} + v)v \right.$$

$$- \frac{\cos\phi\cos\theta}{k_z} (k_y + v)u + \frac{\sin\phi\cos\theta}{k_z} (k_x + u)u$$

$$\left. + \frac{\sin\theta}{k_{tz}k_z} \left[(k_x + u)(k_{ty} + v)u + (k_{ty} + v)(k_y + v)v \right] \right\}$$

$$- \sin\phi_t \cos\theta_t \left\{ \frac{\cos\phi\cos\theta}{k_{tz}} (k_{tx} + u)u + \frac{\sin\phi\cos\theta}{k_{tz}} (k_{tx} + u)v \right.$$

$$- \frac{\sin\phi\cos\theta}{k_z} (k_x + u)v + \frac{\cos\phi\cos\theta}{k_z} (k_y + v)v$$

$$\left. + \frac{\sin\theta}{k_{tz}k_z} \left[(k_{tx} + u)(k_x + u)u + (k_{tx} + u)(k_y + v)v \right] \right\}$$

$$- \frac{\sin\theta_t}{k_{tz}k_z} \left\{ \cos\phi\cos\theta \left[(k_{tx} + u)(k_y + v)u + (k_{ty} + v)(k_y + v)v \right] \right.$$

$$\left. - \sin\phi\cos\theta \left[(k_{tx} + u)(k_x + u)u + (k_x + u)(k_{ty} + v)v \right] \right\} \tag{4D.15}$$

$$B_{t3}(u, v, k_x, k_y, k_{tx}, k_{ty}) = \hat{v}_t \cdot \vec{\hat{n}} \times (\vec{\hat{n}}' \cdot \hat{v})\vec{\hat{g}}$$

$$= - \left[\sin\theta - \frac{\cos\phi\cos\theta}{k_z} (k_x + u) - \frac{\sin\phi\cos\theta}{k_z} (k_y + v) \right]$$

$$\left[\frac{\sin\theta_t}{k_{tz}} (k_{tx}v - k_{ty}u) + \cos\theta_t (v\cos\phi_t - u\sin\phi_t) \right] \tag{4D.16}$$

$$B_{t4}(u, v, k_x, k_y, k_{tx}, k_{ty}) = k\hat{h}_t \cdot \vec{\hat{n}} \times (\vec{\hat{n}}' \times \hat{v})$$

$$= k\cos\phi_t\Bigg[\sin\phi\cos\theta - \frac{\cos\phi\cos\theta}{k_{tz}k_z}(k_{tx}+u)\,(k_y+v)$$

$$+\frac{\sin\phi\cos\theta}{k_{tz}k_z}(k_{tx}+u)\,(k_x+u) + \frac{\sin\theta}{k_z}(k_y+v)\Bigg][$$

$$-k\sin\phi_t\Bigg[\cos\phi\cos\theta + \frac{\cos\phi\cos\theta}{k_{tz}k_z}(k_{ty}+v)\,(k_y+v)$$

$$-\frac{\sin\phi\cos\theta}{k_{tz}k_z}(k_x+u)\,(k_{ty}+v) + \frac{\sin\theta}{k_z}(k_x+u)\Bigg] \tag{4D.17}$$

$$B_{t5}(u, v, k_x, k_y, k_{tx}, k_{ty}) = \hat{h}_t \cdot \vec{n} \times (\vec{n}' \times \hat{h}) \times \vec{g}$$

$$= \cos\phi_t\Bigg[\frac{\cos\phi}{k_{tz}}(k_{tx}+u)\,v - \frac{\sin\phi}{k_{tz}}(k_{tx}+u)\,u$$

$$-\frac{\cos\phi}{k_z}(k_x+u)\,v - \frac{\sin\phi}{k_z}(k_y+v)\,v\Bigg]$$

$$+\sin\phi_t\Bigg[\frac{\cos\phi}{k_{tz}}(k_{ty}+v)\,v - \frac{\sin\phi}{k_{tz}}(k_{ty}+v)\,u$$

$$+\frac{\cos\phi}{k_z}(k_x+u)\,u + \frac{\sin\phi}{k_z}(k_y+v)\,u\Bigg] \tag{4D.18}$$

$$B_{t6}(u, v, k_x, k_y, k_{tx}, k_{ty}) = \hat{h}_t \cdot \vec{n} \times (\vec{n}' \cdot \hat{h})\,\vec{g}$$

$$= [\,(k_y+v)\cos\phi - (k_x+u)\sin\phi]\,[\,(u\cos\phi_t + v\sin\phi_t)/k_z] \tag{4D.19}$$

4D.3 SPECIAL CASES

In this section we consider various special cases of the results from previous section. That is, the C_t and B_t coefficients are evaluated at specific values of u and v and also the resulting expressions of $F_{tqp}(u, v)$.

4D.3.1 Special *B* and *C* Coefficients

The C_t coefficients in (4D.6) through (4D.11) evaluated at $u = -k_x = -k\sin\theta\cos\phi$ and $v = -k_y = -k\sin\theta\sin\phi$ are

$$C_{t1}(-k_x, -k_y, k_x, k_y, k_{tx}, k_{ty}) = k\cos(\phi_t - \phi) \tag{4D.20}$$

$$C_{t2}(-k_x, -k_y, k_x, k_y, k_{tx}, k_{ty}) = \frac{k\sin\theta\cos\theta}{\cos\theta_t}\left[\frac{\sin\theta\cos(\phi_t-\phi)}{\sqrt{\varepsilon_r\mu_r}} - \sin\theta_t\right] \tag{4D.21}$$

$$C_{t3}(-k_x, -k_y, k_x, k_y, k_{tx}, k_{ty}) = k\sin^2\theta\cos(\phi_t-\phi) \tag{4D.22}$$

$$C_{t4}(-k_x, -k_y, k_x, k_y, k_{tx}, k_{ty}) = \frac{k\cos\theta}{\cos\theta_t}\left[\frac{\sin\theta\sin\theta_t}{\sqrt{\varepsilon_r\mu_r}} - \cos(\phi_t-\phi)\right] \tag{4D.23}$$

$$C_{t5}(-k_x, -k_y, k_x, k_y, k_{tx}, k_{ty}) = C_{t6}(-k_x, -k_y, k_x, k_y, k_{tx}, k_{ty}) = 0 \tag{4D.24}$$

Another set of C coefficients needed is evaluated at $u = -k_{tx} = -k\sin\theta_t\cos\phi_t$ and $v = -k_{ty} = -k\sin\theta_t\sin\phi_t$.

$$C_{t1}(-k_{tx}, -k_{ty}, k_x, k_y, k_{tx}, k_{ty}) = k\cos(\phi_t-\phi) \tag{4D.25}$$

$$C_{t2}(-k_{tx}, -k_{ty}, k_x, k_y, k_{tx}, k_{ty}) = 0 \tag{4D.26}$$

$$C_{t3}(-k_{tx}, -k_{ty}, k_x, k_y, k_{tx}, k_{ty}) = k\varepsilon_r\mu_r\sin^2\theta_t\cos(\phi_t-\phi) \tag{4D.27}$$

$$C_{t4}(-k_{tx}, -k_{ty}, k_x, k_y, k_{tx}, k_{ty}) = \frac{k\cos\theta_t[\sqrt{\varepsilon_r\mu_r}\sin\theta\sin\theta_t - \cos(\phi_t-\phi)]}{\cos\theta} \tag{4D.28}$$

$$C_{t5}(-k_{tx}, -k_{ty}, k_x, k_y, k_{tx}, k_{ty}) = \frac{k\sqrt{\varepsilon_r\mu_r}\sin\theta_t\cos\theta_t[\sqrt{\varepsilon_r\mu_r}\sin\theta_t\cos(\phi_t-\phi) - \sin\theta]}{\cos\theta} \tag{4D.29}$$

$$C_{t6}(-k_{tx}, -k_{ty}, k_x, k_y, k_{tx}, k_{ty}) = 0 \tag{4D.30}$$

For cross polarization we need to evaluate the B_t coefficients. For $u = -k_x = -k\sin\theta\cos\phi$ and $v = -k_y = -k\sin\theta\sin\phi$ they are

$$B_{t1}(-k_x, -k_y, k_x, k_y, k_{tx}, k_{ty}) = -k\sin(\phi_t-\phi)/\cos\theta_t \tag{4D.31}$$

$$B_{t2}(-k_x, -k_y, k_x, k_y, k_{tx}, k_{ty}) = \frac{k\sin^2\theta\cos\theta}{\sqrt{\varepsilon_r\mu_r}}\sin(\phi_t-\phi) \tag{4D.32}$$

$$B_{t3}(-k_x, -k_y, k_x, k_y, k_{tx}, k_{ty}) = -k\sin^2\theta\sin(\phi_t-\phi)/\cos\theta_t \tag{4D.33}$$

$$B_{t4}(-k_x, -k_y, k_x, k_y, k_{tx}, k_{ty}) = -k\cos\theta\sin(\phi_t - \phi) \tag{4D.34}$$

$$B_5(-k_x, -k_y, k_x, k_y, k_{tx}, k_{ty}) = B_6(-k_x, -k_y, k_x, k_y, k_{tx}, k_{ty}) = 0 \tag{4D.35}$$

The vanishing of the last two coefficients is similar to the C coefficients. For $u = -k_{tx} = -k\sin\theta_t\cos\phi_t$ and $v = -k_{ty} = -k\sin\theta_t\sin\phi_t$, the other set of the B coefficients are

$$B_{t1}(-k_{tx}, -k_{ty}, k_x, k_y, k_{tx}, k_{ty}) = -k\cos\theta_t\sin(\phi_t - \phi) \tag{4D.36}$$

$$B_{t2}(-k_{tx}, -k_{ty}, k_x, k_y, k_{tx}, k_{ty}) = -k\varepsilon_r\mu_r\sin^2\theta_t\cos\theta_t\sin(\phi_t - \phi) \tag{4D.37}$$

$$B_{t3}(-k_{tx}, -k_{ty}, k_x, k_y, k_{tx}, k_{ty}) = 0 \tag{4D.38}$$

$$B_{t4}(-k_{tx}, -k_{ty}, k_x, k_y, k_{tx}, k_{ty}) = -k\sin(\phi_t - \phi)/\cos\theta \tag{4D.39}$$

$$B_{t5}(-k_{tx}, -k_{ty}, k_x, k_y, k_{tx}, k_{ty}) = 0 \tag{4D.40}$$

$$B_{t6}(-k_{tx}, -k_{ty}, k_x, k_y, k_{tx}, k_{ty}) = k\varepsilon_r\mu_r\sin^2\theta_t\sin(\phi_t - \phi)/\cos\theta \tag{4D.41}$$

The above B and C coefficients are those needed in evaluating single scattering calculations discussed in (5.40) through (5.43). The more general B and C coefficients are required only for multiple scattering terms where integrations over u and v variables have to be carried out.

4D.3.2 Special $F_{qp}(-k_x, -k_y)$ Coefficients

With special C and B coefficients known we can find the corresponding special F_{tqp} coefficients, $F_{tvv}(-k_x, -k_y)$, $F_{thh}(-k_x, -k_y)$, $F_{thv}(-k_{tx}, -k_{ty})$, $F_{tvh}(-k_{tx}, -k_{ty})$. Here, $k_x = k\sin\theta\cos\phi$, $k_y = k\sin\theta\sin\phi$ and $q = k\cos\theta$, $q_t = k(\mu_r\varepsilon_r - \sin^2\theta)^{1/2}$. For simplicity of writing we shall use the following notations in writing the complementary field coefficients:

$$s = \sin\theta,\ st = \sin\theta_t$$

$$cs = \cos\theta,\ cst = \cos\theta_t$$

$$sf = \sin(\phi_t - \phi),\ csf = \cos(\phi_t - \phi)$$

$$sq = (\mu_r\varepsilon_r - \sin^2\theta)^{1/2}$$

$$\mu_r = \mu_t/\mu,\ \ \varepsilon_r = \varepsilon_t/\varepsilon,\ \ rem = \sqrt{\varepsilon_r\mu_r},\ \ \eta_r = \sqrt{\mu_r/\varepsilon_r}$$

$$a_1 = (st\, s - rem\; csf)/(sq\; cst)$$

$$a_2 = s\,(st - s\; csf/rem)/cst$$

$$T_v = 1 + R_{\|},\; T_{vm} = 1 - R_{\|}$$

$$T_h = 1 + R_{\perp},\; T_{hm} = 1 - R_{\perp}$$

$$T_p = 1 + R,\; T_m = 1 - R$$

$$R = (R_{\|} - R_{\perp})/2$$

With the above notations, we can write the $C_{ti}(-k_x, -k_y)$ coefficients as

$$C_{t1}/k = csf$$

$$C_{t2}/k = -\frac{s\; cs}{cst}\left(st - \frac{s\; csf}{rem}\right)$$

$$C_{t3}/k = s^2 csf$$

$$C_{t4}/k = \frac{cs}{cst}\left(\frac{st\; s}{rem} - csf\right)$$

$$C_{t5} = C_{t6} = 0$$

Then, we have

$$\begin{aligned}
F_{tvv}(-k_x, -k_y) &= -\left[\frac{T_{vm}}{cs} - \frac{T_v \mu_r}{sq}\right] T_v C_{t1}(-k_x, -k_y)/k \\
&\quad + \left[\frac{T_{vm}}{cs} - \frac{T_v}{\varepsilon_r sq}\right] T_v C_{t3}(-k_x, -k_y)/k \\
&\quad + \left[\frac{T_{vm}}{cs} - \frac{T_v}{sq}\right] T_{vm} C_{t2}(-k_x, -k_y)/k \\
&\quad + \left[\frac{T_v}{cs} - \frac{T_{vm}\varepsilon_r}{sq}\right] T_{vm} \eta_r C_{t4}(-k_x, -k_y)/k \\
&= -(cs\; T_{vm} - sqT_v/\varepsilon_r)\, T_v csf - (csT_{vm} - sqT_v/\varepsilon_r)\, T_{vm} a_1 \\
&\quad - (T_{vm}^2 - T_v T_{vm} cs/sq)\, s\,(st - s\; csf/rem)/cst \\
&= -(csT_{vm} - sqT_v/\varepsilon_r)\,(T_v csf + T_{vm} a_1) - (T_{vm}^2 - T_v T_{vm} cs/sq)\, a_2
\end{aligned} \tag{4D.42}$$

In the above, we have first combined C_{t1} and C_{t3} terms and then the resultant with the C_{t4} term to obtain the first term in (4D.42). The coefficient $F_{thh}(-k_x, -k_y)$ can be calculated in a similar way. However, since it is the negative dual of $F_{tvv}(-k_x, -k_y)$ multiplied by η_r, we can write it down immediately using (4D.42), i.e.,

$$F_{thh}(-k_x, -k_y) = (csT_{hm} - sqT_h/\mu_r)(T_h csf + T_{hm}a_1)\eta_r$$
$$+ (T_{hm}^2 - T_h T_{hm} cs/sq)\, a_2 \eta_r \tag{4D.43}$$

To find the cross polarized coefficients we list the $B_{ti}(-k_x, -k_y)$ coefficients in simplified notations as follows

$$B_{t1}/k = -sf/cst$$

$$B_{t2}/k = s^2 sf\; cs/rem$$

$$B_{t3}/k = -s^2 sf/cst$$

$$B_{t4}/k = -sf\; cs$$

$$B_{t5} = B_{t6} = 0$$

Then, we have

$$F_{thv}(-k_x, -k_y) = \left[\frac{T_m}{cs} - \frac{T_p\mu_r}{sq}\right] T_p B_{t1}(-k_x, -k_y)/k$$
$$-\left[\frac{T_m}{cs} - \frac{T_p}{sq}\right] T_m B_{t2}(-k_x, -k_y)/k$$
$$-\left[\frac{T_m}{cs} - \frac{T_p}{\varepsilon_r sq}\right] T_m B_{t3}(-k_x, -k_y)/k$$
$$+\left[\frac{T_p}{cs} - \frac{T_m\varepsilon_r}{sq}\right] T_m \eta_r B_{t1}(-k_x, -k_y)/k$$
$$= -(csT_m - sqT_p/\varepsilon_r)\, T_p sf/cst + (csT_m - sqT_p/\varepsilon_r)\, T_m rem\; sf/sq$$
$$-(T_m^2 - T_p T_m cs/sq)\, s^2 sf/rem$$
$$= (csT_m - sqT_p/\varepsilon_r)(remT_m/sq - T_p/cst)\, sf$$
$$-(T_m^2 - T_p T_m cs/sq)\, s^2 sf/rem \tag{4D.44}$$

In the above, we have first combined B_{t1} and B_{t3} terms and then the resultant with the B_{t4} term to obtain the first term in (4D.42). The coefficient $F_{tvh}(-k_x, -k_y)$ can be calculated in a similar way. However, since it is the dual of $F_{thv}(-k_x, -k_y)$ multiplied by η_r, we can write it down immediately using (4D.42), i.e.,

$$F_{tvh}(-k_x, -k_y) = (csT_p - sqT_m/\mu_r)(remT_p/sq - T_m/cst)\,\eta_r sf$$
$$-(T_p^2 - T_p T_m cs/sq)\, s^2 \eta_r sf/rem \tag{4D.45}$$

4D.3.3 Special $F_{tqp}(-k_{tx}, -k_{ty})$ Coefficients

When the arguments of the field coefficients change, a different set of results is expected. This is particularly obvious in view of the changes in the C and B coefficients. Thus, we need to recompute these field coefficients going through a routine similar to that in the previous section. For this section $k_{tx} = k_t \sin\theta_t \cos\phi_t$, $k_{ty} = k_t \sin\theta_t \sin\phi_t$, $q = k(1 - rem^2 st^2)^{1/2}$ and $q_t = krem\ cst$.

Some additional notations analogous to those in the previous section are as follows:

$$a_{1t} = (rem\ st\ s - csf)/(sqt\ cs)$$

$$a_{2t} = st(s - rem\ st\ csf)/cs$$

$$sqt = (1 - rem^2 st^2)^{1/2}$$

The set of $C_{ti}(-k_{tx}, -k_{ty})$ coefficients becomes

$$C_{t1}/k = csf$$

$$C_{t3}/k = rem^2 st^2 csf$$

$$C_{t4}/k = cst(rem\ st\ s - csf)/cs$$

$$C_{t5}/k = rem\ st\ cst(rem\ st\ csf - s)/cs$$

$$C_{t2} = C_{t6} = 0$$

Using the above coefficients, we have

$$F_{vv}(-k_{tx}, -k_{ty}) = -\left[\frac{T_{vm}}{q} - \frac{T_v \mu_r}{q_t}\right] T_v C_{t1}(-k_{tx}, -k_{ty})$$

$$+ \left[\frac{T_{vm}}{q} - \frac{T_v}{\varepsilon_r q_t}\right] T_v C_{s3}(-k_{sx}, -k_{sy})$$

$$+ \left[\frac{T_v}{q} - \frac{T_{vm}\varepsilon_r}{q_t}\right] T_{vm} \eta_r C_{s4}(-k_{sx}, -k_{sy})$$

$$+ \left[\frac{T_v}{q} - \frac{T_{vm}}{q_t}\right] T_v \eta_r C_{s5}(-k_{sx}, -k_{sy})$$

$$= -\left(sqtT_{vm} - \frac{rem\ cstT_v}{\varepsilon_r}\right) T_v csf - \left(sqtT_{vm} - \frac{rem\ cstT_v}{\varepsilon_r}\right) T_{vm} \frac{(rem\ st\ s - csf)}{sqt\ cs}$$

$$-(cstT_v^2/sqt - T_v T_{vm}/rem)\mu_r st(s - rem\ st\ csf)/cs$$

$$= -\left(sqtT_{vm} - \frac{rem\ cstT_v}{\varepsilon_r}\right)(T_v csf + T_{vm}a_{1t}) - \left(\frac{cstT_v^2}{sqt} - \frac{T_v T_{vm}}{rem}\right)\mu_r a_{2t} \quad (4D.46)$$

In the above, we have first combined C_{s1} and C_{s3} terms and then the resultant with the C_{s4} term to obtain the first term in (4D.42). The coefficient $F_{hh}(-k_{sx}, -k_{sy})$ can be calculated in a similar way. However, since it is the negative dual of $F_{vv}(-k_{sx}, -k_{sy})$ multiplied by η_r, we can write it down immediately using (4D.42), i.e.,

$$F_{hh}(-k_{sx}, -k_{sy}) = (sqtT_{hm} - rem\ cstT_h/\mu_r)\eta_r(T_h csf + T_{hm}a_{1t})$$
$$+ (cstT_{hm}^2/sqt - T_h T_{hm}/rem)\ rem\ a_{2t} \quad (4D.47)$$

To find the cross polarized coefficients we list the $B_{ti}(-k_{sx}, -k_{sy})$ coefficients in simplified notations as follows

$$B_{t1}/k = -sf\ cst$$

$$B_{t2}/k = -rem^2 st^2 sf\ cst$$

$$B_{t4}/k = -sf/cs$$

$$B_{t6}/k = rem^2 st^2 sf/cs$$

$$B_{t5} = B_{t3} = 0$$

Then, we have

$$F_{hv}(-k_{tx}, -k_{ty}) = \left[\frac{T_m}{q} - \frac{T_p\mu_r}{q_t}\right]T_p B_{t1}(-k_{tx}, -k_{ty})$$

$$-\left[\frac{T_m}{q} - \frac{T_p}{q_t}\right]T_m B_{t2}(-k_{tx}, -k_{ty})$$

$$+\left[\frac{T_p}{q} - \frac{T_m}{\mu_r q_t}\right]T_m \eta_r B_{t6}(-k_{tx}, -k_{ty})$$

$$+\left[\frac{T_p}{q} - \frac{T_m\varepsilon_r}{q_t}\right]T_m \eta_r B_{s4}(-k_{tx}, -k_{ty})$$

$$= -\left(sqtT_p - \frac{rem\ cstT_m}{\mu_r}\right)\frac{T_m\eta_r sf}{cs} + \left(sqtT_p - \frac{rem\ cstT_m}{\mu_r}\right)\frac{T_p\mu_r sf}{rem\ sqt}$$

$$+ (rem\ cstT_m^2/sqt - T_p T_m)\ rem\ st^2 sf$$

$$= \left(sqtT_p - \frac{rem\ cstT_m}{\mu_r}\right)\left(\frac{T_p}{sqt} - \frac{T_m}{cs}\right)\eta_r sf + \left(\frac{rem\ cstT_m^2}{sqt} - T_p T_m\right)rem\ st^2 sf$$

(4D.48)

In the above, we have first combined B_{t4} and B_{t6} terms and then the resultant with the B_{t1} term to obtain the first term in (4D.42). The coefficient $F_{vh}(-k_{tx}, -k_{ty})$ can be calculated in a similar way. However, since it is the dual of $F_{hv}(-k_{tx}, -k_{ty})$ multiplied by η_r, we can write it down immediately using (4D.42), i.e.,

$$F_{vh}(-k_{tx}, -k_{ty}) = (sqtT_m - rem\ cstT_p/\varepsilon_r)\ (T_m/sqt - T_p/cs)\ sf$$
$$+ (rem\ cstT_p^2/sqt - T_pT_m)\ \mu_r st^2 sf \qquad (4D.49)$$

Chapter 5

Surface Model and Special Cases

5.1 INTRODUCTION

In the previous chapter a surface scattering model was derived in terms of multiple integrals. In this chapter we shall evaluate these integrals and obtain practically useful surface scattering models under various conditions. For convenience in mathematical evaluation we provide two different forms for the surface scattering model depending upon whether the surface height is moderate (Section 5.2) or large (Section 5.3) in terms of the incident wavelength, i.e, in terms of $k\sigma$ values. A large $k\sigma$ may be interpreted as corresponding to high frequency. Thus, the model for large $k\sigma$ is also the high-frequency model. A brief summary of the surface scattering model is given in Section 5.4.

In Section 5.5 we regroup the terms in the scattering model considered in Sections 5.2 and 5.3 into two types: one type representing single scattering and the other representing multiple scattering. This division of the terms in the model makes it convenient for the user to identify whether single or multiple scattering is important in a given application. From user point of view simplicity of the final expression for the scattering model is important. For this reason terms in the scattering coefficient not making significant contributions will be identified and discarded. Usually, we expect single scattering to dominate when the root mean square (rms) surface slope is not large (less than 0.4). This is true irrespective of the surface height. In most applications a moderate surface height usually associates with a small or moderate surface slope. If so, multiple scattering terms in the model are of a higher order compared with single scattering terms. Intuitively, if a surface has many roughness scales, backscattering away from nadir tends to be sensitive to those scales with a small or moderate $k\sigma$. Hence, we expect the scattering model consisting of only single scattering terms for small or moderate surface heights to be most useful in practice. Indeed, this single-scatter scattering coefficient *reduces analytically* to the geometric optics model (Section 5.3) in the high-frequency limit and the first-order small perturbation model in the low-frequency region (Section 5.5.3). The multiple scattering terms may be viewed as corrections to single scattering models in the literature for both the high- and the low-frequency regions. For surfaces with heights large relative to the incident wavelength, the surface rms slope may be small or large. In the latter case multiple scattering terms are important and can be shown to cause enhancement in backscattering (Section 5.3).

Multiple scattering is also important in cross polarized, backscattering calculations because in this special case the single scattering contribution is negligible.

In Section 5.6 the bistatic transmitted scattering coefficients are given in forms similar to those in Section 5.5. Only single-scatter transmitted scattering coefficients and multiple scattering coefficients for small or moderate values of $k\sigma$ are provided. When $k\sigma$ is large we expect backward scattering into the upper medium to dominate and the transmission into the lower medium to be small in most applications.

Special cases illustrated numerically in Section 5.7 are divided into two categories: (1) backscattering and (2) bistatic scattering off the plane of incidence. Within category (1) we show the scattering behaviors of like and cross polarized scattering coefficients as a function of surface roughness and frequency and comparisons between the Kirchhoff model (KM), the first-order perturbation model (SPM) [Ulaby et al., 1982] and the single-scatter scattering coefficient based on the integral equation method (IEM). Within category (2) we shall illustrate polarized and cross polarized bistatic scattering from perfectly conducting surfaces and comparisons of IEM with multi-frequency, multi-angle and multi-polarization measurements from a statistically known surface. Also illustrated are the bistatic transmitted scattering coefficients for a surface with a dielectric constant of the form, $\varepsilon = 4 - j0.5$. To provide a point of reference, the results of the transmitted scattering coefficient from the SPM are also shown. Additional numerical illustrations of backscattering and some of its applications were given in Chapter 2. Special applications to sea surface scattering are given in Chapter 7.

5.2 SCATTERING MODEL FOR SURFACES WITH SMALL OR MODERATE HEIGHTS

For these surfaces multiple scattering is usually of a higher order and the surface rms height normalized to wave number, $k\sigma$, should be smaller than 2.0. More detailed regions of applicability will be given in Chapter 6. To indicate that we are dealing with surface heights smaller than the incident wavelength, we denote the scattering coefficient for this case by

$$\sigma_{qp}^{0}(S) = \sigma_{qp}^{k}(S) + \sigma_{qp}^{kc}(S) + \sigma_{qp}^{c}(S) \tag{5.1}$$

We shall evaluate the terms in (5.1) one by one.

Consider first $\sigma_{qp}^{k}(S)$. The exponential factor involving the surface correlation function in the integrand of (4.140) can be expressed as a sum of two terms,

$$\exp\left[\sigma^2 (k_{sz} + k_z)^2 \rho(\xi, \zeta)\right] = 1 + \sum_{n=1}^{\infty} \frac{\left[\sigma^2 (k_{sz} + k_z)^2\right]^n \rho^n(\xi, \zeta)}{n!} \tag{5.2}$$

Substituting this into the integrand of (4.141) we have

$$\sigma_{qp}^{k}(S) = \frac{k^2}{4\pi}|f_{qp}|^2\exp[-\sigma^2(k_{sz}+k_z)^2]\sum_{n=1}^{\infty}\frac{[\sigma^2(k_{sz}+k_z)^2]^n}{n!}\int\rho^n(\xi,\varsigma)$$

$$\exp[j(k_{sx}-k_x)\xi + j(k_{sy}-k_y)\xi]\,d\xi d\varsigma$$

$$= 0.5k^2|f_{qp}|^2\exp[-\sigma^2(k_{sz}+k_z)^2]\sum_{n=1}^{\infty}\frac{[\sigma^2(k_{sz}+k_z)^2]^n W^{(n)}(k_{sz}-k_x, k_{sy}-k_y)}{n!}$$

(5.3)

where $W^{(n)}(k_{sz}-k_x, k_{sy}-k_y)$ is the roughness spectrum of the surface related to the nth power of the surface correlation function by the Fourier transform as follows:

$$W^{(n)}(k_{sx}-k_x, k_{sy}-k_y) = \frac{1}{2\pi}\int\rho^n(\xi,\varsigma)\,e^{j(k_{sx}-k_x)\xi + j(k_{sy}-k_y)\varsigma}\,d\xi d\varsigma \quad (5.4)$$

Note that (5.3) is the Kirchhoff scattering coefficient or the Kirchhoff model (KM). It accounts for single scattering only. This is evidenced by the fact that, although (5.3) contains an infinite sum, only one pair of surface spectral components, $k_{sx}-k_x, k_{sy}-k_y$, appears in the scattering process.

Next, we consider the cross term, $\sigma_{qp}^{kc}(S)$. From (4.141) the factors that contain the surface-height autocorrelation functions in the integrand can again be written as a sum of two terms as in (5.2). Then, the following factor in the integrand is rewritten as

$$\exp[-\sigma^2 k_z k_{sz}\rho(\xi-\xi',\varsigma-\varsigma')]\,\{\exp[\sigma^2 k_{sz}(k_{sz}+k_z)\rho(\xi,\zeta)]$$

$$\exp[\sigma^2 k_z(k_{sz}+k_z)\rho(\xi',\zeta')] - 1\} \quad (5.5)$$

$$\equiv \sum_{n=0}^{\infty}\frac{(-\sigma^2 k_z k_{sz}\rho_{23})^n}{n!}\left\{\sum_{i=0}^{\infty}\frac{[\sigma^2 k_{sz}(k_{sz}+k_z)\rho_{12}]^i}{i!}\sum_{j=0}^{\infty}\frac{[\sigma^2 k_z(k_{sz}+k_z)\rho_{13}]^j}{j!} - 1\right\}$$

$$= \sum_{i=1}^{\infty}\frac{[\sigma^2 k_{sz}(k_{sz}+k_z)\rho_{12}]^i}{i!} + \sum_{j=1}^{\infty}\frac{[\sigma^2 k_z(k_{sz}+k_z)\rho_{13}]^j}{j!}$$

$$+ \sum_{n=1}^{\infty}\frac{(-\sigma^2 k_z k_{sz}\rho_{23})^n}{n!}\sum_{i=1}^{\infty}\frac{[\sigma^2 k_{sz}(k_{sz}+k_z)\rho_{12}]^i}{i!}$$

$$+ \sum_{n=1}^{\infty}\frac{(-\sigma^2 k_z k_{sz}\rho_{23})^n}{n!}\sum_{j=1}^{\infty}\frac{[\sigma^2 k_z(k_{sz}+k_z)\rho_{13}]^j}{j!}$$

$$+\sum_{i=1}^{\infty}\frac{[\sigma^2 k_{sz}(k_{sz}+k_z)\rho_{12}]^i}{i!}\sum_{j=1}^{\infty}\frac{[\sigma^2 k_z(k_{sz}+k_z)\rho_{13}]^j}{j!}$$

$$+\sum_{n=1}^{\infty}\frac{(-\sigma^2 k_z k_{sz}\rho_{23})^n}{n!}\sum_{i=1}^{\infty}\frac{[\sigma^2 k_{sz}(k_{sz}+k_z)\rho_{12}]^i}{i!}\sum_{j=1}^{\infty}\frac{[\sigma^2 k_z(k_{sz}+k_z)\rho_{13}]^j}{j!}$$

(5.6)

In (5.5) we have omitted the arguments in the surface correlation function by adding subscripts to keep track of changes in arguments. Thus, $\rho_{12} = \rho(\xi, \varsigma)$, $\rho_{13} = \rho(\xi', \varsigma')$ and $\rho_{23} = \rho(\xi-\xi', \varsigma-\varsigma')$. Of the above six terms, we expect the first two and the fifth one to be important since the other three are multiplied by a sum of terms with alternating signs, $\Sigma(-\sigma^2 k_{sz}k_z\rho_{23})^n/n!$, whose magnitude is less than one. The accuracy of the last statement will be investigated by making comparisons of the scattering coefficient with that of the moment method in the next chapter. Thus, we approximate this scattering coefficient by three terms as

$$\sigma_{qp}^{kc}(S) \approx \frac{k^2}{16\pi^3}\exp[-\sigma^2(k_{sz}^2+k_z^2+k_z k_{sz})]\,\mathrm{Re}\{\int F_{qp} f_{qp}{}^*$$

$$\iint\Bigg\{\sum_{n=1}^{\infty}\frac{[\sigma^2 k_{sz}(k_z+k_{sz})\rho(\xi,\varsigma)]^n}{n!}+\sum_{m=1}^{\infty}\frac{[\sigma^2 k_z(k_z+k_{sz})\rho(\xi',\varsigma')]^m}{m!}$$

$$+\sum_{n=1}^{\infty}\frac{[\sigma^2 k_{sz}(k_z+k_{sz})\rho(\xi,\varsigma)]^n}{n!}\sum_{m=1}^{\infty}\frac{[\sigma^2 k_z(k_z+k_{sz})\rho(\xi',\varsigma')]^m}{m!}\Bigg\}$$

$$\exp\{j[(k_{sx}+u)\xi+(k_{sy}+v)\varsigma-(k_x+u)\xi'-(k_y+v)\varsigma']\}\,d\xi d\varsigma d\xi' d\varsigma' du dv\}$$

$$= 0.5k^2\exp[-\sigma^2(k_{sz}^2+k_z^2+k_z k_{sz})]\,\mathrm{Re}\{f_{qp}{}^*$$

$$\Bigg[F_{qp}(-k_x,-k_y)\sum_{n=1}^{\infty}\frac{[\sigma^2 k_{sz}(k_z+k_{sz})]^n}{n!}W^{(n)}(k_{sx}-k_x, k_{sy}-k_y)$$

$$+F_{qp}(-k_{sx},-k_{sy})\sum_{m=1}^{\infty}\frac{[\sigma^2 k_z(k_z+k_{sz})]^m}{m!}W^{(m)}(k_x-k_{sx}, k_y-k_{sy})$$

$$+\frac{1}{2\pi}\sum_{n=1}^{\infty}\frac{[\sigma^2 k_{sz}(k_z+k_{sz})]^n}{n!}\sum_{m=1}^{\infty}\frac{[\sigma^2 k_z(k_z+k_{sz})]^m}{m!}$$

$$\int F_{qp}(u,v)\,W^n(k_{sx}+u, k_{sy}+v)\,W^m(k_x+u, k_y+v)\,du dv\Bigg]\} \qquad (5.7)$$

In the above the single sum terms are similar in character to the Kirchhoff term and are single scattering terms. The double sum term requires integration between pairs of surface spectral components indicating interaction between surface spectral components and hence represents multiple scattering.

Finally, we consider the third term in the scattering coefficient, $\sigma_{qp}^{c}(S)$. To avoid writing the arguments of the surface correlation function we again use subscripts letting $\rho_{12} = \rho(\tau+\xi-\xi', \kappa+\varsigma-\varsigma')$, $\rho_{13} = \rho(\xi, \varsigma)$, $\rho_{34} = \rho(\tau, \kappa)$, $\rho_{24} = \rho(\xi', \varsigma')$, $\rho_{14} = \rho(\tau+\xi, \kappa+\varsigma)$, and $\rho_{23} = \rho(\xi'-\tau, \varsigma'-\kappa)$. To evaluate the integral in it we rewrite the factor in (4.142) which contains the surface-height autocorrelation function as follows:

$$\exp\left[-\sigma^2 k_z k_{sz}(\rho_{12}+\rho_{34})\right]\{\exp\left[\sigma^2(k_{sz}^2\rho_{13}+k_z k_{sz}\rho_{14}+k_z k_{sz}\rho_{23}+k_z^2\rho_{24})\right]-1\}$$

$$= \sum_{p=0}^{\infty}\frac{(-\sigma^2 k_z k_{sz}\rho_{12})^p}{p!}\sum_{q=0}^{\infty}\frac{(-\sigma^2 k_z k_{sz}\rho_{34})^q}{q!}$$

$$\left[\sum_{n=0}^{\infty}\frac{(\sigma^2 k_z k_{sz}\rho_{14})^n}{n!}\sum_{l=0}^{\infty}\frac{(\sigma^2 k_z k_{sz}\rho_{23})^l}{l!}\sum_{i=0}^{\infty}\frac{(\sigma^2 k_z^2\rho_{24})^i}{i!}\sum_{m=0}^{\infty}\frac{(\sigma^2 k_{sz}^2\rho_{13})^m}{m!}-1\right]$$

$$= \sum_{m=1}^{\infty}(\;)+\sum_{n=1}^{\infty}(\;)+\sum_{l=1}^{\infty}(\;)+\sum_{i=1}^{\infty}(\;)$$

$$+\sum_1^{\infty}\sum_1^{\infty}(\;)+\ldots+\sum_1^{\infty}\sum_1^{\infty}\sum_1^{\infty}(\;)+\ldots+\sum_1^{\infty}\sum_1^{\infty}\sum_1^{\infty}\sum_1^{\infty}(\;)+\ldots$$

$$+\sum_1^{\infty}\sum_1^{\infty}\sum_1^{\infty}\sum_1^{\infty}\sum_1^{\infty}(\;)+\ldots+\sum_1^{\infty}\sum_1^{\infty}\sum_1^{\infty}\sum_1^{\infty}\sum_1^{\infty}\sum_1^{\infty}(\;) \tag{5.8}$$

There are sixty terms in the above expression. Terms involving more than one sum represent multiple scattering. These terms should be small if $k\sigma$ is small or the surface slope is small. Under these conditions, only terms involving a single sum are important. If $k\sigma$ becomes large we shall see in the next section that there is a tendency for the multiple scattering terms involving more than two sums to cancel one another. In addition, terms multiplied by sums involving p and q are smaller than others because sums over p or q are smaller than unity. As a result, only terms with a product of two sums involving ρ_{24} and ρ_{13} or ρ_{14} and ρ_{23} are important. Justification for keeping only these two terms plus the single scattering terms will be given in the next chapter. The method of evaluation of this scattering coefficient is similar to the cross term and is left as an exercise. The final result is

$$\sigma^c_{qp}(S) \approx 0.125k^2 \exp[-\sigma^2(k_{sz}^2 + k_z^2)]$$

$$\{|F_{qp}(-k_x, -k_y)|^2 \sum_{m=1}^{\infty} \frac{(\sigma^2 k_{sz}^2)^m}{m!} W^{(m)}(k_{sx} - k_x, k_{sy} - k_y)$$

$$+ F_{qp}(-k_x, -k_y) F_{qp}^*(-k_{sx}, -k_{sy}) \sum_{n=1}^{\infty} \frac{(\sigma^2 k_z k_{sz})^n}{n!} W^{(n)}(k_{sx} - k_x, k_{sy} - k_y)$$

$$+ F_{qp}^*(-k_x, -k_y) F_{qp}(-k_{sx}, -k_{sy}) \sum_{m=1}^{\infty} \frac{(\sigma^2 k_z k_{sz})^m}{m!} W^{(m)}(k_{sx} - k_x, k_{sy} - k_y)$$

$$+ |F_{qp}(-k_{sx}, -k_{sy})|^2 \sum_{n=1}^{\infty} \frac{(\sigma^2 k_z^2)^n}{n!} W^{(n)}(k_{sx} - k_x, k_{sy} - k_y)$$

$$+ \frac{1}{2\pi} \sum_{m=1}^{\infty} \frac{(\sigma^2 k_{sz}^2)^m}{m!} \sum_{n=1}^{\infty} \frac{(\sigma^2 k_z^2)^i}{i!}$$

$$\int |F_{qp}(u, v)|^2 W^{(m)}(k_{sx} + u, k_{sy} + v) W^{(i)}(k_x + u, k_y + v)\, du dv$$

$$+ \frac{1}{2\pi} \sum_{n=1}^{\infty} \sum_{n=1}^{\infty} \frac{(\sigma^2 k_z k_{sz})^{n+m}}{n!m!} \int F_{qp}(u, v) F_{qp}^*[-(u + k_x + k_{sx}), -(v + k_y + k_{sy})]$$

$$W^{(n)}(k_{sx} + u, k_{sy} + v) W^{(m)}(k_x + u, k_y + v)\, du dv\} \qquad (5.9)$$

In (5.9) the single scattering terms are again represented by terms with only one sum and do not involve u, v integration, while terms with more than one sum and u, v integration represent multiple scattering. It is interesting to note that (5.9) is also valid for large $k\sigma$ values, because theoretically the series expansion of exponential factors does not require $k\sigma$ to be small. Thus, if we let $k\sigma$ approach infinity, we see that the single scattering terms vanish and only the two multiple scattering terms remain finite. We shall verify this point in the next section where we seek a representation for the scattering coefficient specifically for large $k\sigma$.

In (5.3), (5.7) and (5.9) the Kirchhoff and the complementary field coefficients f_{qp} and F_{qp} are given in Appendix 4B. The sum of these three equations modified by the shadowing function is the final representation of the scattering coefficient for moderately rough surfaces.

5.3 SCATTERING MODEL FOR SURFACES WITH LARGE HEIGHTS

When the surface height is large, it is not practical to use the series solution as shown in the previous section. Instead, we shall reevaluate the three terms in the scattering coefficient expression defined by (4.139). It is understood that the complete surface scattering model is given by (4.139) plus the shadowing function.

Now we reconsider the Kirchhoff term. Let us rewrite (4.140) in the form,

$$\sigma_{qp}^{k} = \frac{k^2|f_{qp}|^2}{4\pi}\int \{\exp[-\sigma^2(k_{sz}+k_z)^2(1-\rho(\xi,\varsigma))]$$

$$-\exp[-\sigma^2(k_{sz}+k_z)^2]\}\exp[j(k_{sx}-k_x)\xi+j(k_{sy}-k_y)\zeta]\,d\xi d\varsigma \qquad (5.10)$$

Large rms height means the negative exponential factor is negligible compared with unity and hence the coherent term can be ignored. For the integrand to be appreciable the normalized surface correlation must be close to unity. This means that for a sufficiently rough surface we can expand the surface correlation about the origin and use the approximation

$$1-\rho(\xi,\zeta) \approx \frac{1}{2}|\rho_{\xi\xi}(0)|\xi^2+\frac{1}{2}|\rho_{\zeta\zeta}(0)|\zeta^2+|\rho_{\xi\zeta}(0)|\xi\zeta \qquad (5.11)$$

where the subscripts $\xi\xi$, $\varsigma\varsigma$ and $\xi\varsigma$ denote partial derivatives. From the above approximation we obtain

$$\sigma_{qp}^{k}(L) = \frac{k^2|f_{qp}|^2}{4\pi}$$

$$\int_{-\infty}^{\infty}\exp\left[-\frac{\sigma^2}{2}(k_{sz}+k_z)^2(|\rho_{\xi\xi}(0)|\xi^2+|\rho_{\zeta\zeta}(0)|\zeta^2+2|\rho_{\xi\zeta}(0)|\xi\zeta)\right]$$

$$\exp[j(k_{sx}-k_x)\xi+j(k_{sy}-k_y)\zeta]\,d\xi d\zeta \qquad (5.12)$$

In the above integral we have added L in the scattering coefficient notation to emphasize that we are dealing with large surface heights. The integral in (5.12) can be carried out using the following integral formula,

$$\int_{-\infty}^{\infty}\exp(-a^2x^2-bx)\,dx = \frac{\sqrt{\pi}}{a}\exp\left(\frac{b^2}{4a^2}\right) \qquad (5.13)$$

to yield

$$\sigma_{qp}^{k}(L) = \frac{k^2|f_{qp}|^2}{2\sigma^2(k_{sz}+k_z)^2[|\rho_{\xi\xi}(0)||\rho_{\varsigma\varsigma}(0)|-\rho^2_{\xi\varsigma}(0)]^{1/2}}$$

$$\exp\left\{-\frac{|\rho_{\varsigma\varsigma}(0)|(k_{sx}-k_x)^2+|\rho_{\xi\xi}(0)|(k_{sy}-k_y)^2}{2\sigma^2(k_{sz}+k_z)^2\left[|\rho_{\xi\xi}(0)||\rho_{\varsigma\varsigma}(0)|-\rho^2_{\xi\varsigma}(0)\right]}\right\}$$

$$\exp\left\{\frac{2|\rho_{\xi\varsigma}(0)|(k_{sy}-k_y)(k_{sx}-k_x)}{2\sigma^2(k_{sz}+k_z)^2\left[|\rho_{\xi\xi}(0)||\rho_{\varsigma\varsigma}(0)|-\rho^2_{\xi\varsigma}(0)\right]}\right\} \quad (5.14)$$

The above result agrees with the geometric optics solution commonly reported as the limit to the Kirchhoff solution. It shows that the Kirchhoff term is proportional to the surface slope distribution where the quantities, $\sigma^2|\rho_{\xi\xi}(0)|$ and $\sigma^2|\rho_{\varsigma\varsigma}(0)|$, are the variances of the surface slopes along ξ and ς axis, respectively [Ulaby et al., 1982].

Next, let us consider the cross term. We rewrite (4.141) as

$$\sigma_{qp}^{kc}(L) = \frac{k^2}{16\pi^3}\mathrm{Re}\{f_{qp}^*\int F_{qp}(u,v)\,\Im(\xi,\varsigma,\xi',\varsigma')$$

$$\exp\{j[(k_{sx}+u)\xi+(k_{sy}+v)\zeta-(k_x+u)\xi'-(k_y+v)\zeta']\}\,d\xi d\zeta d\xi' d\zeta' du dv \quad (5.15)$$

where

$$\Im_{kc}(\xi,\varsigma,\xi',\varsigma') = \exp\{\sigma^2 k_{sz}k_z[1-\rho(\xi-\xi',\varsigma-\varsigma')]\}$$

$$\{\exp(-\sigma^2\{k_{sz}(k_{sz}+k_z)[1-\rho(\xi,\varsigma)]+k_z(k_{sz}+k_z)[1-\rho(\xi',\varsigma')]\})$$

$$-\exp\left[-\sigma^2(k_{sz}+k_z)^2\right]\} \quad (5.16)$$

Equation (5.16) shows that for large $k\sigma$, $\Im_{kc}(\xi,\varsigma,\xi',\varsigma')$ is very small unless $(\xi,\varsigma,\xi',\varsigma')$ are small enough to force the surface correlation functions to stay close to unity. If so, the factor $\exp\{\sigma^2 k_{sz}k_z[1-\rho(\xi-\xi',\varsigma-\varsigma')]\}$ is close to unity because $\rho(\xi-\xi',\varsigma-\varsigma')\approx 1$. Thus, we can ignore the coherent term and expand the surface correlations as we did in evaluating the Kirchhoff term. That is,

$$1-\rho(\xi,\varsigma) = \frac{1}{2}\left[|\rho_{\xi\xi}(0)|\xi^2+|\rho_{\varsigma\varsigma}(0)|\varsigma^2+2|\rho_{\xi\varsigma}(0)|\xi\varsigma\right] \quad (5.17)$$

$$1-\rho(\xi',\varsigma') = \frac{1}{2}\left[|\rho_{\xi\xi}(0)|\xi'^2+|\rho_{\varsigma\varsigma}(0)|\varsigma'^2+2|\rho_{\xi\varsigma}(0)|\xi'\varsigma'\right] \quad (5.18)$$

Using (5.17), (5.18) and the above stated approximations we can approximate (5.16) as

$$\Im_{kc}(\xi,\varsigma,\xi',\varsigma') \approx \exp\left[-\frac{\sigma^2}{2}k_{sz}(k_{sz}+k_z)\left(|\rho_{\xi\xi}(0)|\xi^2+|\rho_{\varsigma\varsigma}(0)|\varsigma^2+2|\rho_{\xi\varsigma}(0)|\xi\varsigma\right)\right]$$

$$\exp\left[-\frac{\sigma^2}{2}\{k_z(k_{sz}+k_z)\left[|\rho_{\xi\xi}(0)|\xi'^2+|\rho_{\varsigma\varsigma}(0)|\varsigma'^2+2|\rho_{\xi\varsigma}(0)|\xi'\varsigma'\right]\}\right] \qquad (5.19)$$

Substituting (5.19) into (5.15) and then integrating with respect to all spatial variables using the integral identity

$$\int_{-\infty}^{\infty}\int_{-\infty}^{\infty}\exp\{-\frac{A}{2}[a_1\xi^2+a_2\varsigma^2+2a_3\xi\varsigma]+jb_1\xi+jb_2\varsigma\}\,d\xi d\varsigma$$

$$=\frac{2\pi}{A[a_1a_2-a_3^2]^{1/2}}\exp\{-\frac{a_1b_2^2+a_2b_1^2-2a_3b_1b_2}{2A(a_1a_2-a_3^2)}\} \qquad (5.20)$$

we have

$$\sigma_{qp}^{kc}(L) = \frac{k^2}{4\pi\sigma^4k_zk_{sz}(k_{sz}+k_z)^2\left[|\rho_{\xi\xi}(0)||\rho_{\varsigma\varsigma}(0)|-\rho_{\xi\varsigma}^2(0)\right]}\mathrm{Re}\{\int f_{qp}^*$$

$$\exp\left\{-\frac{|\rho_{\varsigma\varsigma}(0)|(k_{sx}+u)^2+|\rho_{\xi\xi}(0)|(k_{sy}+v)^2-2|\rho_{\xi\varsigma}(0)|(k_{sy}+v)(k_{sx}+u)}{2\sigma^2k_{sz}(k_{sz}+k_z)\left[|\rho_{\xi\xi}(0)||\rho_{\varsigma\varsigma}(0)|-\rho_{\xi\varsigma}^2(0)\right]}\right\}$$

$$\exp\left\{-\frac{|\rho_{\varsigma\varsigma}(0)|(k_x+u)^2+|\rho_{\xi\xi}(0)|(k_y+v)^2-2|\rho_{\xi\varsigma}(0)|(k_y+v)(k_x+u)}{2\sigma^2k_z(k_{sz}+k_z)\left[|\rho_{\xi\xi}(0)||\rho_{\varsigma\varsigma}(0)|-\rho_{\xi\varsigma}^2(0)\right]}\right\}$$

$$F_{qp}(u,v)\,dudv\,\} \qquad (5.21)$$

Equation (5.21) is the contribution to the scattering coefficient from the cross term when the surface roughness is large. It corresponds to the multiple scattering term in (5.7). Note that algebraic terms are gone and only the integral term remains. Thus, the single scattering terms in (5.7) are not important when $k\sigma$ is large. If the surface rms slope is small, the exponential factors in (5.21) will make the cross term negligible compared with the Kirchhoff term in (5.14) because the latter has only one exponential factor as opposed to two in (5.21). Thus, even though there is a multiple scattering term in (5.21), it can be ignored relative to the Kirchhoff term, if the surface rms slope is small.

Finally, we want to reevaluate the complementary term in a similar way. Here, the situation is a little more complicated. Let us rewrite (4.142) as follows:

$$\sigma_{qp}^{c}(L) = \left(\frac{k}{16\pi^{2.5}}\right)^2 \iint F_{qp}(u,v) F_{qp}^{*}(u',v') \iiint \Im \exp[j(k_{sx}+u)\xi]$$
$$\exp[j(k_{sy}+v)\varsigma - j(k_x+u)\xi' - j(k_y+v)\varsigma' + j(u-u')\tau + j(v-v')\kappa]$$
$$d\xi d\varsigma d\xi' d\varsigma' d\tau d\kappa du dv du' dv' \qquad (5.22)$$

where

$$\Im = \Im(\xi, \varsigma, \xi', \varsigma', \tau, \kappa) = \exp\{-\sigma^2 k_{sz}^2 [1-\rho(\xi,\varsigma)]\}$$
$$\exp\{-\sigma^2 k_z^2 [1-\rho(\xi',\varsigma')]\} \exp\{\sigma^2 k_{sz} k_z [\rho(\xi+\tau, \varsigma+\kappa) + \rho(\xi'-\tau, \varsigma'-\kappa)]\}$$
$$\exp\{-\sigma^2 k_{sz} k_z [\rho(\xi-\xi'+\tau, \varsigma-\varsigma'+\kappa) + \rho(\tau,\kappa)]\}$$
$$-\exp[-\sigma^2 (k_{sz}^2 + k_z^2)] \exp\{-\sigma^2 k_{sz} k_z [\rho(\xi-\xi'+\tau, \varsigma-\varsigma'+\kappa) + \rho(\tau,\kappa)]\} \qquad (5.23)$$

In view of (5.23) we are dealing with a six-dimensional integral, and we want to consider approximate evaluations of it when $k\sigma$ is large. This means that we must find the regions within the six-dimensional space where the integrand is significant, and then we integrate over these regions. In the evaluations of the Kirchhoff and the cross term, the regions for integration are readily apparent. It is less clear here, and we must rely on our previous series solution to guide us. Recall that the series solution for the complementary term remains valid for large $k\sigma$ although the expression is not practical to use. It indicates to us that there are at least two terms representing multiple scattering. Thus, we must find at least two regions within the six-dimensional space. Let us determine the first region in the next paragraph.

For large $k\sigma$ we see from the first two factors in (5.23) that the variables $\xi, \varsigma, \xi', \varsigma'$, should be near the origin for (5.23) to be significant. The coherent term in (5.23) is again negligible due to large $k\sigma$ and other parts of (5.23) cancel one another, leading to

$$\Im(\xi, \varsigma, \xi', \varsigma', \tau, \kappa) \approx \exp\{-\sigma^2 k_{sz}^2 [1-\rho(\xi,\varsigma)]\} \exp\{-\sigma^2 k_z^2 [1-\rho(\xi',\varsigma')]\}$$
$$= \exp\left[-\frac{\sigma^2}{2} k_{sz}^2 (|\rho_{\xi\xi}(0)|\xi^2 + |\rho_{\varsigma\varsigma}(0)|\varsigma^2 + 2|\rho_{\xi\varsigma}(0)|\xi\varsigma)\right]$$
$$\exp\left[-\frac{\sigma^2}{2} k_z^2 (|\rho_{\xi\xi}(0)|\xi'^2 + |\rho_{\varsigma\varsigma}(0)|\varsigma'^2 + 2|\rho_{\xi\varsigma}(0)|\xi'\varsigma')\right] \qquad (5.24)$$

To find the second region we shall make a change of variables letting $\xi = \xi + \tau$, $\zeta = \varsigma + \kappa$, $\xi' = \xi' - \tau$ and $\zeta' = \varsigma' - \kappa$. Then,

$$\Im(\xi, \varsigma, \xi', \varsigma', \tau, \kappa) = \exp\{-\sigma^2 (k_{sz}^2 + k_z^2) - \sigma^2 k_z k_{sz} [\rho(\xi - \xi' - \tau, \zeta - \zeta' - \kappa)]\}$$
$$\exp\{\sigma^2 k_z k_{sz} [\rho(\xi, \zeta) + \rho(\xi', \zeta') - \rho(\tau, \kappa)]\}$$

$$\exp\{\sigma^2[k_z^2\rho(\xi'+\tau,\varsigma'+\kappa)+k_{sz}^2\rho(\xi-\tau,\varsigma-\kappa)]\} \tag{5.25}$$

In (5.25) we have discarded the coherent term. Next, we shall rewrite (5.25) by completing the square $(k_{sz}-k_z)^2$

$$\Im(\xi,\varsigma,\xi',\varsigma',\tau,\kappa) = \exp\{-\sigma^2(k_{sz}-k_z)^2-\sigma^2k_zk_{sz}[\rho(\xi-\xi'-\tau,\varsigma-\varsigma'-\kappa)]\}$$

$$\exp\{-\sigma^2k_zk_{sz}[1-\rho(\xi,\varsigma)+1-\rho(\xi',\varsigma')+\rho(\tau,\kappa)]\}$$

$$\exp\{\sigma^2[k_z^2\rho(\xi'+\tau,\varsigma'+\kappa)+k_{sz}^2\rho(\xi-\tau,\varsigma-\kappa)]\} \tag{5.26}$$

For large $k\sigma$ we require $\xi,\varsigma,\xi',\varsigma'$ to be small. This leads to

$$\rho(\xi'+\tau,\varsigma'+\kappa)\approx\rho(\xi-\tau,\varsigma-\kappa)\approx\rho(\xi-\xi'-\tau,\varsigma-\varsigma'-\kappa)\approx\rho(\tau,\kappa) \tag{5.27}$$

and hence we have another approximate representation for $\Im(\xi,\varsigma,\xi',\varsigma',\tau,\kappa)$, which is significant in a region of the six-dimensional space different from (5.24),

$$\Im(\xi,\varsigma,\xi',\varsigma',\tau,\kappa)\approx\exp\{-\sigma^2(k_{sz}-k_z)^2[1-\rho(\tau,\kappa)]\}$$

$$\exp\left[-\frac{\sigma^2}{2}k_{sz}k_z(|\rho_{\xi\xi}(0)|\xi^2+|\rho_{\varsigma\varsigma}(0)|\varsigma^2+2|\rho_{\xi\varsigma}(0)|\xi\varsigma)\right]$$

$$\exp\left[-\frac{\sigma^2}{2}k_{sz}k_z(|\rho_{\xi\xi}(0)|\xi'^2+|\rho_{\varsigma\varsigma}(0)|\varsigma'^2+2|\rho_{\xi\varsigma}(0)|\xi'\varsigma')\right] \tag{5.28}$$

In what follows we shall consider the evaluation of the two approximate forms of $\Im(\xi,\varsigma,\xi',\varsigma',\tau,\kappa)$ given by (5.24) and (5.28). Each form contributes to a part of the complementary term. Substitution of (5.24) into (5.22) yields

$$\sigma_{qp}^c(L1) = \left(\frac{k}{16\pi^{2.5}}\right)^2\iint F_{qp}(u,v)\,F_{qp}{}^*(u',v')\iiint\exp[j(k_{sx}+u)\xi]$$

$$\exp\left[-\frac{\sigma^2}{2}k_{sz}^2(|\rho_{\xi\xi}(0)|\xi^2+|\rho_{\varsigma\varsigma}(0)|\varsigma^2+2|\rho_{\xi\varsigma}(0)|\xi\varsigma)\right]$$

$$\exp\left[-\frac{\sigma^2}{2}k_z^2(|\rho_{\xi\xi}(0)|\xi'^2+|\rho_{\varsigma\varsigma}(0)|\varsigma'^2+2|\rho_{\xi\varsigma}(0)|\xi'\varsigma')\right]$$

$$\exp[j(k_{sy}+v)\varsigma-j(k_x+u)\xi'-j(k_y+v)\varsigma'+j(u-u')\tau+j(v-v')\kappa]$$

$$d\xi d\varsigma d\xi' d\varsigma' d\tau d\kappa du dv du' dv' \tag{5.29}$$

In the above integral we can integrate τ,κ first, which gives rise to delta functions, $2\pi\delta(u-u')$, $2\pi\delta(v-v')$. Then, we integrate u',v' to obtain

$$\sigma_{qp}^{c}(L1) = \left(\frac{k}{8\pi^{1.5}}\right)^2 \int |F_{qp}(u,v)|^2 \iint \exp[j(k_{sx}+u)\xi]$$

$$\exp\left[-\frac{\sigma^2}{2}k_{sz}^2\left(|\rho_{\xi\xi}(0)|\xi^2 + |\rho_{\varsigma\varsigma}(0)|\varsigma^2 + 2|\rho_{\xi\varsigma}(0)|\xi\varsigma\right)\right]$$

$$\exp\left[-\frac{\sigma^2}{2}k_z^2\left(|\rho_{\xi\xi}(0)|\xi'^2 + |\rho_{\varsigma\varsigma}(0)|\varsigma'^2 + 2|\rho_{\xi\varsigma}(0)|\xi'\varsigma'\right)\right]$$

$$\exp[j(k_{sy}+v)\varsigma - j(k_x+u)\xi' - j(k_y+v)\varsigma']$$

$$d\xi d\varsigma d\xi' d\varsigma' du dv \tag{5.30}$$

Now, we carry out the integration with respect to ξ, ς, ξ', ς' using the integral formula given by (5.20) to obtain

$$\sigma_{qp}^{c}(L1) = \frac{k^2}{16\pi\sigma^4(k_z^2k_{sz}^2)\left[|\rho_{\xi\xi}(0)||\rho_{\varsigma\varsigma}(0)| - \rho_{\xi\varsigma}^2(0)\right]}\int |F_{qp}(u,v)|^2$$

$$\exp\left\{-\frac{|\rho_{\varsigma\varsigma}(0)|(k_{sx}+u)^2 + |\rho_{\xi\xi}(0)|(k_{sy}+v)^2 - 2|\rho_{\xi\varsigma}(0)|(k_{sx}+u)(k_{sy}+v)}{2\sigma^2 k_{sz}^2\left[|\rho_{\xi\xi}(0)||\rho_{\varsigma\varsigma}(0)| - \rho_{\xi\varsigma}^2(0)\right]}\right\}$$

$$\exp\left\{-\frac{|\rho_{\varsigma\varsigma}(0)|(k_x+u)^2 + |\rho_{\xi\xi}(0)|(k_y+v)^2 - 2|\rho_{\xi\varsigma}(0)|(k_x+u)(k_y+v)}{2\sigma^2 k_z^2\left[|\rho_{\xi\xi}(0)||\rho_{\varsigma\varsigma}(0)| - \rho_{\xi\varsigma}^2(0)\right]}\right\}$$

$$du dv \tag{5.31}$$

The above is the first part of the complementary term. It corresponds to the first multiple scattering term in (5.9). In fact, if we reevaluate this term in (5.9) by replacing its sums by exponentials and then apply a large $k\sigma$ approximation to it, we will obtain (5.31). As in the case of the cross term, the single scattering terms in (5.9) are not significant for very rough surfaces and do not appear in (5.31). Like the cross term in (5.21), if the rms slope of the surface is small, (5.31) will be negligible compared to the Kirchhoff term in (5.14).

Next, we consider the second part of the complementary term by substituting (5.28) into (5.22).

$$\sigma_{qp}^{c}(L2) = \left(\frac{k}{16\pi^{2.5}}\right)^2 \iint F_{qp}(u,v)F_{qp}^*(u',v') \iiint \exp[j(k_{sx}+u)\xi]$$

$$\exp\{-\sigma^2(k_{sz}-k_z)^2[1-\rho(\tau,\kappa)] - j(u+u'+k_{sx}+k_x)\tau - j(v+v'+k_{sy}+k_y)\kappa\}$$

$$\exp\left[-\frac{\sigma^2}{2}k_{sz}k_z\left(|\rho_{\xi\xi}(0)|\xi^2 + |\rho_{\varsigma\varsigma}(0)|\varsigma^2 + 2|\rho_{\xi\varsigma}(0)|\xi\varsigma\right)\right]$$

$$\exp\left[-\frac{\sigma^2}{2}k_{sz}k_z\left(|\rho_{\xi\xi}(0)|\xi'^2+|\rho_{\varsigma\varsigma}(0)|\varsigma'^2+2|\rho_{\xi\varsigma}(0)|\xi'\varsigma'\right)\right]$$

$$\exp\left[j(k_{sy}+v)\,\varsigma-j(k_x+u)\,\xi'-j(k_y+v)\,\varsigma'\right]$$

$$d\xi d\varsigma d\xi' d\varsigma' d\tau d\kappa du dv du' dv' \tag{5.32}$$

Let us integrate first with respect to $\xi, \varsigma, \xi', \varsigma'$. Equation (5.32) becomes

$$\sigma_{qp}^c(L2) = \left(\frac{k}{8\pi^{1.5}}\right)^2\iint F_{qp}(u,v)\,F_{qp}^*(u',v')\int\exp\left[-j(u+u'+k_{sx}+k_x)\,\tau\right]$$

$$\frac{\exp\{-\sigma^2(k_{sz}-k_z)^2[1-\rho(\tau,\kappa)]-j(v+v'+k_{sy}+k_y)\,\kappa\}}{\sigma^4(k_z^2k_{sz}^2)\left[|\rho_{\xi\xi}(0)||\rho_{\varsigma\varsigma}(0)|-\rho_{\xi\varsigma}^2(0)\right]}$$

$$\exp\left\{\frac{|\rho_{\varsigma\varsigma}(0)|(k_{sx}+u)^2+|\rho_{\xi\xi}(0)|(k_{sy}+v)^2-2|\rho_{\xi\varsigma}(0)|(k_{sx}+u)(k_{sy}+v)}{2\sigma^2k_zk_{sz}\left[|\rho_{\xi\xi}(0)||\rho_{\varsigma\varsigma}(0)|-\rho_{\xi\varsigma}^2(0)\right]}\right\}$$

$$\exp\left\{\frac{|\rho_{\varsigma\varsigma}(0)|(k_x+u)^2+|\rho_{\xi\xi}(0)|(k_y+v)^2-2|\rho_{\xi\varsigma}(0)|(k_x+u)(k_y+v)}{2\sigma^2k_zk_{sz}\left[|\rho_{\xi\xi}(0)||\rho_{\varsigma\varsigma}(0)|-\rho_{\xi\varsigma}^2(0)\right]}\right\}$$

$$d\tau d\kappa du dv du' dv' \tag{5.33}$$

Two more integrals can be evaluated in (5.33). The method to be used depends on the magnitude of $\sigma^2(k_{sz}-k_z)^2$. If this quantity is large, we can expand $1-\rho(\tau,\kappa)$. That is,

$$1-\rho(\tau,\kappa)\approx\frac{1}{2}\left[|\rho_{\tau\tau}(0)|\tau^2+|\rho_{\kappa\kappa}(0)|\kappa^2+2|\rho_{\tau\kappa}(0)|\tau\kappa\right] \tag{5.34}$$

Then, (5.33) becomes

$$\sigma_{qp}^c(L2) = \left(\frac{k}{8\pi}\right)^2\iint\frac{2F_{qp}(u,v)\,F_{qp}^*(u',v')}{\left(|\rho_{\tau\tau}(0)||\rho_{\kappa\kappa}(0)|-|\rho_{\tau\kappa}(0)|^2\right)^{1/2}}$$

$$\frac{1}{\sigma^6(k_{sz}-k_z)^2(k_z^2k_{sz}^2)\left[|\rho_{\xi\xi}(0)||\rho_{\varsigma\varsigma}(0)|-\rho_{\xi\varsigma}^2(0)\right]}$$

$$\exp\left\{\frac{|\rho_{\varsigma\varsigma}(0)|(k_{sx}+u)^2+|\rho_{\xi\xi}(0)|(k_{sy}+v)^2-2|\rho_{\xi\varsigma}(0)|(k_{sx}+u)(k_{sy}+v)}{2\sigma^2k_zk_{sz}\left[|\rho_{\xi\xi}(0)||\rho_{\varsigma\varsigma}(0)|-\rho_{\xi\varsigma}^2(0)\right]}\right\}$$

$$\exp\left\{\frac{|\rho_{\varsigma\varsigma}(0)|(k_x+u)^2+|\rho_{\xi\xi}(0)|(k_y+v)^2-2|\rho_{\xi\varsigma}(0)|(k_x+u)(k_y+v)}{2\sigma^2k_zk_{sz}\left[|\rho_{\xi\xi}(0)||\rho_{\varsigma\varsigma}(0)|-\rho_{\xi\varsigma}^2(0)\right]}\right\}$$

$$\exp\left\{-\frac{|\rho_{\tau\tau}(0)|\,(v+v'+k_{sy}+k_y)^2+|\rho_{\kappa\kappa}(0)|\,(u+u'+k_{sx}+k_x)^2}{2\sigma^2(k_{sz}-k_z)^2\left[|\rho_{\tau\tau}(0)||\rho_{\kappa\kappa}(0)|-\rho_{\tau\kappa}^2(0)\right]}\right\}$$

$$\exp\left\{\frac{|\rho_{\xi\varsigma}(0)|\,(u+u'+k_{sx}+k_x)\,(v+v'+k_{sy}+k_y)}{\sigma^2(k_{sz}-k_z)^2\left[|\rho_{\tau\tau}(0)||\rho_{\kappa\kappa}(0)|-\rho_{\tau\kappa}^2(0)\right]}\right\}dudvdu'dv' \tag{5.35}$$

The above equation contains only one multiple scattering term. It has a relatively small value compared to the case when $\sigma^2(k_{sz}-k_z)^2$ is small. For the latter case, we shall use series expansion to deal with the τ, κ integration. Let

$$\exp\{\sigma^2(k_{sz}-k_z)^2\rho(\tau,\kappa)\} = 1+\sum_{n=1}^{\infty}\frac{\left[\sigma^2(k_{sz}-k_z)^2\rho(\tau,\kappa)\right]^n}{n!} \tag{5.36}$$

Then, (5.33) becomes

$$\sigma_{qp}^{c}(L2) = \left(\frac{k}{8\pi^{1.5}}\right)^2\iint F_{qp}(u,v)\,F_{qp}{}^*(u',v')\int\exp[-j(u+u'+k_{sx}+k_x)\tau]$$

$$\frac{\exp\{-\sigma^2(k_{sz}-k_z)^2-j(v+v'+k_{sy}+k_y)\kappa\}}{\sigma^4(k_z^2k_{sz}^2)\left[|\rho_{\xi\xi}(0)||\rho_{\varsigma\varsigma}(0)|-\rho_{\xi\varsigma}^2(0)\right]}\left[1+\sum_{n=1}^{\infty}\frac{\left[\sigma^2(k_{sz}-k_z)^2\rho(\tau,\kappa)\right]^n}{n!}\right]$$

$$\exp\left\{-\frac{|\rho_{\varsigma\varsigma}(0)|\,(k_{sx}+u)^2+|\rho_{\xi\xi}(0)|\,(k_{sy}+v)^2-2|\rho_{\xi\varsigma}(0)|\,(k_{sx}+u)\,(k_{sy}+v)}{2\sigma^2k_zk_{sz}\left[|\rho_{\xi\xi}(0)||\rho_{\varsigma\varsigma}(0)|-\rho_{\xi\varsigma}^2(0)\right]}\right\}$$

$$\exp\left\{-\frac{|\rho_{\varsigma\varsigma}(0)|\,(k_x+u)^2+|\rho_{\xi\xi}(0)|\,(k_y+v)^2-2|\rho_{\xi\varsigma}(0)|\,(k_x+u)\,(k_y+v)}{2\sigma^2k_zk_z{}^2\left[|\rho_{\xi\xi}(0)||\rho_{\varsigma\varsigma}(0)|-\rho_{\xi\varsigma}^2(0)\right]}\right\}$$

$$dudvdu'dv'd\tau d\kappa \tag{5.37}$$

In the above equation the infinite sum in powers of the surface correlation can be represented by its Fourier transform, which is the roughness spectrum of the surface. The term involving unity leads to Dirac delta functions after τ, κ integration. Hence, we can integrate u', v' also. Finally, (5.37) becomes

$$\sigma_{qp}^{c}(L2) = \frac{k^2\exp\left[-\sigma^2(k_{sz}-k_z)^2\right]}{16\pi\sigma^4(k_z^2k_{sz}^2)\left[|\rho_{\xi\xi}(0)||\rho_{\varsigma\varsigma}(0)|-\rho_{\xi\varsigma}^2(0)\right]}$$

$$\left\{\iint\left[F_{qp}(u,v)\,F_{qp}{}^*(-u-k_{sx}-k_x,-v-k_{sy}-k_y)+\left\{\sum_{n=1}^{\infty}\frac{\left[\sigma^2(k_{sz}-k_z)^2\right]^n}{2\pi n!}\right.\right.\right.$$

$$\int F_{qp}(u,v)\,F_{qp}^{*}(u',v')\,W^{(n)}(u+u'+k_{sx}+k_x, v+v'+k_{sy}+k_y)\,du'dv' \} \Bigg]$$

$$\exp\left\{\frac{|\rho_{\zeta\zeta}(0)|(k_{sx}+u)^2+|\rho_{\xi\xi}(0)|(k_{sy}+v)^2-2|\rho_{\xi\zeta}(0)|(k_{sx}+u)(k_{sy}+v)}{2\sigma^2 k_z k_{sz}\left[|\rho_{\xi\xi}(0)||\rho_{\zeta\zeta}(0)|-\rho_{\xi\zeta}^2(0)\right]}\right\}$$

$$\exp\left\{\frac{|\rho_{\zeta\zeta}(0)|(k_x+u)^2+|\rho_{\xi\xi}(0)|(k_y+v)^2-2|\rho_{\xi\zeta}(0)|(k_x+u)(k_y+v)}{2\sigma^2 k_z k_{sz}\left[|\rho_{\xi\xi}(0)||\rho_{\zeta\zeta}(0)|-\rho_{\xi\zeta}^2(0)\right]}\right\}$$

$$du dv \} \tag{5.38}$$

Again, the above equation contains only multiple scattering terms. In particular, if we ignore higher order terms in (5.38) we see that the lowest order term corresponds to the second multiple scattering term in (5.9). This term has a special property that is of interest. Consider the exponential factor $\exp\left[-\sigma^2(k_{sz}-k_z)^2\right]$ and the field coefficient F_{qp}. In the backscattering direction this factor is unity since $k_z = k_{sz}$ and the field coefficients become $|F_{qp}(u,v)|^2$ since $k_{sx}+k_x = k_{sy}+k_y = 0$. Thus, a maximum value for $\sigma_{qp}^{c}(L2)$ should occur in this direction. In addition, we also have (5.31) equal to (5.38) in the backscattering direction! Elsewhere, the exponential factor decreases rather rapidly because $k\sigma$ is assumed large. In the forward direction, the exponential factor is also unity but the field coefficients do not form a squared magnitude, resulting in smaller values when we integrate with respect to u and v. This phenomenon where the scattering coefficient is a maximum in the backscattering direction has been referred to as *backscatter enhancement*.

5.4 SURFACE SCATTERING MODEL SUMMARY

The formulation of the surface scattering model is given by (4.139). After evaluation of the integrals we obtain the model as the sum of (5.3), (5.7) and (5.9) when the surface parameter $k\sigma$ is less than approximately two. In the high-frequency region or for large $k\sigma$ the model is given by (5.14), (5.21), (5.31) and (5.35) or (5.38) depending on the factor $\left[-\sigma^2(k_{sz}-k_z)^2\right]$. If this factor is larger than two, use (5.35); otherwise use (5.38). In applications not all the terms are significant. In the next section we rearrange the terms in the model into a single scattering coefficient and a multiple scattering coefficient and use one or the other or both depending on the problem. For example, the single scattering coefficient is the important term in like polarized backscattering. Although multiple scattering may make a contribution, it is generally small. On the other hand, the multiple scattering coefficient is the only important term in cross polarized backscattering because the single scattering cross polarized coefficient vanishes in the plane of incidence. Bistatic scattering off the plane of incidence is dominated mainly by the single scattering coefficient for both like and cross polarizations except near 90 degrees from the incident plane where the like polarizations reach their minima. As a result the surface scattering model is considerably simplified in each application.

5.5 SINGLE AND MULTIPLE SURFACE SCATTERING MODELS

In this section we want to consider a rearrangement of the terms in the surface scattering model discussed in the previous section. The purpose is to obtain special but simpler models so that physical interpretations and relations to existing special surface scattering models can be established. Where applicable these models are easier to use and understand than the general surface scattering model given in the previous section.

Three general topics are covered in this section: (1) single and multiple scattering coefficients and the associated effects of shadowing, (2) single-scatter backscattering coefficient for a randomly rough surface and its reduction to the first-order perturbation model, and (3) bistatic scattering coefficient for a perfectly conducting rough surface. Model coefficients for different polarizations are given separately and explicitly.

5.5.1 Single and Multiple Scattering Coefficients

We shall first split the scattering coefficient in the previous section into two terms: a scattering coefficient for single scattering and another one for multiple scattering. The former corresponds to terms already integrated, while the latter requires further integration. Most natural terrains have a small rms slope. Hence, we expect to use the single scattering formula more often than the multiple.

Single-Scatter Scattering Coefficients

The scattering coefficient for single scattering from (5.3), (5.7) and (5.9) is

$$\sigma_{qp}^{S} = \frac{k^2}{2}\exp\left[-\sigma^2\left(k_z^2 + k_{sz}^2\right)\right]\sum_{n=1}^{\infty}\sigma^{2n}\left|I_{qp}^{n}\right|^2\frac{W^{(n)}\left(k_{sx}-k_x, k_{sy}-k_y\right)}{n!} \tag{5.39}$$

where

$$I_{qp}^{n} = \left(k_{sz}+k_z\right)^n f_{qp}\exp\left(-\sigma^2 k_z k_{sz}\right) + \frac{\left(k_{sz}\right)^n F_{qp}\left(-k_x, -k_y\right) + \left(k_z\right)^n F_{qp}\left(-k_{sx}, -k_{sy}\right)}{2} \tag{5.40}$$

The above expression is suitable for small to moderate $k\sigma$ calculations. If $k\sigma$ is larger than 1.5, the second term in (5.40) can be ignored. This is because the first term in (5.40) has a large enough growing factor to compensate for the associated exponential decay factor, while the second term does not. Thus, for large $k\sigma$, only the first term in (5.40) (the Kirchhoff term or the KM model) remains, and the series in (5.39) should be rewritten in exponential form. The resulting integral is then evaluated as in (5.12). That is,

$$\sigma_{qp}^{S}(L) \approx \frac{k^2 |f_{qp}|^2}{2\sigma^2 (k_{sz}+k_z)^2 [|\rho_{\xi\xi}(0)||\rho_{\varsigma\varsigma}(0)| - \rho^2_{\xi\varsigma}(0)]^{1/2}}$$

$$\exp\left\{-\frac{|\rho_{\varsigma\varsigma}(0)|(k_{sx}-k_x)^2 + |\rho_{\xi\xi}(0)|(k_{sy}-k_y)^2 - 2|\rho_{\xi\varsigma}(0)|(k_{sy}-k_y)(k_{sx}-k_x)}{2\sigma^2 (k_{sz}+k_z)^2 [|\rho_{\xi\xi}(0)||\rho_{\varsigma\varsigma}(0)| - \rho^2_{\xi\varsigma}(0)]}\right\}$$

(5.41)

The above result indicates that for a large $k\sigma$ the single scattering coefficient is dominated by the Kirchhoff term. Note that a large $k\sigma$ may result from either a surface with a large rms height or when the incident frequency is high. Thus, (5.39) reduces naturally to the Kirchhoff solution in the high-frequency region. It should be noted that the Kirchhoff solution is not necessarily the only solution in the high-frequency region because multiple scattering may occur. If so, multiple scattering terms given by (5.43) must be included in estimating the total scattering coefficient.

Multiple-Scatter Scattering Coefficient

From (5.7) and (5.9) the scattering coefficient representing multiple scattering is given by the following three terms:

$$\sigma_{qp}^{M} = \frac{k^2}{4\pi} e^{-\sigma^2(k_z^2+k_{sz}^2)} \sum_{n=1}^{\infty}\sum_{m=1}^{\infty} \sigma^{2n+2m} \int\Bigg(e^{-\sigma^2 k_z k_{sz}} \{ \frac{[k_{sz}(k_{sz}+k_z)]^n}{n!}$$

$$\frac{[k_z(k_{sz}+k_z)]^m}{m!} \mathrm{Re}[f_{qp}{}^* F_{qp}(u,v)]\ \} + \frac{k_{sz}^{2n} k_z^{2m}}{4n!\, m!} |F_{qp}(u,v)|^2$$

$$+ \frac{(k_z k_{sz})^{m+n}}{4m!n!} F_{qp}(u,v) F_{qp}{}^*(-u-k_x-k_{sx}, -v-k_y-k_{sy}) \Bigg)$$

$$W^{(n)}(k_{sx}+u, k_{sy}+v)\, W^{(m)}(k_x+u, k_y+v)\, du dv \qquad (5.42)$$

Again, due to the use of the series representation, (5.42) is not suitable for large $k\sigma$ or high-frequency calculations. In the latter case the multiple scattering coefficient is given by the sum of (5.21), (5.31) and (5.35) or (5.38). For ease of reference we sum up (5.21), (5.31) and (5.38) below:

$$\sigma_{qp}^{M}(L) = \frac{k^2}{4\pi\sigma^4 k_z k_{sz} [|\rho_{\xi\xi}(0)||\rho_{\varsigma\varsigma}(0)| - \rho^2_{\xi\varsigma}(0)]} \int\Bigg\{\Bigg(\frac{\mathrm{Re}[f_{qp}{}^* F_{qp}(u,v)]}{(k_{sz}+k_z)^2}$$

$$\exp\left[-\frac{SXY}{2\sigma^2 k_{sz}(k_{sz}+k_z)}-\frac{XY}{2\sigma^2 k_z(k_{sz}+k_z)}\right]\Bigg)$$

$$+\frac{|F_{qp}(u,v)|^2}{4k_z k_{sz}}\exp\left(-\frac{SXY}{2\sigma^2 k_{sz}^2}-\frac{XY}{2\sigma^2 k_z^2}\right)$$

$$+\frac{\exp\left[-\sigma^2(k_{sz}-k_z)^2\right]}{4k_z k_{sz}}\left[F_{qp}(u,v)\,F_{qp}{}^*(-u-k_{sx}-k_x,-v-k_{sy}-k_y)\right.$$

$$\left.+\sum_{n=1}^{\infty}\frac{\left[\sigma^2(k_{sz}-k_z)^2\right]^n}{2\pi n!}\int F_{qp}(u,v)\,F_{qp}{}^*(u',v')\,W^{(n)}(\;)\,du'dv'\right]$$

$$\exp\left[-\frac{SXY+XY}{2\sigma^2 k_z k_{sz}}\right]\Big\}\,dudv \qquad (5.43)$$

where $W^{(n)}(\;) = W^{(n)}(u+u'+k_{sx}+k_x, v+v'+k_{sy}+k_y)$ and

$$SXY=\frac{|\rho_{\zeta\zeta}(0)|(k_{sx}+u)^2+|\rho_{\xi\xi}(0)|(k_{sy}+v)^2-2|\rho_{\xi\zeta}(0)|(k_{sx}+u)(k_{sy}+v)}{\left[|\rho_{\xi\xi}(0)||\rho_{\zeta\zeta}(0)|-\rho_{\xi\zeta}^2(0)\right]}$$

$$XY=\frac{|\rho_{\zeta\zeta}(0)|(k_x+u)^2+|\rho_{\xi\xi}(0)|(k_y+v)^2-2|\rho_{\xi\zeta}(0)|(k_x+u)(k_y+v)}{\left[|\rho_{\xi\xi}(0)||\rho_{\zeta\zeta}(0)|-\rho_{\xi\zeta}^2(0)\right]} \qquad (5.44)$$

In the above the field coefficients f_{qp} and F_{qp} are given in Appendix 4B.

Effects of Shadowing

Due to the use of approximate tangential surface fields in the derivation of all the scattering coefficients given in this section, an additional function is needed to account for the effect of shadowing. Some shadowing functions have been derived to correct the single scattering coefficient [see Chapter 4]. The method of correction is to multiply the scattering coefficient by the shadowing function. These shadowing functions are based on the assumption of the geometric optics condition, and hence while the modification is in the right direction, it is not necessarily accurate.

At this time a general shadowing function is not available. For multiple scattering a direct multiplication of the scattering coefficient by a shadowing function cannot represent shadowing of the surface for the rescattered field. An acceptable way to incorporate it into the multiple scattering coefficients is not known. A possible way to include a shadowing function $S(\theta, \sigma_s)$ is suggested below.

Usually, a shadowing function (see Chapter 4 (4.146)) depends on the cotangent of the incident angle. For the scattering coefficient in (5.43), two shadowing functions are needed: one depends on the cotangent of the incident angle, $\cot\theta$, and the other on the cotangent of the incident angle θ^{re} of the *rescattered field*, i.e.,

$$\cot\theta^{re} = (1-u^2-v^2)^{1/2}/(u^2+v^2)^{1/2} \tag{5.45}$$

Thus, the latter shadowing function should be integrated; i.e., the rescattered field along every direction is modified by a shadowing function evaluated along that direction. This means that (5.43) should be multiplied by a shadowing function, say, $R_1(\theta, \sigma_s)$ outside its integrals and by $R_1(\theta^{re}, \sigma_s)$ inside its integrals.

5.5.2 Backscattering from Randomly Rough Surfaces

For backscattering $k_{sz} = k_z$, $k_{sx} = -k_x$ and without loss of generality we can take $k_{sy} = k_y = 0$. Under these conditions, the single scattering coefficient reduces to

$$\sigma_{qp}^{s} = \frac{k^2}{2}\exp(-2\sigma^2 k_z^2)\sum_{n=1}^{\infty}\sigma^{2n}\left|I_{qp}^{n}\right|^2\frac{W^{(n)}(-2k_x, 0)}{n!} \tag{5.46}$$

where $k_z = k\cos\theta$, $k_x = k\sin\theta$ and

$$I_{qp}^{n} = (2k_z)^n f_{qp}\exp(-\sigma^2 k_z^2) + \frac{k_z^n[F_{qp}(-k_x, 0) + F_{qp}(k_x, 0)]}{2} \tag{5.47}$$

For a specific polarization we can find f_{qp}, $F_{qp}(k_x, 0)$ and $F_{qp}(-k_x, 0)$ from Appendix 4B. For ease of reference we list these coefficients below:

$$f_{vv} = \frac{2R_{\|}}{\cos\theta} \tag{5.48}$$

$$f_{hh} = -\frac{2R_{\perp}}{\cos\theta} \tag{5.49}$$

$$F_{vv}(-k_x, 0) + F_{vv}(k_x, 0) = \frac{2\sin^2\theta\,(1+R_{\|})^2}{\cos\theta}\left[\left(1-\frac{1}{\varepsilon_r}\right) + \frac{\mu_r\varepsilon_r - \sin^2\theta - \varepsilon_r\cos^2\theta}{\varepsilon_r^2\cos^2\theta}\right] \tag{5.50}$$

$F_{hh}(-k_x, 0) + F_{hh}(k_x, 0)$

$$= -\frac{2\sin^2\theta\,(1+R_{\perp})^2}{\cos\theta}\left[\left(1-\frac{1}{\mu_r}\right)+\frac{\mu_r\varepsilon_r-\sin^2\theta-\mu_r\cos^2\theta}{\mu_r^{\,2}\cos^2\theta}\right] \tag{5.51}$$

In the above we have assumed that the local incident angle in the Fresnel coefficients in f_{pp} can be approximated by the incident angle. This assumption leads to a restriction on the applicability of (5.46). Such a restriction should be a function of the surface roughness parameters, the shape of the correlation function and the relative permittivity ε_r of the surface (see Chapter 6). The exact range of validity of (5.46) is not known. For surfaces with a Gaussian correlation function, studies in Chapter 6 indicate that it should be restricted to

$$(k\sigma)\,(kL) < 1.2\sqrt{\varepsilon_r} \tag{5.52}$$

where σ is the rms height and L is the surface correlation length. For exponential type correlation function the condition should be much more relaxed (see Chapter 6).

Illustrations of the surface backscattering model given by (5.46) have been given in Chapter 2, Section 2.5. As indicated in Section 2.5.1, we have two model expressions. One expression is as given above; the other is for surfaces satisfying the Kirchhoff approximation where the Fresnel coefficients in (5.48) and (5.49) are evaluated along normal incidence rather than the incident angle. The ranges of validity of the two forms of this model are different and so are their regions of validity (Chapter 6). If the two forms are combined by introducing an appropriate transition of the local angle from the incident to the specular angle [Fung et al., 1992], then the overall model should have a wider region of validity. At this time a completely satisfactory answer is not yet available. More details are given in Chapter 6. An alternative is to evaluate the two model expressions, i.e., evaluate (5.46) in two ways: (1) approximating the local angle in the Fresnel reflection coefficients in f_{pp} by the incident angle and (2) approximating it by the specular angle. The correct backscattering coefficient should be bounded by these two calculations.

For cross polarization, the single-scatter backscattering coefficient is zero. Contributions must come from multiple scattering terms. This backscattering coefficient is a special case of (5.9)

$$\sigma^{M}{}_{qp}(\theta) = \frac{k^2}{16\pi}\exp(-2k_z^2\sigma^2)\sum_{n=1}^{\infty}\sum_{m=1}^{\infty}\frac{(k_z^2\sigma^2)^{n+m}}{n!\,m!}$$

$$\int\left[\left|F_{qp}(u,v)\right|^2 + F_{qp}(u,v)\,F_{qp}{}^{*}(-u,-v)\right]W^{n}(u-k_x,v)\,W^{m}(u+k_x,v)\,du\,dv \tag{5.53}$$

Note that the limits over u, v are symmetric and hence $F_{qp}(u,v)\,F_{qp}{}^{*}(-u,-v)$ must occur in conjugate pairs leading to real integrated values.

5.5.3 Backscattering Coefficients for Slightly Rough Surfaces

Additional special cases of interest are the reduction of I^n_{vv} and I^n_{hh} when $k\sigma$ is much smaller than unity. This may result from either a low-frequency condition or a surface with a small rms height. Let us further assume that $\mu_r = 1$ and consider

$$I^1_{vv} \approx 2k_z\left(\frac{2R_{\|}}{\cos\theta}\right) + \frac{k_z[F_{vv}(-k_x, 0) + F_{vv}(k_x, 0)]}{2}$$

$$= 4k\left\{R_{\|} + \frac{\sin^2\theta\,(1+R_{\|})^2}{4}\left[\left(1-\frac{1}{\varepsilon_r}\right) + \left(1-\frac{1}{\varepsilon_r}\right)\frac{\tan^2\theta}{\varepsilon_r}\right]\right\}$$

$$= 4k\left[R_{\|}\cos^2\theta + \frac{\sin^2\theta\,(1+R_{\|})^2}{2}\left(1-\frac{1}{\varepsilon_r}\right)\right]$$

$$+ 4k\left[R_{\|}\sin^2\theta - \frac{\sin^2\theta\,(1+R_{\|})^2}{4}\left[\left(1-\frac{1}{\varepsilon_r}\right) - \left(1-\frac{1}{\varepsilon_r}\right)\frac{\tan^2\theta}{\varepsilon_r}\right]\right] \quad (5.54)$$

The first term in (5.54) is identical to the coefficient needed in the first-order scattering coefficient derived from the small perturbation method [Ulaby et al., 1982, p. 965]. We want to show that the second term in (5.54) is zero. That is,

$$R_{\|}\sin^2\theta - \frac{\sin^2\theta\,(1+R_{\|})^2}{4}\left[\left(1-\frac{1}{\varepsilon_r}\right) - \left(1-\frac{1}{\varepsilon_r}\right)\frac{\tan^2\theta}{\varepsilon_r}\right]$$

$$= -\frac{\sin^2\theta\,(1-R_{\|})^2}{4} + \frac{\sin^2\theta\,(1+R_{\|})^2}{4\varepsilon_r}\left[1 + \left(1-\frac{1}{\varepsilon_r}\right)\tan^2\theta\right]$$

$$= -\frac{\sin^2\theta\,(1-R_{\|})^2}{4} + \frac{\sin^2\theta\cos^2\theta\,(1-R_{\|})^2}{4\,(\varepsilon_r - \sin^2\theta)}[\varepsilon_r + (\varepsilon_r - 1)\tan^2\theta]$$

$$= -\frac{\sin^2\theta\,(1-R_{\|})^2}{4} + \frac{\sin^2\theta\,(1-R_{\|})^2}{4\,(\varepsilon_r - \sin^2\theta)}[\varepsilon_r\cos^2\theta + (\varepsilon_r - 1)\sin^2\theta] = 0 \quad (5.55)$$

Hence, the final form for I^1_{vv} is

$$I^1_{vv} = 4k\left[R_{\|}\cos^2\theta + \frac{\sin^2\theta\,(1+R_{\|})^2}{2}\left(1-\frac{1}{\varepsilon_r}\right)\right] \quad (5.56)$$

and the vertically polarized scattering coefficient for small $k\sigma$ is obtained by substituting (5.56) into (5.46) and dropping higher order terms as

$$\sigma_{vv}^{1} = 8k^4\sigma^2\left|R_{\|}\cos^2\theta + \frac{\sin^2\theta\,(1+R_{\|})^2}{2}\left(1-\frac{1}{\varepsilon_r}\right)\right|^2 W(-2k_x, 0) \tag{5.57}$$

Equation (5.57) is identical to the scattering coefficient derived from the first-order small perturbation method given in Ulaby et al. [1982]. It is known as the small perturbation model (SPM) in the literature.

For horizontal polarization the reduction of the scattering coefficient for small $k\sigma$ and $\mu_r = 1$ is much simpler. Upon substituting (5.49) and (5.51) into (5.47) we have

$$\begin{aligned} I_{hh}^{1} &\approx -\left\{2k_z\left(\frac{2R_\perp}{\cos\theta}\right) + \frac{k_z\,[F_{hh}(-k_x, 0) + F_{hh}(k_x, 0)]}{2}\right\} \\ &= -\left\{2k_z\left(\frac{2R_\perp}{\cos\theta}\right) + \left[k\sin^2\theta\,(1+R_\perp)^2\right]\left[\frac{\varepsilon_r - \sin^2\theta - \cos^2\theta}{\cos^2\theta}\right]\right\} \\ &= -\{4kR_\perp - 4k\sin^2\theta R_\perp\} = -4kR_\perp\cos^2\theta \end{aligned} \tag{5.58}$$

Then, substituting (5.58) into (5.46) and dropping higher order terms, we have the scattering coefficient for horizontal polarization under small $k\sigma$ condition as

$$\sigma_{hh}^{1} = 8k^4\sigma^2\left|R_\perp\cos^2\theta\right|^2 W(-2k_x, 0) \tag{5.59}$$

Equations (5.57) and (5.59) are also the scattering coefficients in the low-frequency region where $k\sigma$ is small. They are referred to as the SPM in numerical illustrations given in Section 5.6.

5.5.4 Scattering from Perfectly Conducting Surfaces

For perfectly conducting surfaces, all the scattering coefficient expressions in the previous subsections apply. The only difference lies in the field coefficients, which are now much simpler. For ease of reference they are listed below for bistatic scattering:

$$f_{vv} = f_{hh} = \frac{2}{\cos\theta + \cos\theta_s}\,[\sin\theta\sin\theta_s - (1+\cos\theta\cos\theta_s)\cos(\phi_s - \phi)] \tag{5.60}$$

$$f_{hv} = f_{vh} = 2\sin(\phi - \phi_s) \tag{5.61}$$

$$F_{vv}(u, v) = \left[\frac{4}{\sqrt{k^2 - u^2 - v^2}}\right] C_5(u, v) \tag{5.62}$$

$$F_{vv}(-k\sin\theta\cos\phi, -k\sin\theta\sin\phi) = 0 \tag{5.63}$$

$$F_{vv}(-k\sin\theta_s\cos\phi_s, -k\sin\theta_s\sin\phi_s) = \frac{4\sin\theta_s\cos\theta_s[\sin\theta - \sin\theta_s\cos(\phi_s - \phi)]}{\cos\theta\cos\theta_s} \tag{5.64}$$

$$F_{hh}(u, v) = -\left[\frac{4}{\sqrt{k^2 - u^2 - v^2}}\right] C_2(u, v) \tag{5.65}$$

$$F_{hh}(-k\sin\theta\cos\phi, -k\sin\theta\sin\phi) = -\frac{4\sin\theta\cos\theta[\sin\theta_s - \sin\theta\cos(\phi_s - \phi)]}{\cos\theta\cos\theta_s} \tag{5.66}$$

$$F_{hh}(-k\sin\theta_s\cos\phi_s, -k\sin\theta_s\sin\phi_s) = 0 \tag{5.67}$$

$$F_{hv}(u, v) = \left[\frac{4}{\sqrt{k^2 - u^2 - v^2}}\right] B_5(u, v) \tag{5.68}$$

$$F_{vh}(u, v) = -\left[\frac{4}{\sqrt{k^2 - u^2 - v^2}}\right] B_2(u, v) \tag{5.69}$$

The C and B coefficients are given in Appendix 4B. The above field coefficients reduce to much simpler forms in backscattering because the C and B coefficients are much simplified, yielding

$$f_{vv} = f_{hh} = \frac{2}{\cos\theta} \tag{5.70}$$

$$f_{hv} = f_{vh} = 0 \tag{5.71}$$

$$F_{vv} = \frac{4}{\sqrt{k^2 - u^2 - v^2}}\left[\frac{\sin\theta}{k^2\cos^2\theta}(uk\sin\theta - u^2 - v^2)(k\sin\theta + u) + u\sin\theta + \frac{u^2 - v^2}{k}\right] \tag{5.72}$$

$$F_{hh} = \frac{4}{\sqrt{k^2 - u^2 - v^2}}\left[\frac{\sin\theta}{k^2\cos^2\theta}(uk\sin\theta + u^2 + v^2)(k\sin\theta - u) + u\sin\theta + \frac{v^2 - u^2}{k}\right] \tag{5.73}$$

$$F_{hv} = F_{vh} = \left[\frac{8}{\sqrt{k^2 - u^2 - v^2}}\right]\frac{uv}{k\cos\theta} \tag{5.74}$$

Furthermore, the field coefficients needed for the single-scatter scattering coefficient are

$$F_{vv}(k\sin\theta, 0) = \frac{8\sin^2\theta}{\cos\theta} \tag{5.75}$$

$$F_{vv}(-k\sin\theta, 0) = 0 \tag{5.76}$$

$$F_{hh}(k\sin\theta, 0) = 0 \tag{5.77}$$

$$F_{hh}(-k\sin\theta, 0) = -\frac{8\sin^2\theta}{\cos\theta} \tag{5.78}$$

5.6 TRANSMITTED SCATTERING COEFFICIENTS

Due to the similarity in the forms of the scattered and transmitted field expressions given by (4.68), (4.69) and (4.120), (4.121), it follows that the scattering coefficients given by (5.79), (5.81), (5.82) and (5.43) have corresponding transmitted scattering coefficients, which can be obtained by making the following changes in these quantities:

1. C becomes C_t.
2. $\vec{k}_s$ becomes $\vec{k}_t$.
3. f_{pq} becomes f_{pqt}.
4. F_{pq} becomes F_{pqt}.

As a result, the single-scatter transmitted scattering coefficients become

$$\sigma^t_{qp} \approx \frac{|k_t|^2}{2} exp\left[-\sigma^2(k_z^2 + k_{tzr}^2)\right] \sum_{n=1}^{\infty} \sigma^{2n} \left|I^n_{qpt}\right|^2 \frac{W^{(n)}(k_{tx} - k_x, k_{ty} - k_y)}{n!} \tag{5.79}$$

where k_{tzr} is the real part of k_{tz} and the approximating sign results from ignoring the imaginary part of k_{tz}.

$$I^n_{qpt} = (k_z - k_{tzr})^n f_{qpt} \exp(\sigma^2 k_z k_{tzr}) + \frac{(k_{tzr})^n F_{qpt}(-k_x, -k_y) + (k_z)^n F_{qpt}(-k_{tx}, -k_{ty})}{2} \tag{5.80}$$

The above expression is suitable for small to moderate $k\sigma$ calculations. If $k\sigma$ is larger than two, the second term in (5.80) can be ignored. This is because the first term in (5.80) has a large enough growing factor to compensate for the associated exponential decay factor, while the second term does not. Thus, for large $k\sigma$, only the first term in (5.80) (the Kirchhoff term or the KM model) remains and the series in (5.79) should be rewritten in exponential form. The resulting integral is then evaluated as in (5.12). That is, for a sufficiently large $k\sigma$, we have another approximate expression for the transmitted scattering coefficients as

$$\sigma^t_{qp}(L) \approx \frac{|k_t|^2 |f_{qpt}|^2}{2\sigma^2 (k_z - k_{tzr})^2 [|\rho_{\xi\xi}(0)||\rho_{\varsigma\varsigma}(0)| - \rho^2_{\xi\varsigma}(0)]^{1/2}}$$

$$\exp\left\{ -\frac{|\rho_{\varsigma\varsigma}(0)|(k_{tx} - k_x)^2 + |\rho_{\xi\xi}(0)|(k_{ty} - k_y)^2 - 2|\rho_{\xi\varsigma}(0)|(k_{ty} - k_y)(k_{tx} - k_x)}{2\sigma^2 (k_z - k_{tzr})^2 [|\rho_{\xi\xi}(0)||\rho_{\varsigma\varsigma}(0)| - \rho^2_{\xi\varsigma}(0)]} \right\}$$

(5.81)

In both (5.79) and (5.81) we have ignored a decay factor of the form $\exp(-2k_{ti}R)$ in C_t, where k_{ti} is the imaginary part of k_t and R is the range from the center of the illuminated area to the point of observation. In most applications, we are interested in the transmitted scattering coefficient at the surface boundary, and this is why this factor is not needed.

Similarly, the transmitted scattering coefficients representing multiple scattering can be obtained from (5.82) as

$$\sigma^M_{qpt} = \frac{|k_t|^2}{4\pi} \exp[-\sigma^2 (k_z^2 + k_{tz}^2)] \sum_{n=1}^{\infty} \sum_{m=1}^{\infty} \sigma^{2n+2m} \iint \left(e^{\sigma^2 k_z k_{tz}} \left\{ \frac{[k_{tz}(k_{tzr} - k_z)]^n}{n!} \right.\right.$$

$$\left. \frac{[k_z(k_z - k_{tzr})]^m}{m!} \mathrm{Re}[f_{qpt}^* F_{qpt}(u, v)] \right\} + \frac{k_{tz}^{2n} k_z^{2m}}{4n!\,m!} |F_{qpt}(u, v)|^2$$

$$\left. + \frac{(-k_z k_{sz})^{m+n}}{4m!n!} F_{qpt}(u, v) F_{qpt}(-u - k_x - k_{tx}, -v - k_y - k_{ty}) \right)$$

$$W^{(n)}(k_{tx} + u, k_{ty} + v)\, W^{(m)}(k_x + u, k_y + v)\, du dv \qquad (5.82)$$

In the above, the field coefficients f_{qpt} and F_{qpt} are given in Appendix 4D.

In applications, both the scattering coefficients for the upper medium and the transmitted scattering coefficients for the lower medium need to be modified by an appropriate amount of shadowing. This point is particularly important for surfaces with large slopes. Unfortunately, a general shadowing function is yet to be found. In addition, we used an approximate surface current in the derivation of the scattered field. While this scattered field is shown to be quite accurate in backward scattering into the upper medium, it is less accurate in transmitted scattering into the lower medium especially for highly lossy medium and at large angles of scattering, which is around 55 degrees or more. As a compromise we suggest multiplying the single-scatter transmitted scattering coefficients given by (5.46) by an empirical function of the form

$$\exp(-0.5 \tan\theta_s) \qquad (5.83)$$

when the scattering angle exceeds 55 degrees. Furthermore, (5.46) must use a complex dielectric constant to avoid a singularity, which will result from nonphysical dielectric constant. An illustration of the behaviors of these coefficients will be given in the next section.

As mentioned in Chapter 2, most natural surfaces are multi-scaled so that as the exploring wavelength gets shorter, the wave begins to see smaller scales instead of approaching the geometric optics limit. When this happens, we do not need to use the high-frequency expression given by (5.41) since the effective surface scattering parameters will remain small. This is another reason why the single-scatter scattering coefficients given by (5.39) and (5.46) are the most useful in practice.

5.7 NUMERICAL ILLUSTRATIONS

To illustrate the IEM model behavior, we shall consider the dependence of its polarized and cross polarized backscattering coefficients on the normalized rms height $k\sigma$, correlation length kL and frequency in Figures 5.1 through 5.9 for perfectly conducting, Gaussian correlated surfaces. Then, we shall illustrate like and cross polarized bistatic backward and forward transmitted scattering and comparisons with laboratory controlled backward scattering measurements from a known, randomly rough surface.

5.7.1 Backscattering Model Behavior

We shall use (5.46) for like polarization and (5.53) for cross polarization. The like polarized field coefficients are defined by (5.70) and (5.75) through (5.78), and the cross polarized field coefficients are given by (5.71) and (5.74). In Figures 5.1 through 5.3 we show the dependence of the like and cross polarized backscattering coefficients on the roughness parameter $k\sigma$ between 0.25 and 0.8 while kL is kept constant at 3.14. We purposely select an intermediate kL value since in the low- and high-frequency regions, scattering characteristics can be easily determined from the small perturbation (SPM) and the Kirchhoff models (KM). For the purpose of obtaining some reference the predictions of the Kirchhoff and the first-order small perturbation models in like polarization and the second-order perturbation in cross polarization are also plotted on the same graphs. At $k\sigma = 0.25$ the IEM predictions agree well with the small perturbation results in cross polarization but deviate at larger angles of incidence in like polarizations, because the kL value is larger than what is acceptable by the small perturbation model. The Kirchhoff predictions are correct only at near normal incidence where the difference between VV and HH polarizations is small. This is not because the Kirchhoff model is valid for the surface at this frequency but because a large roughness scale tends to dominate scattering at small angles of incidence thus allowing closer agreement to the Kirchhoff model. The reason why the cross polarized prediction from the small perturbation model agrees well with the IEM model is because the calculation for cross polarization is calculated to the second-order. Hence, in applying the small perturbation model its cross polarized component always has a wider range of applicability.

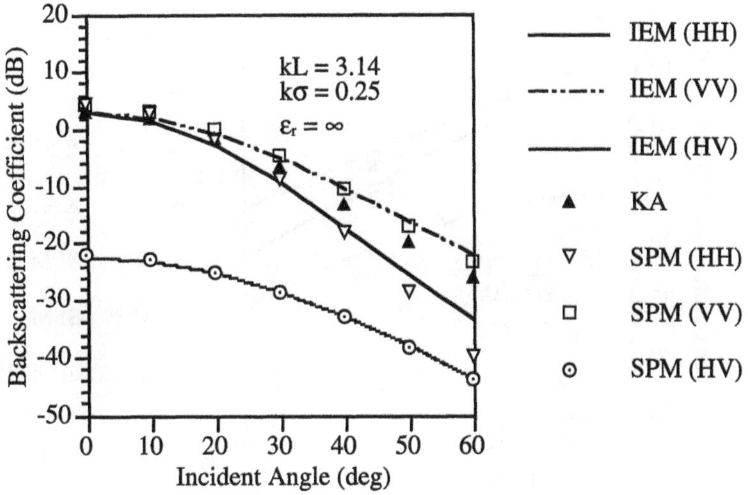

Figure 5.1 Comparisons of backscattering coefficients between the KM, SPM and IEM for $k\sigma = 0.25$.

As $k\sigma$ increases to 0.5, IEM predictions differ significantly from the small perturbation results in like polarizations, because the latter is no longer valid (Figure 5.2). The difference occurs in both angular trends and levels for vertical and horizontal polarizations. The Kirchhoff model continues to lie between the vertically and horizontally polarized scattering predicted by the IEM, and the agreement seems to be better and over a wider range of incident angles. The cross polarized returns between the IEM and the small perturbation model have also begun to differ but not significantly. This shows that the cross polarized coefficient given by the second-order perturbation method has a much wider range of validity in kL and $k\sigma$ values than the first-order like polarized coefficients. As $k\sigma$ increases further to 0.8, the differences in cross polarized scattering become more significant, and it is clear that the second-order small perturbation model predicts a faster drop-off angular trend than the IEM model (Figure 5.3). In all cases shown the Kirchhoff model predictions for like polarizations continue to lie between VV and HH predictions of the IEM. Furthermore, the spacing between VV and HH polarizations of the IEM model narrows, indicating that they are approaching the Kirchhoff solution for larger roughness conditions. The overall angular trends for both like and cross polarizations are to drop-off slower as $k\sigma$ increases. The increase in the level with $k\sigma$ is faster in cross polarization than like polarization. Again, this is because we calculate cross polarization up to the second order.

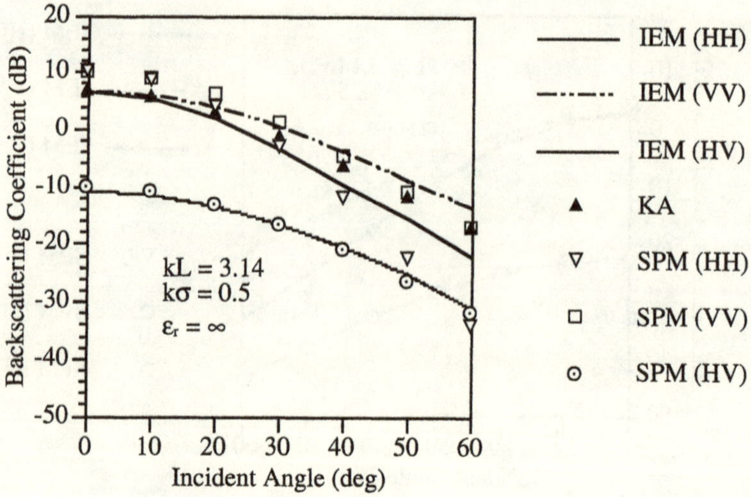

Figure 5.2 Comparisons of backscattering coefficients between the KM, SPM and IEM for $k\sigma = 0.5$.

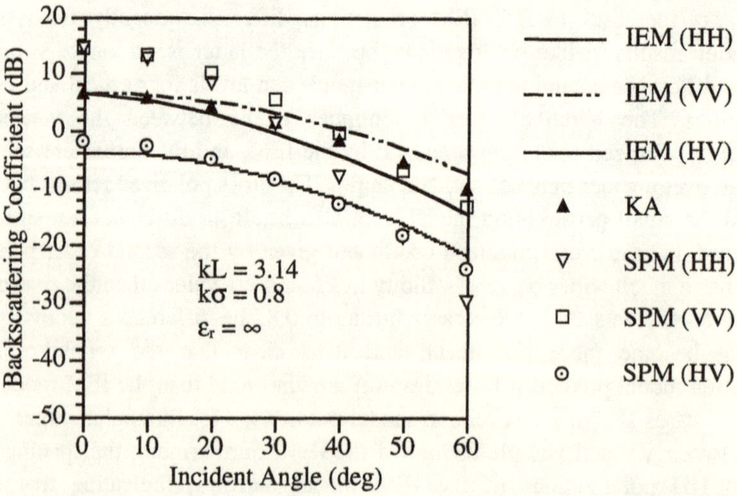

Figure 5.3 Comparisons of backscattering coefficients between the KM, SPM and IEM for $k\sigma = 0.8$.

Figure 5.4 shows the IEM model behavior when the roughness parameter kL is varied from 1 to 4.5 while $k\sigma$ is kept constant at 0.2. In these graphs it is seen that for $k\sigma = 0.2$ and $kL = 1$, IEM agrees very well with the small perturbation model in both like and cross polarizations. As kL increases, the separation between VV and HH polarizations decreases. The angular trends for both like and cross polarizations drop off faster (Figure 5.5) and the IEM model approaches the Kirchhoff model in like polarization (Figure 5.6). Due to the choice of the small $k\sigma$ value the difference between the cross polarized coefficients of the perturbation and IEM models is not very significant even when kL is 4.5 (Figure 5.6). This again shows that the second-order perturbation model for the cross polarization has a wider range of validity than the first-order like polarized model. This behavior is consistent with theoretical predictions. It is specially interesting to note that the level of the cross polarized returns rises as kL increases from 1 to 3 but then it decreases for further increases in kL to a lower level than when kL was 1. The initial increase in level is due to having a larger scale of roughness, while the decrease at still larger kL values is due to having a significantly smaller slope. This decrease is consistent with the fact that a Kirchhoff type surface has a low level of depolarization.

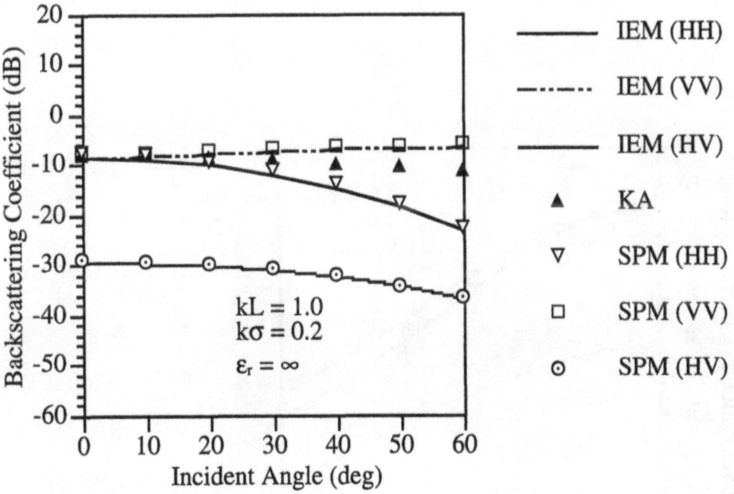

Figure 5.4 Comparisons of backscattering coefficients between KM, SPM and IEM when $kL = 1.0$.

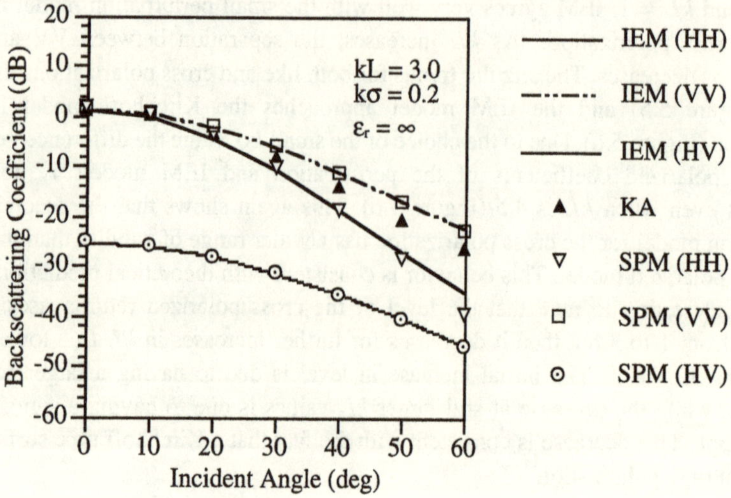

Figure 5.5 Comparisons of backscattering coefficients between KM, SPM and IEM when $kL = 3.0$.

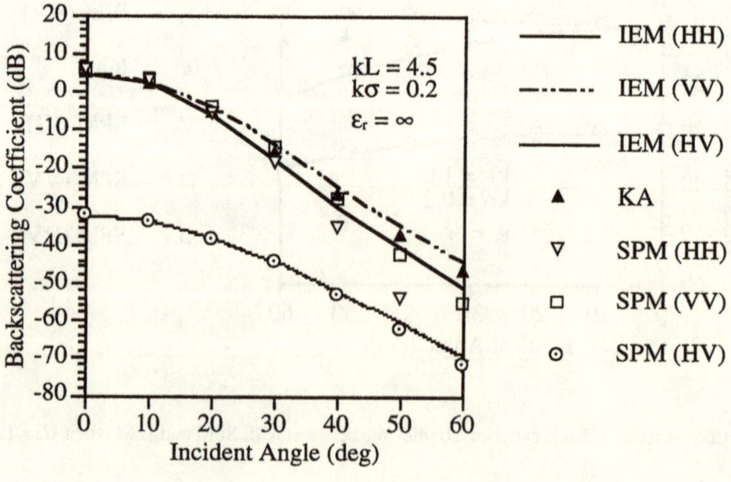

Figure 5.6 Comparisons of backscattering coefficients between KM, SPM and IEM when $kL = 4.5$.

To show frequency dependence we illustrate in Figures 5.7 through 5.9 the like and cross polarized scattering properties by varying $k\sigma$ from 0.125 to 1 and kL from 0.75 to 6, a change in frequency by a factor of 8. We see a gradual decrease in the separation between the VV and HH polarizations, a fast increase in the level of the backscattering coefficients in the low-frequency region, a gradually faster angular drop-off and the approach to the geometric optics solution at the high frequency end.

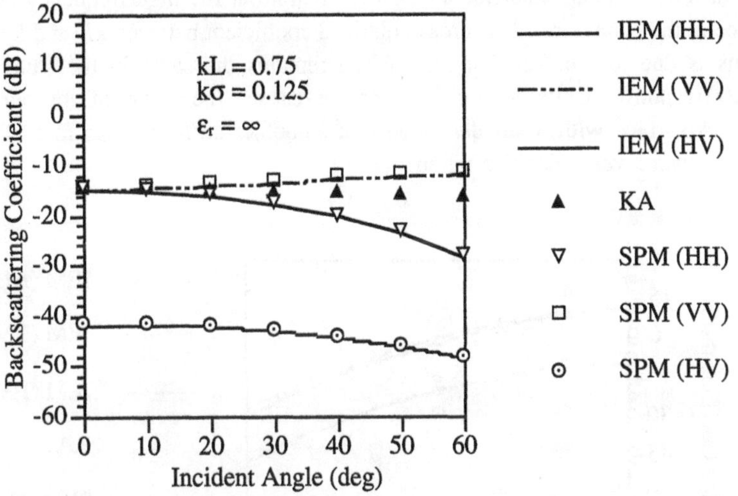

Figure 5.7 Comparisons of backscattering coefficients between KM, SPM and IEM when frequency is low.

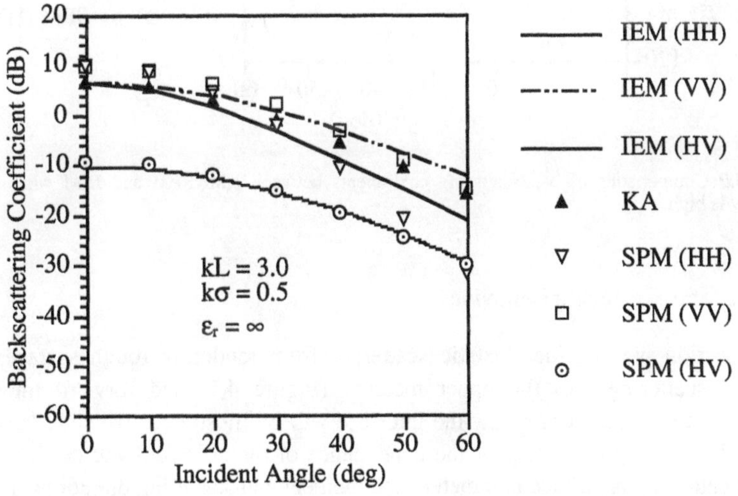

Figure 5.8 Comparisons of backscattering coefficients between KM, SPM and IEM when frequency is at intermediate level.

From Figure 5.7, both like and cross polarized returns based on the IEM model agree with the perturbation solution at the low-frequency end. Then, as frequency increases, the like polarized coefficients begin to deviate from the perturbation solution (Figure 5.8) and approach the Kirchhoff solution. This is then followed by a deviation in both the like and the cross polarized coefficients at a higher frequency (Figure 5.9), where the IEM model in like polarization is now in agreement with the Kirchhoff solution. Note that, while the polarized coefficients approach the Kirchhoff solution at the high-frequency end similar to the case when kL increases (Figure 5.6), the cross polarized coefficient near vertical incidence appears to saturate at the high-frequency end. The difference in the behaviors of the cross polarized coefficient between kL and frequency variations is due to the fact that the surface remains the same in the study of the frequency response, while letting kL increase causes the slope of the surface to decrease. A surface with a smaller slope is a smoother surface. Therefore, it should generate a lower level of depolarization.

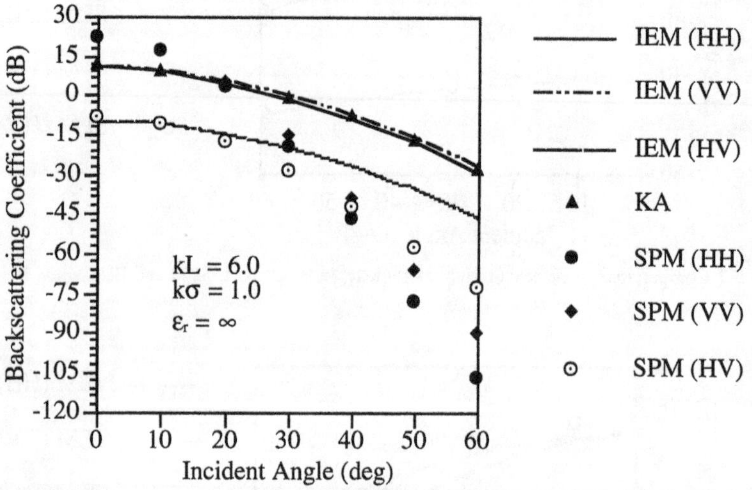

Figure 5.9 Comparisons of backscattering coefficients between KM, SPM and IEM when incident frequency is high.

5.7.2 Bistatic Scattering Behavior

In this section we consider bistatic scattering from randomly rough surfaces. Both backward scattering into the upper medium (Figure 4.1) and forward transmitted scattering into the medium below the interface will be illustrated. We shall begin with backward scattering and consider the dependence of the polarized and cross polarized bistatic scattering on surface parameters and incident and scattering directions. In all the numerical illustrations, we use the 1.5-power correlation function given in Appendix 2B.

Backward Scattering into the Upper Medium

First, we show the effects of surface parameters. In Figure 5.10 we see that an increase in the rms height of the surface causes an increase of all the scattering coefficients shown. Unlike the case of backscattering the increase is uniform over all the scattering angles and for all four scattering coefficients, VV, HH, VH and HV. As a function of the scattering angle θ_s there is a crossover between VV and HH and also between VH and HV. All four scattering coefficients peak at around 40 degrees in θ_s. This is because the incident angle is at 40 degrees and the azimuthal angle is 20 degrees, which is very close to the forward scattering direction. In like polarized scattering HH is lower than VV at small angles of scattering (about 12 degrees) and becomes significantly higher at large angles. Similar changes take place between VH and HV except the crossover occurs at a much larger scattering angle (about 40 degrees) and the difference between the pair of scattering coefficients is much smaller. For the case shown like polarized scattering coefficients are higher than those of the cross polarized coefficients. This is because the azimuthal angle used in the calculation is close to the plane of incidence (20 degrees). If the selected azimuthal direction is orthogonal to the incident plane we expect the reverse to be true.

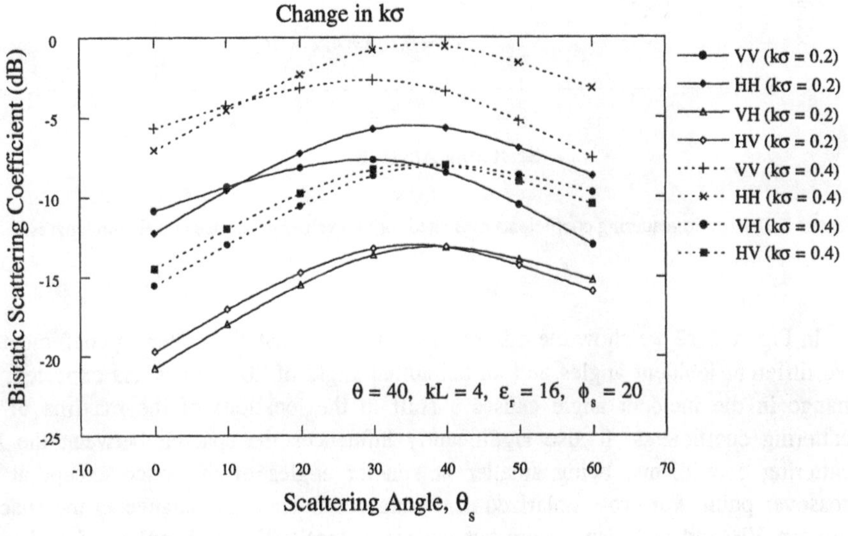

Figure 5.10 Bistatic scattering coefficients evaluated for two values of surface rms heights.

Next, we consider how a change in kL affects bistatic scattering. In Figure 5.11 we show the four scattering coefficients plotted versus the scattering angle for two different values of kL. A large value of kL gives a higher scattering coefficient. The increase in level, however, is changing with the scattering angle; first it increases to a maximum as the scattering angle increases to about 40 degrees and then it declines somewhat as the scattering angle increases further. The crossover points for VV, HH and VH, HV remain the same as in Figure 5.10. Unlike in backscattering the change in like and cross polarizations due to a change in *kL* is similar in both level and angular trend. The reason for this is because in bistatic scattering both like and cross scattering are dominated by single scattering, whereas in backscattering the cross polarization component is due mainly to multiple scattering.

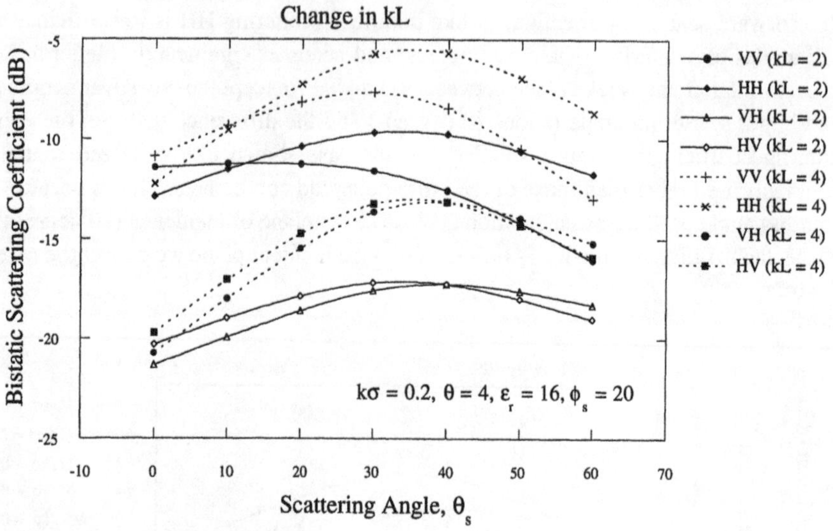

Figure 5.11 Bistatic scattering coefficients evaluated for two values of surface correlation lengths.

In Figure 5.12 we show the differences in the four bistatic scattering coefficients at two different incident angles and an azimuthal angle of 20 degrees. As expected, the change in the incident angle causes a shift in the locations of the maxima of the scattering coefficients. It also significantly influences the spacing between the like scattering coefficients, being smaller at smaller angles of incidence except at the crossover point. For cross polarized scattering coefficients, the change in the spacing between VH and HV also occurs but not significantly. For both polarized and cross polarized scattering coefficients, if they peak at smaller angles of scattering, they also decline to smaller values at large scattering angles. Thus, even though the coefficients for 20 degree incidence have higher peaks at around 20 degree scattering angle, they are lower than the coefficients at 40 degree incidence, when the scattering angle exceeds 40 degrees except for VV. This exception occurs because the set of surface parameters used leads to a large spacing between VV and HH scattering coefficients at 40 degrees incidence.

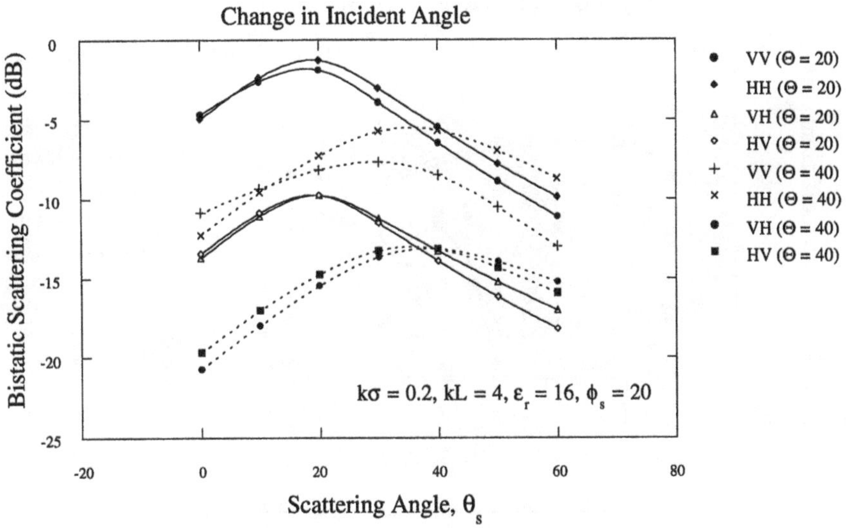

Figure 5.12 Bistatic scattering coefficients evaluated at incident angles of 20 and 40 degrees.

The above three parameter effects are shown at an azimuthal angle of 20 degrees. Clearly, azimuthal angle variations can change the relative behavior of the scattering coefficients or even reverse the observations made above. To get an idea of this angular effect, we show in Figures 5.13 and 5.14 plots of the scattering coefficients versus the azimuthal angle. In Figure 5.13 we fix the scattering angle at 20 degrees and show scattering curves at incident angles of 40 and 60 degrees. It is clear that the cross polarized scattering coefficients are lower than those of the like only for azimuthal angles within approximately 45 degrees of the plane of incidence, and this range is dependent upon the incident and scattering angles. It is also interesting to note that the relative levels of VV and HH switches over near the azimuthal angular region orthogonal to the plane of incidence. Around the backscattering direction (azimuthal angle = 180 degrees), VV is higher than HH and the reverse is true around the forward scattering direction (azimuthal angle = 0 degree). Such behavior is not shared by the cross polarized scattering coefficients. They appear parallel over the entire range of the azimuthal angle. While the dips of the polarized scattering coefficients are between 50 to 100 degrees in azimuthal angle, the dips of the cross polarized coefficients are always in the plane of incidence. These latter dips, however, are not shown in Figure 5.13 or Figure 5.14.

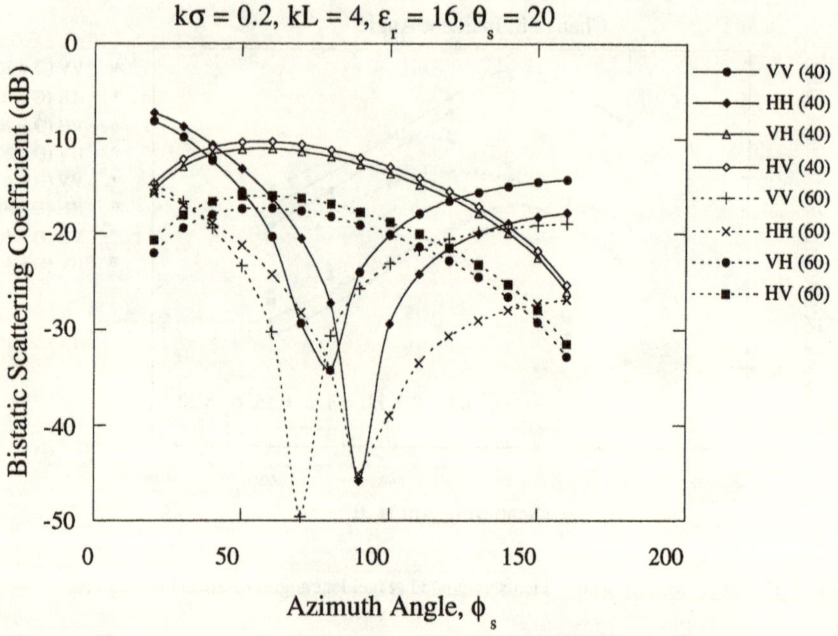

Figure 5.13 Bistatic scattering coefficients evaluated at incident angles of 40 and 60 degrees.

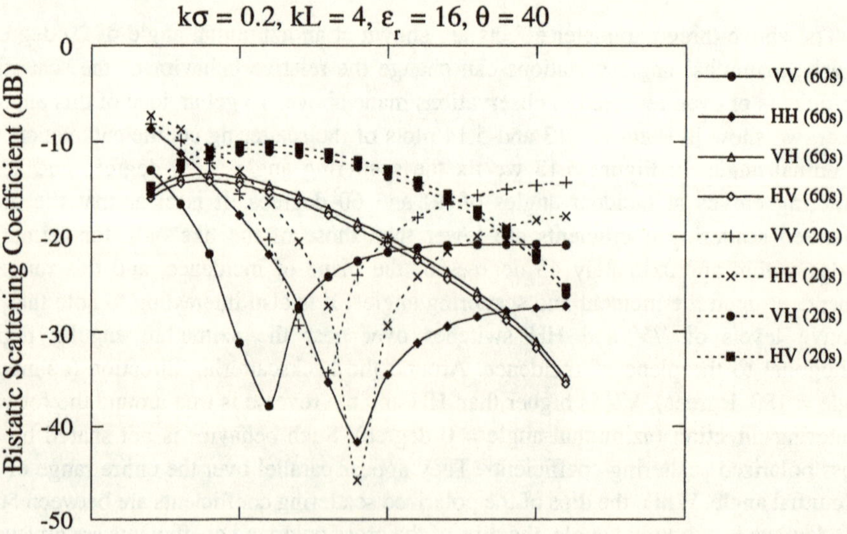

Figure 5.14 Bistatic scattering coefficients evaluated at scattering angles of 20 and 60 degrees.

Transmitted Scattering into the Lower Medium

The incoherent transmitted scattering into the lower medium below the rough interface is generally not measured and therefore of less interest. However, in modeling multi-layered media an appropriate representation of the transmitted signal is needed. To understand the behavior of the transmitted scattering coefficients we shall plot (5.39) and the corresponding scattering coefficients based on the small perturbation model (SPM) from Chapter 12 of Ulaby et al. [1982] as a function of the scattering angle for incident angles of 20, 40 and 60 degrees near the plane of incidence (ϕ_s = 20). The dependence on the incident angle θ is given in Figures 5.15 through 5.17. Then, we shall consider their behavior when the azimuth direction is nearly orthogonal to the incident plane (Figure 5.18). Finally, we show the effect of the dielectric constant on these coefficients. From these figures it is clear that the general angular trends between the predictions of the two scattering models are quite similar.

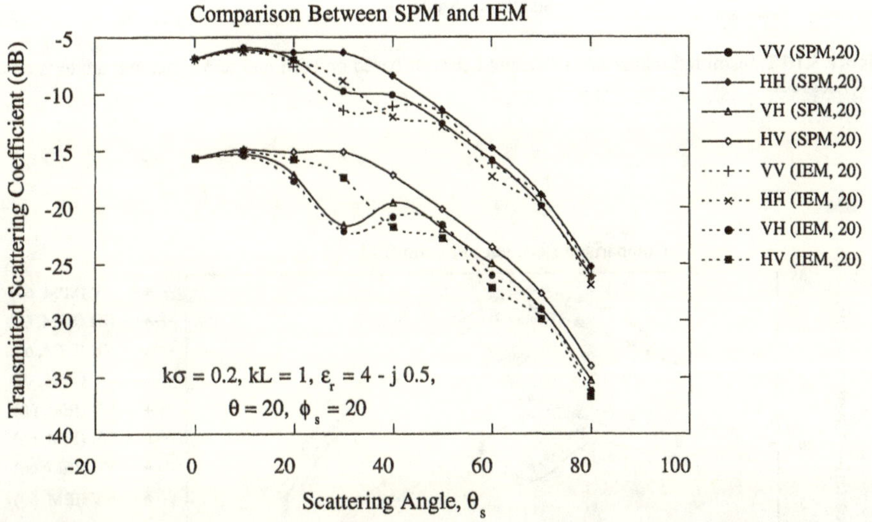

Figure 5.15 Transmitted scattering coefficient behavior based on IEM and SPM. Incident angle is equal to 20 degrees.

As the incident angle increases the peak region of the transmitted scattering curves continues to shift towards larger angles (Figure 5.16 and Figure 5.17). Although the shift is not large because the lower medium is the denser medium, a dip gradually appears in the small angular region that confirms the shift in the peak. Unlike the bistatic scattering coefficients studied earlier, there are some unexpected dips in the VH coefficients appearing in both the IEM and the SPM model. Another point to note is that the dip in VH occurs approximately at the scattering angle where other transmitted scattering coefficients are near their peaks.

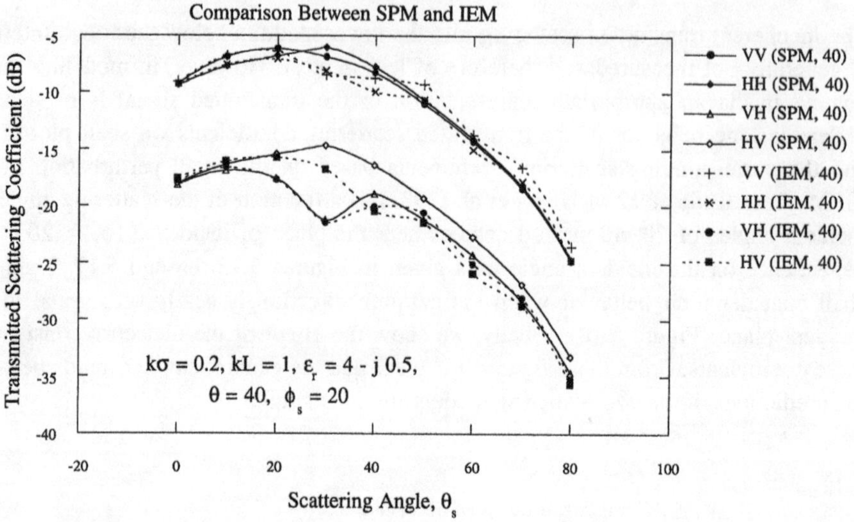

Figure 5.16 Transmitted scattering coefficient behavior based on IEM and SPM. Incident angle is equal to 40 degrees.

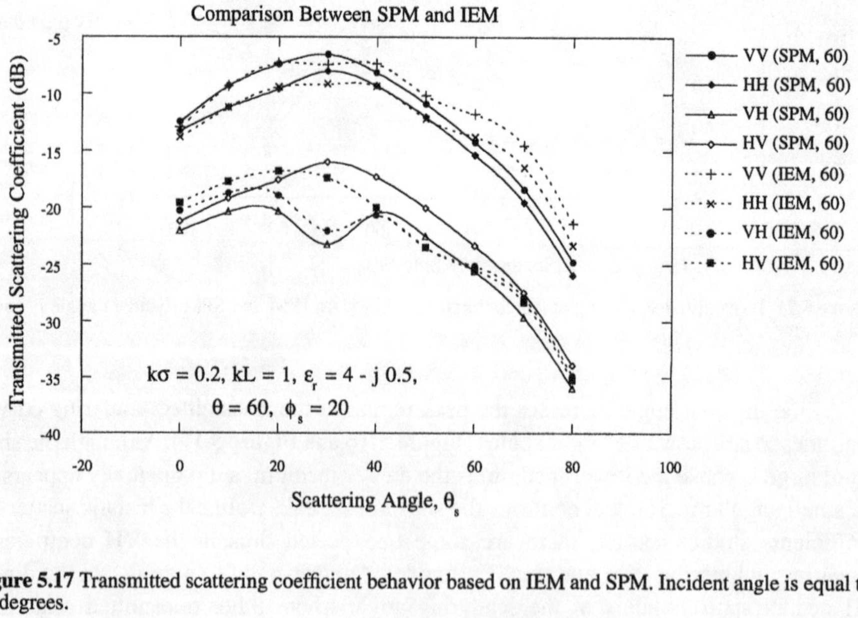

Figure 5.17 Transmitted scattering coefficient behavior based on IEM and SPM. Incident angle is equal to 60 degrees.

Other than some differences in level, one major difference between SPM and IEM is that IEM has oscillations not present in the SPM model. This is because the F_{pqt} coefficients of the IEM model has a singular point caused by the quantity denoted as sqt when the dielectric constant is real (shown in Appendix 4D). For a physical dielectric constant this singularity is removed but there are oscillations in the scattering coefficients that are dependent on the imaginary part of the dielectric constant. To demonstrate this effect we show in Figures 5.18 and 5.19 how the scattering curves change when the imaginary part is increased from 0.5 to 3.5. There is a very significant reduction in the oscillatory portions of the scattering curves in both the IEM and SPM. The larger imaginary part leads to higher scattering curves because we are showing incoherent scattering. Another point shown in Figure 5.18 is that the relative level between like and cross scattering coefficients has interchanged when we change the azimuthal angle from 20 degrees in Figure 5.17 to 80 degrees in Figure 5.18.

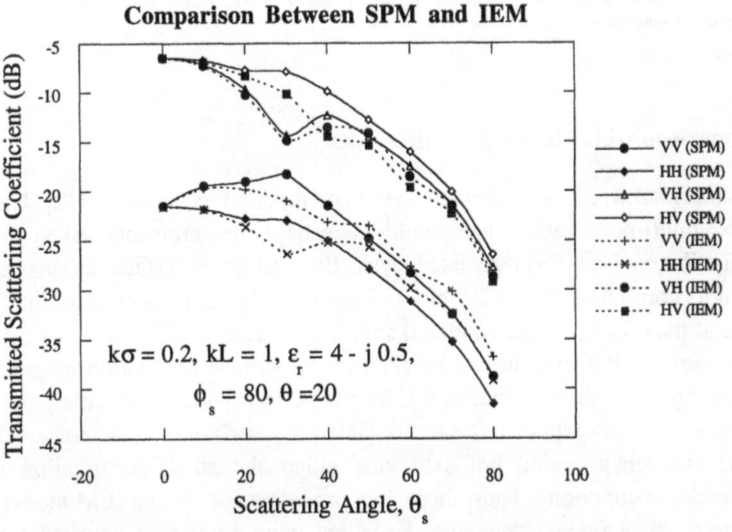

Figure 5.18 Transmitted scattering coefficient behavior based on IEM and SPM. Azimuthal angle is equal to 80 degrees and the dielectric constant is 4 - j 0.5.

For the scattering curves shown in Figures 5.15 through 5.19, we have purposely chosen small surface parameters so that we can compare the general angular trends between the IEM and the SPM models. Note that the surface parameters listed in these figures are for a relative dielectric constant of unity. The surface parameters in the lower medium are larger by a factor proportional to the square root of the dielectric constant.

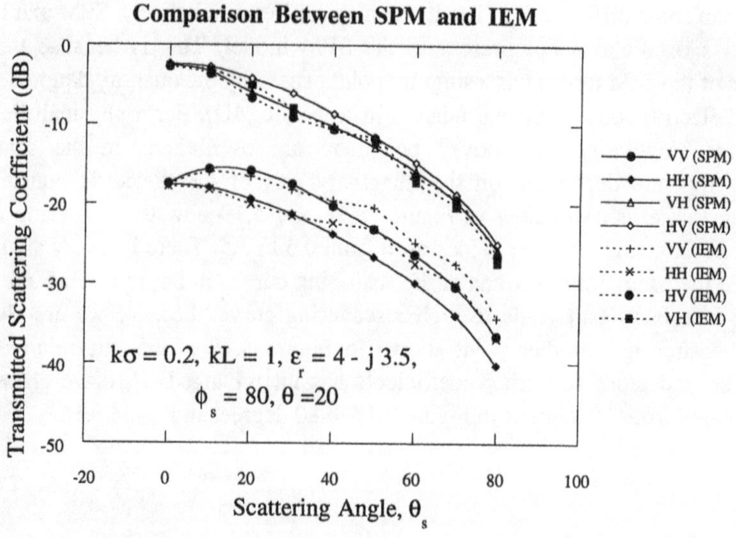

Figure 5.19 Transmitted scattering coefficient behavior based on IEM and SPM. Azimuthal angle is equal to 80 degrees and the dielectric constant is 4 - j 3.5.

5.7.3 Comparison with Bistatic Measurements

A set of bistatic backward measurements have been reported by Nance [1992]. These are multi-angle, multi-polarization and multi-frequency measurements taken from a Gaussian distributed, Gaussian correlated, perfectly conducting surface. Its rms height is 0.25 cm and its correlation length is 2 cm. In Figure 5.20 we show a comparison for VV polarization at fixed incident and scattered angles over a range of frequencies from 5 to 10 GHz. In view of the rms surface height and the surface correlation length values shown in the figure, it follows that at 5 GHz the surface parameters satisfy the small perturbation model assumptions. Near 10 GHz, the surface parameters are in the intermediate frequency region not satisfying either the small perturbation or the Kirchhoff model assumptions. Thus, these data provide a test for the IEM model in the low and intermediate frequency region. Excellent trend agreement between the IEM model and measurements is obtained at VV polarization over most part of the frequency range shown in Figure 5.20 except at the low-frequency end, where there is a difference of about 1 dB. In HH polarization (Figure 5.21) only a point at the low-frequency end lies outside the confidence interval by a small fraction of a dB. We know that at 5 GHz, the IEM model agrees with the small perturbation model. Hence, the predictions should be reliable. Excellent agreements are obtained at VH (Figure 5.22) and HV (Figure 5.23) polarizations over all frequencies. This lends further confidence to the IEM model.

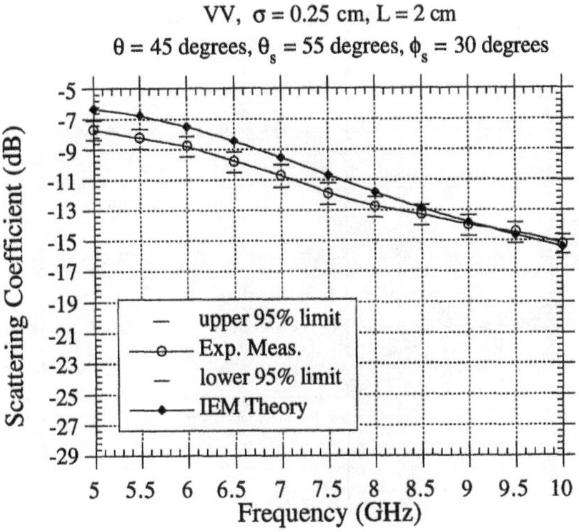

Figure 5.20 Comparisons of IEM model with bistatic VV measurements at a scattering angle of 55 degrees.

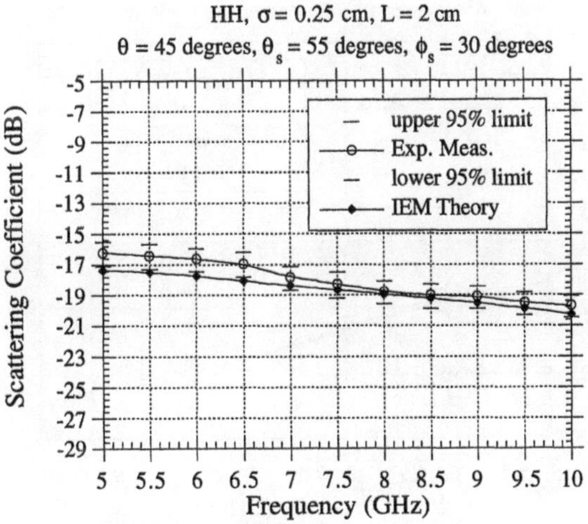

Figure 5.21 Comparisons of IEM model with bistatic HH measurements at a scattering angle of 55 degrees.

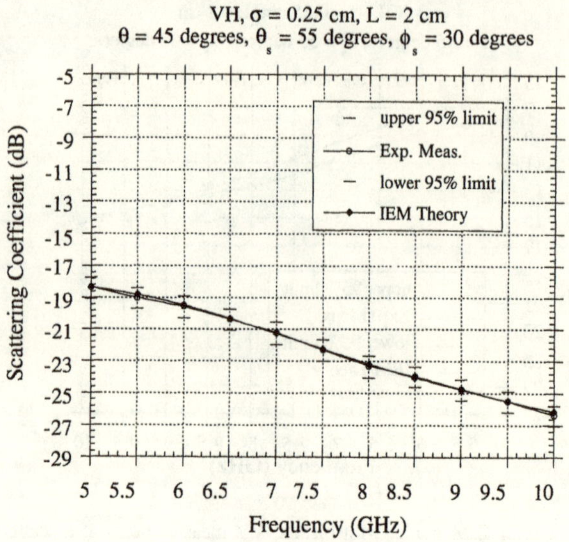

Figure 5.22 Comparisons of IEM model with bistatic VH measurements at a scattering angle of 55 degrees.

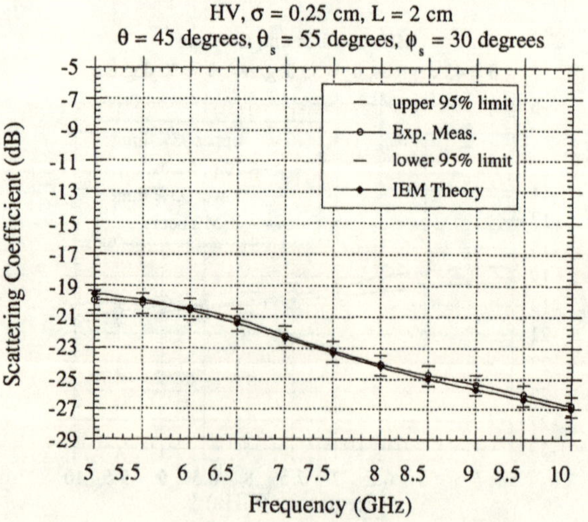

Figure 5.23 Comparisons of IEM model with bistatic HV measurements at a scattering angle of 55 degrees.

Next, we consider similar comparisons at a scattering angle of 70 degrees. Although some variations exist between data and model prediction, the difference is within the 95% confidence interval for VV polarization (Figure 5.24). For HH polarization substantial difference in trend and level appear as shown in Figure 5.25 over the mid-frequency range. This is attributed to the constructed surface being of finite size generated edge effects at large angles of scattering in horizontal polarization. When we compare IEM with the HV polarization (Figure 5.26), the agreement is very good except at the low-frequency end where the prediction is higher than the data by more than 1 dB. The angular trend, however, is very similar. Finally, we compare the model prediction with VH polarization. In this case it is the high-frequency end that deviates from the mean value of the data, but all points are within the confidence interval as shown in Figure 5.27. Thus, we conclude that the IEM model is very accurate up to scattering angles as large as 70 degrees.

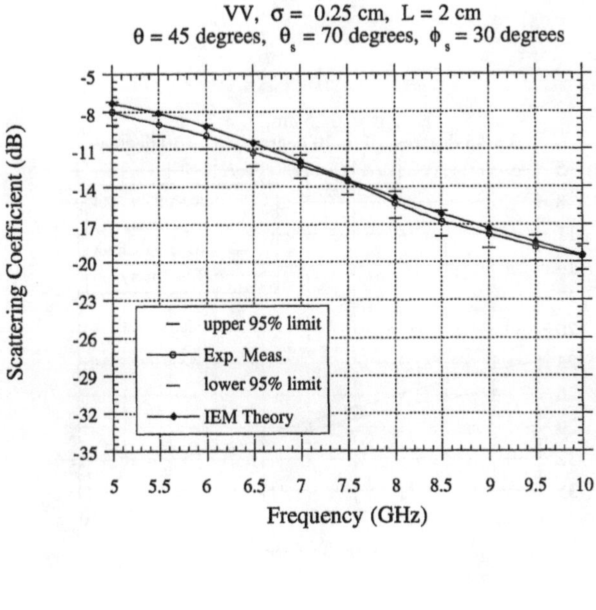

Figure 5.24 Comparisons of IEM model with bistatic VV measurements at a scattering angle of 70 degrees.

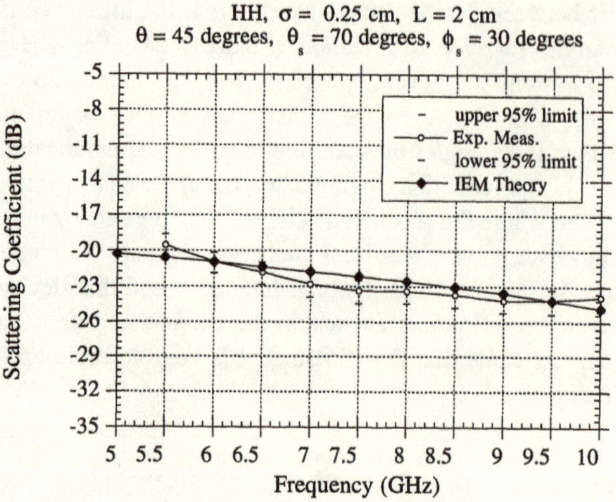

Figure 5.25 Comparisons of IEM model with bistatic HH measurements at a scattering angle of 70 degrees.

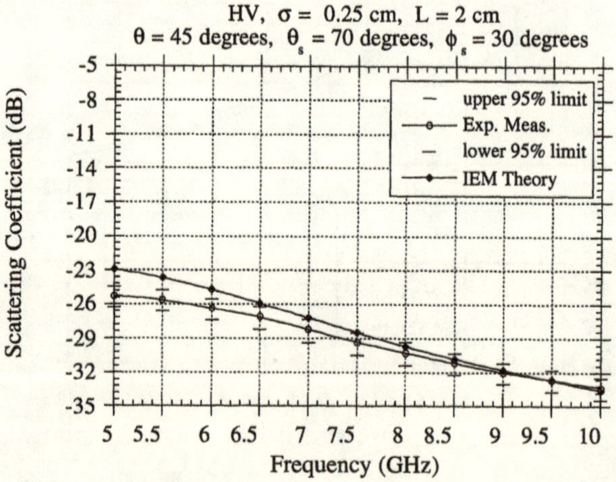

Figure 5.26 Comparisons of IEM model with bistatic HV measurements at a scattering angle of 70 degrees.

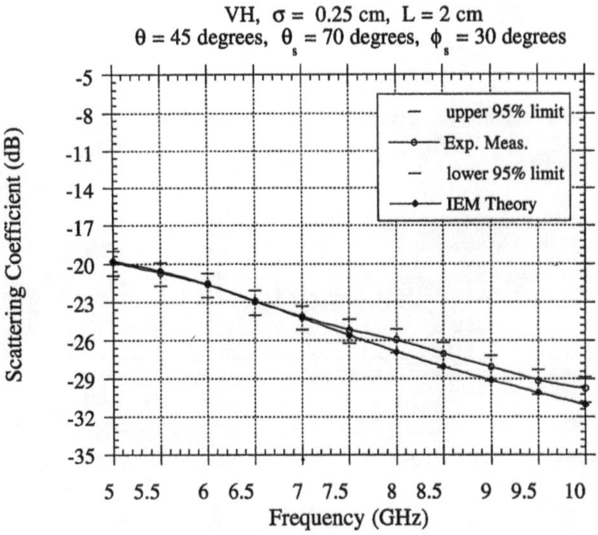

Figure 5.27 Comparisons of IEM model with bistatic VH measurements at a scattering angle of 70 degrees.

REFERENCES

Fung, A.K., Z. Li and K.S. Chen, "Backscattering from a Randomly Rough Dielectric Surface," *IEEE Tran. Geoscience Remote Sensing*, Vol. 30, no. 2, 1992, pp. 356–369.

Nance, C.E., "Scattering and Image Analysis of Conducting Rough Surfaces," Ph.D. Dissertation, E.E. Department, University of Texas at Arlington, 1992.

Ulaby, F.T., R.K. Moore and A.K. Fung, *Microwave Remote Sensing*, Vol. 2, Artech House, Dedham; MA, 1982, Chapter 12.

Chapter 6

Ranges of Validity of the IEM Model

6.1 INTRODUCTION

The surface scattering model developed in Chapters 4 and 5 has made one major assumption and several simplifying approximations that require further justification. The major assumption is that the unknown tangential field in the integral terms of the expressions for the total tangential surface fields given by (4.36) and (4.37) can be estimated by the corresponding Kirchhoff tangential fields. The additional simplifying approximations are

1. Apply integration by parts and discard the edge terms in Section 4.4.1.
2. Ignore the absolute phase factor in the Green's function in Section 4.4.2.
3. Discard terms in the tangential field expression involving the sum of $R_\perp$ and $R_\parallel$ in Section 4.3.
4. Approximate the local incident angle in $R_\perp$ and $R_\parallel$ by the incident angle for surfaces with small scale roughness and by the specular angle for surfaces with large scale roughness.

The first two simplifying approximations affect both perfectly conducting and dielectric surfaces, and the last two affect surfaces with small or moderate values of the dielectric constant. If the dielectric value is large, the last two approximations become very good. They are completely correct if the surface is perfectly conducting. The study of the validity of these approximations will be carried out in this chapter. We shall first verify that the simplifying approximations are valid for perfectly conducting surfaces by comparing numerical calculations without using these approximations with the surface scattering model predictions that depend on these approximations in Section 6.2. These calculations are carried out for a three-dimensional problem. In Section 6.3 we shall investigate the validity of the major assumption for perfectly conducting surfaces by comparing the model predictions with exact numerical calculations by the moment method in two dimensional problems. Then, we repeat the same investigation as in Section 6.3 but for dielectric surfaces in Section 6.4. Finally, we verify the model prediction against multifrequency measurements from a statistically known perfectly conducting surface.

6.2 INVESTIGATION OF SIMPLIFYING ASSUMPTIONS

From Section 4.2 we can estimate the tangential surface fields by substituting the Kirchhoff tangential fields into the unknown surface fields under the integral signs on the right-hand sides of (4.36) and (4.37) and then use these tangential fields to calculate the scattered fields. Without further simplification such calculations can be done only numerically because the tangential field expressions are very complicated. These numerical calculations can be carried out for three-dimensional scattering problems [Fung, 1990] and offer the opportunity to calculate both like and cross polarized scattering and to verify the effect of the additional simplifying assumptions used in theoretical expressions given by (4.139), (5.46) and (5.54).

To verify the effects of the simplifying assumptions stated in Section 6.1 we generate on a digital computer a rough surface in three dimensions with Gaussian height distribution and a Gaussian correlation function [Rochier et al., 1989]. Using such a surface we can calculate approximate tangential surface fields using (4.36) and (4.37) as described in the previous paragraph. Then, we calculate a scattered field sample using (4.60) and subsequently a scattered power sample. We repeat the process to obtain enough samples to calculate the average scattered power and the backscattering coefficient. The result obtained uses only the assumption that the Kirchhoff tangential fields can be used in the terms under the integral signs in (4.36) and (4.37) to estimate the total tangential fields. On the other hand, additional simplifying approximations are used in the scattering models given by (5.39) and (5.42). Thus, by comparing numerically simulated result with the model result we can assess the effects of the additional approximations.

In Figures 6.1 through 6.3 we show evaluations of (5.39) and (5.42) for like and cross polarized backscattering coefficients, respectively, and their comparisons with numerical simulation results. In these figures the notation *IEM* denotes the model with simplifying approximations and the word *Simulation* refers to numerical simulation calculations. The surface is assumed perfectly conducting. Excellent agreement in level and trend is seen between the two types of calculations for both like and cross polarizations. In Figure 6.1 the surface parameters are for a slightly rough surface. After increasing the incident frequency by a factor of two we obtain a set of normalized surface parameters $k\sigma$ and kL that fall into the intermediate frequency region in which neither a small perturbation model nor a Kirchhoff model would apply. The comparison between model and numerical simulation results for this case is shown in Figure 6.2. Further increase in frequency by another factor of two leads to another set of normalized surface parameters that do not satisfy the assumptions of either the Kirchhoff or the perturbation model. In all cases the IEM gives excellent agreement indicating that the use of the additional simplifying approximations does not make an appreciable difference. Note that we purposely select normalized roughness parameters that fall into the intermediate frequency range because in the low- and high-frequency regions the IEM reduces analytically to the first-order small perturbation model and the Kirchhoff model, respectively (see Chapter 5). Hence, there is no need to do further comparisons in those regions. Furthermore, such reductions also indicate that the additional simplifying approximations used in the IEM do not exceed those used in these standard models in the respective frequency regions. This is so because the IEM model has the same analytical expression as the perturbation model in the low-frequency region and as the Kirchhoff model in the high-frequency region.

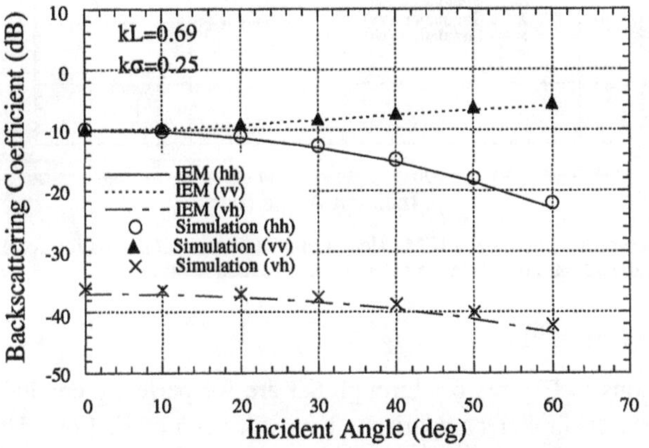

Figure 6.1 A comparison between IEM calculation and numerical simulation for a slightly rough surface.

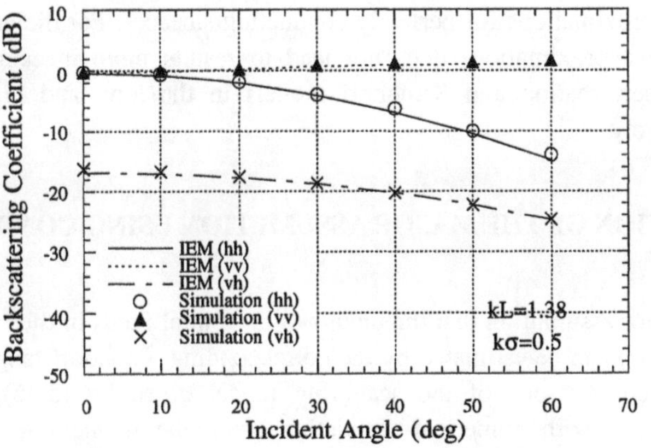

Figure 6.2 A comparison between IEM calculation and numerical simulation for a rough surface with normalized roughness parameters in the intermediate frequency range.

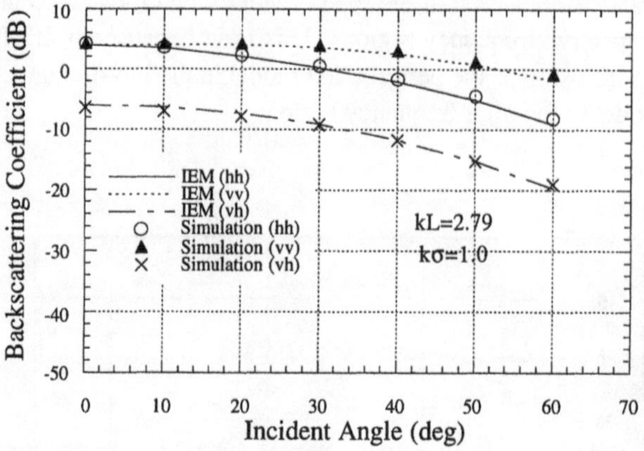

Figure 6.3 Another comparison between IEM calculation and numerical simulation for a rougher surface with normalized roughness parameters in the intermediate frequency range.

The comparisons in Figures 6.1 through 6.3 are for perfectly conducting surfaces. For dielectric surfaces similar comparisons have not been carried out. However, since the model reduces analytically to the first-order perturbation model in the low-frequency region for dielectric surfaces, the additional simplifying assumptions did not cause more inaccuracy than the standard perturbation model. Similar remarks are applicable to the high-frequency region where the IEM reduces to the Kirchhoff model.

In conclusion, the additional simplifying approximations used to derive (5.39) and (5.42) do not cause appreciable error when compared with calculations performed without these approximations for perfectly conducting surfaces. For dielectric surfaces the use of these approximations does not lead to results more inaccurate than the standard small perturbation and Kirchhoff models in the low- and high-frequency regions, respectively.

6.3 VERIFICATION OF THE MAJOR ASSUMPTION USING CONDUCTING SURFACES

To verify the major assumption that the unknown tangential fields in the integral terms of (4.36) and (4.37) can be estimated by the corresponding Kirchhoff tangential fields, we shall make comparisons of the scattering model given by (5.46), where this assumption is made with numerical simulation using the moment method in two dimensions [Chen et al., 1989]. The two-dimensional version of (5.46) is obtained by dividing (5.46) by $2k$ and using a two-dimensional surface spectrum. The explicit expression of the model and the standard first-order perturbation and Kirchhoff models in two dimensions are given in the next subsection. Then, comparisons will be carried out between the model and numerical simulations in the subsection to follow. All calculations are done for perfectly conducting surfaces. Dielectric surfaces are considered in Section 6.4.

6.3.1 Backscattering Models in Two Dimensions

In addition to the integral equation model three other surface scattering models are given in this section to provide additional references for different types of surface. They are the well-known Kirchhoff model, the first-order perturbation model and the two-scale model. All these models assume Gaussian height distribution.

Kirchhoff Model (KM)

The Kirchhoff surface scattering model is known to be valid in the high-frequency region, and it has several versions depending on the approximations used. The one given below is for surfaces with small rms slopes. We shall let σ denote the surface height standard deviation, L represent correlation length, k be the wave number and θ be the angle of incidence. In this case the backscattering coefficient under the scalar approximation is [Ulaby et al., 1982]

$$\sigma^K(\theta) = \frac{k}{\cos^2\theta} e^{-4s^2} \sum_{n=1}^{\infty} \frac{(4s^2)^n}{n!} W^{(n)}(2k\sin\theta) \tag{6.1}$$

where the surface spectrum $W^{(n)}(2k\sin\theta)$ is the cosine transform of the nth power of the surface correlation coefficient and $s = k\sigma\cos\theta$. For Gaussian correlation

$$W^{(n)}(2k\sin\theta) = \int \rho^n(x)\cos(2kx\sin\theta)\,dx = \sqrt{\pi/n}\,L e^{-(kL\sin\theta)^2/n} \tag{6.2}$$

For more complex correlation the surface spectrum may have to be evaluated numerically. Note that for a perfectly conducting surface, the Kirchhoff model does not differentiate vertical and horizontal polarization.

First-Order Small Perturbation Model (SPM)

This model is valid for surfaces with small roughness parameters. This means that both the surface standard deviation and its correlation length should be small compared with the incident wavelength [Chen and Fung, 1988]. Its backscattering coefficients are

$$\sigma^S_{pp}(\theta) = 4k|\alpha_{pp}|^2 (k\sigma)^2 \cos^4\theta\, W(2k\sin\theta) \tag{6.3}$$

where $\alpha_{hh} = -1$ and $\alpha_{vv} = -(1+\sin^2\theta)/\cos^2\theta$

Two-Scale Model (TSM)

There are several forms of the two-scale surface scattering model in the literature. The most common is the one which is the sum of the Kirchhoff model and the first-order perturbation model averaged over the slope distribution of the large scale roughness. The Kirchhoff model has several forms depending on the simplifying assumptions used in addition to the tangent plane approximation. The averaging operation of the small scale roughness also varies due to the way shadowing is accounted for. For the purpose of providing a reference, it is sufficient to use a much simplified version, namely, the direct sum of the Kirchhoff and the small perturbation model given in the above two subsections. Thus, we use

$$\sigma^{T}(\theta) = \sigma^{K}(\theta) + \sigma^{S}_{pp}(\theta) \tag{6.4}$$

Integral Equation Model (IEM)

The two-dimensional version of surface scattering model given in (5.46) differs from the three-dimensional problem in two respects: its roughness spectrum is the one- instead of two-dimensional Fourier transform and its coefficient is $1/(2k)$ times that of the three-dimensional scattering coefficient. That is,

$$\sigma^{0}_{pp}(\theta) = \frac{k}{4}e^{-2s^2}\sum_{n=1}^{\infty} s^{2n}\left[2^n f_{qp} e^{-s^2} + \frac{F_{qp}(-k\sin\theta, 0) + F_{qp}(k\sin\theta, 0)}{2}\right]^2 \frac{W^{(n)}(2k\sin\theta)}{n!} \tag{6.5}$$

where $s = k\sigma\cos\theta$; the field coefficients, f_{qp} and F_{qp}, are given by (5.48) through (5.51). For perfectly conducting surfaces the use of (5.70) and (5.75) through (5.78) reduces (6.5) to

$$\sigma^{0}_{pp}(\theta) = \frac{k}{\cos^2\theta}e^{-2s^2}\sum_{n=1}^{\infty} s^{2n}\left(2^n e^{-s^2} \pm 2\sin^2\theta\right)^2 \frac{W^{(n)}(2k\sin\theta)}{n!} \tag{6.6}$$

where $s = k\sigma\cos\theta$. Again (6.6) differs from the three-dimensional problem in two respects: its roughness spectrum is the one- instead of two-dimensional Fourier transform and its coefficient is $1/(2k)$ times that of the three dimensional scattering coefficient. Equations (6.5) and (6.6) will be used for comparison with the moment method calculations in the subsections to follow.

6.3.2 Comparisons Between Models and Moment Method Simulation

Moment method simulation is an approach to compute the surface current numerically without making any simplifying assumptions in the formulation of the problem. Thus, its result is regarded as the exact solution subject only to numerical inaccuracy.

In the comparisons to follow we also show the small perturbation model (SPM) and the Kirchhoff model (KM) to provide additional reference. The two-scale model is not plotted but the applicability of its concept can be seen from the plots given. Gaussian distributed, perfectly conducting surfaces with the following correlation functions are considered in this section:

1. Gaussian correlated surfaces.
2. Non-Gaussian correlated surfaces.
3. Two-parameter, Gaussian correlated surface where the parameter sizes differ by a factor of eight.
4. Two-parameter, Gaussian correlated surface where the parameter sizes differ by a factor of two.

In Figure 6.4 we show comparisons of the IEM with the moment method solution for a Gaussian correlated surface. The IEM follows the polarization, frequency and angular variations very well, beginning at a frequency where the small perturbation model is applicable, i.e., the low-frequency region. After the frequency increases by a factor of two, only the IEM agrees with the moment method simulation. Here, the normalized surface roughness parameters fall into the intermediate frequency region. A further increase in frequency by another factor of two causes the simulated and the IEM results to approach the Kirchhoff model, which is valid in the high-frequency region. Thus, the IEM is applicable over the entire frequency band.

Next, we show comparisons of the IEM with the simulation for a non-Gaussian correlated surface. The correlation function is taken to be

$$\rho(\xi) = \exp\left[-\xi^2/(l^4+\xi^2L^2)^{1/2}\right] \tag{6.7}$$

The comparisons are shown in Figures 6.5 and 6.6. At a wavelength of 18 units in simulation, the IEM and the perturbation model results are in agreement with the moment method simulation (MM). As the wavelength decreases to 9 units, the perturbation model results (SPM) begin to deviate from the simulation especially in HH polarization which is higher than the moment method calculations in the small incident angle region ($0 \text{ degree} \le \theta < 20 \text{ degrees}$) and the converse is true in the large incident angle region ($40 \text{ degrees} < \theta$). The VV polarization of the SPM is higher in the small incident angle region but is able to follow the simulation results at large incident angles. The IEM continues to follow the simulation results for all cases in Figure 6.5. Further decrease in wavelength by another factor of two is shown in Figure 6.6, where the deviation of the vertically polarized predictions of the SPM from the simulated results has become apparent. The differences in HH polarizations between the SPM and MM are similar to but larger than those in Figure 6.5. Due to the properties of the non-Gaussian surface correlation the angular behavior of the scattering coefficients in Figures 6.5 and 6.6 is quite different from those in Figure 6.4. In particular, as frequency increases the convergence of the vertical and horizontal polarizations to the Kirchhoff model (KM) is much slower. However, the IEM model continues to track the simulation results in all cases.

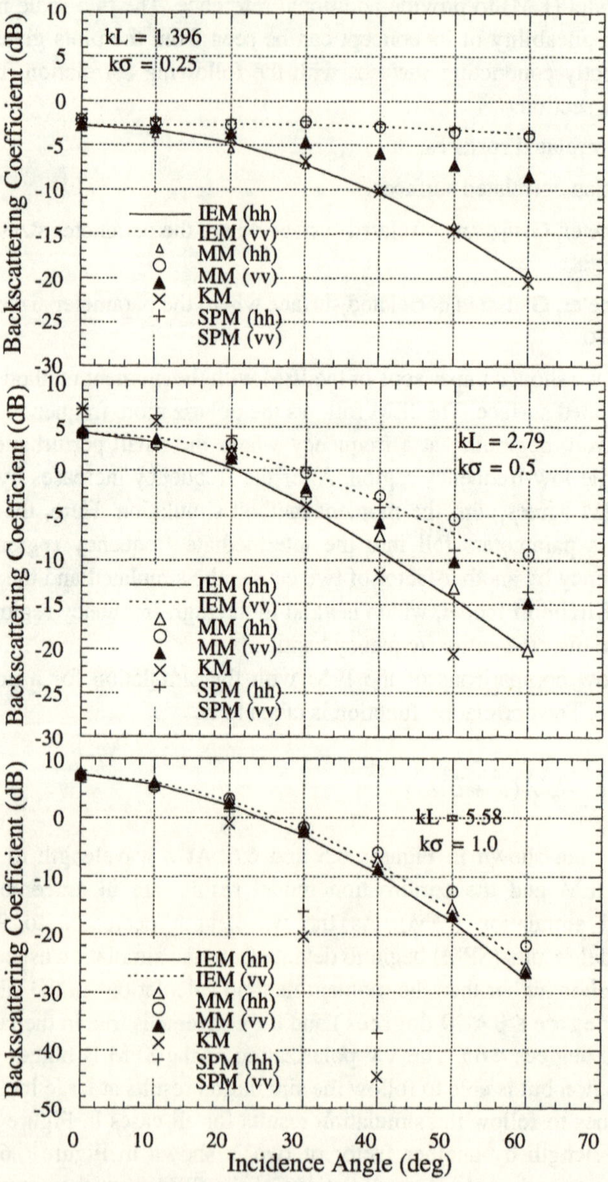

Figure 6.4 Comparisons of the IEM with moment method simulations from a Gaussian correlated surface under three different frequency conditions.

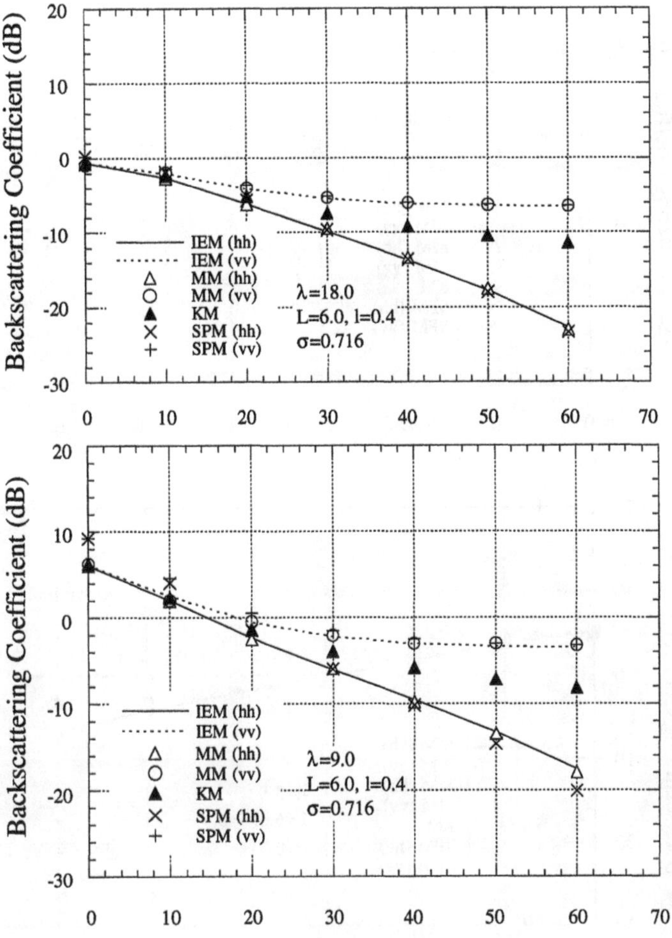

Figure 6.5 Comparisons of the IEM with moment method simulations from a non-Gaussian correlated surface at wavelengths of 18 and 9.

Next, we consider randomly rough surfaces characterized by a correlation function that is the sum of two Gaussian functions given by (2B.5) in Appendix 2B. This function has two correlation parameters and is repeated here for convenience:

$$\rho(\xi) = a\exp(-\xi^2/L_1^2) + b\exp(-\xi^2/L_2^2) \tag{6.8}$$

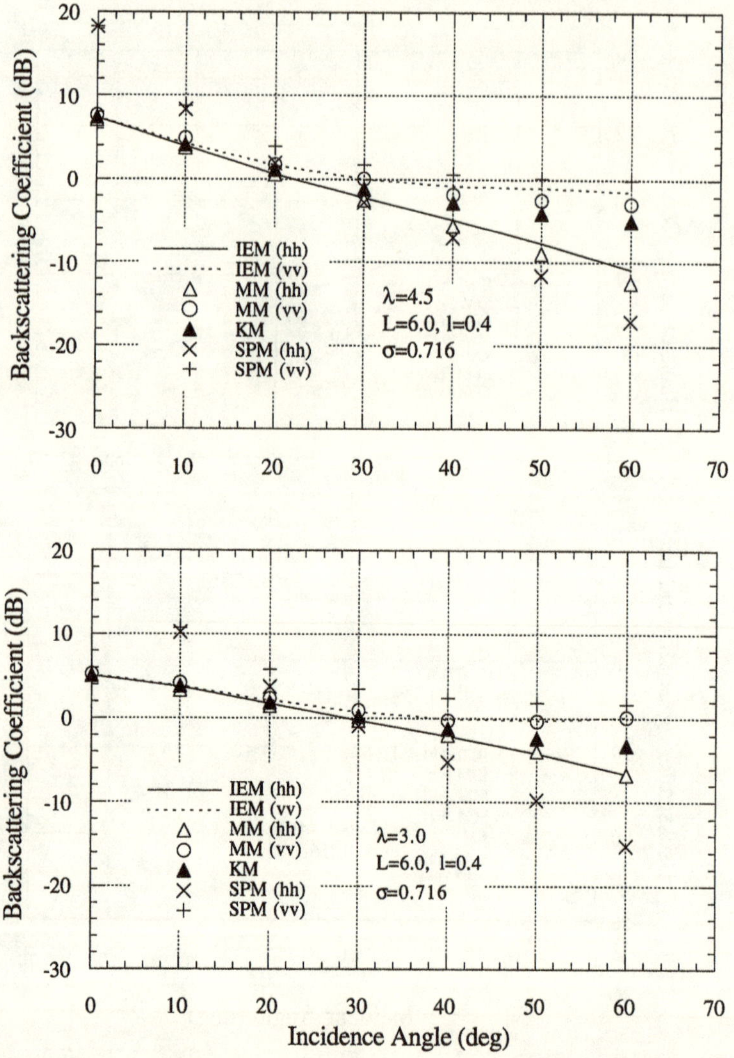

Figure 6.6 Comparisons of the IEM with moment method simulations from a non-Gaussian correlated surface at wavelengths of 4.5 and 3.

where $a = \sigma_1^2/\sigma^2$, $b = \sigma_2^2/\sigma^2$ and σ_1^2, σ_2^2 are the variances of the two scales while σ^2 is the total variance of the surface. We shall first consider the case where the correlation lengths of the two scales differ by a factor of eight in Figure 6.7. This is the condition under which we expect the two-scale theory to apply.

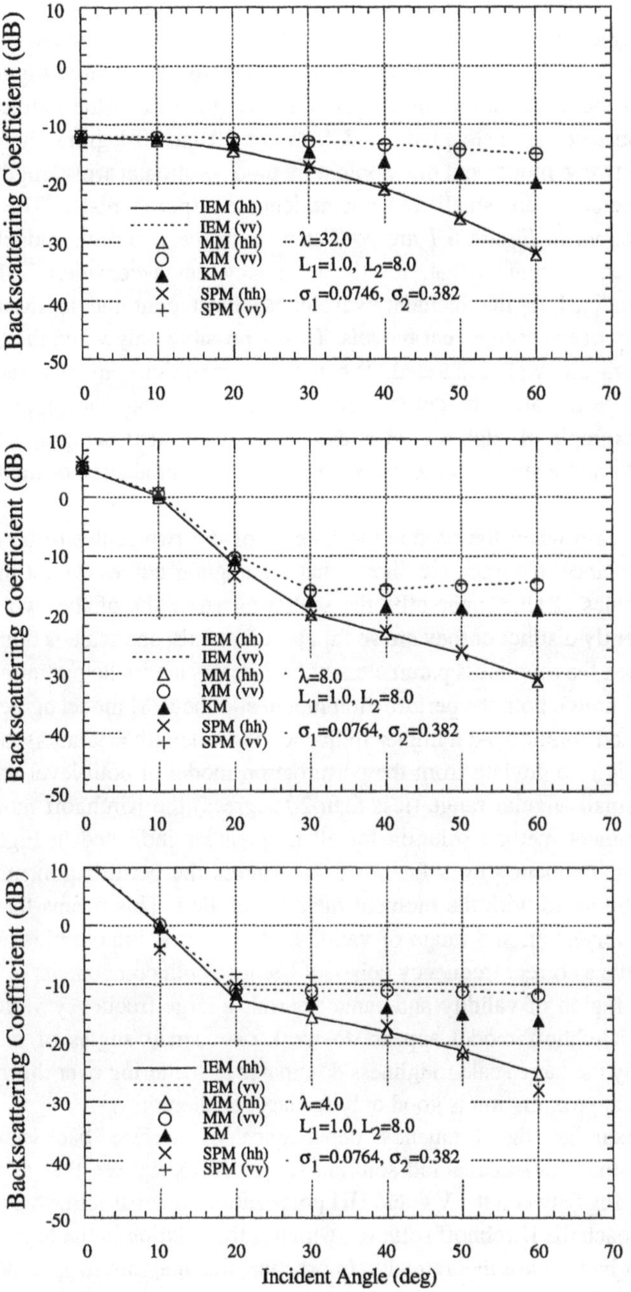

Figure 6.7 Comparisons of the IEM with moment method simulations from a two-scale surface where the correlation lengths of the scales differ by a factor of eight.

When the small scale variance is within the perturbation region and the large scale correlation length falls into the Kirchhoff region, we see from Figure 6.7 that in the small incident angle region ($\theta < 20$ degrees) the Kirchhoff model is in close agreement with the moment method simulation at all three wavelengths. Over large angular regions ($\theta > 40$ degrees) the perturbation model gives a very good fit to the moment method solution for the surface parameters indicated. Between 20 and 40 degrees, there are some disagreements between model and the moment method solution at wavelengths of 4 and 8. These disagreements are small and not noticeable experimentally. The scattering characteristics shown in Figure 6.7 are consistent with the two-scale model given by (6.4). The key requirement is that, for the frequency range considered, the surface parameters normalized to the incident wavelength must continue to lie within the regions of validity of the component models. This is possible only when the two sets of surface parameters are well separated. When these conditions are not satisfied, we expect the small perturbation model to differ substantially from the moment method simulation. An example of such a case is discussed in the next paragraph, where we choose a larger σ_1 and a smaller ratio between the correlation lengths of the two scales of roughness.

Consider the case when the correlation lengths of the two scales differ only by a factor of two. A glance of Figure 6.8 shows that the angular curves do not indicate the effects of two scales at all. Apparently, the scattering strengths of the two scales are either not sufficiently distinct or they are so far apart that only one scale is dominating at a given time. When the roughness parameters of both scales are in the perturbation range (wavelength = 18 units), both the perturbation model and the IEM model agree well with the moment method solution. At a higher frequency (wavelength = 9 units) the moment method results begin to deviate from the perturbation model in both level and angular trend. Over the small angular range (less than 20 degrees) the Kirchhoff model agrees well with the moment method solution for all frequencies indicated in Figure 6.8. A further increase in frequency by a factor of two causes the perturbation model to be completely off compared with the moment method solution. This means that the two-scale idea has a very restricted range of validity. Indeed, it is impossible for the two-scale model to have a correct frequency behavior because both components of the model have a restricted region of validity and cannot permit a large frequency variation. The reason why the Kirchhoff model appears to work over small angles of incidence is because physically the large scale roughness dominates in scattering over this region and the tangent plane approximation is good only for large scatterers.

When we examine the frequency dependence of surface backscattering by considering only the moment method solution (Figure 6.8), we see that as frequency increases the spacing between the VV and HH polarizations decrease. Furthermore, both VV and HH approach the Kirchhoff solution, which is the solution in the high-frequency region. This must be the case theoretically. In practice, this may not happen because the roughness scales of a natural surface are not restricted to only two. As frequency increases, smaller roughness scales become effective scatterers because they are now comparable to the incident wavelength.

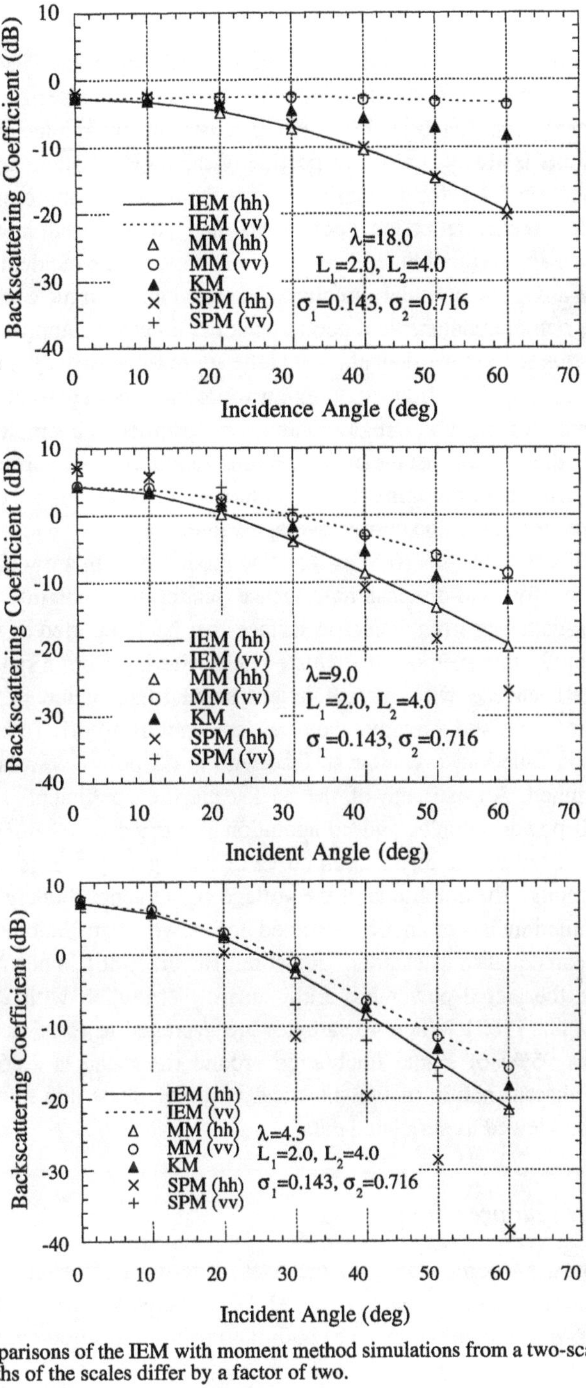

Figure 6.8 Comparisons of the IEM with moment method simulations from a two-scale surface where the correlation lengths of the scales differ by a factor of two.

6.4 VERIFICATION OF THE MAJOR ASSUMPTION USING DIELECTRIC SURFACES

In the previous section we have demonstrated the applicability of the IEM for Gaussian and non-Gaussian correlated, perfectly conducting surfaces characterized by a one- or two-parameter correlation function. For such surfaces the magnitude of the Fresnel reflection coefficients is always unity irrespective of the incident angle and local slope. As explained in Section 4.4.1, for dielectric surfaces there are two approximations to the local angle in the Fresnel reflection coefficients $R_{\perp,\parallel}$ to be used in the Kirchhoff coefficient f_{qp}: one approximation replaces the local angle by the incident angle and the other by the angle along the specular direction. The local angle in the Fresnel reflection coefficients in the complementary field coefficients F_{qp} is always approximated by the incident angle. In this section we shall show that the approximation by the incident angle is good for low to intermediate frequency region while the other approximation is good in the high frequency region. These regions cannot be specified in a simple way because they depend on the dielectric constant and correlation function of the surface. In general, in the region where neither approximation is applicable, these two approximations give the upper and lower bounds of the correct average value.

The study in this section uses (6.5) as the IEM model to compare with the moment method calculations for two-dimensional surface scattering problems. The moment method applied to scattering from dielectric surfaces has been reported by Chen and Bai [1990]. We shall apply this method to evaluate backscattering from a randomly rough, Gaussian correlated surface with surface height standard deviation of 0.429 cm, a correlation length of 3 cm and a relative dielectric constant of 3-j 0.1. To understand the effect of correlation function, a similar surface with a 1.5-power correlation function will also be examined. An estimate of the backscattering coefficient is obtained by averaging over 40 power samples and computation is carried out from 1 to 10 GHz. From Chapter 1, the tail of the distribution curve of the signal power is approximately exponential even though the amplitude of the voltage signal is not Rayleigh. As a result, the exponential function is commonly assumed for power distribution. This density function has its mean equal to its standard deviation [Beyer, 1986]. When N samples are averaged together, the signal power has a chi-square distribution with $2N$ degrees of freedom [Ulaby et al., 1982]. When 40 samples are averaged to form an estimate, the 90% range (5% to 95%) of signal fluctuation around the mean is 2.26 dB. Such a fluctuation is an inherent nature of the statistical problem. Thus, the moment method calculations may be viewed as simulated data.

Gaussian Correlated Surface

We shall begin showing comparisons of the backscattering coefficients at 10 degree incidence over the frequency range, 1 to 10 GHz, between the moment method calculations and IEM model given by (6.5) with the Fresnel reflection coefficient in f_{qp}

evaluated either at the incident angle or normal for vertical polarization. Results are shown in Figure 6.9.

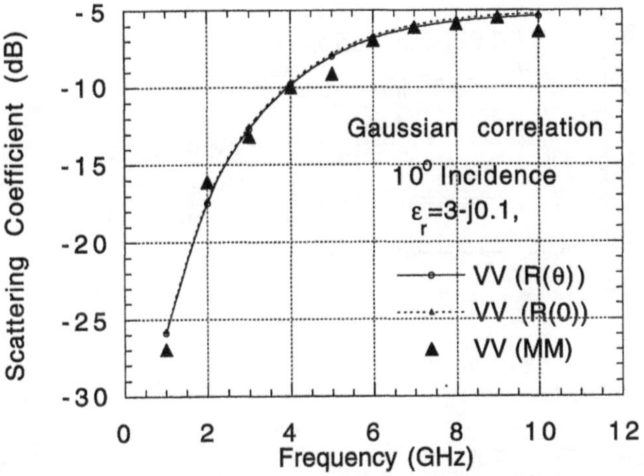

Figure 6.9 Comparisons between the moment method calculations and IEM in backscattering plotted versus frequency.

Due to the small incident angle used in Figure 6.9 there is a negligible difference between the two forms of the IEM model. The difference between VV and HH polarizations should also be insignificant. Hence, no further comparisons with HH polarization at 10 degrees incidence will be shown. Despite the expected fluctuations in the moment method calculation (MM) the agreement between the frequency trends of IEM and MM is excellent. Next, we show similar comparisons at 30 and 50 degree incidence in Figure 6.10 for vertically polarized wave. At the lower frequency region (1 to 5 GHz in Figure 6.10) the IEM with $R(\theta)$ gives better agreement with the MM calculations; in the higher frequency region (7 to 10 GHz in Figure 6.10), it is the IEM with $R(0)$ that agrees better with MM results. It appears that there is a transition region between 5 and 7 GHz, which may be characterized by the normalized surface parameters, $k\sigma = 0.54$ and $kL = 3.77$ at 6 GHz. It is clear that the two forms of the IEM model cover most of the frequency region but not the whole region, when we are dealing with a dielectric surface. This is because the local angle inside the Fresnel reflection coefficients cannot be approximated by either of the two forms when the surface roughness scales lie in the intermediate region.

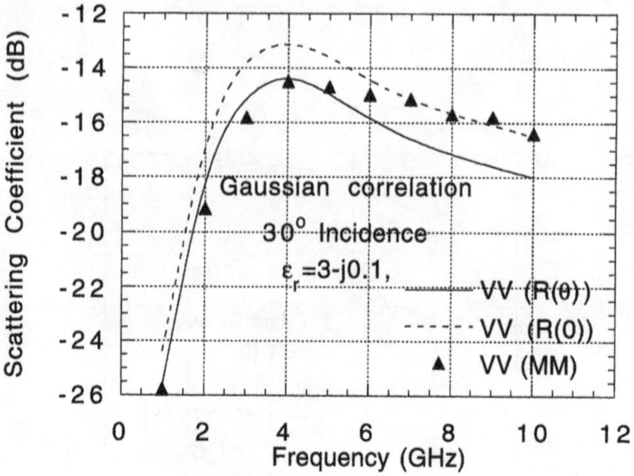

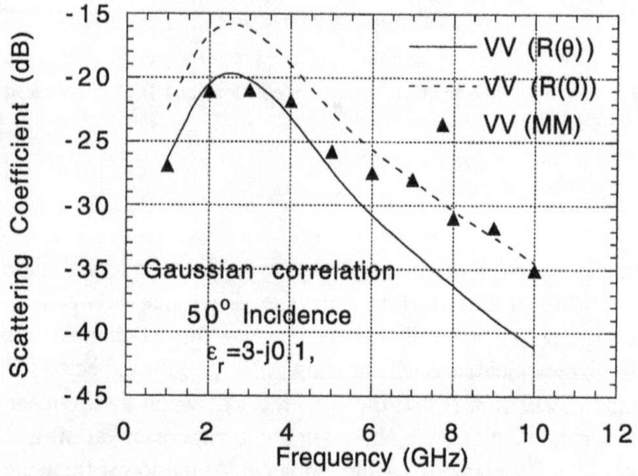

Figure 6.10 Comparisons between the moment method calculations and IEM in backscattering plotted versus frequency for 30 and 50 degree incidence and vertical polarization.

The corresponding comparisons for HH polarization are shown in Figure 6.11. Here, the transition region appears to be between 6 and 8 GHz at 30 degree and 5 and 8 GHz at 50 degree incidence. It appears that the fluctuation in MM is more severe in horizontal polarization. If we recognize that a low-probability sample, which deviates significantly from the mean, can and does occur in practice to distort an average value, we can formulate a rule of thumb for Gaussian correlated surfaces as described below.

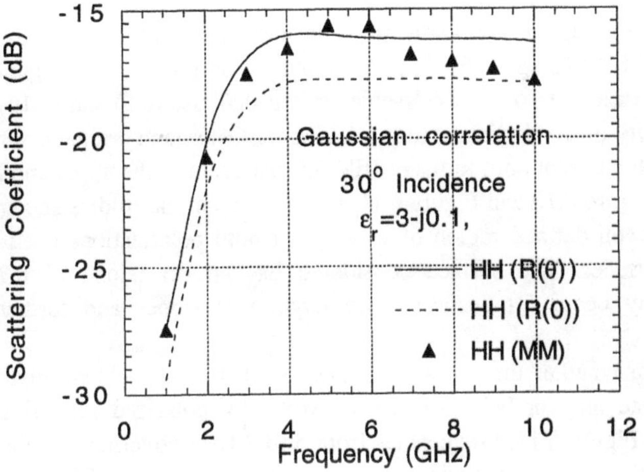

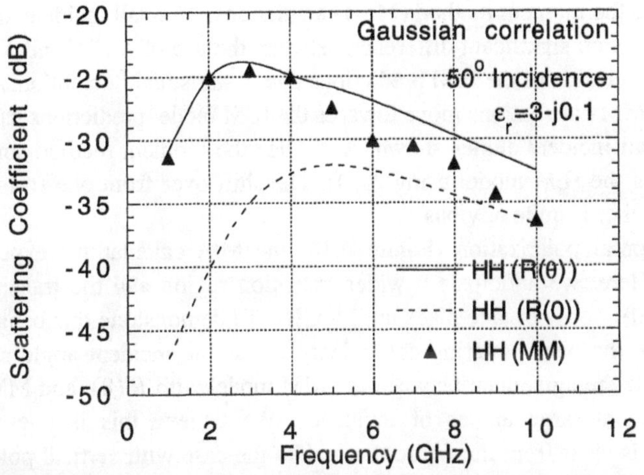

Figure 6.11 Comparisons between the moment method calculations and IEM in backscattering plotted versus frequency for 30 and 50 degree incidence and horizontal polarization.

By assuming that the use of $R(\theta)$ is valid at 5 GHz but the frequency must be less than 6 GHz for the surface under consideration, at 6 GHz with $\varepsilon_r = 3$ we obtain

$$(k\sigma)(kL) \approx 1.2\sqrt{\varepsilon_r}$$

Thus, we say that the form of the IEM model which uses $R(\theta)$ is applicable when

$$k^2\sigma L < 1.2\sqrt{\varepsilon_r} \tag{6.9}$$

At 8 GHz, $kL \approx 5$. The use of $R(0)$ for both polarizations requires

$$kL > 5 \tag{6.10}$$

In the transition region, we know that the correct average value lies between the predictions of IEM using $R(\theta)$ and $R(0)$ but we do not have a simple expression to cover this transition region. For convenience, we shall use (6.9) and (6.10) to decide on the applicability of the particular form of IEM for Gaussian correlated surfaces. It may also be desirable to plot both forms of IEM and use them as the upper and lower bounds as shown in Figure 6.10 and Figure 6.11. It is clear from the studies shown that there is no clear-cut, well-defined region of validity for both polarizations even for Gaussian correlated surfaces. The surface considered has an rms slope of 0.2. Additional variations may occur for surfaces with larger rms slopes and further studies are necessary.

To confirm whether the above guidelines work for all incident angles we plot in Figure 6.12 the angular behaviors of the vertically polarized scattering coefficients considered in Figure 6.11. Frequencies from 5 to 7 GHz covering the transition region shown by vertically polarized waves are selected. At 5 GHz the IEM with $R(\theta)$ gives the better fit to the moment method (MM) calculations. At small incident angles ($\theta < 20$ degrees) there is no significant difference between the use of $R(\theta)$ and $R(0)$ and the MM results lie outside of the IEM predictions due to statistical fluctuation. As frequency increases the MM calculations move towards the IEM model predictions using $R(0)$ as expected for all incident angles shown. At 7 GHz the moment method solution agrees very well with the IEM model using $R(0)$. The shift over from one form of the IEM model to the other is quite obvious.

For horizontal polarization (Figure 6.13) the MM calculations experience more fluctuations. Here, we anticipate a wider transition region and the transition towards IEM with $R(0)$ does not take place until 8 GHz. To demonstrate this behavior we plot in Figure 6.13 the backscattering coefficients versus the incident angle at 5, 7 and 8 GHz. At 5 GHz the agreement between the IEM model with $R(\theta)$ and MM solution is not very good at large angles of incidence. We believe this is due to statistical fluctuation. It is clear from the figure that unlike the case with vertical polarization the MM results at 7 GHz still lie between the predictions of the two forms of IEM at large angles of incidence. At 8 GHz the angular trend of IEM with $R(0)$ gets very close to the MM calculations. It is also interesting to note that for the cases shown in vertical polarization the IEM with $R(0)$ is higher than the IEM with $R(\theta)$ (Figure 6.12) and the reverse is true in horizontal polarization (Figure 6.13). This is because up to and around the Brewster angle region, $|R_{\parallel}(\theta)| \le |R(0)|$, while $|R_{\perp}(\theta)| \ge |R(0)|$.

In all the above calculations the permittivity of the surface is 3 - j0.1. This small value is purposely selected to illustrate one of the worst cases where the local angle approximation used in IEM is likely to break down. Recall from Section 6.3 that IEM works well for perfectly conducting surfaces because in this case the local angle approximation in the Fresnel reflection coefficient is not an issue. Larger values of the surface permittivity permit less variations in the Fresnel reflection coefficients. Hence, the IEM model should have larger range of validity for such surfaces.

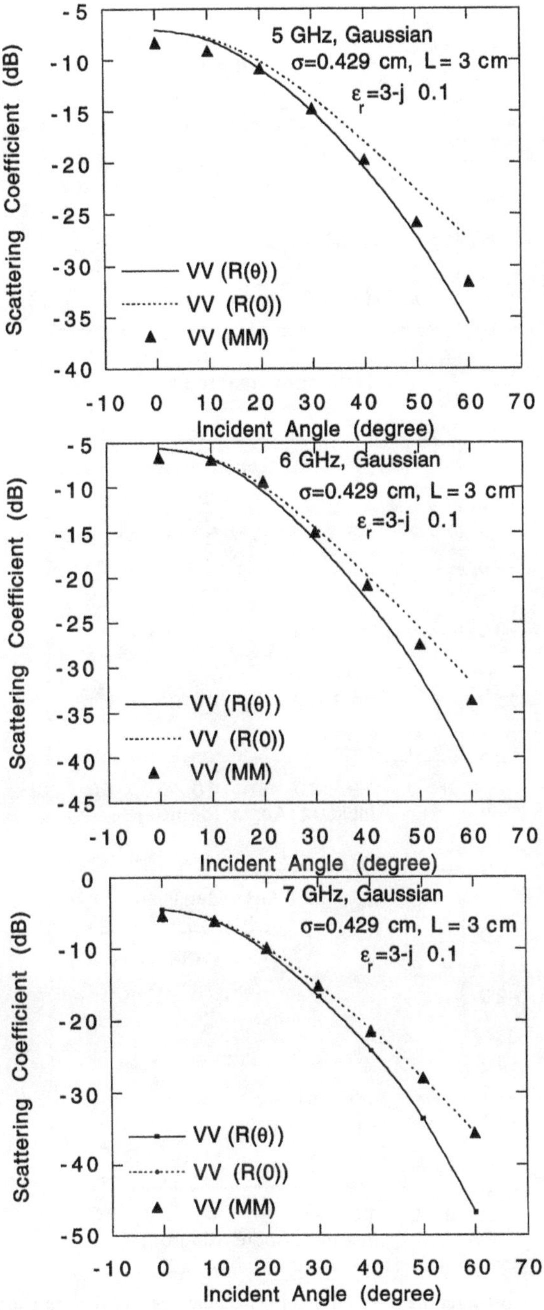

Figure 6.12 Comparisons between the moment method calculations and IEM in backscattering plotted versus angle at 5, 6 and 7 GHz for vertical polarization.

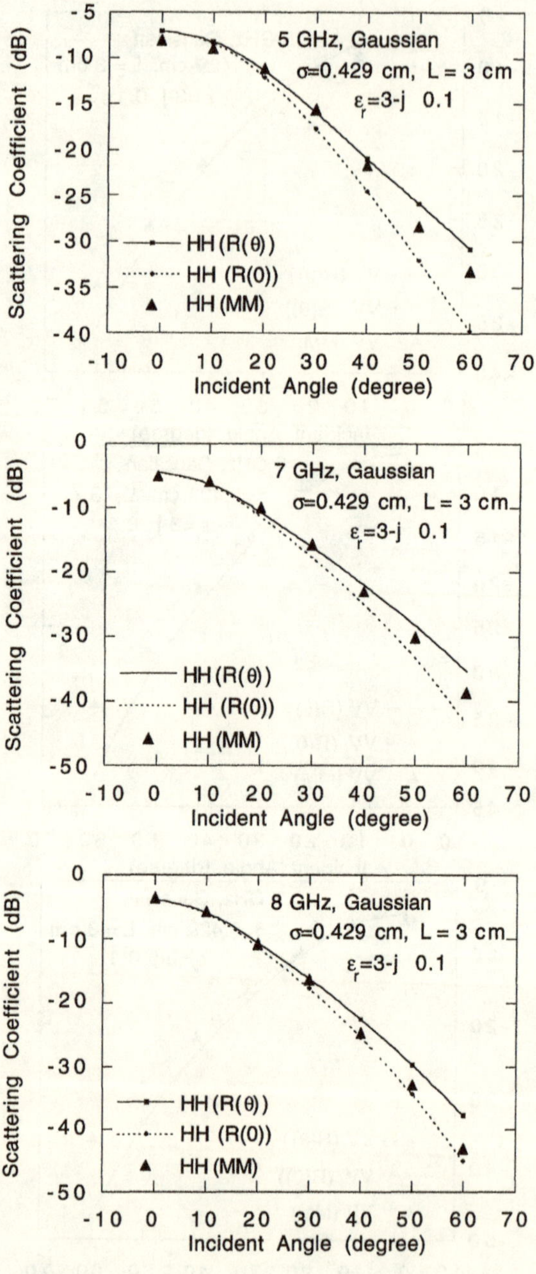

Figure 6.13 Comparisons between the moment method calculations and IEM in backscattering plotted versus angle at 5, 7 and 8 GHz for horizontal polarization.

To illustrate the effect of larger permittivity we plot the backscattering coefficient in Figure 6.14 using the same surface parameter as in Figure 6.12 and Figure 6.13 but with a relative permittivity of 36 - j 0.1. This larger permittivity should give closer agreement between the IEM and MM calculations because the spread between the two forms of IEM should be smaller. This effect is illustrated in Figure 6.14. Note that at 6 GHz the agreement between IEM and MM results were poor at large angles of incidence (see Figure 6.12). However, due to this large permittivity value much better agreement appears for both VV and HH polarization in Figure 6.14. All the results in this section is based on a Gaussian correlated surface. A non-Gaussian correlated surface will be considered in the next section.

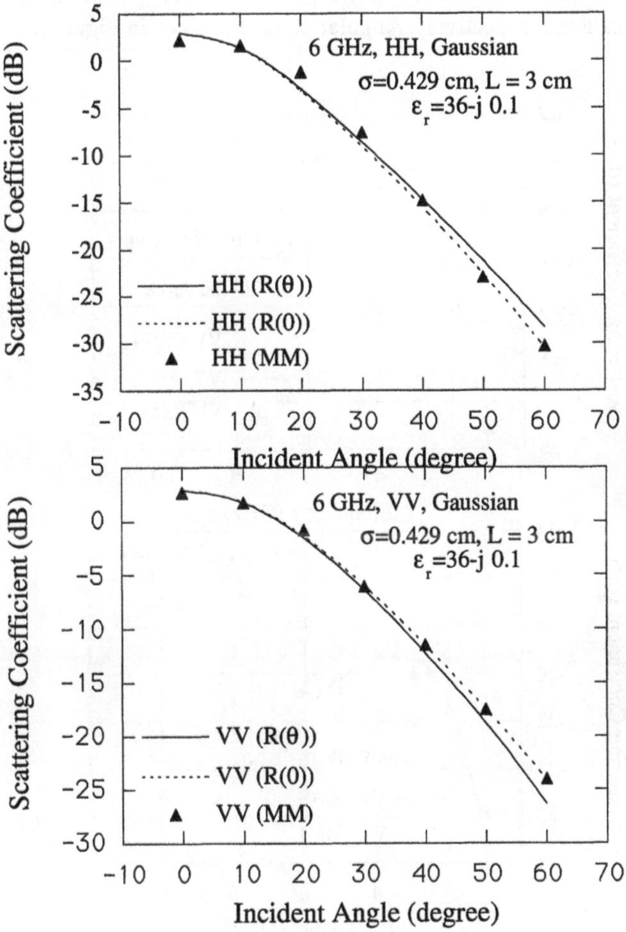

Figure 6.14 Comparisons between the moment method calculations and IEM in backscattering plotted versus angle at 6 GHz for vertical and horizontal polarization with a high permittivity of 36 -j0.1.

1.5-Power Correlated Surfaces

In this section we consider a non-Gaussian correlated surface with the same surface height standard deviation, σ = 0.429 cm, and correlation length, L = 3 cm, as the Gaussian surface in the previous section. The normalized correlation function has the form

$$\rho(\xi) = (1+\xi^2/L^2)^{-1.5} \tag{6.11}$$

As in the previous section we shall examine frequency and angular plots of the backscattering coefficients based on the moment method and the IEM model. Results plotted versus frequency are summarized in Figure 6.15 and Figure 6.16 for vertical and horizontal polarizations, respectively. Angular plots are shown in Figure 6.17 and 6.18.

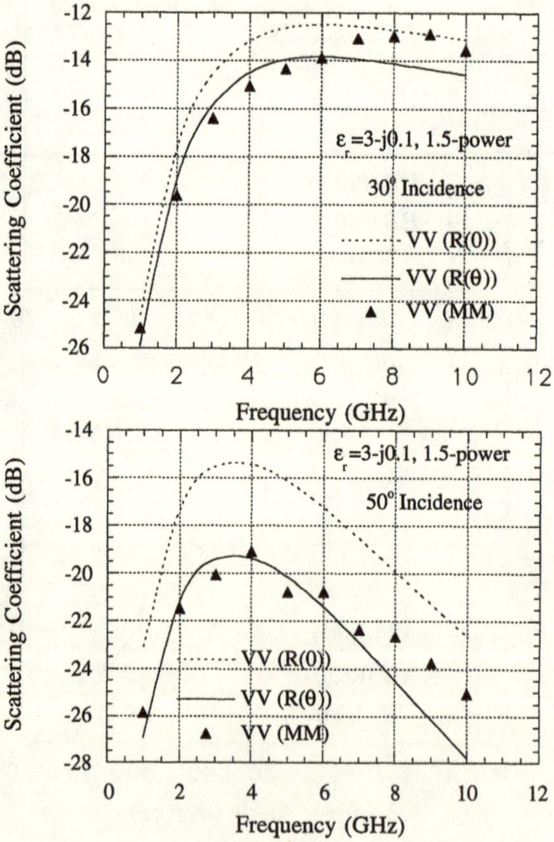

Figure 6.15 Comparisons between the moment method calculations and IEM in backscattering plotted versus frequency for a non-Gaussian correlated surface, 30 and 50 degree incidence and vertical polarization.

Figure 6.15 shows plots of the backscattering coefficients versus frequency for vertical polarization. It appears that the IEM with $R(\theta)$ is applicable to the lower portion of the frequency axis up to 6 GHz at 30 degrees and to 7 GHz at 50 degrees. The other form of the IEM model is applicable after 8 GHz at 30 degrees incidence but it is not clear what is the applicable frequency at 50 degrees incidence due to fluctuations in the MM calculations. In Figure 6.16 we show similar plots for horizontal polarization. Here, at 30 degrees incidence the IEM with $R(\theta)$ seems to be valid till 9 GHz but at 50 degree incidence it appears to be applicable only to 6 GHz. This lack of a clearer region is again the result of statistical fluctuation.

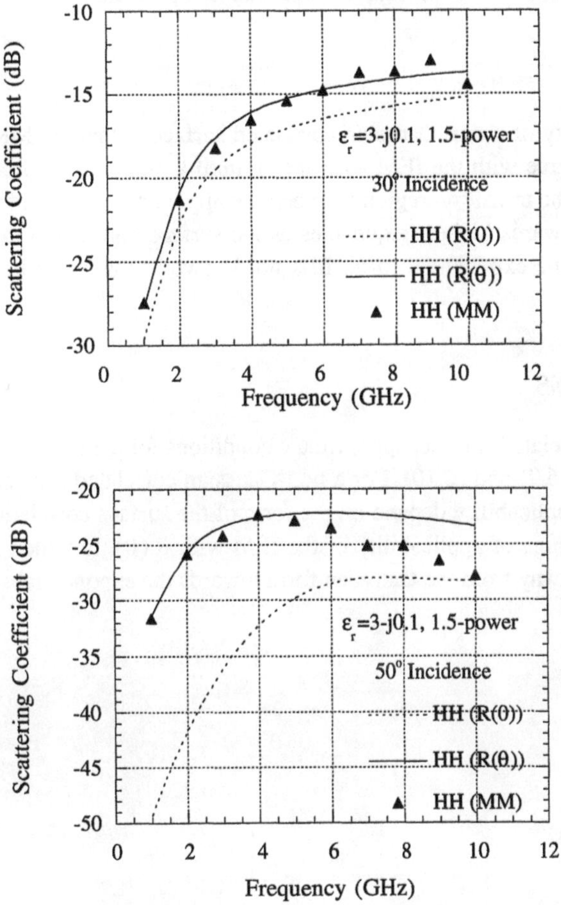

Figure 6.16 Comparisons between the moment method calculations and IEM in backscattering plotted versus frequency for 30 and 50 degree incidence and horizontal polarization.

In all cases, it is safe to conclude that the IEM with $R(\theta)$ is valid to a higher frequency (between 6 and 7 GHz) for non-Gaussian correlation than for Gaussian correlation. To confirm the last statement we plot in Figure 6.17 and Figure 6.18 the backscattering coefficients versus the incident angle for vertical and horizontal polarizations, respectively, at 6, 7 and 8 GHz. It is seen that good agreement is obtained at 6 and 7 GHz for vertical polarization and some deviation occurs only at 8 GHz, indicating that the range is now wider as compared with the Gaussian correlated case given in Figure 6.12. For horizontal polarization (Figure 6.18) the deviation of IEM with $R(\theta)$ from the MM calculations occurs at 7 GHz especially at large angles of incidence. This leads to the following rule of thumb for IEM with $R(\theta)$ when applied to the non-Gaussian correlated surface as described in this section:

$$k^2\sigma L < 1.6\sqrt{\varepsilon_r} \tag{6.12}$$

Another property of a non-Gaussian correlated surface is that the MM calculations generally do not agree with the IEM with $R(0)$ until a much higher frequency. Thus, we anticipate that the transition region between the applicable regions of the two forms of IEM to move towards higher frequencies as the surface correlation moves from the Gaussian towards the exponential form. It is not known at this point how wide is the transition region.

6.5 CONCLUSIONS

For a Gaussian correlated surface, approximate conditions for applying the two forms of IEM are given by (6.9) and (6.10). For a non-Gaussian correlated surface we find that the conditions of applicability depend on the form of the surface correlation. Generally, we anticipate the range of applicability of the IEM with $R(\theta)$ to widen as the surface correlation moves away from the Gaussian form towards the exponential form.

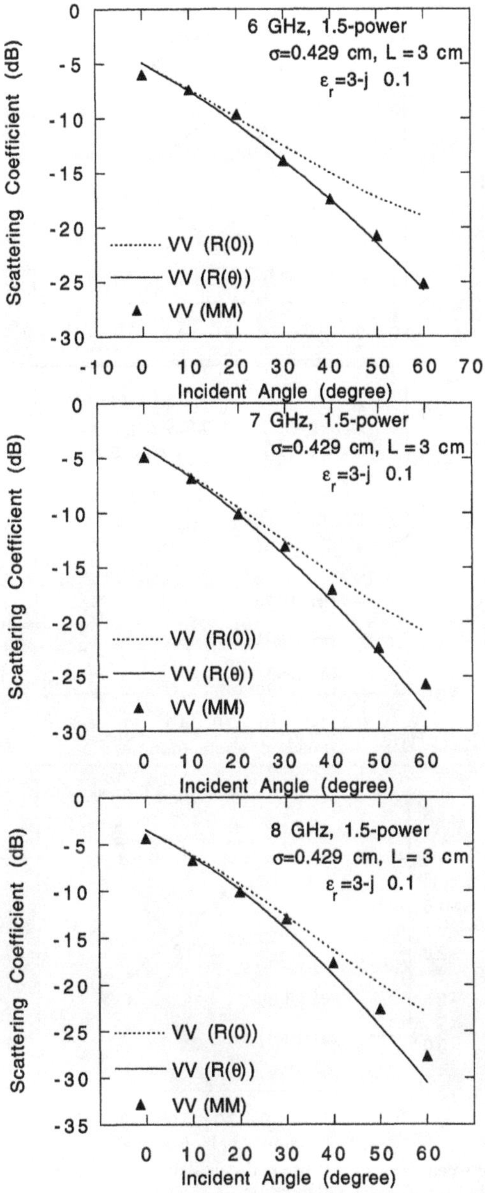

Figure 6.17 Comparisons between the moment method calculations and IEM in backscattering plotted versus angle at 6, 7 and 8 GHz for vertical polarization.

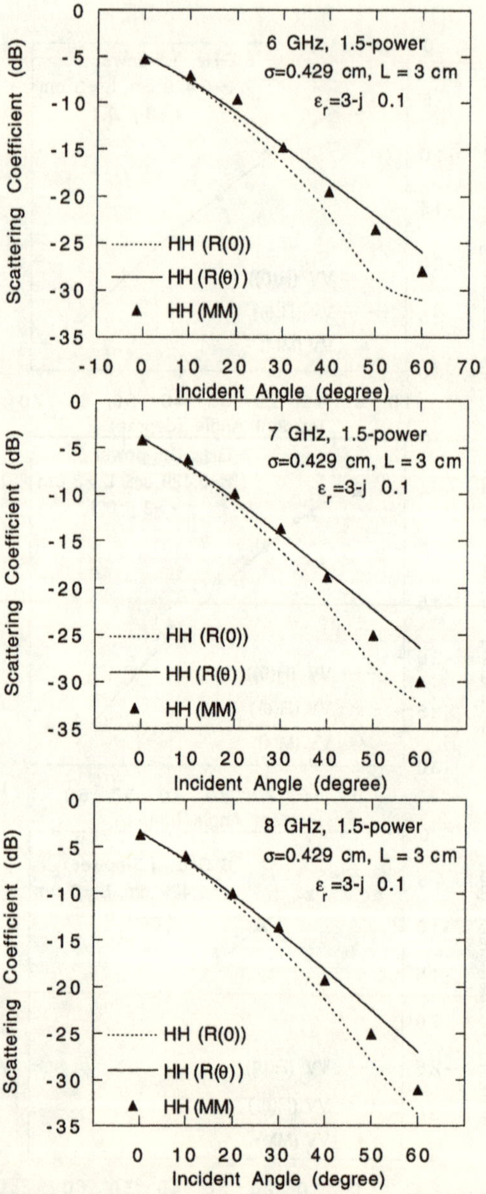

Figure 6.18 Comparisons between the moment method calculations and IEM in backscattering plotted versus angle at 6, 7 and 8 GHz for horizontal polarization

REFERENCES

Beyer, W.H., *CRC Standard Mathematical Tables*, CRC Press, Boca Raton, FL, 1986, p. 509.

Chen, M.F., and S.Y. Bai, "Computer Simulation of Wave Scattering from a Dielectric Random Surface in Two Dimensions—Cylindrical Case," *J. Electromagnetic Waves and Applications*, Vol. 4, no. 10, 1990, pp. 963–982.

Chen, M.F., and A.K. Fung, "A Numerical Study of the Regions of Validity of the Kirchhoff and Small Perturbation Rough Surface Scattering Models," *Radio Science*, Vol. 23, no. 2, 1988, pp. 163–170.

Chen, M.F., K.S. Chen and A.K. Fung, "A Study of the Validity of the Integral Equation Model by Moment Method Simulation—Cylindrical Case," *Remote Sensing of Environment*, Vol. 29, 1989, pp. 217–228.

Fung, A.K., "Effect of Cell Size on Radar Clutter Statistics, *RADC Tech. Report* , 1990, pp. 90–235.

Fung, A.K., and K.S. Chen, "Numerical Simulation of Scattering from Simple and Composite Random Surfaces," *J. Opt. Soc. Am.* A/Vol. 2, no. 12, 1985, pp. 2274–2284.

Fung, A.K., Z. Li and K.S. Chen, "Backscattering from a Randomly Rough Dielectric Surface," *IEEE Trans. Geoscience and Remote Sensing*, Vol., 30, no. 2, 1992, pp. 356–369.

Rochier, J.D., A.J. Blanchard and M.F. Chen, The Generation of Surface Targets with Specified Surface Statistics," *International Journal of Remote Sensing,* Vol. 10, no. 7, 1989, pp. 1155–1174

Ulaby, F. T., R.K. Moore and A.K. Fung, *Microwave Remote Sensing*, Vol. 2, Artech House, MA, 1982, Chapter 7.

Chapter 7

Scattering from Oceanlike Surfaces

7.1 INTRODUCTION

The objective of this chapter is to extend the surface scattering model in Chapter 4 to include non-Gaussian distributed surfaces. Such a model will be shown to consist of two parts: one part is proportional to the surface roughness spectrum, and the other to the surface bispectrum. The bispectrum part comes into the model when the third-order surface statistics are included. While the second-order statistics account for the wind directional dependence, the third-order statistics account for the dependence on the sense of direction of the wind. Thus, it is the critical part for explaining the difference between upwind and downwind observations. In general, the bispectrum is a complex quantity and the asymmetric effect of the sea surface is represented by its imaginary part. The model characteristics such as polarization and azimuthal dependence are illustrated through numerical calculations. The predictions of the model are compared with field measurements.

Over the last twenty years considerable effort has been devoted to the development of scattering models to support data interpretation from ocean surface. Among them, the two-scale model has been widely used [Fung and Lee, 1982; Plant, 1986], although it is really a very restrictive model [Chen and Fung, 1988]. The integral equation model developed by Fung and Pan [1987] has shown a wide range of validity [Chen et al., 1989; Fung and Chen, 1992]. The same technique has been extended to dielectric surfaces as shown in Chapter 4. In this chapter further extension is made of the integral equation model to include the asymmetric surface properties by carrying the surface statistics up to the third order in scattering calculations.

The description of a random surface including asymmetric surface effects has been studied both theoretically and experimentally by Longuet-Higgins [1963].The skewness of the surface height related to the significant slope of the waves was also discussed by Srokosz and Longuet-Higgins [1980]. The bispectrum of the ocean waves was studied by many researchers [Hasselmann et al., 1963] and has been applied to study the nonlinear interactions of the random wave. Roden and Bendliner [1973] used cross-bispectra to investigate profiles of oceanographic variables, such as density and salinity. A numerical computation of bispectra arising from weak nonlinear resonant interactions of internal waves was reported by McComas and Briscoe [1980]. German et al. [1980] used bispectral technique to analyze the sea-level variations. Elgar and Guza [1985] used the bispectrum to study the nonlinear dynamics of waves shoaling between 9 and 1 m water depths. More details on bispectrum estimation have been given by Nikias and Raghuveer [1987]. Despite these various applications of the bispectral analysis, the surface bispectrum was not included in most of the sea surface scattering models, possibly because they represent higher order effects in scattering. The surface scattering model under the geometric optics approximation [Barrick, 1968] shows that backscattering is in direct proportion to surface slope distribution. Thus, skewness in surface slope that appears through the surface slope distribution function [Longuet-Higgins, 1982] is included in the geometric optics model. However, such a model cannot adequately account for the upwind/downwind difference [Cox and Munk, 1954]. In 1990, Chen and Fung extended the standard small perturbation model to include the third-order surface statistics. They [Fung and Chen, 1991] also modified the Kirchhoff model by including the third-order statistics of the random surface to account for both skewness in surface heights and surface slopes. They demonstrated that such a Kirchhoff model reduced to the geometric optics model in the high-frequency limit containing a slope distribution function with skewness as given in Longuet-Higgins [1982] and Cox and Munk [1954]. Although these efforts have extended the model, the basic restrictions on the Kirchhoff model to the high-frequency region and on the small perturbation model to the low-frequency region remain unchanged. Therefore, it is desirable to incorporate the bispectrum into a scattering model with a wider range of applicability. In particular, the model given in Chapter 4 is extended in this chapter to include the third-order surface statistics represented by the surface bispectrum or the skewness function.

In the next section, the definition of the surface roughness spectrum and bispectrum along with their basic properties are summarized. The model development is given in Section 7.3. In Section 7.4, numerical calculations of the backscattering coefficient are given to illustrate the effect of surface skewness on backscattering. Comparisons with measurements are provided to illustrate the model predictions. The concluding remarks are presented in Section 7.5.

7.2 SURFACE PROPERTIES

This section gives the definitions of the surface roughness spectrum and bispectrum. Then, their basic properties are discussed. Let the surface function $z(x, y)$ be a real stationary random process with zero mean, $\langle z \rangle = 0$, variance $\langle z^2 \rangle = \sigma^2$ and third-order moment $\langle z^3 \rangle = \mu_3$. Then, the moments of the process are related to its second-order cumulant function (correlation function) $\rho(\tau_x, \tau_y)$ and its third-order cumulant function (bicorrelation function) $C(\tau_x, \tau_y; \varsigma_x, \varsigma_y)$ as

$$\langle z(x, y) z(x+\tau_x, y+\tau_y) \rangle = \sigma^2 \rho(\tau_x, \tau_y) \tag{7.1}$$

and

$$\langle z(x,y)\,z(x+\tau_x, y+\tau_y)\,z(x+\varsigma_x, y+\varsigma_y)\rangle = \mu_3 C(\tau_x, \tau_y; \varsigma_x, \varsigma_y)$$
$$= \sigma^3 S(\tau_x, \tau_y; \varsigma_x, \varsigma_y) \qquad (7.2)$$

In the above, $S(\tau_x, \tau_y; \varsigma_x, \varsigma_y)$ is referred to as the *skewness function* because it represents the distribution of the surface skewness coefficient μ_3/σ^3 as given in Longuet-Higgins [1963]. This function is introduced because the skewness coefficient rather than the third moment is the quantity commonly used to measure departure from symmetry.

The sea surface roughness spectrum is defined as the Fourier transform of the surface correlation:

$$W(k_x, k_y) = \frac{1}{2\pi}\int \sigma^2 \rho(\tau_x, \tau_y) \exp[-jk_x\tau_x - jk_y\tau_y]\, d\tau_x d\tau_y \qquad (7.3)$$

Note that this definition of surface spectrum does not include normalization by the variance as in earlier chapters because most reported sea spectra are not normalized. Since the surface correlation is a centro-symmetric function, so is the surface roughness spectrum. As a result, it is also real and does not carry any information regarding surface asymmetry. Additional basic properties of the directional sea spectrum are summarized in Fung et al. [1989].

The bispectrum is the Fourier transform of the surface bicorrelation function:

$$B(k_x, k_y; \bar{k}_x, \bar{k}_y) = \frac{1}{(2\pi)^2}\int \sigma^3 S(\tau_x, \tau_y; \varsigma_x, \varsigma_y)$$
$$\exp[-jk_x\tau_x - jk_y\tau_y - j\bar{k}_x\varsigma_x - j\bar{k}_y\varsigma_y]\, d\tau_x d\tau_y d\varsigma_x d\varsigma_y \qquad (7.4)$$

In general, the bispectrum B is complex and is a function of four variables. In modeling calculations two special cases of the skewness function are of interest. When $\tau_x = \varsigma_x$ and $\tau_y = \varsigma_y$, we have

$$\langle z(x,y)\, z^2(x+\tau_x, y+\tau_y)\rangle = \sigma^3 S(\tau_x, \tau_y) \qquad (7.5)$$

and when $\varsigma_x = \varsigma_y = 0$, we have

$$\langle z^2(x,y)\, z(x+\tau_x, y+\tau_y)\rangle = \langle z^2(x'-\tau_x, y'-\tau_y)\, z(x', y')\rangle = \sigma^3 S(-\tau_x, -\tau_y) \qquad (7.6)$$

Note that the variables have reduced from four to two. For a real process *z(x, y)*, it follows that $B(k_x, k_y) = B^*(-k_x, -k_y)$ where * is the complex conjugate operator. The results in (7.5) and (7.6) suggest that to understand the impact of the negative sign in the arguments we should decompose the skewness function into a symmetric part, $S_s(\tau_x, \tau_y)$, and an asymmetric part, $S_a(\tau_x, \tau_y)$, as

$$S_s(\tau_x, \tau_y) = \frac{S(\tau_x, \tau_y) + S(-\tau_x, -\tau_y)}{2} \quad (7.7)$$

$$S_a(\tau_x, \tau_y) = \frac{S(\tau_x, \tau_y) - S(-\tau_x, -\tau_y)}{2} \quad (7.8)$$

such that the skewness function can be written as

$$S(\tau_x, \tau_y) = S_s(\tau_x, \tau_y) + S_a(\tau_x, \tau_y) \quad (7.9)$$

Since the bispectrum is the Fourier transform of $S(\tau_x, \tau_y)$, we can use the properties of $S_s(\tau_x, \tau_y)$ and $S_a(\tau_x, \tau_y)$ to write

$$B(k_x, k_y) = B_s(k_x, k_y) + jB_a(k_x, k_y) \quad (7.10)$$

or in polar form

$$B(K, \phi) = B_s(K, \phi) + jB_a(K, \phi) \quad (7.11)$$

where

$$B_s(k_x, k_y) = \frac{1}{2\pi}\int \sigma^3 S_s(\tau_x, \tau_y)\, e^{-j(k_x\tau_x + k_y\tau_y)}\, d\tau_x d\tau_y \quad (7.12)$$

$$jB_a(k_x, k_y) = \frac{1}{2\pi}\int \sigma^3 S_a(\tau_x, \tau_y)\, e^{-j(k_x\tau_x + k_y\tau_y)}\, d\tau_x d\tau_y \quad (7.13)$$

Thus, the real part of the bispectrum relates to the symmetric property of the random surface, while the imaginary part represents its asymmetric property [Masuda and Kuo, 1981; Elgar, 1987]. In another words,

$$B_s(K, \phi) = B_s(K, \phi + \pi)$$

$$B_a(K, \phi) = -B_a(K, \phi + \pi)$$

Finally, the surface height variance and the third order moment can be retrieved from the following expressions, respectively:

$$\sigma^2 = \int_{-\infty}^{\infty} W(k_x, k_y)\, dk_x dk_y \quad (7.14)$$

$$\mu_3 = \int_{-\infty}^{\infty} B(k_x, k_y)\, dk_x dk_y = \int_{-\infty}^{\infty} B_s(k_x, k_y)\, dk_x dk_y \tag{7.15}$$

Note that in (7.15), only the real part of the bispectrum remains because the imaginary part integrates to zero. This is consistent with the fact that the third moment is real for real random process, $z(x, y)$ and it does not bear information on surface asymmetry. Additional details about the bispectrum and its comparisons with the spectrum is given in Appendix 7A.

7.3 MODEL DEVELOPMENT

A detailed development of the surface scattering model based on the integral equation method has been given in Chapter 4. However, only second-order surface statistics are included. To provide the reader with a brief summary of that development we show in the following subsections the estimates of the surface tangential fields needed to calculate the scattered far-zone field, the far-zone scattered field expressions and the average power expression that includes up to the third-order surface statistics.

7.3.1 Surface Tangential Fields

Based on the procedures given in Chapter 4 and equations (4.43), (4.44), (4.51) and (4.55), the approximate vertically polarized tangential surface fields for finitely conducting surface are given by

$$(\hat{n} \times \vec{E})_v \approx (1 - R_{\parallel})\, (\hat{n} \times \hat{v} E^i) - \frac{1}{4\pi} (1 - R_{\parallel})\, \hat{n} \times \int \vec{\mathcal{E}}_v\, ds$$
$$- \frac{1}{4\pi} (1 + R_{\parallel})\, \hat{n} \times \int \vec{\mathcal{E}}_{vt}\, ds' \tag{7.16}$$

$$(\hat{n} \times \vec{H})_v \approx (1 + R_{\parallel})\, \hat{n} \times (\hat{k}_i \times \hat{v} E^i / \eta) + \frac{1}{4\pi} (1 + R_{\parallel})\, \hat{n} \times \int \vec{\mathcal{H}}_v\, ds$$
$$+ \frac{1}{4\pi} (1 - R_{\parallel})\, \hat{n} \times \int \vec{\mathcal{H}}_{vt}\, ds' \tag{7.17}$$

where $\hat{n}$ is the unit normal vector to the random surface pointing into the upper medium; E^i denotes the incident electric field; η is the intrinsic impedance in the upper medium; $\hat{k}_i$ is the incident unit vector and $R_{\parallel}$ is the Fresnel reflection coefficient for vertical polarization. From (4.7)–(4.10) and the approximate tangential field expressions in Section 4.3, the contents of the integrands in (7.16) and (7.17) are

$$\vec{\mathcal{E}}_v = jk\eta\, (1 + R_{\parallel})\, (\hat{n}' \times \vec{H}_v^i)\, G - (1 - R_{\parallel})\, (\hat{n}' \times \vec{E}_v^i) \times \nabla' G$$

$$-(1+R_{\|})\,(\hat{n}'\cdot\vec{E}_v^i)\,\nabla' G \tag{7.18}$$

$$\vec{\mathcal{E}}_{vt} = jk_t\eta_t(1+R_{\|})\,(\hat{n}'\times\vec{H}_v^i)\,G_t-(1-R_{\|})\,(\hat{n}'\times\vec{E}_v^i)\times\nabla' G_t$$
$$-\frac{1}{\varepsilon_r}(1+R_{\|})\,(\hat{n}'\cdot\vec{E}_v)\,\nabla' G_t \tag{7.19}$$

$$\vec{\mathcal{H}}_v = j\frac{k}{\eta}(1-R_{\|})\,(\hat{n}'\times\vec{E}_v^i)\,G+(1+R_{\|})\,(\hat{n}'\times\vec{H}_v^i)\times\nabla' G$$
$$+(1-R_{\|})\,(\hat{n}'\cdot\vec{H}_v^i)\,\nabla' G \tag{7.20}$$

$$\vec{\mathcal{H}}_{vt} = j\frac{k_t}{\eta_t}(1-R_{\|})\,(\hat{n}'\times\vec{E}_v^i)\,G_t+(1+R_{\|})\,(\hat{n}'\times\vec{H}_v^i)\times\nabla' G_t$$
$$+\frac{1}{\mu_r}(1-R_{\|})\,(\hat{n}'\cdot\vec{H}_v)\,\nabla' G_t \tag{7.21}$$

where $\hat{n}'$ has the same meaning as $\hat{n}$ but it is located at integration points on the surface, $\vec{n} = \hat{n}(1+Z_x^2+Z_y^2)^{1/2}$, Z_x, Z_y are the surface slopes and G, G_t are the Green's functions in the upper and lower medium, respectively. The symbols ε_r, μ_r are the permittivity and permeability of the lower medium relative to the upper medium as illustrated in Figure 7.1.

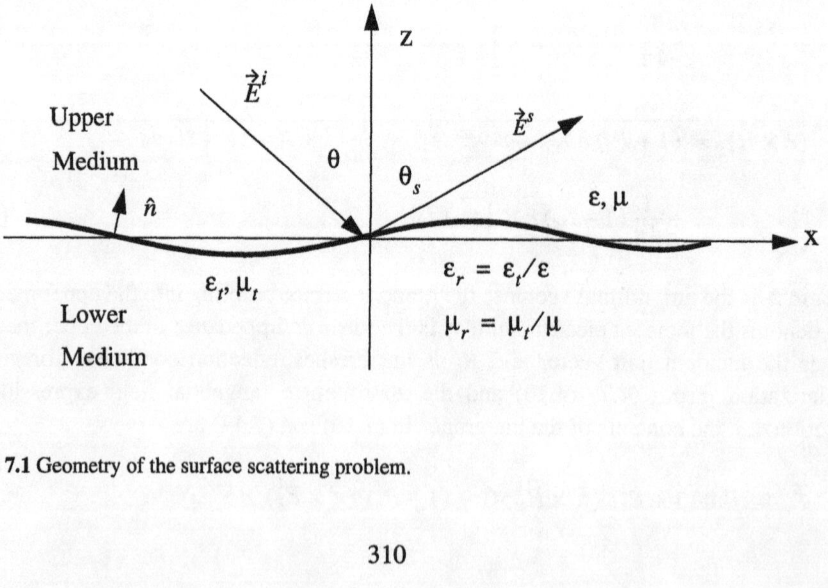

Figure 7.1 Geometry of the surface scattering problem.

Similarly, we have the approximate magnetic and electric tangential fields for horizontal polarization from (4.45), (4.46), (4.56) and (4.57) as

$$(\hat{n}\times\vec{E})_h \approx (1+R_\perp)\,(\hat{n}\times\vec{E}_h^i) - \frac{1}{4\pi}(1+R_\perp)\,\hat{n}\times\int\vec{\mathcal{E}}_h dS'$$

$$+\frac{1}{4\pi}(1-R_\perp)\,\hat{n}_1\times\int\vec{\mathcal{E}}_{ht}\,dS' \tag{7.22}$$

$$(\hat{n}\times\vec{H})_h \approx (1-R_\perp)\,(\hat{n}\times\vec{H}_h^i) + \frac{1}{4\pi}(1-R_\perp)\,\hat{n}\times\int\vec{\mathcal{H}}_h dS'$$

$$-\frac{1}{4\pi}(1+R_\perp)\,\hat{n}\times\int\vec{\mathcal{H}}_{ht}\,dS' \tag{7.23}$$

Analogous to (7.18)-(7.21), the integrands in (7.22) and (7.23) are

$$\vec{\mathcal{E}}_h = jk\eta\,(1-R_\perp)\,(\hat{n}'\times\vec{H}_h^i)\,G - (1+R_\perp)\,(\hat{n}'\times\vec{E}_h^i)\times\nabla'G$$

$$-(1-R_\perp)\,(\hat{n}'\cdot\vec{E}_h^i)\,\nabla'G \tag{7.24}$$

$$\vec{\mathcal{E}}_{ht} = jk_t\eta_t(1-R_\perp)\,(\hat{n}'\times\vec{H}_h^i)\,G_t - (1+R_\perp)\,(\hat{n}'\times\vec{E}_h^i)\times\nabla'G_t$$

$$-\frac{1}{\varepsilon_r}(1-R_\perp)\,(\hat{n}'\cdot\vec{E}_h^i)\,\nabla'G_t \tag{7.25}$$

$$\vec{\mathcal{H}}_h = j\frac{k}{\eta}(1+R_\perp)\,(\hat{n}'\times\vec{E}_h^i)\,G + (1-R_\perp)\,(\hat{n}'\times\vec{H}_h^i)\times\nabla'G$$

$$+\,(1+R_\perp)\,(\hat{n}'\cdot\vec{H}_h^i)\,\nabla'G \tag{7.26}$$

$$\vec{\mathcal{H}}_{th} = j\frac{k_t}{\eta_t}(1+R_\perp)\,(\hat{n}'\times\vec{E}_h^i)\,G_t + (1-R_\perp)\,(\hat{n}'\times\vec{H}_h^i)\times\nabla'G_t$$

$$+\frac{1}{\mu_r}(1+R_\perp)\,(\hat{n}'\cdot\vec{H}_h^i)\,\nabla'G_t \tag{7.27}$$

where $R_\perp$ is the Fresnel reflection coefficient for horizontal polarization. In view of (7.16), (7.17) and (7.22), (7.23), the approximate tangential surface fields, in general, consist of two components: the one without the integral sign is recognized as the tangential Kirchhoff fields, and the other with the integral sign is a tangential surface field complementary to the Kirchhoff field. Next, we calculate the far-zone scattered fields from the approximate tangential surface fields.

7.3.2 Far-Zone Scattered Fields

Making use of the Stratton-Chu integral [Ulaby et al., 1982], we can write the far- zone scattered fields in the upper medium as

$$\vec{E}_p^s = -\frac{jke^{-jkR}}{4\pi R}\hat{k}_s \times \int [(\hat{n}\times\vec{E}_p) - \eta\hat{k}_s \times (\hat{n}\times\vec{H}_p)] \exp[j(\vec{k}_s\cdot\vec{r})]\, dS \tag{7.28}$$

where p denotes the incident polarization equal to $\hat{v}$ (vertical) or $\hat{h}$ (horizontal).

$$\hat{h} = \hat{y} \tag{7.29}$$

$$\hat{v} = \hat{x}\cos\theta + \hat{z}\sin\theta \tag{7.30}$$

and $\vec{k}_s = k\hat{k}_s$. In backscattering, $\hat{k}_s = -\hat{x}\sin\theta + \hat{z}\cos\theta = -\hat{k}_i$.

Since there are two components in the tangential surface field, the far-zone backscattered field can be similarly separated into two corresponding components: one is the Kirchhoff scattered field and the other is its complementary field denoted as

$$E^s_{pp} = E^k_{pp} + E^c_{pp} \tag{7.31}$$

where

$$E^k_{pp} = CE_0 \int f_{pp} \exp\left[-j2\vec{k}_i\cdot\vec{r}\right] dxdy \tag{7.32}$$

$$E^c_{pp} = \frac{CE_0}{8\pi^2}\int F_{pp} \exp\left[-j\vec{k}_i\cdot\vec{r} - j\vec{k}_i\cdot\vec{r}' + ju(x-x') + jv(y-y')\right] dxdydx'dy'dudv \tag{7.33}$$

where $C = -(jke^{-jkR})/(4\pi R)$.

The bistatic expressions for the field coefficients f_{pp}, F_{pp} have been given in Appendix 4B. In backscattering under single scattering condition these coefficients simplify to

$$f_{vv} = (2R_{\parallel})/\cos\theta \tag{7.34}$$

$$f_{hh} = -(2R_{\perp})/\cos\theta \tag{7.35}$$

$$F_{vv} = \frac{2\sin^2\theta}{\cos\theta}\left[\left(1 - \frac{\varepsilon_r\cos^2\theta}{\varepsilon_r - \sin^2\theta}\right)(1-R_{\parallel})^2 + \left(1-\frac{1}{\varepsilon_r}\right)(1+R_{\parallel})^2\right] \tag{7.36}$$

$$F_{hh} = \frac{2\sin^2\theta}{\cos\theta}\left[4R_{\perp} - \left(1 - \frac{1}{\varepsilon_r}\right)(1+R_{\perp})^2\right] \tag{7.37}$$

7.3.3 Scattered Power and Backscattering Coefficient

Once the scattered field is found, we can follow the procedure in Chapter 4 and carry our computation up to the third-order surface statistics to obtain the incoherently scattered power defined as

$$\begin{aligned} P_{pp} &= \langle E^s{}_{pp} E^s{}_{pp}{}^* \rangle - \langle E^s{}_{pp} \rangle \langle E^s{}_{pp}{}^* \rangle \\ &= \langle E^k{}_{pp} E^k_{pp}{}^* \rangle - \langle E^k{}_{pp} \rangle \langle E^k{}_{pp}{}^* \rangle + 2\mathrm{Re}\left[\langle E^k{}_{pp}{}^* E^c_{pp} \rangle - \langle E^k{}_{pp}{}^* \rangle \langle E^c{}_{pp} \rangle\right] \\ &\quad + \langle E^c{}_{pp} E^c{}_{pp}{}^* \rangle - \langle E^c{}_{pp} \rangle \langle E^c{}_{pp}{}^* \rangle \end{aligned} \tag{7.38}$$

where < > means ensemble average and * is the complex conjugate operator. Each term on the right-hand side of (7.38) can be evaluated as follows:

$$\begin{aligned} P^k_{pp} &= \langle E^k{}_{pp} E^k_{pp}{}^* \rangle - \langle E^k{}_{pp} \rangle \langle E^k{}_{pp}{}^* \rangle \\ &= |CE_0 f_{pp}|^2 \{ \langle \iint \exp\left[-j\vec{k}_i \cdot (\vec{r} - \vec{r}')\right] dx'dy'dxdy \rangle \\ &\quad - \left| \langle \int \exp(-j\vec{k}_i \cdot \vec{r})\, dxdy \rangle \right|^2 \} \end{aligned} \tag{7.39}$$

where the ensemble averages from Appendix 4C are

$$\langle e^{-j2k_z z} \rangle = \exp\left[-2k_z^2\sigma^2 + (j4k_z\sigma^3)/3\right] \tag{7.40}$$

$$\langle e^{-j2k_z(z-z')} \rangle = \exp\{4k_z^2\sigma^2[\rho(\zeta,\varsigma) - 1] + jk_z^3\sigma^3 s_a(\zeta,\varsigma)\} \tag{7.41}$$

and $\xi = x - x', \varsigma = y - y'$.

Next, we consider the cross term

$$P^{kc}_{pp} = 2\mathrm{Re}\left[\langle E^c_{pp} E^k_{pp}{}^* \rangle - \langle E^c_{pp} \rangle \langle E^k_{pp}{}^* \rangle\right] = \frac{1}{(2\pi)^2} |CE_0|^2 \mathrm{Re}\{ \int F_{pp} f_{pp}{}^*$$

$$\iint \langle \exp\left[-j\vec{k}_i \cdot (\vec{r} + \vec{r}' - 2\vec{r}'') + ju(x - x') + jv(y - y')\right] dxdydx'dy'dx''dy''dudv \rangle$$

$$- \langle \iint \exp\left[-j\vec{k}_i \cdot (\vec{r} + \vec{r}') + ju(x - x') + jv(y - y')\right] \rangle$$

$$\langle \int \exp(j2\vec{k}_i \cdot \vec{r}'') \rangle dxdydx'dy'dx''dy''dudv \} \tag{7.42}$$

The following averages from Appendix 4C are needed:

$$\langle e^{-jk_z(z+z')} \rangle = e^{-k_z^2\sigma^2[1+\rho(\xi-\xi',\varsigma-\varsigma')] + (jk_z^3\sigma^3)/3 + jk_z^3\sigma^3 S_s(\xi-\xi',\varsigma-\varsigma')} \tag{7.43}$$

$$\langle e^{jk_z(-z-z'+2z'')} \rangle = e^{k_z^2\sigma^2[2\rho(\xi,\varsigma)+2\rho(\xi',\varsigma')-\rho(\zeta,\iota)-3]}$$

$$e^{-jk_z^3\sigma^3[2S_a(\xi,\varsigma)+2S_a(\xi',\varsigma')+S(\xi,\varsigma)+S(\xi',\varsigma')-S_s(\zeta,\iota)+1]} \tag{7.44}$$

where $\xi = x - x''$, $\varsigma = y - y''$, $\xi' = x' - x''$, $\varsigma' = y' - y''$, $\zeta = x - x'$, $\iota = y - y'$.

Finally, the complementary term is

$$P_{pp}^c = \langle E_{pp}^c E_{pp}^{c\,*} \rangle - \langle E_{pp}^c \rangle \langle E_{pp}^{c\,*} \rangle = \frac{1}{(8\pi^2)^2} |CE_0|^2 \{ \int\int F_{pp} F_{pp}^*$$

$$\int\int\int\int \langle \exp[-j\vec{k}_i \cdot (\vec{r} + \vec{r}' - \vec{r}'' - \vec{r}''') + ju(x-x') + ju'(x''-x''')] \rangle$$

$$\exp[jv(y-y') - jv'(y''-y''')] dxdydx'dy'dx''dy''dx'''dy'''dudvdu'dv'$$

$$-\left| \langle \int\int\int F_{pp} \exp[-j\vec{k}_i \cdot (\vec{r} + \vec{r}') + ju(x-x') + jv(y-y')] dxdydx'dy'dudv \rangle \right|^2 \}$$

$$\tag{7.45}$$

The averages needed here from Appendix 4C are

$$\langle e^{jk_z(-z-z'+z''+z''')} \rangle = \exp\{k_z^2\sigma^2[\rho(\xi+\tau,\varsigma+\kappa) + \rho(\xi'-\tau,\varsigma'-\kappa)]\}$$

$$\exp\{k_z^2\sigma^2[-2-\rho(\xi-\xi'+\tau,\varsigma-\varsigma'+\kappa) - \rho(\tau,\kappa) + \rho(\xi,\varsigma) + \rho(\xi',\varsigma')]\}$$

$$\exp\{-jk_z^3\sigma^3[S_s(\xi,\varsigma) + S_s(\xi',\varsigma') + S_a(\xi-\xi'+\tau,\varsigma-\varsigma'+\kappa)]\}$$

$$\exp\{-jk_z^3\sigma^3[S_a(\xi+\tau,\varsigma+\kappa) + S_a(\xi'-\tau,\varsigma'-\kappa) + S_a(\tau,\kappa)]\} \tag{7.46}$$

where

$$\xi = x - x'', \varsigma = y - y'', \xi' = x' - x''', \varsigma' = y' - y''', \tau = x'' - x''', \kappa = y'' - y'''$$

After a lengthy but straightforward manipulation, we obtain the expression for the average scattered power as

$$P^{k}_{pp} = 2\pi A_0 E^2{}_0 |C|^2 |f_{pp}|^2 e^{-4k^2\sigma^2\cos^2\theta}$$

$$\cdot \sum_{n=1}^{\infty} \left[\frac{(4k^2\cos^2\theta)^n}{n!} W^{(n)}(K,\phi) + \frac{(-8k^3\cos^3\theta)^n}{n!} B_a^{(n)}(K,\phi) \right] \qquad (7.47)$$

$$P^{kc}_{pp} = \pi A_0 E^2{}_0 |C|^2 f_{pp}{}^* F_{pp} e^{-3k^2\sigma^2\cos\theta}$$

$$\cdot \sum_{n=1}^{\infty} \left[\frac{(2k^2\cos^2\theta)^n}{n!} W^{(n)}(K,\phi) + \frac{(-3k^3\cos^3\theta)^n}{n!} B_a^{(n)}(K,\phi) \right.$$

$$\left. + (-1)^{n+1} \frac{(3k^3\cos^3\theta)^{2n}}{(2n)!} B_s^{(2n)}(K,\phi) \right] \qquad (7.48)$$

$$P^{c}_{pp} = 0.5\pi A_0 E^2{}_0 |C|^2 |F_{pp}|^2 e^{-2k^2\sigma^2\cos^2\theta}$$

$$\cdot \sum_{n=1}^{\infty} \left[\frac{(k^2\cos^2\theta)^n}{n!} W^{(n)}(K,\phi) + \frac{(-k^3\cos^3\theta)^n}{n!} B_a^{(n)}(K,\phi) \right] \qquad (7.49)$$

The above result shows that the scattered power consists of two types of terms: one associated with the roughness spectrum; the other with the bispectrum. Next, we write the backscattering coefficient in two parts each involving one type of terms. From Chapter 4, the backscattering coefficient is defined in terms of the average power expression as

$$\sigma^0_{pp} = 4\pi R^2 \frac{P_{pp}}{E^2{}_0 A_0} \qquad (7.50)$$

Thus, the backscattering coefficient can be written as

$$\sigma^0{}_{pp} = \sigma^0{}_{pp}(N) + \sigma^0{}_{pp}(S) \qquad (7.51)$$

where

$$\sigma^0{}_{pp}(N) = \frac{k^2}{2} exp(-2k^2\sigma^2\cos^2\theta) \sum_{n=1}^{\infty} \frac{|I^n{}_{pp}|^2}{n!} W^{(n)}(K,\phi) \qquad (7.52)$$

with

$$I^n = (2k\cos\theta)^n f_{pp} \exp(-k^2\sigma^2\cos^2\theta) + \frac{(k\cos\theta)^n}{2} F_{pp}$$

and

$$\sigma^0_{pp}(S) = \frac{k^2}{2}|f_{pp}|^2 \exp(-4k^2\sigma^2\cos^2\theta) \sum_{n=1}^{\infty} \frac{(-8k^3\cos^3\theta)^n}{n!} B_a^{(n)}(K, \phi)$$

$$+ \frac{k^2}{2}\mathrm{Re}(f_{pp}{}^*F_{pp}) \exp(-3k^2\sigma^2\cos^2\theta) \sum_{n=1}^{\infty} \left[\frac{(-3k^3\cos^3\theta)^n}{n!} B_a^{(n)}(K, \phi)\right.$$

$$\left. + (-1)^{n+1}\frac{(k^3\cos^3\theta)^{2n}}{(2n)!} B_s^{(2n)}(K, \phi)\right]$$

$$+ \frac{k^2}{8}|F_{pp}|^2 \exp(-2k^2\sigma^2\cos^2\theta) \sum_{n=1}^{\infty} \frac{(-k^3\cos^3\theta)^n}{n!} B_a^{(n)}(K, \phi) \tag{7.53}$$

In the above, both the real and imaginary parts of the surface bispectrum appear in the backscattering coefficient expression. It is seen that, when $n = 1$, only the imaginary part appears, while for $n \geq 2$, both parts appear. The $n \geq 2$ terms are of higher order compared to the $n = 1$ terms. In addition, B_a represents asymmetry while B_s is symmetric, just like the surface spectrum. Hence, it is practical to ignore B_s in scattering calculations that illustrate azimuthal dependence. Clearly, the difference between the upwind and downwind observations can only be attributed to B_a.

7.4 ILLUSTRATIONS AND COMPARISONS WITH MEASUREMENTS

To illustrate the model behavior, we adopt the sea spectrum given in Pierson and Moskowitz [1964] and Fung and Lee [1982]. The directional sea spectrum is taken to be

$$W(K, \phi) = \frac{1}{2\pi}(1 + r\cos 2\phi)\, W(K) \tag{7.54}$$

where

$$W(K) = 0.875K^{-1}(2\pi)^{p-1}\left(1 + 3\frac{K^2}{K_m^2}\right) g^{\frac{(1-p)}{2}} \left[K\left(1 + \frac{K^2}{K_m^2}\right)\right]^{-\frac{(1+p)}{2}} \tag{7.55}$$

and

$$K_m^2 = \frac{g\rho}{\tau} = (3.63)^2 cm^{-2}$$

$g = 981$ cm/s/s, ρ = seawater density, and τ = sea surface tension. The symbol r in (7.54) is defined as

$$r = 2\frac{(1-\nu)}{(1+\nu)}$$

where ν is the ratio of the slope variances in crosswind and upwind directions given by Cox and Munk [1954] as

$$\nu = \frac{\sigma_c^2}{\sigma_u^2} = \frac{0.003 + 1.92 \times 10^{-3} U}{3.16 \times 10^{-3} U} \tag{7.56}$$

The U in (7.56) is the windspeed in meters per second at an altitude of 12.5 m above the sea horizon. In (7.55) $p = 5 - \log U^*$ and the friction velocity U^* in cm/s is related to the windspeed U at altitude z in centimeters above the sea level by [Fung and Lee, 1982]

$$U = \left(\frac{U^*}{0.4}\right) \ln\left(\frac{z}{Z_0}\right)$$

and $Z_0 = 0.684/U^* + 4.28 \times 10^{-5} U^{*2} - 0.0443$.

To find the imaginary part of the bispectrum, we shall use the Fourier transform of the asymmetric part of the skewness function given in Appendix 7A as

$$\sigma^3 S_a(\xi, \varphi) = \sigma^3 (\xi \cos\varphi / s_0)^3 \exp(-\xi^2/s_0^2) \tag{7.57}$$

Then, the imaginary part of the bispectrum becomes

$$B_a(K, \phi) = -\frac{K\sigma^3 s_0^3 (6 - K^2 s_0^2 \cos^2\phi) \cos\phi}{16} \exp\left(-\frac{K^2 s_0^2}{4}\right) \tag{7.58}$$

Note that in (7.58) the bispectrum changes sign when the wind direction changes 180 degrees and becomes zero when $\phi = 90$ degrees. This means that there is no skewness effect in the crosswind direction.

In Figure 7.2(a), a comparison of the backscattering coefficient calculated from the present model with and without the bispectrum is plotted to illustrate the asymmetric properties of the backscattering coefficient. The incident angle is 50 degrees. When skewness is set to zero, the azimuthal curve is symmetric with respect to the upwind or downwind direction, and there is no difference in scattering along the upwind ($\phi = 0$ degree) and downwind ($\phi = 180$ degrees) directions. After the skewness is included, it is seen that not only there is a level difference between upwind and downwind directions, but also the minima of the azimuthal curve are shifted from the crosswind direction toward the downwind direction. After increasing the incident angle to 70 degrees, similar observations can be drawn from Figure 7.2(b). Furthermore, the level difference is now larger than at 50 degree incidence and the minima of the azimuthal curve are further shifted toward the downwind direction.

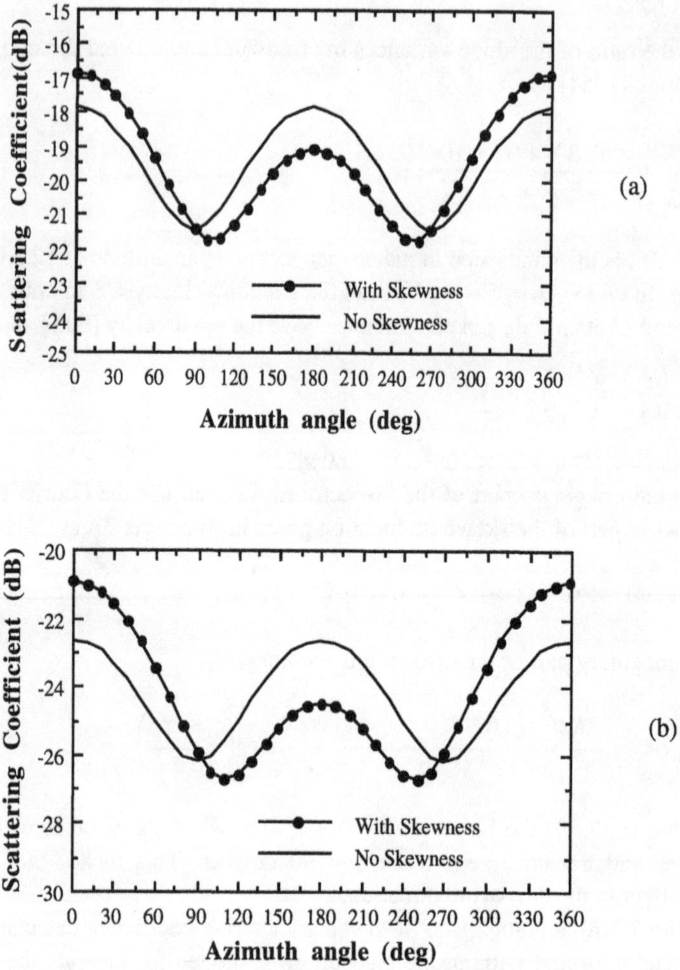

Figure 7.2 Illustration of the surface skewness effect: (a) incident angle equal to 50 degrees, (b) incident angle equal to 70 degrees.

As an application, we compare the model predictions and scattering measurements from the sea surface. The above selection of the sea spectrum and bispectrum fixes all parameters at a given wind speed except the parameter s_0 in the bispectrum. We assume that s_0 is also wind speed dependent so that for a given wind speed it will take on one value and will remain unchanged as polarization, azimuthal angle and incidence angle change. The first set of data was acquired during the Frontal Air-Sea Interaction

Experiment (FASINEX) from the NASA-JPL Ku band (14.6 GHz) scatterometer mounted on a C130 aircraft [Weissman, 1990; Li et al., 1989]. The windspeed is 7 m/s. At this windspeed, we choose ks_0 equal to 5.99. In Figure 7.3, comparisons of the backscattering coefficient between the model and data are shown versus the incident angle. The wind direction in Figure 7.3(a) is upwind. The model matches well with the data over the angular range from 20 to 60 degrees for both HH and VV polarizations. When the wind direction is changed to downwind, again, the agreement between the model and measurement is very good as shown in Figure 7.3(b). From Figures 7.3(a) and 7.3(b), the model is able to account for the upwind/downwind difference for all incident angles. It also predicts correctly the change in the spacing between HH and VV polarizations. An additional comparison is shown in Figure 7.3(c) for the crosswind direction. It is seen that the model correctly predicts the level and change in the spacing between HH and VV polarizations in the crosswind direction. From Figure 7.3, the trend behaviors indicate that both the spacing between VV and HH polarizations and the difference between upwind and downwind scattering coefficients increase with the incident angle and the spacing between VV and HH is larger at downwind than at either upwind or crosswind at large incident angles.

As an additional verification of the bispectrum model given by (7A.21), we compare the model predictions with field measurements taken from AAFE RADSCAT (13.9 GHz) reported in Donelan and Pierson [1987] at a higher wind speed, 7.5 m/s, and an incident angle of 40.9 degrees along the azimuthal direction in Figure 7.4. The skewness parameter used is $ks_0 = 5.5$. It is seen that excellent agreement is obtained over all azithmual angles. The minima in the crosswind are shifted toward the downwind direction same as in the measurements. Another significant feature is that the spacing between levels of VV and HH polarizations is clearly larger near downwind than at other azimuthal angles. This apparently is because the incidence angle is large and HH is lower at downwind than upwind. Another point to note is that the up/downwind difference in HH polarization is larger than that in VV polarization.

For further comparisons another set of data is taken from Masuko et al. [1986] at Ka band (34.43 GHz). The windspeed is 14 m/s. Figure 7.5 shows the backscattering coefficients according to the present model with ks_0 chosen to be 0.288. Figure 7.5(a) shows the comparison at 32 degrees incident angle. A close agreement between the model and data is seen for both HH and VV polarizations. Figure 7.5(b) and Figure 7.5(c) are comparisons at 52 degrees and 70 degrees incident angles, respectively. At 32 degrees the spacing between VV and HH is fairly uniform over all azimuthal angles. At larger incident angles the spacing around the downwind region is again larger than at upwind and crosswind.

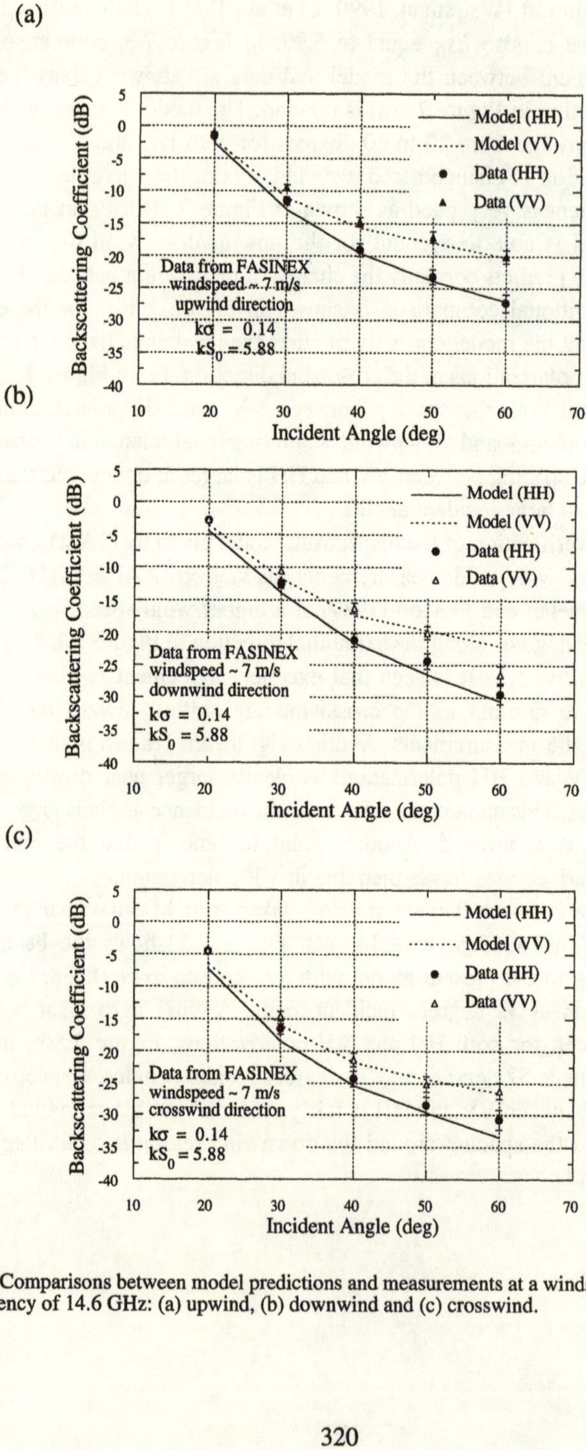

Figure 7.3 Comparisons between model predictions and measurements at a windspeed of 7 m/s and a radar frequency of 14.6 GHz: (a) upwind, (b) downwind and (c) crosswind.

7.5 SUMMARY

The scattering model in Chapter 4 has been generalized to include the third-order surface statistics to model sea surface scattering. The model consists of two types of terms: one type involves the surface roughness spectrum; the other the surface bispectrum. Since the roughness spectrum is centro-symmetric, it cannot account for the change in the sense of direction. The main contribution from the asymmetric surface property is given by the imaginary part of the bispectrum (which measures asymmetry). It is the cause of the difference between upwind and downwind observations. This interpretation is confirmed by our comparisons between model and measurements. Good agreements are obtained for three different windspeeds, three frequencies, a range of incident angles, all azimuthal angles and for both HH and VV polarizations using the sea spectrum proposed by Pierson and an empirical bispectrum given in Appendix 7A that involves one unknown parameter.

All comparisons are carried out for angles of incidence greater than or equal to 30 degrees. A different bispectrum may be needed to account for scattering over smaller angles of incidence from a skewed sea surface.

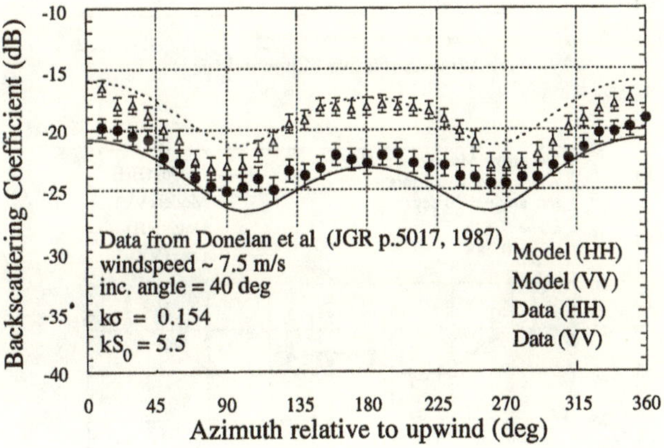

Figure 7.4 Model predictions using proposed bispectrum vs. measured data from AAFE experiment.

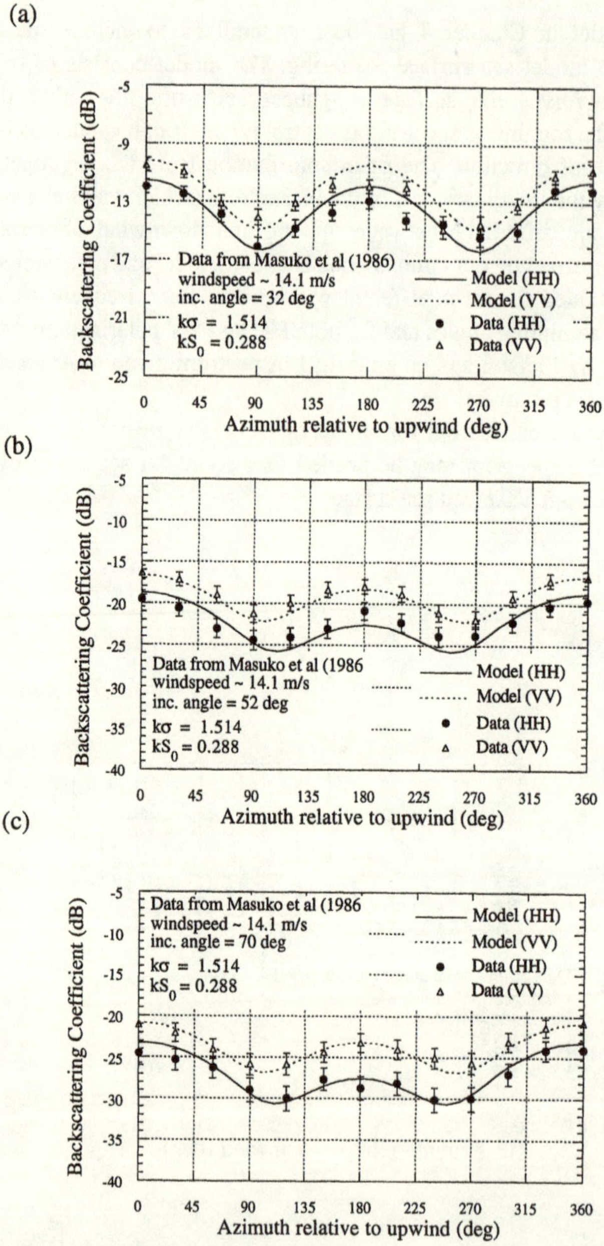

Figure 7.5 Comparisons of azimuthal behaviors between model predictions and measurements at a windspeed of 14 m/s and 34.43 GHz for incident angles of (a) 32 deg, (b) 52 deg and (c) 70 deg.

REFERENCES

Amar, F., "Directional Random Sea Surface Generation," M.S. Thesis, University of Texas at Arlington, 1989.

Barrick, D.E., "Relationship Between Slope Probability Density Function and the Physical Optics Integral in Rough Surface Scattering," *Proc. IEEE*, Oct. 1968, pp. 1728–1729.

Chen, M.F., K.S. Chen and A.K. Fung, "A Study of the Validity of the Integral Equation Model by Moment Method Simulation—Cylindrical Case," *Remote Sensing of Environment*, Vol. 29, 1989, pp. 217–228.

Chen, M.F., and A.K. Fung, "A Numerical Study of the Regions of Validity of the Kirchhoff and Small Perturbation Rough Surface Scattering Models," *Radio Science*, Vol. 23, no. 2, 1988, pp. 163–170.

Chen, K.S., and A.K. Fung, "A Bragg Scattering Model for the Sea Surface," Ocean 90, Conference Proc., 1990, pp. 249–252.

Chen, K.S., A.K. Fung and D.E. Weissman, "A Backscattering Model for Ocean Surface," *IEEE Trans. Geosci. and Remote Sensing*, Vol. 30, no. 4, 1992, pp. 811-817.

Cox, C., and W. Munk, "Measurement of the Roughness of the Sea Surface from Photographs of the Sun Glitter," *J. Opt. Soc. Am.*, Vol. 44, no. 11, 1954, pp. 838–850.

Donelan, M.A., and W.J. Pierson, Jr., "Radar Scattering and Equilibrium Ranges in Wind-Generated Waves with Application to Scatterometry," *J. Geophysical Res.*, Vol. 92, no. C5, 1987, pp. 4971–5029.

Elgar, S., "Relationships Involving Third Moments and Bispectra of a Harmonic Process, " *IEEE Trans. Acoustics, Speech, and Signal Proc.*, Vol. ASSP-35, no. 12, 1987, pp. 1725–1726.

Elgar, S., and R.T. Guza, "Observations of Bispectra of Shoaling Surface Gravity Waves," *J. Fluid Mech.*, Vol. 161, 1985, pp. 425–448.

Fung, A.K., and K.S. Chen, "Kirchhoff Model for a Skewed Random Surface," *J. of Electromagnetic Waves and Applications*, Vol. 5, no. 2, 1991, pp. 205–216.

Fung, A.K., and K.S. Chen, "Dependence of the Surface Backscattering Coefficients on Roughness, Frequency and Polarization States," *Intl. J. Remote Sensing*, 1992, pp. 1663–1680.

Fung, A.K., K.S. Chen and M.F. Chen, "A Note on the Directional Sea Spectrum," *Remote Sensing of Environment*, Vol. 30, 1989, pp. 96–106.

Fung, A.K, and K.K. Lee, "A Semi-Empirical Sea-Spectrum Model for Scattering Coefficient Estimation," *IEEE J. Ocean Eng.*, Vol. OE-7, no. 4, 1982, pp. 166–176.

Fung, A.K., Z. Li, and K.S. Chen, "Backscattering from a Randomly Rough Dielectric Surface" *IEEE Trans. Geosci. and Remote Sensing*, Vol. 30, no. 2, 1992, pp. 356–369.

Fung, A.K., and G.W. Pan, "A Scattering Model for Perfectly Conducting Random Surface: I. Model Development," *Intl. J. Remote Sensing*, Vol. 8, no. 11, 1987, pp. 1579–1605.

German, V. K., S.P. Levikov and A.S. Tsvetsinskii, "Bispectral Analysis of Sea-Level Variations," *Sov. meteorol. Hydrol.*, no. 11, 1980, pp. 50–57.

Guissard, A., C. Baufays and P. Sobieski, "Sea Surface Description Requirements for Electromagnetic Scattering Calculations," *J. Geophys. Res.*, Vol. 91, no. C2, 1986, pp. 2477–2492.

Hasselmann, K., W. Munk and G. MacDonald, "Bispectra of Ocean Waves," in *Time Series Analysis* (ed. M. Rosenblatt), John Wiley & Sons, N.Y., 1963, pp. 126–139.

Li, Z., and A.K. Fung, "A Reformulation of the Surface Field Integral Equations," *J. Electromagnetic Waves and Applications*, Vol. 5, no. 2, 1991, pp. 195–203.

Li, F., W. Large, W. Shaw, E.J. Walsh and K. Davidson, "Ocean Radar Backscatter Relationship with Near-Surface Winds: A Case Study During FASINEX," *J. Phys. Oceanogr.*, Vol. 12, 1989, pp. 342–353.

Longuet-Higgins, M.S., "The Effect of Non-Linearities on Statistical Distribution in the Theory of Sea Waves," *J. Fluid Mech.*, Vol. 17, 1963, pp. 459–480.

Longuet-Higgins, M.S., "On the Skewness of Sea Surface Slopes, "*J. Phys. Ocean*, Vol. 12, 1982, pp. 1283–1291.

Masuda, A., and Y. Kuo, "A Note on the Imaginary Part of Bispectra," *Deep-Sea Research*, Vol. 28A, no. 3, 1981, pp. 213–222.

Masuko, H., K. Okamoto, M. Shimada and S. Niwa, "Measurement of Microwave Backscattering Signatures of the Ocean Surface Using X Band and Ka Band Airborne Scatterometers," *J. Geophysical Res.*, Vol. 91, no. C11, 1986, pp. 13605–13083.

McComas, C.H., and M.G. Briscoe, "Bispectra of Internal Waves," *J. Fluid Mech.*, Vol. 97, part 1, 1980, pp. 205–213.

Nikias, C.L., and M.R. Raghuveer, "Bispectrum Estimation," *Proc. IEEE*, Vol. 75, no. 7, 1987, pp. 869–891.

Pierson, W.J., and L. Moskowitz, "A Proposed Spectral Form of Fully Developed Seas Based on the Similarity Theory of S.A. Kitaigorodskii," *J. Geophys. Res.*, Vol. 69, no. 24, 1964, pp. 5181–5190.

Plant, W. J., "A Two-Scale Model for Short Wind-Generated Waves and Scatteometry," *J. Geophys. Res.*, Vol. 91, no. C9, 1986, pp. 10735–10749.

Roden, G.I., and D.J. Bendliner, "Bispectra and Cross-Bispectra of Temperature, Salinity, Sound Velocity and Density Fluctuations with Depth off Northeastern Japan," *J. Phys. Ocean*, Vol. 3, no. 3, 1973, pp. 308–317.

Srokosz, M.A., and M.S. Longuet-Higgins, "On the Skewness of Sea-Surface Elevation," *J. Fluid Mech.*, Vol. 164, 1980, pp. 487–497.

Stogryn, A., and G.J. Desargant, "The Dielectric Properties of Brine in Sea Ice at Microwave Frequencies," *IEEE Trans. Ant. and Prop.*, Vol. 33, 1985, pp. 523-532.

Tayfun, A.M. , "On Narrow-Band Representation of Ocean Waves; 1: Theory," *J. Geophys. Res.*, Vol. 91, no. C6, 1986, pp. 7743–7752.

Ulaby, F.T., R.K. Moore and A.K. Fung, *Microwave Remote Sensing*, Vol. II, Artech House, Norwood, MA, 1982.

Weissman, D.E., "Dependence of the Radar Cross Section on Ocean Surface Variables: Comparison of Measurements and Theory Using Data from the Frontal Air-Sea Interaction Experiment," *J. Geophys. Res.*, Vol. 95, no. C3, 1990, pp. 3387–3398.

Wentz, F.J., S. Peteherych and L.A. Thomas, "A Model Function for Ocean Radar Cross Section at 14.6 GHz," *J. Geophys. Res.*, Vol. 89, 1984, pp. 3689–3704.

Appendix 7A
Bicoherence Function and Bispectrum

7A.1 INTRODUCTION

The properties of a skewed surface spectrum and bispectrum are found by generating such a surface on a digital computer and then evaluating its correlation function, bicoherence function, power spectrum and bispectrum. The bispectrum is defined to be the Fourier transform of the bicoherence function. It is found that the surface bicoherence function, its first and second derivatives must all vanish at the origin. In general, the surface bispectrum is a complex function. Its real part is centro-symmetric, the same as the surface spectrum but much smaller in magnitude, and its imaginary part is asymmetric. A function with the above stated properties is introduced to represent the imaginary part of the sea surface bispectrum. The unknown parameter in this function is calibrated using a data set from the FASINEX experiment. In the next section we will describe an approach to obtain the bicoherence characteristics of a skewed surface. Then, an empirical bispectrum model is given. In Section 7A.3, an illustration of the effects of skewness is carried out.

7A.2 BICOHERENCE FUNCTION AND BISPECTRUM

Direct evaluation of the sea bispectrum from a sea surface profile is difficult because its magnitude is small and usually a time rather than a spatial profile is measured. Also, it requires a high spatial resolution, and this is difficult to achieve in the presence of viscosity, surface tension and capillary effects. To investigate the properties of the bispectrum, the approach here is to generate a skewed surface profile numerically in one dimension [Tayfun, 1986; Amar, 1989]. The surface bicoherence function and bispectrum are then calculated from such a profile. For comparison, the correlation function and surface spectrum are also calculated.

7A.2.1 Generation of a Skewed Surface

It is well known that the ocean surface exhibits asymmetry due to the influence of the wind. In particular, the crests of the waves are tilted towards the wind direction. This loss of symmetry causes the surface height distribution to deviate from the usual Gaussian form and assumes the Gram-Charlier distribution [Cox and Munk, 1954].

In order to provide a mechanism for surface generation, a perturbation expansion is used to write the surface equation as the superposition of several profiles representing different perturbation orders of the Gaussian model [Longuet-Higgins, 1963]. For a one-dimensional surface, the perturbation expansion is given by

$$z(x, y) = z_1 + z_2 + z_3 + \ldots \tag{7A.1}$$

The first-order perturbation or Gaussian approximation, z_1, is a summation of random amplitude and phase as given in (7A.2)

$$z_1(x, y) = \sum_m a_m \cos(\psi_m) \tag{7A.2}$$

where $\psi_m = k_m x - \omega_m t + \varepsilon_m$ is the phase term with k_m being the physical wavenumber in the x-direction, ω_m is the frequency such that $\omega_m^2 = g k_m$, g is the gravitational acceleration, ε_m is a random phase distributed uniformly over $(-\pi, \pi)$ and a_m is a random amplitude chosen such that the random variables $a_m \cos(\varepsilon_m)$ and $a_m \sin(\varepsilon_m)$ are normally distributed. In addition, the mean-square of the random amplitude is related to the roughness spectrum, $W(k)$, as

$$\sum_{k_m} \overline{\frac{1}{2} a_m^2(k_m)} = W(k)\,\Delta k \tag{7A.3}$$

where $k_m \in [k - 0.5\Delta k, k + 0.5\Delta k]$. If each differential grid sample, Δk, contains one and only one wavenumber component, then the amplitude becomes

$$a_m = \sqrt{2W(k_m)\,\Delta k_m} \tag{7A.4}$$

For a narrow-band spectrum, Tayfun [1986] proposed a model for the surface including a part of the second-order expansion. The result is a random profile whose crests are narrow and peaked and whose troughs are long and flat. Such asymmetry in the vertical direction is referred to as *vertical skewness*. The expansion, however, does not include any asymmetry due to tilted crests in the horizontal direction, which is referred to as *horizontal skewness* [Amar, 1989]. In what follows, we shall modify Tayfun's model for nonlinear surfaces to accommodate both vertical and horizontal skewness. Since horizontal skewness is asymmetric with respect to a vertical axis, it is the component that influences the difference in the scattering coefficient of the ocean surface in the upwind/downwind direction. The modified surface model is

$$z(x) = z_1 + \frac{1}{2}\tilde{k}\left[(z_1^2 - \tilde{z}_1^2) + 2 z_1 \tilde{z}_1\right] \tag{7A.5}$$

where $\tilde{k}$ is the mean wavenumber,

$$\tilde{k} = m_1 / m_0 \tag{7A.6}$$

$$\tilde{z}_1(x) = \sum_m a_m \sin\psi_m \tag{7A.7}$$

$$m_j = \int k^j W(k)\, dk \tag{7A.8}$$

In (7A.5) the first term on the right-hand side is given by (7A.2) and represents the linear component of the random wave. This is also the component of the surface that is Gaussian distributed. The second term represents the nonlinear component. It is the surface component that contains the skewness effect. The average wavenumber, $\bar{k}$, is defined in terms of the zeroth and first moments as shown in (7A.6). The nonlinear component of the wave profile contains two terms. The vertical skewness is due to the $(z_1^2 - \tilde{z}_1^2)$ term, which contains $\cos(2\psi_m)$ coefficients. The horizontal skewness is due to the $(z_1\tilde{z}_1)$ term, which contains $\sin(2\psi_m)$ coefficients. It is worth noting that the cosine is an even function, while sine is an odd function. This is a key issue since the Fourier transformation of a real, even function is a real, even function, and the Fourier transform of a real, odd function is an imaginary, odd function.

Figure 7A.1 is a sample of a computer generated profile where horizontally skewed and vertically skewed surfaces are plotted separately for the purpose of illustrating the effects of the different skewness types. In addition, the linear component of the surface which is Gaussian distributed, is also plotted on the same figure. For an actual sea surface, both horizontal and vertical skewness may be present at the same time, and it may not be possible to isolate them. From Figure 7A.1, the main differences between the skewed and nonskewed profiles are in the crests and troughs. The height probability density functions (pdf) of the skewed surfaces are shown in Figure 7A.2 along with the height pdf of the linear surface term. From Figure 7A.1 it is clear that vertically skewed profile has flattened troughs and narrower and larger peaks. Hence, there are more points just below the mean sea surface (taken to be the origin), and its pdf is more peaked just below the origin (Figure 7A.1). For a horizontally skewed surface, the crests are more peaked and tilted in one direction and the troughs are tilted in the opposite direction, generating some flat spots (Figure 7A.1). Thus, there are more points near the origin and at large deviations from the mean, and fewer points at intermediate deviations relative to the Gaussian distribution (Figure 7A.2). However, the height distribution is still evenly distributed about the mean sea surface.

7A.2.2 Surface Functions

Consider a one-dimensional random surface $z(x)$ with zero mean, $\langle z(x) \rangle = 0$, a variance of $\langle z^2(x) \rangle = \sigma^2$, and a third moment, $\langle z^3(x) \rangle = \mu_3$. It is assumed that the surface is generated by a stationary random process.

In second-order statistics we have the correlation (or coherence) function defined as

$$c(\tau) = \langle z(x) z(x+\tau) \rangle \tag{7A.9}$$

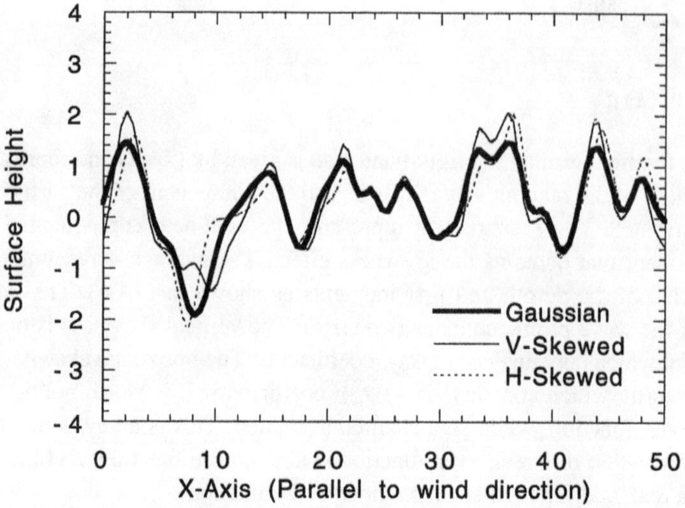

Figure 7A.1 Sample profiles of computer generated random surface.

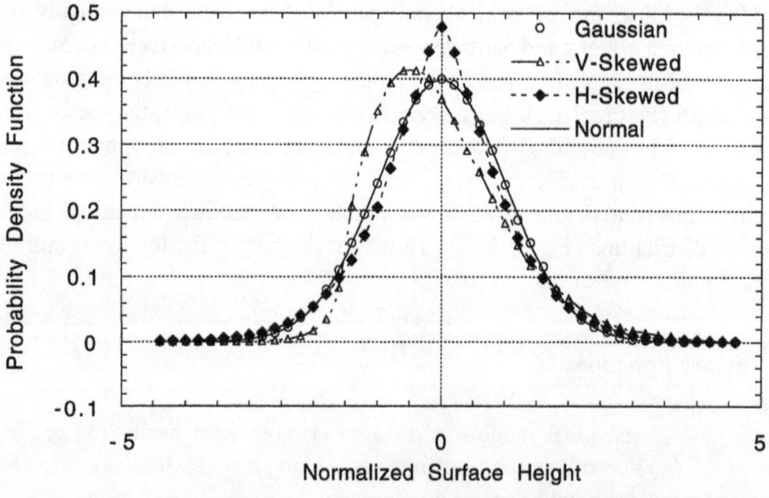

Figure 7A.2 Surface height distribution functions computed from profiles for a surface with no skewness (Gaussian), a surface with horizontal skewness and a surface with vertical skewness. The normal curve is a theoretical plot. Results indicate excellent agreement between theory and profile.

The Fourier transform of it is the surface spectrum. Since the correlation function is a real, even function, the corresponding surface spectrum is also a real, even function. The variance of the surface function, σ^2, is equal to the correlation function evaluated at the origin, i.e., $\sigma^2 = c(0)$.

In third-order statistics we have the bicoherence function defined as [Chen et al., 1993]:

$$S(\tau, \varsigma) = \sigma^3 \langle z(x) z(x+\tau) z(x+\varsigma) \rangle \tag{7A.10}$$

The surface bispectrum is the Fourier transform of the bicoherence function. The skewness coefficient, μ_3/σ^3, is the normalized bicoherence function evaluated at the origin

$$\lambda_3 = S(0, 0)/\sigma^3 = \mu_3/\sigma^3 \tag{7A.11}$$

The bicoherence function is neither even nor odd. Its Fourier transform, the bispectrum, is in general complex:

$$B(k, k') = \frac{1}{2\pi}\iint S(\tau, \varsigma) e^{-jk\tau - jk'\varsigma} d\tau d\varsigma = B_s(k, k') + jB_a(k, k') \tag{7A.12}$$

Since $S(\tau, \varsigma)$ is a real-valued function, $B_s(k, k')$ is even and $B_a(k, k')$ is odd. When extended to two-dimensional surface, $B_s(k, k')$ is centro-symmetric and $B_a(k, k')$ is antisymmetric. In modeling calculations a special case of the bicoherence function is of interest. It occurs when $\tau = \varsigma$,

$$S_1(\tau) = \langle z(x) z^2(x+\tau) \rangle \tag{7A.13}$$

For this special case, the corresponding bispectrum reduces to

$$B(k) = \frac{1}{\sqrt{2\pi}} \int S_1(\tau) e^{-jk\tau} d\tau = B_s(k) + jB_a(k) \tag{7A.14}$$

Furthermore, the bicoherence function may be written as the sum of two functions, one even, $S_s(\tau)$, and one odd, $S_a(\tau)$:

$$S_1(\tau) = S_s(\tau) + S_a(\tau) \tag{7A.15}$$

where

$$\sigma^3 S_s(\tau) = \frac{\langle z(x) z^2(x+\tau) \rangle + \langle z^2(x) z(x+\tau) \rangle}{2} = \sigma^3 \frac{S_1(\tau) + S_2(\tau)}{2} \tag{7A.16}$$

$$\sigma^3 S_a(\tau) = \frac{\langle z(x) z^2(x+\tau) \rangle - \langle z^2(x) z(x+\tau) \rangle}{2} = \sigma^3 \frac{S_1(\tau) - S_2(\tau)}{2} \tag{7A.17}$$

Substituting (7A.15) into (7A.14) we can relate the real and imaginary parts of the bispectrum to the even and odd components of $S_1(\tau)$ as

$$B_s(k) = \frac{1}{\sqrt{2\pi}}\int \sigma^3 S_s(\tau)\, e^{-jk\tau} d\tau \tag{7A.18}$$

$$jB_a(k) = \frac{1}{\sqrt{2\pi}}\int \sigma^3 S_a(\tau)\, e^{-jk\tau} d\tau \tag{7A.19}$$

In view of (7A.18) and (7A.19), it can be deduced that the real part of the bispectrum, $B_s(k)$, is due to the vertical skewness only and that the imaginary part, $B_a(k)$, is due to the horizontal skewness only. This becomes clear when we recall that the term containing vertical skewness is made up of $\cos(2\psi_m)$ coefficients. Taking the Fourier transform of such coefficients results in a real-valued, even function, i.e., $B_s(\tau)$. On the other hand, the term containing horizontal skewness is composed of $\sin(2\psi_m)$ coefficients. Taking the Fourier transform of such coefficients results in an imaginary-valued, odd function, i.e., $jB_a(\tau)$.

The estimated spatial surface correlation function is shown in Figure 7A.3 and its corresponding spectrum is shown in Figure 7A.4. As expected, the correlation function is symmetric about the origin. The sea spectrum is a positive, even function that starts at zero and reaches a peak before tapering off to zero again.

Similarly, the estimated spatial surface bicoherence function and its components are shown in Figure 7A.5. The real and imaginary parts of its bispectrum are shown in Figure 7A.6. *It is important to note that the bicoherence function, its first and second derivatives are equal to zero at the origin.* Since the special bicoherence functions S_1 and S_2 are neither even nor odd, the bispectrum is a complex function whose real part is even. The imaginary part, on the other hand, is odd and is zero at the origin. Both the real and imaginary parts of the bispectrum become negative before tapering off to zero. In addition, the maxima of the real and imaginary parts are about one order of magnitude smaller than the maximum of the surface spectrum.

Although the real and imaginary parts of the bispectrum are of the same order of magnitude, their effect on the radar cross section is very different. As discussed previously, the real part, which stems from vertical skewness, preserves the same symmetry properties as the surface spectrum but is much smaller in magnitude. Therefore, it does not have much effect on the scattering coefficient [Chen et al., 1992]. On the other hand, the imaginary part, which stems from the horizontal skewness, has asymmetry with respect to the look direction especially at large incident angles. As a result, in practice we only need to introduce a function corresponding to $S_a(\tau)$ which we define in polar coordinates as

$$\sigma^3 S_a(\zeta, \varphi) = \sigma^3 (\zeta \cos\varphi / s_0)^3 \exp(-\zeta^2 / s_0^2) \tag{7A.20}$$

where s_0 is the skewness parameter. It is important to note that, while the above suggested function has the desired properties of a skewness function, it is not adequate for a real sea surface where a more complex form is expected. In applications, this form has been found to give good results when the angle of incidence exceeds 30 degrees.

It is worth noting that $S_a(0) = S'_a(0) = S''_a(0) = 0$. This is required by the simulated results in Figure 7A.5. The corresponding asymmetrical part of the bispectrum is readily computed as

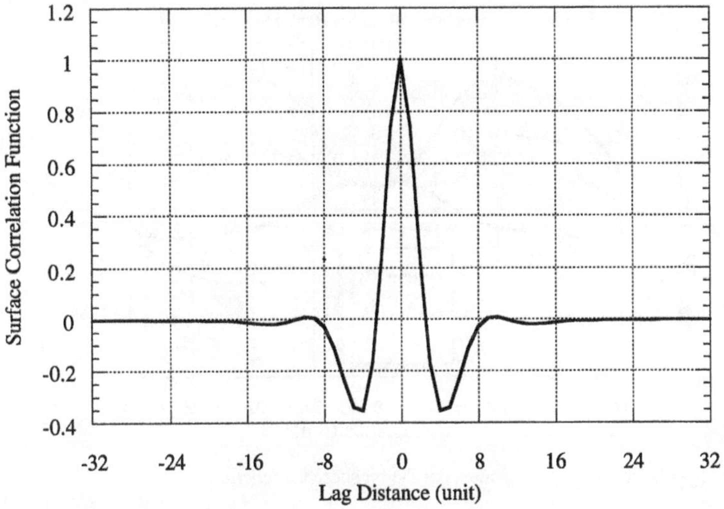

Figure 7A.3 Surface correlation function calculated from generated surface.

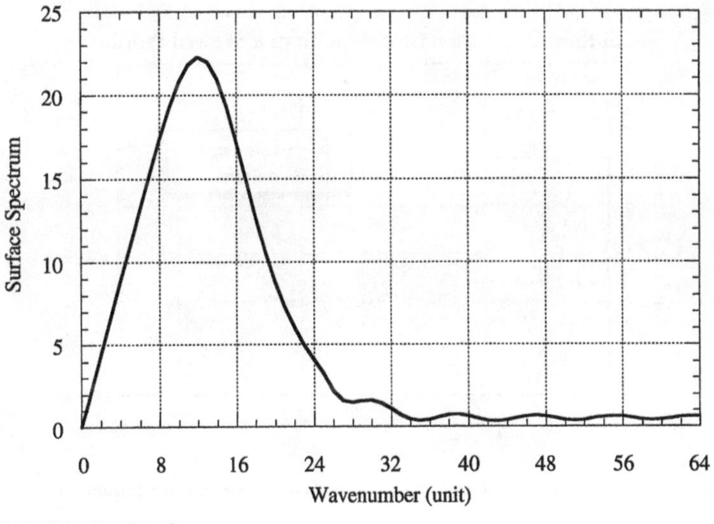

Figure 7A.4 Calculated surface spectrum.

$$B_a(K, \phi) = -\frac{K\sigma^3 s^3{}_0 (6 - K^2 s_0^2 \cos^2\phi) \cos\phi}{16} \exp(-K^2 s^2{}_0/4) \tag{7A.21}$$

where ϕ is the azimuthal angle and $\phi = 0$ is in the upwind direction.

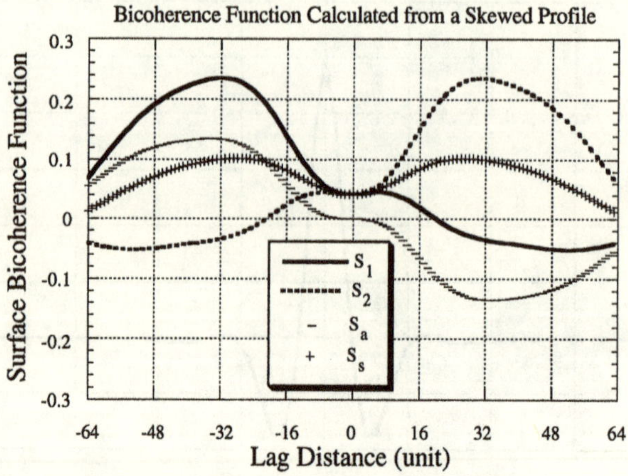

Figure 7A.5 Special forms and components of the bicoherence function.

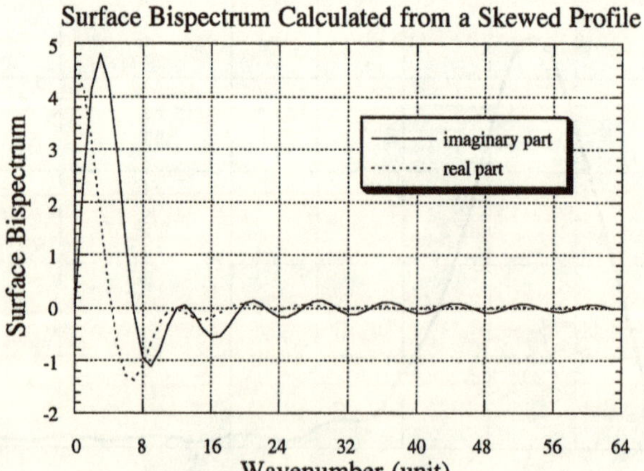

Figure 7A.6 An illustration of the real and imaginary parts of the surface bispectrum.

From (7A.21), the bispectrum vanishes in the crosswind direction ($\phi = 90$), and it changes sign as the wind direction changes 180 degrees, i.e., a sign difference between upwind and downwind direction. Moreover, the presence of the bispectrum will shift the minima (around the crosswind directions) of the scattering curve toward the downwind direction, as will be shown in the next section.

7A.3 EFFECTS OF SKEWNESS

For the purpose of illustration, we shall use the backscattering model in Chapter 7 and adopt the surface spectrum given by Fung and Lee [1982] and the bispectrum given in (7A.21). We first demonstrate the variations of the backscattering coefficient as a function of the skewness constant s_0 along the upwind direction as shown in Figure 7A.7. As the skewness constant increases, the radar return is higher in the upwind direction and consequently lower in the downwind direction due to the sign change mentioned in the previous section. Note that if the sense of the skewness function is reversed, mathematically it can also make downwind higher than upwind. Experimentally, such a situation may occur in the angular region when the incident angle is between 0 and 25 degrees. This implies that the complete skewness function $S_a(\tau)$ for the sea surface is more involved than the one given by (7A.20).

To show the influence of surface skewness in the azimuthal direction, we plot the part of the scattering coefficient due to the addition of skewness versus the azimuthal angle in Figure 7A.8 at an incident angle of 30 degrees. A change of sign is observed in moving from upwind to downwind direction. In the crosswind direction, the contribution due to skewness is zero.

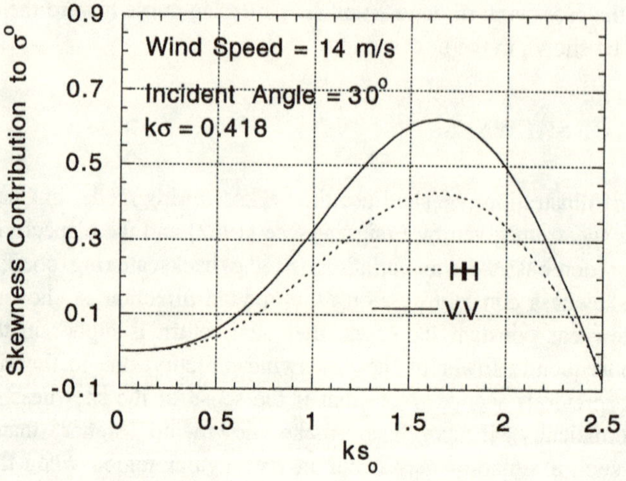

Figure 7A.7 Surface skewness effects on the backscattering coefficient in the upwind direction.

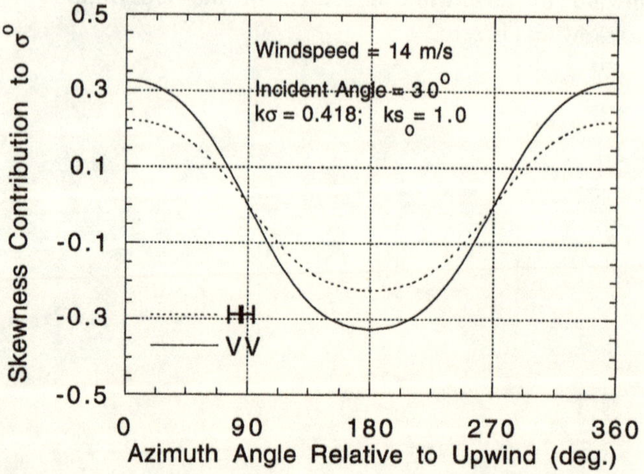

Figure 7A.8 Effect of the surface bispectrum along the azimuthal direction

Appendix 7B

High-Frequency Limit of the Kirchhoff Model for a Skewed Surface

7B.1 INTRODUCTION

It is well known that the Kirchhoff model (KM) should reduce to the geometric optics model in the high-frequency limit. For a skewed surface, we expect the skewness in slope to appear naturally as the limit of the terms involving skewness in surface height. To demonstrate this point, we will consider a one-dimensional case to avoid mathematical complication.

In the KM , the scattering coefficient is of the form [Fung and Chen, 1991]:

$$\sigma_{pp}^{0} = \frac{kq^2|U_{pp}|^2}{4q_z^2}\int \exp\{jq_x x - \sigma^2 q_z^2 (1-\rho) + j(q_z\sigma)^3 (S/2)\}\,dx \tag{7B.1}$$

where q_x, q_z, U_{pp} are functions of angle; $q^2 = q_x^2 + q_z^2$; k is the wave number; σ is the rms surface height; ρ is the surface correlation function and S is the skewness function normalized by σ^3. In the high-frequency limit, we will show that (7B.1) reduces to the geometric optics solution:

$$\sigma_{geo}^{0} = \frac{2\pi kq^2|U_{pp}|^2}{4q_z^3}\,p\left(-\frac{q_x}{q_z}\right) \tag{7B.2}$$

where $p(-q_x/q_z)$ is the probability density function of the surface slope evaluated at $Z_x = -q_x/q_z$. The explicit form of the probability density function including up to the third order is given by [Longuet-Higgins, 1982]

$$p(Z_x) = \frac{1}{\sigma_s\sqrt{2\pi}}\left[1 - Z_x\frac{\lambda_3}{2\sigma_s} + Z_x^3\frac{\lambda_3}{6\sigma_s^3}\right]e^{-Z_x^2/(2\sigma_s^2)} \tag{7B.3}$$

where

$\langle Z_x\rangle = 0;$

$\lambda_3 = \kappa_3/\sigma_s^3 = \mu_3/\sigma_s^3;$

μ_m = mth moment of the slope distribution and

κ_m = mth cumulant of the slope distribution.

7B.2 REDUCTION OF (7B.1) IN THE HIGH-FREQUENCY LIMIT

Consider the integral in (7B.1),

$$I = \int \exp\{jq_x x - \sigma^2 q_z^2 (1-\rho) + j(q_z\sigma)^3 (S/2)\}\, dx$$
$$= \int \left\{ \exp\left[jq_x x - \sigma^2 q_z^2 (1-\rho)\right] \sum_{n=0}^{\infty} \left[j(q_z\sigma)^3 S/2\right]^n / n! \right\} dx \tag{7B.4}$$

In the high-frequency limit, only the region of x around the origin is important. Hence, we can expand ρ and S about $x = 0$. Note that $S(0) = S'(0) = S''(0) = 0$ and

$$I = \int \left\{ \exp\left[jq_x x - \sigma^2 q_z^2 |\rho''(0)| \frac{x^2}{2}\right] \sum_{n=0}^{\infty} \left[j(q_z\sigma)^3 S'''(0) \frac{x^3}{12}\right]^n / n! \right\} dx \tag{7B.5}$$

For small skewness, the terms in the sum are getting successively smaller. Hence, we shall retain only the first two terms as

$$I = \int \exp\left[jq_x x - \sigma^2 q_z^2 |\rho''(0)| \frac{x^2}{2}\right] \left[1 + j(q_z\sigma)^3 S'''(0) \frac{x^3}{12}\right] dx$$
$$= \int \exp\left[-\sigma^2 q_z^2 |\rho''(0)| \frac{x^2}{2}\right] \left[\cos(q_x x) - (q_z\sigma)^3 S'''(0) \frac{x^3 \sin(q_x x)}{12}\right] dx \tag{7B.6}$$

where $\sigma^2 |\rho''(0)|$ is recognized as the variance of surface slope denoted by σ_s^2. The integral in (7B.6) can be carried out yielding

$$I = \frac{\sqrt{2\pi}}{\sigma_s q_z} \left[1 - \frac{q_x \sigma^3 S'''(0)}{4 q_z \sigma_s^4} + \frac{q_x^3 \sigma^3 S'''(0)}{12 q_z^3 \sigma_s^6}\right] e^{-\frac{1}{2}\left(\frac{q_x}{\sigma_s q_z}\right)^2} \tag{7B.7}$$

Substitution of (7B.7) into (7B.1), we reach the final form of the scattering coefficient in the high-frequency limit as

$$\sigma_{pp}^0 = \frac{\sqrt{2\pi} k q^2 |U_{pp}|^2}{4 \sigma_s q_z^3} \left[1 - \frac{q_x \sigma^3 S'''(0)}{4 q_z \sigma_s^4} + \frac{q_x^3 \sigma^3 S'''(0)}{12 q_z^3 \sigma_s^6}\right] e^{-\frac{1}{2}\left(\frac{q_x}{\sigma_s q_z}\right)^2} \tag{7B.8}$$

In order to show that the above equation is the same as (7B.2) we need to find the relation between λ_3 or μ_3 and the third derivative of the skewness function evaluated at the origin. This is carried out in the next section.

7B.3 RELATION BETWEEN μ_3 AND S

The third moment of the surface slope can be related to the skewness function as follows:

$$\mu_3 = \langle Z_x^3 \rangle = \lim_{\Delta x \to 0} \left\langle \left(\frac{z - z'}{\Delta x} \right)^3 \right\rangle = \lim_{\Delta x \to 0} \left\langle \frac{z^3 - 3z^2 z' + 3zz'^2 - z'^3}{(\Delta x)^3} \right\rangle$$

$$= \lim_{\Delta x \to 0} \frac{\sigma^3 + \langle 3z'^2 z - 3z' z^2 \rangle + 3z' z^2 - \sigma^3}{(\Delta x)^3}$$

$$= \lim_{\Delta x \to 0} \frac{\langle 3zz'^2 - 3z^2 z' \rangle}{(\Delta x)^3} = \lim_{\Delta x \to 0} \frac{-3\sigma^3 S}{\Delta x^3}$$

$$= \lim_{\Delta x \to 0} \frac{-3\sigma^3}{\Delta x^3} \left[S(0) + S'(0)\,\Delta x + \frac{S''(0)}{2} \Delta x^2 + \frac{S'''(0)}{2} \Delta x^2 + \ldots \right]$$

$$= \frac{-\sigma^3 S'''(0)}{2} \tag{7B.9}$$

Thus, the λ_3 in (7B.3) is given by

$$\lambda_3 = \mu_3 / \sigma_s^3 = \frac{-\sigma^3 S'''(0)}{2\sigma_s^3} \tag{7B.10}$$

7B.4 COMPARISON BETWEEN (7B.8) AND (7B.2)

After replacing λ_3 and setting $Z_x = q_x / q_z$ in (7B.3), we can write (7B.2) explicitly as

$$\sigma_{geo}^0 = \frac{\sqrt{2}\pi k q^2 |U_{pp}|^2}{4\sigma_s q_z^3} \left[1 - \frac{q_x \sigma^3 S'''(0)}{4 q_z \sigma_s^4} + \frac{q_x^3 \sigma^3 S'''(0)}{12 q_z^3 \sigma_s^6} \right] e^{-\frac{1}{2}\left(\frac{q_x}{\sigma_s q_z}\right)^2} \tag{7B.11}$$

which is identical to (7B.8).

Appendix 7C

Dielectric Constant of Saline Water

7C.1 INTRODUCTION

The dielectric constant of saline water has been studied by many investigators [Ulaby et al., 1986, Appendix E]. In the next section we summarize a dielectric model developed by Stogryn and Desargant [1985].

7C.2 DIELECTRIC CONSTANT OF SALINE WATER

The complex dielectric constant of saline water or brine can be expressed in terms of the operating frequency f and the following quantities:

1. Conductivity,

$$\sigma = -t \exp[0.5193 + 0.08755\,\mathrm{t}], \quad \text{if } \mathrm{t} \geq -22.9°C$$
$$= -t \exp[1.0334 + 0.11\,\mathrm{t}], \quad \text{if } \mathrm{t} < -22.9°C$$

2. Staticdielectric constant,

$$\varepsilon_s = (939.66 - 19.068t) / (10.737 - t)$$

3. High-frequency limit of the real part of the dielectric constant,

$$\varepsilon_\infty = (82.79 + 8.19t^2) / (15.68 + t^2)$$

4. Relaxation time τ multiplied by 2π,

$$2\pi\tau = 0.1099 + 0.00136t + 0.00020894t^2 + 2.82\times10^{-6}t^3$$

in the form,

$$\varepsilon = \varepsilon_\infty + \frac{\varepsilon_s - \varepsilon_s}{1 - j2\pi f\tau} + j\frac{\sigma}{2\pi\varepsilon_0 f} \tag{7C.1}$$

In the above equation frequency is in Hertz and ε_0 is the permittivity of free space equal to $10^{-9} / (36\pi)$ F/m.

Chapter 8

Matrix Doubling Formulation

8.1 INTRODUCTION

In Chapter 2 a first-order radiative transfer solution was obtained that is effective for scattering from an inhomogeneous layer with a small albedo. When multiple scattering due to volume scattering or interaction between volume inhomogeneity and boundary is important, an exact solution of the radiative transfer equation is needed. This can be done numerically using the eigen-analysis technique [Tsang et al., 1985; Ulaby et al, 1986] or the matrix doubling method [Ulaby et al., 1986]. In terms of computation, matrix doubling is a more intuitive and efficient method especially for cases where the layer optical thickness is large.

The matrix doubling method was originally developed for the calculation of incoherent multiple scattering in the atmosphere. It was subsequently modified by Leader [1975] to include the plane boundary effects. Fung and Eom [1981b] generalized the method further to include an inhomogeneous layer with irregular boundaries and the effects of close spacing between adjacent scatterers that are small. Further extensions of the matrix doubling method to model both scattering and emission from an irregular multilayered, inhomogeneous medium have been carried out by Tjuatja [1992] and Tjuatja and Fung [1992]. Furthermore, their models permit large spherical scatterers with close spacing between adjacent scatterers.

In this chapter the basic concept and implementation techniques of the matrix doubling method is discussed and applied to a multilayered inhomogeneous medium with irregular interfaces. The inhomogeneous medium is assumed to consist of sublayers with different physical properties for both the scatterers and the host layer. Each sublayer is assumed to be statistically homogeneous. The basic concept of the matrix doubling method is explained in Section 8.2. Section 8.3 provides a solution method using Fourier decomposition with respect to the azimuthal angle in the phase matrix elements and intensity vectors. The use of this decomposition allows matrix elements to depend only on the incident and scattered polar angles. The boundary conditions at the interface between two adjacent media and the interactions between boundary and media are derived in Section 8.4. The scattering coefficient for an irregular layer is given in Section 8.5 and and that for a multi-layered inhomogeneous medium in Section 8.6. Similar development for the emission problem is given in Section 8.7.

8.2 BASIC CONCEPT OF MATRIX DOUBLING METHOD

Consider an irregular inhomogeneous layer shown in Figure 8.1. The scattered intensity $\mathbf{I}^s$ due to the layer is related to the incident intensity $\mathbf{I}$ by

$$\mathbf{I}^s(\theta_s, \phi_s) = \frac{1}{4\pi}\int_{4\pi} \mathbf{S}_{T1}(\theta_s, \theta; \phi_s - \phi)\mathbf{I}(\theta, \phi)d\Omega \tag{8.1}$$

where $\mathbf{S}_{T1}(\theta_s, \theta; \phi_s - \phi)$ is the total scattering phase matrix of the irregular layer. $\mathbf{S}_{T1}$ accounts for effects due to the interface scattering, volume scattering, volume-interface interactions and multiple scattering within the layer. Using the matrix doubling method, $\mathbf{S}_{T1}$ can be expressed in terms of the phase matrix of an infinitesimal layer and the phase matrices of the boundaries.

For an infinitesimally thin layer of the medium with thickness $\Delta\tau$ (see Figure 8.2), its single scattering phase matrices in backward direction $\mathbf{S}$ and forward direction $\mathbf{F}$ are related to the single scattering phase matrix, $\mathbf{P}(\theta_s, \theta, \phi_s - \phi)$, of the scatterer in the medium by [Leader, 1975; Twomey et al., 1966, 1975]

$$\mathbf{S}(\theta_s, \theta, \phi_s - \phi) = a\mathbf{U}^{-1}\mathbf{P}(\theta_s, \pi - \theta, \phi_s - \phi)\Delta\tau \tag{8.2}$$

$$\mathbf{F}(\theta_t, \theta, \phi_t - \phi) = a\mathbf{U}^{-1}\mathbf{P}(\pi - \theta_t, \pi - \theta, \phi_t - \phi)\Delta\tau \tag{8.3}$$

where $\mathbf{U}$ is the diagonal matrix containing the directional cosines of the scattered angles, and a is the single scattering albedo of the medium. When the incidence direction is reversed, the single scattering phase matrices for an infinitesimally thin layer are

$$\mathbf{S}^*(\theta_s, \theta, \phi_s - \phi) = a\mathbf{U}^{-1}\mathbf{P}(\pi - \theta_s, \theta, \phi_s - \phi)\Delta\tau$$

$$\mathbf{F}^*(\theta_t, \theta, \phi_t - \phi) = a\mathbf{U}^{-1}\mathbf{P}(\theta_t, \theta, \phi_t - \phi)\Delta\tau$$

When two adjacent layers of optical thickness $\Delta\tau_1$ and Dt2 are combined into one layer with optical thickness $\Delta\tau_1 + \Delta\tau_2$, the scattering process due to a unit incident intensity is detailed in Figure 8.3. Note that it is the combination of two infinitesimal layers that generates multiple scattering. The phase matrices of the combined layer including multiple scattering can be found by summing up the terms indicated in Figure 8.3 [Ulaby et al., 1986] to yield

$$\mathbf{S} = \mathbf{S}_1 + \mathbf{T}^*{}_1\mathbf{S}_2(\mathbf{I} - \mathbf{S}^*{}_1\mathbf{S}_2)^{-1}\mathbf{T}_1 \tag{8.4}$$

$$\mathbf{T} = \mathbf{T}_2(\mathbf{I} - \mathbf{S}_1^*\mathbf{S}_2)^{-1}\mathbf{T}_1 \tag{8.5}$$

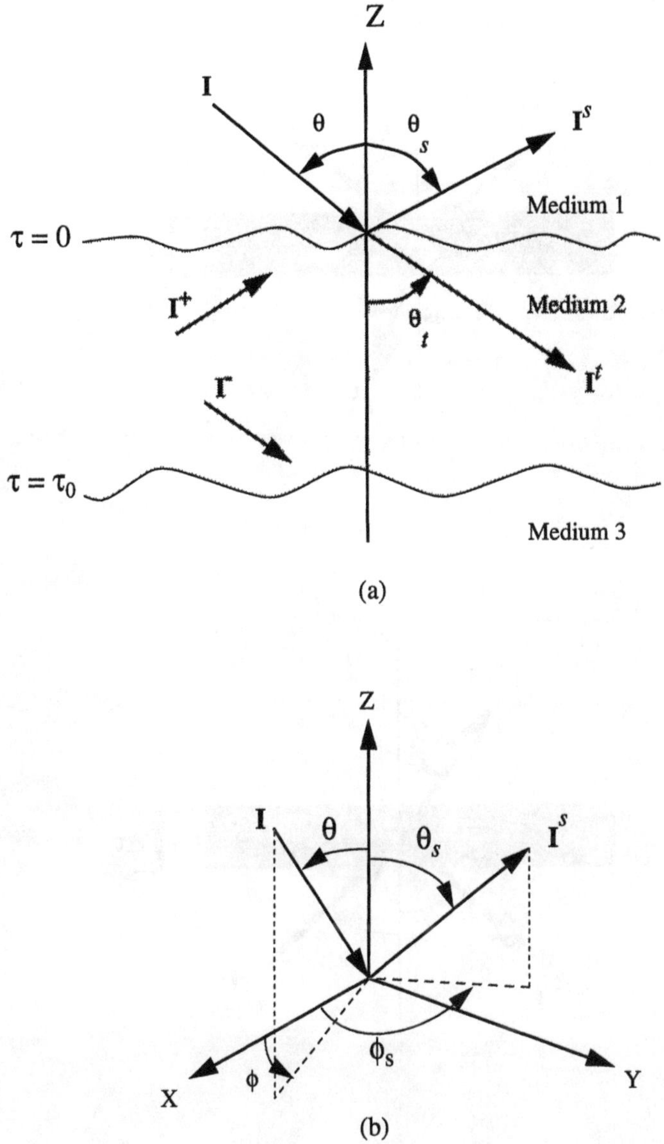

Figure 8.1 (a) Geometry of the single layer scatter problem. (b) The coordinate system.

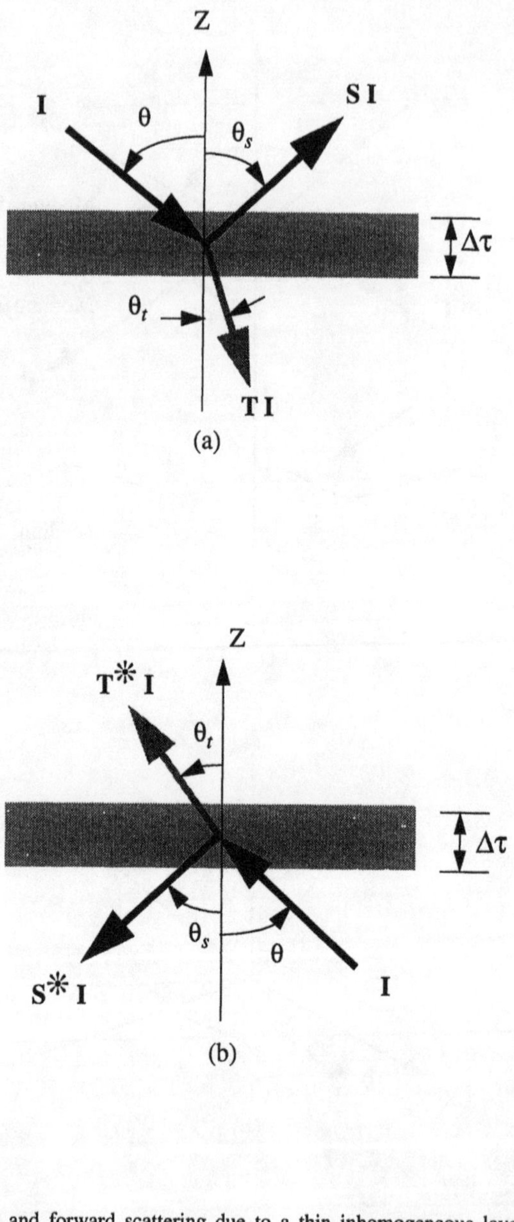

Figure 8.2 Backward and forward scattering due to a thin inhomogeneous layer. (a) Downward (-z) incidence. (b) Upward (+z) incidence.

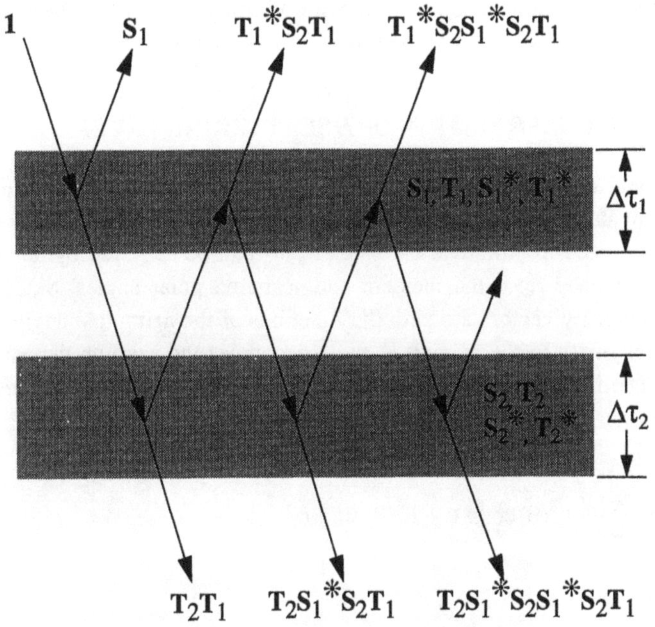

Figure 8.3 The scattering process due to two adjacent inhomogeneous infinitesimal layers for a unit incident intensity.

$$\mathbf{S}^* = \mathbf{S}_1^* + \mathbf{T}_1\mathbf{S}_2^*(\mathbf{I} - \mathbf{S}_1\mathbf{S}_2^*)^{-1}\mathbf{T}_1^* \tag{8.6}$$

$$\mathbf{T}^* = \mathbf{T}_2^*(\mathbf{I} - \mathbf{S}_1\mathbf{S}_2^*)^{-1}\mathbf{T}_1^* \tag{8.7}$$

where **I** is the identity matrix. The starred quantities in the above equations are scattering phase matrices for incidence in the $+z$ direction and are defined in a similar way as the unstarred quantities. The incident intensity traversing the layer undergoes both absorption and scattering due to medium inhomogeneity. Hence, the total forward intensity inside a layer consists of a diffuse component and a direct component. The direct intensity originates from the incident intensity with reduced strength due to both scattering and absorption. Thus, the single scattering phase matrix, **T**, for an infinitesimal layer in the forward direction always consists of two parts:

$$\mathbf{T} = \mathbf{F} + \mathbf{E}$$

where **E** is the extinction matrix taken to be diagonal for isotropic media. Its diagonal elements are $\exp(-\Delta\tau/\mu_i)$, where μ_i is the directional cosine. The multiple scattering phase matrix of an inhomogeneous layer with arbitrary optical thickness can thus be obtained by repeated application of the process shown in Figure 8.3.

8.3 HARMONIC PHASE MATRIX OF A SCATTERING LAYER

In scattering and emission calculations, the intensity vectors and phase matrices are functions of incident and scattered polar angles and azimuthal angles. To use matrix computation we need to eliminate one angle. This can be achieved by decoupling the azimuthal dependence from the incident and scattered polar angles. Since the phase matrices and intensity vectors are periodic functions of the azimuthal angle, expanding them in Fourier series in the azimuthal angle decouples the azimuth dependence from polar angles. The incident and scattered intensities expanded in Fourier series are given by

$$\mathbf{I}(\theta, \phi) = \sum_{m=0}^{\infty} \left[\mathbf{I}_e^m(\theta)\cos m\phi + \mathbf{I}_o^m(\theta)\sin m\phi\right]$$

$$\mathbf{I}^s(\theta_s, \phi_s) = \sum_{m=0}^{\infty} \left[\mathbf{I}_{es}^m(\theta_s)\cos m\phi_s + \mathbf{I}_{os}^m(\theta_s)\sin m\phi_s\right]$$

where the superscript m denotes the mth Fourier coefficient, and the subscripts e and o denote the Fourier cosine (even) and sine (odd) coefficients. The backward and forward scattering phase matrices expanded in Fourier series are given by

$$\mathbf{S}(\theta_s, \theta, \phi_s - \phi) = \sum_{m=0}^{\infty} \left[\mathbf{S}_e^m(\theta_s, \theta)\cos m\,(\phi_s - \phi) + \mathbf{S}_o^m(\theta_s, \theta)\sin m\,(\phi_s - \phi)\right]$$

$$\mathbf{F}(\theta_t, \theta, \phi_t - \phi) = \sum_{m=0}^{\infty} \left[\mathbf{F}_e^m(\theta_t, \theta)\cos m\,(\phi_t - \phi) + \mathbf{F}_o^m(\theta_t, \theta)\sin m\,(\phi_t - \phi)\right]$$

The doubling equations given in (8.4) through (8.7) can be expressed in terms of Fourier coefficients. The harmonic multiple scattering phase matrices for the mth Fourier coefficient are then given by

$$\bar{\mathbf{S}}^m = \bar{\mathbf{S}}_1^m + \bar{\mathbf{T}}_1^{m*}\bar{\mathbf{S}}_2^m\left(\mathbf{I} - f_m^2\bar{\mathbf{S}}_1^{m*}\bar{\mathbf{S}}_2^m\right)^{-1}\bar{\mathbf{T}}_1^m \tag{8.8}$$

$$\bar{\mathbf{T}}^m = \bar{\mathbf{T}}_2^m\left(\mathbf{I} - f_m^2\bar{\mathbf{S}}_1^{m*}\bar{\mathbf{S}}_2^m\right)^{-1}\bar{\mathbf{T}}_1^m \tag{8.9}$$

$$\bar{\mathbf{S}}^{m*} = \bar{\mathbf{S}}_1^{m*} + \bar{\mathbf{T}}_1^{m}\bar{\mathbf{S}}_2^{m*}\left(\mathbf{I} - f_m^2\bar{\mathbf{S}}_1^{m}\bar{\mathbf{S}}_2^{m*}\right)^{-1}\bar{\mathbf{T}}_1^{m*} \tag{8.10}$$

$$\bar{\mathbf{T}}^{m*} = \bar{\mathbf{T}}_2^{m*}\left(\mathbf{I} - f_m^2\bar{\mathbf{S}}_1^{m}\bar{\mathbf{S}}_2^{m*}\right)^{-1}\bar{\mathbf{T}}_1^{m*} \tag{8.11}$$

where

$$\bar{\mathbf{S}}^{m} = \begin{bmatrix} \mathbf{S}_e^m & -\mathbf{S}_o^m \\ \mathbf{S}_o^m & \mathbf{S}_e^m \end{bmatrix}$$

$$\bar{\mathbf{T}}^{m} = f_m\bar{\mathbf{F}}^{m} + \bar{\mathbf{E}} = f_m\begin{bmatrix} \mathbf{F}_e^m & -\mathbf{F}_o^m \\ \mathbf{F}_o^m & \mathbf{F}_e^m \end{bmatrix} + \begin{bmatrix} \mathbf{E} & 0 \\ 0 & \mathbf{E} \end{bmatrix}$$

$$f_m = \left\{ \begin{matrix} 1/2 & m = 0 \\ 1/4 & m > 0 \end{matrix} \right.$$

To account for full polarization effect, $\mathbf{S}_e^m$, $\mathbf{S}_o^m$, $\mathbf{T}_e^m$ and $\mathbf{T}_o^m$ are 4×4 matrices that relate the four incident to the four scattered Stokes parameters' Fourier coefficients. Thus, $\bar{\mathbf{S}}^m$ and $\bar{\mathbf{T}}^m$ are 8×8 matrices. For special cases where the phase matrix is symmetric with respect to the plane of incidence, the first two Stokes parameters are azimuthally even and the other two are odd. Under these conditions, $\bar{\mathbf{E}}$ reduces to $\mathbf{E}$, and both $\bar{\mathbf{S}}^m$ and $\bar{\mathbf{T}}^m$ reduce to 4×4 matrices of the forms

$$\mathbf{S}^m = \begin{bmatrix} S_{e11}^m & S_{e12}^m & S_{o13}^m & S_{o14}^m \\ S_{e21}^m & S_{e22}^m & S_{o23}^m & S_{o24}^m \\ S_{o31}^m & S_{o32}^m & S_{e33}^m & S_{e34}^m \\ S_{o41}^m & S_{o42}^m & S_{e43}^m & S_{e44}^m \end{bmatrix} \text{ and } \mathbf{T}^m = \begin{bmatrix} T_{e11}^m & T_{e12}^m & T_{o13}^m & T_{o14}^m \\ T_{e21}^m & T_{e22}^m & T_{o23}^m & T_{o24}^m \\ T_{o31}^m & T_{o32}^m & T_{e33}^m & T_{e34}^m \\ T_{o41}^m & T_{o42}^m & T_{e43}^m & T_{e44}^m \end{bmatrix}$$

where the first and second numeric subscripts denote scattered and incident Stokes parameters.

Note that there is no difference in the formulation of the problem whether we are dealing with $\mathbf{S}^m$ or $\bar{\mathbf{S}}^m$. The difference between the quantities with or without a bar only lies in the contents of these quantities. *Hence, for simplicity we shall use the simpler symbol without the bar in the problem formulation to follow.* It is understood that if a problem does not have the necessary symmetry, we should replace all relevant quantities with bars.

Let *N*-incident polar angles be chosen for calculations. To find the multiple scattering phase matrix using the doubling technique requires the calculations of N-scattered polar angles corresponding to the N-incident angles. Hence, S^m_{epq}, S^m_{opq}, T^m_{epq}, and T^m_{opq} are $N \times N$ while $\mathbf{S}^m$ and $\mathbf{T}^m$ are $4N \times 4N$ matrices.

If the numerical integration is carried out using the Gaussian quadrature method with directional cosine μ as the integration variable, $s^m_{\alpha pq}$, the elements of $S^m_{\alpha pq}$ with ith scattered and jth incident polar angles for an infinitesimally thin layer are given by

$$s^m_{\alpha pq}(\mu_i, \mu_j) = aS^m_{\alpha pq}(\mu_i, \mu_j)\frac{\Delta\tau}{\mu_i}w_j \tag{8.12}$$

where μ_i and μ_j are the Gaussian quadrature zeros, w_i is the weight at μ_i and $1 < i, j < N$. The subscript α denotes either e or o. Similarly, $t^m_{\alpha pq}$, the elements of $T^m_{\alpha pq}$, can be expressed as

$$t^m_{\alpha pq}(\mu_i, \mu_j) = aF^m_{\alpha pq}(\mu_i, \mu_j)\frac{\Delta\tau}{\mu_i}w_i + \delta(\mu_i, \mu_j)\exp(\frac{-\Delta\tau}{\mu_i}) \tag{8.13}$$

where $\delta(\ )$ is the Kronecker delta function. Now $s^m_{\alpha pq}$ and $t^m_{\alpha pq}$ of (8.12) and (8.13) are then substituted into (8.8) and (8.11) to generate the harmonic phase matrix for an inhomogeneous layer with arbitrary thickness.

8.4 BOUNDARY MEDIUM INTERACTION

The scattering at the boundary between two media are affected by the boundary roughness, the permittivity difference, and the inhomogeneities of the media. In the radiative transfer formulation, interactions between the interface and the inhomogeneities in the media are solved by imposing the boundary conditions. In the matrix doubling method, the boundary medium interactions are determined graphically using a ray tracing technique. Four types of interfaces are commonly encountered in the modeling of natural terrain, namely, interfaces between (1) two homogeneous media, (2) upper homogeneous and lower inhomogeneous media, (3) upper inhomogeneous and lower homogeneous media, and (4) two inhomogeneous media. The boundary relations of the first type of interface are discussed in Section 8.4.1, the second and third types are derived in Section 8.4.2, and the fourth type is developed in Section 8.4.3.

8.4.1 Interface Between Two Homogeneous Mdia

At the interface between two homogeneous media, the scattering characteristics are determined by the interface roughness and the discontinuity between the media. When the interface is between two homogeneous half spaces, as shown in Figure 8.4, the problem reduces to a classical surface scattering formulation. Several surface scattering models have been developed in the last three decades [Ulaby et al., 1982]. They are the small perturbation model (SPM), Kirchhoff model (KM), phase perturbation model (PPM) [Winebrenner and Ishimaru, 1985], full wave model (FWM) [Bahar, 1991], and the integral equation model (IEM) developed in Chapter 4. The small perturbation model is applicable only to surfaces with small roughnesses relative to the incident wavelength, whereas the Kirchhoff model is valid for surfaces with large scale roughnesses. The PPM, FWM, and IEM surface models are expected to have wider ranges of validity than either SPM or KM. The range of validity of the IEM model has been discussed in Chapter 6. The model was also verified by laboratory measurements of bistatic scattering from surfaces with small, intermediate and large scale roughnesses [Sections 2.5.3 and 5.7.3; Nance et al., 1990; Nance, 1992].

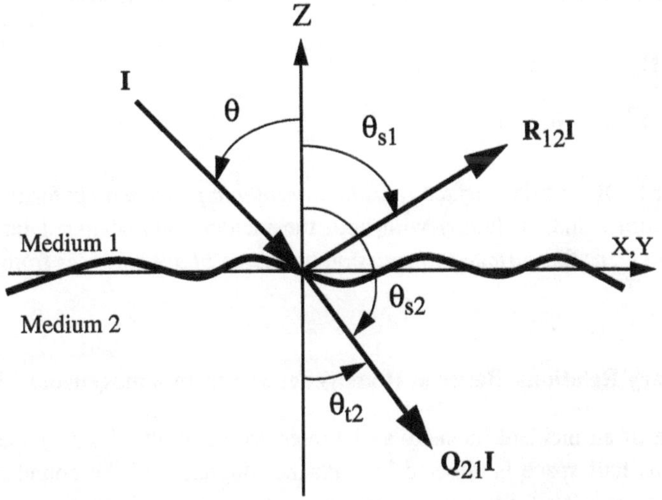

Figure 8.4 The scattering process at the interface between two homogeneous media.

The surface scattering model is incorporated into the matrix doubling method through its phase matrix. It was shown that the scattered intensity due to a rough surface is related to the incident intensity by [Fung and Eom, 1981a]

$$\mathbf{I}^s(\theta_s, \phi_s) = \frac{1}{4\pi}\int_0^{2\pi}\int_0^{\pi/2}\left[\frac{\sigma^\circ(\theta_s, \theta, \phi_s - \phi)}{\cos\theta_s}\right]\mathbf{I}(\theta, \phi)\sin\theta d\theta d\phi \tag{8.14}$$

where the quantity inside the bracket is the phase function, and the scattered direction can be either backward, $\theta_s \le 90°$, or forward, $\theta_s > 90°$. Equation (8.14) can be expressed in Fourier component form as

$$\mathbf{I}^{sm}(\theta_s) = f_m \int_0^{\pi/2} \left[\frac{\sigma^m(\theta_s, \theta)}{\cos\theta_s}\right] \mathbf{I}^m(\theta)\sin\theta d\theta \tag{8.15}$$

By applying the N-point Gaussian quadrature technique to (8.15) and computing the scattered intensity in N directions in accordance with the quadrature zeros as described in Section 8.3, the following matrix equation is obtained:

$$\mathbf{I}^{sm} = f_m \Gamma^m \mathbf{I}^m, \quad f_m = \begin{cases} 1/2 & m = 0 \\ 1/4 & m > 0 \end{cases} \tag{8.16}$$

where $\mathbf{I}^{sm}$ and $\mathbf{I}^m$ are $4N$ column vectors, and Γ^m is a $4N \times 4N$ matrix,

$$\Gamma^m = \begin{cases} \mathbf{R}_{ij}^m & ,\theta_s \le 90° \\ \mathbf{Q}_{ji}^m & ,\theta_s > 90° \end{cases} = \frac{\sigma^m(\theta_s, \theta)}{\cos\theta_s}$$

We shall refer to $\mathbf{R}_{ij}^m$ as the *surface reflection harmonic phase matrix* for the interface between medium i and medium j with both the incident and scattered intensities in medium i and $\mathbf{Q}_{ji}^m$ as the *surface transmission harmonic phase matrices* from medium i to medium j.

8.4.2 Boundary Relations Between Homogeneous and Inhomogeneous Media

The scattering of an incident intensity at the interface between a homogeneous and an inhomogeneous half space is affected by both the roughness of the boundary and the medium inhomogeneity. Consider the case where the upper half space (medium 1) is homogeneous and the lower half space (medium 2) is inhomogeneous. When an incident intensity in medium 1 impinges upon medium 2, the effective reflection and transmission phase matrices, $\tilde{\mathbf{R}}_{12}$ and $\tilde{\mathbf{Q}}_{21}$, are given by summing the appropriate terms as shown in Figure 8.5(a):

$$\tilde{\mathbf{R}}_{12} = \mathbf{R}_{12} + \mathbf{Q}_{12}\mathbf{S}_2(\mathbf{I} - \mathbf{R}_{21}\mathbf{S}_2)^{-1}\mathbf{Q}_{21} \tag{8.17}$$

$$\tilde{\mathbf{Q}}_{21} = (\mathbf{I} - \mathbf{R}_{21}\mathbf{S}_2)^{-1}\mathbf{Q}_{21} \tag{8.18}$$

where $\mathbf{I}$ is the identity matrix, $\mathbf{S}_2$ is the backward scattering phase matrix of medium 2, and $\mathbf{R}_{ij}$, $\mathbf{Q}_{ji}$ are the surface reflection and transmission phase matrices respectively given in Section 8.4.1.

When the direction of the incident intensity is reversed as shown in Figure 8.5(b), the effective reflection and transmission phase matrices, $\tilde{\mathbf{R}}_{21}$ and $\tilde{\mathbf{Q}}_{12}$, are

$$\tilde{\mathbf{R}}_{21} = \mathbf{R}_{21}(\mathbf{I} - \mathbf{S}_2\mathbf{R}_{21})^{-1} \tag{8.19}$$

$$\tilde{\mathbf{Q}}_{12} = \mathbf{Q}_{12}(\mathbf{I} - \mathbf{S}_2\mathbf{R}_{21})^{-1} \tag{8.20}$$

The effective reflection and transmission phase matrices given by (8.17) through (8.20), can be expressed in terms of their Fourier coefficients as

$$\tilde{\mathbf{R}}_{12}^{m} = \mathbf{R}_{12}^{m} + f_m^2 \mathbf{Q}_{12}^{m}\mathbf{S}_2^{m}(\mathbf{I} - f_m^2 \mathbf{R}_{21}^{m}\mathbf{S}_2^{m})^{-1}\mathbf{Q}_{21}^{m} \tag{8.21}$$

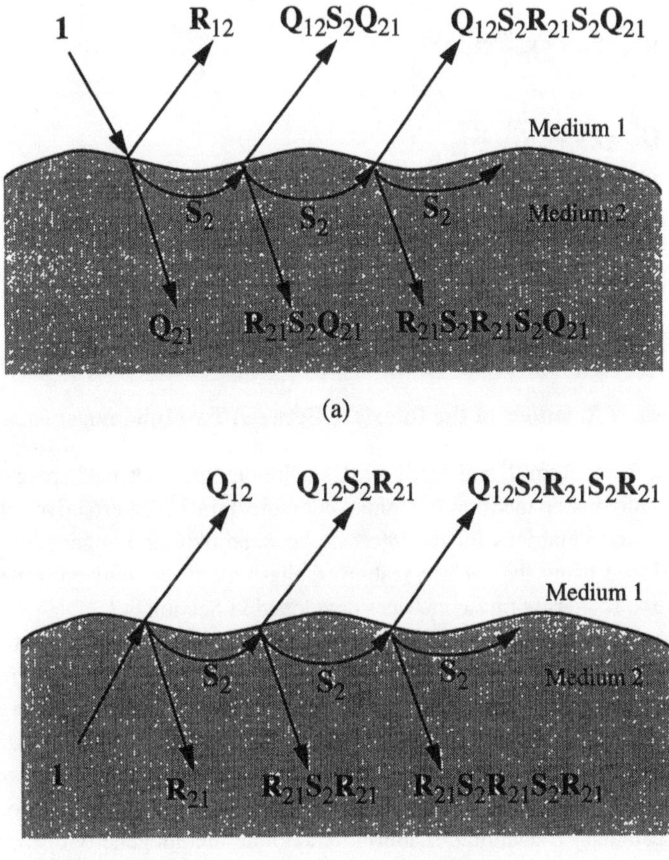

Figure 8.5 The scattering processes at the interface between homogeneous (medium 1) and inhomogeneous (medium 2) half spaces due to a unit incident intensity. $\mathbf{S}_2$ is the backward scattering phase matrix of medium 2.

$$\tilde{\mathbf{Q}}_{21}^{m} = (\mathbf{I} - f_m^2 \mathbf{R}_{21}^m \mathbf{S}_2^m)^{-1} \mathbf{Q}_{21}^m \tag{8.22}$$

$$\tilde{\mathbf{R}}_{21}^{m} = \mathbf{R}_{21}^m (\mathbf{I} - f_m^2 \mathbf{S}_2^m \mathbf{R}_{21}^m)^{-1} \tag{8.23}$$

$$\tilde{\mathbf{Q}}_{12}^{m} = \mathbf{Q}_{12}^m (\mathbf{I} - f_m^2 \mathbf{S}_2^m \mathbf{R}_{21}^m)^{-1} \tag{8.24}$$

Similarly, when the upper half space (medium 1) is inhomogeneous and lower half space (medium 2) is homogeneous, the effective reflection and transmission harmonic phase matrices are given by (see Figure 8.6)

$$\underset{\sim}{\mathbf{R}}_{12}^{m} = \mathbf{R}_{12}^m (\mathbf{I} - f_m^2 \mathbf{S}_1^{m*} \mathbf{R}_{12}^m)^{-1} \tag{8.25}$$

$$\underset{\sim}{\mathbf{Q}}_{21}^{m} = \mathbf{Q}_{21}^m (\mathbf{I} - f_m^2 \mathbf{S}_1^{m*} \mathbf{R}_{12}^m)^{-1} \tag{8.26}$$

$$\underset{\sim}{\mathbf{R}}_{21}^{m} = \mathbf{R}_{21}^m + f_m^2 \mathbf{Q}_{21}^m \mathbf{S}_1^{m*} (\mathbf{I} - f_m^2 \mathbf{R}_{12}^m \mathbf{S}_1^{m*})^{-1} \mathbf{Q}_{12}^m \tag{8.27}$$

$$\underset{\sim}{\mathbf{Q}}_{12}^{m} = (\mathbf{I} - f_m^2 \mathbf{R}_{12}^m \mathbf{S}_1^{m*})^{-1} \mathbf{Q}_{12}^m \tag{8.28}$$

8.4.3 Boundary Relations at the Interface Between Two Inhomogeneous Media

Assuming both medium 1 and medium 2 are inhomogeneous half spaces, let a unit incident intensity inside medium 1 impinge upon medium 2. The effective reflection and transmission phase matrices for the interface between two inhomogeneous half spaces can be developed using the surface scattering phase matrices obtained in Section 8.4.1 and the volume scattering phase matrices developed in Section 8.2.

To derive the effective reflection and transmission phase matrices at the interface between two inhomogeneous half spaces, consider four basic surface-volume interactions shown in Figures 8.7 and 8.8. In each basic surface-volume interaction process, a unit ray undergoes the multiple scattering process (primary scattering) that generates secondary rays that have not undergone the multiple scattering processes. For a unit incident ray in medium 1, the primary reflected multiple scattering process and its associated secondary transmitted scattering due to the inhomogeneity of medium 1 and the interface is shown Figure 8.7(a). The primary reflection phase matrix for the incident ray, $\mathbf{R}_{ip}$, is given by

$$\mathbf{R}_{ip} = (\mathbf{I} - \mathbf{R}_{12} \mathbf{S}_1^{*})^{-1} \mathbf{R}_{12} \tag{8.29}$$

and the associated secondary transmission phase matrix for the incident ray, $\mathbf{Q}_{is}$, is

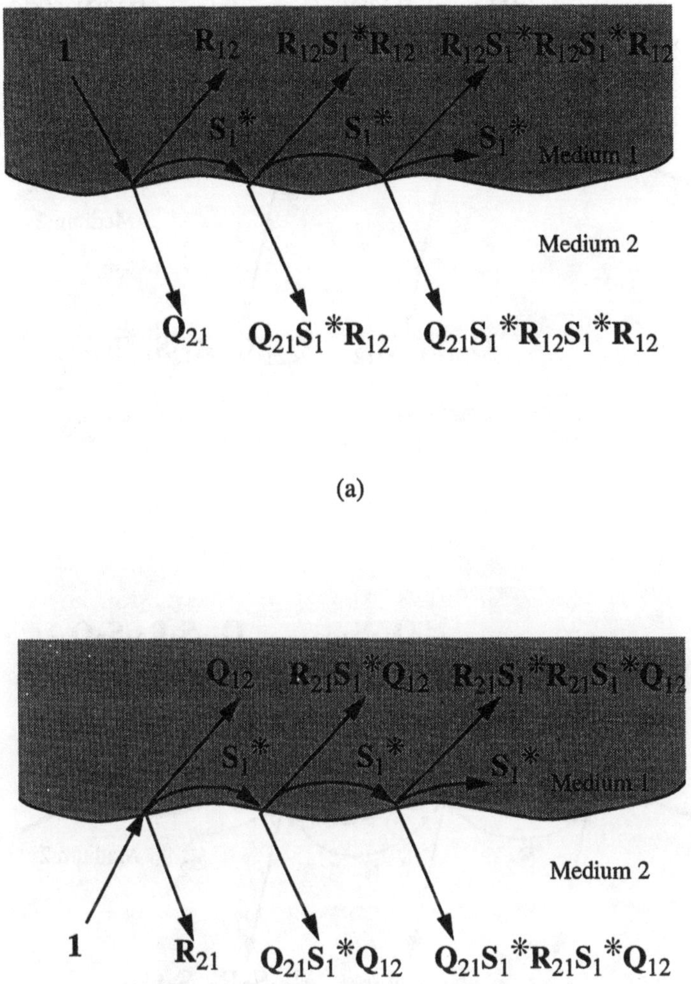

(a)

(b)

Figure 8.6 The scattering processes at the interface between inhomogeneous (medium 1) and homogeneous (medium 2) half spaces due to a unit incidenct intensity. $\mathbf{S}_1^*$ is medium 1 backward scattering phase matrix for incidence in $+z$ direction.

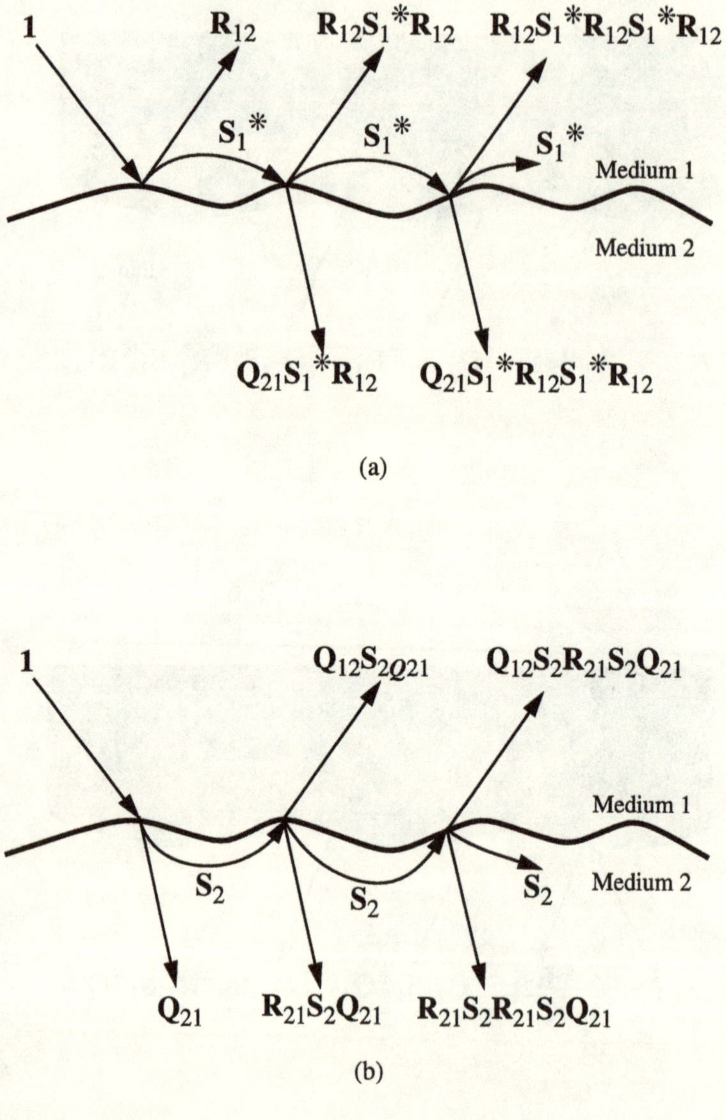

Figure 8.7 Basic scattering processes at the interface between two inhomogeneous media due to a unit incident intensity from medium 1. (a) Reflected multiple (primary) scattering in medium 1. (b) Transmitted multiple (primary) scattering in medium 2.

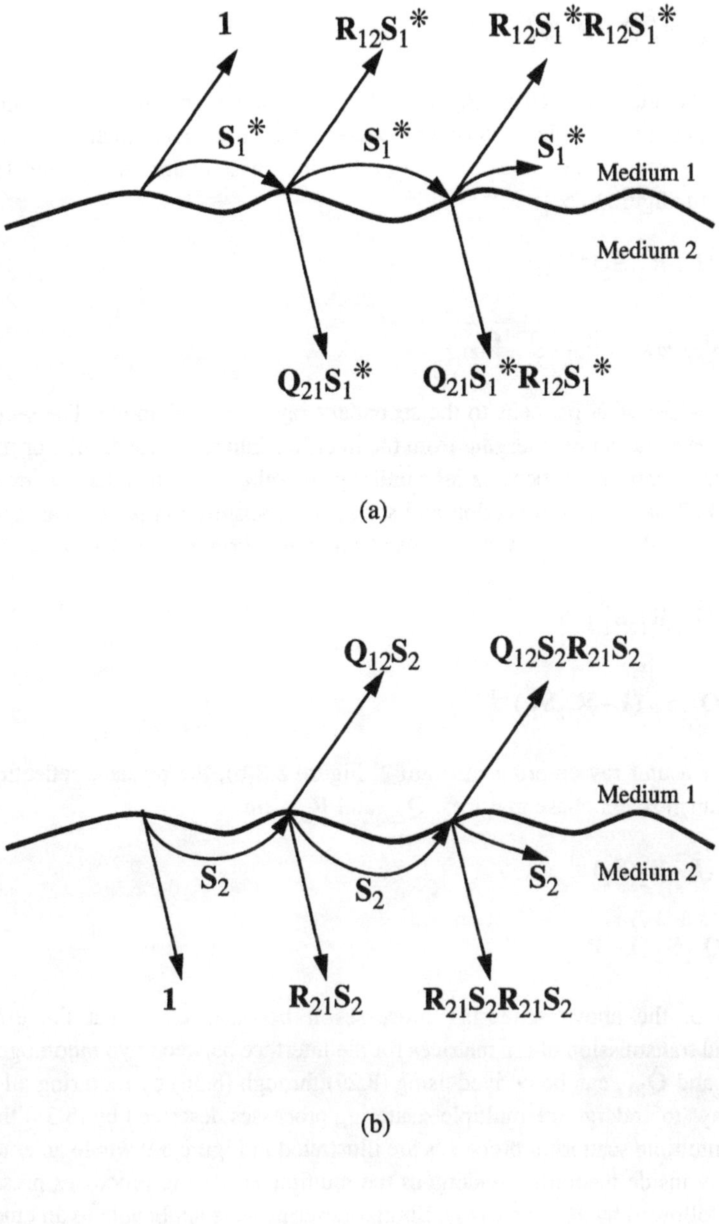

Figure 8.8 Basic scattering processes at the interface between two inhomogeneous media due to a unit intensity emerging upward from the interface. (a) Reflected multiple (primary) scattering in medium 1. (b) Reflected multiple (primary) scattering in medium 2.

$$\mathbf{Q}_{is} = \mathbf{Q}_{21}\mathbf{S}_1^* (\mathbf{I} - \mathbf{R}_{12}\mathbf{S}_1^*)^{-1}\mathbf{R}_{12} \tag{8.30}$$

where $\mathbf{I}$ is the identity matrix. Figure 8.7(b) shows the primary transmission and secondary reflection multiple scattering processes for a unit incident intensity in medium 1. The primary transmission and secondary reflection phase matrices, $\mathbf{Q}_{ip}$ and $\mathbf{R}_{is}$, are shown from the figure to be

$$\mathbf{Q}_{ip} = (\mathbf{I} - \mathbf{R}_{21}\mathbf{S}_2)^{-1}\mathbf{Q}_{21} \tag{8.31}$$

$$\mathbf{R}_{is} = \mathbf{Q}_{12}\mathbf{S}_2 (\mathbf{I} - \mathbf{R}_{21}\mathbf{S}_2)^{-1}\mathbf{Q}_{21} \tag{8.32}$$

Next, consider what happens to the secondary ray discussed above. The secondary ray can be viewed as a ray emerging from the interface into either medium 1 or medium 2. The multiple scattering processes for a unit ray emerging from the interface are shown in Figure 8.8. The primary reflection and secondary transmission phase matrices for a unit ray inside medium 1, $\mathbf{R}_{up}$ and $\mathbf{Q}_{ds}$, can be obtained from Figure 8.8(a) and they are

$$\mathbf{R}_{up} = (\mathbf{I} - \mathbf{R}_{12}\mathbf{S}_1^*)^{-1} \tag{8.33}$$

$$\mathbf{Q}_{ds} = \mathbf{Q}_{21}\mathbf{S}_1^* (\mathbf{I} - \mathbf{R}_{12}\mathbf{S}_1^*)^{-1} \tag{8.34}$$

Similarly, for a unit ray entering medium 2, Figure 8.8(b), the primary reflection and secondary transmission phase matrices, $\mathbf{Q}_{dp}$ and $\mathbf{R}_{us}$, are

$$\mathbf{Q}_{dp} = (\mathbf{I} - \mathbf{R}_{21}\mathbf{S}_2)^{-1} \tag{8.35}$$

$$\mathbf{R}_{us} = \mathbf{Q}_{12}\mathbf{S}_2 (\mathbf{I} - \mathbf{R}_{21}\mathbf{S}_2)^{-1} \tag{8.36}$$

In view of the above scattering processes it becomes clear that the effective reflection and transmission phase matrices for the interface between two inhomogeneous media, $\hat{\mathbf{R}}_{12}$ and $\hat{\mathbf{Q}}_{21}$, can be derived using (8.29) through (8.36) by requiring all of the secondary rays to undergo the multiple scattering processes described by (8.33) through (8.36). The multiple scattering processes are illustrated in Figure 8.9 where an emerging secondary ray inside medium 1 undergoes the multiple scattering processes prescribed by $\mathbf{R}_{is}$ and followed by $\mathbf{R}_{up}$ and $\mathbf{Q}_{ds}$. Similar description is applicable to an emerging secondary ray inside medium 2 which undergoes the multiple scattering process prescribed by $\mathbf{Q}_{is}$. As a result, the following relations are obtained:

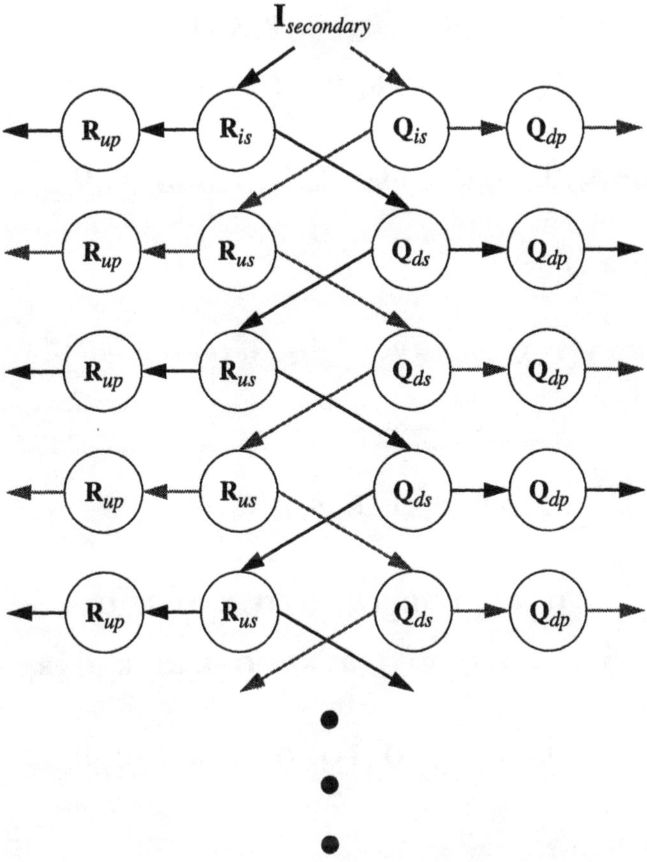

Figure 8.9 The reflected and transmitted scattered intensities resulted from multiple scattering of the secondary rays.

$$\hat{\mathbf{R}}_{12} = \mathbf{R}_{ip} + \mathbf{R}_{up}\mathbf{R}_{is} + \mathbf{R}_{up}\mathbf{R}_{us}\mathbf{Q}_{ds}\mathbf{R}_{is} + \mathbf{R}_{up}\mathbf{R}_{us}\mathbf{Q}_{ds}\mathbf{R}_{us}\mathbf{Q}_{ds}\mathbf{R}_{is} + \ldots$$

$$+ \mathbf{R}_{up}\mathbf{R}_{us}\mathbf{Q}_{is} + \mathbf{R}_{up}\mathbf{R}_{us}\mathbf{Q}_{ds}\mathbf{R}_{us}\mathbf{Q}_{is} + \mathbf{R}_{up}\mathbf{R}_{us}\mathbf{Q}_{ds}\mathbf{R}_{us}\mathbf{Q}_{ds}\mathbf{R}_{us}\mathbf{Q}_{is} + \ldots$$

$$= \mathbf{R}_{ip} + \mathbf{R}_{up}(\mathbf{I} - \mathbf{R}_{us}\mathbf{Q}_{ds})^{-1}\mathbf{R}_{is} + \mathbf{R}_{up}(\mathbf{I} - \mathbf{R}_{us}\mathbf{Q}_{ds})^{-1}\mathbf{R}_{us}\mathbf{Q}_{is}$$

$$= (\mathbf{I} - \mathbf{R}_{12}\mathbf{S}_1^*)^{-1}$$

$$\{\mathbf{R}_{12} + \left[\mathbf{I} - \mathbf{Q}_{12}\mathbf{S}_2(\mathbf{I} - \mathbf{R}_{21}\mathbf{S}_2)^{-1}\mathbf{Q}_{21}\mathbf{S}_1^*(\mathbf{I} - \mathbf{R}_{12}\mathbf{S}_1^*)^{-1}\right]^{-1}$$

$$\mathbf{Q}_{12}\mathbf{S}_2(\mathbf{I} - \mathbf{R}_{21}\mathbf{S}_2)^{-1}$$

$$\left[\mathbf{Q}_{21} + \mathbf{Q}_{21}\mathbf{S}_1^*(\mathbf{I} - \mathbf{R}_{12}\mathbf{S}_1^*)^{-1}\mathbf{R}_{12}\right]\} \tag{8.37}$$

$$\hat{\mathbf{Q}}_{21} = \mathbf{Q}_{ip} + \mathbf{Q}_{dp}\mathbf{Q}_{is} + \mathbf{Q}_{dp}\mathbf{Q}_{ds}\mathbf{R}_{us}\mathbf{Q}_{is} + \mathbf{Q}_{dp}\mathbf{Q}_{ds}\mathbf{R}_{us}\mathbf{Q}_{ds}\mathbf{R}_{us}\mathbf{Q}_{is} + \ldots$$

$$+ \mathbf{Q}_{dp}\mathbf{Q}_{ds}\mathbf{R}_{is} + \mathbf{Q}_{dp}\mathbf{Q}_{ds}\mathbf{R}_{us}\mathbf{Q}_{ds}\mathbf{R}_{is} + \mathbf{Q}_{dp}\mathbf{Q}_{ds}\mathbf{R}_{us}\mathbf{Q}_{ds}\mathbf{R}_{us}\mathbf{Q}_{ds}\mathbf{R}_{is} + \ldots$$

$$= \mathbf{Q}_{ip} + \mathbf{Q}_{dp}(\mathbf{I} - \mathbf{Q}_{ds}\mathbf{R}_{us})^{-1}\mathbf{Q}_{is} + \mathbf{Q}_{dp}(\mathbf{I} - \mathbf{Q}_{ds}\mathbf{R}_{us})^{-1}\mathbf{Q}_{us}\mathbf{R}_{is}$$

$$= (\mathbf{I} - \mathbf{R}_{21}\mathbf{S}_2)^{-1}$$

$$\{\mathbf{Q}_{21} + \left[\mathbf{I} - \mathbf{Q}_{21}\mathbf{S}_1^*(\mathbf{I} - \mathbf{R}_{12}\mathbf{S}_1^*)^{-1}\mathbf{Q}_{12}\mathbf{S}_2(\mathbf{I} - \mathbf{R}_{21}\mathbf{S}_2)^{-1}\right]^{-1}$$

$$\mathbf{Q}_{21}\mathbf{S}_1^*(\mathbf{I} - \mathbf{R}_{12}\mathbf{S}_1^*)^{-1}$$

$$[\mathbf{R}_{12} + \mathbf{Q}_{12}\mathbf{S}_2(\mathbf{I} - \mathbf{R}_{21}\mathbf{S}_2)^{-1}\mathbf{Q}_{21}]\} \tag{8.38}$$

The corresponding harmonic forms of (8.37) and (8.38) are

$$\hat{\mathbf{R}}_{12}^m = (\mathbf{I} - f_m^2\mathbf{R}_{12}^m\mathbf{S}_1^{m*})^{-1}$$

$$\{\mathbf{R}_{12}^m + \left[\mathbf{I} - f_m^4\mathbf{Q}_{12}^m\mathbf{S}_2^m(\mathbf{I} - f_m^2\mathbf{R}_{21}^m\mathbf{S}_2^m)^{-1}\mathbf{Q}_{21}^m\mathbf{S}_1^{m*}(\mathbf{I} - f_m^2\mathbf{R}_{12}^m\mathbf{S}_1^{m*})^{-1}\right]^{-1}$$

$$f_m^2\mathbf{Q}_{12}^m\mathbf{S}_2^m(\mathbf{I} - f_m^2\mathbf{R}_{21}^m\mathbf{S}_2^m)^{-1}\left[\mathbf{Q}_{21}^m + f_m^2\mathbf{Q}_{21}^m\mathbf{S}_1^{m*}(\mathbf{I} - f_m^2\mathbf{R}_{12}^m\mathbf{S}_1^{m*})^{-1}\mathbf{R}_{12}^m\right]\} \tag{8.39}$$

$$\hat{\mathbf{Q}}_{21}^{m} = (\mathbf{I} - f_m^2 \mathbf{R}_{21}^m \mathbf{S}_2^m)^{-1}$$

$$\{\mathbf{Q}_{21}^m + \left[\mathbf{I} - f_m^4 \mathbf{Q}_{21}^m \mathbf{S}_1^{m*} (\mathbf{I} - f_m^2 \mathbf{R}_{12}^m \mathbf{S}_1^{m*})^{-1} \mathbf{Q}_{12}^m \mathbf{S}_2^m (\mathbf{I} - f_m^2 \mathbf{R}_{21}^m \mathbf{S}_2^m)^{-1}\right]^{-1}$$

$$f_m^2 \mathbf{Q}_{21}^m \mathbf{S}_1^{m*} (\mathbf{I} - f_m^2 \mathbf{R}_{12}^m \mathbf{S}_1^{m*})^{-1}$$

$$\left[\mathbf{R}_{12}^m + f_m^2 \mathbf{Q}_{12}^m \mathbf{S}_2^m (\mathbf{I} - f_m^2 \mathbf{R}_{21}^m \mathbf{S}_2^m)^{-1} \mathbf{Q}_{21}^m\right]\}$$

(8.40)

Note that $\hat{\mathbf{R}}_{12}^m$ and $\hat{\mathbf{Q}}_{21}^m$ reduce to $\tilde{\mathbf{R}}_{12}^m$ and $\tilde{\mathbf{Q}}_{21}^m$ when medium 1 is homogeneous, i.e. $\mathbf{S}_1^m = \mathbf{S}_1^{m*} = \mathbf{0}$. Similarly, they reduce to $\underset{\sim}{\mathbf{R}}_{12}^m$ and $\underset{\sim}{\mathbf{Q}}_{21}^m$ when medium 2 is homogeneous. They also reduce to $\mathbf{R}_{12}^m$ and $\mathbf{Q}_{21}^m$ when the interface is between two homogeneous half spaces. Thus, the boundary conditions at two inhomogeneous half spaces can be viewed as generalized boundary conditions for an interface between two half spaces.

When the incident direction is reversed, i.e., changed to $+z$ direction, the same arguments apply in the derivation of the effective reflection and transmission phase matrices, $\hat{\mathbf{R}}_{21}^m$ and $\hat{\mathbf{Q}}_{12}^m$, and are given by

$$\hat{\mathbf{R}}_{21}^{m} = (\mathbf{I} - f_m^2 \mathbf{R}_{21}^m \mathbf{S}_2^m)^{-1}$$

$$\{\mathbf{R}_{21}^m + \left[\mathbf{I} - f_m^4 \mathbf{Q}_{21}^m \mathbf{S}_1^{m*} (\mathbf{I} - f_m^2 \mathbf{R}_{12}^m \mathbf{S}_1^{m*})^{-1} \mathbf{Q}_{12}^m \mathbf{S}_2^m (\mathbf{I} - f_m^2 \mathbf{R}_{21}^m \mathbf{S}_2^m)^{-1}\right]^{-1}$$

$$f_m^2 \mathbf{Q}_{21}^m \mathbf{S}_1^{m*} (\mathbf{I} - f_m^2 \mathbf{R}_{12}^m \mathbf{S}_1^{m*})^{-1}$$

$$\left[\mathbf{Q}_{12}^m + f_m^2 \mathbf{Q}_{12}^m \mathbf{S}_2^m (\mathbf{I} - f_m^2 \mathbf{R}_{21}^m \mathbf{S}_2^m)^{-1} \mathbf{R}_{21}^m\right]\}$$

(8.41)

$$\hat{\mathbf{Q}}_{12}^{m} = (\mathbf{I} - f_m^2 \mathbf{R}_{12}^m \mathbf{S}_1^{m*})^{-1}$$

$$\{\mathbf{Q}_{12}^m + \left[\mathbf{I} - f_m^4 \mathbf{Q}_{12}^m \mathbf{S}_2^m (\mathbf{I} - f_m^2 \mathbf{R}_{21}^m \mathbf{S}_2^m)^{-1} \mathbf{Q}_{21}^m \mathbf{S}_1^{m*} (\mathbf{I} - f_m^2 \mathbf{R}_{12}^m \mathbf{S}_1^{m*})^{-1}\right]^{-1}$$

$$f_m^2 \mathbf{Q}_{12}^m \mathbf{S}_2^m (\mathbf{I} - f_m^2 \mathbf{R}_{21}^m \mathbf{S}_2^m)^{-1} \left[\mathbf{R}_{21}^m + f_m^2 \mathbf{Q}_{21}^m \mathbf{S}_1^{m*} (\mathbf{I} - f_m^2 \mathbf{R}_{12}^m \mathbf{S}_1^{m*})^{-1} \mathbf{Q}_{12}^m\right]\} \quad (8.42)$$

8.5 SCATTERING COEFFICIENTS FOR AN INHOMOGENEOUS LAYER WITH IRREGULAR BOUNDARIES

The scattering process incurred on a unit incident intensity by an inhomogeneous layer with irregular boundaries is shown in Figure 8.1. The inhomogeneous medium is characterized by backward volume-scattering phase matrices **S** and $\mathbf{S}^*$ and forward volume-scattering phase matrices **T** and $\mathbf{T}^*$. The reflection and transmission phase matrices at the layer's boundaries are derived in Section 8.4.2. Thus, the total reflection and transmission scattering phase matrices of the irregular inhomogeneous layer, $\mathbf{S}_{T1}$ and $\mathbf{T}_{T1}$, can be derived using the ray tracing method shown in Figure 8.10 and are given by

$$\mathbf{S}_{T1} = \tilde{\mathbf{R}}_{12} + \tilde{\mathbf{Q}}_{12}(\mathbf{I} - \mathbf{T}^* \underset{\sim}{\mathbf{R}}_{23} \mathbf{T} \tilde{\mathbf{R}}_{21})^{-1} \mathbf{T}^* \underset{\sim}{\mathbf{R}}_{23} \mathbf{T} \tilde{\mathbf{Q}}_{21} \tag{8.43}$$

$$\mathbf{T}_{T1} = \underset{\sim}{\mathbf{Q}}_{32}(\mathbf{I} - \mathbf{T}\tilde{\mathbf{R}}_{21}\mathbf{T}^* \underset{\sim}{\mathbf{R}}_{23})^{-1} \mathbf{T}\tilde{\mathbf{Q}}_{21} \tag{8.44}$$

The Fourier components of $\mathbf{S}_{T1}$ and $\mathbf{T}_{T1}$ are

$$\mathbf{S}_{T1}^m = \tilde{\mathbf{R}}_{12}^m + f_m^2 \tilde{\mathbf{Q}}_{12}^m (\mathbf{I} - f_m^2 \mathbf{T}^{m*} \underset{\sim}{\mathbf{R}}_{23}^m \mathbf{T}^m \tilde{\mathbf{R}}_{21}^m)^{-1} \mathbf{T}^{m*} \underset{\sim}{\mathbf{R}}_{23}^m \bar{\mathbf{T}}^m \tilde{\mathbf{Q}}_{21}^m \tag{8.45}$$

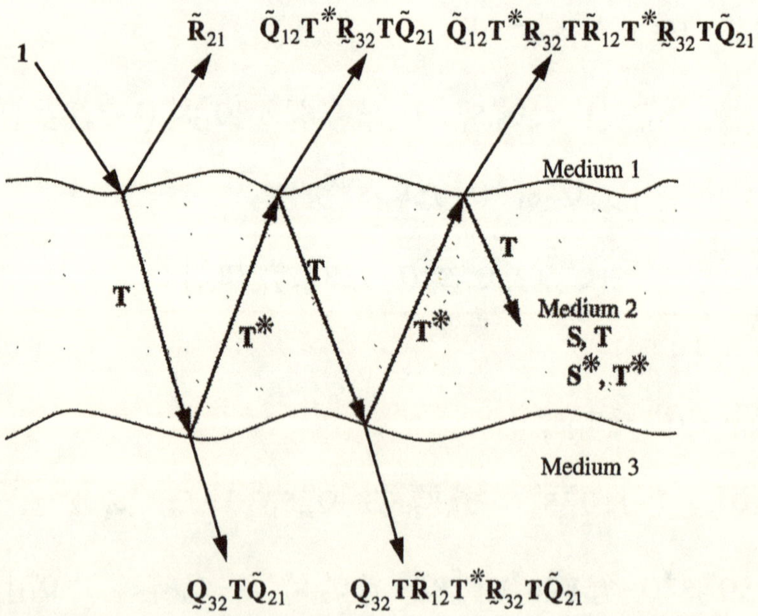

Figure 8.10 The scattering process due to an irregular inhomogeneous layer (medium 2) characterized b the volume scattering phase matrices **S**, $\mathbf{S}^*$, **T** and $\mathbf{T}^*$.

$$\mathbf{T}_{T1}^{m} = f_m^2 \underset{\sim}{\mathbf{Q}}_{32}^{m}(\mathbf{I} - f_m^2 \mathbf{T}^{m}\tilde{\mathbf{R}}_{21}^{m}\mathbf{T}^{m*}\underset{\sim}{\mathbf{R}}_{23}^{m})^{-1}\mathbf{T}^{m}\tilde{\mathbf{Q}}_{21}^{m} \tag{8.46}$$

In Fourier component form, the scattered intensity from an irregular inhomogeneous layer is related to the incident intensity by

$$\mathbf{I}^{sm} = f_m \mathbf{S}_{T1}^{m}\mathbf{I}^{m} \tag{8.47}$$

The harmonic scattering coefficient of the irregular inhomogeneous layer can be obtained by comparing (8.47) to (8.15) and (8.16), and it is given by

$$\sigma_{\alpha pq}^{m}(\theta_i, \theta_j) = 4\pi\cos\theta_i\left[\mathbf{S}_{T1}^{m}(\theta_i, \theta_j)\right]_{\alpha pq}$$

where α is either a for a cosine series' coefficient or b for a sine series' coefficient; p and q denote the scattered and incident polarization states, respectively. The total scattering coefficient of an irregular inhomogeneous layer, $\sigma_{pq}^{o}(\theta_i, \theta_j; \phi_s - \phi)$, is thus given by

$$\sigma_{pq}^{o}(\theta_i, \theta_j; \phi_s - \phi) = \sum_{m=0}^{\infty} 4\pi\cos\theta_i\Big[\left[\mathbf{S}_{T1}^{m}(\theta_i, \theta_j)\right]_{epq}\cos m(\phi_s - \phi) + \left[\mathbf{S}_{T1}^{m}(\theta_i, \theta_j)\right]_{opq}\sin m(\phi_s - \phi)\ )\Big] \tag{8.48}$$

When the scattering layer is isotropic, such as a Rayleigh or Mie medium with isotropically rough boundaries, its total scattering coefficient simplifies to

$$\sigma_{pq}^{o}(\theta_i, \theta_j; \phi_s - \phi) = \sum_{m=0}^{\infty} 4\pi\cos\theta_i\left[\mathbf{S}_{T1}^{m}(\theta_i, \theta_j)\right]_{epq}\cos m(\phi_s - \phi) \tag{8.49}$$

8.6 EXTENSION TO A MULTILAYERED INHOMOGENEOUS MEDIUM

To analyze the scattering processes for a multilayered inhomogeneous medium, consider the scattering geometry of an N-layered inhomogeneous medium shown in Fig. 8.11. The uppermost and lowermost interfaces are between homogeneous and inhomogeneous media and their boundary-volume interactions are described in Section 8.4.2. The couplings between two adjacent inhomogeneous layers are determined using the generalized boundary relations derived in Section 8.4.4. The scattering process within each layer, with appropriate boundary conditions, is described in Section 8.5.

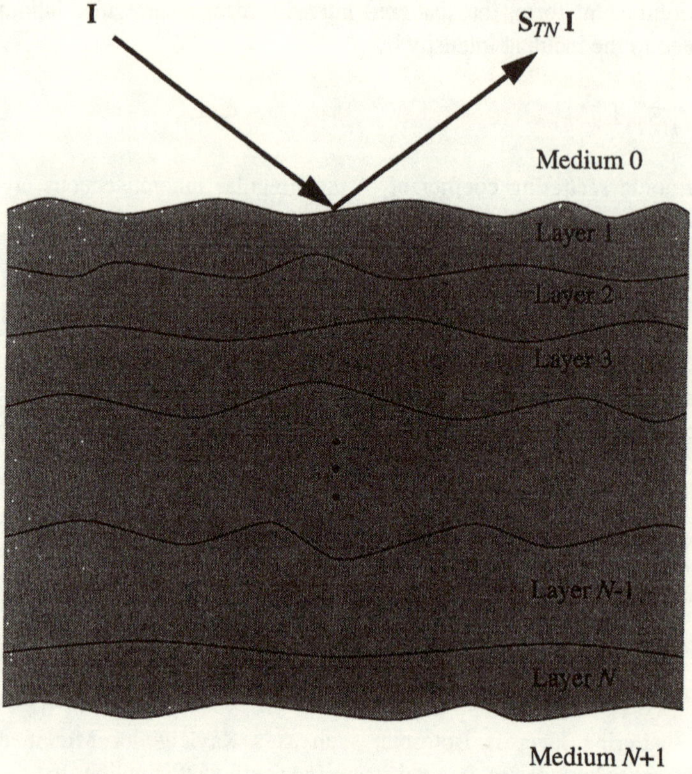

Figure 8.11 The scattering geometry for an N-layered inhomogeneous medium with irregular interfaces. Medium 0 and medium $N + 1$ are homogeneous.

The total reflection phase matrix of the N-layered media, $\mathbf{S}_{TN}$, can be computed using the following algorithm:

1. Determine the volume scattering phase matrices for each inhomogeneous layer, $\mathbf{S}_i$ and $\mathbf{T}_i$, $1 < i < N$, using the matrix doubling method described in Section 8.2 and Section 8.3. These phase matrices have not taken into account the layer boundary effects.
2. Determine the reflection and transmission phase matrices, $\hat{\mathbf{R}}_{ji}$ and $\hat{\mathbf{Q}}_{ji}$ for all of the interfaces assuming the interfaces are between two half spaces (see Section 8.4). $\hat{\mathbf{R}}_{ji}$ represents scattering into medium j by the interface between layers i and j, and $\hat{\mathbf{Q}}_{ji}$ represents transmission from medium i to medium j.
3. Starting with the lowest inhomogeneous layer, layer N, determine its total reflection phase matrix observed in layer N - 1, $\mathbf{R}_N^T$. In this derivation assume that

medium N - 1 is a half space above the Nth layer. The scattering phase matrix $\mathbf{R}_N^T$ for the layer is given by

$$\mathbf{R}_N^T = \hat{\mathbf{R}}_{N-1,N} + \hat{\mathbf{Q}}_{N-1,N}(\mathbf{I} - \mathbf{T}_N^* \underset{\sim}{\mathbf{R}}_{N,N+1}\mathbf{T}_N \hat{\mathbf{R}}_{N,N-1})^{-1}\mathbf{T}_N^* \underset{\sim}{\mathbf{R}}_{N,N+1}\mathbf{T}_N \hat{\mathbf{Q}}_{N,N-1} \tag{8.50}$$

As illustrated in section , the phase matrix $\mathbf{R}_N^T$ accounts for scattering effects due to the volume, boundaries, boundary-volume interactions, and coupling between layer N and medium N - 1 (inhomogeneous upper half space).

4. Similarly, the total reflection phase matrix for layer N - 1, $\mathbf{R}_{N-1}^T$, is obtained by assuming layer N - 2 as the upper half space and layer N as the homogeneous lower half space with effective reflection phase matrix $\mathbf{R}_N^T$. The phase matrix $\mathbf{R}_{N-1}^T$ is given by

$$\mathbf{R}_{N-1}^T = \hat{\mathbf{R}}_{N-2,N-1}$$
$$+ \hat{\mathbf{Q}}_{N-2,N-1}(\mathbf{I} - \mathbf{T}_{N-1}^* \mathbf{R}_N^T \mathbf{T}_{N-1} \hat{\mathbf{R}}_{N-1,N-2})^{-1}\mathbf{T}_{N-1}^* \mathbf{R}_N^T \mathbf{T}_{N-1} \hat{\mathbf{Q}}_{N-1,N-2} \tag{8.51}$$

Note that the phase matrix $\mathbf{R}_{N-1}^T$ accounts not only for the scattering effects due to layer N - 1 but also the effects due to all layers beneath it.

5. Repeat the steps progressively from layer N - 2 to layer 1. The total reflected scattering phase matrix for an N-layered inhomogeneous medium is given by

$$\mathbf{S}_{TN} = \mathbf{R}_1^T = \hat{\mathbf{R}}_{01} + \underset{\sim}{\mathbf{Q}}_{01}(\mathbf{I} - \mathbf{T}_1^* \mathbf{R}_2^T \mathbf{T}_1 \hat{\mathbf{R}}_{10})^{-1}\mathbf{T}_1^* \mathbf{R}_2^T \mathbf{T}_1 \hat{\mathbf{Q}}_{10} \tag{8.52}$$

and its Fourier component form is

$$\mathbf{S}_{TN}^m = \tilde{\mathbf{R}}_{01}^m + f_m^2 \tilde{\mathbf{Q}}_{01}^m (\mathbf{I} - f_m^2 \mathbf{T}_1^{m*} \mathbf{R}_2^{Tm} \mathbf{T}_1^m \tilde{\mathbf{R}}_{10}^m)^{-1} \mathbf{T}_1^{m*} \mathbf{R}_2^{Tm} \mathbf{T}_1^m \tilde{\mathbf{Q}}_{10}^m \tag{8.53}$$

The scattering coefficients for the N-layered inhomogeneous medium can be obtained from the total reflected scattering phase matrix, $\mathbf{S}_{TN}$, using the method described in Section 8.5 and (8.48) or (8.49).

8.7 MATRIX DOUBLING FORMULATION FOR EMISSION

The matrix doubling method has been utilized in modeling the emission from a sparse irregular inhomogeneous layer. It was applied in the computation of emission from an inhomogeneous layer with no boundaries by Twomey [1975, 1979] and the derivation is summarized in Section 8.7.1. Fung and Eom [1981c] extended the doubling technique to incorporate the rough boundary effects on emission from an irregular Rayleigh layer, and the summary of their work is provided in Section 8.7.2. For a dense layer, instead of using the Rayleigh phase matrix given in Appendix 2C, the volume scattering phase matrices for a closely packed medium developed in Chapter 9 are used in the emission model. Further extension of the matrix doubling method to model the emission from a multilayered irregular inhomogeneous medium is presented in Section 8.7.3.

8.7.1 Emission from an Inhomogeneous Medium Without Boundaries

Consider the emission-scattering processes in an inhomogeneous layer without boundaries depicted in Figure 8.12. The emission source is from an infinitesimally thin layer with optical thickness $\Delta\tau$ at vertical distance τ from the top. Under local thermodynamic equilibrium, the upwelling and downwelling emitted intensities, ΔI_u and ΔI_d, at wavelength λ from a thin layer with thickness $\Delta\tau$ at vertical level τ are given by the Kirchhoff-Planck Law:

$$\Delta I_u(\tau;\,\theta,\phi) = \Delta I_d(\tau;\,\theta,\phi) = \frac{K}{\lambda^2}\varepsilon_r T(\tau)\frac{\{1-\omega(\theta,\phi)\}}{\cos\theta}\Delta\tau \tag{8.54}$$

where θ and ϕ are the polar and azimuthal angles defining the propagation direction, ε_r is the relative permittivity of the layer, ω is the single scattering albedo, $T(\tau)$ is the temperature of the thin layer, and the layer relative permeability μ_r is assumed to be 1. The emitted intensities ΔI_u and ΔI_d undergo multiple scattering and absorption processes due to the inhomogeneities in the media above and below the thin layer at level τ. The effects of the multiple scattering and absorption are accounted for through the matrix doubling process.

For a statistically isotropic medium the emitted intensities are independent of the azimuthal angle. When the intensity is expanded in Fourier components with respect to the azimuthal angle, integration over the azimuthal angle eliminates all Fourier components except the zeroth order. The sizes of the intensity vectors and phase matrices used in the matrix doubling formulation are dependent on the number of incident and scattering angles chosen and the number of Stokes parameters used in the computation. In natural emission, we assume that there is no correlation between the horizontally and vertically emitted field components. Hence, only the first two modified Stokes parameters, the vertically and horizontally polarized intensities, are nonzero. Therefore, if N polar angles are chosen in accordance with the Gaussian quadrature technique, the intensity vectors will be $2N$-column vectors and the phase matrices will be $2N\times 2N$ matrices.

The zeroth-order Fourier components of the backward- and forward-diffused volume multiple-scattering phase matrices, $\mathbf{S}^0$ and $\mathbf{F}^0$, for an infinitesimally thin layer with thickness $\Delta\tau$ are

$$\mathbf{S}^0(\theta_s,\theta;\,\Delta\tau) = \omega\mathbf{U}^{-1}\mathbf{P}^0(\theta_s,\pi-\theta)\,(\Delta\tau)$$

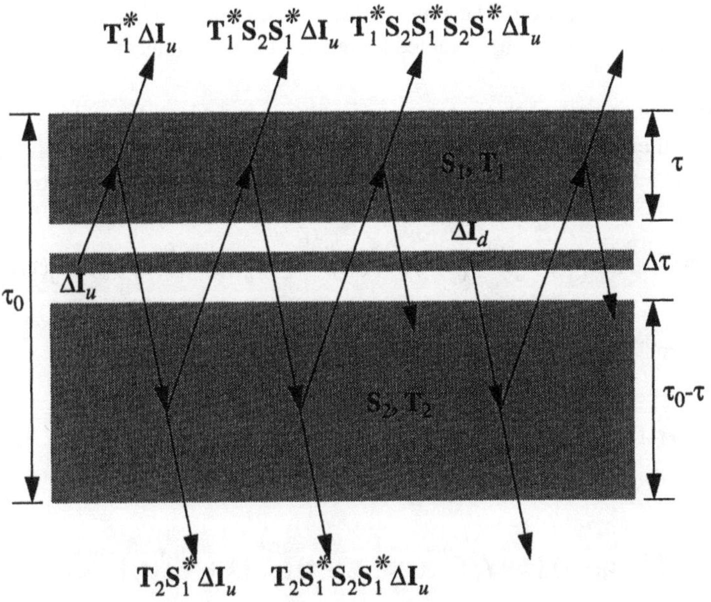

Figure 8.12 The scattering processes undergone by the emissions from an infinitesimally thin layer at level τ.

$$\mathbf{F}^0(\theta_t, \theta; \Delta\tau) = \omega\mathbf{U}^{-1}\mathbf{P}^0(\pi - \theta_t, \pi - \theta)\Delta\tau$$

where ω is the single scattering albedo, $\mathbf{U}$ is the diagonal matrix of the cosine of the scattered angles and $\mathbf{P}^0$ is the $2N\times 2N$ zeroth-order Fourier component of the single scattering phase matrix for the first two Stokes parameters. Note that the zeroth harmonic total forward scattering phase matrix, $\mathbf{T}^0$, is given by

$$\mathbf{T}^0(\Delta\tau) = f_0\mathbf{F}^0(\Delta\tau) + \mathbf{E}(\Delta\tau)$$

where f_0 is 1/2, and $\mathbf{E}(\Delta\tau)$ is a diagonal matrix with elements $\exp(-\Delta\tau/\cos\theta_i)$.

When the thickness of the layer is doubled, i.e., to $2\Delta\tau$, its multiple scattering phase matrices can be obtained by combining $\mathbf{S}^0(\Delta\tau)$ and $\mathbf{T}^0(\Delta\tau)$ using the doubling process described in Section 8.2. They are given by

$$\mathbf{S}^0(2\Delta\tau) = \mathbf{S}^0 + \mathbf{T}^{0*}\mathbf{S}^0(\mathbf{I} - f_0^2\mathbf{S}^{0*}\mathbf{S}^0)^{-1}\mathbf{T}^0 \tag{8.55}$$

$$\mathbf{T}^0(2\Delta\tau) = \mathbf{T}^0(\mathbf{I} - f_0^2\mathbf{S}^{0*}\mathbf{S}^0)^{-1}\mathbf{T}^0 \tag{8.56}$$

where **I** is the identity matrix, and the superscript $*$ denotes scattering phase matrix for incidence in the $+z$ direction. The total multiple scattering phase matrices for a layer with thickness $2^n\Delta\tau$ can thus be obtained by repeated applications of (8.55) and (8.56) n times.

For a layer with optical thickness τ_0 and without boundary, the net upwelling and downwelling intensities, $\Delta\mathbf{u}_u(\tau)$ and $\Delta\mathbf{u}_d(\tau)$, due to emission from an infinitesimally thin sublayer at level τ sandwiched between two scattering layers are given by (see 8.12)

$$\Delta\mathbf{u}_u(\tau) = \mathbf{T}^{0*}(\tau)\{\mathbf{I} - f_0^2\mathbf{S}^0(\tau_0-\tau)\mathbf{S}^{0*}(\tau)\}^{-1}\{\Delta\mathbf{I}_u(\tau) + \mathbf{S}^0(\tau_0-\tau)\Delta\mathbf{I}_d(\tau)\}$$

$$= \mathbf{T}^{0*}(\tau)\{\mathbf{I} - f_0^2\mathbf{S}^0(\tau_0-\tau)\mathbf{S}^{0*}(\tau)\}^{-1}\{\mathbf{I} + \mathbf{S}^0(\tau_0-\tau)\}$$

$$\mathbf{U}^{-1}\left[\{1-\omega\}\frac{K}{\lambda^2}\varepsilon_r T(\tau)\Delta\tau\right] \tag{8.57}$$

$$\Delta\mathbf{u}_d(\tau) = \mathbf{T}^{0*}(\tau_0-\tau)\{\mathbf{I} - f_0^2\mathbf{S}^{0*}(\tau)\mathbf{S}^0(\tau_0-\tau)\}^{-1}\{\mathbf{I} + \mathbf{S}^{0*}(\tau)\}$$

$$\mathbf{U}^{-1}\left[\{1-\omega\}\frac{K}{\lambda^2}\varepsilon_r T(\tau)\Delta\tau\right] \tag{8.58}$$

where ω is the single scattering albedo and $T(\tau)$ is layer temperature at level τ. The total upward emission $\mathbf{u}_u$ from such a layer with thickness τ_0 is obtained by integrating over the thickness of the layer, i.e.,

$$\mathbf{u}_u(\tau_0) = \int_0^{\tau_0}\mathbf{T}^{0*}(\tau)\{\mathbf{I} - f_0^2\mathbf{S}^0(\tau_0-\tau)\mathbf{S}^{0*}(\tau)\}^{-1}\{\mathbf{I} + \mathbf{S}^0(\tau_0-\tau)\}$$

$$\{1-\omega\}\frac{K}{\lambda^2}\varepsilon_r\mathbf{U}^{-1}\mathbf{T}(\tau)d\tau \tag{8.59}$$

where $\mathbf{u}_u$ and $\mathbf{T}$ are $2N$ column vectors, and $\mathbf{T}$ is the layer temperature profile. Similarly, the total downward emission from the layer is given by

$$\mathbf{u}_d(\tau_0) = \int_0^{\tau_0}\mathbf{T}^{0*}(\tau_0-\tau)\{\mathbf{I} - f_0^2\mathbf{S}^{0*}(\tau)\mathbf{S}^0(\tau_0-\tau)\}^{-1}\{\mathbf{I} + \mathbf{S}^{0*}(\tau)\}$$

$$\{1-\omega\}\frac{K}{\lambda^2}\varepsilon_r\mathbf{U}^{-1}\mathbf{T}(\tau)d\tau \tag{8.60}$$

The scattering phase matrices $\mathbf{S}^0(\tau)$ and $\mathbf{T}^0(\tau)$ are computed using either (8.55)–(8.56) or the generalized doubling method given in (8.4)–(8.7). The generalized doubling method must be employed if the computations of scattering phase matrices for arbitrary thicknesses τ and $\tau_0 - \tau$ are needed. However, considerable saving of computational time can be achieved if the optical thickness τ is chosen such that [Twomey, 1979]

$$\tau = 2^{-N-1}\tau_0, \ldots, \frac{\tau_0}{4}, \frac{\tau_0}{2}, \tau_0$$

where $\mathbf{S}^0(\tau)$ and $\mathbf{T}^0(\tau)$ can be calculated using (8.55)–(8.56). In this case, $\mathbf{S}^0(\tau_0 - \tau)$ and $\mathbf{T}^0(\tau_0 - \tau)$ can be determined from $\mathbf{S}^0(\tau)$ and $\mathbf{T}^0(\tau)$ without carrying out a full sequence of doublings for each $\tau_0 - \tau$, and the computation procedure is as follows:

$$\{\mathbf{S}^0(\tau_0 - \tau), \mathbf{T}^0(\tau_0 - \tau)\}\Big|_{\tau = \tau_0} = \{\mathbf{0}, \mathbf{I}\}$$

$$\{\mathbf{S}^0(\tau_0 - \tau), \mathbf{T}^0(\tau_0 - \tau)\}\Big|_{\tau = \tau_0/2} = \{\mathbf{S}^0(0), \mathbf{T}^0(0)\} \otimes \{\mathbf{S}^0(\tau_0/2), \mathbf{T}^0(\tau_0/2)\}$$

$$\{\mathbf{S}^0(\tau_0 - \tau), \mathbf{T}^0(\tau_0 - \tau)\}\Big|_{\tau = \tau_0/4}$$

$$= \{\mathbf{S}^0(\tau_0/2), \mathbf{T}^0(\tau_0/2)\} \otimes \{\mathbf{S}^0(\tau_0/4), \mathbf{T}^0(\tau_0/4)\}$$

$$\{\mathbf{S}^0(\tau_0 - \tau), \mathbf{T}^0(\tau_0 - \tau)\}\Big|_{\tau = \tau_0/8} = \{\mathbf{S}^0(3\tau_0/4), \mathbf{T}^0(3\tau_0/4)\}$$

$$\otimes \{\mathbf{S}^0(\tau_0/8), \mathbf{T}^0(\tau_0/8)\}$$

.

.

.

where $\otimes$ denotes the doubling process given in (8.8)–(8.11).

8.7.2 Emission from an Inhomogeneous Layer with Irregular Boundaries

The geometry of the emission problem for an inhomogeneous layer (medium 2) with irregular boundaries is shown in Figure 8.1, where medium 1 and medium 3 are assumed to be homogeneous half spaces and the layer backward and forward volume scattering phase matrices are $\mathbf{S}$ and $\mathbf{T}$. The upward and downward emission sources in the layer, $\mathbf{u}_u$

and $\mathbf{u}_d$, are calculated using (8.61) and (8.63), and the emission from the lower homogeneous half space, $\mathbf{u}_l$, can be determined using Rayleigh-Jean's law and the lower boundary emissivity **e** as

$$\mathbf{u}_l = \mathbf{e}\,(KT/\lambda^2)\,[\mu\varepsilon/(\mu_0\varepsilon_0)]$$

where K is the Boltzmann constant, T is the physical temperature in Kelvins, μ is the permeability, λ is the operating wavelength and ε is the permittivity of the half space.

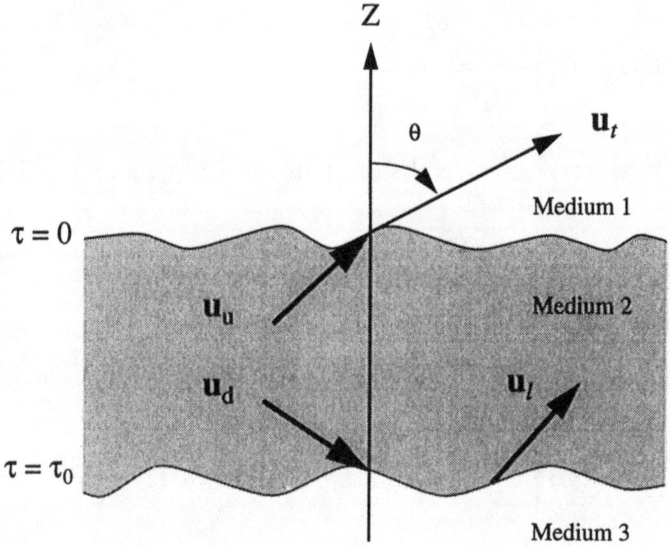

Figure 8.13 Geometry of the single layer emission problem.

Because of the inhomogeneities in the layer and the discontinuity at the interfaces, emissions from the three sources $\mathbf{u}_u$, $\mathbf{u}_d$ and $\mathbf{u}_l$ will further undergo the multiple scattering processes. The total emission $\mathbf{u}_t$ into medium 1 due to $\mathbf{u}_u$, $\mathbf{u}_d$ and $\mathbf{u}_l$ may be written as

$$\mathbf{u}_t = \boldsymbol{L}_u\mathbf{u}_u + \boldsymbol{L}_d\mathbf{u}_d + \boldsymbol{L}_l\mathbf{u}_l \tag{8.61}$$

where $\boldsymbol{L}$'s are the multiple scattering operators.

The multiple scattering operators can be obtained by applying the ray tracing methods shown in Figures 8.14–8.16, and utilizing the effective reflected and transmitted scattering phase matrices for the interface between homogeneous and inhomogeneous half spaces developed in Section 8.4.2. Note that in the emission calculations for statistically isotropic media only the zeroth-order Fourier components are needed. When the incident intensity is in medium 2, the zeroth harmonic of the effective interface scattering matrices for the upper boundary are given by

$$\tilde{\mathbf{R}}_{21}^{0} = \mathbf{R}_{21}^{0}(\mathbf{I} - f_0^2\mathbf{S}^0\mathbf{R}_{21}^{0})^{-1} \tag{8.62}$$

$$\tilde{\mathbf{Q}}_{12}^{0} = \mathbf{Q}_{12}^{0}(\mathbf{I} - f_0^2\mathbf{S}^0\mathbf{R}_{21}^{0})^{-1} \tag{8.63}$$

and the effective reflected scattering phase matrix at the lower boundary is

$$\underset{\sim}{\mathbf{R}}_{23}^{0} = \mathbf{R}_{23}^{0}(\mathbf{I} - f_0^2\mathbf{S}^0\mathbf{R}_{23}^{0})^{-1} \tag{8.64}$$

Consider the multiple scattering processes for an upward emission source $\mathbf{u}_u$ shown in Figure 8.14. The effective emitted intensity in medium 1 due to $\mathbf{u}_u$ is obtained by summing all the upwardly transmitted rays from the layer, and the multiple scattering operator L_u is thus given by

$$\begin{aligned} L_u &= f_0\tilde{\mathbf{Q}}_{12}^{0} + f_0^3\tilde{\mathbf{Q}}_{12}^{0}\mathbf{T}^{0*}\underset{\sim}{\mathbf{R}}_{23}^{0}\mathbf{T}^{0}\tilde{\mathbf{R}}_{21}^{0} + f_0^5\tilde{\mathbf{Q}}_{12}^{0}\mathbf{T}^{0*}\underset{\sim}{\mathbf{R}}_{23}^{0}\mathbf{T}^{0}\tilde{\mathbf{R}}_{21}^{0}\mathbf{T}^{0*}\underset{\sim}{\mathbf{R}}_{23}^{0}\mathbf{T}^{0}\tilde{\mathbf{R}}_{21}^{0} + \dots \\ &= f_0\tilde{\mathbf{Q}}_{12}^{0}(\mathbf{I} - f_0^2\mathbf{T}^{0*}\underset{\sim}{\mathbf{R}}_{23}^{0}\mathbf{T}^{0}\tilde{\mathbf{R}}_{21}^{0})^{-1} \end{aligned} \tag{8.65}$$

Similarly, from Figure 8.1 and Figure 8.1, the operators L_d and L_l are given by

$$L_d = f_0^2\tilde{\mathbf{Q}}_{12}^{0}(\mathbf{I} - f_0^2\mathbf{T}^{0*}\underset{\sim}{\mathbf{R}}_{23}^{0}\mathbf{T}^{0}\tilde{\mathbf{R}}_{21}^{0})^{-1}\mathbf{T}^{0*}\underset{\sim}{\mathbf{R}}_{23}^{0} \tag{8.66}$$

$$L_l = f_0\tilde{\mathbf{Q}}_{12}^{0}(\mathbf{I} - f_0^2\mathbf{T}^{0*}\underset{\sim}{\mathbf{R}}_{23}^{0}\mathbf{T}^{0}\tilde{\mathbf{R}}_{21}^{0})^{-1}\mathbf{T}^{0*} \tag{8.67}$$

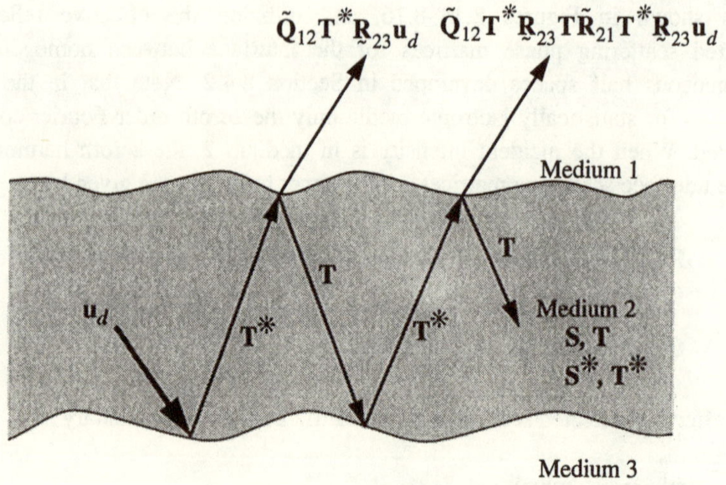

Figure 8.14 The emission-scattering processes due to a downwelling emission source, $\mathbf{u}_d$, in the irregular inhomogeneous layer (medium 2) characterized by the volume scattering phase matrices $\mathbf{S}$, $\mathbf{S}^*$, $\mathbf{T}$ and $\mathbf{T}^*$.

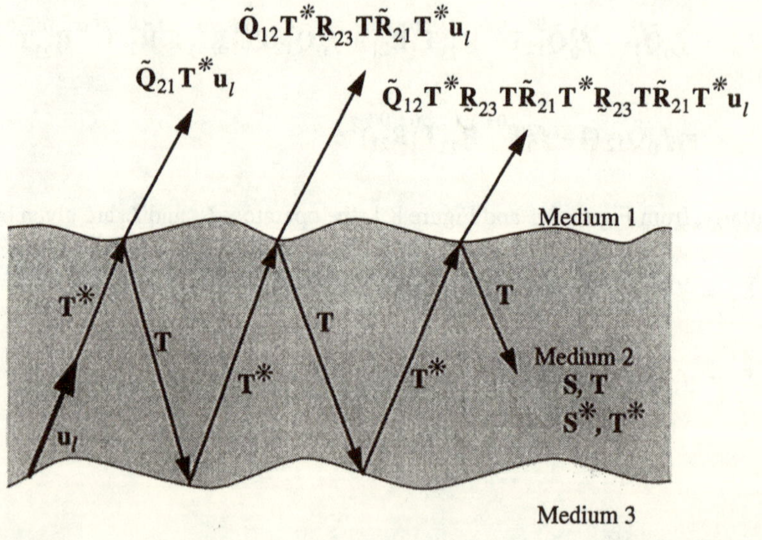

Figure 8.15 The emission-scattering processes due to an upwelling emission source from medium 3, $\mathbf{u}$ in the irregular inhomogeneous layer (medium 2) characterized by the volume scattering phase matrice $\mathbf{S}$, $\mathbf{S}^*$, $\mathbf{T}$ and $\mathbf{T}^*$.

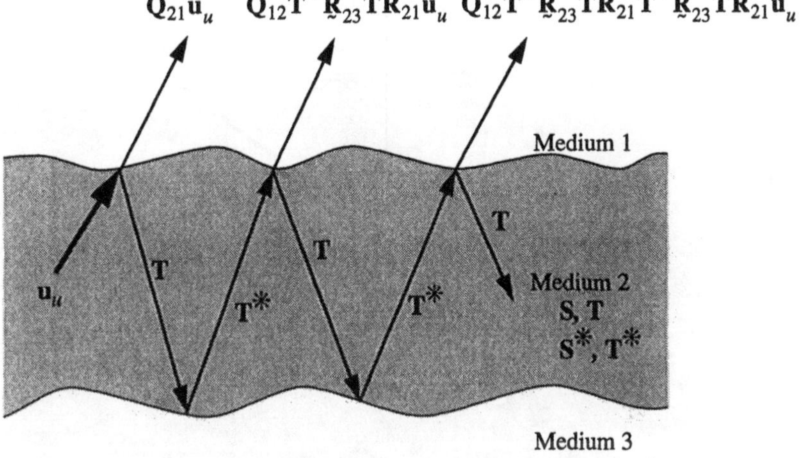

Figure 8.16 The emission-scattering processes due to an upwelling emission source, $\mathbf{u}_u$, in the irregular inhomogeneous layer (medium 2) characterized by the volume scattering phase matrices $\mathbf{S}$, $\mathbf{S}^*$, $\mathbf{T}$ and $\mathbf{T}^*$.

8.7.3 Emission from a Multilayered Irregular Inhomogeneous Medium

Consider the multilayer emission problem shown in Figure 8.17 where medium 0 and medium $N + 1$ are homogeneous half spaces. Unlike the multilayer scattering problem where only one energy source (dominant source) is present, emission sources from each layer are contributing to the total emission. The emitted intensities from each layer not only undergo the multiple scattering process inside the layer, as illustrated in the previous section, but must also go through additional multiple scattering processes inside other layers. The generalized boundary conditions and the total reflected scattering phase matrix for N - layer developed in Sections 8.4.3 and 8.6 are utilized in the multilayer emission model.

Assuming that the measurements are performed in medium 0 (upper half space), the multilayer emission calculations are carried out using the following algorithm:

1. Determine the upward and downward emission sources, $\mathbf{u}_u$ and $\mathbf{u}_d$, inside each of the inhomogeneous layers using (8.61) and (8.63).
2. Start with the lowest inhomogeneous layer, layer N, compute the total emission into layer N - 1, $\mathbf{u}_{t,N}$, using

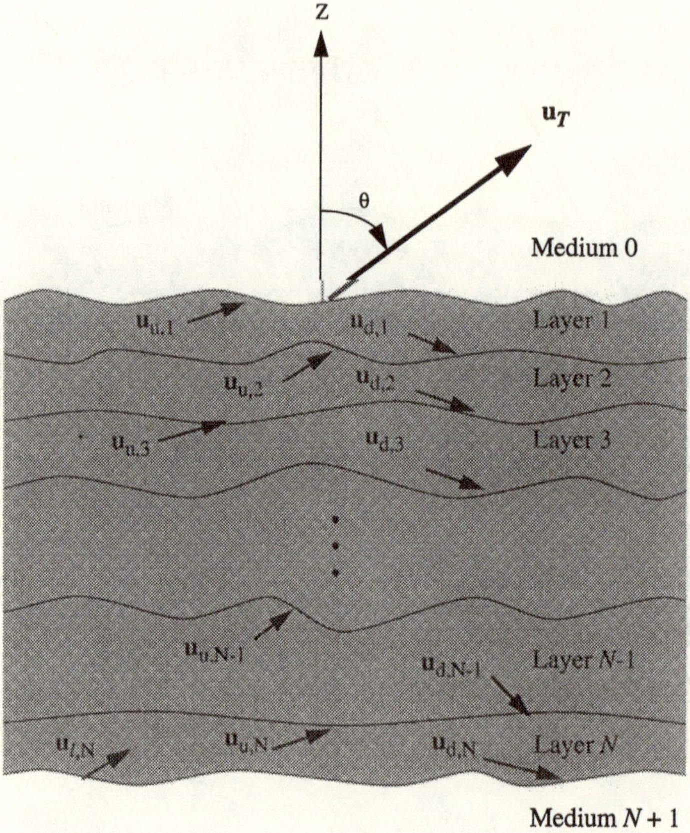

Figure 8.17 The emission geometry for an N-layered inhomogeneous medium with irregular interfaces. Medium 0 and medium $N+1$ are homogeneous and $\mathbf{u}$'s are the emission sources.

$$\mathbf{u}_{t,N} = \boldsymbol{L}_{u,N}\mathbf{u}_{u,N} + \boldsymbol{L}_{d,N}\mathbf{u}_{d,N} + \boldsymbol{L}_{l,N}\mathbf{u}_{l,N} \tag{8.68}$$

where

$$\boldsymbol{L}_{u,N} = f_0 \hat{\mathbf{Q}}^0_{N-1,N}(\mathbf{I} - f_0^2 \mathbf{T}^{0*}_N \underset{\sim}{\mathbf{R}}^0_{N,N+1} \mathbf{T}^0_N \hat{\mathbf{R}}^0_{N,N-1})^{-1}$$

$$\boldsymbol{L}_{d,N} = f_0^2 \hat{\mathbf{Q}}^0_{N-1,N}(\mathbf{I} - f_0^2 \mathbf{T}^{0*}_N \underset{\sim}{\mathbf{R}}^0_{N,N+1} \mathbf{T}^0_N \hat{\mathbf{R}}^0_{N,N-1})^{-1} \mathbf{T}^{0*}_N \underset{\sim}{\mathbf{R}}^0_{N,N+1}$$

$$L_{l,N} = f_0\hat{\mathbf{Q}}^0_{N-1,N}(\mathbf{I} - f_0^2\mathbf{T}^{0*}_N\mathbf{R}^0_{N,N+1}\mathbf{T}^0_N\hat{\mathbf{R}}^0_{N,N-1})^{-1}\mathbf{T}^{0*}_N$$

Note that the generalized boundary conditions are utilized at the interfaces between two inhomogeneous layers. The source $\mathbf{u}_{l,N}$ is the emission from homogeneous half space with rough boundary and can be determined from the boundary emissivity.

3. The total emission from layer N, $\mathbf{u}_{t,N}$, is an emission source for layer N - 1. However, all of the emission sources in layer N - 1 must undergo the multiple scattering processes associated with the inhomogeneities and discontinuities underneath the layer. This is achieved through the use of total reflected scattering phase matrix $\mathbf{R}^T_N$ (see (8.50) and (8.51)) at the lower interface of layer N - 1. The cumulated emission out of layer N - 1 is thus given by

$$\mathbf{u}_{t,N-1} = L_{u,N-1}\mathbf{u}_{u,N-1} + L_{d,N-1}\mathbf{u}_{d,N-1} + L_{l,N-1}\mathbf{u}_{t,N} \tag{8.69}$$

where the multiple scattering operators are

$$L_{u,N-1} = f_0\hat{\mathbf{Q}}^0_{N-2,N-1}(\mathbf{I} - f_0^2\mathbf{T}^{0*}_{N-1}\mathbf{R}^{T0}_N\mathbf{T}^0_{N-1}\hat{\mathbf{R}}^0_{N-1,N-2})^{-1}$$

$$L_{d,N-1} = f_0^2\hat{\mathbf{Q}}^0_{N-2,N-1}(\mathbf{I} - f_0^2\mathbf{T}^{0*}_{N-1}\mathbf{R}^{T0}_N\mathbf{T}^0_{N-1}\hat{\mathbf{R}}^0_{N-1,N-2})^{-1}\mathbf{T}^{0*}_{N-1}\mathbf{R}^{T0}_N$$

$$L_{l,N-1} = f_0\hat{\mathbf{Q}}^0_{N-2,N-1}(\mathbf{I} - f_0^2\mathbf{T}^{0*}_{N-1}\mathbf{R}^{T0}_N\mathbf{T}^0_{N-1}\hat{\mathbf{R}}^0_{N-1,N-2})^{-1}\mathbf{T}^{0*}_{N-1}$$

4. Repeat the steps progressively for layer N - 2 to layer 1. The total emission from an N-layered inhomogeneous medium is given by

$$\mathbf{u}_T = \mathbf{u}_{t,1} = L_{u,1}\mathbf{u}_{u,1} + L_{d,1}\mathbf{u}_{d,1} + L_{l,1}\mathbf{u}_{t,2} \tag{8.70}$$

and the multiple scattering operators are

$$L_{u,1} = f_0\hat{\mathbf{Q}}^0_{01}(\mathbf{I} - f_0^2\mathbf{T}^{0*}_1\mathbf{R}^{T0}_2\mathbf{T}^0_1\hat{\mathbf{R}}^0_{10})^{-1}$$

$$L_{d,1} = f_0^2\hat{\mathbf{Q}}^0_{01}(\mathbf{I} - f_0^2\mathbf{T}^{0*}_1\mathbf{R}^{T0}_2\mathbf{T}^0_1\hat{\mathbf{R}}^0_{10})^{-1}\mathbf{T}^{0*}_1\mathbf{R}^{T0}_2$$

$$L_{l,1} = f_0\hat{\mathbf{Q}}^0_{01}(\mathbf{I} - f_0^2\mathbf{T}^{0*}_1\mathbf{R}^{T0}_2\mathbf{T}^0_1\hat{\mathbf{R}}^0_{10})^{-1}\mathbf{T}^{0*}_1$$

REFERENCES

Bahar, E., "Full Wave Analysis for Rough Surface Difuse, Incoherent Radar Cross Section with Height-Slope Correlations Included," *IEEE Trans. Ant. Prop.*, Vol. 39, no. 9, 1991, pp. 1293–1304.

Fung, A.K., and H.J. Eom, "Multiple Scattering and Depolarization by a Randomly Rough Kirchhoff Surface," *IEEE Trans. Ant. Prop.*, Vol. 29, 1981a, pp. 463–471.

Fung, A.K., and H.J. Eom, "A Theory of Wave Scattering from an Inhomogeneous Layer with Irregular Interface," *IEEE Trans. Ant. Propagation*, Vol. AP-29, no. 6, 1981b, pp. 899–910.

Fung A.K., and H.J. Eom, "Emission from Rayleigh Layer with Irregular Boundaries," *J. Quart. Spectrosc. Rad. Transfer*, Vol. 26, 1981c, pp. 397–409.

Leader, J.C., "Polarization Dependence in EM Scattering from Rayleigh Scatterers Embedded in a Dielectric Slab," *J. Appl. Phys.*, Vol. 46, no. 10, October 1975, pp. 4371–4385.

Nance, C.E., "Microwave Imaging Technique," Ph.D. Dissertation, Electrical Engineering Department, The University of Texas at Arlington, 1992.

Nance, C.E., A.K. Fung and J.W. Bredow, "Comparison of Integral Equation Method Predictions and Experimental Backscatter Measurements from Random Conducting Rough Surfaces," *Proc. IGARSS'90 Symp.*, May 20–24, 1990, College Park, MD, pp. 477–480.

Tjuatja, S., "Theoretical Scatter and Emission Models for Inhomogeneous Layers with Application to Snow and Sea Ice," Ph.D. Dissertation, Electrical Engineering Department, The University of Texas at Arlington, 1992.

Tjuatja, S., and A.K. Fung, "Doubling Method Applied to Multi-Layered Dense Inhomogeneous Medium with Irregular Interfaces," *IEEE, IGARSS'92,* Vol. 2, 1992, pp. 1141–1143.

Tsang, L., J.A. Kong and R.T. Shin, *Theory of Microwave Remote Sensing*, John Wiley & Sons, 1985.

Twomey, S., "Explicit Calculation of Layer Absorption in Radiative Transfer Computations by Doubling Methods," *J. Quart. Spectrosc. Rad. Transfer*, Vol. 15, 1975, pp. 775–778.

Twomey, S., "Doubling and Superposition Methods in the Presence of Thermal Emission," *J. Quart. Spectrosc. Rad. Transfer*, Vol. 22, 1979, pp. 355–363.

Twomey, S., H. Jacobowitz and H.B. Howell, "Matrix Methods for Multiple-Scattering Problems," *J. Atmos. Sci.*, Vol. 23, 1966, pp. 289–296.

Ulaby, F.T., and C. Elachi, *Radar Polarimetry for Geoscience Applications*, Artech House, Norwood, MA, 1990.

Ulaby, F.T., R.K. Moore and A.K. Fung, *Microwave Remote Sensing*, Vol. 2, Artech House, Norwood, MA, 1982, Chapter 12.

Ulaby, F.T., R.K. Moore and A.K. Fung, *Microwave Remote Sensing*, Vol. 3, Artech House, Norwood, MA, 1986, Appendix E and Chapter 13.

Winebrenner, D.P., and A. Ishimaru, "Application of Phase Perturbation Technique to Randomly Rough Surface," *J. Opt. Soc. Am. A.*, Vol. 2, no. 12, 1985, pp. 2285–2293.

Chapter 9

Scattering and Emission Models for Snow and Sea Ice

9.1 INTRODUCTION

In Chapter 8 a method of solving the radiative transfer equation is given for a multilayered medium with irregular boundaries. A sea ice medium with snow cover and a layered snow medium above a rough ground surface are multilayered media. The difference in modeling between these two types of media lies in the physical and geometric properties of the layers and scatterers within them. In general, the scatterers are nonspherical in shape but randomly oriented. As a result, good estimates can be obtained by modeling the scatterers by spheres. When an incident wave impinges upon an irregular inhomogeneous layer, some of its power is absorbed and the rest is scattered by the layer. In the radiative transfer formulation the scattering and absorption effects inside an inhomogeneous layer are accounted for in the volume scattering phase matrix and the extinction matrix. Thus, it is of interest to derive the scattering phase matrix and the extinction matrix of a sphere.

In the early development of the radiative transfer method only media with sparsely populated scatterers were treated. Hence, scatterers interacted in the far field of one another, and it was sufficient to define the phase matrix elements in terms of scattered fields in the far zone. For a snow medium the scatterers are the ice needles, which are closely spaced occupying typically 10% to 40% of the volume. A sea ice medium may have either brine pockets or air bubbles as scatterers, whose volume fraction may range from 5% to 20% of the sea ice volume. Thus, the adjacent scatterers are not always in the far field of one another, and it is necessary to use the exact scattered field without the far-zone approximation in the calculation of the scattering phase matrix elements. A modified Mie phase matrix that accounts for the effects of close spacing between spherical scatterers with arbitrary size is developed in Section 9.2.1. A coordinate transformation needed in phase matrix development is given in Section 9.2.2. The single scattering volume extinction and absorption coefficients are presented in Section 9.2.4. The volume scattering phase matrix is then implemented in the matrix doubling method to model scattering and emission from snow and sea ice layers as described in Chapter 8, where the effects of interface roughness and discontinuity are accounted for in the scattering model through some generalized boundary relations. The particular theoretical scattering characteristics of snow and sea ice are presented in Section 9.3 and

Section 9.4, respectively. Theoretical emission characteristics of snow and sea ice are presented in Section 9.5 and Section 9.6.

9.2 SCATTERING PHASE MATRIX AND EXTINCTION COEFFICIENT OF A MIE SPHERE

In a random inhomogeneous medium there may be more than one type of scatterer, each with a size and an orientation distribution. In modeling scattering from snow and ice media the scatterers are assumed to be randomly oriented and hence they are replaced by spheres. In addition, instead of a size distribution usually only an effective size is used to simplify the problem. Despite these approximations such a modeling approach has been shown to provide satisfactory predictions [Fung and Eom, 1985; Tjuatja et al., 1992, 1993]. Other modeling approaches using a strong fluctuation theory and a size distribution in the radiative transfer theory for snow and ice have been considered by Stogryn [1985, 1986], Nghiem [1991] and Wen et al. [1990].

The Mie solution of the field scattered from a sphere is utilized in Section 9.2.1 to derive the volume scattering phase matrix for a closely packed medium. The coordinate transformation method to obtain various incident and scattered directions is available from Fung and Eom [1985].

9.2.1 Mie Phase Matrix for Closely Spaced Scatterers

Consider the scattering geometry for a single sphere shown in Figure 9.1, where a time harmonic plane wave is incident on the sphere along the z-axis. The sphere has a radius r_0, a relative permittivity, $\varepsilon_r = \varepsilon' + j\varepsilon''$, and a permeability μ equal to that of the background medium. The incident electric and magnetic field components (with $\exp[-j\omega t]$ suppressed) are

$$\vec{E}^i = \hat{x}E_0 e^{jkz}$$

$$\vec{H}^i = \hat{y}\frac{E_0}{\eta}e^{jkz} \tag{9.1}$$

where

$$k = \omega\sqrt{\mu\varepsilon} \text{ and } \eta = (\mu/\varepsilon)^{1/2}$$

The scattered fields due to the sphere are [Stratton, 1941]

$$\vec{E}^s = E_0 \sum_{n=1}^{\infty} j^n \frac{(2n+1)}{n(n+1)} \{-b_n \vec{m}^{(3)}_{o1n} + ja_n \vec{n}^{(3)}_{e1n}\} \tag{9.2a}$$

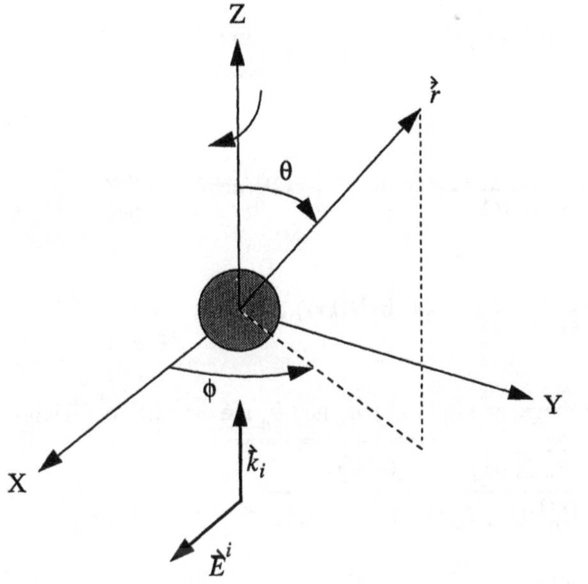

Figure 9.1 Scattering geometry for a single sphere.

$$\vec{H}^s = \frac{E_0}{\eta}\sum_{n=1}^{\infty} j^n \frac{(2n+1)}{n(n+1)}\{a_n \vec{m}_{e1n}^{(3)} + jb_n \vec{n}_{o1n}^{(3)}\} \tag{9.2b}$$

where the subscripts o and e of the spherical vector wave function $\vec{m}_{o1n}^{(3)}$, $\vec{n}_{e1n}^{(3)}$ denote odd and even. The vector wave functions are expressed in terms of the spherical Hankel function of the first kind, $h_n^{(1)}(kr)$, and the associated Legendre polynomials $P_n^1(\cos\theta)$ as

$$\vec{m}_{\substack{o\\e}1n}^{(3)} = \pm\frac{\hat{\theta}}{\sin\theta}h_n^{(1)}(kr)\ P_n^1(\cos\theta)\ \begin{matrix}\sin\phi\\ \cos\phi\end{matrix}$$

$$-\hat{\phi}h_n^{(1)}(kr)\frac{\partial}{\partial\theta}\{P_n^1(\cos\theta)\}\ \begin{matrix}\sin\phi\\ \cos\phi\end{matrix}$$

$$\vec{n}^{(3)}_{\substack{o\\e}1n} = \hat{r}\frac{n(n+1)}{kr}\mathrm{h}_n^{(1)}(kr)\,\mathrm{P}_n^1(\cos\theta)\begin{matrix}\sin\phi\\ \cos\phi\end{matrix}$$

$$+\frac{\hat{\theta}}{kr}\frac{d}{d(kr)}\{kr\mathrm{h}_n^{(1)}(kr)\}\frac{\partial}{\partial\theta}\{\mathrm{P}_n^1(\cos\theta)\}\begin{matrix}\sin\phi\\ \cos\phi\end{matrix}$$

$$\pm\frac{\hat{\phi}}{kr\sin\theta}\frac{d}{d(kr)}\{kr\mathrm{h}_n^{(1)}(kr)\}\,\mathrm{P}_n^1(\cos\theta)\begin{matrix}\cos\phi\\ \sin\phi\end{matrix} \tag{9.3}$$

In (9.3) r is the range, and in (9.2b) a_n and b_n are the Mie coefficients given by

$$a_n = \frac{\sqrt{\varepsilon_r}\hat{\mathrm{J}}_n'(u)\hat{\mathrm{J}}_n(v) - \hat{\mathrm{J}}_n(u)\hat{\mathrm{J}}_n'(v)}{\sqrt{\varepsilon_r}[\hat{\mathrm{H}}_n^{(1)}(u)]'\hat{\mathrm{J}}_n(v) - \hat{\mathrm{H}}_n^{(1)}(u)\hat{\mathrm{J}}_n'(v)}$$

$$b_n = \frac{\sqrt{\varepsilon_r}\hat{\mathrm{J}}_n(u)\hat{\mathrm{J}}_n'(v) - \hat{\mathrm{J}}_n'(u)\hat{\mathrm{J}}_n(v)}{\sqrt{\varepsilon_r}\hat{\mathrm{H}}_n^{(1)}(u)\hat{\mathrm{J}}_n'(v) - [\hat{\mathrm{H}}_n^{(1)}(u)]'\hat{\mathrm{J}}_n(v)} \tag{9.4}$$

where

$$u = kr_0$$

$$v = k\sqrt{\varepsilon_r}r_0 = k_s r_0$$

In (9.4), $\hat{\mathrm{J}}_n(v)$ is the Ricatti-Bessel function, $\hat{\mathrm{H}}_n^{(1)}(u)$ is the Ricatti-Hankel function of the first kind, and $[\ ]'$ denotes differentiation with respect to the argument. Note that the Mie coefficients are functions of the sphere radius, whereas the spherical vector wave functions are functions of the spacing between scatterers, incident angle and scattering angle. In the microwave region the average size of the scatterers is small (radius < 1 mm) compared to centimeter scale, $k_s r_0 < 0.8$. Hence, only the first two terms of the spherical harmonics are needed. The corresponding terms of the spherical vector wave functions are

$$\vec{m}^{(3)}_{\substack{o\\e}11} = \frac{e^{jkr}}{kr}\alpha\left[\pm\hat{\theta}\begin{matrix}\cos\phi\\ \sin\phi\end{matrix} - \hat{\phi}\cos\theta\begin{matrix}\sin\phi\\ \cos\phi\end{matrix}\right]$$

$$\vec{m}^{3}_{\substack{o\\e}12} = \frac{e^{jkr}}{kr}\beta\left[\pm\hat{\theta}\cos\theta\begin{matrix}\cos\phi\\ \sin\phi\end{matrix} - \hat{\phi}(-1+2\cos^2\theta)\begin{matrix}\sin\phi\\ \cos\phi\end{matrix}\right]$$

$$\vec{n}^{(3)}_{\substack{o\\e}11} = \frac{e^{jkr}}{kr}\left[\hat{r}\gamma\sin\theta \begin{matrix}\sin\phi\\ \cos\phi\end{matrix} + \delta\left(\hat{\theta}\cos\theta \begin{matrix}\sin\phi\\ \cos\phi\end{matrix} \pm \hat{\phi} \begin{matrix}\cos\phi\\ \sin\phi\end{matrix}\right)\right]$$

$$\vec{n}^{(3)}_{\substack{o\\e}12} = \frac{e^{jkr}}{kr}\left[\hat{r}\kappa\sin\theta\cos\theta \begin{matrix}\sin\phi\\ \cos\phi\end{matrix} + \zeta\left(\hat{\theta}\,(-1+2\cos^2\theta) \begin{matrix}\sin\phi\\ \cos\phi\end{matrix} \pm \hat{\phi}\cos\theta \begin{matrix}\cos\phi\\ \sin\phi\end{matrix}\right)\right] \tag{9.5}$$

where

$$\alpha = -1 - j\frac{1}{w}$$

$$\beta = 3j\left(1 - \frac{3}{w^2} + j\frac{3}{w}\right)$$

$$\gamma = -2\left(\frac{1}{w} + j\frac{1}{w^2}\right)$$

$$\delta = \frac{1}{w} + j\left(-1 + \frac{1}{w^2}\right)$$

$$\kappa = 18j\left(\frac{1}{w} - \frac{3}{w^3} + j\frac{3}{w^2}\right)$$

$$\zeta = 3\left[-1 + \frac{6}{w^2} + j\left(-\frac{3}{w} + \frac{6}{w^3}\right)\right] \tag{9.6}$$

and

$$w = kr$$

The above results are for the geometry shown in Figure 9.1, where the incident direction is along the z-axis. To construct the phase matrix (defined in Chapter 1), we need scattered field components for arbitrary incident angle. Hence, a coordinate transformation is required to convert (9.2b) to permit arbitrary incident angles for both vertical and horizontal polarizations. This transformation is given in the next section.

9.2.2 Transformation Between Field Components

A description of the transformation of the scattered field components from one coordinate system to the other has been given in Fung and Eom [1985] and is summarized below. The purpose of this transformation is to convert the scattered field components in the xyz-system given in the previous section, which is for normal incidence along the z-axis, into those in the primed system for arbitrary incidence.

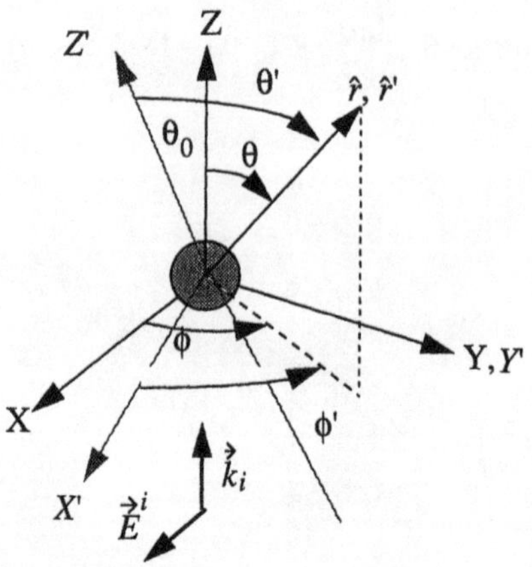

Figure 9.2 Geometry showing the incident field to be vertically polarized in the primed system incident at an angle θ_0. In the unprimed coordinate the incident field is parallel to the x-axis.

Vertically Polarized Incident Wave

Starting with the problem described in the previous section we can convert the incident field given by (9.1) into a vertically polarized incident field at an angle q_0 with respect to a primed coordinate system generated by a rotation around the y-axis (see Figure 9.2). The field components of these two coordinate systems are related by

$$\begin{bmatrix} E_{x'} \\ E_{y'} \\ E_{z'} \end{bmatrix} = \begin{bmatrix} \cos\theta_0 & 0 & \sin\theta_0 \\ 0 & 1 & 0 \\ -\sin\theta_0 & 0 & \cos\theta_0 \end{bmatrix} \begin{bmatrix} E_x \\ E_y \\ E_z \end{bmatrix} \equiv T_v E(\vec{x}) \equiv E(\vec{x}') \qquad (9.7)$$

The transformation between the field components in rectangular and spherical coordinate systems is

$$\begin{bmatrix} E_{x'} \\ E_{y'} \\ E_{z'} \end{bmatrix} = \begin{bmatrix} \sin\theta'\cos\phi' & \cos\theta'\cos\phi' & -\sin\phi' \\ \sin\theta'\sin\phi' & \cos\theta'\sin\phi' & \cos\phi' \\ \cos\theta' & -\sin\theta' & 0 \end{bmatrix} \begin{bmatrix} E_{r'} \\ E_{\theta'} \\ E_{\phi'} \end{bmatrix} \equiv T_{RS} E(\vec{r}') \equiv E(\vec{x}') \qquad (9.8)$$

Since the matrix $T_{R'S}$ is unitary, its inverse is equal to its transpose, i.e.,

$$E(\vec{r}') = T'_{SR}E(\vec{x}') = \begin{bmatrix} E_{r'} \\ E_{\theta'} \\ E_{\phi'} \end{bmatrix} = \begin{bmatrix} \sin\theta'\cos\phi' & \sin\theta'\sin\phi' & \cos\theta' \\ \cos\theta'\cos\phi' & \cos\theta'\sin\phi' & -\sin\theta' \\ -\sin\phi' & \cos\phi' & 0 \end{bmatrix} \begin{bmatrix} E_{x'} \\ E_{y'} \\ E_{z'} \end{bmatrix} \tag{9.9}$$

For the unprimed system we have a similar relation,

$$\begin{bmatrix} E_x \\ E_y \\ E_z \end{bmatrix} = \begin{bmatrix} \sin\theta\cos\phi & \cos\theta\cos\phi & -\sin\phi \\ \sin\theta\sin\phi & \cos\theta\sin\phi & \cos\phi \\ \cos\theta & -\sin\theta & 0 \end{bmatrix} \begin{bmatrix} E_r \\ E_\theta \\ E_\phi \end{bmatrix} \equiv T_{RS}E(\vec{r}) \equiv E(\vec{x}) \tag{9.10}$$

Using the above relations we can relate the scattered field components in the primed coordinate system to the unprimed system as

$$\begin{bmatrix} E^s_{r'} \\ E^s_{\theta'} \\ E^s_{\phi'} \end{bmatrix} = T'_{SR}T_v T_{RS} \begin{bmatrix} E^s_r \\ E^s_\theta \\ E^s_\phi \end{bmatrix} \tag{9.11}$$

Thus, for an arbitrary incident angle, we have found the scattered field in any direction in terms of the fields given by (9.2b). In numerical evaluation the known quantities are the incident angle θ_0 and angles in the primed coordinates, θ', ϕ'. We need to express all θ, ϕ in terms of them. From Figure 9.2, the observation point is physically the same in either coordinate system. Hence, we can use the following relations to determine θ, ϕ:

$$\cos\theta = \hat{r}\cdot\hat{z} = \hat{r}'\cdot\hat{z} = \sin\theta_0\sin\theta'\cos\phi' + \cos\theta_0\cos\theta' \tag{9.12}$$

$$\sin\theta\cos\phi = \hat{r}\cdot\hat{x} = \hat{r}'\cdot\hat{x} = \cos\theta_0\sin\theta'\cos\phi' - \sin\theta_0\cos\theta' \tag{9.13}$$

$$\sin\theta\sin\phi = \hat{r}\cdot\hat{y} = \hat{r}'\cdot\hat{y} = \sin\theta'\sin\phi' \tag{9.14}$$

Horizontally Polarized Incident Wave

To find a coordinate system in which the incident field given by (9.1) appears as a horizontally polarized field, we first generate an intermediate coordinate system by rotating the *xyz*-system as shown in 9.1 90 degrees around the *z*-axis. Then, we rotate the intermediate system around its *y*-axis by an angle θ_0 to obtain a primed system. The field components in the primed and unprimed systems are related as follows:

$$\begin{bmatrix} E_{x'} \\ E_{y'} \\ E_{z'} \end{bmatrix} = \begin{bmatrix} \cos\theta_0 & 0 & \sin\theta_0 \\ 0 & 1 & 0 \\ -\sin\theta_0 & 0 & \cos\theta_0 \end{bmatrix} \begin{bmatrix} 0 & -1 & 0 \\ 1 & 0 & 0 \\ 0 & 0 & 1 \end{bmatrix} \begin{bmatrix} E_x \\ E_y \\ E_z \end{bmatrix} = \begin{bmatrix} 0 & \cos\theta_0 & \sin\theta_0 \\ 1 & 0 & 0 \\ 0 & \sin\theta_0 & \cos\theta_0 \end{bmatrix} \begin{bmatrix} E_x \\ E_y \\ E_z \end{bmatrix} \equiv T_h E(\vec{x}) \equiv E(\vec{x}') \tag{9.15}$$

The transformation from rectangular to spherical or spherical to rectangular has been given in the previous section. From (9.9), (9.10) and (9.15) we can write

$$\begin{bmatrix} E^s_{r'} \\ E^s_{\theta'} \\ E^s_{\phi'} \end{bmatrix} = T_{SR} T_h T_{RS} \begin{bmatrix} E^s_r \\ E^s_\theta \\ E^s_\phi \end{bmatrix} \tag{9.16}$$

which are the scattered field components in the θ', ϕ' direction responding to a horizontally polarized incident wave at an angle θ_0. Here again, the values of θ, ϕ can be found in terms of θ_0, θ', ϕ' by the following relations:

$$\cos\theta = \hat{r}\cdot\hat{z} = \hat{r}'\cdot\hat{z} = \sin\theta_0\sin\theta'\cos\phi' + \cos\theta_0\cos\theta' \tag{9.17}$$

$$\sin\theta\cos\phi = \hat{r}\cdot\hat{x} = \hat{r}'\cdot\hat{x} = \sin\theta'\sin\phi' \tag{9.18}$$

$$\sin\theta\sin\phi = \hat{r}\cdot\hat{y} = \hat{r}'\cdot\hat{y} = -\cos\theta_0\sin\theta'\cos\phi' + \sin\theta_0\cos\theta' \tag{9.19}$$

9.2.3 Phase Matrix for the First Two Stokes Parameters

With the scattered field components known, all the elements of the phase matrix defined in (1.44) of Chapter 1 can be computed. For many applications only like and cross polarizations are of interest. For such cases we need only the first four elements of the phase matrix as

$$\mathbf{P}_s = \frac{4\pi r^2 \eta n_o}{\kappa_s |E_0|^2} \mathrm{Re} \begin{bmatrix} (E^s_{\theta'} H^{s\,*}_{\phi'})_{v\text{-}inc} & (E^s_{\theta'} H^{s\,*}_{\phi'})_{h\text{-}inc} \\ (-E^s_{\phi'} H^{s\,*}_{\theta'})_{v\text{-}inc} & (-E^s_{\phi'} H^{s\,*}_{\theta'})_{h\text{-}inc} \end{bmatrix} \tag{9.20}$$

where κ_s is the effective volume-scattering coefficient and n_o is the number density of scatterers. Other parameters have been defined in (9.1). The range r is related to the volume fraction v_f of the spherical scatterer by [Goedecke,1977]

$$r = (v_o / v_f)^{1/3} \tag{9.21}$$

where

$$v_o = \frac{4}{3}\pi a^3$$

Note that (9.20) is a phase matrix for single scattering calculations. It corresponds to the scattering phase matrix for an infinitesimally thin layer in the matrix doubling method given in Chapter 8. Since the full scattered field expression is used in (9.20), there is no restriction on the spacing between scatterers in carrying out the doubling process described in Chapter 8.

9.2.4 Extinction and Absorption Coefficients

The extinction and scattering cross sections of a Mie sphere have been calculated using the forward scattering theorem [Ishimaru, 1978; Karam and Fung, 1982] and Poynting's theorem [Karam and Fung, 1987]. In terms of the Mie coefficients a_n, b_n defined in (9.4), the extinction and scattering cross sections are given by [Stratton, 1941],

$$Q_e = \frac{2\pi}{k^2}\sum_{n=1}^{\infty}(2n+1)\,\mathrm{Re}(a_n+b_n) \tag{9.22}$$

$$Q_s = \frac{2\pi}{k^2}\sum_{n=1}^{\infty}(2n+1)\,(|a_n|^2+|b_n|^2) \tag{9.23}$$

where k is the propagation constant in the host medium. The volume extinction and scattering coefficients, κ_e and κ_s, for an infinitesimally thin dense medium can be approximated by

$$\kappa_e = n_0 Q_e,\ \kappa_s = n_0 Q_s \tag{9.24}$$

In modeling natural terrain such as snow or sea ice, the medium effective relative permittivity, $\varepsilon_l = \varepsilon_l' + j\varepsilon_l''$, is usually known from measurements or can be estimated using empirical or semi-empirical formulas. In which case a better approximation for the volume extinction coefficient can be obtained using

$$\kappa_e = \kappa_s + \kappa_a = \kappa_s + 2k_0\left|\mathrm{Im}(\sqrt{\varepsilon_l})\right| \tag{9.25}$$

9.3 A SCATTERING MODEL FOR A SNOW LAYER

A snow layer can be viewed as a homogeneous medium embedded densely with needle-shaped ice particles that may have aggregated into many shapes and sizes after a period of metamorphism. These ice particles are randomly oriented and hence can be approximated by a collection of spheres with an effective radius. Due to the small spacing between the ice particles the fractional volume of snow is usually between 10% and 20% when fresh and may reach 40% or more after aging. Depending on the liquid water content, a snow layer can be classified as dry or wet snow. In a completely dry snow layer, the host medium is air and the scatterers are ice particles. For a wet snow layer there are water droplets which reside between ice particles [Colbeck, 1979]. The average droplet size is much smaller than the ice particle size and does not make a significant contribution to scattering. Thus, the water inclusion in a wet snow layer may be considered as part of the host medium and contributes to absorption. In this study, a snow layer is modeled as a volume of Mie scatterers (ice particles) that are closely packed and bounded by irregular boundaries [Fung and Tjuatja, 1991]. The average spacing of the scatterers is a function of the volume fraction of the snow layer. The boundaries of the snow layer are modeled using the surface scattering model (IEM) described in Chapter 4 with a correlation function given by

$$\rho(\xi) = \left[1 + \left(\frac{\xi}{L}\right)^2\right]^{-1.5} \tag{9.26}$$

where ξ is the spacing between two points and L is the surface correlation length.

The effects of volume and surface scattering are integrated by applying the matrix doubling method given in Chapter 8. The mixture of liquid water inclusion and air is considered as the host medium. An irregular dry snow layer is chosen as the reference, and its parameters are listed in Table 9.1.P1111

Table 9.1
Snow Layer Reference Parameters

Snow parameters	*Reference values*
Frequency	17 GHz
Layer thickness (optical depth)	0.5 m, (0.13)
Scatterer volume fraction	30%
Scatterer radius (albedo)	0.5 mm, (0.8)
Scatterer permittivity	$3.15 + j0.015$
Liquid water content	<0.1% (dry snow)
Layer permittivity, ε_r (Formula in Appendix 9A)	$1.482 + j0.00017$
Lower half-space permittivity	5.0
Top surface rms height ($k\sigma$)	0.14 cm, ($k\sigma = 0.5$)
Top surface correlation length (kL)	0.7 cm, ($kL = 2.5$)
Bottom surface rms height, ($k\sigma$)	0.056 cm, ($k\sigma = 0.2$)
Bottom surface correlation length (kL)	0.28 cm, ($kL = 1.0$)

When a parameter variation is studied, all other parameters are kept as shown in Table 9.1, while the layer permittivity is calculated using the formula given in Appendix 9A. Note that layer permittivity is dependent on frequency, wetness and the volume fraction of the scatterer. The effects of snow parameter variations on the backscattering coefficients are shown in Figure 9.3 through Figure 9.11.

9.3.1 Theoretical Studies on Effects of Snow Parameters

In Figure 9.3 a variation of the ice particle radius is shown for both like and cross polarizations. The layer permittivity relative to vacuum is around 1.48 so that scattering from the air-snow interface is very small. As expected larger particles cause greater backscattering in like and cross polarization. Although the snow layer is 50 cm thick, loss through the layer is not very high because it is dry. The extinction losses at radii of 0.5, 0.75 and 1 mm are 0.26, 0.77 and 1.78 nepers/m respectively. Thus, at small angles of incidence and for the small particle size (radius of 0.5 mm) we see a faster drop-off of the backscattering coefficient in like polarization due to surface scattering effects. In addition, at larger angles of incidence, HH is higher than VV, indicating that there is a contribution from snow layer interacting with the lower boundary (snow-ground interface). For the largest particle size (radius of 1 mm) considered, the loss in the snow layer is high enough so that both direct surface contribution at small angles of incidence and snow-ground interaction are negligible. The change in the level of cross polarized scattering in 9.2 is close to 18 dB while the corresponding change for the like polarizations is around 6 dB indicating that the former is dominated by multiple scattering while the latter is by single scattering.

In Figure 9.4 backscattering coefficients versus frequency are shown. At 5 GHz the albedo and optical depth are 0.21 and 0.00375, respectively. Since the layer optical thickness is very small, cross polarized scattering resulting from multiple scattering is small at small angles of incidence and rises as the incident angle increases because the effective depth seen by the incident wave increases. For polarized scattering there is no rise of scattering with the incident angle because the dominant mechanism is single scattering. Although the snow-ground interface is seen, there is no significant contribution due to surface-volume interaction, because the albedo is small. Thus, HH is not higher than VV. At 17 GHz the albedo is 0.81 while the optical depth is 0.13. The interaction term becomes appreciable and HH is higher than VV at large angles of incidence. At 35 GHz albedo and optical depth are 0.97 and 2. The snow-ground interface is not contributing and VV is higher than HH at large angles of incidence due to pure volume scattering in polarized scattering.

Next, we consider layer thickness effect in Figure 9.5. Here, layer thickness changes from an optical depth of 1.3 at 5 m to 0.026 at 0.1 m. The snow-ground interface has roughness characterized by $k\sigma = 0.2$ and $kL = 1.0$ at 17 GHz, where k is the wavenumber in air. With snow cover these roughness parameters should be larger by a factor of $\sqrt{\varepsilon_r}$. Clearly, the ground effect is seen at a thin optical depth in like polarization causing a faster drop off at small angles of incidence. At intermediate thickness, the surface-volume interaction term is also observable by having HH returns higher than VV. The overall effect of an increase in thickness is to permit more volume scattering to occur leading to higher like and cross backscattering. The corresponding increase in cross polarization is larger than the like because the latter is dominated by single scattering while the former is by multiple scattering.

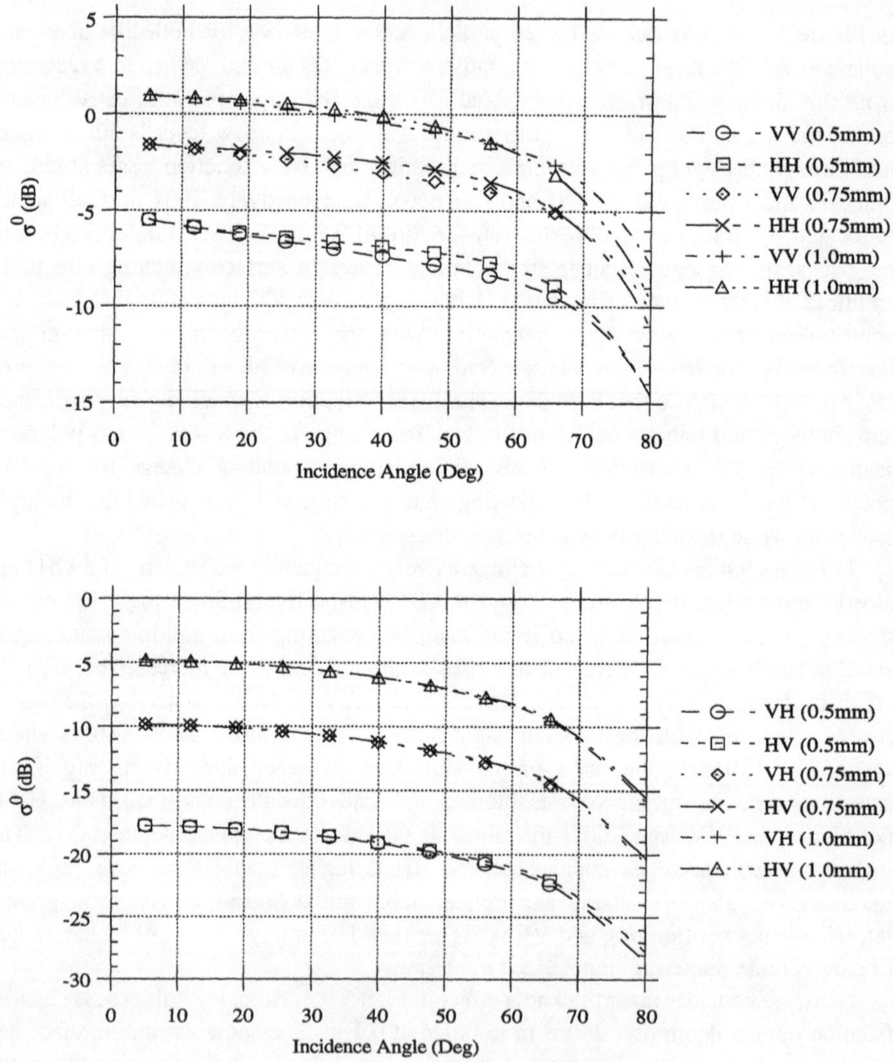

Figure 9.3 Effect of ice particle radius, *a*, on snow backscattering coefficients.

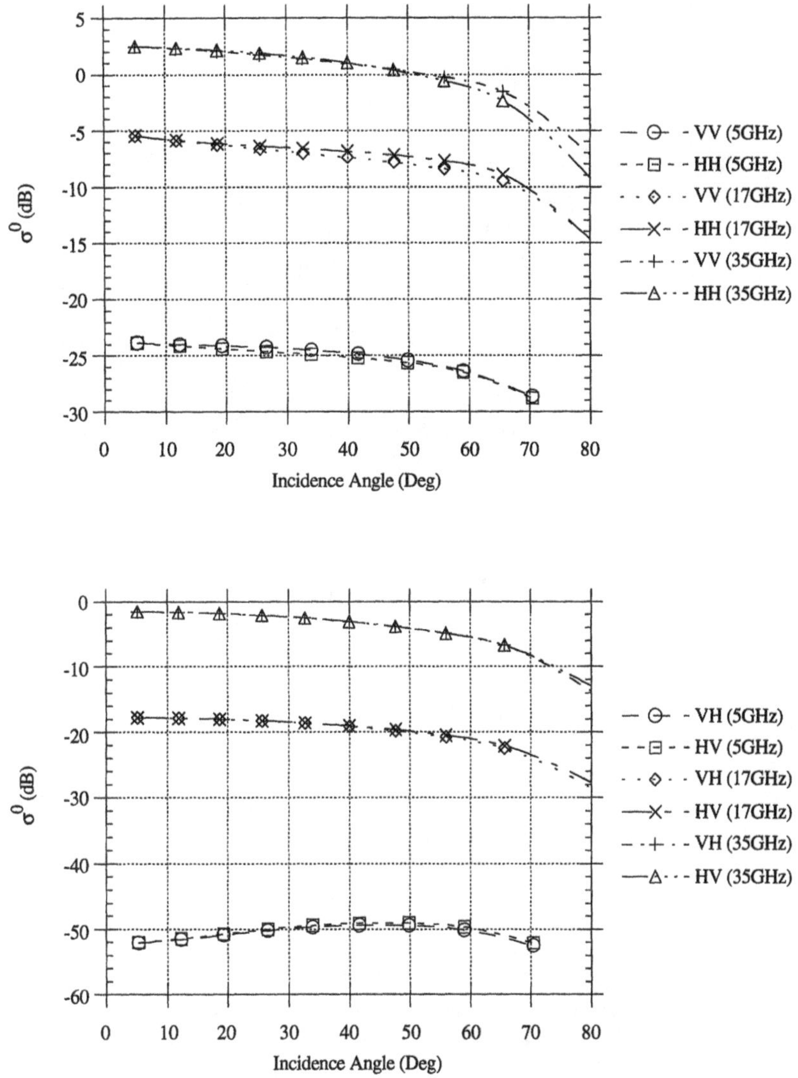

Figure 9.4 Effects of signal frequency variation on snow backscattering coefficients.

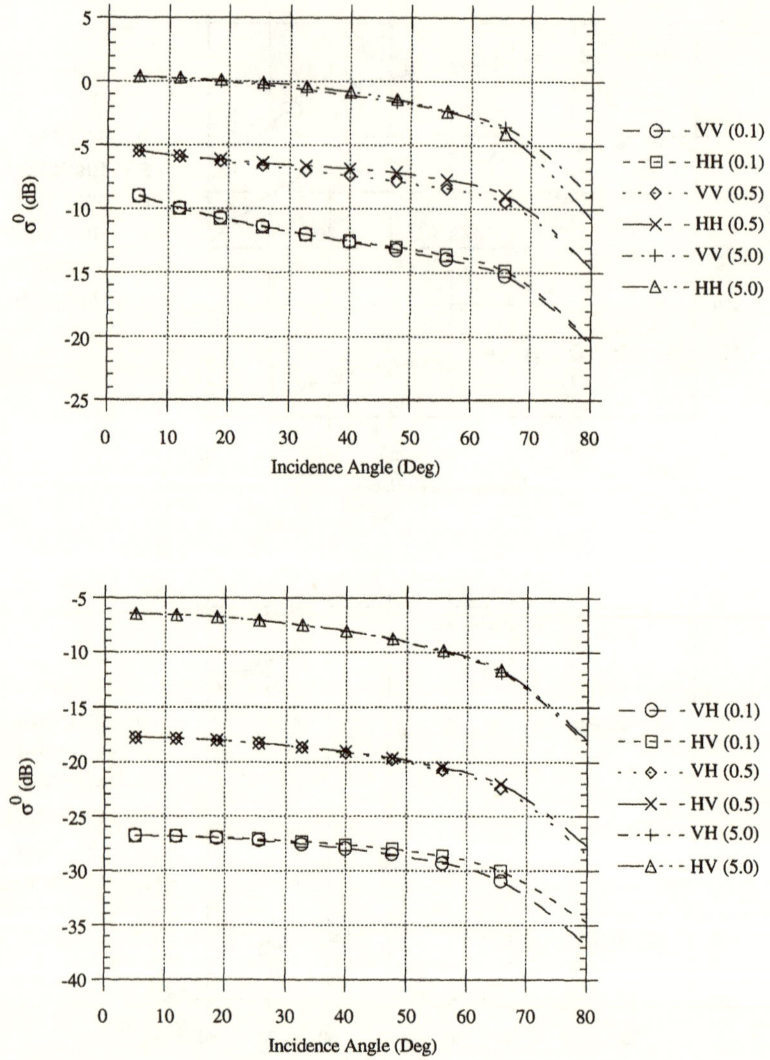

Figure 9.5 Effects of layer thickness *d* (in meters) on snow backscattering coefficients.

The effect of volume fraction of scatterers is shown in Figure 9.6. Over the range of values, 20% to 40%, the backscattering coefficient is seen to increase with the volume fraction for both like and cross polarizations. The reason for this change is because the optical depth has increased from 0.086 to 0.176, which is more than what is needed to offset the decrease in albedo from 0.82 to 0.79. It is conceivable that at much larger volume fractions, it is possible for the albedo to become very small so that the effect of the decrease in albedo dominates over the increase in optical depth. In the limiting case, when the layer volume is completely filled with scatterers, the albedo should be zero because the medium has become homogeneous. Thus, it appears that as the volume fraction of a dense medium increases, incoherent scattering must be increasing initially and then decreasing towards zero at large volume fractions. Another way to view the problem is that as the volume fraction exceeds 50% the roles of the scatterer and the host medium are interchanged. Thus, the reason applicable to explain the increase in scattering in the small volume fraction region is also applicable to explain the decrease in scattering in the large volume fraction region. In like polarized scattering HH is higher than VV because the optical depths of the three cases in Figure 9.6 are small.

The presence of liquid water in a snow medium is to increase the absorption of the medium and causes the albedo to decrease. Thus, we anticipate a much lower scattering strength when the snow is wet. Figure 9.7 shows wet snow calculations relative to dry snow. For cross polarization the wet snow cases have a similar angular variation as the dry but their levels are much lower because the albedo is very small (0.01 and 0.0041, respectively, for 2% and 4% wetness). The ground surface is not seen by the incident wave because the optical depths for the wet cases are 11.0 and 25.6. In like polarization the angular curves for wet snow are substantially different from the dry. They drop off very quickly with the incident angle between 5 degrees and 50 degrees. More so for 4% wetness because its layer permittivity is $1.68 + j0.186$, while that at 2% is $1.58 + j\,0.075$. At larger angles the angular trend is not entirely due to scattering from the top surface boundary, because there is no significant difference between VV and HH polarizations. This lack of separation between the polarizations is due to volume scattering contributions from the snow layer.

In Figure 9.8 we examine the effect of changing the rms height ($k\sigma$ varies from 0.01 to 0.7, with kL fixed at 2.5) of the top boundary. For dry snow we expect the direct surface scattering contribution to be very small (1 dB or less) because the dielectric discontinuity is small (1.48 versus 1). It is observable only between 5 to 30 degrees. In like polarization at large angles of incidence, the difference between the three cases shown is negligible in VV due to the Brewster angle effect, while HH polarization increases with the roughness although the change is about 0.5 of a dB. There is about 1 dB separation between VV and HH polarizations. This is due to volume scattering interacting with the snow-ground boundary because HH is higher than VV. In cross polarization we know it should be dominated by multiple scattering generated within the snow layer. However, some separation between the three cases are observable. This is due to surface-volume interaction leading to a larger scattering coefficient for the rougher surface, i.e., larger rms height. Note that the maximum change between the three cases is less than 1.5 dB.

Next, we consider the effect of surface correlation of the top snow layer boundary by varying kL from 2.5 to 12.5 and keeping $k\sigma$ fixed at 0.5 as shown in Figure 9.9. In like polarization the influence of direct surface scattering contribution appears only at small angles of incidence, 5 to 20 degrees. Apparently, volume scattering is dominating at large angles of incidence and the effect of kL is small compared with it. In cross polarized scattering volume scattering is again dominating. Here, there is no direct surface scattering contribution, and surface-volume interaction is the only mechanism that causes the change among the cases. The maximum variation appears to be within a dB. All cases studied indicate that due to the small dielectric discontinuity at the air-snow interface, the influence of this boundary is generally quite small except possibly at small angles of incidence.

The effect of the boundary between snow and ground is generally larger than the air-snow boundary unless the snow layer is so lossy that this boundary is not seen by the incident wave. In Figure 9.10 $k\sigma$ is varied from 0.01 to 0.28. Since the ground is covered with snow, the value of k used in calculations is larger by a factor of $\sqrt{\varepsilon_r}$, where ε_r is the snow layer permittivity. We give the $k\sigma$ in vacuum instead of the layer so that when the layer permittivity changes this value is not affected. Although the change in $k\sigma$ is smaller than that considered for the air-snow boundary, the corresponding change in the backscattering coefficient in like polarization is just as large. This is because the dielectric discontinuity here is larger. Note that larger $k\sigma$ leads to larger VV coefficients at large angles of incidence, but the corresponding changes for HH coefficients are negligible. This result appears to be the reverse of the one in Figure 9.8. As in Figure 9.8 this phenomenon can be explained in terms of the Brewster angle. Here, more VV than HH polarized energy has been transmitted and propagated towards the snow-ground interface. Thus, because of the Brewster angle effect VV is responding more to the lower interface while HH is to the upper interface. The optical depth for this case is 0.13 and does not cause much loss. For cross polarization the change in the backscattering coefficient is comparable to that for the top boundary because cross polarization depends mainly on multiple volume scattering and surface-volume interaction and does not depend on direct surface scattering contribution in the backscattering direction.

Finally, we show the effect of changing kL of the snow-ground interface from 1 to 7. In like polarization there are large changes over small incident angles between 5 and 20 degrees. This is due to direct surface scattering. At large angles of incidence, a much smaller change in the scattering coefficients is observed indicating a dominance by volume scattering. The optical depth is 0.13 and hence surface-volume interactions between the layer and the ground are significant resulting in HH coefficients higher than those of VV. In cross polarized scattering less than 1 dB variation occurs. Again, this is because cross polarization is dominated by volume multiple scattering in the backscattering direction. There is no direct surface scattering contribution, and the small change is the result of surface-volume interactions.

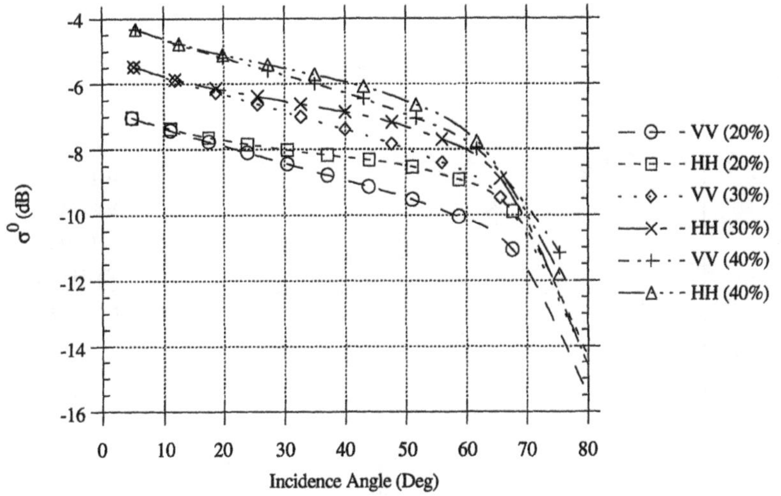

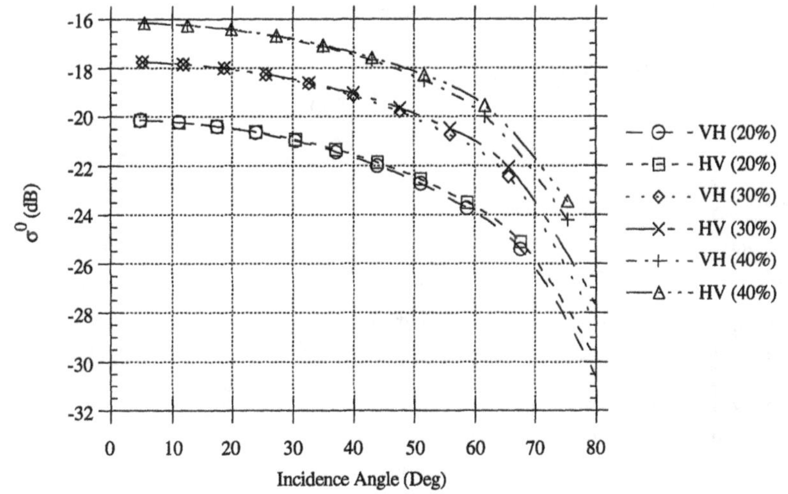

Figure 9.6 Effect of scatterer volume fraction on snow backscattering.

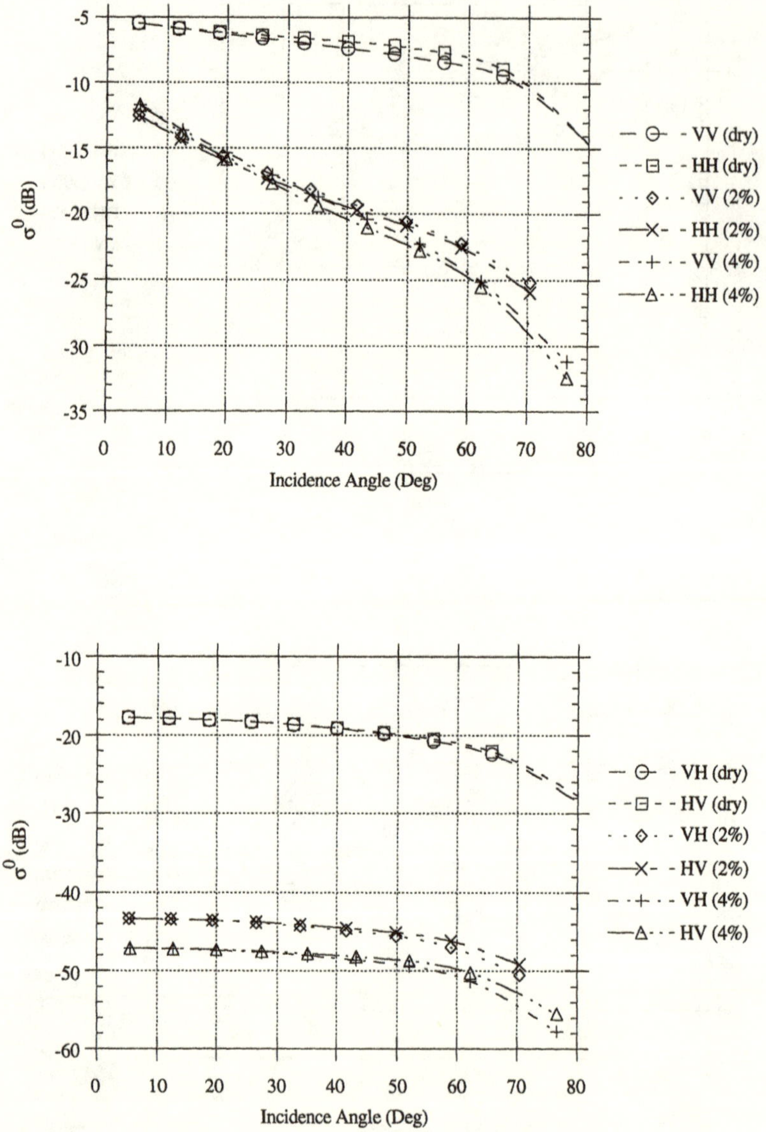

Figure 9.7 Effect of volumetric liquid water content on snow backscattering.

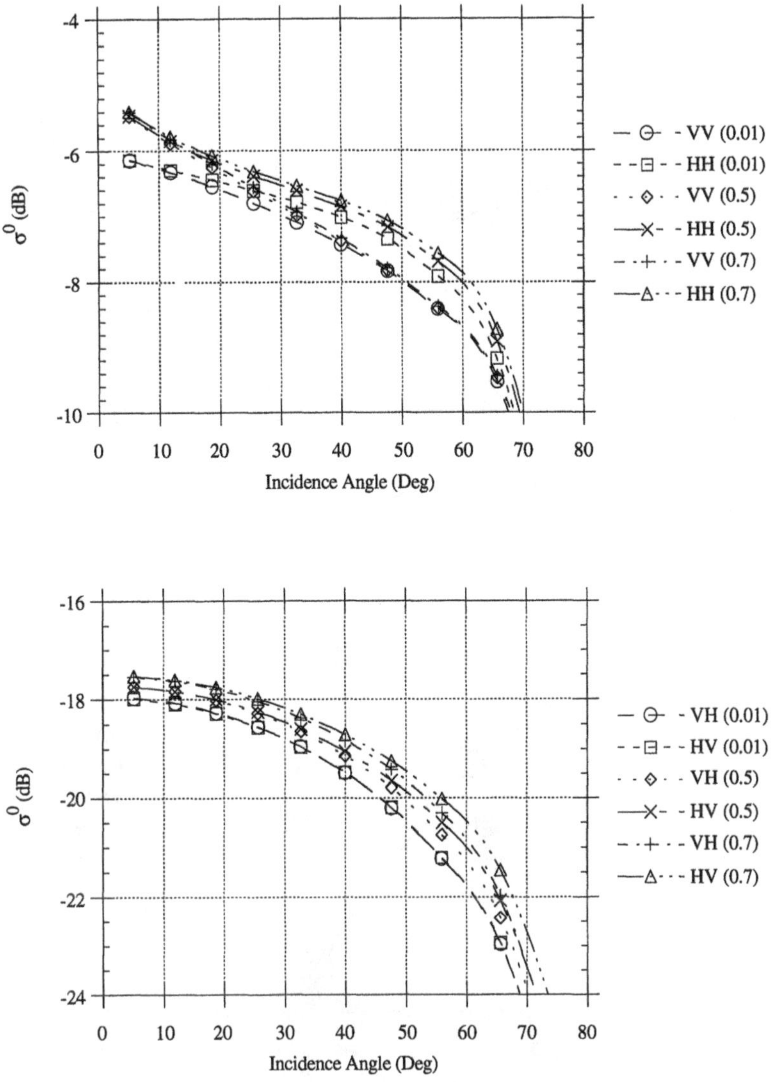

Figure 9.8 Effect of top surface $k\sigma$ variation on snow backscattering coefficients.

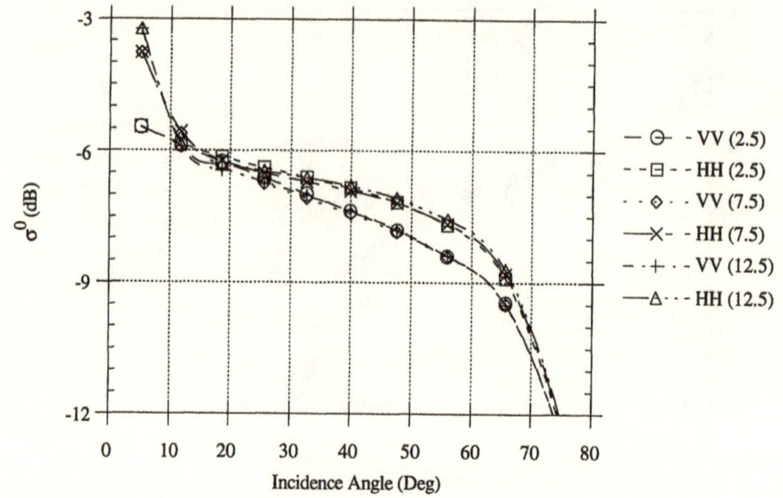

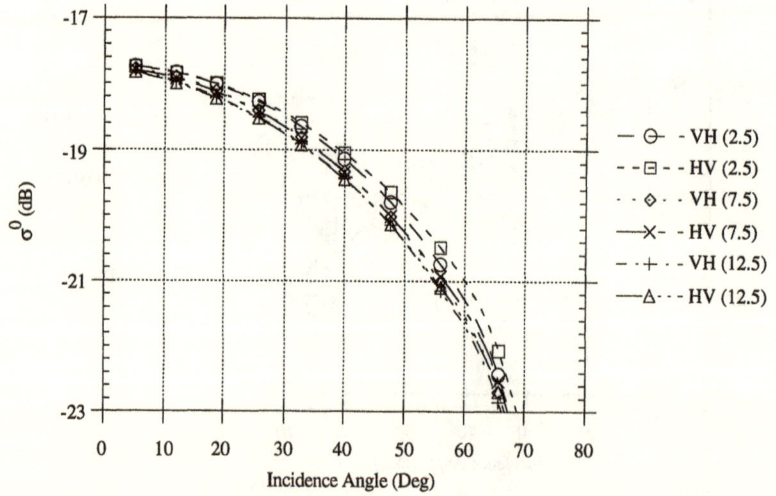

Figure 9.9 Effect of top surface kL variation on snow backscattering coefficients.

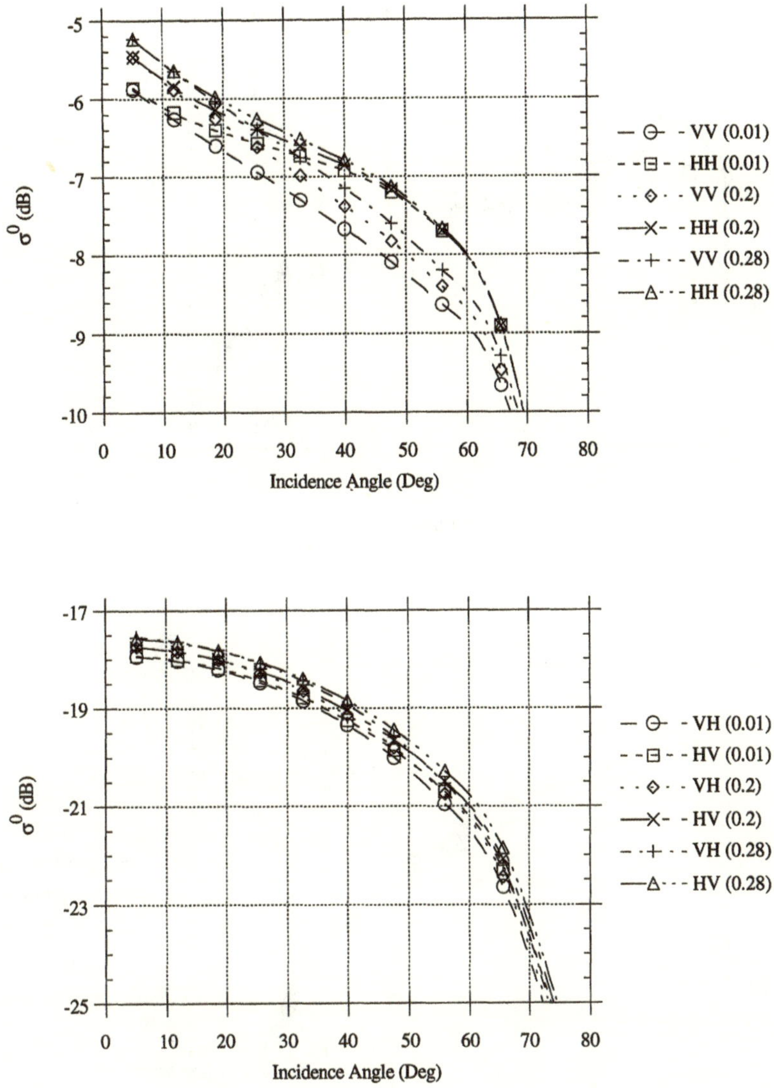

Figure 9.10 Effect of bottom surface $k\sigma$ variation on snow backscattering coefficients.

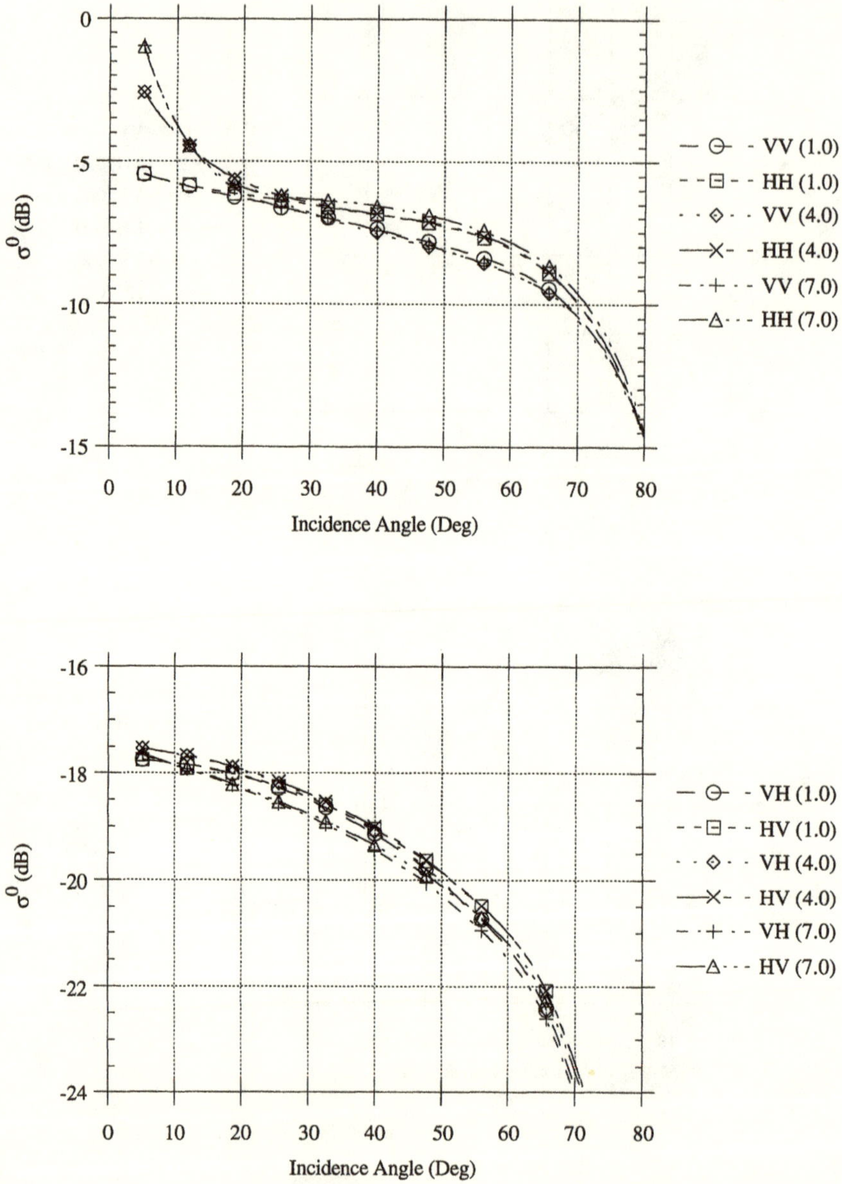

Figure 9.11 Effect of bottom surface *kL* variation on snow layer backscattering coefficients.

9.4 A SCATTERING MODEL FOR A SEA ICE LAYER

It is a common practice to separate sea ice into salinated ice (which includes young ice less than 30 cm thick and first year sea ice which is between 30 cm and 2 m thick) and desalinated ice which is multiyear sea ice generally over 2 m thick. The salinated sea ice in the Arctic Ocean typically has a salinity profile of 5-16 ‰ near the surface, about 4-5 ‰ in the bulk of the ice and around 30 ‰ near the ice-water interface. This high salinity is due to brine (saline water) inclusions interspersed within the ice medium. Sea water itself has salinity around 32‰ but is much lower close to land, where it is mixed with fresh water from rivers. Saline ice is highly absorptive and hence has a low albedo. It should be high in emission and low in volume scattering relative to desalinated ice. The desalinated sea ice has a salinity less than 1 ‰ near the surface and 2-3 ‰ in the bulk of the ice. This is because the highly absorptive brine inclusions have been drained and replaced by air pockets.

In modeling we expect volume scattering from a sea ice layer to be dominated by either the brine inclusions (around 0.25 mm in radius) in saline ice or air pockets (1-2 mm in radius) in desalinated ice. It has been shown that the volume scattering contribution from saline ice is quite small [Fung et al., 1992; Tjuatja et al., 1992, 1993] because brine inclusions are small and their volume fraction is around 5%. As a result the total backscattering is dominated by the air-ice boundary. Thus, the problem becomes a surface scattering problem, if ice is not covered with snow, or a snow layer problem, if the ice is covered with snow. To examine the combined surface-volume scattering effects of ice, we shall consider, in the next section, parameter studies of a multiyear (desalinated) ice layer above a thick saline ice. The desalinated ice layer is modeled as an irregular layer of pure ice embedded with air bubbles, and its boundaries are modeled using the IEM surface model with the 1.5-power correlation (Appendix 2B). The thick saline ice underneath the desalinated ice layer is treated as a homogeneous half space. The reference parameters for a desalinated ice sheet are listed in Table 9.2. The effects of the sea ice parameter variations on the backscattering coefficients are shown in Figure 9.12 through Figure 9.19.

9.4.1 Theoretical Studies on Effects of Sea Ice Parameters

In Figure 9.12 a variation of the air bubble radius is shown for both like and cross polarizations. The layer permittivity relative to vacuum is around 2.7 so that scattering from air-ice interface makes a contribution. The discontinuity at the lower boundary is small because the permittivity in the lower half space is around 3.5. As expected larger particles cause greater backscattering in like and cross polarization. Although the ice layer is 50 cm thick, loss through the layer is not high with zero salinity. The optical depths at radii of 0.5, 0.75 and 1 mm are 0.11, 0.21 and 0.39 nepers/m respectively. Thus, the backscattering from this sea ice layer is dominated by the top surface and the ice layer. This leads to VV polarization higher than HH because the top surface parameters are in the small roughness range, $k\sigma = 0.3$ and $kL = 1.5$, and albedo in the layer is 0.84 at 1 mm scatterer radius. At 10 degree incidence top surface contribution is around -13.6 dB. For the small scatterer size of 0.5 mm, surface scattering contribution is evident in the angular curve, which drops 6 dB between 5 and 60 degrees in HH polarization, whereas the corresponding angular drop-off due to volume scattering alone for larger scatterers is only about 4 dB. The change in the levels of cross polarized scattering is larger than the like indicating that multiple scattering is dominating in cross polarization. A small separation between VH and HV at large angles of incidence may be due to surface-volume interaction where bistatic surface scattering is present.

Table 9.2
Sea Ice Layer Reference Parameters

Sea ice parameters	*Reference values*
Frequency	10 GHz
Layer thickness (optical depth)	0.5 m, (0.108)
Scatterer volume fraction	20%
Scatterer radius (albedo)	0.5 mm, (0.41)
Scatterer relative permittivity	1.0 (air bubbles)
Salinity	0%
Layer permittivity	$2.7 + j0.0001$
Lower half space permittivity	$3.5 + j0.25$
Top boundary rms height ($k\sigma$)	0.143 cm, (0.3)
Top boundary correlation length (kL)	0.716 cm, (1.5)
Bottom boundary rms height ($k\sigma$)	0.05 cm, (0.1)
Bottom boundary correlation length (kL)	0.29 cm, (0.6)

In Figure 9.13 backscattering coefficients versus frequency are shown. At 5 GHz the albedo and optical depth are 0.08 and 0.035, respectively, and the top surface contribution is about -23.6 dB at 10 degrees. Thus, there is some surface contribution making the HH polarization drop off by about 2 dB faster than if there is only volume scattering. This effect exists at all frequencies because both surface and volume parameters are scaled by frequency. For example, surface contribution is -13.6 at 10 degrees and 10 GHz. Again, HH is lower than VV because there is no significant contribution from the lower boundary. At 14 GHz the albedo is 0.65 while the optical depth is 0.255. The corresponding surface contribution goes up to -9.8 dB at 10 degrees. The cross polarized backscattering is due to multiple volume scattering enhanced by surface-volume interaction. Thus, its level increases faster with frequency than the like polarized scattering as expected.

We consider layer thickness effect in Figure 9.14 with thicknesses of 0.1, 0.5 and 5 m. Here, the top surface contribution is dominating at small angles of incidence (less than 20 degrees) for the 0.1 m layer. At 10 degrees incidence surface contribution is –13.6 dB. At larger angles of incidence volume scattering becomes increasingly more important. The volume scattering contribution is also more important as layer thickness increases. Thus, at 5 m thickness, direct surface scattering is negligible. Since volume scattering increases with the thickness of the layer, the level of backscattering increases with the layer thickness especially at large angles of incidence where surface scattering is less important. In cross polarized scattering the mechanism remains to be multiple volume scattering enhanced by surface-volume interaction.

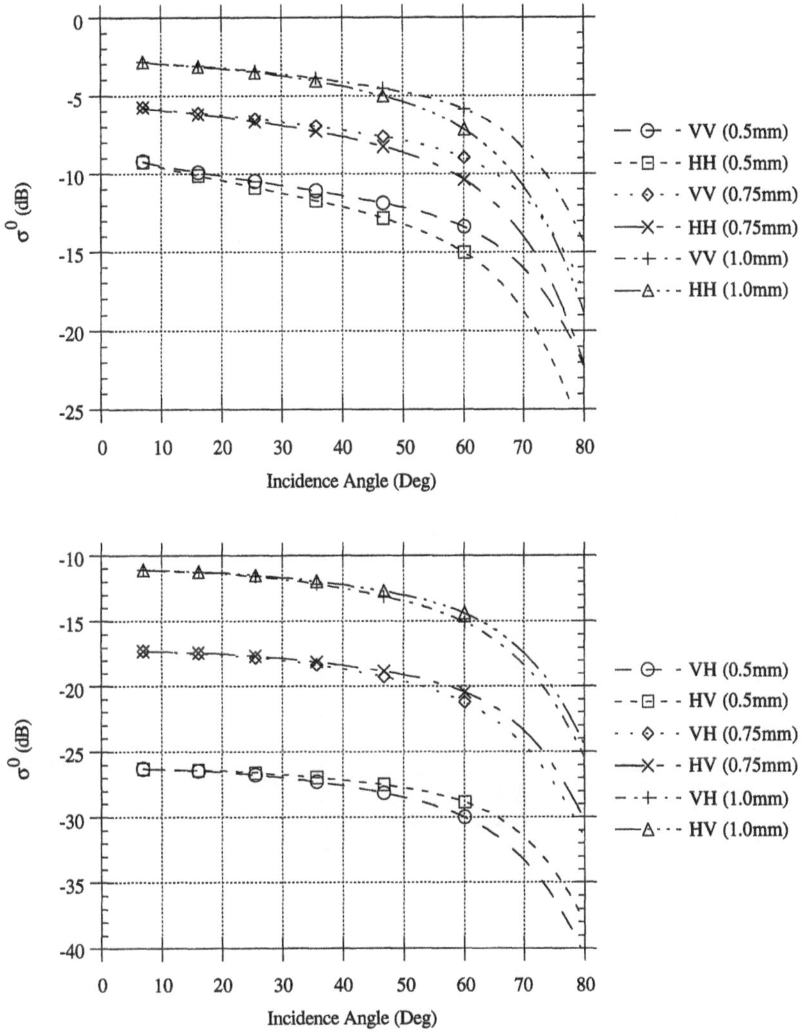

Figure 9.12 Effect of scatterer (air bubble) size on multiyear sea ice backscattering coefficients.

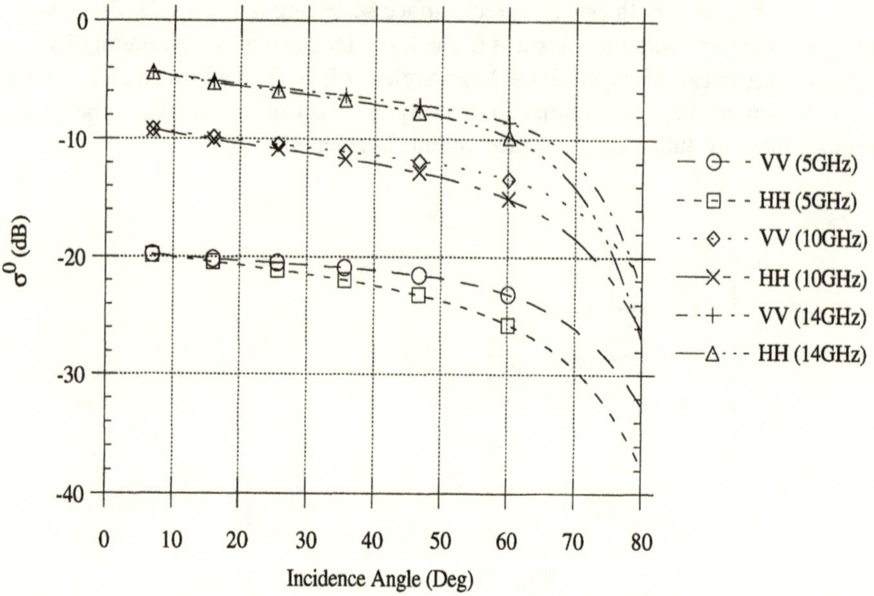

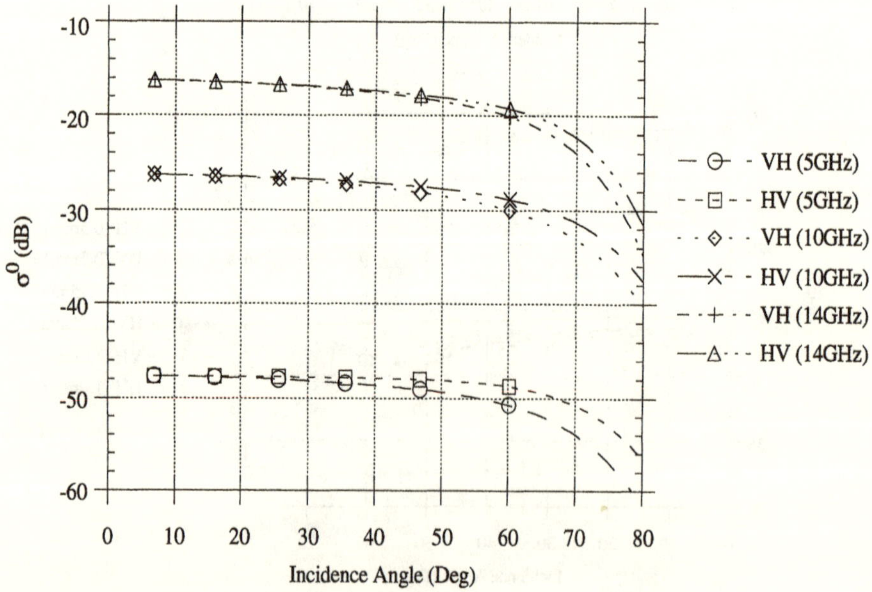

Figure 9.13 Effect of signal frequency on multiyear sea ice backscattering coefficients.

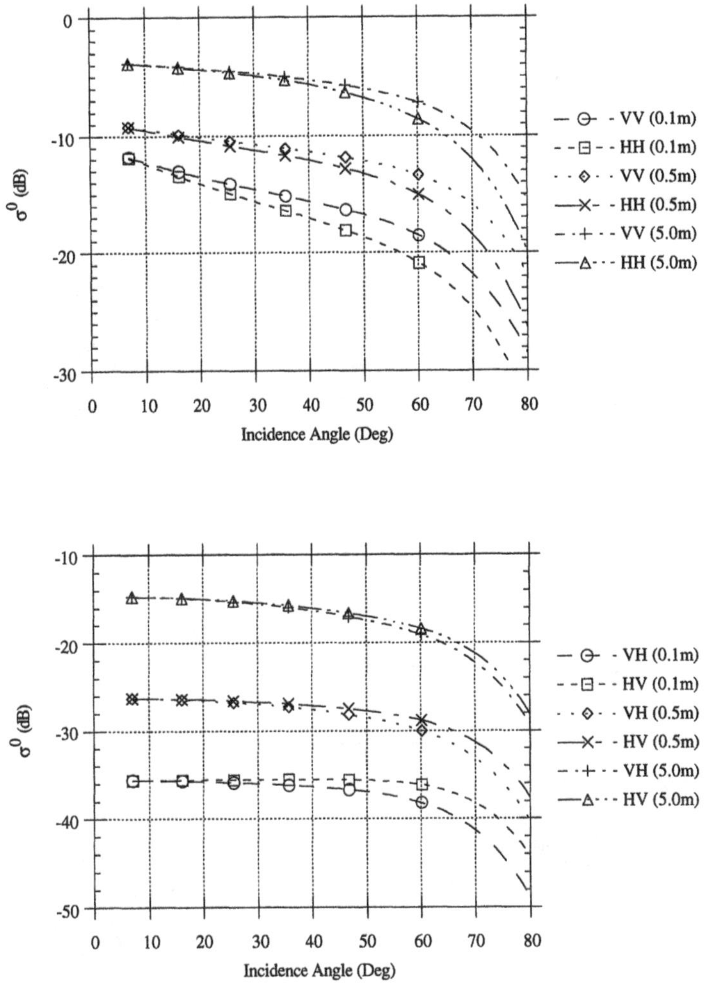

Figure 9.14 Effect of layer thickness on multiyear sea ice backscattering coefficients.

In Figure 9.15 we consider the effect of the volume fraction of scatterers. As the volume fraction changes, it affects the imaginary part of the layer permittivity and hence the albedo of the layer medium. On the other hand, its impact on optical depth is quite small, because the imaginary part is small. In fact, optical depth is around 0.1 for all three volume fractions. As volume fraction increases from 10% to 30%, albedo increases from 0.18 to 0.51 causing the backscattering coefficient to increase along with it. Note that, as volume fraction increases, the spacing between HH and VV narrows because volume scattering becomes more important. In the meantime, the angular drop-off decreases from about 8 dB at 10% to about 4.5 dB at 30% between 5 degrees and 60 degrees in like polarization. There is some sign of saturation in the sense that the percentage of increase in backscattering from 10% to 20% volume fraction is larger than that from 20% to 30%. Such a saturation effect is much more pronounced in cross polarization.

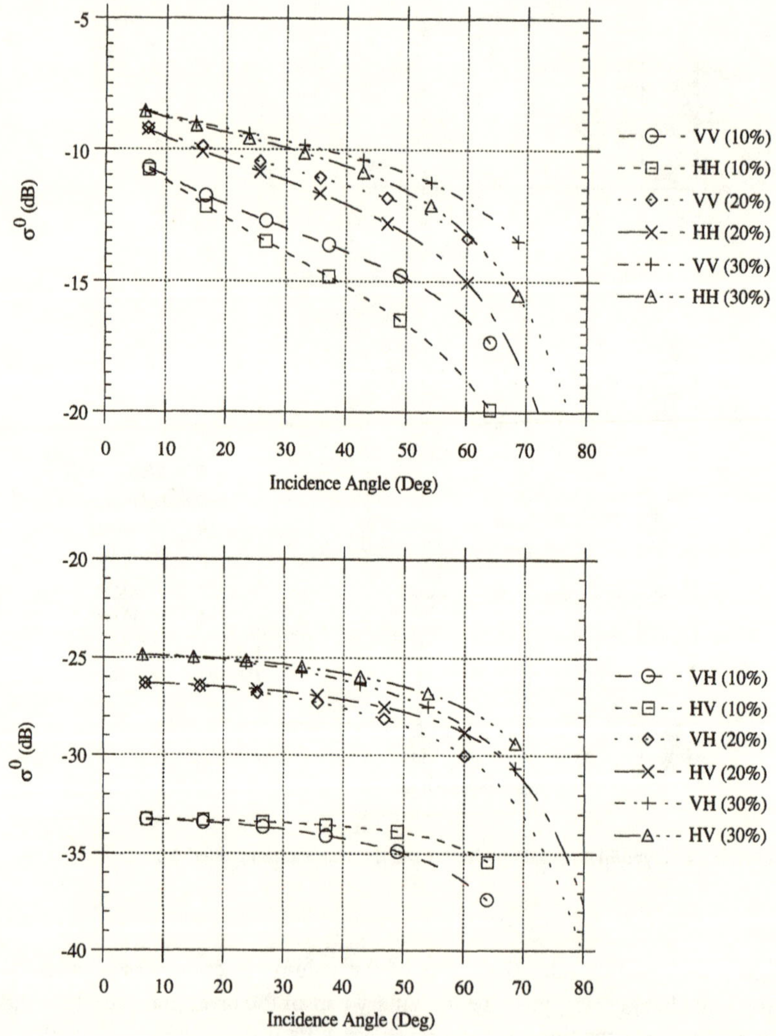

Figure 9.15 Effect of the air bubble volume fraction on multiyear sea ice backscattering coefficients.

In summary, an increase in any one of the four parameters, scatterer radius, frequency, layer thickness or volume fraction, will cause

1. an increase in polarized and cross polarized backscattering;
2. a slower angular drop-off of the polarized backscattering coefficient;
3. an increase in the relative importance of volume versus direct surface scattering from the air-ice boundary; and
4. a narrower spacing between HH and VV polarization.

Next, we consider the effects of surface parameters. We shall begin with a change in $k\sigma$ of the air-ice boundary in Figure 9.16. When $k\sigma = 0.03$, incoherent surface backscattering is very small and volume scattering is dominating. As $k\sigma$ increase to 0.3, the backscattering coefficient rises over the small angular range (5 degrees to 35 degrees) and surface scattering appears to dominate between 5 degrees to 35 degrees. Further increase of $k\sigma$ to 0.42 causes the backscattering coefficient to rise some more and surface scattering to dominate over a still wider range of angles. Beyond 55 degrees, the backscattering coefficients for VV and HH do not change much, indicating that volume scattering is dominating. In cross polarized scattering, there is a continued rise of the backscattering coefficients with the $k\sigma$ fairly uniformly over all incident angles. This is caused by the surface-volume interaction. Note that when $k\sigma$ is small, VH and HV are the same over all angles satisfying the reciprocity theorem in electromagnetic theory. However, when $k\sigma$ is equal to 0.3, VH and HV are not the same after about 30 degrees. This effect increases with further increase in $k\sigma$ to 0.42. The reason for this behavior appears to indicate that surface-volume interaction is dominated by bistatic cross polarized surface scattering to which reciprocity does not apply. In this case HV is consistently higher than VH.

In Figure 9.17 we examine the effect of kL at the air-ice boundary. Since kL controls the rate of change of the angular backscattering curve and a larger kL causes a faster drop-off, it follows that an increase in kL results in a larger and steeper backscattering coefficient at small angles of incidence and allows volume scattering to be dominating over a wider range of incident angles. Thus, while larger variations of the backscattering coefficients appear over small incident angles, there is not much change at large angles of incidence because of the dominance by volume scattering. Although surface-volume interaction should also have an impact, it is of higher order compared with the direct surface or volume scattering and hence is not causing much change. In addition, significant changes are confined to the first 15 degrees. Thus, it requires antenna with a much narrow beam to observe these changes. Experimentally, it is also difficult to differentiate incoherent and coherent scattering that may result when kL is large. In cross polarized scattering surface-volume interaction is more important because cross polarized backscattering is due to multiple scattering, a higher order effect. Here, the change due to surface roughness appears throughout all incident angles. Although the change in Figure 9.17 is only about 1 dB at 5 degrees it increases to about 2 dB at 60 degrees. This indicates that surface-volume interaction is an important mechanism for cross polarized scattering.

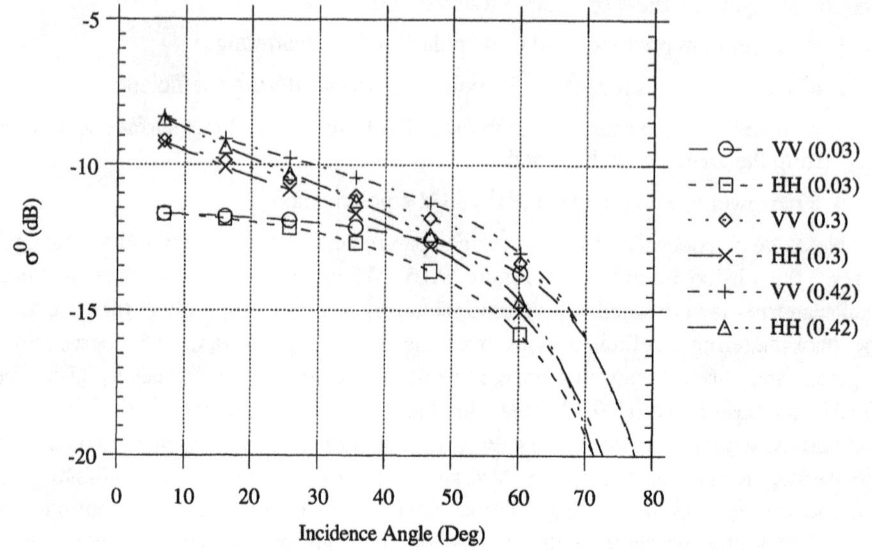

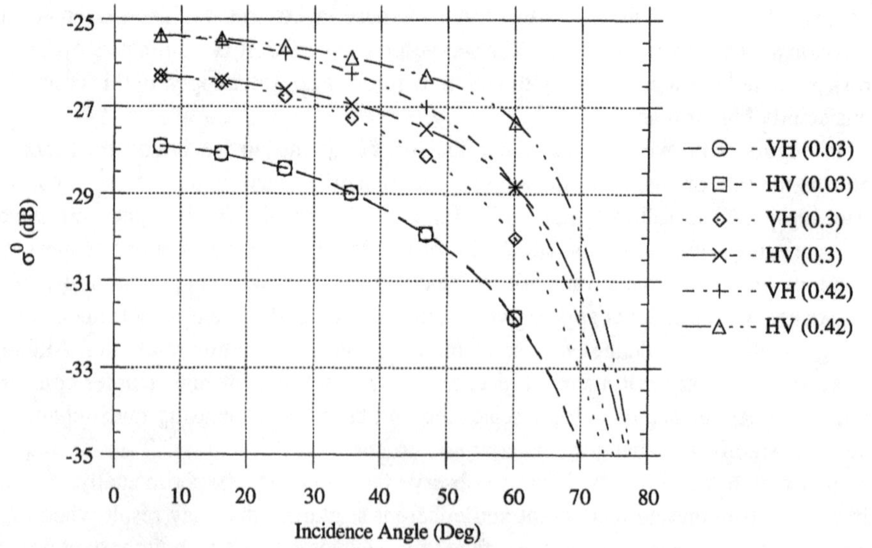

Figure 9.16 Effect of top surface $k\sigma$ on multiyear sea ice backscattering coefficients.

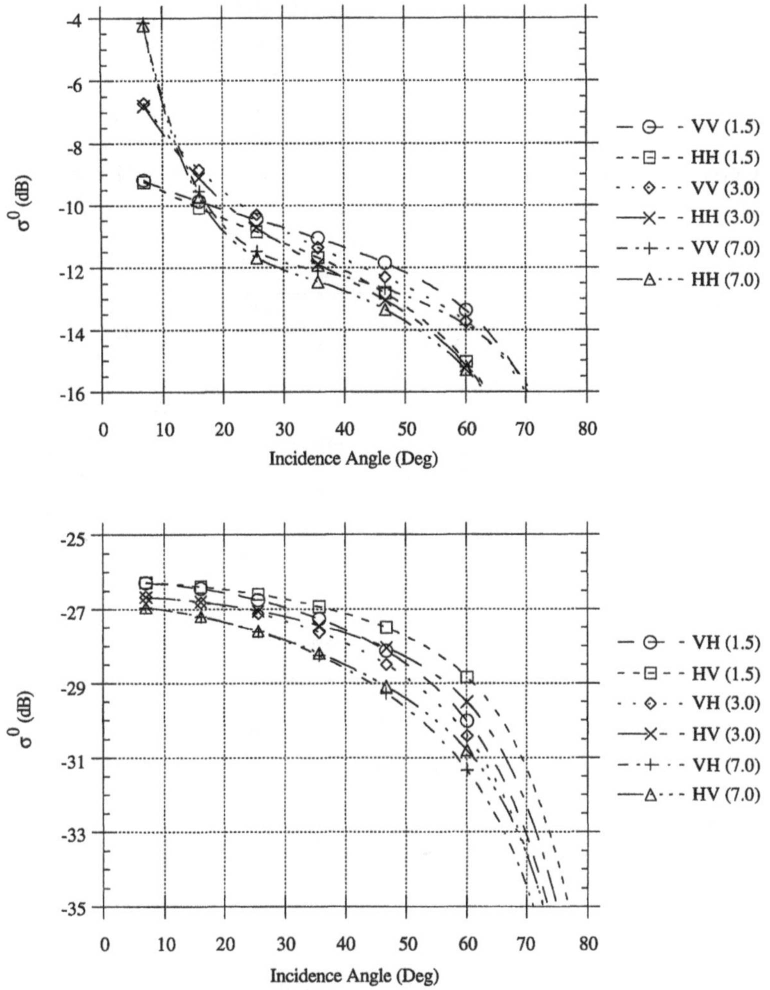

Figure 9.17 Effect of top surface kL on multiyear sea ice backscattering coefficients.

As we indicated earlier, the discontinuity at the lower boundary between desalinated ice and salinated ice is rather small. Hence, a change in the roughness condition of this boundary should have a very little impact on backscattering. This point is illustrated in Figures 9.18 and 9.19, where $k\sigma$ and kL variations are shown. The modeling assumption made here is that the total ice thickness is large and the saline portion of the ice is thick. Thus, the loss over the saline portion of the ice is such that we can treat it as a half space. It is clear from Figure 9.18 that changes in $k\sigma$ have no appreciable influence on backscattering for both like and cross polarizations. The major mechanisms remain to be direct surface scattering from the air-ice boundary and volume scattering from the ice layer for like polarization. For cross polarization, the mechanisms are still multiple volume scattering and interaction between the top surface boundary and the ice volume. When kL of the lower boundary is changed from 2 to 6, there a rise of about 0.5 dB at 5 degrees incidence, and the increase becomes insignificant after 10 degrees. Experimentally, this will be difficult to recognize because, at small angles of incidence, there are effects of antenna pattern and coherent contributions. For cross polarization kL has more influence on backscattering. This is because the overall level of scattering due to multiple volume scattering is low.

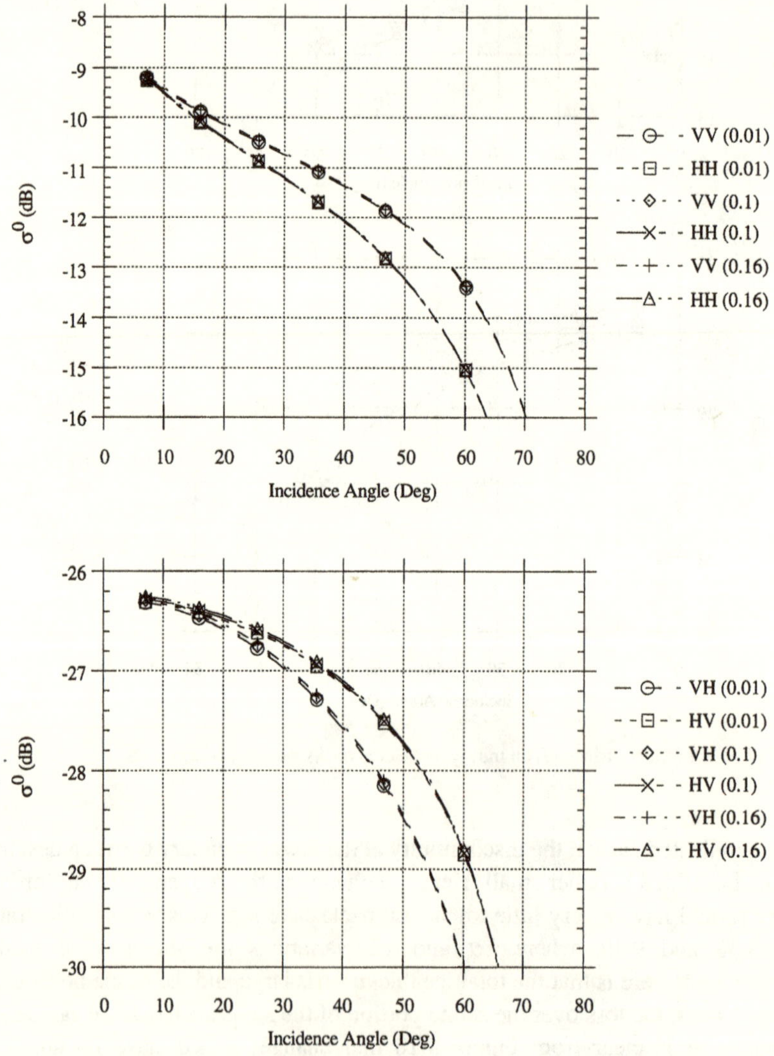

Figure 9.18 Effect of bottom surface $k\sigma$ on multiyear sea ice backscattering coefficients.

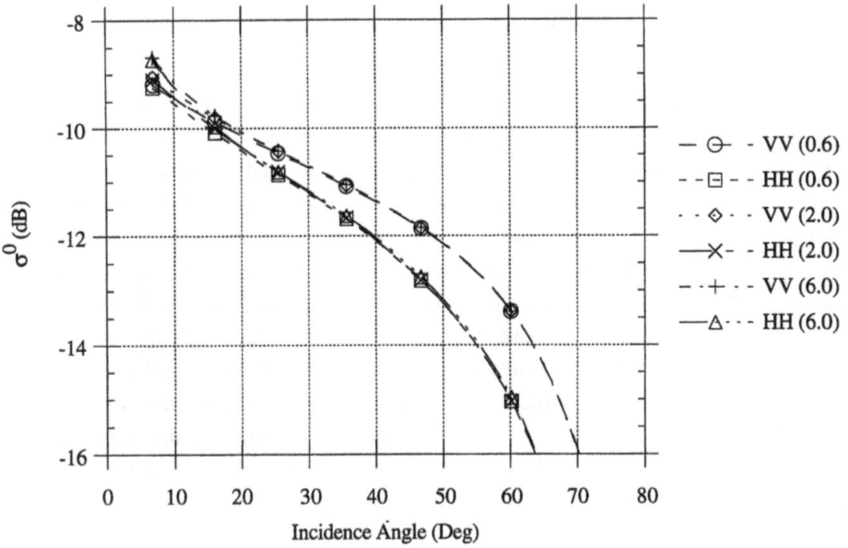

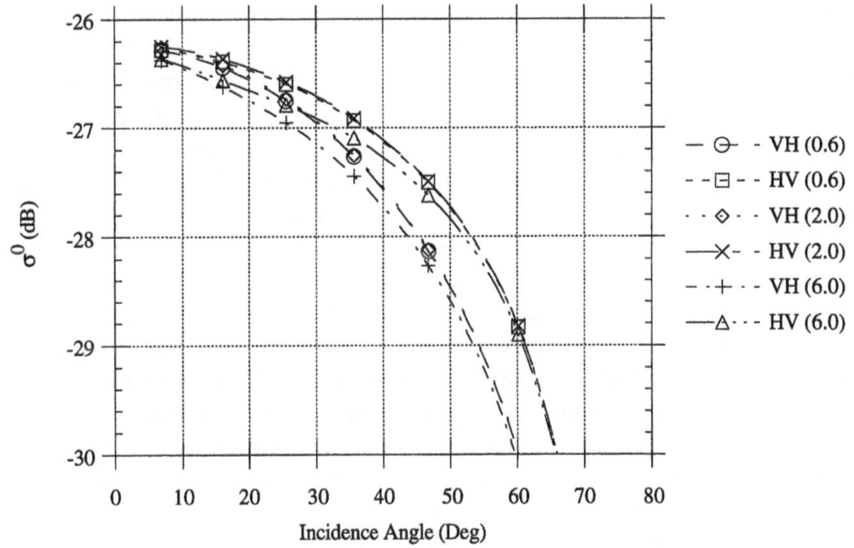

Figure 9.19 Effect of bottom surface *kL* on multiyear sea ice backscattering coefficients.

9.5 AN EMISSION MODEL FOR A SNOW LAYER

The snow layer is modeled as a closely packed irregular inhomogeneous layer above a homogeneous half space. The temperature at the air-snow boundary is taken to be 262 K and at the lower boundary it is 272 K. A linear temperature profile is assumed in the calculations of the snow layer emission. Except for the difference in the nature of the signal sources, the emitted waves experience scattering and attenuation processes similar to the incident wave in active scattering problems. The snow layer parameters to be used for the passive emission calculations are the same as those for the active problem given in Table 9.1. The brightness temperature characteristics of snow due to variations of snow parameters are shown in Figure 9.20 through Figure 9.26. In general, there are three major emission sources: upwelling and downwelling of emitted intensities within the snow layer and upward emission from the lower half space. These intensities are expected to go through some or all of the following processes, volume scattering, surface scattering, attenuation, surface-volume interactions and transmission across irregular boundary or boundaries, before they arrive at the receiver. A detailed description of these processes have been given in Chapter 8.

9.5.1 Theoretical Emission Characteristics of a Snow Layer

In this section we shall show the angular dependence of snow emission on various system, volume and surface parameters. These include scatterer size, frequency, layer thickness, volume fraction of scatterers, liquid water content, and surface roughness parameters. Other parameters are kept the same as in Table 9.1. As in scattering models, ice particles are modeled as spheres because they are randomly oriented.

The variations of the brightness temperature with the nadir angle denoted as theta is shown in Figure 9.20 for three different scatterer sizes. When the radius of the scatterer is changed from 0.5 to 0.75 and 1.0 mm, the optical depth changes from 0.13 to 0.38 and 0.89, respectively. Due to relatively small optical depths, emissions from the layer, either directly or through lower boundary reflection, are much smaller than those from the lower half space. The contributions from both upward and downward layer emissions, and the lower half space are 7.9 K, 0.97 K and 236 K, respectively, for a scatterer radius of 0.5 mm. As the scatterer size increases, the snow layer is optically thicker and hence there is more emissions from the layer. However, these increases are not larger than the decrease in contributions from the lower half space. Thus, the snow layer acts as an attenuator to the emission from lower half space under these conditions. The total emitted brightness temperature decreases with an increase in scatterer size. The layer permittivity is around 1.48 and so the Brewster angle effect is small. The contributions from both the upward layer emission and the lower half space increase with the nadir angle from 5 degrees to about 66 degrees in vertical polarization, while that from the downward emission decreases over the same angular region, when the radius is 0.5 mm. As the scatterer radius increases to 0.75 mm, the emission from the lower half space becomes flat, while the upward layer emission remains increasing.

Further increase in scatterer size to 1 mm radius results in a decreasing angular trend for the emission from the lower half space. This change in angular trend is due to attenuation of the snow layer which is effectively thicker for emission at larger nadir angles. This is the reason why the overall angular trend for vertical polarization becomes flat and slightly decreasing when the scatterer radius is 1 mm. This change in angular trend also causes the spacing between vertically and horizontally polarized emissions to narrow.

To examine the dependence of layer emission on frequency we plot in Figure 9.21 brightness temperatures at 5, 17 and 35 GHz. The optical depths of the snow layer at these frequencies are 0.008, 0.13 and 2. At the lower frequencies, the brightness temperature is dominated by the lower half space. At 5 degrees the upward layer emission is 7.92 degrees while the lower half space contribution is 236 degrees. The attenuation through the snow layer is small and the brightness temperature rises with the nadir angle up to 65 degrees. The spacing between the two polarized temperatures is large because the vertically polarized temperature is rising, while the horizontally polarized temperature is falling between 5 degrees and 65 degrees. At 35 GHz, the temperatures from the upward and downward layer emissions at the 5 degrees nadir angle are 24.7 degrees and 1.27 degrees, while the lower half space emission is 130 degrees. The attenuation of the snow layer is much larger, causing the half-space emission to decrease with the nadir angle. As a result, the spacing between the two polarization components is much smaller compared with lower frequency cases and there is no peaking in vertical polarization. In general, as the influence of the snow layer increases, the level of total emission and the spacing between polarization components decrease. This behavior is true for a dry snow layer.

In Figure 9.22 we show the effect of layer thicknesses of 0.1, 0.5 and 5 m. These depths correspond to optical depths of 0.026, 0.13 and 1.3. Thus, the effect on brightness temperature is similar to increasing frequency, which also leads to increases in optical depth and therefore attenuation of the contribution from lower half space. However, unlike frequency variation, the albedo of the layer and boundary roughness are not affected by a change in physical depth. Thus, the overall change is less drastic compared with a change in frequency. The general change in the level and spacing between vertically and horizontally polarized brightness temperatures is very similar to that in Figure 9.21 indicating that the dominant contributor is the emission from the lower half space attenuated by the snow layer.

In a closely packed medium one effect of interest is the volume fraction of scatterers. This is illustrated in Figure 9.24. A change in volume fraction from 20% to 40% causes a small decline in albedo from 0.82 to 0.8 and a more significant increase in optical depth from 0.085 to 0.176, resulting in higher emissions. The reason for having higher emissions is because the increase in optical depth generates more emission from the snow layer than the loss of emission from the lower half space due to attenuation by the layer. This happens when the albedo of the layer is smaller than its absorption coefficent. Note that this is an interesting situation where both backscattering (see Figure 9.6) and emission increase with the volume fraction. Intuitively, one expects emission and backscattering to go in opposite directions, which would happen if the layer albedo were large.

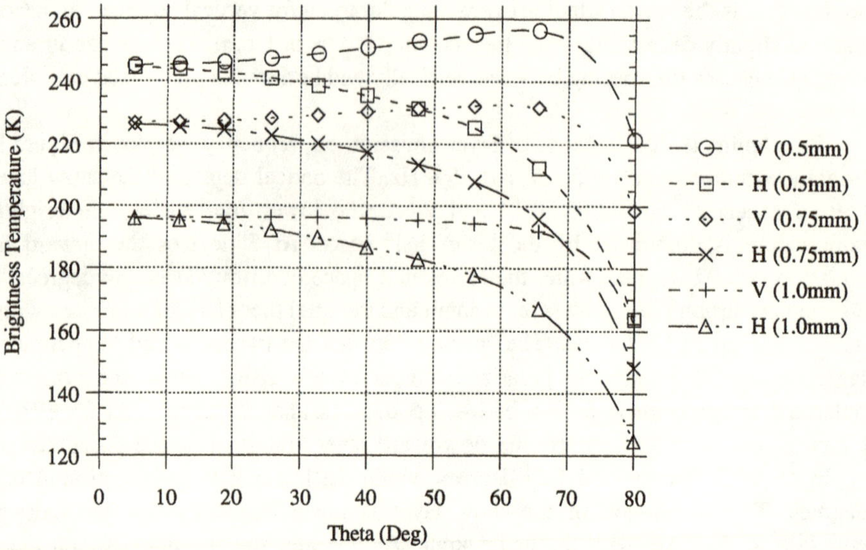

Figure 9.20 Effect of scatterer (ice particle) size on emission from a snow layer.

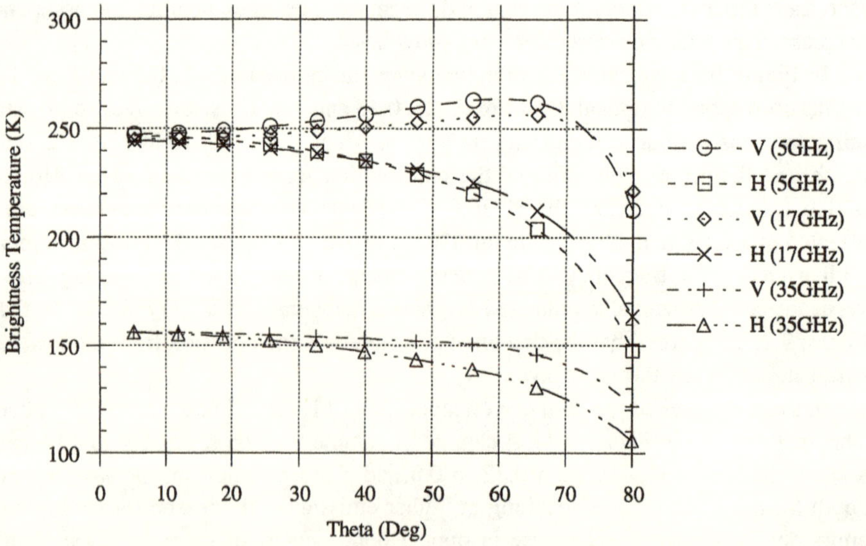

Figure 9.21 Effect of signal frequency on emission from a snow layer.

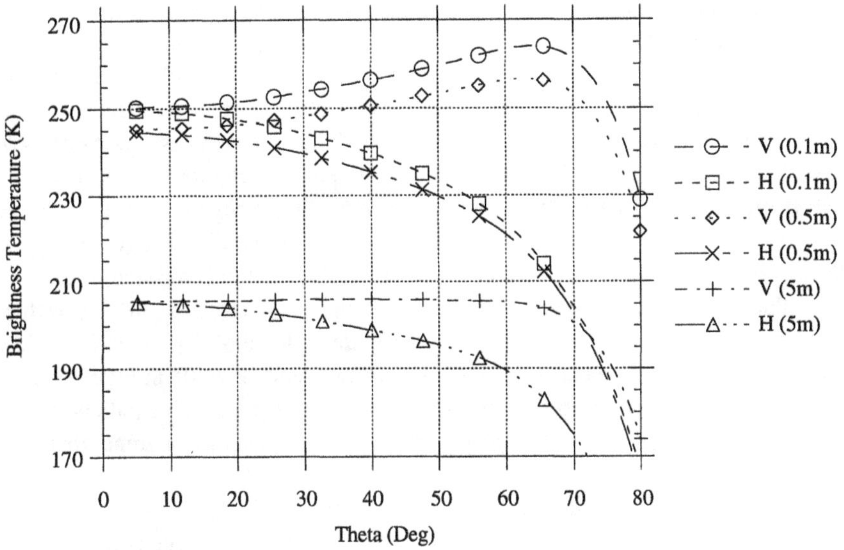

Figure 9.22 Effect of layer thickness on emission from a snow layer.

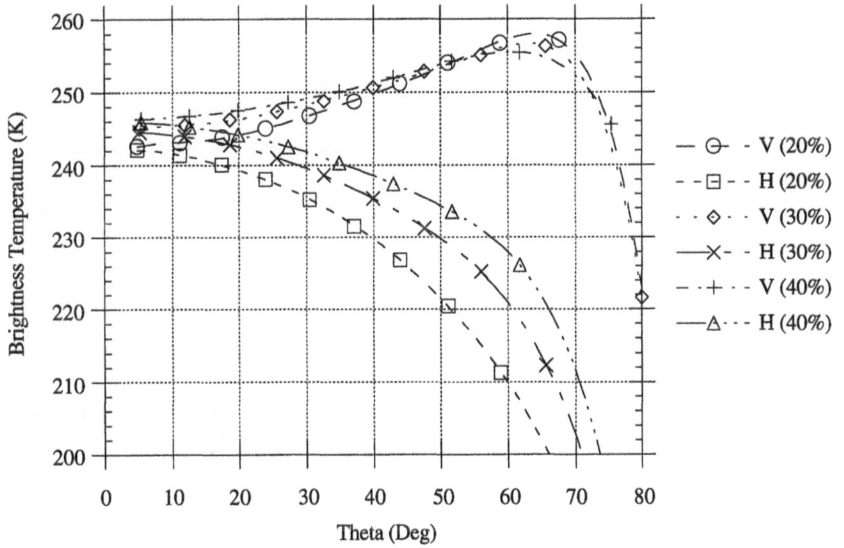

Figure 9.23 Effect of ice particle volume fraction on emission from a snow layer.

The effect of wetness is less straightforward. When compared with dry snow, it is clear that there should be more absorption and therefore more emission from the snow layer. However, as wetness increases the effective layer permittivity also increases causing a larger discontinuity at the air-snow boundary. Thus, the total emission should increase initially with wetness but it will decrease when wetness is too large. In the Figure 9.26 we see that emission continues to increase from dry to 1% wetness. Upon comparing the emission at 2% (Figure 9.24a) with the emission at 1% (Figure 9.26b) the emission at 2% is only slightly higher than that at 1% indicating a region where the increase in emission due to wetness is largely offset by a decrease due to dielectric discontinuity at the air-snow boundary. At 3% wetness (Figure 9.24b) the total emission is clearly lower than that at 2%. This decline continues through 4% and 5%. At 5% wetness the brightness temperature is approaching the brightness level of dry snow. Thus, in Figure 9.24a we see both a rise and a fall with increasing wetness, while in Figure 9.24b, there is a continued decline of brightness temperature with wetness. This indicates that if not enough wetness cases are available, one can come up with a trend that could be misleading.

Next, we consider surface roughness effect by changing $k\sigma$ of the air-snow boundary from 0.05 to 0.7 and keeping $kL = 2.5$ in Figure 9.25. The volume parameters, albedo and optical depth, are fixed at 0.81 and 0.13, respectively, at 17 GHz. In general, boundary roughness tends to slow down angular variations in the emission curves. Thus, the emission increases with an increase in the nadir angle for horizontal polarization. In vertical polarization there is both an increase and a decrease around the Brewster angle region to slow down the angular variations in the brightness temperature curves. It is anticipated that because the permittivity of the snow layer is small, the surface roughness effect at the air-snow interface is also small. For an inhomogeneous layer with larger permittivity, there should be a similar but larger changes over all nadir angles for both polarizations.

We now consider the roughness of the bottom boundary of the snow layer by varying $k\sigma$ from 0.01 to 0.28 in Figure 9.26. Other snow layer parameters are kept the same as in Table 9.1. Unlike the air-snow boundary the discontinuity in permittivity is much larger here, resulting in a greater surface effect at small nadir angles. An increase in roughness causes a corresponding increase in brightness temperature, which decreases with an increase in nadir angle. The reason for having a greater effect at small nadir angles than at large angles is due to the attenuation of the snow layer, which is greater for larger nadir angles. Thus, the $k\sigma$ of the air-snow boundary has more influence on emission at larger nadir angles, while that of the snow-ground boundary affects emission more at smaller nadir angles. Recall that noncoherent backscattering increases with surface roughness. Thus, with respect to surface rms height both emission and backscattering increase with surface roughness. It was explained in Chapter 3 that with respect to a change in the dielectric constant the emission and backscattering must move in opposite directions.

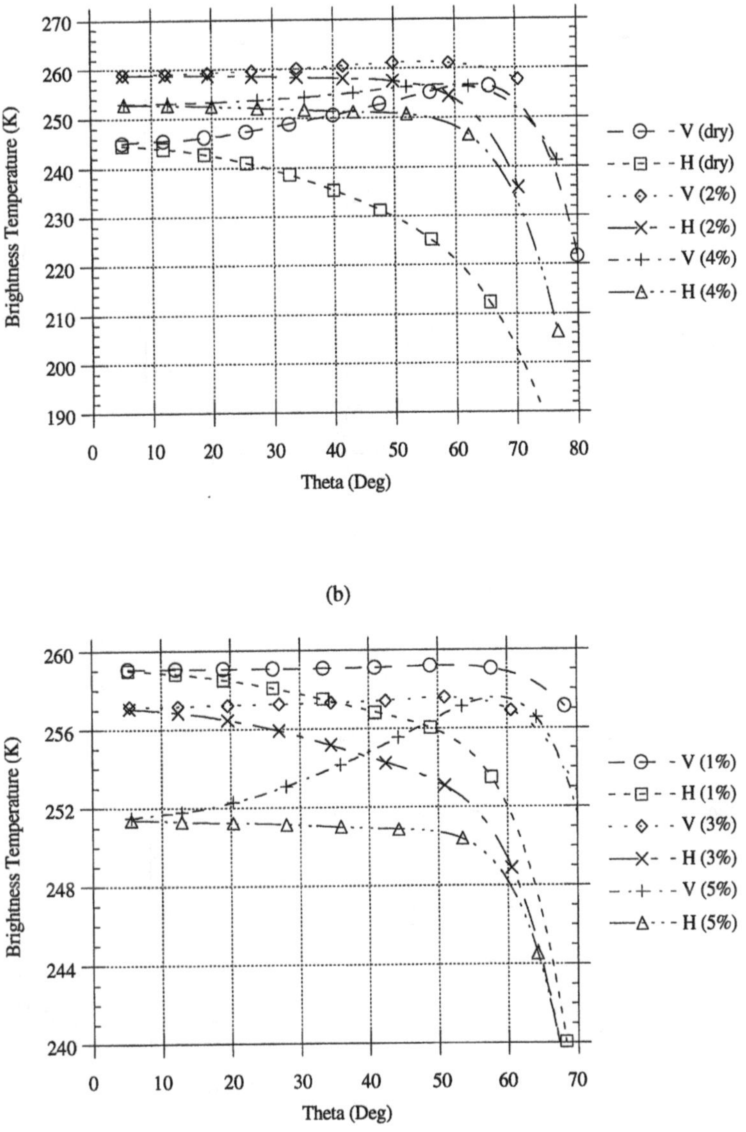

Figure 9.24 (a) Effect of volumetric liquid water content at 2% and 4% on emission from a snow layer. (b) Effect of volumetric liquid water content at 1%, 3% and 5% on emission from a snow layer.

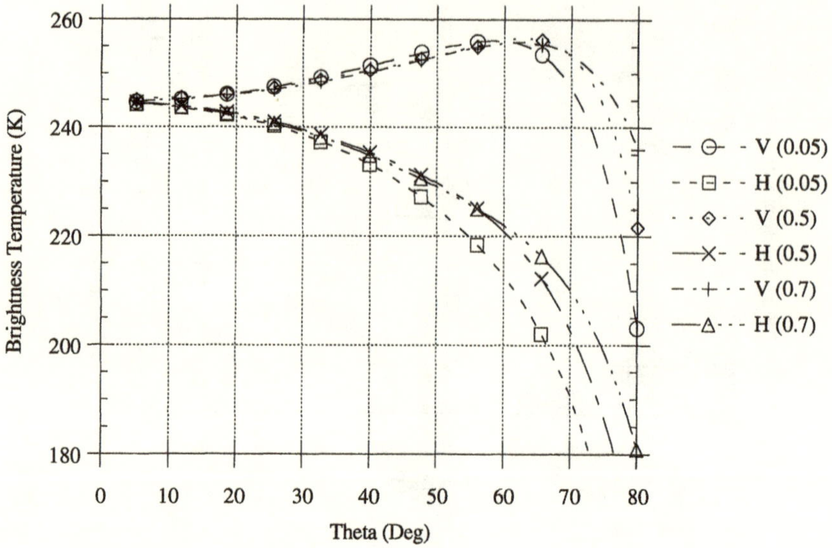

Figure 9.25 Effect of top surface $k\sigma$ variation on emission from a snow layer.

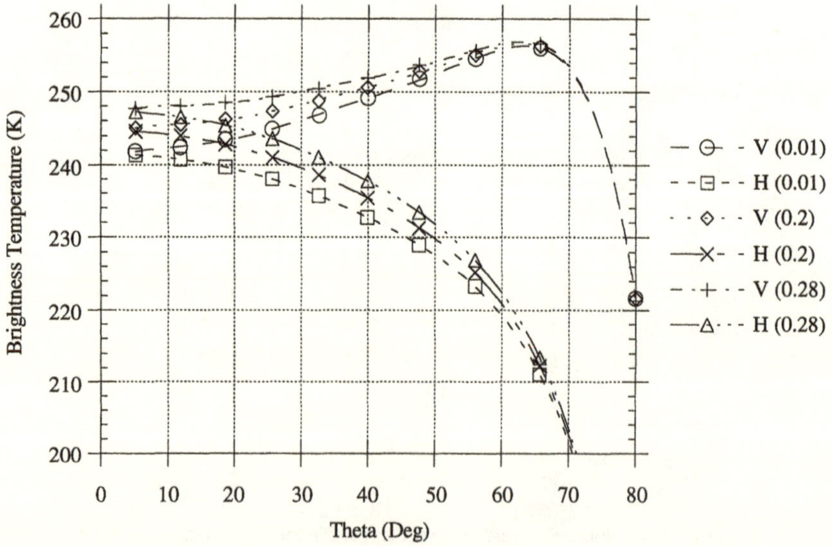

Figure 9.26 Effect of bottom surface $k\sigma$ variation on emission from a snow layer.

9.6 AN EMISSION MODEL FOR A SEA ICE LAYER

The multiyear (desalinated) sea ice layer with the same reference parameters as listed in Table 9.2 is utilized in the theoretical emission calculations. The ice layer is assumed to have a linear temperature profile with an upper interface temperature of 262 K and a lower interface temperature of 272 K. The variations of the parameters (air bubble size, frequency, layer thickness, volume fraction of scatterers and boundary roughness) on emission from such a desalinated sea ice layer are shown in Figures 9.27 through 9.32. The general structure of the ice is a layer of desalinated ice embedded with air bubbles sitting above a half space of saline ice. The reason for assuming a half space of saline ice is because generally the total ice layer is over 2 m thick and the desalinated portion is around 0.5 of a meter. Thus, the sea water below the ice is not likely to be seen. In addition, the interface between desalinated and saline ice does not have a large dielectric discontinuity. Hence, we expect the major sources of scattering to come from the air-ice interface and the air bubbles within the desalinated ice layer.

9.6.1 Theoretical Emission Characteristics of a Sea Ice Layer

The effect of the size of the air bubbles (scatterers) is shown in Figure 9.27. As the size increases from 0.5 to 1 mm, albedo increases from 0.41 to 0.836 and optical depth from 0.108 to 0.39. The air bubbles are embedded within a desalinated ice layer. This implies that a thicker layer has a larger volume scattering coefficient than absorption coefficient. Although the increase in optical depth generates more emission from the layer, it also attenuates more the emission from the lower half space. On balance, it acts as an attenuator to the emission from the lower half space, which has a much larger absorption coefficient because it is a saline ice. The general trend is a decrease in brightness as the scatterer size increases.

When frequency changes from 5 to 10 and 14 GHz, albedo changes from 0.08 to 0.41 and 0.65; optical depth from 0.0347 to 0.108 and 0.255; $k\sigma$ of the air-ice interface from 0.15 to 0.3 and 0.42; and kL from 0.75 to 1.5 and 2.1. These changes are illustrated in Figure 9.28. The initial increase in optical depth by a factor of about three, when frequency changes from 5 to 10 GHz, raises emission at small nadir angles. At larger nadir angles, the effect of surface boundary roughness is significant and causes the spacing between vertical and horizontal polarization to narrow. Further increase in frequency to 14 GHz generates more scattering than emission resulting in a lower emission curve near nadir. However, boundary roughness has increased further, causing additional narrowing of the spacing between the two polarizations. Overall, there is less change in vertical polarization than horizontal polarization. In fact, the change in vertical polarization is negligible when frequency changes from 10 to 14GHz.

Figure 9.29 illustrates the effect of layer thickness. Generally, a thicker layer emits more. However, because total emission is dominated by the lower half space, the thicker ice layer causes more attenuation on the contribution from lower half space than it emits. As a result, the total emission is lowered as ice layer thickness increases.

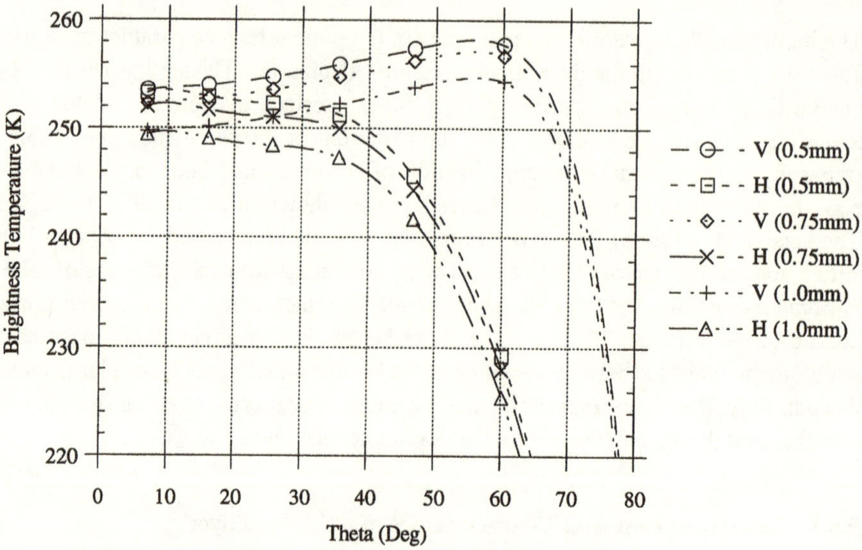

Figure 9.27 Effect of air bubble size on emission from a multiyear sea ice layer.

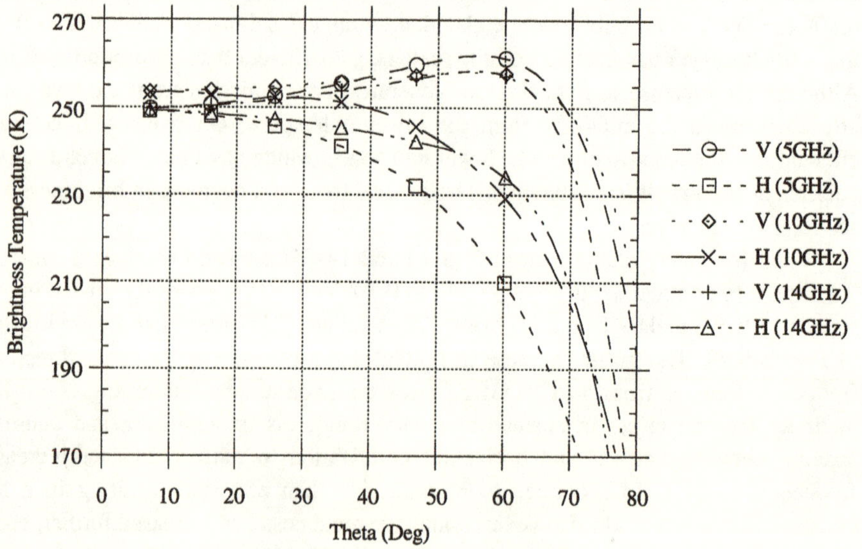

Figure 9.28 Effect of signal frequency on emission from a multiyear sea ice layer.

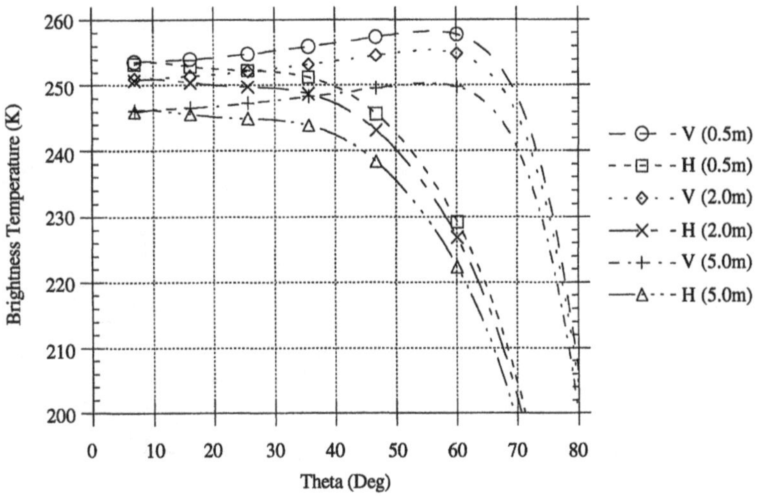

Figure 9.29 Effect of layer thickness on emission from a multiyear sea ice layer.

When volume fraction increases, albedo increases and absorption decreases with the result that optical depth remains almost constant. At 10% volume fraction albedo is 0.18 and layer dielectric is $2.9 + j0.0015$. The main contributions are from the layer and lower half space. After volume fraction increases to 20% albedo increases to 0.41 and layer dielectric drops to $2.7 + j0.001$. This increase in albedo causes layer emission to go down, but the lower half-space contribution goes up because of somewhat lower attenuation by the layer and higher overall transmission through the layer (Figure 9.30). The total emission is slightly higher than that at 10% volume fraction. Further increase of volume fraction to 30% leads to an albedo of 0.51 and a further drop in layer emission. The layer dielectric constant is $2.3 + j0.0008$, which gives a higher transmission at the air-ice interface but a lower transmission at the interface between desalinated and saline ice. The result is that there is negligible change in emission by the half space as compared to that at 20%. The total emission is the lowest because of the decrease in layer emission.

Finally, we consider surface roughness effects in Figures 9.31 and 9.32. In Figure 9.31 we change the $k\sigma$ of the air-ice boundary from 0.03 to 0.3 and then 0.42. The major effect appears to be slower varying angular curves for nadir angles between 5 and 60 degrees at larger $k\sigma$ values. In addition, the spacing between vertical and horizontal polarizations becomes narrower over the same angular region. These changes in angular trends make the Brewster angle effect on vertical polarization less pronounced. When a similar change is made of the $k\sigma$ at the interface between desalinated and saline ice, a negligible change in emission is observed in Figure 9.32. This is due to the small dielectric discontinuity at such an interface.

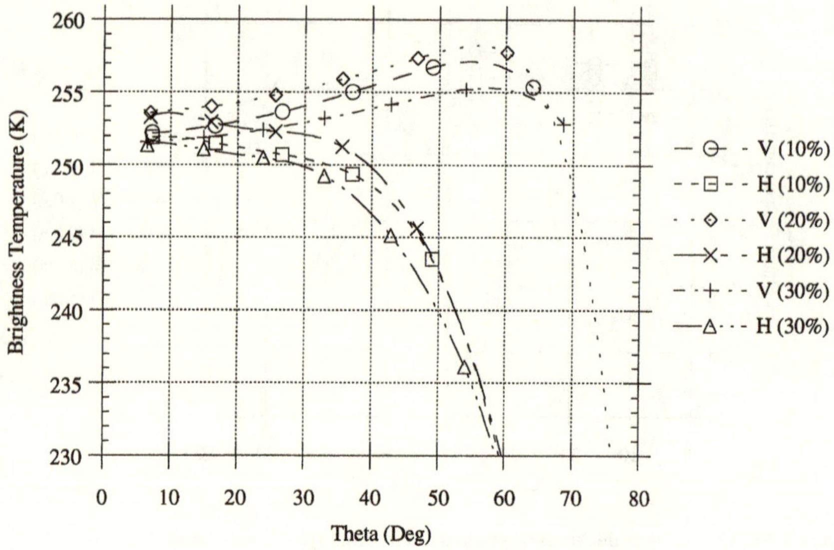

Figure 9.30 Effect of air bubble volume fraction on emission from a multiyear sea ice layer.

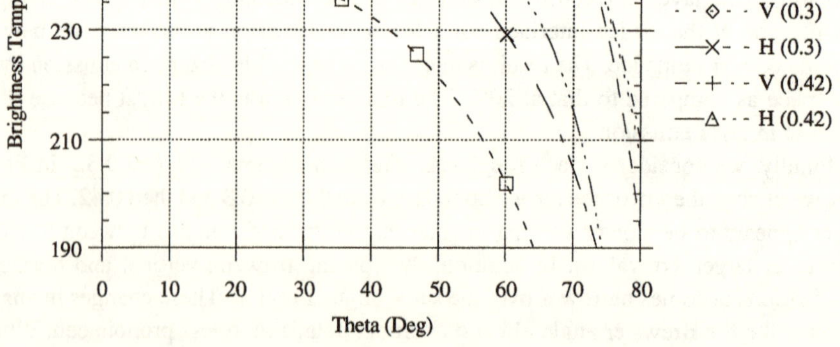

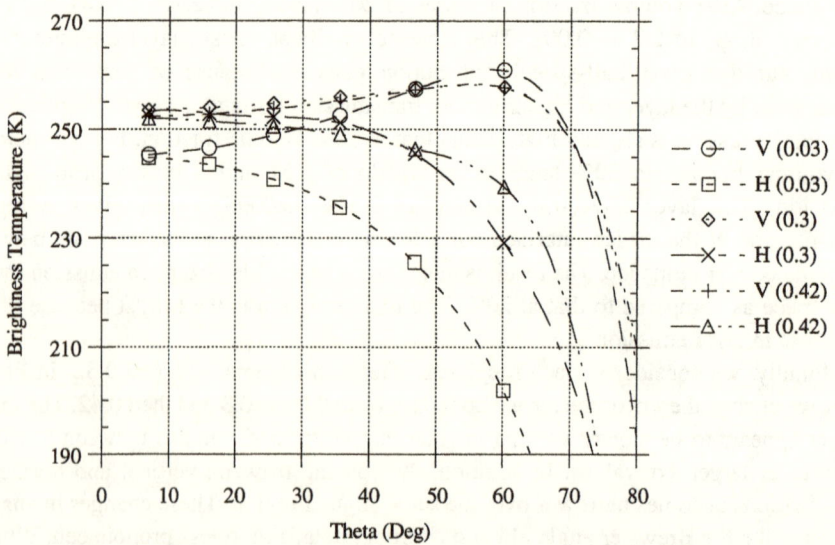

Figure 9.31 Effect of top surface $k\sigma$ on emission from a multiyear sea ice layer.

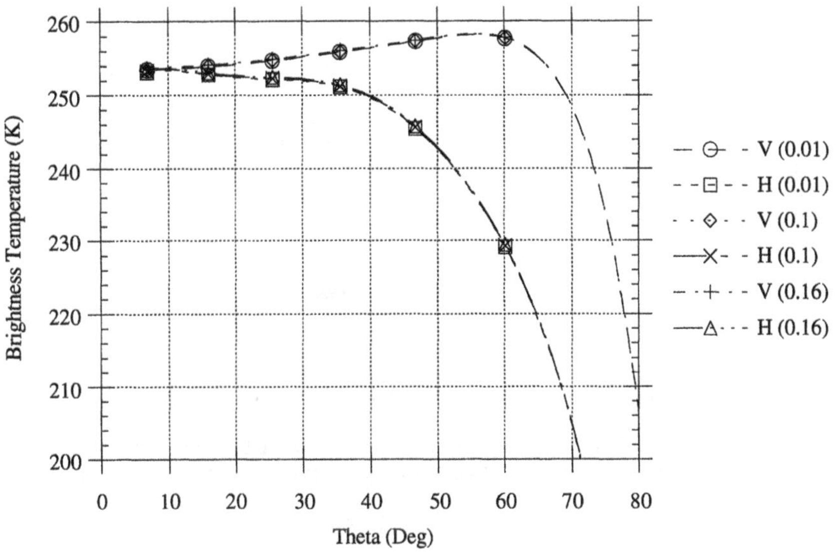

Figure 9.32 Effect of bottom surface $k\sigma$ on emission from a multiyear sea ice layer.

REFERENCES

Colbeck, S.C., "Grain Cluster in Wet Snow," *J. Colloid and Interface Science*, Vol. 72, no. 3, December 1979, pp. 371–384.

Frankenstein, G., and R. Garner, "Equations for Determining the Brine Volume of Sea Ice from -0.5°C to -22.9°C," *J. Glaciol.*, Vol. 6, 1967, pp. 943–947.

Fung, A.K., and H.J. Eom, "A Study of Backscattering and Emission from Closely Packed Inhomogeneous Media," *IEEE Trans. Geosci. Remote Sens.*, Vol. GE-23, no. 5, 1985, pp. 761-767.

Fung, A.K., and S. Tjuatja, "Snow Parameter Effects in Scattering and Emission," *Proc. IGARSS'91 Symp.*, June 3–6, Helsinki, Finland, 1991.

Fung, A.K., M. Dawson and S. Tjuatja, "An Analysis of Scattering from a Thin Saline Ice Layer," *Proc. IGARSS'92 Symp.*, Houston, TX, May 25–29, 1992.

Goedecke, G.H., "Radiative Transfer in Closely Packed Media," *GOSA*, Vol. 67, no. 10, October 1977, pp. 1339–1348.

Ishimaru, A., *Wave Propagation and Scattering in Random Media*, Vols. 1 and 2, Academic Press, New York, 1978.

Hallikainen, M., F.T. Ulaby and M. Abdelrazik, "Dielectric Properties of Snow," *IEEE Trans. Ant. Prop.*, Vol. 34, no. 11, 1986, pp. 1329–1340.

Hallikainen, M., F.T. Ulaby and T.E. Van Deventer, "Extinction Behavior of Dry Snow in the 18 to 90 GHz Range," *IEEE Trans. on Geoscience and Remote Sensing*, Vol. 25, no. 6, 1987, pp. 737–745.

Karam, M.A., and A.K. Fung, "Vector Forward Scattering Theorem," *Radio Science*, Vol. 17, 1982, pp. 752–756.

Karam, M.A., and A.K. Fung, "An Investigation of the Forward Scattering Theorem," *Proc. IGARSS'87 Symp.*, 1987, pp. 1007–1012.

Linlor, W.I., "Permittivity and Attenuation of Wet Snow Between 4 and 12 GHz," *J. Appl. Physics*, Vol. 51, 1980, pp. 2811–2816.

Matzler, C., "Applications of the Interaction of Microwaves with the Natural Snow Cover," *Remote Sensing Reviews*, Vol. 2, 1987, pp. 259–387.

Matzler, C., and U. Wegmuller, "Dielectric Properties of Fresh-Water Ice at Microwave Frequencies," *J. Appl. Phys.*, Vol. 20, 1987, pp. 1623–1630.

Nghiem, S., "The Electromagnetic Wave Model for Polarimetric Remote Sensing of Geophysical Media, Ph.D. Dissertation, Massachusetts Institute of Technology, Cambridge, 1991.

Poe, G., A. Stogryn and A. T. Edgerton, "A Study of the Microwave Emission Characteristics of Sea Ice," Final Tech. Rept. 1749R-2, Contract no. 2-35340, Aerjet Electrosystems Co., Azusa, CA., 1972.

Stratton, J.A., *Electromagnetic Theory*, McGraw-Hill, New York, 1941.

Stogryn, A., Strong Fluctuation Theory for Moist Granular Media," *IEEE Trans. GRS*, Vol. 23, no. 2, 1985, pp. 78–83.

Stogryn, A., "A Study of the Microwave Brightness Temperature of Snow from the Point of View of Strong Fluctuation Theory," *IEEE Trans. GRS,* Vol. 24, no. 2, 1986, pp. 220–231.

Stogryn, A., and G.J. Desargent, "The Dielectric Properties of Brine in Sea Ice at Microwave Frequencies," *IEEE Trans. on Ant. and Prop.*, Vol. 33, 1985, pp. 523–532.

Tiuri, M., A.H. Sihvola, E.G. Nyfors and M.T. Hallikainen, "The Complex Dielectric Constant of Snow at Microwave Frequencies," *IEEE J. Oceanic Engineering*, Vol. 9, 1984, pp. 377–382.

Tjuatja, S., A.K. Fung and J.W. Bredow, "A Scattering Model for Snow-Covered Sea Ice," *IEEE Trans. Geosci. Remote Sensing*, Vol. 30, no. 4, 1992, pp. 804–810.

Tjuatja, S., A.K. Fung and M.K. Dawson, "An Analysis of Scattering and Emission from Sea Ice," *Remote Sensing Reviews*, Vol. 7, 1993, pp. 83–106.

Wen, B.H., L. Tsang, D.P. Winebrenner and A. Ishimaru, "Dense Medium Radiativetransfer Theory: Comparison with Experiment and Application to Microwave Remote Sensing and Polarimetry," *IEEE Trans. Geosci. Remote Sensing*, Vol. 28, no. 1, 1990, pp. 46–59.

Ulaby, F.T., R.K. Moore, and A.K. Fung, *Microwave Remote Sensing*, Vol. 3, Appendix E, Artech House, 1986, Appendix E.

Vant, M.R., R.O. Ramseier and V. Makios, "The Complex Dielectric Constant of Sea Ice at Frequencies in the Range 0.1–40 GHz," *J. Appl. Physics*, Vol. 49, 1978, pp. 1264–1280.

Appendix 9A

Microwave Dielectric Properties of Snow and Sea Ice

9A.1 INTRODUCTION

Extensive reviews of the microwave dielectric and extinction properties of snow and sea ice have been provided by Hallikainen [1986, 1987], Matzler [1987] and Ulaby et al. [1986]. Over the past two decades, many experiments and measurements were carried out to determine the empirical permittivity models for snow and sea ice. It was found that the snow permittivity is sensitive to the following parameters: (1) snow liquid water content, (2) volume fraction of ice and (3) operating frequency; while the sea ice permittivity is sensitive to another set of parameters: (1) sea ice salinity, (2) volume fractions of brine and air pockets and (3) operating frequency. Some common empirical relations for snow and sea ice permittivities are provided in the following sections. It is worth noting that in actual applications snow layer permittivity is usually estimated from some of the empirical formulas, while sea ice permittivity is measured and its component, brine permittivity is estimated by formula.

9A.2 SNOW RELATIVE PERMITTIVITY

Snow is an inhomogeneous medium consisting of ice particles, liquid water and air. Based on the quantity of liquid water in the mixture, snow is categorized into two types: (1) dry snow, which is a mixture of ice and air containing no liquid (free) water; and (2) wet snow, which is a mixture of dry snow and liquid water. The relative permittivity of dry snow is, therefore, governed by the dielectric properties of ice and the snow density (or ice volume fraction). The relative permittivity of wet snow, on the other hand, depends on the dielectric properties of ice, liquid water and the volume fraction of ice and liquid water. The empirical formulas for relative permittivities of dry and wet snow are given in the following subsections.

9A.2.1 Dry Snow

The empirical formulas for the dielectric constant, ε_{ds}', of dry snow can be found in Ulaby et al. [1986]. One formula for the real part of the snow dielectric constant has the form

$$\varepsilon_{ds}' = (1 + 0.51\rho_s)^3 \tag{9A.1}$$

where ρ_s is the snow density in g cm^{-3}. Under natural conditions, the dry snow density, ρ_s, rarely exceeds 0.5 g cm^{-3}. The following linear relations, though less accurate, can also be used to compute the real part of the dielectric constant for dry snow:

$$\varepsilon'_{ds} = \begin{cases} 1.0 + 1.9\rho_s, & \text{for } \rho_s \leq 0.5 \text{ g cm}^{-3} \\ 0.51 + 2.88\rho_s, & \text{for } \rho_s \geq 0.5 \text{ g cm}^{-3} \end{cases} \tag{9A.2}$$

Both formulas have been shown to agree very well with experimental results [Hallikainen, 1987; Ulaby et al. 1986].

To date, experimental data for the imaginary part of the dielectric constant or the dielectric loss factor of dry snow are limited to frequencies below 13 GHz. Furthermore, discrepancies still exist among the empirical formulas for computing the dielectric loss factor of dry snow. If ice particles in the dry snow are assumed to be spherical, the following relation, which is derived from Polder-Van-Santen's mixing formula, can be used to estimate the dielectric loss factor of dry snow:

$$\varepsilon''_{ds} = 3v_i\varepsilon''_i \frac{\varepsilon'^{2}_{ds}(2\varepsilon'_{ds} + 1)}{(\varepsilon'_i + 2\varepsilon'_{ds})(\varepsilon'_i + 2\varepsilon'^{2}_{ds})} \tag{9A.3}$$

where ε'_i and ε''_i are, respectively, the real and imaginary parts of the dielectric constant of pure-water ice, and v_i is the volume fraction of ice in the dry snow. In the microwave region ε'_i is set to 3.15, and ε''_i is estimated from experimental measurements shown on p. 2027 of Ulaby et al. (1986); ε'_{ds} is calculated using (9A.1) or (9A.2).

9A.2.2 Wet Snow

Wet snow is a mixture of ice particles, air and liquid water. Due to the high relative permittivity of liquid water compared to that of ice and air, the presence of liquid water strongly affects the relative permittivity of the snow mixture. The dielectric behavior of wet snow has been investigated in the frequency range of 1 to 37 GHz [Linlor, 1980; Tiuri et al., 1984; Hallikainen, 1986; Ulaby et al., 1986]. For a frequency range of 3 to 37 GHz and snow density of 0.09 to 0.38 g cm^{-3}, empirical relations, modified Debye-like models for computing the real part of the dielectric constant, ε'_{ws}, and its imaginary part or dielectric loss factor, ε''_{ws}, of wet snow are given by

$$\varepsilon'_{ws} = A + \frac{Bm_v^x}{1 + (f/f_0)^2} \tag{9A.4}$$

$$\varepsilon''_{ws} = \frac{C(f/f_0)^2 m_v^x}{1 + (f/f_0)^2} \tag{9A.5}$$

where f is the operating frequency in GHz, and $m_v(\%)$ is the volume fraction of liquid water expressed as a percentage which is restricted to a range of 1% to 12%. The constants A, B, C, x and f_0 are given below,

$$A = 1.0 + 1.83\rho_s + 0.02m_v^{1.015}$$

$$B = 0.073, \quad C = 0.073$$

$$x = 1.31$$

$$f_0 = 9.07 \text{ GHz}$$

9A.3 SEA ICE RELATIVE PERMITTIVITY

Sea ice is an inhomogeneous medium consisting of fresh-water ice, liquid brine (a mixture of salt and water) and air pockets. Depending on the volume fraction of brine inclusion, sea ice is classified into two major categories: (1) saline ice such as young ice (less than 30 cm thick) and first year sea ice (between 30 cm and 2 m); and (2) desalinated ice such as multiyear sea ice (usually more than 2 m). The scattering and extinction coefficients of saline ice are dominated by the brine pockets in the ice, whereas those of desalinated ice are dominated by the air bubble. The dielectric properties of saline ice, desalinated ice and brine are given in the following subsections.

9A.3.1 Relative Permittivity of Saline Ice

Since the dielectric constant of brine is very large at microwave frequencies, even a small brine volume will have a significant impact on the overall sea ice dielectric properties. The relative permittivity of sea ice can be estimated using a simple empirical model proposed by Vant et al. (1978) and given by

$$\varepsilon_{si}' = 3.05 + 0.0072 v_b, \tag{9A.6}$$

$$\varepsilon_{si}'' = 0.024 + 0.0033 v_b \tag{9A.7}$$

where v_b is the brine volume fraction which is a function of sea ice salinity S_i and temperature T. The volume fraction of brine in saline ice can be estimated using the empirical expression provided by Frankenstein and Garner [1967] as follows:

$$v_b = \begin{cases} 10^{-3} S_i\left(\dfrac{-52.56}{T} - 2.28\right), & -2.06 \le T \le -0.5°\text{C} \\ 10^{-3} S_i\left(\dfrac{-45.917}{T} - 0.930\right), & -8.2 \le T \le -2.06°\text{C} \\ 10^{-3} S_i\left(\dfrac{-43.795}{T} - 1.189\right), & -22.9 \le T \le -8.2°\text{C} \end{cases} \tag{9A.8}$$

where S_i is in ‰. This quantity is usually measured.

Under thermal equilibrium, the brine salinity S_b (‰) in sea ice is related to the ice temperature T (°C) by the following empirical relations [Poe et al., 1972]:

$$S_b = \begin{cases} 1.725 - 18.756T - 0.3964T^2, & -8.2 \le T \le -2°\text{C} \\ 57.014 - 9.929T - 0.16204T^2 - 0.002396T^3, & -22.9 \le T \le -8.2°\text{C} \\ 242.94 + 1.5299T + 0.0429T^2, & -36.8 \le T \le -22.9°\text{C} \\ 508.18 + 14.535T + 0.2018T^2, & -43.2 \le T \le -36.8°\text{C} \end{cases} \quad (9A.9)$$

When the density of the brine ρ_b is known, the volume fraction of brine is also given by

$$v_b = (S_i\rho_i)/(S_b\rho_b) \quad (9A.10)$$

where ρ_i is the density of pure ice.

9A.3.2 Relative Permittivity of Desalinated Ice

Desalinated ice can be modeled as homogeneous fresh-water ice embedded with air bubbles. In the microwave region, the relative permittivity of fresh-water ice, ε_i, is practically the same and independent of temperature. The relative permittivity of desalinated ice ε_{di} is, therefore, only a function of volume fraction of air bubbles, v_a. The relative permittivity of desalinated ice can be estimated using the following mixing formula:

$$\varepsilon_{di} = \varepsilon_i\left(\frac{1 + 2v_a y}{1 - v_a y}\right) \quad (9A.11)$$

where

$$y = \frac{1 - \varepsilon_i}{1 + 2\varepsilon_i}$$

9A.3.3 Relative Permittivity of Brine in Sea Ice

In the microwave region, the real and imaginary parts of the dielectric constant of brine in sea ice, ε_b' and ε_b'', respectively, can be estimated using the following empirical Debye type relaxation formulas [Stogryn and Desargant, 1985]

$$\varepsilon_b' = \text{Re}(K) \text{ and } \varepsilon_b'' = \text{Im}(K) \quad (9A.12)$$

where

$$K = \varepsilon_\infty + \frac{\varepsilon_s - \varepsilon_\infty}{1 - j2\pi f\tau} + j\frac{\sigma}{2\pi\varepsilon_0 f} \quad (9A.13)$$

and

$$j = \sqrt{-1}$$

In (9A.13), ε_s and ε_∞ are the limiting static and high frequency values of ε_b', τ is the relaxation time, f is the operating frequency in GHz, σ is the ionic conductivity of the dissolved salts, and ε_0 is the free space permittivity. Then, ε_s, ε_∞ and τ are calculated using the following equations:

$$\varepsilon_s = \frac{939.66 - 19.068T}{10.737 - T} \tag{9A.14}$$

$$\varepsilon_\infty = \frac{82.79 + 8.19T^2}{15.68 + T^2} \tag{9A.15}$$

$$2\pi\tau = 0.10990 + 0.13603\times10^{-2}T + 0.20894\times10^{-3}T^2 + 0.28167\times10^{-5}T^3 \tag{9A.16}$$

where T is the temperature of sea ice in °C and has a range of 0°C to -25°C. The ionic conductivity σ is calculated using

$$\sigma = \begin{cases} -T\exp(0.5193 + 0.08755\times10^{-1}T), & T \geq -22.9°\text{'C} \\ -T\exp(1.0334 + 0.1100T), & T < -22.9°\text{'C} \end{cases} \tag{9A.17}$$

Chapter 10

Comparisons of Model Predictions with Backscattering and Emission Measurements from Snow and Ice

10.1 INTRODUCTION

The scattering and emission models developed in Chapter 8 and illustrated in Chapter 9 are applied to the interpretation of scattering and emission measurements in this chapter. The radiometric data sets selected are those for which properties of the snow field or ice condition have been measured independently or can be estimated by existing formulas based on inputs from direct measurements of relevant snow or ice parameters. Thus, the input parameters to the models are well bounded and the data sets can also serve to validate the scattering and emission models.

Scattering and emission measurements from two types of ice are considered: (i) thin saline ice and (ii) multiyear ice. The data for thin saline ice were obtained from the CRREL'88 experiment conducted at the U.S. Army Cold Region Research and Engineering Laboratory (CRREL) during the winter of 1988 [Onstott, 1990]. The measurements for multiyear ice were conducted during the winter of 1988 and 1989 segments of the Coordinated Eastern Arctic Experiment (CEAREX) in the Arctic Ocean [Onstott, 1991]. For the snow case, comparisons are performed using the measured data reported by Kuga et al. [1991], Matzler et al. [1980], and Stiles and Ulaby [1980]. Descriptions of the ice condition and comparisons of data with model predictions for the two ice types are provided in Section 10.2 and Section 10.3, respectively. Comparisons between the models and scattering and emission measurements from snow fields are provided in Section 10.4.

10.2 COMPARISONS WITH MEASUREMENTS FROM A THIN SALINE ICE

Saline ice is a very lossy ice at microwave frequencies. It is also highly inhomogeneous because of the presence of brine pockets. As a result, it is generally thought that scattering should be due mainly to the air-ice boundary and to a much lesser extent the inhomogeneities within the layer volume because the absorption due to brine inclusions is high and albedo has to be low. This is definitely the case for thick saline ice. For thin ice, questions exist as to whether or not there is some contribution from the bottom interface of the ice. The reason why the lower interface is of interest is because the dielectric discontinuity is much larger at this boundary than at the air-ice boundary. Furthermore, the roughness at this interface is magnified by the square root of the ice layer permittivity. For similar reasons, emission should be dominated by the layer volume because of the large absorption and the contribution from the lower half space is reduced by the large dielectric discontinuity at the lower (ice-water) boundary and attenuation through the layer. To better understand the sources and mechanisms of scattering and emission, controlled experiments were conducted at CRREL in 1988 where saline ice was grown and ice properties were measured. In addition, active and passive microwave measurements were carried out at several frequencies on the same ice layer. The current study is aimed at interpreting four-frequency, two-polarization, active and passive measurements taken on the same ice layer using a combined surface and volume scattering model. It is anticipated that joint interpretations of active and passive measurements can better determine the medium parameters and sources of contribution.

The geometry of the thin ice problem is depicted in Figure 10.1 in which it is noted that scatterers are brine pockets and hence the ice medium is very lossy.

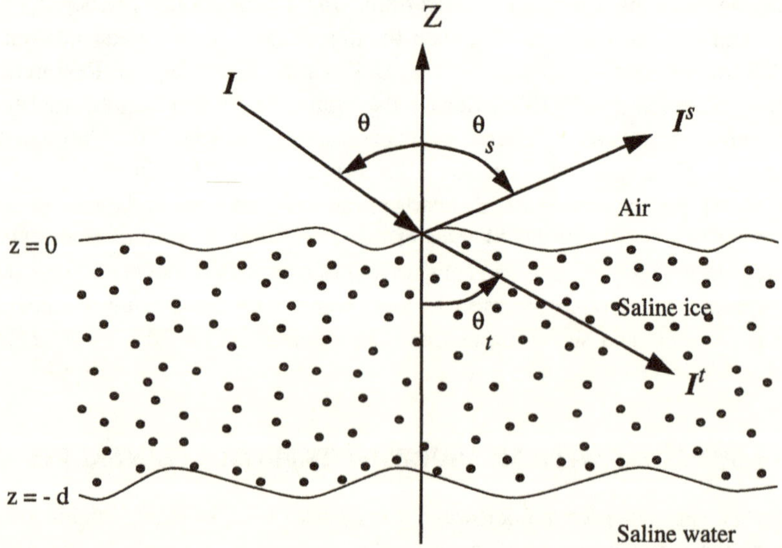

Figure 10.1 Geometry of the problem of a thin saline ice layer above saline water. The scatterers in the ice layer are brine pockets.

10.2.1 Modeling Considerations for a Thin Saline Ice Sheet

A saline ice layer above water is modeled as an irregular inhomogeneous layer above a half space of saline water. The formulation of the problem based on the radiative transfer theory was provided in Chapter 1 and reformulated using the matrix doubling method in Chapter 8. The phase function used is the modified Mie phase function developed without making the far-field simplifying assumption (see Chapter 8). This is done so that the spacing between adjacent scatterers does not have to be large compared with the operating wavelength. The surface phase function is derived from an integral equation model (IEM) given in Chapter 5 and summarized in Appendix 2A. The matrix doubling method can be used in numerical evaluation for both the active and passive models (see Chapter 8).

A complete model of wave scattering or emission from an inhomogeneous layer consists of two components: (a) modeling of electromagnetic wave interaction with the inhomogeneous layer and (b) modeling of the electrical parameters of the inhomogeneous layer in terms of the physical and geometric properties of the scatterer and the host medium. In general, the modeling of the electrical parameters is quite difficult and the model may depend on temperature, frequency, properties of the scatterer, properties of the background medium etc. There is as much a question about the range of validity of the electromagnetic wave interaction model as about the electrical parameter model. Only in the event when both component models are correct, will the complete model be correct. Since theoretical modeling must involve some simplifying assumptions, whenever the electrical parameters of a medium are known from direct measurements, these values are preferred to theoretically or empirically estimated values.

10.2.2 Model Comparisons with CRRELEX Measurements

In this study backscattering and emission from a thin saline ice above a water surface is considered. Unlike comparisons of theoretical models with field measurements where ice parameters are not quantified, in the CRREL experiments all parameters needed in modeling are either known, well bounded or can be estimated by existing formulas in the literature (see Appendix 9A). Directly measured ice parameters are listed in Table 10.1 [Bredow, 1989; Onstott, 1990]. Frequency dependent parameters and parameters estimated from formulas are given in Table 10.2. The purpose of this study is twofold: (a) to find out how well the models agree with measurements using the predetermined parameter values at four different frequencies, two polarizations and incident and nadir angles within the range, 10 to 70 degrees, and (b) based on the agreement or disagreement between model and measurements infer the mechanisms for scattering and emission. Since saline ice is a highly lossy medium, its lower boundary and the medium below it were not expected to be important contributors to scattering or emission. One major finding in the study to follow is that when the saline ice is thin, the ice-water boundary can be a significant contributor to scattering.

Table 10.1
CRREL'88 Saline Ice Physical Parameters

Ice parameters	*Measured/observed values*
Thickness	6.5 – 8 cm
Air-ice interface rms height (σ)	0.02 – 0.048 cm
Air-ice interface correlation length (L)	0.669 – 1.77 cm
Air-ice interface temperature	-16 °C
Ice-water interface temperature	-1°C
Ice salinity	7 – 11.5 ‰
Water salinity	24 ‰

Table 10.2
Estimated Parameters in Modeling the Thin Saline Ice

Parameters	*Estimated values*
Brine volume fraction	5%
Scatterer (brine) radius	0.25 mm
Scatterer permittivity [Stogryn and Desargent, 1985]	
5 GHz	50.4 - *j*40.3
6.7 GHz	50.4 - *j*40.3
10 GHz	33.5 - *j*38.0
18.7 GHz	18.4 - *j*28.2
Ice permittivity [Ulaby et al., 1986]	
5 GHz	3.4 - *j*0.22
6.7 GHz	3.4 - *j*0.22
10 GHz	3.35 - *j*0.23
18.7 GHz	3.3 - *j*0.24
Saline water permittivity [Stogryn and Desargent, 1985]	
5 GHz	61.4 - *j*39.4
6.7 GHz	61.4 - *j*39.4
10 GHz	35.0 - *j*38.0
18.7 GHz	18.4 - *j*30.2
Air-ice interface rms height	0.048 cm
Air-ice interface correlation length	0.669 cm
Ice-water interface rms height	0.028 cm
Ice-water interface correlation length	2.1 cm

Using the parameters from the tables in the backscattering model for one irregular inhomogeneous layer given in Chapter 8, we show comparisons with backscattering measurements at 5 GHz in Figure 10.2. First, we note that the data show an angular drop-off and a separation between VV and HH polarization, which are characteristic of surface backscattering. However, when we calculate backscattering from the top surface boundary alone the predicted level is about 7 dB lower than the measurement at 20 degrees and the angular drop-off for VV is slower than the measurement. When we consider volume scattering by itself, its level at 20 degrees is even lower because the medium has a large absorption coefficient, which leads to a very small albedo (order of 10^{-4}). Thus, the contribution from the lower boundary between ice and water must be significant even in the presence of a very lossy but thin ice layer. This is possible because the dielectric discontinuity at the ice-water interface is much larger than that at the air-ice interface. When an irregular layer model is used, a very good agreement is obtained over all angles as shown in Figure 10.2. This means that contributions from the entire layer including the interactions between volume and boundary surface scattering are needed. A breakdown of the contributions from different parts of the layer using a first-order layer model was given in Figure 2.40 of Chapter 2. It is clear from Figure 2.40 that the dominant contribution does come from the bottom interface at small angles of incidence. At larger angles contributions from volume and the air-ice interface are significant because their angular scattering curves are not as steep as those from the bottom layer interface. However, the simple model does not calculate surface-volume interaction accurately and the fit there with data is not as good as the one shown in Figure 10.2 using the more complete model.

When the incident frequency is changed to 10 GHz, all model parameters remain the same except for those that are frequency dependent. The comparison between backscattering measurements and the model predictions is shown in Figure 10.3. Here again a very good agreement is obtained in level, angular trends and polarization. In this case contribution from the upper interface is more important than that from the lower interface of the ice layer, when the incident angle exceeds 10 degrees. This is because the attenuation by the ice layer increases very quickly with the incident angle and hence there is a corresponding reduction in scattering from the lower interface making it unimportant at larger incident angles. Volume scattering has a relatively flat angular response and becomes appreciable at large angles of incidence approximately after 40 degrees. Its level of scattering approaches that from the upper ice layer interface when the incident angle reaches 60 degrees. The fact that the level difference between VV and HH polarization follows the surface scattering model fairly closely over the incident angle from near normal to 60 degrees indicates that surface scattering is still the dominant mechanism over all the angular range shown in Figure 10.3. A similar comparison was given in Figure 2.41 in Chapter 2 with a first-order model. As expected, the agreement is not as good as the one in Figure 10.3. However, the first-order model is easier to use and the contributions from different parts of the layer are easily computed and interpreted as shown in Figure 2.41.

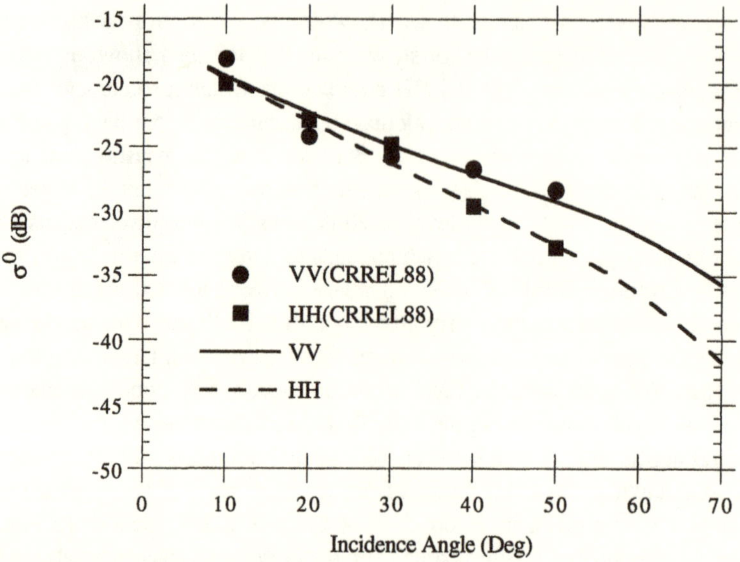

Figure 10.2 Comparison of model predictions with backscattering data from thin saline ice at 5 GHz in CRREL'88. A scatterer radius of 0.25 mm is used in the model.

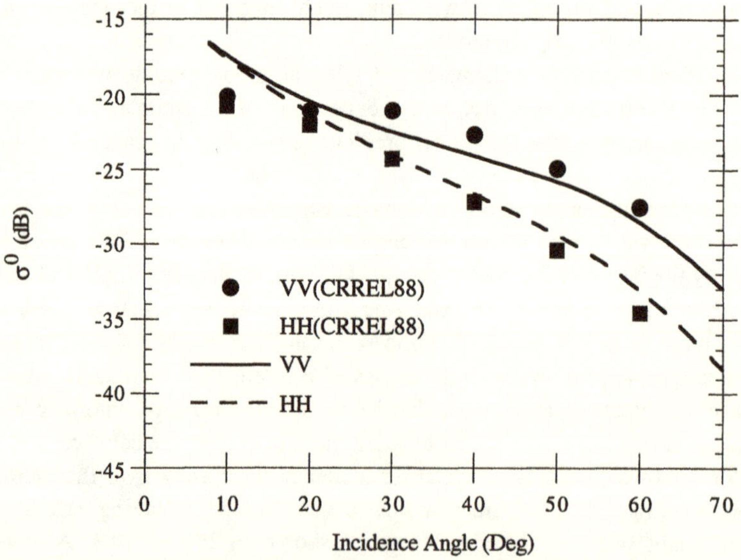

Figure 10.3 Comparison of model predictions with backscattering data from thin saline ice at 10 GHz in CRREL'88. A scatterer radius of 0.25 mm is used in the model.

Next, we consider passive measurements at 6.7 and 18.7 GHz in Figures 10.4 and 10.5 using the same set of physical layer parameters as in Figures 10.2 and 10.3. It is seen that good agreements are obtained in Figure 10.8 for both vertical and horizontal polarization. The Brewster angle effect in the theoretical prediction, however, does not appear in the data. Realizing that a consistent set of model parameters have been used to fit both active and passive measurements, two polarizations, four frequencies and many observation angles in all four figures, Figures 10.2 through 10.5, we can say that the agreements in both level and trend in Figures 10.2 through 10.4 are excellent. The predicted emission curve in Figure 10.5 is clearly higher than the measured and the predicted spacing between vertical and horizontal polarization is narrower than the measured. A possible reason for a higher emission is that there may be more volume scattering than assumed by the current set of parameters. If we increase, for example, the size of the scatterer, there will be more volume scattering and less emission resulting in a lower emission over all nadir angles. To test this idea we increase the albedo of the layer by using a larger scatterer radius and recompute both backscattering at 10 GHz and emission at 18.7 GHz as shown in Figures 10.6 and 10.7, respectively. As expected the predicted emission curves are lower and agree better with measurements. There is no improvement in the spacing between polarizations in the emission curves. On the other hand, the backscattering curve in HH polarization is higher than the measured values (see Figure 10.6). That is, when there is an improvement in agreement between model and passive measurements, there is a corresponding deterioration in active comparisons. This means that the model parameters should not be changed much from what they were. It is clear that if we consider backscattering and emission separately and only one frequency and one polarization at a time, it is possible to realize much better agreement between model predictions and measurements. However, we may have inconsistency in the choice of model parameters between active and passive comparisons. Clearly, a good agreement at one polarization and frequency between theoretical model predictions and either passive or active measurements does not provide as tight a bound on the medium parameter as when both active and passive measurements are considered together. Similarly, when only active or passive comparisons are made, a tighter bound on the model parameters can be realized with multifrequency data as opposed to single frequency data. From the point of view of using theoretical predictions to guide measurements, it appears that when we can predict three out of four sets of data taken at four frequencies on the same ice layer with a well-defined set of model parameters, the predictions for the fourth frequency are likely to be reliable and useful for indicating the correct data trend. The above analysis also indicates that in retrieval of medium parameters through remote sensing, better defined parameters can be obtained through active and passive measurements than is possible with active or passive measurements alone. In parameter retrieval, it is particularly important to have data over several frequencies, incident angles and polarizations, because there is always some uncertainty associated with each data set. Similarly, there is uncertainty associated with theoretical model predictions due to uncertainty in its range of validity and its choice of model parameters.

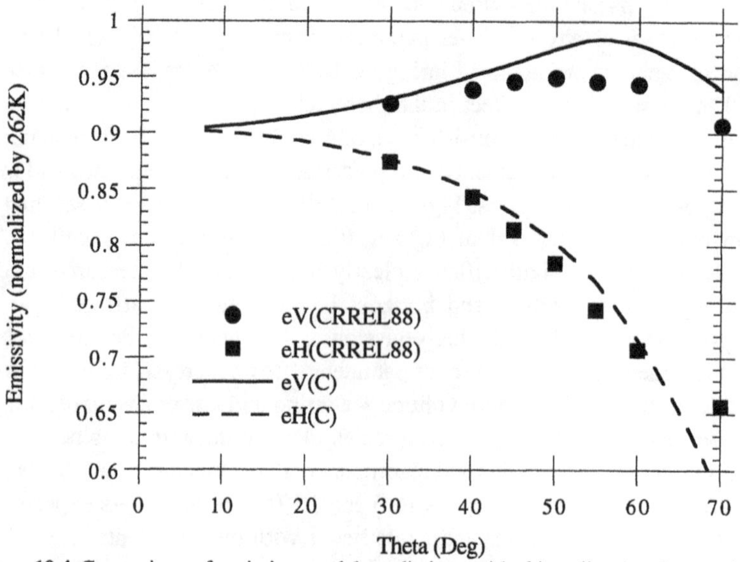

Figure 10.4 Comparison of emission model predictions with thin saline ice data at 6.7 GHz from CRREL'88. A scatterer radius of 0.25 mm is used in the model.

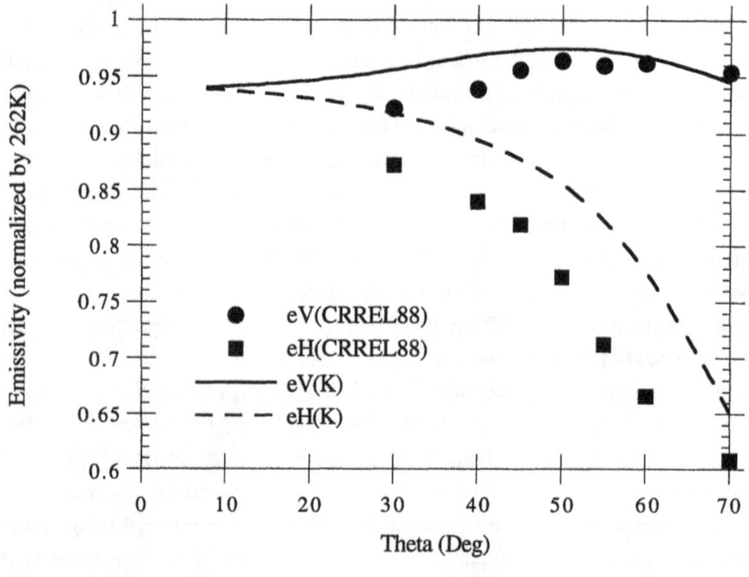

Figure 10.5 Comparison of emission model predictions with thin saline ice data at 18.7 GHz from CRREL'88. A scatterer radius of 0.25 mm is used in the model.

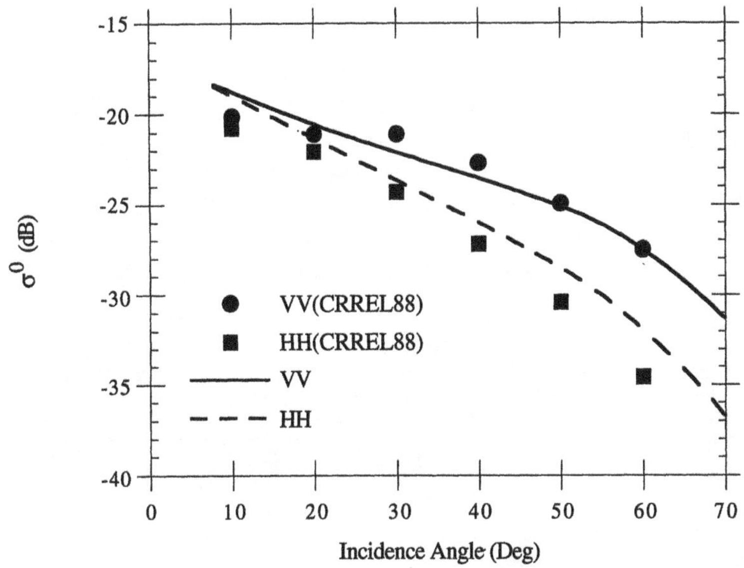

Figure 10.6 Comparison of backscattering model predictions with thin saline ice data at 10 GHz from CRREL'88. A scatterer radius of 0.35 mm is used in the model.

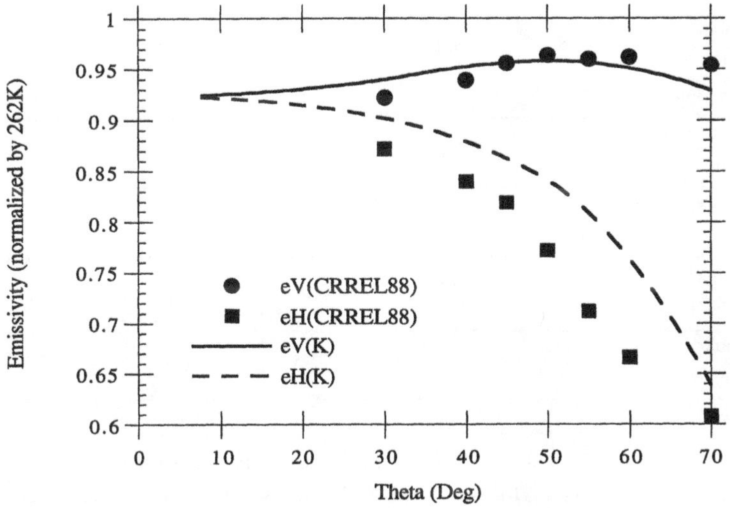

Figure 10.7 Comparison of emission model predictions with thin saline ice data at 18.7 GHz from CRREL'88. A scatterer radius of 0.35 mm is used in the model.

10.3 COMPARISONS WITH MEASUREMENTS FROM MULTIYEAR SEA ICE

Backscattering and emission measurements were performed on multiyear sea ice during the winters of 1988 and 1989 segments of the Coordinated Eastern Arctic Experiment (CEAREX) in regions located to the North and East of Svalbard [Grenfell, 1992; Onstott, 1991]. Five multiyear ice sites were selected as representative of the area.Four of these sites are from hummocks (denoted as Alpha-35, DS-7, DS-9 and DS-13) and one is from a fresh water meltpond (denoted as DS-MP). Their observed physical properties are summarized in Table 10.3 in which all radii are given in millimeters.

Table 10.3
CEAREX' 89 Upper Ice Sheet Characterizations [Onstott, 1991] (LDI = Low density ice)

Parameters		*Alpha-35*	*DS-7*	*DS-9*	*DS-13*	*DS-MP*
σ rms (cm)		0.33 ± 0.2	0.14 ± 0.02	0.77 ± 0.12	1.01 ± 0.31	0.08 ± 0.03
L (cm)		4.7 ± 0.21	2.0 ± 1.3	4.6 ± 1.2	3.2 ± 1.2	2.8 ± 1.9
LDI-1	Thickness (cm)	14	5.0 ± 0.6	4.9 ± 1.0	3.0 ± 1.3	5.2 ± 0.3
	Salinity (‰)	0.0	0.0	0.0	0.0	0.0
	Density (kg/m^3)	815	457	513	513	914
	Bubble radius	0.8	1.25	1.15	1.65	0.65
	Void radius	0.25	4	3.5	2.5	0
LDI-2	Thickness (cm)	–	3.5 ± 0.4	4.9 ± 0.4	13.6 ± 0.4	4.8 ± 0.3
	Salinity (‰)	–	0	0	0	0
	Density (kg/m^3)	–	728	728	929	919
	Bubble radius	–	2	2.15	0.5	0
	Void radius	–	1	0	0	0

One interesting point about these ice types is that there is a significant variation in ice density from a low of 457 kg/m^3 to a high of 914 kg/m^3. This change in density is due to the percent of air bubbles embedded in the ice. The corresponding data set is unique in that it can indicate scattering variations versus ice density or we can convert ice density to volume fraction of the scatterer (air bubbles) using 916 kg/m^3 as the density for pure ice. Therefore, this data set provides the dense medium effect on backscattering based on field measurements. Note that the higher is the density, the smaller the volume fraction of scatterers and vice versa. Another point to note about the ice properties is that there are two sites, DS-7 and DS-9, with two inhomogeneous layers of ice of significantly different air bubble sizes. In particular, the lower layers have larger air bubbles. Thus, a two-layer model is needed for these sites (see Chapter 8 for multilayer models). The site, DS-13, also has two ice layers but the air bubble in the lower layer is less than a third of that of the upper layer. Hence, the lower layer does not contribute much to scattering and can be treated as part of the half space below. The other two sites, Alpha-35 and DS-MP, have only one inhomogeneous ice layer. A cross-section view of the upper portion of the ice structure is shown in Figure 10.8. From this figure it is seen that there is a layer of low density snow above the ice. Thus, in general we have a multi-layered structure.

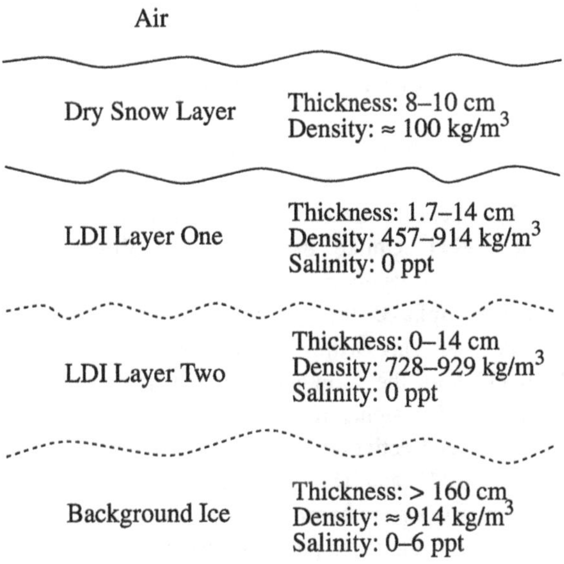

Figure 10.8 The cross-section view of the upper portion of the multiyear ice sheet (CEAREX).

10.3.1 Scattering and Emission Modeling Considerations for Multiyear Sea Ice

From Figure 10.8 and Table 10.3 the multiyear sea ice layers measured during CEAREX can be modeled either as an irregular inhomogeneous layer over a homogeneous half space or as a two-layer irregular inhomogeneous medium over a homogeneous half space. Although the measurement sites are covered by 8–10 cm of low-density dry snow, the effects of the snow layer on backscattering measurements at 10 GHz are negligible compared with scattering from the ice (the snow layer optical thickness is on the order of

10^{-2}) and, therefore, not included in modeling. The only exception occurs in modeling scattering and emission from the meltpond where a layer of snow is included because there is a set of passive measurements taken at 18.7 GHz where the snow layer does have an effect on emission. The model formulations for active and passive problems involving either single or multilayer structures have been given in Chapter 8.

In summary, each of the ice layers at sites Alpha-35 and DS-13 is modeled as an irregular inhomogeneous layer above a homogeneous half space of desalinated ice. The low-density snow layer above the ice is ignored at 10 GHz because it is nearly transparent. For sites DS-7 and DS-9 each ice type is modeled as two layers of low-density bubbly ice above a homogeneous half space of desalinated ice. The two-layer model is also used for the site DS-MP in both active and passive problems. Here, the top layer is a low-density snow and the lower layer is ice embedded with small air bubbles. To maintain consistency identical physical structure and associated model parameters are used in both active and passive calculations.

10.3.2 Model Comparisons with CEAREX Measurements

The parameters used in the computations of emission and backscattering from multi-year sea ice are listed in Table 10.4, where the layer permittivity given is at 10 GHz. Note that the selected physical parameter values are within the range reported in the field experiment [Onstott, 1991]. Unlike the saline ice observed in the CRREL'88 experiment, the multi-year ice layers measured in CEAREX have very low absorption loss. Except for the site Alpha-35 the backscattered signals from the hummocks are dominated by volume scattering within the ice layers, where air bubbles are the scatterers. For Alpha-35 both the snow-ice interface and the volume inhomogeneity contribute to the backscattering signature.

The comparisons between active and passive models with measurements from the meltpond (site DS-MP) at 10 and 18.7 GHz are shown in Figures 10.9 and 10.10, respectively. These comparisons are carried out using the same physical parameters for active and passive cases and the layer permittivities are calculated by the formulas in Appendix 9A that account for the difference in frequency. Very good agreement between model predictions and data is obtained in both level and angular trends. The applicability of the models is, therefore, justified. Based upon the model calculation, backscattering from meltpond (DS-MP) at 10 GHz is due primarily to the snow-ice interface until the incident angle is over 45 degrees, beyond which some volume scattering contribution is observable. In particular, bistatic scattering from the snow-ice boundary causes horizontal polarization to be higher than the vertical at large angles (Figure 10.9). The reason for the low-volume scattering is because the scatterer is small leading to an albedo of the ice layer of 0.0072 and an optical thickness of 0.062. The snow layer albedo is much higher. It is at 0.49 but the optical depth is quite small, only 0.0015. Thus, volume scattering from both the snow and the ice layer is low, permitting scattering from the snow ice interface to dominate. In Figure 10.10 the reported emissivity is an effective quantity not accounting for temperature variations within or across the layers. The main contributions to the level of emissivity are from the ice layer and the lower half space. The former can contribute because it has a significant absorption coefficient and a low albedo, while the latter is homogeneous ice with some salinity. The snow layer acts mostly as an attenuator because its absorption coefficient is very low and its albedo is relatively high.

Table 10.4
Estimated Parameters in the Multiyear Sea Ice Model

Parameters	*Alpha-35*	*DS-7*	*DS-9*	*DS-13*	*MP*
Layer #1:					
Thickness	14 cm	5 cm	3.9 cm	4.3 cm	8 cm
Scatterer volume fraction	11%	50%	44%	44%	11%
Scatterer radius	0.8 mm	1.75 mm	1.15 mm	2.0 mm	0.25 mm
Scatterer ε_r	1.0	1.0	1.0	1.0	3.2-j0.0015
Layer, ε_r	2.86 - j0.009	1.95 - j0.002	2.1 - j0.002	2.1 - j0.002	$1.2 - j\,5.10^{-5}$
Layer #2:					
Thickness	–	3.5 cm	4.5 cm	–	5.2 cm
Scatterer volume fraction	–	22%	22%	–	1%
Scatterer radius	–	2.0 mm	2.15 mm	–	0.65 mm
Scatterer ε_r	–	1.0	1.0	–	1.0
Layer ε_r	–	2.65 - j0.008	2.65 - j0.008	–	3.12 - j0.01
Interface roughness (σ = rms height, L = correlation length):					
Top σ	0.35 cm	0.16 cm	0.65 cm	0.70 cm	0.06 cm
Top L	4.56 cm	1.76 cm	5.79 cm	4.40 cm	0.89 cm
Middle σ	–	0.16 cm	0.89 cm	–	0.06 cm
Middle L	–	1.76 cm	3.4 cm	–	1.2 cm
Bottom σ	0.35 cm	0.16 cm	0.89 cm	0.70 cm	0.11 cm
Bottom L	4.56 cm	1.76 cm	3.4 cm	4.40 cm	0.94 cm

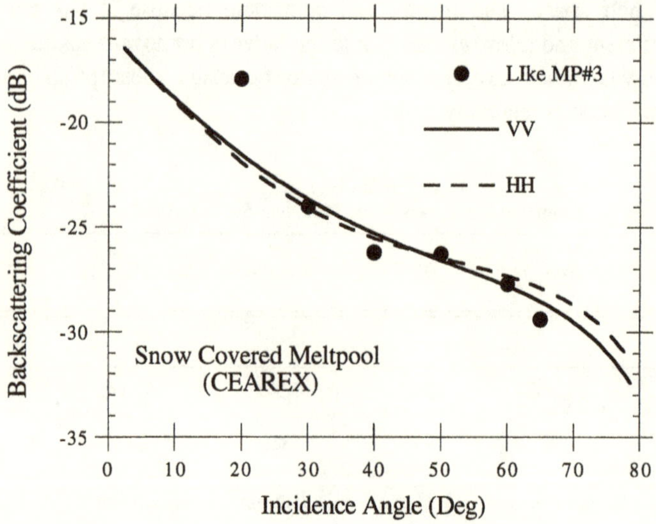

Figure 10.9 Comparisons of the two-layer backscattering model predictions with meltpond site data at 10 GHz from CEAREX.

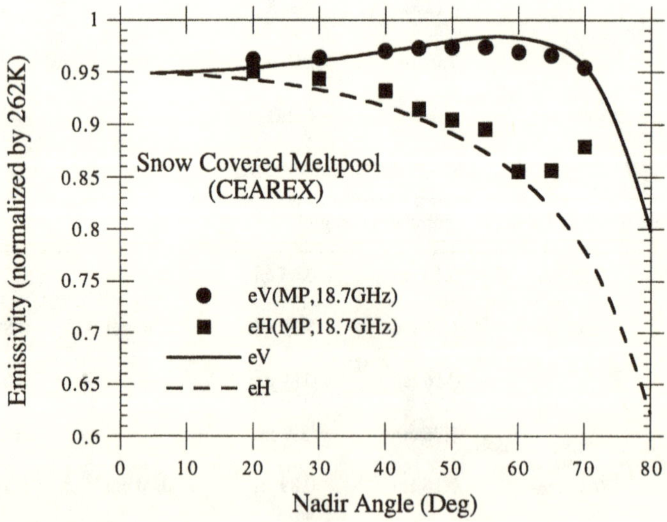

Figure 10.10 Comparisons of the two-layer emission model predictions with meltpond site data at 18.7 GHz from CEAREX.

Each of the sites, Alpha-35 and DS-13, is modeled as a single inhomogeneous layer above a homogeneous half space, as shown in Figures 10.11 and 10.12, respectively. Recall from the physical properties of the ice layer that Alpha-35 is a weakly inhomogeneous layer, while DS-13 is a strongly scattering layer. In Figure 10.11 the agreement in like polarized scattering is quite good. It appears from like polarized scattering that between 10 and 30 degrees the effect of surface scattering is dominating and volume scattering takes over after 40 degrees. The general level for the cross polarized scattering is correct except for angles between 10 and 40 degrees. Cross polarized scattering is due to multiple surface and volume scattering and surface-volume interaction. All these mechanisms contribute to a slow decaying scattering curve versus the incident angle. The fast drop-off at small incident angles in the measured cross polarized scattering is usually due to antenna pattern effect and its convolution with coherent surface scattering, which is not modeled. Note that coherent surface scattering is dependent upon system beamwidth and the size of the illuminated area [Fung and Eom, 1983]. For this reason its coefficient is not chosen to represent terrain properties. The DS-13 site has much larger air bubbles than the Alpha-35. From Figure 10.12 scattering is mainly dominated by volume scattering except in the angular region between 10 and 20 degrees, where some surface effect can be observed. The theoretical curve in like polarization appears to be in good agreement with the data. The cross polarized angular curve drops with the incident angle, while the data has an opposite trend. It is believed that such a data behavior is not correct because physically only an optically thin, weakly inhomogeneous layer can have such a behavior. On the other hand, DS-13 is a strongly scattering medium. A possible cause for the increase in cross polarized scattering with angle is that the antenna was not looking at a statistically homogeneous medium as the incident angle was changed.

The theoretical model for sites DS-7 and DS-9 is a two-layered low density bubbly ice above a homogeneous half space of desalinated ice. The comparisons between model and data are shown in Figures 10.13 and 10.14, respectively, for the sites DS-7 and DS-9. The theoretical curve for like polarization in Figure 10.13 has the correct level and angular trend and agrees well with the data. The cross polarized curve drops slowly with the incident angle, while the data has a dip at 30 degrees and then rises with the incident angle. Such a data trend is again not meaningful physically for a statistically homogeneous ice layer. It appears that either the special properties of the ice condition are not recorded or the antenna side lobe effect is not fully accounted for in data reduction. For DS-9 site the comparisons between model and data in Figure 10.14 show that there is a very good agreement in the angular trend in like polarization. However, the level of the theoretical curve is a bit high relative to the data. In cross polarization the angular trend of the data again increases with the incident angle. As noted earlier, such a data behavior is not consistent with a statistically homogeneous ice layer. Another indication of problem in this set of data is the spacing between the like and cross polarization. A less than 5 dB spacing is very suspicious in microwave measurements. More knowledge of the ice geometry and its statistics is needed to resolve the difference between model and data.

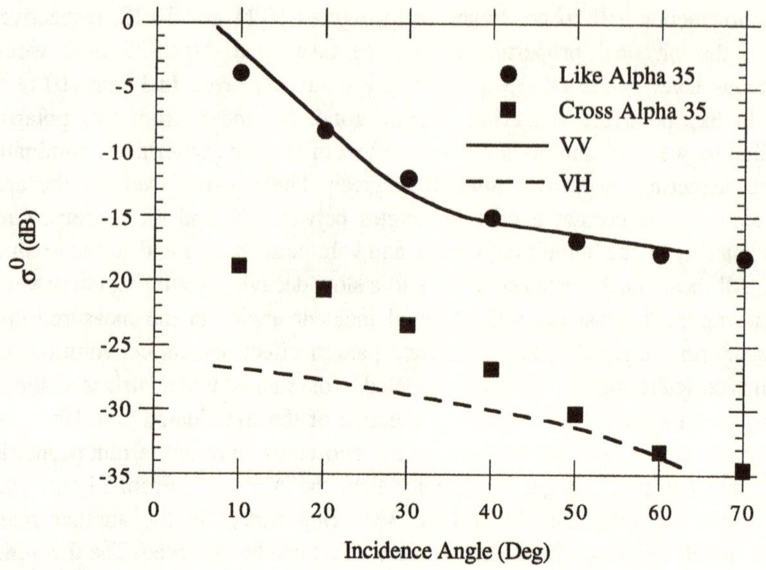

Figure 10.11 Comparison of a single-layer model predictions with backscattering data at 10 GHz from site Alpha-35 in CEAREX.

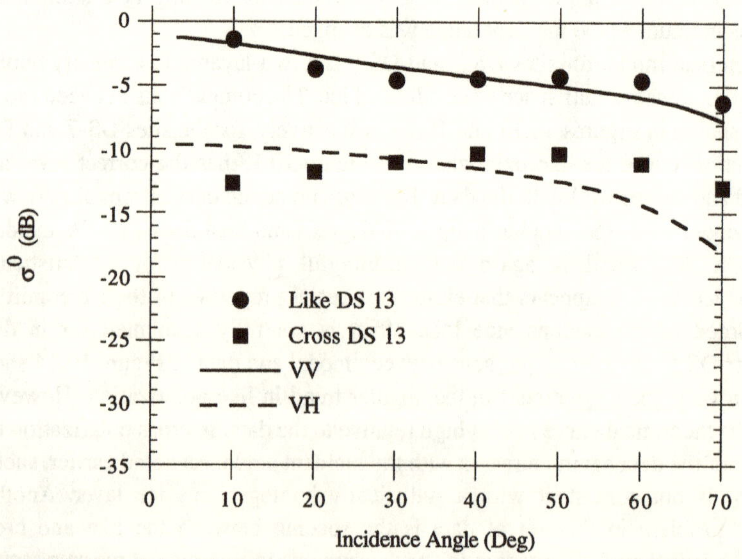

Figure 10.12 Comparison of single-layer backscattering model prediction with CEAREX site DS-13 data at 10 GHz.

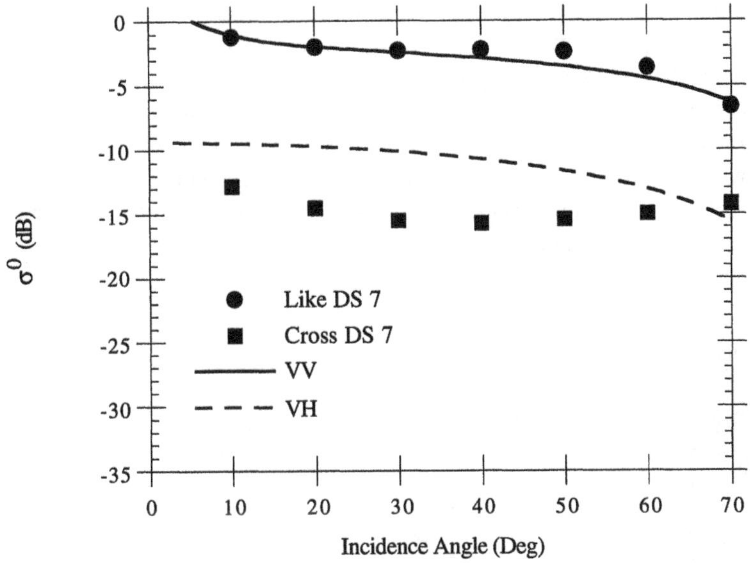

Figure 10.13 Comparison of two-layer backscattering model prediction with CEAREX site DS-7 data at 10 GHz.

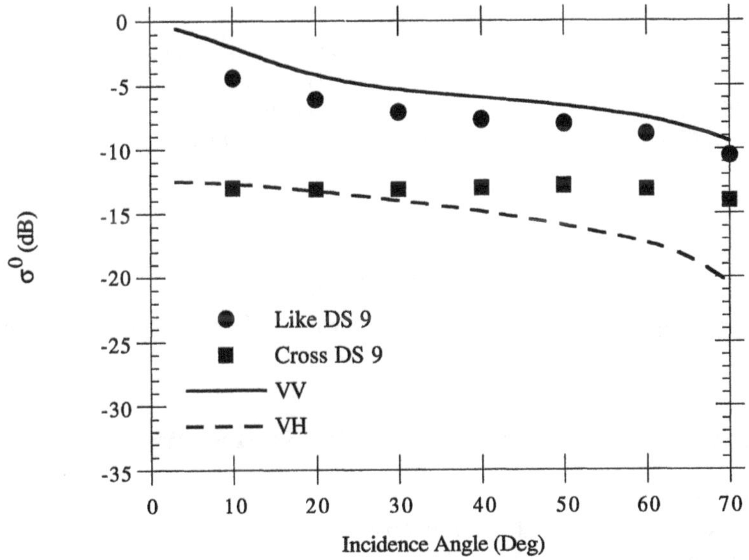

Figure 10.14 Comparison of two-layer backscattering model prediction with CEAREX site DS-9 data at 10 GHz.

Figure 10.15 shows the relationship between the backscattering coefficient and the volume fraction of the scatterer for multiyear sea ice at an incident angle of 50 degrees. Over the available region of the volume fraction, both the data and the model predictions show an increase in backscattering with the volume fraction of the scatterer. The calculations were carried out for both the one-layer model and the two-layer model. In both cases the rate of increase in backscattering with the volume fraction is seen to be in agreement with the measurements. It is anticipated that for larger volume fractions incoherent backscattering must gradually decrease because the roles of the host medium and embedding scatterers are now interchanged.

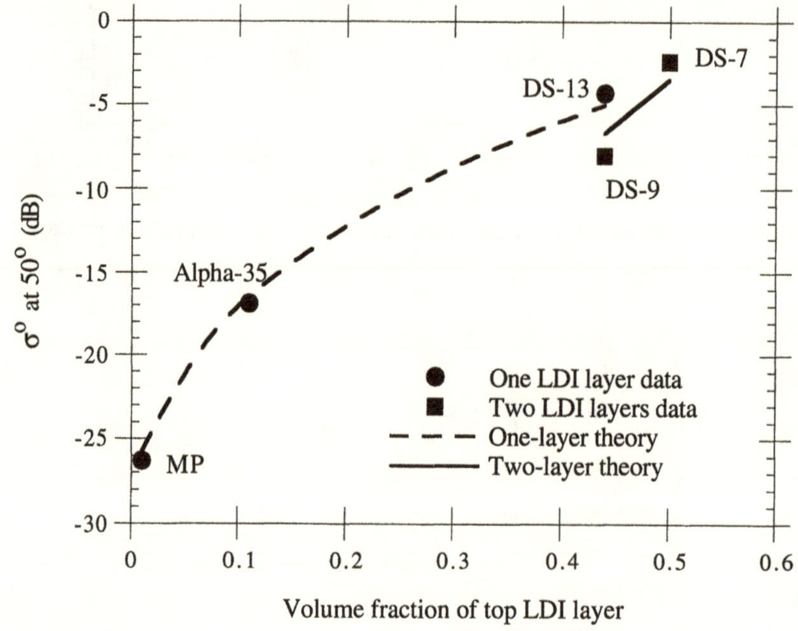

Figure 10.15 Backscattering coefficients of multiyear ice sheets (CEAREX) versus the scatterer volume fractions.

10.4 COMPARISONS WITH MEASUREMENTS FROM SNOW

In this section we consider model comparisons with two sets of measurements from snow fields. One set is a combined active and passive measurements of a dry snow field reported by Matzler et al. [1980]. The other set is a multifrequency backscattering measurement of a dry snowpack acquired by Ulaby et al [1991].

Consider first the backscattering measurements by Ulaby et al. [1991]. The measured physical parameters of the snow layer are listed in Table 10.5.

Table 10.5
Physical Parameters of the Dry Snow Layer

Parameter	*Measured/observed value*
Layer thickness	12 cm
Snow density	0.2 gr/cm^3
Mean crystal diameter	1 mm
Air-snow surface rms height (σ)	0.14 cm

The snow layer is modeled as an irregular inhomogeneous layer above a homogeneous ground (a half space). The permittivity of the ground is not given and is taken to be 5.0. The scatterers are assumed to be spherical ice crystals with an effective radius a. The effective value is used because in reality the snow particles near the top are generally smaller than those near the bottom of a snow layer [Stiles and Ulaby, 1980]. This is because of the effect of metamorphism. In addition, a difference in frequency means different penetration depths. Thus, we select a higher effective size for the 35 GHz and a smaller effective size at 94 GHz as shown in Table 10.6. Although the snow is said to be dry, there could be a very small percent of moisture present that cannot be measured directly. For modeling purposes 0.1% of moisture is added to generate a larger value for the imaginary part of the layer permittivity. This is found to be necessary to keep the albedo small enough to achieve the scattering level shown in the measurement at 94 GHz. From Ulaby et al. [1986, Appendix E, Table E.1 and Figure E.3] the permittivity of ice is $2.92 - j\,0.0032$ at 35 GHz and $3.08 - j\,0.0041$ at 94 GHz. From these numbers the layer permittivities are estimated by the empirical formula given in Appendix 9A and shown in Table 10.6.

Table 10.6
Estimated Snow Layer Parameters

Parameter	*Estimated value*
Layer thickness	12 cm
Snow density	0.2 gr/cm^3
Scatterer diameter	1.1 mm (35 GHz) 0.5 mm (94 GHz)
Ground permittivity	5.0 (35 GHz) 5.0 (94 GHz)
Snow layer permittivity	$1.37 - j0.00177$ (35 GHz) $1.38 - j0.00172$ (95 GHz)
Air-snow surface rms height (σ)	0.14 cm
Air-snow surface correlation length (L)	2.66 cm
Snow-ground interface rms height (σ)	0.14 cm
Snow-ground interface correlation length (L)	1.54 cm

The IEM surface model is utilized for both the upper and lower snow layer boundaries. All the estimated parameters used in the model calculations are listed in Table 10.6. Note that except for the changes noted above all the measured/observed physical parameters are used in the calculations without change. The comparisons between model predictions and measurements at 35 GHz and 95 GHz are shown in Figures 10.16 and 10.17, respectively

At 35 GHz, the major contribution is due to volume scattering from the snow layer and the scattering from the snow-ground interface, which significantly affects the signature at small angles of incidence (less than 30 degrees). The backscattering contribution from the air-snow boundary is negligible since the discontinuity in permittivity at this interface is small. The calculated value at this boundary is at least 10 dB lower than the combined volume and snow-ground interface contribution. The surface scattering contribution from the lower boundary is visible in Figure 10.16 at small angles of incidence (θ less than 30 degrees). The volume scattering albedo is 0.75 and the total extinction is 0.53 for the 12 cm thick snow at 35 GHz. This value is not large even though we added some moisture to the dry snow. At large angles of incidence (over 40 degrees) both like and cross polarized scattering agree well with measurements. At smaller angles of incidence only the like polarized scattering is in good agreement with data. The cross polarized data appear to have a peak around 25 degrees, which is uncharacteristic of normal cross polarized scattering from a snow field.

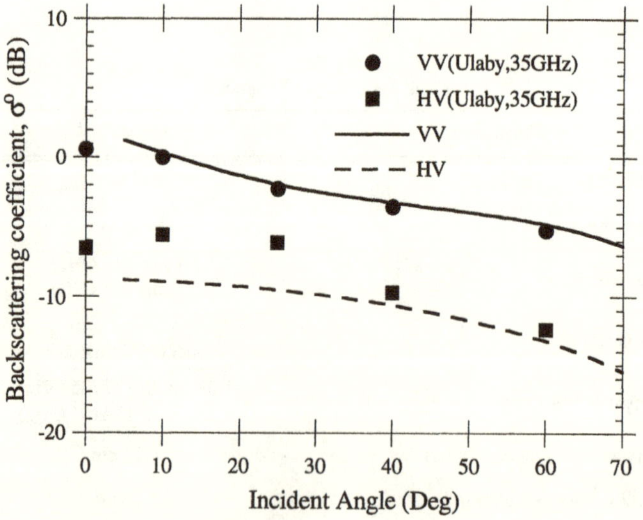

Figure 10.16 Dry snow backscattering coefficients at 35 GHz.

It is evident from the model computations that the backscattering signature of the dry snow layer at 94 GHz is dominated by volume scattering (albedo ≈ 0.86, optical thickness ≈ 2.5). The effect of the snow-ground interface is negligible due to the large optical thickness of the snow layer. The choice of the amount of moisture and the change in the size of the scatterer lead to very good agreements at both 35 and 94 GHz (Figures 10.16 and 10.17). There is the possibility that another choice may be possible.

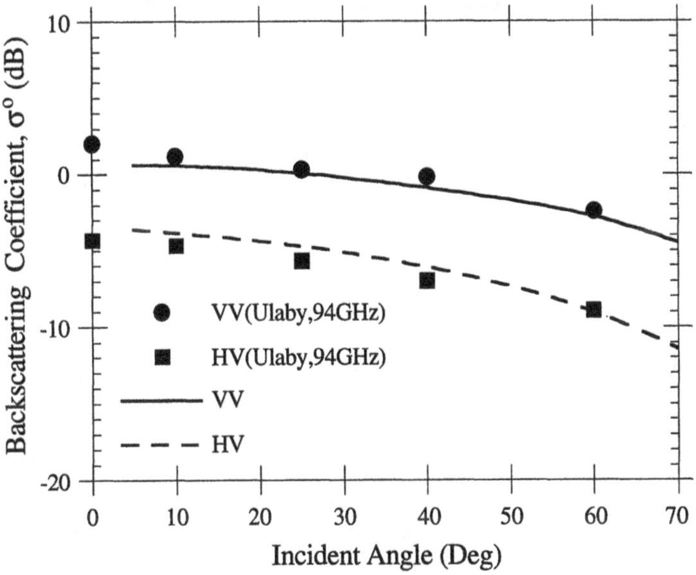

Figure 10.17 Dry snow backscattering coefficients at 94 GHz.

In Figure 10. 18 we show a case where the radius of the scatterer is chosen to be 0.3 mm. This causes the albedo to increase to 0.91 and the optical depth to 4.2. The increase in albedo raises the level of the cross polarized return much more than the like, resulting in a clearly poorer agreement between model and measurements than those shown in Figure 10.17. If we try to lower the albedo by increasing the absorption or the imaginary part of the layer permittivity, this will cause a corresponding increase in the absorption at 35 GHz resulting in a lower model prediction than those shown in Figure 10.16. Thus, this other choice may give better agreement at 94 GHz but will not provide a better result for both 35 and 94 GHz cases.

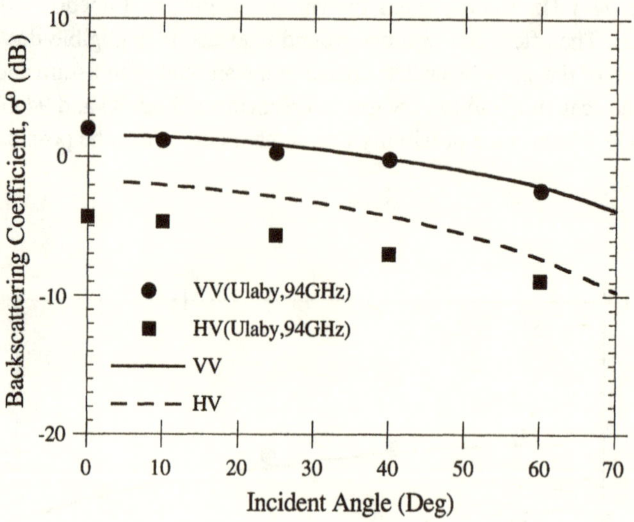

Figure 10.18 Backscattering from dry snow when the radius of the scatterer is increased to 0.3 mm.

Next, we want to compare the model in Chapter 8 with the measurements by Matzler et al. [1980]. This set of measurements is special in that both active and passive measurements of a dry snow field were taken at the same frequency, 10.4 GHz. Hence, all snow parameters are the same for both types of measurements. The computed albedo and optical depth values will also be the same; they are equal to 0.34 and 0.053, respectively. The temperature for the snow layer and the half space below was 268 K. The parameter values used in modeling are listed in Table 10. 7.

Table 10.7
Snow Parameters used for Data from Matzler et al.

Parameters	*Estimated value*
Layer thickness	1.5 m
Volume fraction	0.24
Scatterer diameter	0.8 mm
Ground permittivity	6.5
Snow layer permittivity	1.375 - j0.000125
Air-snow surface rms height (σ)	0.23 mm
Air-snow surface correlation length (L)	6.9 mm
Snow-ground interface rms height (σ)	1.8 mm
Snow-ground interface correlation length (L)	12.4 mm

In Figure 10.19 we show a comparison between model and backscattering measurements for both vertical and horizontal polarizations. The computed results indicate that for incident angles up to 40 degrees, scattering is dominated by the ground. Over this region vertical polarization is higher than horizontal polarization but the separation is small and indistinguishable experimentally. At larger angles volume scattering by the snow layer becomes significant and the ground-volume interaction causes horizontal polarization to be higher than the vertical by a small amount. Experimentally such a change is not seen but the separation between vertical and horizontal polarization remains small.

In Figure 10.20 the emission from the same snow layer as the one in Figure 10.19 is shown. An excellent agreement between model and measurements is obtained between 10 and 60 degrees. Beyond 60 degrees the data points become flat with angle in horizontal polarization possibly because of sky emission reflected into the receiver. For vertical polarization the agreement between model and data is excellent for all angles. The level of vertically polarized emission is much higher than the horizontal especially at large nadir angles. This is because of the Brewster angle effect. For the same reason a much smaller amount of sky emission is reflected into the antenna for vertical polarization and hence has no appreciable effect on data comparison.

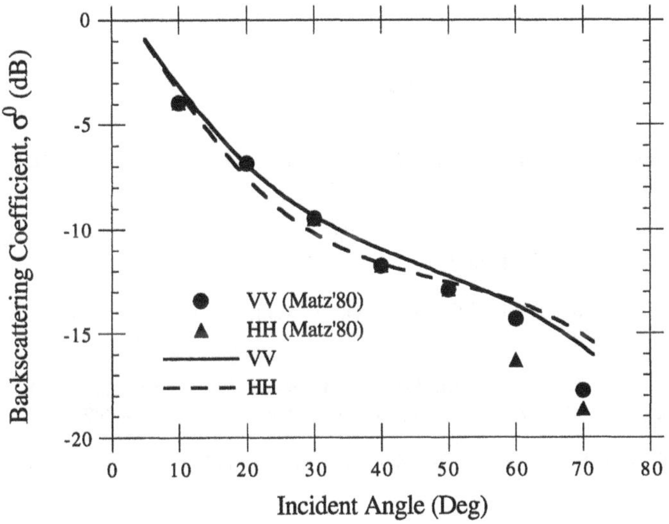

Figure 10.19 Comparison with active measurements by Matzler et al. [1980] from a dry snow layer.

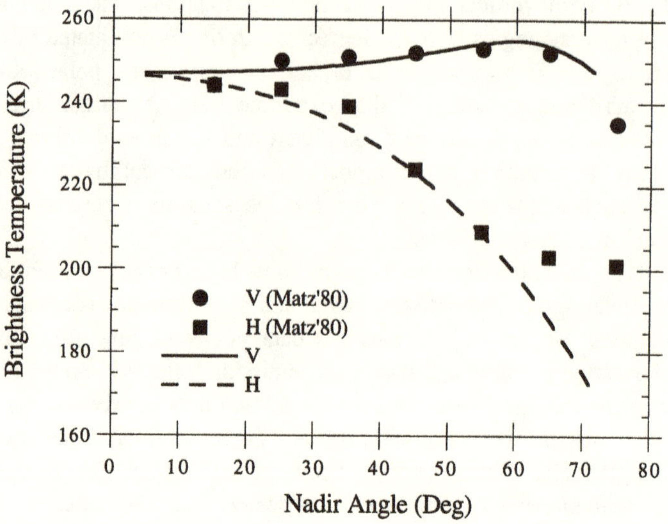

Figure 10.20 Comparison with passive measurements by Matzler et al. [1980] from a dry snow layer.

REFERENCES

Bredow, J.W., *A Laboratory Investigation into Microwave Backscattering from Sea Ice*, RSL Tech. Report 8240-1, August 1989.

Fung, A.K., and H.J. Eom, "Coherent Scattering of a Spherical Wave from an Irregular Surface," *IEEE Trans. Ant. Propagation*, Vol. AP-31, no. 1, 1983, pp. 68–72.

Grenfell, T.C., "Surface-Based Passive Microwave Study of Multiyear Sea Ice," *J. Geophysical Res.*, Vol. 86, C9, 1992, pp. 3485–3501.

Kuga, Y., F.T. Ulaby, T.F. Haddock, and R.D. DeRoo, "Millimeter-Wave Radar Scattering from Snow: Radiative Transfer Model," *Radio Science*, Vol. 26, March–April, 1991, pp. 329–342.

Matzler, C., and E. Schanda, "Snow Mapping with Active Microwave Sensors," *Int. J. Remote Sensing*, Vol. 5, 1984, pp. 409–450.

Matzler, C, Schanda, E., Hofer, R., and W. Good, "Microwave Signatures of the Natural Snow Cover at Weissfluhjoch," *NASA Conference Publication 2153.*, 1980, pp. 203-223.

Onstott, R.G., *Polarimetric Radar Measurements of Artificial Sea Ice During CRRELEX'88*, ERIM Tech. Report 196100-23-T, April 1990.

Onstott, R.G., "Active Microwave Observations of Arctic Sea Ice During the fall Freeze-Up," *Proc. IGARSS'91 Symp.*, June 3–6, Helsinki, Finland, 1991.

Stiles, W.H., and F.T. Ulaby, "The Active and Passive Microwave Response to Snow Parameters: Part 1—Wetness," *J. Geophysics. Res.*, Vol. 85, 1980, pp. 1037–1044.

Stogryn, A., and G.J. Desargent, "The Dielectric Properties of Brine in Sea Ice at Microwave Frequencies," *IEEE Trans. Antenna Prop.*, Vol. AP-33, no. 5, May 1985, pp. 523–532.

Ulaby, F.T., T.F. Haddock, R.T. Austin, and Y. Kuga," Millimeter-Wave Radar Scattering from Snow: Comparison of Theory with Experimental Observation," *Radio Science*, Vol. 26, March-April, 1991, pp. 343–351.

Ulaby, F.T., R.K. Moore, and A.K. Fung, *Microwave Remote Sensing*, Vol. 3, Appendix E, Artech House, Dedham, MA, 1986.

Chapter 11

Scattering and Emission Models for Vegetation

11.1 INTRODUCTION

In general we may consider a vegetated medium as a multi-layered medium above an irregular ground surface. For example, when we model a forest medium, we can choose its crown region as one layer and its trunk region as another layer. For deciduous forest we can separate its crown region further into two layers with the upper layer consisting mostly of leaves and some small twigs and the lower layer containing mostly branches and some leaves. This is a special multilayer medium in that there is no discontinuity in permittivity between the layers inside the vegetation. A discontinuity in permittivity occurs only at the vegetation-ground interface. Furthermore, the medium generally contains more than one type of scatterer. For instance, a branch is a scatterer and a leaf is another scatterer. We can apply the multilayer scattering and emission models developed in Chapter 8 to a vegetated medium, provided appropriate phase functions can be found for the vegetation components. The most often encountered components in a vegetated medium are cylindrical structures such as a stem, branch or trunk and disc or needle structures such as disc- or needle-shaped leaves [Karam et al., 1992; Ulaby et al., 1990; Durden et al., 1989; Fung et al., 1987; Seker, 1986; Tsang, et al., 1984; Lang and Sidhu, 1983; Karam and Fung, 1983]. A long deciduous leaf may curve and bend. From scattering viewpoint such a leaf may be modeled by several small discs each positioned with a different slope to simulate the long leaf. An element of the phase matrix in the scattering or emission models becomes a weighted sum of the phase functions for the vegetation components. Therefore, there is *no change in the framework of the model* but different phase functions for the vegetation components are needed. The simple models developed in Chapter 2 and 3 are specially useful when we wish to examine contributions from each scatterer. An extension of these models to two- and three-layer vegetated medium have been reported by Karam et al [1991, 1992]. Thus, what is needed is the phase matrix for the vegetation components, and we shall develop the elements of the scattering phase matrices for disc- and needle-shaped scatterers in Section 11.2 and for a dielectric cylinder of finite length in Section 11.3. A scattering model for a two-layer medium is described in Section 11.4. Theoretical scattering behaviors of a two-layer and a three-layer canopy model are presented in Section 11.5. This is followed by comparisons with radar measurements in Section 11.6. Finally, we show the effect of a slanted terrain in 11.7.

Because the average relative permittivity of a canopy is slightly larger than one and that of the ground is usually larger than three, generally the emission from a forest canopy dominates over that of the ground and is quite high [Barton, 1978; Ulaby et al., 1986, Chapter 19]. As a result the emission from a forest canopy is not expected to be very sensitive to forest parameters. Hence, no illustrations will be given in this chapter about emission. For agricultural fields the canopy height is much smaller, and there is row-directional dependence in emission due to the ground. A detailed treatment of this problem can be found in Ulaby et al. [1986, Chapter 19].

11.2 PHASE FUNCTIONS OF DISC- AND NEEDLE-SHAPED SCATTERERS

Both circular disc- and needle-shaped scatterers can be viewed as special cases of circular cylinders characterized by a radius and a length or thickness. For a disc its thickness is assumed to be very small compared with its radius, and for a needle the reverse is true, i.e., its radius must be small compared with its length. When the small dimension of the scatterer is small compared with the exploring wavelength, we can assume that the electromagnetic field is a constant across that dimension. The development of the phase function in this section makes use of this basic assumption.

11.2.1 The Scattered Fields

From Appendix 11A the $\hat{p}_s$ polarized component of the scattered field, where $\hat{p}_s$ is transverse to the direction of observation, can be expressed in terms the field inside the scatterer $\vec{E}$ as

$$\hat{p}_s \cdot \vec{E}^s(\vec{r}) = \frac{k^2(\varepsilon_r - 1)}{4\pi} \int_V \frac{\exp(-jk|\vec{r} - \vec{r}'|)}{|\vec{r} - \vec{r}'|} \left(\hat{p}_s \cdot \vec{E}\right) dV \tag{11.1}$$

where k is the wavenumber in air; ε_r is the relative permittivity of the scatterer to air and integration is over the volume of the scatterer defined in terms of the prime variables. Clearly, the scattered field can be found if the field inside the scatterer is known or can be estimated.

To permit arbitrary orientation of the scatterer with respect to a reference frame (x, y, z), we need to introduce a local frame which is the principal frame of the scatterer (x'', y'', z''). The orientation of a symmetric scatterer such as a needle or a disc can be specified by a polar angle β and an azimuthal angle α in the reference coordinates. From Figure 11.1, the two coordinate systems are related as follows:

$$\begin{bmatrix} x'' \\ y'' \\ z'' \end{bmatrix} = \begin{bmatrix} \cos\beta\cos\alpha & \cos\beta\sin\alpha & -\sin\beta \\ -\sin\alpha & \cos\alpha & 0 \\ \sin\beta\cos\alpha & \sin\beta\sin\alpha & \cos\beta \end{bmatrix} \begin{bmatrix} x \\ y \\ z \end{bmatrix} \equiv \mathbf{U} \begin{bmatrix} x \\ y \\ z \end{bmatrix} \tag{11.2}$$

where $\mathbf{U}$ is a unitary matrix and known as Euler's transformation. Its inverse is equal to its transpose yielding the relation,

$$\begin{bmatrix} x \\ y \\ z \end{bmatrix} = \mathbf{U}^{-1} \begin{bmatrix} x'' \\ y'' \\ z'' \end{bmatrix} = \mathbf{U}^{T} \begin{bmatrix} x'' \\ y'' \\ z'' \end{bmatrix} \tag{11.3}$$

The unitary matrix is needed not only to permit arbitrary orientation of the scatterer but also to transform integration variables in the reference coordinates to the principal coordinates of the scatterer in which integration over the volume of the scatterer is to be carried out. This is necessary to simplify integration in the scattered field expression. Note that the desired expression for the scattered field is in the reference coordinates, while the integration in it is in the local coordinates.

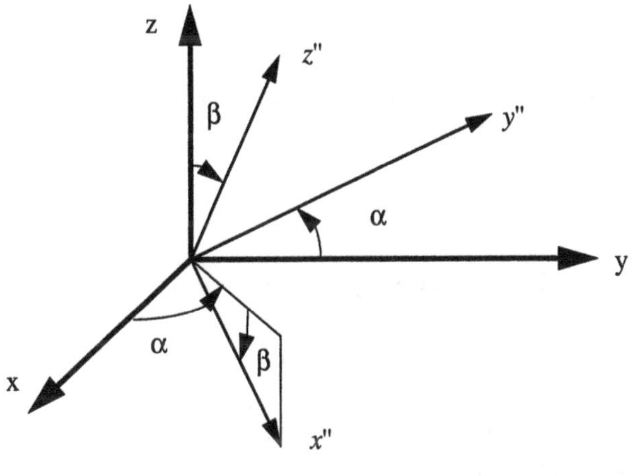

Figure 11.1 Illustration of principal frame of the scatterer relative to the reference frame.

Estimate of the Field Inside an Ellipsoid

Both the disc- and needle-shaped scatterers can be treated as special cases of an ellipsoid. For an ellipsoid the field inside the scatterer is related to the incident field, $\vec{E}_i = \vec{E}_0 \exp(j\vec{k}_i \cdot \vec{r})$ (where $\vec{k} = \hat{i}k$; $\hat{i}$ is the unit vector in the direction of propagation and $\vec{r}$ is the displacement vector in the reference frame), by [Stratton, 1941]

$$\vec{E} = \mathbf{U}^{-1}\begin{bmatrix} 1/a_1 & 0 & 0 \\ 0 & 1/a_2 & 0 \\ 0 & 0 & 1/a_3 \end{bmatrix}\mathbf{U}\cdot\vec{E}_i \equiv \mathbf{A}\cdot\vec{E}_i \tag{11.4}$$

where *a vector that appears with a matrix is understood to be a column matrix* and

$$\begin{aligned} a_1 &= 1 + 0.5abc\,(\varepsilon_r - 1)\,B_1 \\ a_2 &= 1 + 0.5abc\,(\varepsilon_r - 1)\,B_2 \\ a_3 &= 1 + 0.5abc\,(\varepsilon_r - 1)\,B_3 \end{aligned} \tag{11.5}$$

and a, b, c are the semi-axes of the ellipsoid with volume $v_0 = 4\pi abc/3$. These relations are valid under static conditions. It has been shown by Stratton [1941] that

$$B_1 + B_2 + B_3 = 2/(abc) \tag{11.6}$$

The contents of B_n have closed form solutions for spheroids where ($a = b$), $B_1 = B_2$. For oblate spheroids ($a > c$), we have

$$B_3 = 2\,[\,(rt/c) - \tan^{-1}(rt/c)\,]\,/\,(rt)^3 \tag{11.7}$$

where $rt = (a^2 - c^2)^{0.5}$, and for prolate spheroids ($a < c$), we have

$$B_3 = -2\,[\,(\overline{rt}/c) + 0.5\ln\,[\,(c - \overline{rt})\,/\,(c + \overline{rt})\,]\,]\,/\,(\overline{rt})^3 \tag{11.8}$$

where $\overline{rt} = (c^2 - a^2)^{0.5}$. For both types of spheroids B_1 can be found from (11.6) and B_3. For our applications only spheroids are of interest. In this case the elements of the symmetric matrix A are listed below with $a_1 = a_2$,

$$A_{11} = (\cos^2\alpha\cos^2\beta + \sin^2\alpha)\,/\,a_1 + \cos^2\alpha\sin^2\beta/a_3$$

$$A_{12} = \sin\alpha\cos\alpha\,[\,(\cos^2\beta - 1)\,/\,a_1 + \sin^2\beta/a_3]$$

$$A_{13} = \cos\alpha\sin\beta\cos\beta\,(1/a_3 - 1/a_1)$$

$$A_{21} = A_{12}$$

$$A_{22} = (\sin^2\alpha\cos^2\beta + \cos^2\alpha)\,/\,a_1 + \sin^2\alpha\sin^2\beta/a_3$$

$$A_{23} = \sin\alpha\sin\beta\cos\beta\,(1/a_3 - 1/a_1)$$

$$A_{31} = A_{13}$$

$$A_{32} = A_{23}$$

$$A_{33} = \sin^2\beta/a_1 + \cos^2\beta/a_3$$

11.2.2 Scattered Fresnel Field and Scattering Amplitude of a Circular Disc

When an ellipsoid is specialized to a circular disc of equivalent volume with radius $a_0 = a$ and thickness $t = 4c/3$, we have $a = b = a_0 \gg t$ and B_n's are given by

$$B_3 = 2\left(\frac{4}{3t}\right)^3\left[\left[\left(\frac{4a_0}{3t}\right)^2 - 1\right]^{0.5} - \tan^{-1}\left[\left(\frac{4a_0}{3t}\right)^2 - 1\right]^{0.5}\right] \Big/ \left[\left(\frac{4a_0}{3t}\right)^2 - 1\right]^{1.5} \tag{11.9}$$

$$B_1 = \frac{4}{3ta_0^2} - \frac{B_3}{2} \tag{11.10}$$

The substitution of (11.30) and (11.31) into (11.4) gives an estimate of the field inside the circular disc denoted as $\vec{E} = \mathbf{A}_d \cdot \vec{E}_0$. From (11.1) we can write the $\hat{p}_s$ polarized scattered field component for a circular disc as

$$\begin{aligned} \hat{p}_s \cdot \vec{E}^s(\vec{r}) &\approx (\varepsilon_r - 1)\, k^2 \left(\frac{\hat{p}_s \cdot \mathbf{A}_d \cdot \vec{E}_0}{4\pi r}\right) \exp(-jkr)\, I_d \\ &= \{\hat{p}_s \cdot \left[k^2(\varepsilon_r - 1)\frac{\mathbf{A}_d}{4\pi} I_d\right]\} \cdot \vec{E}_0 \frac{\exp(-jkr)}{r} \\ &\equiv \hat{p}_s \cdot \boldsymbol{f}_d(\vec{k}_s, \vec{k}_i) \cdot \hat{p}_i E_0 \frac{\exp(-jkr)}{r} \end{aligned} \tag{11.11}$$

where $\boldsymbol{f}_d(\vec{k}_s, \vec{k}_i)$ is the scattering amplitude matrix for a disc; $\vec{k}_s, \vec{k}_i$ are the scattering and incident unit propagation vectors, $\hat{s}$ and $\hat{i}$, multiplied by the wave number and *calculations involving vector and matrix dot products are carried out from right to left* using matrix convention with column matrices treated as vectors. *Vector products follow the conventional definition for vectors instead of matrices.* The elements of this matrix defined in accordance with vertical and horizontal polarizations can be written as

$$\boldsymbol{f}_d(\vec{k}_s, \vec{k}_i) = \frac{k^2(\varepsilon_r - 1)\, I_d}{4\pi}\begin{bmatrix} (\hat{v}_s \cdot \mathbf{A}_d \cdot \hat{v}_i) & (\hat{v}_s \cdot \mathbf{A}_d \cdot \hat{h}_i) \\ (\hat{h}_s \cdot \mathbf{A}_d \cdot \hat{v}_i) & (\hat{h}_s \cdot \mathbf{A}_d \cdot \hat{h}_i) \end{bmatrix} \equiv \begin{bmatrix} f_{vv} & f_{hv} \\ f_{vh} & f_{vh} \end{bmatrix} \tag{11.12}$$

where the vertical and horizontal polarization unit vectors in (11.12) are chosen to agree with $\hat{\theta}$ and $\hat{\phi}$ unit vectors of a standard spherical coordinate system and form an orthogonal set with $\hat{s}$ as

$$\hat{s} = \hat{x}\sin\theta_s\cos\phi_s + \hat{y}\sin\theta_s\sin\phi_s + \hat{z}\cos\theta_s$$

$$\hat{v}_s = \hat{x}\cos\theta_s\cos\phi_s + \hat{y}\cos\theta_s\sin\phi_s - \hat{z}\sin\theta_s$$

$$\hat{h}_s = -\hat{x}\sin\phi_s + \hat{y}\cos\phi_s \tag{11.13}$$

Similarly, the unit polarization vectors associated with the incident direction are

$$\hat{i} = \hat{x}\sin\theta_i\cos\phi_i + \hat{y}\sin\theta_i\sin\phi_i + \hat{z}\cos\theta_i$$

$$\hat{v}_i = \hat{x}\cos\theta_i\cos\phi_i + \hat{y}\cos\theta_i\sin\phi_i - \hat{z}\sin\theta_i$$

$$\hat{h}_i = -\hat{x}\sin\phi_i + \hat{y}\cos\phi_i \tag{11.14}$$

The definitions of θ_s and θ follow the standard spherical coordinate system and are measured with respect to the positive *z*-axis.

To specialize the scattered field expression to the Fresnel zone we use the expansion $(1+x)^{0.5} \approx 1 + x - x^2/8$ and approximate $|\vec{r} - \vec{r}'|$ by $r - \hat{s}\cdot\vec{r}' + 0.5r'^2[1 - (\hat{s}\cdot\hat{r}')^2]$ in the expression for I_d as,

$$I_d = \int_{v'} \left[1 + \frac{\hat{s}\cdot\vec{r}'}{r}\right] \exp\{j\vec{q}\cdot\vec{r}' - j\frac{kr'^2}{2r}[1 - (\hat{s}\cdot\hat{r}')^2]\}\, d\vec{r}' \tag{11.15}$$

In (11.34) $\vec{q} = \vec{k}_s - \vec{k}_i = \hat{x}q_x + \hat{y}q_y + \hat{z}q_z$ and $\hat{s} = \hat{x}s_x + \hat{y}s_y + \hat{z}s_z$, where

$$s_x = \sin\theta_s\cos\phi_s$$

$$s_y = \sin\theta_s\sin\phi_s$$

$$s_z = \cos\theta_s \tag{11.16}$$

$$q_x = k(s_x - \sin\theta_i\cos\phi_i)$$

$$q_y = k(s_y - \sin\theta_i\sin\phi_i)$$

$$q_z = k(s_z - \cos\theta_i) \tag{11.17}$$

The use of the static approximation for the inner field of the scatterer is valid under the conditions,

$$t \ll a_0 \text{ and } kt(\varepsilon_r)^{0.5} \ll 1 \tag{11.18}$$

Partial Evaluation of I_d

Due to the symmetry property of the disc in its principal frame, the integral I_d can be partially carried out in this frame (the local frame). Conversions of $\vec{r}'$ into local frame $\vec{r}''$ are carried out using the Euler's transformation $\mathbf{U}$:

$$\vec{q}\cdot\vec{r}' = (\mathbf{U}\cdot\vec{q})\cdot(\mathbf{U}\cdot\vec{r}') = \vec{q}''\cdot\vec{r}'' \tag{11.19}$$

$$\hat{s}\cdot\vec{r}' = (\mathbf{U}\cdot\hat{s})\cdot(\mathbf{U}\cdot\vec{r}') = \hat{s}''\cdot\vec{r}'' \tag{11.20}$$

where

$$\begin{aligned}\vec{q}'' = \mathbf{U}\cdot\vec{q} &= \hat{x}''(q_x\cos\alpha\cos\beta + q_y\sin\alpha\cos\beta - q_z\sin\beta)\\ &\quad + \hat{y}''(q_y\cos\alpha - q_x\sin\alpha) + \hat{z}''(q_x\cos\alpha\sin\beta + q_y\sin\alpha\sin\beta + q_z\cos\beta)\\ &\equiv \hat{x}''q''_x + \hat{y}''q''_y + \hat{z}''q''_z\end{aligned} \tag{11.21}$$

$$\begin{aligned}\hat{s}'' = \mathbf{U}\cdot\hat{s} &= \hat{x}''(s_x\cos\alpha\cos\beta + s_y\sin\alpha\cos\beta - s_z\sin\beta)\\ &\quad + \hat{y}''(s_y\cos\alpha - s_x\sin\alpha) + \hat{z}''(s_x\cos\alpha\sin\beta + s_y\sin\alpha\sin\beta + s_z\cos\beta)\\ &\equiv \hat{x}''s''_x + \hat{y}''s''_y + \hat{z}''s''_z\end{aligned} \tag{11.22}$$

With the above relation the integral term can be expressed in the local frame as

$$I_d = \int_{v''}\left[1 + \frac{r''}{r}(\hat{s}''\cdot\hat{r}'')\right]\exp\{j\vec{q}''\cdot\vec{r}'' - j\frac{kr''^2}{2r}[1-(\hat{s}''\cdot\hat{r}'')^2]\}\,d\vec{r}'' \equiv I_{d1} + I_{d2} \tag{11.23}$$

where

$$I_{d1} = \int_{v''}\exp\{j\vec{q}''\cdot\vec{r}'' - j\frac{kr''^2}{2r}[1-(\hat{s}''\cdot\hat{r}'')^2]\}\,d\vec{r}'' \tag{11.24}$$

and

$$I_{d2} = \int_{v''}\frac{r''}{r}(\hat{s}''\cdot\hat{r}'')\exp\{j\vec{q}''\cdot\vec{r}'' - j\frac{kr''^2}{2r}[1-(\hat{s}''\cdot\hat{r}'')^2]\}\,d\vec{r}'' \tag{11.25}$$

If the disk is very thin, i.e., t is very small, we can ignore the z-component of $\vec{r}''$ in the phase and approximate $\vec{r}''$ by

$$\vec{r}'' \approx \rho'' (\cos\phi''\hat{x}' + \sin\phi''\hat{y}')$$

Then,

$$\vec{q}'' \cdot \vec{r}'' = \rho'' (q''_x \cos\phi'' + q''_y \sin\phi'')$$

$$\hat{s}'' \cdot \hat{r}'' = s''_x \cos\phi'' + s''_y \sin\phi''$$

After integration over the thickness of the disc, we have

$$I_{d1} = t\int_0^{2\pi} \int_0^{a_0} \rho''g(\rho'', \phi'')d\rho''d\phi'' \tag{11.26}$$

$$I_{d2} = t\int_0^{2\pi} \frac{1}{r}(s''_x \cos\phi'' + s''_y \sin\phi'') \int_0^{a_0} \rho''^{2}g(\rho'', \phi'')d\rho''d\phi'' \tag{11.27}$$

where

$$g(\rho'', \phi'') = \exp\{j\rho'' (q''_x \cos\phi'' + q''_y \sin\phi'') - \frac{k\rho''^2}{2r}[1 - (s''_x \cos\phi'' + s''_y \sin\phi'')^2]\}$$

$$= \exp[j\rho''q_d - j(m_d\rho''^2/2)]$$

$$q_d = q''_x \cos\phi'' + q''_y \sin\phi''$$

and

$$m_d = k[1 - (s''_x \cos\phi'' + s''_y \sin\phi'')^2]/r$$

where r is related to the volume of the scatterer v_0 and its volume fraction vf by $r = [v_0/(vf)]^{1/3}$ [Goedecke, 1977].

Consider the ρ'' integration for $m_d = 0$ and $m_d \neq 0$ cases, and let

$$g_1(\phi'') = \int_0^{a_0} \rho''g(\rho'', \phi'')d\rho'' = \int_0^{a_0} \rho''\exp[j\rho''q_d - j(m_d\rho''^2/2)]d\rho''$$

$$g_2(\phi'') = \int_0^{a_0} \rho''^2g(\rho'', \phi'')d\rho'' = \int_0^{a_0} \rho''^2\exp[j\rho''q_d - j(m_d\rho''^2/2)]d\rho''$$

If $m_d = 0$,the integration over the variable ρ'' for either $g_1(\phi'')$ or $g_2(\phi'')$ has a closed form representation as follows:

$$g_1(\phi'') = \int_0^{a_0} \rho'' \exp(j\rho'' q_d) d\rho'' = [\exp(ja_0 q_d)(1 - ja_0 q_d) - 1]/q_d^2$$

$$g_2(\phi'') = \int_0^{a_0} \rho''^2 \exp(j\rho'' q_d) d\rho'' = [\exp(ja_0 q_d)(2a_0 q_d - ja_0^2 q_d^2 + j2) - j2]/q_d^3$$

If $m_d \neq 0$, we have the following evaluations for $g_1(\phi'')$ and $g_2(\phi'')$,

$$g_1(\phi'') = \int_0^{a_0} \rho'' \exp\left(j\rho'' q_d - j\frac{m_d \rho''^2}{2}\right) d\rho''$$

$$= \frac{q_d}{m_d}\sqrt{\frac{\pi}{2jm_d}} \exp\left(\frac{jq_d^2}{2m_d}\right)\left\{\mathrm{erf}\left(\sqrt{\frac{jm_d}{2}}\left[a_0 - \frac{q_d}{m_d}\right]\right) - \mathrm{erf}\left(q_d\sqrt{\frac{j}{2m_d}}\right)\right\}$$

$$+ \frac{j}{m_d}\exp\left(j\frac{q_d^2}{2m_d}\right)\left\{\exp\left[\frac{m_d}{2j}\left(a_0 - \frac{q_d}{m_d}\right)^2\right] - \exp\left(j\frac{q_d}{2m_d}\right)\right\}$$

$$= \frac{q_d}{m_d}\sqrt{\frac{\pi}{2jm_d}} \exp\left(\frac{jq_d^2}{2m_d}\right)\left\{\mathrm{erf}\left(\sqrt{\frac{jm_d}{2}}\left[a_0 - \frac{q_d}{m_d}\right]\right) - \mathrm{erf}\left(q_d\sqrt{\frac{j}{2m_d}}\right)\right\}$$

$$+ \frac{j}{m_d}\left\{\exp\left[\frac{m_d a_0^2}{2j} + jq_d a_0\right] - 1\right\}$$

$$g_2(\phi'') = \int_0^{a_0} \rho''^2 \exp\left(j\rho'' q_d - j\frac{m_d \rho''^2}{2}\right) d\rho''$$

$$= j\frac{m_d a_0 + q_d}{m_d^2}\exp\left(ja_0 q_d - j\frac{m_d a_0^2}{2}\right) - \frac{jq_d}{m_d^2} + \frac{q_d^2 - jm_d}{m_d^2}\exp\left(\frac{jq_d^2}{2m_d}\right) \times$$

$$\left[\sqrt{\frac{\pi}{2jm_d}}\left\{\mathrm{erf}\left(\sqrt{\frac{jm_d}{2}}\left[a_0 - \frac{q_d}{m_d}\right]\right) - \mathrm{erf}\left(q_d\sqrt{\frac{j}{2m_d}}\right)\right\}\right]$$

Substituting $g_1(\psi)$ and $g_2(\psi)$ into (11.39) and (11.40), respectively, we obtain

$$I_{d1} = t\int_{0}^{2\pi} g_1(\phi'')d\phi'' \tag{11.28}$$

$$I_{d2} = t\int_{0}^{2\pi} \frac{1}{r}(s''_x\cos\phi'' + s''_y\sin\phi'')\, g_2(\phi'')d\phi'' \tag{11.29}$$

11.2.3 Scattered Fresnel Field and Scattering Amplitude of a Needle

When an ellipsoid is specialized to a needle of equivalent volume with radius $r_0 = a$ and length $L = 4c/3$, we have $a = b = r_0 \ll L$ and $B_n's$ are given by

$$B_3 = -2\,[\,(\overline{rt}/c) + 0.5\ln[\,(c-\overline{rt})/(c+\overline{rt})\,]\,]/(\overline{rt})^3 \tag{11.30}$$

$$B_1 = \frac{4}{3Lr_0^2} - \frac{B_3}{2} \tag{11.31}$$

$$\frac{\overline{rt}}{c} = \left[1-\left(\frac{4r_0}{3L}\right)^2\right]^{0.5},\; c = \frac{3L}{4}$$

The substitution of (11.30) and (11.31) into (11.4) gives an estimate of the field inside the needle denoted as $\vec{E} = \mathbf{A}_n \cdot \vec{E}_0$. From (11.1) we can write the $\hat{p}_s$ polarized scattered field for a needle as

$$\hat{p}_s \cdot \vec{E}^s(\hat{r}) \approx (\varepsilon_r - 1)\,k^2\left(\frac{\hat{p}_s \cdot \mathbf{A}_n \cdot \vec{E}_0}{4\pi r}\right)\exp(-jkr)I_n$$

$$= \{\hat{p}_s \cdot \left[k^2(\varepsilon_r - 1)\frac{\mathbf{A}_n}{4\pi}I_n\right]\} \cdot \vec{E}_0\frac{\exp(-jkr)}{r}$$

$$\equiv \hat{p}_s \cdot f_n(\vec{k}_s, \vec{k}_i) \cdot \vec{E}_0\frac{\exp(-jkr)}{r} \tag{11.32}$$

where $f_n(\vec{k}_s, \vec{k}_i)$ is the scattering amplitude matrix in meter for a needle; $\vec{k}_s, \vec{k}_i$ are the scattering and incident unit propagation vectors multiplied by the wave number. The elements of this matrix defined in accordance with vertical and horizontal polarizations can be written as

$$f_n(\vec{k}_s, \vec{k}_i) = \frac{k^2(\varepsilon_r - 1) I_n}{4\pi} \begin{bmatrix} (\hat{v}_s \cdot \mathbf{A}_n \cdot \hat{v}_i) & (\hat{v}_s \cdot \mathbf{A}_n \cdot \hat{h}_i) \\ (\hat{h}_s \cdot \mathbf{A}_n \cdot \hat{v}_i) & (\hat{h}_s \cdot \mathbf{A}_n \cdot \hat{h}_i) \end{bmatrix} \equiv \begin{bmatrix} f_{vv} & f_{hv} \\ f_{vh} & f_{hh} \end{bmatrix} \tag{11.33}$$

where the incident and scattered vertical and horizontal polarization unit vectors have been defined by (11.13) and (11.14).

To specialize the scattered field to the Fresnel zone we use the expansion $(1+x)^{0.5} \approx 1 + x - x^2/8$ and approximate $|\vec{r} - \vec{r}'|$ by $r - \hat{s} \cdot \vec{r}' + 0.5 r'^2 [1 - (\hat{s} \cdot \hat{r}')^2]$ to simplify the expression for I_n as

$$I_n = \int_{V'} \left(1 + \frac{\hat{s} \cdot \vec{r}'}{r}\right) \exp\{j\vec{q} \cdot \vec{r}' - j\frac{kr'^2}{2r}[1 - (\hat{s} \cdot \hat{r}')^2]\} d\vec{r}' \tag{11.34}$$

In (11.34), $\vec{q} = \vec{k}_s - \vec{k}_i$. and the dimension of I_n is in meters cubed.

The use of the static approximation for the inner field of the scatterer is valid under the conditions

$$r_0 \ll L \text{ and } kr_0 (\varepsilon_r)^{0.5} \ll 1 \tag{11.35}$$

Evaluation of I_n

Similar to the disc scatterer it is simpler to evaluate the integral given by (11.34) in the principal coordinates of the needle. Here, the axis of the needle is chosen to agree with the *z*-axis in the local frame. The starting equation is the same as (11.23) and so are the definitions of $\hat{s}''$ and $\vec{q}''$:

$$\begin{aligned} I_n &= \int_{v''} \left[1 + \frac{r''}{r}(\hat{s}'' \cdot \hat{r}'')\right] \exp\{j\vec{q}'' \cdot \vec{r}'' - j\frac{kr''^2}{2r}[1 - (\hat{s}'' \cdot \hat{r}'')^2]\} d\vec{r}'' \\ &\equiv I_{n1} + I_{n2} \end{aligned} \tag{11.36}$$

where

$$I_{n1} = \int_{v''} \exp\{j\vec{q}'' \cdot \vec{r}'' - j\frac{kr''^2}{2r}[1 - (\hat{s}'' \cdot \hat{r}'')^2]\} d\vec{r}'' \tag{11.37}$$

and

$$I_{n2} = \int_{v''} \frac{r''}{r}(\hat{s}'' \cdot \hat{r}'') \exp\{j\vec{q}'' \cdot \vec{r}'' - j\frac{kr''^2}{2r}[1 - (\hat{s}'' \cdot \hat{r}'')^2]\} d\vec{r}'' \tag{11.38}$$

If the needle is very thin, i.e., r_0 is very small, we can replace the integration over azimuth and radial distance by πr_0^2 so that $\vec{q}'' \cdot \vec{r}'' = q''_z z''$, $\hat{s}'' \cdot \hat{r}'' = s''_z$ and

$$I_{n1} = (\pi r_0^2) \int_0^L g(z'')\, dz'' \tag{11.39}$$

$$I_{n2} = (\pi r_0^2) \frac{s''_z}{r} \int_0^L z'' g(z'')\, dz'' \tag{11.40}$$

where

$$g(z'') = \exp\left[jz''q''_z - j\frac{kz''^2}{2r}(1 - s''^2_z) \right] = \exp(jz''q''_z - j\frac{m_n z''^2}{2})$$

and

$$m_n = k(1 - s''^2_z)/r$$

Consider the z'' integration for $m_n = 0$ and $m_n \neq 0$ cases:
If $m_n = 0$, the integration can be carried out yielding

$$I_{n1} = (\pi r_0^2) \int_0^L \exp(jz''q''_z) dz'' = \{\pi r_0^2 [\exp(jq''_z L) - 1]\} / (jq''_z) \tag{11.41}$$

$$I_{n2} = \frac{\pi r_0^2}{r} \int_0^L z'' \exp(jz''q''_z) dz'' = \frac{\pi r_0^2}{r q''^2_z} [(1 - jq''_z L) \exp(jq''_z L) - 1] \tag{11.42}$$

If $m_n \neq 0$,the integration can also be carried out in terms of special functions as

$$I_{n1} = \pi r_0^2 \int_0^L \exp(jz''q''_z - j\frac{m_n z''^2}{2}) dz''$$

$$= \pi r_0^2 \int_0^L \exp\left\{ \frac{m_n}{2j} \left[\left(z'' - \frac{q''_z}{m_n} \right)^2 - \frac{q''^2_z}{m_n^2} \right] \right\} dz''$$

$$= \pi r_0^2 \exp\left(j\frac{q''^2_z}{2m_n} \right) \int_{-q''_z/m_n}^{L - q''_z/m_n} \exp\{ \frac{m_n}{2j} z^2 \}\, dz$$

$$= \pi r_0^2 \sqrt{\frac{\pi}{m_n}} \exp\left(j \frac{q''^2_z}{2m_n} \right) \{ fc(b_1) + fc(b_2) + j[fs(b_1) + fs(b_2)] \} \tag{11.43}$$

where $fc(\ )$ and $fs(\ )$ are the Fresnel cosine and sine integral functions defined as

$$fc(x) = \left(\frac{2}{\pi}\right)^{0.5} \int_0^x \cos\left(t^2\right) dt \text{ and } fs(x) = \left(\frac{2}{\pi}\right)^{0.5} \int_0^x \sin\left(t^2\right) dt \tag{11.44}$$

$$b_1 = q''_z / m_n \text{ and } b_2 = L - q''_z / m_n \tag{11.45}$$

Next, we evaluate the second integral,

$$I_{n2} = \pi r_0^2 \frac{s''_z}{r} \int_0^L z'' \exp(jz''q''_z - j\frac{m_n z''^2}{2}) dz''$$

$$= \pi r_0^2 \frac{s''_z}{r} \exp\left(j \frac{q''^2_z}{2m_n} \right) \int_{-q''_z/m_n}^{L - q''_z/m_n} \left(z + \frac{q''_z}{m_n} \right) \exp\left(\frac{m_n}{2j} z^2 \right) dz$$

$$= \frac{s''_z q''_z}{r m_n} I_{n1} + \pi r_0^2 \frac{s''_z}{r} \exp\left(j \frac{q''^2_z}{2m_n} \right) \int_{-q''_z/m_n}^{L - q''_z/m_n} z \exp\left(\frac{m_n}{2j} z^2 \right) dz$$

$$= \frac{s''_z q''_z}{r m_n} I_{n1} + \pi r_0^2 \frac{j s''_z}{r m_n} \left[\exp\left(\frac{m_n L^2}{2j} + j q''_z L \right) - 1 \right] \tag{11.46}$$

11.2.4 Absorption and Extinction Coefficients and Phase Matrices

Once the elements of the scattering amplitude matrix are known from (11.12) or (11.33), the corresponding elements of the phase matrix **P** can be found in terms of them [Section 1.6 of Chapter 1]:

$$\mathbf{P} = \mathbf{K}_e^{-1} n_0 \langle \sigma \rangle \tag{11.47}$$

where σ is the Stokes matrix and its elements are expressible directly in terms of the elements of the scattering amplitude matrix $\mathbf{f}(\vec{k}_s, \vec{k}_i)$. In (11.47) n_0 is the number density of the scatterers, $\mathbf{K}_e^{-1}$ is the inverse of the extinction matrix, and $\langle\ \rangle$ denotes ensemble averaging over the orientation angles, α and β.

For practical computation it is convenient to assume a diagonal extinction matrix as

$$\mathbf{K}_e = \text{diag}\,[K_v, K_h, K_3, K_4] \tag{11.48}$$

where

$$K_v = K_{av} + K_{sv}$$

$$K_h = K_{ah} + K_{sh}$$

$$K_3 = K_4 = 0.5\,(K_v + K_h)$$

and K_{ap} and K_{sp} are the absorption and volume scattering coefficients for p polarization. A complete form of the extinction coefficient has been given in Ishimaru and Cheung [1980] and Karam and Fung [1982]. For most practical applications the contribution of the off-diagonal terms of the extinction matrix is small and can be neglected. The volume scattering coefficients for vertical and horizontal polarizations can be calculated from the elements of the scattering amplitude matrix as

$$K_{sv}(\theta) = n_0 \int \langle |f_{vv}|^2 + |f_{hv}|^2 \rangle d\Omega\,(\theta_s, \phi_s) \tag{11.49}$$

$$K_{sh}(\theta) = n_0 \int \langle |f_{vh}|^2 + |f_{hh}|^2 \rangle d\Omega\,(\theta_s, \phi_s) \tag{11.50}$$

where the integration is over the 4π solid angle with respect to the scattering angles. These integrations plus averaging over the orientation angles constitute a fourfold integral. A simpler way to compute extinction is to use the forward scattering theorem as given later by (11.84).The absorption cross section σ_a for a single object may be written in terms of the field inside the scatterer $\vec{E}$ [Ishimaru, 1978, p. 17] as

$$\sigma_a = \int k\varepsilon'' |\vec{E}(\vec{r}')|^2 dV' \tag{11.51}$$

where ε'' is the imaginary part of the permittivity ε of the scatterer. After substitution of the inner fields into (11.51) for a unit incident field amplitude and multiplication by the number density, we find the absorption coefficient for a disc scatterer as

$$K_{ap} = n_0 k\varepsilon'' \pi a_0^2 t |\mathbf{A}_d \cdot \hat{p}_i|^2 \tag{11.52}$$

and for a needle scatterer as

$$K_{ap} = n_0 k\varepsilon'' \pi r_0^2 L |\mathbf{A}_n \cdot \hat{p}_i|^2 \tag{11.53}$$

In practice, it is simpler to calculate the extinction of a scatterer by the forward scattering theorem, because (11.49) and (11.50) involve fourfold integration. This is explained further in Section 11.3.4.

11.3 PHASE MATRIX OF A DIELECTRIC CYLINDER

To find the phase function of a finite-length dielectric cylinder the procedure is the same as in the previous section except for the determination of the field inside the cylinder. There is as yet no exact solution for the field inside a finite-length dielectric cylinder. The approximation to be used here is to use the field inside a similar cylinder of infinite length. Then, an approximate scattered field is found by integrating the inner field over the finite-length of the cylinder of interest. Such an approach requires the length of the cylinder to be much larger than its diameter. This condition is usually satisfied in modeling the branch or the trunk of a tree.

The field inside of and scattered by an infinitely long dielectric cylinder responding to an incident plane wave of either vertical or horizontal polarization have been found by Wait [1955]. To permit arbitrary rotation of the cylinder with respect to a reference frame, Wait's results should be treated as derived in the principal frame of the cylinder. This local frame is related to the reference frame by the Euler transformation, (11.2), shown in the previous section. Hence, we must find the relation between the incident polarization vectors in the local and reference frame in order to recognize the field inside the cylinder due to an incident plane wave of either vertically or horizontally polarized plane wave in the reference frame. This is done in the next subsection. Then, the field inside the cylinder in the local frame is transformed to the reference frame using (11.2) and substituted into (11.1) to find the scattered field.

11.3.1 Relation Between Polarization Vectors in Local and Reference Frames

Consider a cylinder illuminated by a time-harmonic plane wave in $\hat{i}$ direction in the reference frame denoted by

$$\vec{E}^i(\vec{r}) = \hat{p}_i E_0 \exp(-j\vec{k}\cdot\vec{r}) \tag{11.54}$$

where the time dependence $\exp(j\omega t)$ is understood and the polarization vector $\hat{p}_i$ is equal to either $\hat{v}_i$ or $\hat{h}_i$. The polarization vectors and the incident direction $\hat{i}$ are defined in the reference frame by (11.14) in which θ_i and ϕ_i are, respectively, the incident polar and azimuth angles defined in a standard spherical coordinate system. This differs from the work by Wait [1955] in the sign of $\cos\theta_i$ because he defined θ_i with respect to the negative instead of positive z-axis. In the reference frame, the direction of the scattered field $\hat{s}$ and its associated polarization vectors $\hat{v}_s$ and $\hat{h}_s$ have been defined by (11.13).

In the local frame of the cylinder, the incident polarization vectors $\hat{v}_{il}$ and $\hat{h}_{il}$ can be obtained from (11.14) by replacing θ_i and ϕ_i by θ_{il} and ϕ_{il}, respectively,

$$\hat{h}_{il} = \frac{\hat{z}'' \times \hat{i}}{|\hat{z}'' \times \hat{i}|} = -\hat{x}''\sin\phi_{il} + \hat{y}''\cos\phi_{il}$$

$$\hat{v}_{il} = \hat{h}_{il} \times \hat{i} = \cos\theta_{il}(\hat{x}''\cos\phi_{il} + \hat{y}''\sin\phi_{il}) - \hat{z}''\sin\theta_{il} \tag{11.55}$$

where (θ_{il}, ϕ_{il}) is the incident pair of angles in the local frame. Note that the incident and scattering directions do not change but their components are different in different coordinate systems. The local angles θ_{il} and ϕ_{il} are governed by the following relations:

$$\cos\theta_{il} = \hat{z}'' \cdot \hat{i} = \cos\theta_i\cos\beta + \sin\theta_i\sin\beta\cos(\alpha - \phi_i)$$

$$\cos\phi_{il} = (\hat{x}'' \cdot \hat{i}) / [1 - (\hat{z}'' \cdot \hat{i})^2]^{0.5}$$

$$= [\cos\beta\sin\theta_i\cos(\alpha - \phi_i) - \sin\beta\cos\theta_i] / [1 - (\hat{z}'' \cdot \hat{i})^2]^{0.5} \tag{11.56}$$

To find the polarization vectors in the local frame in terms of the polarization vectors in the reference frame, we consider the relation between $\hat{z}''$ and $(\hat{i}, \hat{v}_{il_i}, \hat{h}_{il})$ defined as

$$\hat{z}'' = \cos\theta_{il}\hat{i} + c_{vi}\hat{v}_i + c_{hi}\hat{h}_i \tag{11.57}$$

where using (11.2), we find

$$c_{vi} = \hat{z}'' \cdot \hat{v}_i = \sin\beta\cos\theta_i\cos(\alpha - \phi_i) - \cos\beta\sin\theta_i \tag{11.58}$$

$$c_{hi} = \hat{z}'' \cdot \hat{h}_i = \sin\beta\sin(\alpha - \phi_i) \tag{11.59}$$

Note that

$$c_{vi}^2 + c_{hi}^2 = 1 - (\hat{z}'' \cdot \hat{i})^2 \tag{11.60}$$

To establish the relations between $(\hat{v}_{il}, \hat{h}_{il})$ and $(\hat{v}_i, \hat{h}_i)$ requires the following four dot products,

$$\hat{v}_i \cdot \hat{v}_{il} = \hat{v}_i \cdot \frac{(\hat{z}'' \cdot \hat{i})[(\hat{x}'' \cdot \hat{i})\hat{x}'' + (\hat{y}'' \cdot \hat{i})\hat{y}''] - \hat{z}''[1 - (\hat{z}'' \cdot \hat{i})^2]}{\sqrt{1 - (\hat{z}'' \cdot \hat{i})^2}}$$

$$= \hat{v}_i \cdot \frac{(\hat{z}'' \cdot \hat{i})\,\hat{i} - \hat{z}''}{\sqrt{(c_{vi}^2 + c_{hi}^2)}}$$

$$= \frac{-\hat{v}_i \cdot \hat{z}''}{\sqrt{(c_{vi}^2 + c_{hi}^2)}}$$

$$= \frac{-c_{vi}}{\sqrt{(c_{vi}^2 + c_{hi}^2)}} \tag{11.61}$$

$$\hat{h}_i \cdot \hat{v}_{il} = \hat{h}_i \cdot \frac{(\hat{z}'' \cdot \hat{i})\,\hat{i} - \hat{z}''}{\sqrt{(c_{vi}^2 + c_{hi}^2)}} = \frac{c_{hi}}{\sqrt{(c_{vi}^2 + c_{hi}^2)}} \tag{11.62}$$

$$\hat{v}_i \cdot \hat{h}_{il} = \hat{v}_i \cdot \frac{\hat{z}'' \times \hat{i}}{|\hat{z}'' \times \hat{i}|} = \frac{\hat{z}'' \cdot \hat{i} \times \hat{v}_i}{\sqrt{1 - (\hat{z}'' \cdot \hat{i})^2}}$$

$$= \frac{\hat{z}'' \cdot \hat{h}_i}{\sqrt{c_{vi}^2 + c_{hi}^2}} = \frac{c_{hi}}{\sqrt{c_{vi}^2 + c_{hi}^2}} \tag{11.63}$$

and

$$\hat{h}_i \cdot \hat{h}_{il} = \frac{\hat{z}'' \cdot \hat{i} \times \hat{h}_i}{\sqrt{1 - (\hat{z}'' \cdot \hat{i})^2}} = \frac{c_{vi}}{\sqrt{c_{vi}^2 + c_{hi}^2}} \tag{11.64}$$

The incident polarization vectors in the local frame ($\hat{v}_{il}$, $\hat{h}_{il}$) can be related to those in the reference frame ($\hat{v}_i$, $\hat{h}_i$) using (11.61) through (11.64) and are given by

$$\begin{bmatrix} \hat{v}_{il} \\ \hat{h}_{il} \end{bmatrix} = (c_{vi}^2 + c_{hi}^2)^{-0.5} \begin{bmatrix} c_{vi} & c_{hi} \\ -c_{hi} & c_{vi} \end{bmatrix} \begin{bmatrix} \hat{v}_i \\ \hat{h}_i \end{bmatrix} \equiv \begin{bmatrix} T_{vv} & T_{vh} \\ T_{hv} & T_{hh} \end{bmatrix} \begin{bmatrix} \hat{v}_i \\ \hat{h}_i \end{bmatrix} \tag{11.65}$$

which we can redefine as

$$\vec{p}_{il} = \mathbf{T} \cdot \vec{p}_i \tag{11.66}$$

where **T** is the transformation matrix for the polarization vectors to go from reference frame to local frame.

11.3.2 The Field Inside a Dielectric Cylinder

The field inside a dielectric cylinder of infinite length was obtained by Wait [1955] and adapted by Karam et al. [1988] to scattering from an arbitrarily oriented finite length cylinder. We shall first summarize the field inside an infinitely long dielectric cylinder in the principal frame of the cylinder. Then, we shall convert the local inner field to the reference frame and use it to calculate the scattered field in the reference frame from which we find the scattering amplitude. If we find the local scattering amplitude first and then convert it to the reference frame, the result should be equivalent, if the phase of the incident field is properly accounted for.

Field Inside an Infinite Cylinder in the Local Frame

From Wait [1955] the $\hat{q}$ polarized field inside an infinite dielectric cylinder due to a locally incident plane wave of amplitude, $\vec{E}_l$, along the direction specified by θ_{il} and ϕ_{il} in the local frame defined by (ρ'', ϕ'', z'') is [Karam et al., 1988]

$$E_{xq} = \sum_{n=-\infty}^{\infty} \{c_{nq} J_{n+1}(\lambda_i \rho'')\exp(j\phi'') + d_{nq} J_{n-1}(\lambda_i \rho'')\exp(-j\phi'')\}\ (\hat{q}\cdot\vec{E}_l)\, F_n$$

$$\equiv E_{xq}(\hat{q}\cdot\vec{E}_l)\exp[-j(kz''\cos\theta_{il})]$$

$$E_{yq} = -j\sum_{n=-\infty}^{\infty} \{c_{nq} J_{n+1}(\lambda_i \rho'')\exp(j\phi'') - d_{nq} J_{n-1}(\lambda_i \rho'')\exp(-j\phi'')\}\, F_n$$

$$\equiv E_{yq}(\hat{q}\cdot\vec{E}_l)\exp[-j(kz''\cos\theta_{il})]$$

$$E_{zq} = \sum_{n=-\infty}^{\infty} e_{nq} J_n(\lambda_i \rho'') F_n \equiv E_{zq}(\hat{q}\cdot\vec{E}_l)\exp[-j(kz''\cos\theta_{il})] \qquad (11.67)$$

where $J_n(\ )$ is the Bessel function; $\hat{q} = \hat{v}_{il}$ or $\hat{h}_{il}$.; and

$$F_n = j^{-n}\exp[jn(\phi'' - \phi_{il}) - jkz''\cos\theta_{il}]$$

$$\lambda_i = k(\varepsilon_r - \cos^2\theta_{il})^{0.5}$$

$$c_{nq} = 0.5k(\eta h_{nq} + je_{nq}\cos\theta_{il})/\lambda_i$$

$$d_{nq} = 0.5k(\eta h_{nq} - je_{nq}\cos\theta_{il})/\lambda_i$$

$$e_{nv} = \frac{j\sin\theta_{il}}{J_n(u)R_n}\left[\frac{H_n^{(2)\prime}(w)}{wH_n^{(2)}(w)} - \frac{\mu_r J_n'(u)}{uJ_n(u)}\right]$$

$$h_{nv} = \frac{n\cos\theta_{il}\sin\theta_{il}}{\eta J_n(u)R_n}\left[\frac{1}{w^2} - \frac{1}{u^2}\right]$$

$$R_n = \frac{\pi w^2}{2}H_n^{(2)}(w)\left\{\left[\frac{H_n^{(2)\prime}(w)}{wH_n^{(2)}(w)} - \frac{\varepsilon_r J_n'(u)}{uJ_n(u)}\right]\left[\frac{H_n^{(2)\prime}(w)}{wH_n^{(2)}(w)} - \frac{\mu_r J_n'(u)}{uJ_n(u)}\right]\right.$$

$$\left. -n^2\cos^2\theta_{il}\left(\frac{1}{w^2} - \frac{1}{u^2}\right)^2\right\}$$

$$e_{nh} = \frac{n\cos\theta_{il}\sin\theta_{il}}{J_n(u)R_n}\left[\frac{1}{w^2} - \frac{1}{u^2}\right]$$

$$h_{nh} = \frac{-j\sin\theta_{il}}{J_n(u)R_n}\left[\frac{H_n^{(2)\prime}(w)}{wH_n^{(2)}(w)} - \frac{\varepsilon_r J_n'(u)}{uJ_n(u)}\right]$$

$$u = \lambda_i r_0$$

$$w = \lambda r_0$$

$$\lambda = k\sin\theta_{il}$$

Note that in (11.67), ϕ'', z'' appear only in the phase and the prime on the Bessel and Hankel functions, $J_n'(u)$ and $H_n^{(2)\prime}(w)$, indicates differentiation with respect to the argument of the function; η is the intrinsic impedance in free space; ε_r and μ_r are the relative permittivity and permeability of the cylinder.

In the above relations $\vec{E}_l$ is the local incident field amplitude and part of its propagation phase is $\exp(-jkz''\cos\theta_{il})$ in the local frame. Note that, while the incident field amplitude decomposes into locally vertical and horizontal components, the incident propagation phase is the same whether it is expressed in the local or the reference frame. For a $\hat{p}_i$ polarized incident plane wave, from (11.66), $\vec{E}_l$ is given by

$$\vec{E}_l = \mathbf{T}\cdot\hat{p}_i \tag{11.68}$$

From (11.68) and (11.67) we can relate the field inside the cylinder in the local frame to the incident field vector in the reference frame as follows

$$\vec{E}_{in} = [(\hat{x}''E_{xq} + \hat{y}''E_{yq} + \hat{z}''E_{zq})\hat{q}] \cdot (\mathbf{T} \cdot \hat{p}_i) \exp(-jkz''\cos\theta_{il})$$

$$= (\vec{E}_q\hat{q}) \cdot (\mathbf{T} \cdot \hat{p}_i) \exp(-jkz''\cos\theta_{il}) \tag{11.69}$$

Note that $\vec{E}_q\hat{q}$ is a dyad and $(\mathbf{T} \cdot \hat{p}_i)$ gives the two components of the local incident field, $T_{vp}\hat{v}_{il}$ and $T_{hp}\hat{h}_{il}$ in the form of a column matrix. The amplitude product in (11.69) yields a sum of vectors and a more explicit form of the field inside the cylinder as

$$\vec{E}_{in} = (\vec{E}_v T_{vp} + \vec{E}_h T_{hp}) \exp(-jkz''\cos\theta_{il}) \tag{11.70}$$

11.3.3 Scattered Field and Scattering Amplitude of a Circular Cylinder

In this section we shall first calculate the scattered field and then derive the scattering amplitude from it. By applying the inverse Euler transformation from (11.3) to (11.69) we can write the inner field in the reference frame due to a $\hat{p}_i$ polarized incident plane wave as

$$\vec{E}_{inp} = \mathbf{U}^{-1} \cdot (\vec{E}_q\hat{q}) \cdot (\mathbf{T} \cdot \hat{p}_i) \exp(-jkz''\cos\theta_{il}) \tag{11.71}$$

Note that although $\vec{E}_{inp}$ is in the reference frame, $\vec{E}_q$ is still a function of variables in the principal frame of the cylinder. Hence, the integration to be carried out to find the scattered field is in the local coordinate system.

From (11.1) the $\hat{p}_s$ polarized component of the scattered field in the reference frame is

$$\hat{p}_s \cdot \vec{E}^s(\vec{r}) = \frac{k^2(\varepsilon_r - 1)}{4\pi} \int_V \frac{\exp(-jk|\vec{r} - \vec{r}'|)}{|\vec{r} - \vec{r}'|} (\hat{p}_s \cdot \vec{E}_{inp})\, dV \tag{11.72}$$

The scattered field component in the far zone is

$$\hat{p}_s \cdot \vec{E}^s(\vec{r}) \approx \frac{k^2(\varepsilon_r - 1)e^{-jkr}}{4\pi r} \hat{p}_s \cdot \mathbf{U}^{-1} \cdot \int_V \vec{E}_q\hat{q} \cdot (\mathbf{T} \cdot \hat{p}_i) \exp\left[j\vec{k}_s \cdot \vec{r}' - jkz''\cos\theta_{il}\right] dV$$

$$\equiv \hat{p}_s \cdot \left[\frac{k^2(\varepsilon_r - 1)}{4\pi} \mathbf{U}^{-1} \cdot \vec{I}_c\hat{q} \cdot \mathbf{T}\right] \cdot \hat{p}_i \frac{\exp(-jkr)}{r}$$

$$\equiv \hat{p}_s \cdot \mathbf{f}_c(\vec{k}_s, \vec{k}_i) \cdot \hat{p}_i \frac{\exp(-jkr)}{r} \tag{11.73}$$

where $f_c(\vec{k}_s, \vec{k}_i)$ is the quantity in the bracket and is the scattering amplitude matrix of the dielectric cylinder; and

$$\vec{I}_c = \int_V (\vec{E}_v T_{vp} + \vec{E}_h T_{hp}) \exp\left[j\vec{k}_s \cdot \vec{r}' - jkz''\cos\theta_{il}\right] dV \tag{11.74}$$

where the subscript p represents incident polarization in reference frame and the subscript in $\vec{E}_v$ denotes polarization in local frame. The T_{qp}'s in (11.74) are not functions of integration variables. From (11.67) the mathematical forms of $\vec{E}_v$ and $\vec{E}_h$ are the same. Hence, same types of integrals appear in both terms of (11.74).

Evaluation of Integrals in I_c

As in the previous sections on disc and needle scatterers, it is more convenient to evaluate the integrals in the local frame. Therefore, the phase factor in (11.74) is rewritten using (11.21) and (11.22) as follows:

$$\begin{aligned}
\exp\left[j\vec{k}_s \cdot \vec{r}' - jkz''\cos\theta_{il}\right] &= \exp\{jk[\hat{s}'' \cdot \vec{r}'' - z''\cos\theta_{il}]\} \\
&= \exp[jk(s''_x\rho''\cos\phi'' + s''_y\rho''\sin\phi'') + jk(s''_z - \cos\theta_{il})z''] \\
&\equiv \exp[j\lambda_s\rho''\cos(\phi'' - \phi_0) + jq''_z z'']
\end{aligned} \tag{11.75}$$

where $\hat{s}''$ is defined in (11.22), $\lambda_s = k(s''^2_x + s''^2_y)^{0.5}$ and ϕ_0 is the arc tangent of s''_y/s''_x. Substituting the above phase factor into (11.74) we can carry out the integration over the volume of the cylinder of length, L. The integral involving z'' is of the form

$$\begin{aligned}
I_z(\hat{s}, \hat{i}, \alpha, \beta) &= \int_{-L/2}^{L/2} \exp[jk(s''_z - \cos\theta_{il})z'']\,dz'' \\
&= L\sin[k(s''_z - \cos\theta_{il})L/2] / [k(s''_z - \cos\theta_{il})L/2]
\end{aligned} \tag{11.76}$$

The oscillations in the sine function above are not likely to occur in measurements. It could be advantageous to add a window function to reduce these oscillations.

From (11.67) and (11.75) the integration over ϕ'' takes the form

$$I_\phi(n \pm 1) = \int_0^{2\pi} \exp[j(n \pm 1)\phi''] \exp[j\lambda_q\rho''\cos(\phi'' - \phi_0)]\,d\phi'' \tag{11.77}$$

Note the relation for Bessel functions $J_n(-x) = (-1)^n J_n(x) = J_{-n}(x)$. Then, applying the formula for plane wave representation in terms of cylindrical wave functions [Balanis, 1989]

$$\exp[-j\beta\rho\cos\phi] = \sum_{m=-\infty}^{\infty} j^{-m} J_m(\beta\rho)\exp(jm\phi) \tag{11.78}$$

we have the following relation,

$$\begin{aligned} I_\phi &= \sum_{m=-\infty}^{\infty} \int_0^{2\pi} \exp[j(n\pm1)\phi''] j^m J_m(\lambda_s\rho'')\exp[-jm(\phi''-\phi_0)]\,d\phi'' \\ &= 2\pi j^{n\pm1} J_{n\pm1}(\lambda_s\rho'')\exp[j(n\pm1)\phi_0] \\ &\equiv I_\phi(n\pm1)\,J_{n\pm1}(\lambda_s\rho'') \end{aligned} \tag{11.79}$$

when $m = n\pm1$. Otherwise, I_ϕ is equal to zero. Note that because the integrand of ϕ'' contains ρ'' which will be integrated out later, we define the integrated result of ϕ'' excluding the ρ'' dependence part as $I_\phi(n\pm1)$. Finally, let r_0 be the radius of the cylinder. From (11.67) and (11.79), the integration over ρ'' takes the form [Karam et al., 1988],

$$\begin{aligned} I_\rho(n) &= \int_0^{r_0} J_n(\lambda_i\rho'')J_n(\lambda_s\rho'')\,\rho''d\rho'' \\ &= [r_0/(\lambda_i^2-\lambda_s^2)]\,[\lambda_i J_n(\lambda_s r_0)\,J_{n+1}(\lambda_i r_0) - \lambda_s J_n(\lambda_i r_0)\,J_{n+1}(\lambda_s r_0)] \end{aligned} \tag{11.80}$$

From (11.76), (11.79) and (11.80) we see that the integration over the volume of the cylinder comes to a closed form. The use of these integrated results allows us to write the component of a term in $\vec{I}_c$ of (11.74) as follows:

$$I_{cx} = \sum_{n=-\infty}^{\infty} \{c_{nq}I_\phi(n+1)\,I_\rho(n+1) + d_{nq}I_\phi(n-1)\,I_\rho(n-1)\}\,Z_{qp} \tag{11.81}$$

$$I_{cy} = \sum_{n=-\infty}^{\infty} \{d_{nq}I_\phi(n-1)\,I_\rho(n-1) - c_{nq}I_\phi(n+1)\,I_\rho(n+1)\}\,jZ_{qp} \tag{11.82}$$

$$I_{cz} = \sum_{n=-\infty}^{\infty} \{e_{nq} I_{\phi}(n) I_{\rho}(n)\} Z_{qp} \tag{11.83}$$

where $Z_{qp} = j^{-n} I_z(\hat{s}, \hat{\imath}, \alpha, \beta) T_{qp}$

11.3.4 Extinction Cross Section of a Randomly Oriented Cylinder

According to the forward scattering theorem, the extinction cross section of a scatterer can be written in terms of its scattering amplitude for a $\hat{p}_i$ polarized incident wave as [Karam and Fung, 1982]

$$\sigma_{ep}(\hat{\imath}) = (4\pi/k) < \mathrm{Im}[f_{pp}(\hat{\imath}, \hat{\imath})] > \tag{11.84}$$

where Im( ) denotes the operation for taking the imaginary part and $\hat{\imath}$ represents the direction of propagation of the incident wave. In (11.84), $\hat{s} = \hat{\imath}$ and from (11.76) $I_z = L$. The dimension of $\sigma_{ep}(\hat{\imath})$ is in meters squared. The extinction coefficient K_{ep} in the radiative transfer theory can be found by multiplying $\sigma_{ep}(\hat{\imath})$ by the number density n_0. This is the more practical way to compute the extinction coefficient needed in the radiative transfer equation than the use of (11.49) through (11.53).

11.4 SINGLE AND DOUBLE SCATTERING FROM A TWO-LAYER MEDIUM

In Chapter 2 we considered single scattering solutions to a one-layer problem, which is useful for many applications. For a forested medium generally a two-layer model is needed because of the presence of a crown and a trunk layer. Furthermore, double scattering should be included to obtain a better estimate of cross polarization. If we use the formulation in Chapter 8, scattering of all orders are included but it is not possible to identify the contributions of individual scatterers. For this reason we summarize the scattering model reported in Karam et al. [1992] in this section. Here, a vegetated medium is modelled as a crown layer and a trunk layer situated above a rough ground surface as illustrated in Figure 11.2, where single scatterings from the ground, the trunk layer and the crown layer are designated as 1, 2 and 3 and double scatterings from trunk-ground, crown-ground and crown-crown are indicated by 4, 5 and 6. Note that in Karam et al. [1992] the incident and scattered angles are kept less than or equal to 90 degrees. When they exceed 90 degrees, they are replaced by $\pi - \theta_i$ or $\pi - \theta_s$ accordingly. Such a change is not included in the previous sections, where we derived the phase matrix elements and we stayed with the standard spherical coordinates throughout.

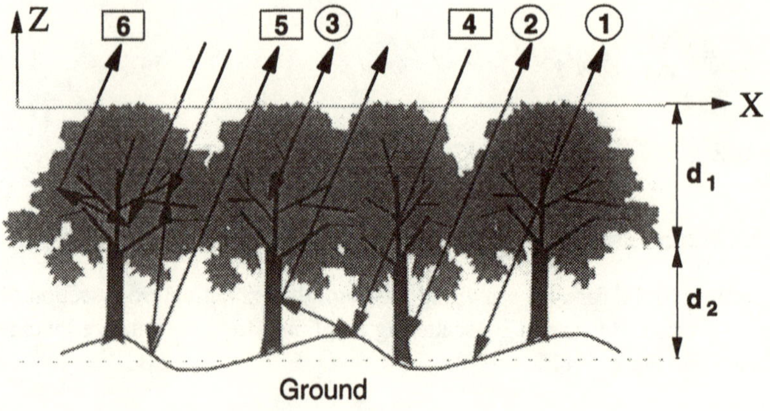

Figure 11.2 Geometry of a two-layer vegetated medium.

11.4.1 Single Scattering Terms

There are three single scattering terms as indicated in Figure 11.2. One comes from soil surface scattering σ^s_{pq}, one from the crown layer σ^c_{pq} and another from the trunk layer σ^t_{pq}. Generally, one type of scatterer is involved in σ^s_{pq} and σ^t_{pq}. However, several types should appear in σ^c_{pq}, corresponding to leaves and branches of different sizes. Thus, the total single scattering coefficient with incident field polarization q and scattered field polarization p may be written as

$$\sigma^0_{spq} = \sigma^s_{pq} + \sigma^c_{pq} + \sigma^t_{pq} \tag{11.85}$$

More explicit forms of the terms in (11.85) are given in the subsections to follow. We give bistatic rather than backscattering expressions because they are needed later for the double scattering problem.

Surface Scattering Term

The surface scattering coefficient in (11.85) may be represented by the surface scattering coefficient given in Chapter 2 multiplied by the attenuation due to passage through the trunk and the crown layers as

$$\sigma^s_{pq} = L_{1p}(\theta_s)\, L_{2p}(\theta_s)\, \sigma_{pq}(S, \theta_s, \phi_s; \pi - \theta_i, \phi_i)\, L_{2q}(\theta_s)\, L_{1q}(\theta_s) \tag{11.86}$$

where $\sigma_{pq}(S, \theta_s, \phi_s; \pi - \theta_i, \phi_i)$ is the pq element of the surface bistatic scattering coefficient matrix given in Appendix 2A. For a nearly plane surface this quantity may be replaced by the power Fresnel reflection coefficient [Ulaby et al., 1982, Appendix 12D of Chapter 12] modified by $\exp\left[-k^2\sigma^2(\cos\theta_s + \cos\theta_i)^2\right]$. The form of the attenuation factors is

$$L_{lq}(\theta_m) = \exp[-k_{lq}(\theta_m) d_l \sec\theta_m], \qquad l = 1, 2 \tag{11.87}$$

where $k_{nq}(\theta_m)$ is the extinction coefficient of layer l; d_n is the height of layer l. In general, the extinction coefficient for a layer may be written as the sum of the products of the number density times the extinction coefficient of each type of scatterers:

$$k_{lq}(\theta_m) = \sum_{j=1}^{N_l} n_j \langle \kappa_{jq}(\theta_m) \rangle \tag{11.88}$$

where N_l is the number of types of scatterers within layer l; n_j is the number density and $\kappa_{jq}(\theta_m)$ is the extinction cross section of the jth type of scatterer and $\langle\ \rangle$ denotes averaging over the orientation angles α and β of the scatterer.

Crown Scattering Term

The crown scattering term consists of contributions by several types of scatterers modified by an average attenuation. This attenuation is found by integrating the loss factors of the incident and scattered intensities, $\exp[k_{1q}(\theta_i) z \sec\theta_i]$ and $\exp[k_{1p}(\theta_s) z \sec\theta_s]$ from $-d_1$ to 0. The bistatic scattering coefficient for the crown layer may be written as

$$\sigma^c_{pq} = 4\pi Q_{1pq}(\theta_s, \phi_s; \pi - \theta_i, \phi_i) \frac{1 - L_{1p}(\theta_s) L_{1q}(\theta_i)}{k_{1q}(\theta_i)\sec\theta_i + k_{1p}(\theta_s)\sec\theta_s} \tag{11.89}$$

where

$$Q_{1pq}(\theta_s, \phi_s; \pi - \theta_i, \phi_i) = \sum_{j=1}^{N_1} n_j \langle \left| f_{jpq}(\vec{k}_s, \vec{k}_i) \right|^2 \rangle \tag{11.90}$$

and $f_{jpq}(\vec{k}_s, \vec{k}_i)$ is the pq element of the scattering amplitude matrix of the jth type scatterer within the crown layer discussed in the previous two sections. The product $4\pi \left| f_{jpq}(\vec{k}_s, \vec{k}_i) \right|^2$ is the cross section of the scatterer.

Trunk Scattering Term

This term is mathematically similar to the crown scattering term except for the additional attenuation factor needed to pass through the crown layer:

$$\sigma^t_{pq} = 4\pi L_{2p}(\theta_s)\, L_{2q}(\theta_i)\, Q_{2pq}(\theta_s, \phi_s; \pi - \theta_i, \phi_i) \frac{1 - L_{2p}(\theta_s)\, L_{2q}(\theta_i)}{k_{2q}(\theta_i)\sec\theta_i + k_{2p}(\theta_s)\sec\theta_s} \tag{11.91}$$

where $Q_{2pq}(\theta_s, \phi_s; \pi - \theta_i, \phi_i)$ is defined in the same way as (11.90) except the scattering amplitudes and number densities are for the scatterers in the trunk layer. Generally, only one size trunk is assumed and hence the summation is not necessary.

11.4.2 Double Scattering Terms

As illustrated in Figure 11.2 there are three types of double scattering terms: crown-ground, trunk-ground and crown-crown interactions. That is,

$$\sigma^0_{dpq} = \sigma_{pq}(c \leftrightarrow g) + \sigma_{pq}(t \leftrightarrow g) + \sigma_{pq}(c \leftrightarrow c) \tag{11.92}$$

The first two terms are important at low frequencies when signal easily reaches the ground. This is particularly important for horizontal polarization because it is not affected by the Brewster angle, which reduces the magnitude of such interactions. The crown-crown interaction term is small at low frequencies due to small albedo, but it is important at higher frequencies when the volume scattering albedo is significant and when we are interested in the cross polarized backscattering.

Crown-Ground Interaction Term

This term originates from two mechanisms: incident signal scattered by the ground first and then scattered into the direction of interest by the crown layer or vice versa. That is,

$$\sigma_{pq}(c \leftrightarrow g) = \sigma_{pq}(c \rightarrow g) + \sigma_{pq}(g \rightarrow c) \tag{11.93}$$

Consider the first term in (11.93). For the incident signal to reach a volume scatterer represented by $Q_{1uq}(\pi - \theta, \phi; \pi - \theta_i, \phi_i)$ located at a point z inside the crown region, it must be attenuated by $\exp[k_{1q}(\theta_i)\, z\sec\theta_i]$. Then, as it is scattered to the crown-trunk interface, it is attenuated further by the factor $\exp[-k_{1u}(\theta)\,(d_1 + z)\sec\theta]$. The integration of the product of these two factors with respect to z from $-d$ to 0 yields the overall attenuation factor in the crown region. Finally, this incident signal must be attenuated by the trunk layer, scattered by the ground and attenuated by both trunk and crown layers before it comes back out. Mathematically, we have

$$\sigma_{pq}(c \to g) = L_{1p}(\theta_s) L_{2p}(\theta_s) \int_0^{2\pi} d\phi \int_0^{\pi/2} \sin\theta \sec\theta_s d\theta$$

$$\sum_{u = v, h} \sigma^s_{pu}(\theta_s, \phi_s; \pi - \theta, \phi) Q_{1uq}(\pi - \theta, \phi; \pi - \theta_i, \phi_i)$$

$$L_{2u}(\theta) \frac{L_{1u}(\theta) - L_{1q}(\theta_i)}{k_{1q}(\theta_i) \sec\theta_i - k_{1u}(\theta) \sec\theta} \tag{11.94}$$

In (11.94) the $\sec\theta_s$ combines with the surface scattering coefficient to form the phase function for surface. The 4π associated with the cross section of the scatterer cancels with the 4π associated with the integration over the solid angle. Similarly, the other term is given by

$$\sigma_{pq}(g \to c) = L_{1q}(\theta_i) L_{2q}(\theta_i) \int_0^{2\pi} d\phi \int_0^{\pi/2} \sin\theta \sec\theta_i d\theta$$

$$\sum_{u = v, h} Q_{1pu}(\theta_s, \phi_s; \theta, \phi) \sigma^s_{uq}(\theta, \phi; \pi - \theta_i, \phi_i) L_{2u}(\theta) \frac{L_{1p}(\theta_s) - L_{1u}(\theta)}{k_{1u}(\theta) \sec\theta - k_{1p}(\theta_s) \sec\theta_s} \tag{11.95}$$

A case of special interest is when we consider backscattering and when the ground surface appears to be slightly rough under the exploring wavelength. Under these conditions reciprocity holds, and only the coherent scattering coefficient is important. This scattering coefficient is given by

$$\sigma^c_{qq}(\theta_i, \phi_i; \pi - \theta_i, \phi_i) = 4\pi \cos\theta_i R_{qq}(\theta_i) \exp\left[-(2k\sigma\cos\theta_i)^2\right]$$

$$\delta(\cos\theta - \cos\theta_i)\, \delta(\phi - \phi_i) \tag{11.96}$$

The integration over the surface scattering coefficient reduces to the power Fresnel reflection coefficient modified by the surface rms height $R_{qq}(\theta_i) \exp\left[-(2k\sigma\cos\theta_i)^2\right]$ [Ulaby et al., 1982, Chapter 12, Appendix 12D]. Hence, (11.93) becomes

$$\sigma_{pq}(c \leftrightarrow g) = 8\pi \cos\theta_i L_{1q}(\theta_i) L^2_{2q}(\theta_i) R_{qq}(\theta_i) \exp\left[-(2k\sigma\cos\theta_i)^2\right]$$

$$Q_{1pq}(\theta_i, \phi_i + \pi; \theta_i, \phi_i) \frac{L_{1p}(\theta_i) - L_{1q}(\theta_i)}{k_{1q}(\theta_i) - k_{1p}(\theta_i)} \tag{11.97}$$

When the incident and scattered polarizations are the same, the ratio in (11.97) is singular and it can be replaced by its limiting value as

$$L_{1p}(\theta_i)\frac{1-\exp\{-[k_{1q}(\theta_i)-k_{1p}(\theta_i)]d_1\sec\theta_i\}}{k_{1q}(\theta_i)-k_{1p}(\theta_i)} = d_1\sec\theta_i L_{1p}(\theta_i)$$

This leads to a scattering for like polarization of the form

$$\sigma_{qq}(c\leftrightarrow g) = 8\pi d_1 L_{1q}^2(\theta_i)L_{2q}^2(\theta_i)R_{qq}(\theta_i)\exp\left[-(2k\sigma\cos\theta_i)^2\right]$$
$$Q_{1pq}(\theta_i,\phi_i+\pi;\theta_i,\phi_i) \tag{11.98}$$

Trunk-Ground Interaction Term

Similar to the crown-ground interaction term the trunk-ground interaction also consists of two terms as

$$\sigma_{pq}(t\leftrightarrow g) = \sigma_{pq}(t\to g)+\sigma_{pq}(g\to t) \tag{11.99}$$

The first term represents scattering by the trunk followed by scattering by the ground, and the second term has a reverse order relative to the first term. The development of these terms is similar to the crown-ground interaction term. The major change is a switching from layer 1 to layer 2 by changing subscripts and vice versa. Note also that the loss in layer 1 involves only one polarization at a time. The final results are

$$\sigma_{pq}(t\to g) = L_{1p}(\theta_s)L_{2p}(\theta_s)L_{1q}(\theta_i)\int_0^{2\pi}d\phi\int_0^{\pi/2}\sin\theta\sec\theta_s d\theta$$
$$\sum_{u=v,h}\sigma_{pu}^s(\theta_s,\phi_s;\pi-\theta,\phi)Q_{2uq}(\pi-\theta,\phi;\pi-\theta_i,\phi_i)\frac{L_{2u}(\theta)-L_{2q}(\theta_i)}{k_{2q}(\theta_i)\sec\theta_i-k_{2u}(\theta)\sec\theta}$$
$$\tag{11.100}$$

$$\sigma_{pq}(g\to t) = L_{1q}(\theta_i)L_{2q}(\theta_i)L_{1p}(\theta_s)\int_0^{2\pi}d\phi\int_0^{\pi/2}\sin\theta\sec\theta_i d\theta$$
$$\sum_{u=v,h}Q_{2pu}(\theta_s,\phi_s;\theta,\phi)\sigma_{uq}^s(\theta,\phi;\pi-\theta_i,\phi_i)\frac{L_{2p}(\theta_s)-L_{2u}(\theta)}{k_{2u}(\theta)\sec\theta-k_{2p}(\theta_s)\sec\theta_s} \tag{11.101}$$

Again a case of special interest is in backscattering and when the surface is slightly rough. In this case we have

$$\sigma_{pq}(t\leftrightarrow g) = 8\pi\cos\theta_i L_{1q}(\theta_i)L_{1p}(\theta_i)L_{2p}(\theta_i)R_{qq}(\theta_i)\exp\left[-(2k\sigma\cos\theta_i)^2\right]$$

$$Q_{2pq}(\pi-\theta_i,\phi_i+\pi;\pi-\theta_i,\phi_i)\frac{L_{2p}(\theta_i)-L_{2q}(\theta_i)}{k_{2q}(\theta_i)-k_{2p}(\theta_i)} \tag{11.102}$$

In like polarization we should replace the ratio in (11.102) by its limiting form yielding

$$\sigma_{qq}(t\leftrightarrow g) = 8\pi d_2 L_{1q}^2(\theta_i)L_{2q}^2(\theta_i)R_{qq}(\theta_i)\exp\left[-(2k\sigma\cos\theta_i)^2\right] \\ Q_{2qq}(\pi-\theta_i,\phi_i+\pi;\pi-\theta_i,\phi_i) \tag{11.103}$$

Crown-Crown Interaction Term

In surface scattering the cross polarized component is generally much smaller than the like in the backscattering direction and the like and cross polarized fields are not strongly correlated. There is also no correlation between the volume scattered field and the surface scattered field. However, in crown-crown interaction simplification due to the above stated properties do not exist. In fact, we need to include all four Stokes parameters due to the nonzero correlation between like and cross polarized scattered fields.

Consider case (a) in Figure 11.3. We denote it by the symbol $\sigma_{pq}(u,u,d)$, which represents a downward incident signal scattered by a unit volume upward and then scattered again upward into the direction of observation. For the first two Stokes parameters these scattering processes are represented by $Q_{1pu}(\theta_s,\phi_s;\theta,\phi)$ times $Q_{1uq}(\theta,\phi;\theta_i,\phi_i)$ and for the last two Stokes parameters by $Q_{1p3}(\theta_s,\phi_s;\theta,\phi)$ times $Q_{13q}(\theta,\phi;\theta_i,\phi_i)$. To reach the first scattering volume the incident signal is attenuated by $\exp[k_{1q}(\theta_i)z'\sec\theta_i]$. In going from the first to the second scattering volume, the signal is further attenuated by $\exp[k_{1u}(\theta)(z'-z)\sec\theta]$. After the second scattering the scattered signal with p polarization is attenuated by $\exp[k_{1p}(\theta_s)z\sec\theta_s]$ before it reaches the crown-air interface. The integration of the products of these attenuation factors over dz' $(-d_1\le z'\le z)$ and dz $(-d_1\le z\le 0)$ gives the loss factors associated with the above stated scattering processes. Mathematically, we have

$$\sigma_{pq}(u,u,d) = 4\pi\int_0^{2\pi}d\phi\int_0^{\pi/2}\sin\theta d\theta$$

$$\left\{\sum_{u=v,h}\frac{\sec\theta_s Q_{1pu}(\theta_s,\phi_s;\theta,\phi)Q_{1uq}(\theta,\phi;\pi-\theta_i,\phi_i)}{k_{1q}(\theta_i)\sec\theta_i+k_{1u}(\theta)\sec\theta}\right.$$

$$\left[\frac{1-L_{1p}(\theta_s)L_{1q}(\theta_i)}{k_{1q}(\theta_i)\sec\theta_i+k_{1p}(\theta_s)\sec\theta_s}+\frac{L_{1q}(\theta_i)[L_{1u}(\theta)-L_{1p}(\theta_s)]}{k_{1u}(\theta)\sec\theta-k_{1p}(\theta_s)\sec\theta_s}\right]$$

$$+2\text{Re}\left(\frac{\sec\theta_s Q_{1p3}(\theta_s,\phi_s;\theta,\phi)\,Q_{13q}(\theta,\phi;\pi-\theta_i,\phi_i)}{k_{1q}(\theta_i)\sec\theta_i+k_{13}(\theta)\sec\theta}\right.$$

$$\left.\left.\left[\frac{1-L_{1p}(\theta_s)L_{1q}(\theta_i)}{k_{1q}(\theta_i)\sec\theta_i+k_{1p}(\theta_s)\sec\theta_s}+\frac{L_{1q}(\theta_i)\,[L_{13}(\theta)-L_{1p}(\theta_s)]}{k_{13}(\theta)\sec\theta-k_{1p}(\theta_s)\sec\theta_s}\right]\right)\right\} \quad (11.104)$$

The summation in (11.104) is over the first two Stokes parameters and the Re( ) operator comes from summing the last two Stokes parameters. The integration over $d\theta$ and $d\phi$ is to include all possible scattering directions through which the signal scattered from location z' towards location z. The contents of Q_{1u3} and Q_{13u} are as follows:

$$Q_{1u3}(\theta_s,\phi_s;\theta,\phi) = \sum_{j=1}^{N_1} n_j\langle f_{juv}(\vec{k}_s,\vec{k}_i)\,f_{juh}{}^*(\vec{k}_s,\vec{k})\rangle$$

$$Q_{13u}(\theta_s,\phi_s;\theta,\phi) = \sum_{j=1}^{N_1} n_j\langle f_{jvu}(\vec{k}_s,\vec{k}_i)\,f_{jhu}{}^*(\vec{k}_s,\vec{k})\rangle \quad (11.105)$$

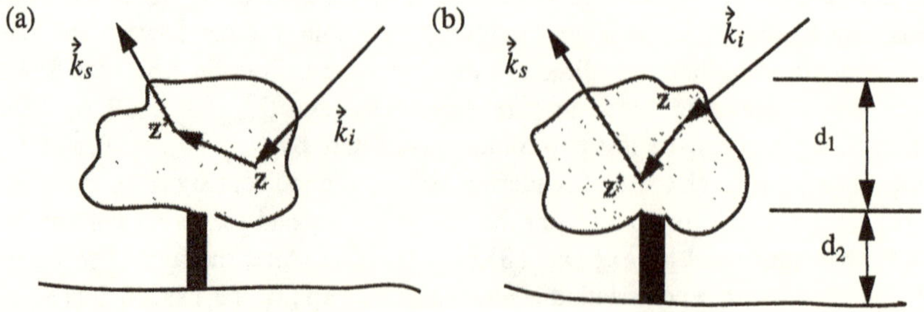

Figure 11.3 Illustration of major terms in the crown-crown interaction: (a) $\sigma(u, u, d)$ and (b) $\sigma(u, d, d)$.

From part (b) of Figure 11.3 we can obtain a similar expression as (11.104) to represent the scattering process where the incident signal is first scattered downward and then it is scattered upward into the direction of observation as

$$\sigma_{pq}(u,d,d) = 4\pi\int_0^{2\pi}d\phi\int_0^{\pi/2}\sin\theta d\theta$$

$$\left\{\sum_{u=v,h}\frac{\sec\theta_s Q_{1pu}(\theta_s,\phi_s;\pi-\theta,\phi)\,Q_{1uq}(\pi-\theta,\phi;\pi-\theta_i,\phi_i)}{k_{1p}(\theta_s)\sec\theta_s+k_{1u}(\theta)\sec\theta}\right.$$

$$\left[\frac{1-L_{1p}(\theta_s)L_{1q}(\theta_i)}{k_{1q}(\theta_i)\sec\theta_i+k_{1p}(\theta_s)\sec\theta_s}+\frac{L_{1p}(\theta_s)\left[L_{1u}(\theta)-L_{1q}(\theta_i)\right]}{k_{1u}(\theta)\sec\theta-k_{1q}(\theta_i)\sec\theta_i}\right]$$

$$+2Re\left(\frac{\sec\theta_s Q_{1p3}(\theta_s,\phi_s;\pi-\theta,\phi)\,Q_{13q}(\pi-\theta,\phi;\pi-\theta_i,\phi_i)}{k_{1p}(\theta_s)\sec\theta_s+k_{13}(\theta)\sec\theta}\right.$$

$$\left.\left.\left[\frac{1-L_{1p}(\theta_s)L_{1q}(\theta_i)}{k_{1q}(\theta_i)\sec\theta_i+k_{1p}(\theta_s)\sec\theta_s}+\frac{L_{1p}(\theta_s)\left[L_{13}(\theta)-L_{1q}(\theta_i)\right]}{k_{13}(\theta)\sec\theta-k_{1q}(\theta_i)\sec\theta_i}\right]\right)\right\} \quad (11.106)$$

Note that the form of the terms from summing the first two Stokes parameters is analogous to that of the sum of the last two Stokes parameters. The sum of $\sigma_{pq}(u, u, d)$ and $\sigma_{pq}(u, d, d)$ is expected to be important for cross polarization especially when the albedo of the crown layer is significant.

The extinction coefficient for the third Stokes parameter and the associated attenuation factor that appear in (11.104) and (11.104) are defined as [Karam et al., 1992]

$$k_{13}(\theta)=0.5\left[k_{1v}(\theta)+k_{1h}(\theta)\right]+\frac{2\pi j}{k}\mathrm{Re}\left[\sum_{m=1}^{N_1} n_m\langle f_{mvv}(\vec{k}_i,\vec{k}_i)-f_{mhh}(\vec{k}_i,\vec{k}_i)\rangle\right] \quad (11.107)$$

$$L_{13}(\theta) = \exp\left[-k_{13}(\theta)\,d_1\sec\theta\right] \quad (11.108)$$

11.5 THEORETICAL MODEL CHARACTERISTICS

In this section we consider the theoretical backscattering characteristics of coniferous and deciduous trees. We use three specific sites where significant ground truth has been gathered. The three sites are the Freiburg, the Landes and a walnut orchard. The first two sites have coniferous trees with average needle lengths of 1.5 and 15 cm, respectively, and the last site has deciduous trees with an average leaf radius of 3.6 cm. Synthetic aperture radar (SAR) measurements are available from the first two sites at 0.44 GHz (P-band), 1.22 GHz (L-band), and 5.3 GHz (C-band), while truck measurements at L- and X-bands were taken over the walnut orchard. To give a more complete picture of the backscattering properties of trees, model calculations at 9.6 GHz in addition to the P-, L- and C-bands are performed for coniferous trees to show theoretical characteristics over incident angles from 20 to 65 degrees. This is followed by comparisons with backscattering measurements at an incident angle of nominally 45 degrees as a function of age and biomass. Incident angular comparisons of the backscattering coefficient with orchard cypress trees are also shown in Section 11.6.

11.5.1 Coniferous Vegetation

The Freiburg forest is of coniferous type situated in the southwestern part of Germany. The major tree species for this area are homogeneous stands of Picea abies, Abies alba and Pinus sylvestris. They range in age from 80 to 100 years and height from 30 to 40 meters. The understory of the stands is composed of young blueberry, spruce and fir. Its needles are short typically 1.5 cm in length. Additional ground truth information is summarized in Tables 11.1 and 11.2.

Table 11.1
General Ground Truth

Ground truth parameters	*Values or range*	
Frequency range	P- through X-bands	Simulation only
Angular range	20° to 65°	Simulation only
Surface roughness	rms height = 2.0 cm	Correlation length = 15 cm
Surface moisture content (SMC)	0.3	
Top layer height	1.50 m	
Middle layer height	15.00 m	
Lower layer height	10. m	

Table 11.2
Needle and Branch Parameters

	Top layer		*Middle layer*				*Bottom*
	Needle 1	*Sm brch*	*Needle 2*	*Sm brch*	*Md brch*	*Lg brch*	*Trunk*
Radius, cm	0.1	0.2	0.1	0.2	0.35	0.85	15.0
1/2 length, cm	0.75	10.0	0.75	10.0	65.0	110.00	575.0
Density, m-3	1000.0	1.60	2000.0	5.80	2.90	0.89	0.005
Gravimetric moisture content	0.65	0.55	0.65	0.55	0.55	0.55	0.50
β_0	0.00	0.00	0.00	0.00	20.0	30.0	0.00
β_1	0.00	20.0	0.00	20.0	20.0	30.0	0.00
β_2	90.0	90.0	90.0	90.0	90.0	90.0	10.0
β_m	90.0	55.0	90.0	55.0	65.0	70.0	5.00
Exponent, n	0.00	1.00	0.00	1.00	2.00	1.00	6.00

Theoretical studies showing the contributions of different forest components are calculated using model parameter values in accordance with Tables 11.1 and 11.2. The reported ground truth assumes that the forest canopy can be separated into three layers: a top layer of 1.5 m containing mainly needles and twigs, a middle layer (10 m) of short and long branches, and a bottom layer of approximately vertical trunks (11.5 m). The size of the needles are the same in different layers but the branch sizes vary in radius and length between and within the layers. Information regarding the angular distributions of branches, leaves, and trunk is not available from measurements. The results in Table 11.2 are based on visual observations described by the angular distribution function,

$$p(\beta) = A\cos^n\left[\frac{\pi}{2}\left(\frac{\beta-\beta_m}{\beta_0-\beta_m}\right)\right], \quad (\beta_1 \le \beta \le \beta_2) \tag{11.109}$$

The dielectric constant is estimated from empirical formula given in Appendix 11B.

The ground truth given in the tables uses a three-layer arrangement. The top layer, about 1.5 m high, contained needles and small branches. The middle layer, about 15 m high, contained needles, small branches, mid-size branches and large branches. The lower layer, about 10 m high, contained the trunk for a total of seven types of scatterers. As it turns out, the top layer has very little effect at P- and L-bands because its scatterers are small. In the study to follow contribution by ground alone is not considered.

To illustrate backscattering as a function of frequency, polarization, incident angle, and scatterers in different layers we use 3D bar charts and the scattering model reported by Karam et al. [1991], which is an extension of the model given in Section 11.4. *Except for a difference in the amount of attenuation, single and double scattering from three layers uses the same formulae as those given in Section 11.4.* The vegetation components along the scatterer-axis in all the 3D bar charts for the Freiburg site are, from left to right, needles and small branches in the top layer, needles, small branches, medium branches and large branches in the middle layer, and trunk in the bottom layer. The other two axes are for the incident angle and the backscattering coefficient. Volume scattering contributions and volume-ground interactions are shown separately at P-band for both vertical and horizontal polarizations because at this frequency the effect of interaction is very significant (Figures 11.4 and 11.5). At L-band and higher frequencies attenuation is much larger so that the interaction terms become much less important even for HH polarization. Thus, only the total contribution by individual scatterers at VV and HH polarizations are shown (Figures 11.6 to 11.7).

From Figures 11.4 and 11.5 we see that at P-band volume scattering is dominated by the large branches, while volume-ground interaction is dominated by the trunk-ground component. For VV polarization the volume scattering term by large branches is about 10 dB higher than the interaction term at large angles of incidence but they are comparable in size at small angles of incidence. For HH polarization the difference between these terms is about 1 dB at large angles of incidence, and the interaction term becomes the largest contributor at small angles of incidence. Thus, the interaction term is very important for HH polarization. The reason why it is not as important for VV is because of the Brewster angle effect. All contributions from other scatterers are more than 20 dB lower for all angles of incidence between 20 and 65 degrees.

At L-band (Figure 11.6) the major contributors to scattering are again the large branches. For VV polarization all other contributors are at least 10 dB lower, but for HH polarization there is some contribution from volume-ground interaction, which is around

6 dB lower than that from the large branches at small angles of incidence. Very significant changes take place at C-band (Figure 11.7). For VV polarization at large angles of incidence needles and large branches are the most important followed by mid-size branches. At small angles of incidence the contributions of the medium and large branches are slightly larger than the needles. For HH polarization important contributors are the same as VV except at large angles of incidence where the mid-size branches contribute more than the needle and the large branches. The reason why the needles are important is because its number density is more than two orders of magnitude larger.

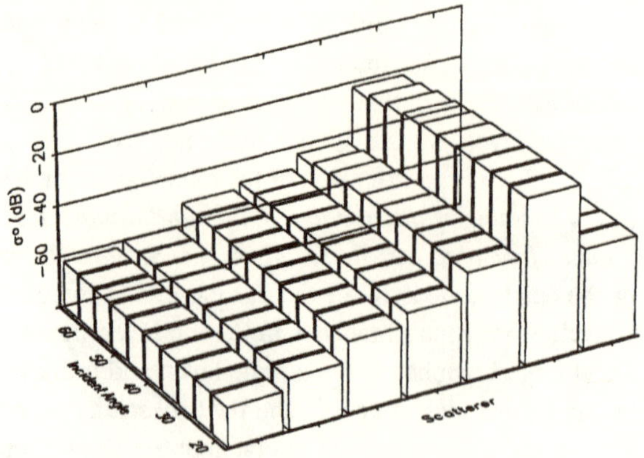

(a) volume scattering with VV polarization

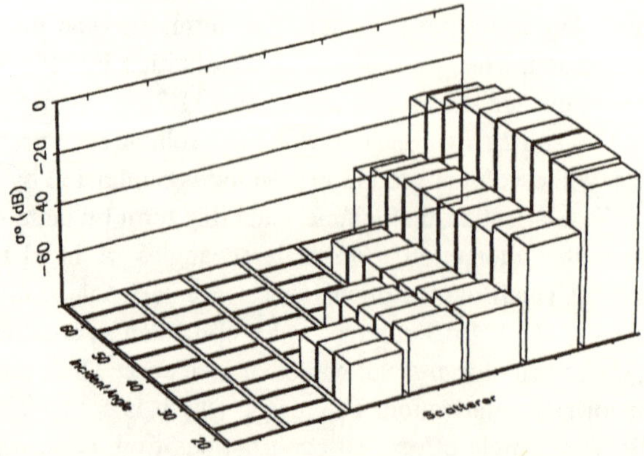

(b) volume-ground interaction with VV polarization

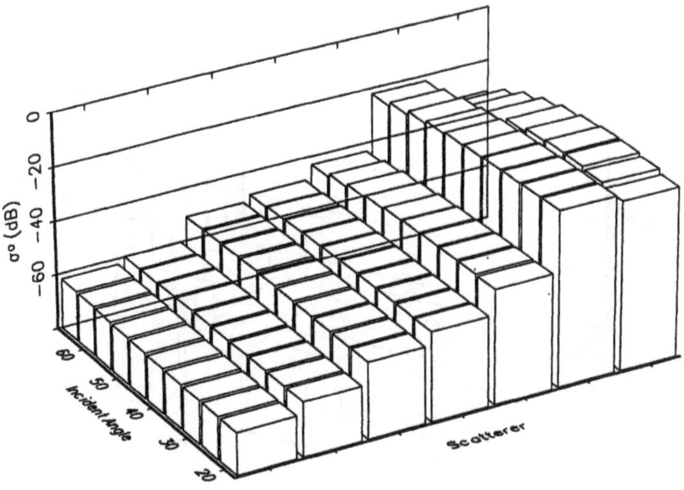

(c) sum of (a) and (b)

Figure 11.4 Freiburg site: Scattering contributions of tree components at P-band with VV polarization. (a) vloume scattering, (b) volume-ground interaction and (c) sum of parts (a) and (b).

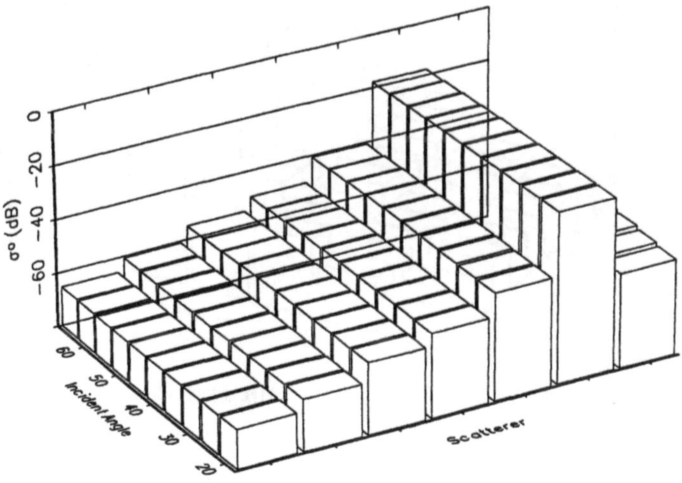

(a) volume scattering with HH polarization

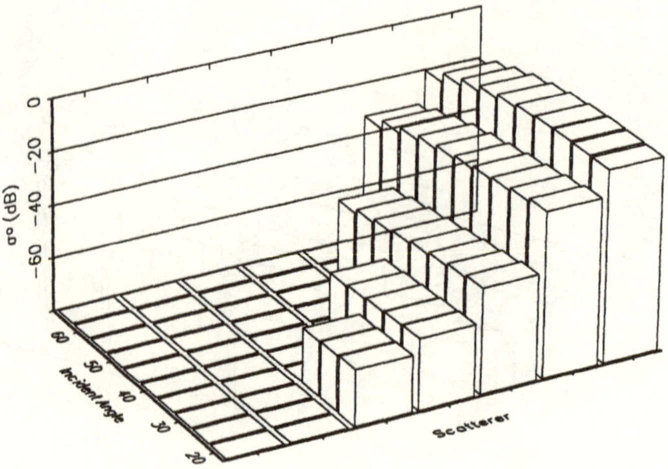

(b) volume-ground interaction with HH polarization

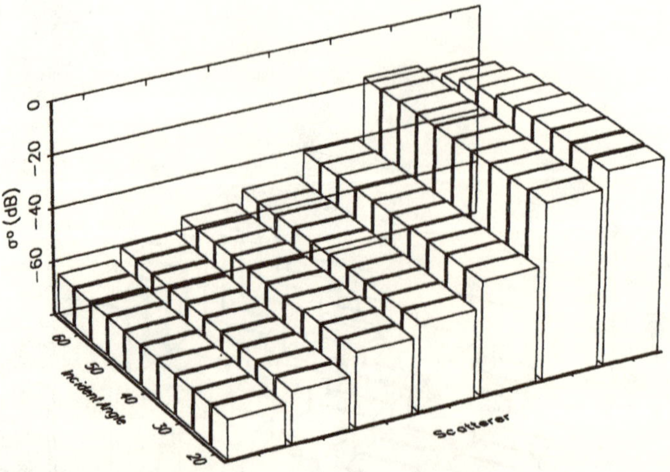

(c) sum of parts (a) and (b) with HH polarization

Figure 11.5 Freiburg site: Scattering contributions of tree components at P-band with HH polarization. (a) volume scattering, (b) volume-ground interaction and (c) sum of parts (a) and (b).

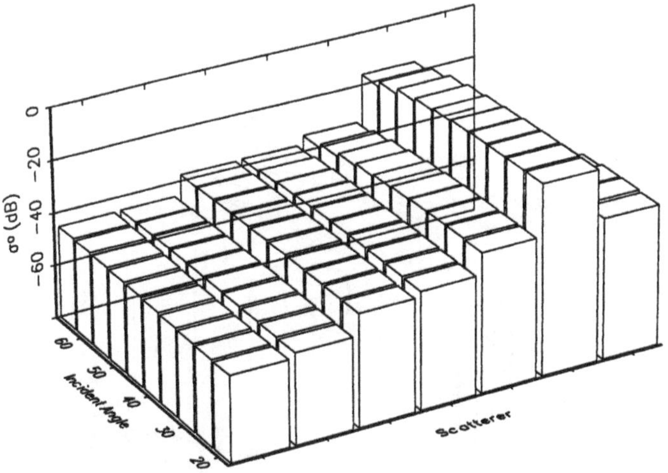

(a) L-band VV polarization

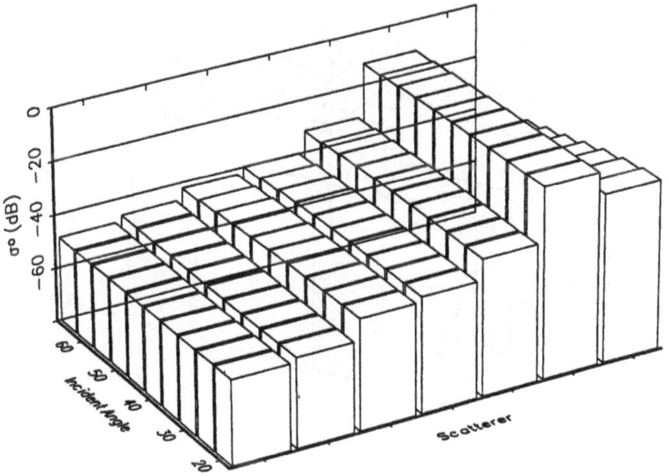

(b) L-band HH polarization

Figure 11.6 Freiburg site: Scattering contributions of tree components at L-band. (a) VV polarization, and (b) HH polarization.

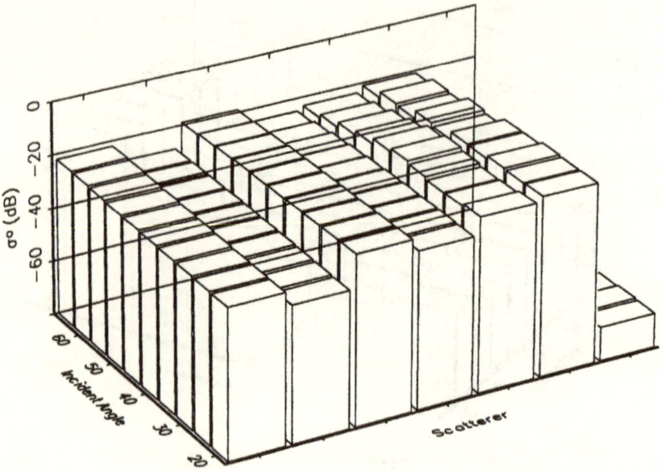

(a) C-band VV polarization

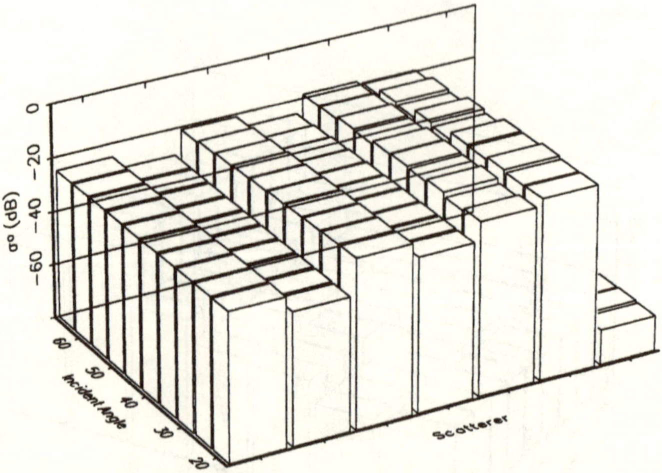

(b) C-band HH polarization

Figure 11.7 Freiburg site: Scattering contributions of tree components at C-band. (a) VV polarization, and (b) HH polarization.

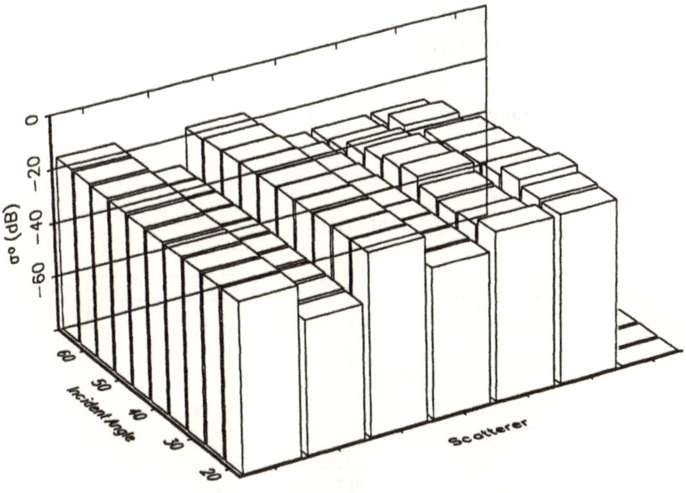

(a) X-band VV polarization

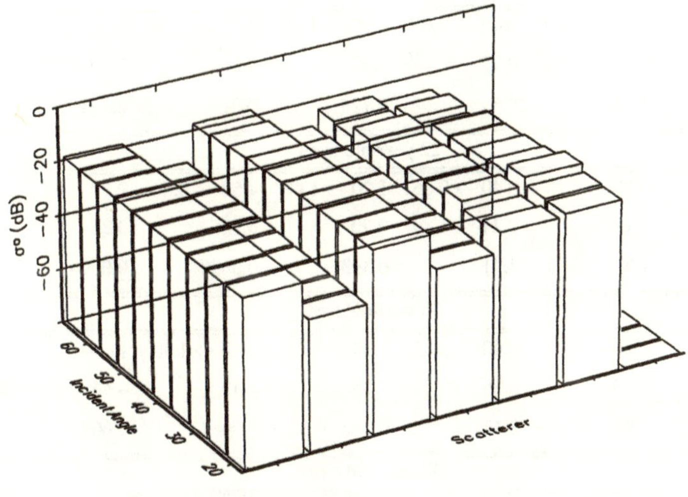

(b) X-band HH polarization

Figure 11.8 Freiburg site: Scattering contributions of tree components at X-band. (a) VV polarization, and (b) HH polarization.

When frequency is increased further to X-band, needles are the only dominant scatterer. This is true for all angles of incidence for the Freiburg site (Figure 11.8).

Next, we consider the Landes site where the needles are a factor of ten longer. Coniferous trees of various ages exist in this site. In particular, ground truths on trees of

6, 14, 22, 30, 38 and 46 years old have been gathered. The information is summarized in Tables 11.3 through 11.7 regarding soil conditions, leaves, branch and trunk dimensions, gravimetric moisture and dielectric values of the tree components, and observed angular distributions of tree components.

Table 11.3
Soil Parameters

Rms height (cm)	*Corr. length (cm)*	*% Moisture content*	*Dielectric constant*
1.5	10.0 → 15.0	10.00	8.0 - j1.0

Table 11.4
Vegetation Components of the Landes Forest

	Needles	*Md brch*	*Lg brch*	*Trunk (t)*	*Trunk(b)*	*Shrubs*
			6 year-old trees			
Radius, cm	0.10 → 0.17	0.40 → 0.65	0.75 → 0.90	3.60 → 4.40	5.50	0.15
1/2 length, m	0.15	0.20	0.63	2.35	0.25	0.075
Density, m^{-3}	1800	7.60	0.868	0.024	0.228	4000
			14 year-old trees			
Radius, cm	0.10 → 0.17	0.40 → 0.65	0.90 → 1.20	4.00 → 4.20	7.80	0.15
1/2 length, m	0.15	0.22	0.70	2.80	2.00	0.075
Density, m^{-3}	1500	5.69	0.728	0.015	0.021	3200
			22 year-old trees			
Radius, cm	0.10 → 0.17	0.40 → 0.65	1.20 → 1.40	4.50 → 5.00	9.70	0.15
1/2 length, m	0.15	0.24	0.90	3.10	3.25	0.075
Density, m^{-3}	1200	4.41	0.393	0.0098	0.0094	2200
			30 year-old trees			
Radius, cm	0.10 → 0.17	0.40 → 0.65	1.30 → 1.50	4.80 → 4.80	11.30	0.15
1/2 length, m	0.15	0.26	1.05	3.30	4.45	0.075
Density, m^{-3}	1000	3.48	0.272	0.0068	0.00505	2400
			38 year-old trees			
Radius, cm	0.10 → 0.17	0.40 → 0.65	1.43 → 1.85	5.00 → 5.80	12.60	0.15
1/2 length, m	0.15	0.28	1.15	3.50	5.65	0.075
Density, m^{-3}	800	2.64	0.189	0.0047	0.00292	1800
			46 year-old trees			
Radius, cm	0.10 → 0.17	0.40 → 0.65	1.30 → 2.40	5.50 → 5.70	13.90	0.15
1/2 length, m	0.15	0.30	1.30	3.65	6.90	0.075
Density, m^{-3}	640	2.00	0.133	0.0033	0.00174	1400

Table 11.5
Layer Heights in Meters

Age	*Top layer*	*Bottom layer*
6 years old	4.70	0.50
14 years old	5.60	4.00
22 years old	6.20	6.50
30 years old	6.60	8.90
38 years old	7.00	11.30
46 years old	7.30	13.80

Table 11.6
Gravimetric Moisture Content in the Canopy Constituents and Their Dielectric Constants

	GMC	*P-band*	*L-band*	*C-band*
Soil	0.10	$8.00 - j1.00$	$8.00 - j1.00$	$6.65 - j0.88$
Needles	0.55	$26.28 - j8.67$	$29.43 - j4.38$	$25.43 - j7.94$
Mid-size branch	0.55	$29.34 - j7.50$	$25.92 - j3.20$	$21.89 - j6.91$
Large branch	0.55	$29.34 - j7.50$	$25.92 - j3.20$	$21.89 - j6.91$
Trunk (t)	0.50	$27.24 - j7.60$	$20.10 - j3.50$	$17.40 - j5.58$
Trunk (b)	0.50	$27.24 - j7.60$	$20.10 - j3.50$	$17.40 - j5.58$
Shrubs	0.60	$33.37 - j11.43$	$29.50 - j4.50$	$18.70 - j5.97$

Table 11.7
Angular Distribution

	Needles	*Md brch*	*Lg brch*	*Trunk (t)*	*Trunk (b)*	*Shrubs*
β_0, deg	20.00	0.00	30.00	0.00	0.00	0.00
β_1, deg	20.00	20.00	30.00	0.00	0.00	0.00
β_2, deg	150.00	180.00	110.00	10.00	10.00	60.00
β_m, deg	90.00	90.00	70.00	5.00	5.00	30.00
Power, n	1.00	2.00	1.00	4.00	4.00	1.00

The labeling of branch sizes is consistent with those for the Freiburg site. The trunk sizes vary because of the many age groups. There is no branch size in the Landes site corresponding to the small branch in the Freiburg site. We use 3D bar charts to show the incident angle, vegetation components and the backscattering coefficients. The six vegetation components along the scatterer-axis from left to right are needles, medium branch, large branch, portions of the trunk in the top and bottom layer and shrubs in the bottom layer. For the 38 year old trees we show in Figures 11.9 and 11.10 volume scattering and volume-ground interaction separately. In VV polarization the large branch clearly dominates in single scattering (Figure 11.9). Although the trunk-ground interaction is dominating in double scattering, it is over 10 dB smaller in total scattering.

Les Landes Site

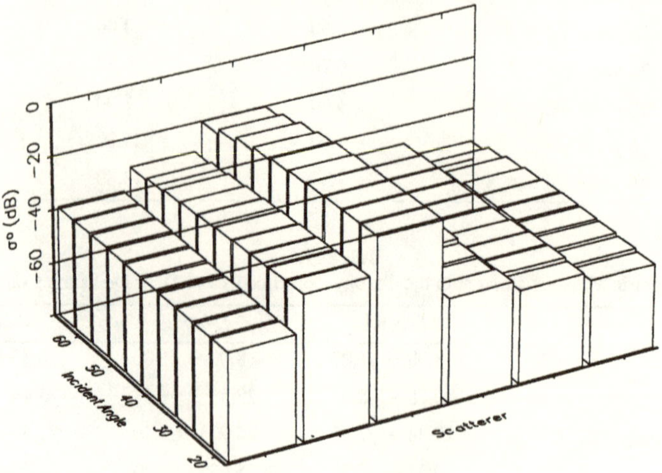

(a) volume scattering VV polarization

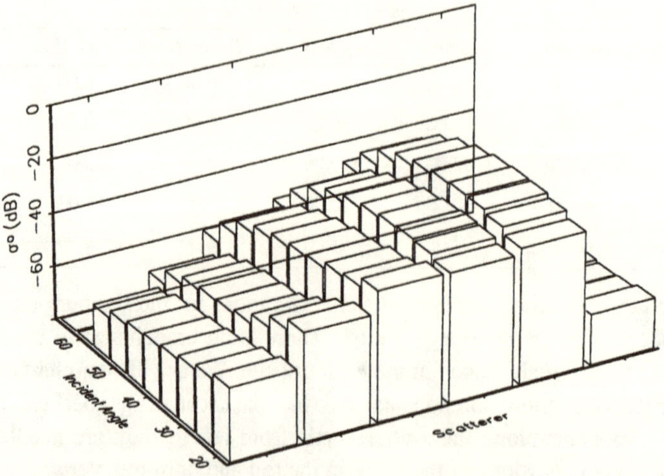

(b) volume-ground interation, VV polarization

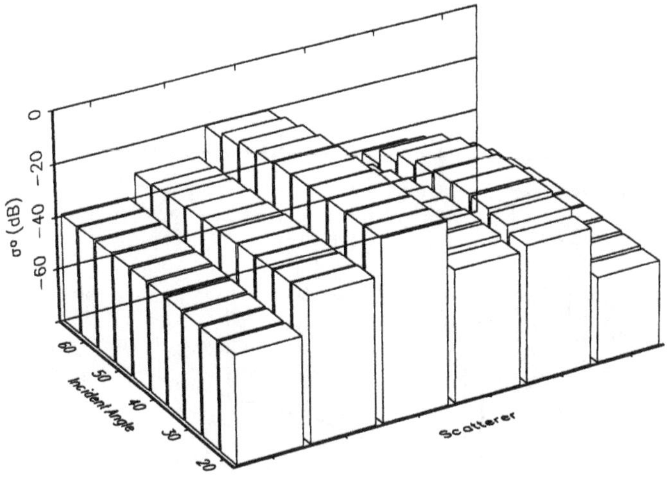

(c) sum of parts (a) and (b)

Figure 11.9 Les Landes site: Scattering contributions of tree components at P-band with VV polarization. (a) volume scattering, (b) volume-ground interaction, and (c) sum of parts (a) and (b).

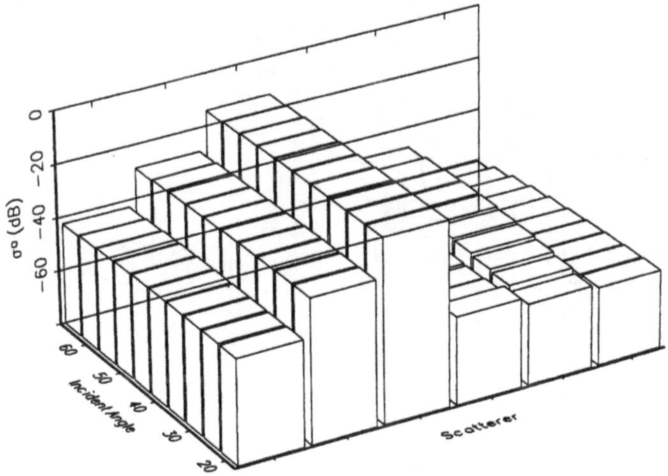

(a) volume scattering HH polarization

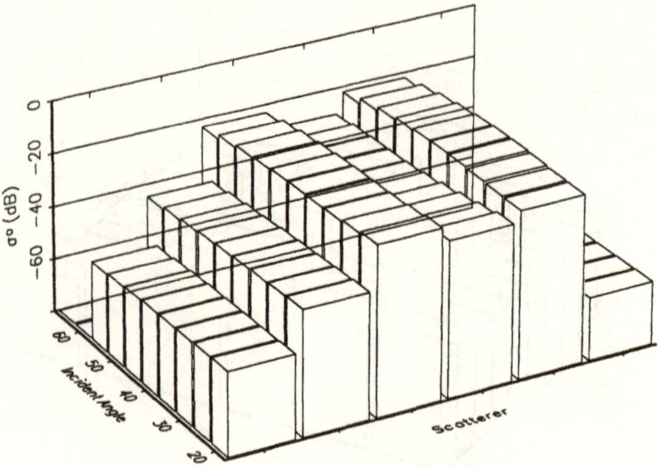

(b) volume-ground interaction HH polarization

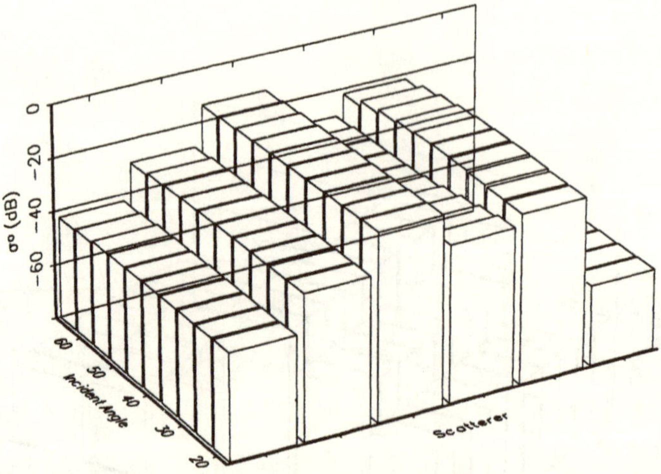

(c) sum of parts (a) and (b)

Figure 11.10 Les Landes site: Scattering contributions of tree components at P-band with HH polarization. (a) volume scattering, (b) volume-ground interaction, and (c) sum of parts (a) and (b).

Figure 11.10(a) shows that in HH polarization the large branch is the dominant scatterer in single scattering same as in VV polarization, and Figure 11.10(b) indicates that in double scattering the trunk-ground interaction is the highest for HH polarization. It is the portion of the trunk in the bottom layer that is important because of the size of the trunk. The large branch-ground interaction is slightly smaller than the trunk-ground term. The effect of the shrub is always small because it is situated at the lowest level in the canopy and its height is not more than a meter. It appears that vegetation growing under the trees will not be important scatterers although they may cause some attenuation to the incident signal and reduce the volume-ground interaction terms. Overall, the large branch contributes the most to scattering (Figure 11.10(c)) and the next in line is the trunk-ground interaction. Thus, the relative importance of the contributions from the large branch and the trunk-ground interaction is reversed when compared to the Freiburg site. This is because of the large number of long needles, which generate more attenuation to the incident signal in the Landes site.

In summary, the large branch is the important scatterer at P-band for both VV and HH polarization for the Landes site. This change in the dominance of scatterers in HH polarization for the Landes site is due to mainly to the attenuation by the crown layer. Because of the trunk-ground interaction, HH polarization is generally higher than VV polarization at P-band. Contribution to backscattering by ground alone is not considered here.

The scattering at L-band is illustrated in Figure 11.11. Here, only the total scattering by both volume and volume-ground interaction is shown. Important contributors are now the branches of different sizes. Apparently, the amount of attenuation by the crown layer is so high that the effect of the trunk-ground interaction is of less importance even for HH polarization where it is the third most important contributor.

At C-band (Figure 11.12) the needles contribute the most to scattering. The mid-size branch is the second in line followed by scattering due to the large branch for both VV and HH polarizations. Scattering by all other scatterers is negligible. When we compare scattering at the Landes site with that at the Freiburg site (Figure 11.7) we see that the needles are not the most important scatterer there in HH polarization and their strength of scattering is comparable to the branches in VV scattering. This is clearly due to the fact that the needles at the Freiburg site are much shorter and the attenuation caused by them is also smaller. Another interesting point to note in Figure 11.12 is the drastic reduction of the contribution by the trunk-ground interaction term. Relative to the contribution by the shrubs, the trunk-ground term is now much smaller. This apparently is because the attenuation by the shrubs has become quite significant. Note also that the plot for the contribution by large branches is not smooth versus the incident angle. This is due to the oscillations in the I_z integral for the dielectric cylinder. These oscillations can be removed by adding an appropriate window function to the integration. Its effect is not obvious at low frequencies but is becoming more and more apparent at higher frequencies.

Scattering at X-band is quite similar to that at the Freiburg site in that the dominant scatterers are the needles for both VV and HH polarization (Figure 11.13). Second in line are the contributions from the branches. They are smaller because there is no significant change in their strength of scattering but they are attenuated more at higher frequencies. On the other hand, the needles are stronger scatterers at higher frequencies. Figure 11.13 also shows that not only the trunk-ground interaction in the bottom layer is negligible but also the effects of shrubs are significantly smaller than at C-band. This indicates a significant increase in the attenuation of the crown layer. As in Figure 11.12, there are oscillations in the angular behaviors of the branches.

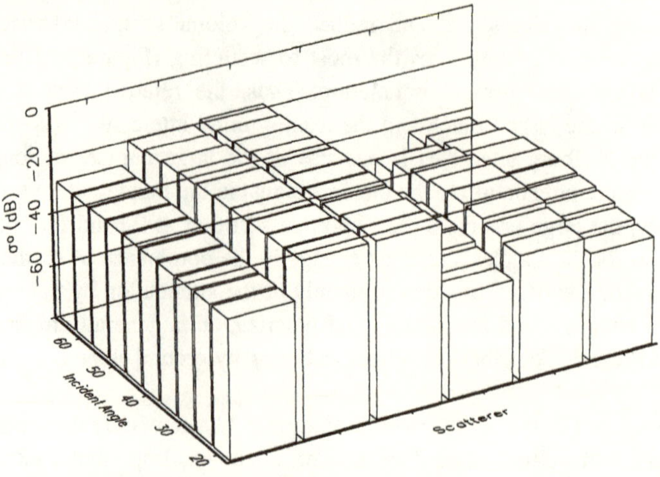

(a) L-band VV polarization

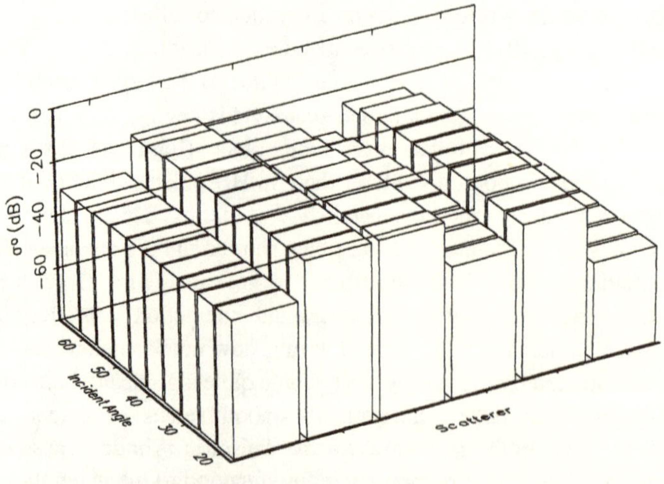

(b) L-band HH polarization

Figure 11.11 Les Landes site: Scattering contributions of tree components at L-band. (a) VV polarization, and (b) HH polarization.

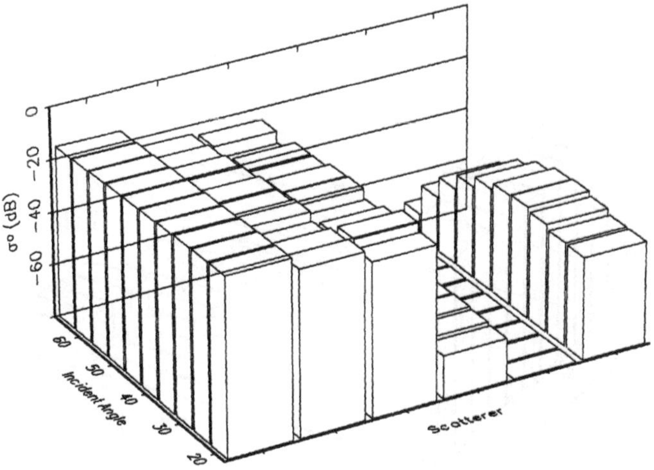

(a) C-band VV polarization

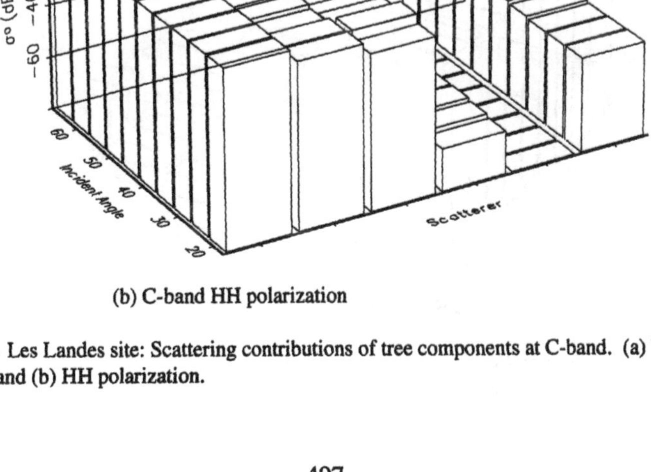

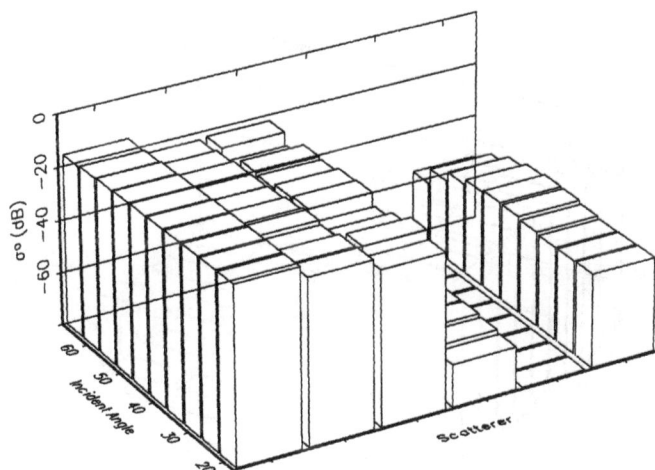

(b) C-band HH polarization

Figure 11.12 Les Landes site: Scattering contributions of tree components at C-band. (a) VV polarization, and (b) HH polarization.

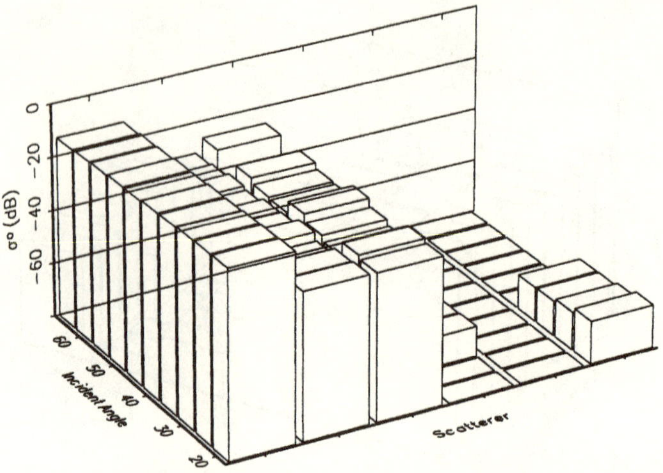

(a) X-band VV polarization

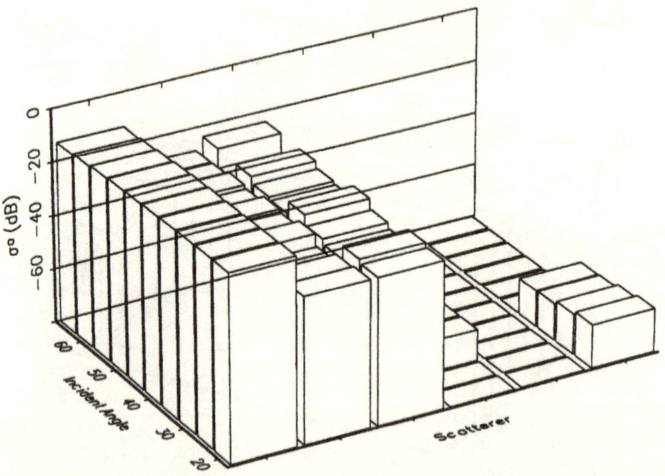

(b) X-band HH polarization

Figure 11.13 Les Landes site: Scattering contributions of tree components at X-band. (a) VV polarization, and (b) HH polarization.

11.5.2 Deciduous Vegetation

A special experimental investigation was conducted by a group of investigators to acquire scatterometer measurements from deciduous trees with detailed ground truth for the purpose of supporting satellite measurements and verifying scattering models [Cimino et al., 1988]. These measurements were taken over a walnut orchard at 1.25 and 9.6 GHz. A description of the characteristics of the orchard and ground truth gathering effort is given in the following paragraphs.

The canopy consists of 6-year-old black walnut trees. The trees have an average height of 4.8 m. The data on tree geometry was collected in two parts. Measurements involving branches with diameter greater than 4 cm were termed *skeleton geometry measurements* and the rest *higher order measurements*. A group of sixteen walnut trees was chosen for the canopy geometry and ground truth measurements. Their heights, width across the row, and the length down the row were measured. The skeleton branches which terminated into a successively smaller diameter branch were physically sampled for their length, diameter, and inclination angle for all the sixteen trees. Small branches that grew along the skeleton tend to fill the interior of the canopy. Such branches with diameter less than 4 cm were sampled only for a couple of trees. The thinnest branches with diameter less than 1 cm and length less than 30 cm, were not sampled for their inclination orientations. The branches were grouped into four different groups according to their radius, and for each group an average length of the branch was computed. Besides these four branch groups, there were green stems that had an external covering of green bark and were located just below the juncture with the petioles. For modeling purposes we shall consider the stems as a group of branches. The stem group will be labeled as group # 1 among the other branch groups. Table 11.8 sums up the relevant parameters for each branch group.

Table 11.8
Stem and Branch Sizes.

Branch group	*Diameter range (cms)*	*Average diameter (cms)*	*Ave. length (cms)*	*Number density (m^{-3})*
1/stems	0.0–0.40	0.1	18	250
2	0.5–1.90	1.28	14	11.4
3	2.0–2.90	2.60	32	0.43
4	3.0–6.90	5.00	58	0.33
5/trunk	7.0–17.1	9.00	76	0.14

From the inclination angle measurements of the branches, the inclination angle probabilities for all branch groups were calculated. Figure 11.14 presents the histograms of the inclination angle probabilities for different radius groups. The data show that as the diameter of the branch increases, it tends to become more vertical. The thin branches do not show any preferred inclination.

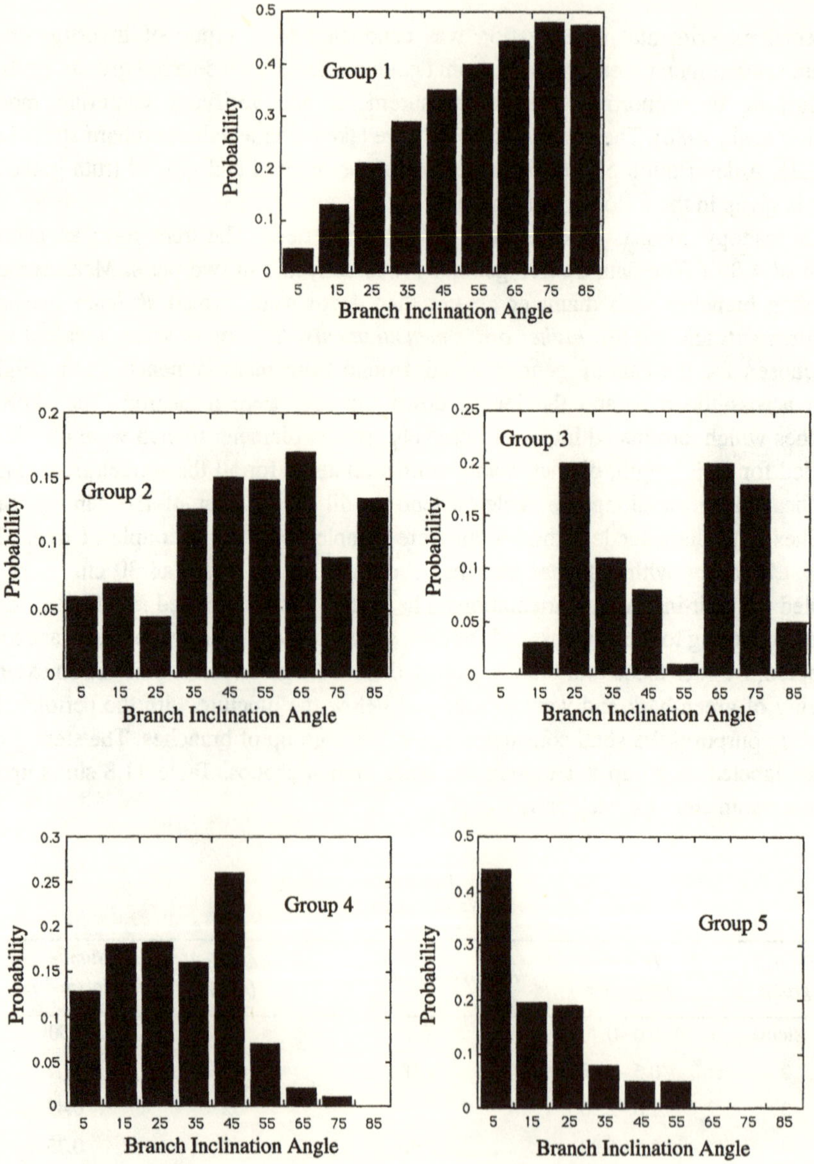

Figure 11.14 Probability distributions of the inclination angles of branch groups.

The leaves were found to be growing only on the branches in groups 1 and 2. The leaves on group 2 branches were determined from the routine sampling of higher order canopy geometry measurements. However, due to the large number of branches in group 1, an exclusive sampling was done to estimate the number of leaves per branch in the group 1 category. These leaf data were extrapolated to cover the whole canopy. The average density of the leaves was 250 m^{-3} [McDonald et al., 1991]. Each leaf had an average leaf area of 254 square cm, and on the average there were five leaflets on one leaf. Assuming the leaves to be circular, the leaf area results gave the average disc radius of approximately 3.6 cm and a thickness equal to 0.1 mm. The leaf inclination angle has a probability distribution function equal to $\sin\beta$ ($0^\circ < \beta < 90^\circ$).

In-situ measurements of dielectric constant of stems, branches and leaves were made by a team from the University of Michigan at L-band ($f = 1.2$ GHz). An electric probe was inserted into a hole drilled into a branch to find the dielectric constant [Cimino et al., 1988, Vol. II, Sec. XIII]. The leaves were stacked in layers upon a flat piece of wood. For each stack, the probe reading was noted at three separate locations on the stack. The behavior of the dielectric constant with depth inside a branch or on a stack of leaves was found from probe readings. Figure 11.15 shows the relative dielectric constant as a function of depth into walnut bole. It is clear that the real part of the dielectric constant has values between 4 and 45. The imaginary part varies between 1 and 30. For stem dielectric constant a representative value at the L-band is found to be $27.3 - j8.4$. As this value for the dielectric constant is a mean value for the bole dielectric constant, it will be taken as a representative value for the branch dielectric constant. The leaf dielectric value varies from $8.77 - j2.88$ to $19.58 - j5.54$ according to the number of the leaves stacked in layers upon a flat piece of wood to measure the dielectric constant. Consequently, an average value of $19.58 - j5.54$ will be used to represent the leaf dielectric constant. There is no independent confirmation of the leaf and branch dielectric constant values as high as those used in McDonald et al. [1991].

The soil relative dielectric constant measurements were repeated on an hourly basis. Each observation sequence consisted of three types of data, designated as wet, dry or mix. The wet and dry consisted of separate samples of the soil surface regions, which were always wet or always dry, respectively. A transect sampling was used to evaluate the spatial average of the dielectric constant over the three moisture regions. The dielectric transect data consisted of twenty-two samples spaced 0.3 m apart and extending from center of one row to the center of the next. The location of the transect along a row with respect to the tree and sprinkler location was randomized. The sprinklers were placed along the rows of the trees to irrigate the trees. The water from sprinklers was sprayed in an approximately 3 ft. wide strip along the rows of the trees. There was no sprinkler in between the tree rows. Thus approximately 70% of the area was not irrigated and can be classified as "dry soil." It was found that inter-row dielectric constant has values between $18.48 - j1.6$ and $2.79 - j0.16$. Furthermore, ten soil samples taken from an area partially covered by organic leaf litter showed an average dielectric constant value of $2.96 - j0.49$. In addition, the spatial averaging for backscatter as seen by the radar was done by rotating the boom across the rows in an arc, and it was not done along the tree rows. Accordingly, it is believed that the ground dielectric constant should have a small average value, which is taken to be 5.0 - j0.7 in the subsequent study.

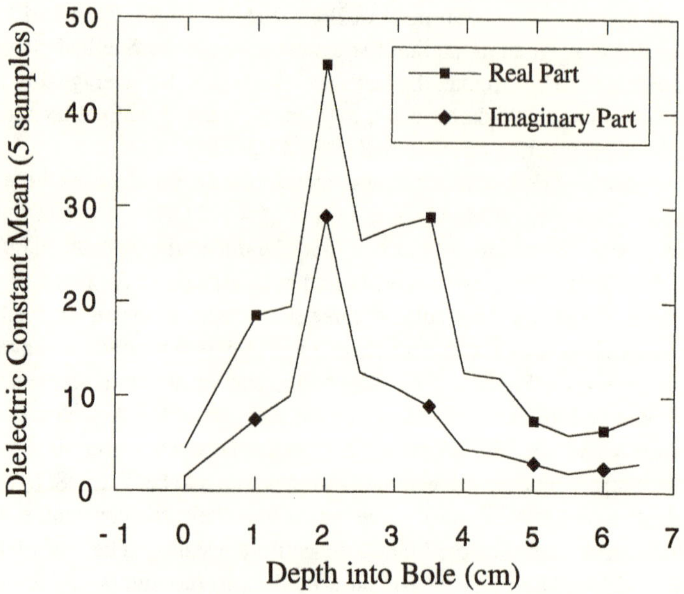

Figure 11.15 Relative dieletric constant as a function depth into the bole of a walnut tree.

To find the values of the leaf and branch dielectric constant at X-band (f = 9.6 GHz), the corresponding values at L-Band are incorporated in Ulaby and El Rayes' dielectric constant formula [1987] to obtain the leaf and branch gravimetric moisture contents (0.55 for leaves and 0.65 for branches). By substituting these values for the moisture contents along with the X-band frequency into Ulaby and El Rayes' formula, we obtain the values of the dielectric constant at X-band (14.9 - j4.9 for leaves and 20 - j9.7 for branches). Since the soil effect at X-band is unimportant in the backscattering calculation, its dielectric constant is not estimated at X-band. Table 11.9 sums up the dielectric constant values used for the leaves and stems and branches

Table 11.9
Dielectric Values

Frequency band	*L*	*X*
Leaves	19.58 - j5.54	14.9 - j4.9
Branch / stems	27.3 - j8.4	20.0 - j9.7
Soil	5.00 - j0.7	5.00 - j0.7

For the purpose of modeling we divide the canopy into two layers above a rough interface. The upper layer with a depth of 3 m is the crown layer and the lower layer with a depth of 1.7 m is for the trunk layer. The crown layer contains leaves and the first four branch groups. The fifth branch group is in the trunk layer. The soil-canopy interface roughness is represented by a Gaussian correlation function with a rms height σ and a correlation length L given by 0.021 m and 0.25 m, respectively. In calculating the crown- and trunk-ground interaction terms the soil scattering coefficient is the sum of its coherent and noncoherent components.

Based upon the above information we can now show backscattering calculations as a function of the incident angle using the scattering model given in the previous section. In illustrating the contributions by different vegetation components we exclude those components whose contributions are more than 10 dB lower than the final backscattering curve. In Figure 11.16 we show the scattering contributions by different branch groups at 1.25 GHz for VV, HH and VH polarizations. In this figure the crown-crown interaction term is not included. It is seen that the major contributor for all polarizations is group 2, because its branch size is large enough at this frequency and its number density is close to a factor of 3 or more greater than the larger branches. The next largest contributor varies depending on polarization.

The secondary contributors to VV polarization are from branch groups 3 and 4, and the interaction of group 4 with the ground. They are more than 5 dB lower than the level of scattering by branch group 2. Both groups 3 and 4 have a broad angular distribution but the size of group 4 is almost a factor of two larger than group 3. Thus, group 4 is the more effective contributor in secondary contribution. While the interaction term is not very important in VV polarization because it decreases very quickly with the incident angle due to Brewster angle effect, it is important in HH polarization where the secondary contributors are from branch-ground interactions of branch groups 2 and 4. Direct contributions from branch group 4 are of the same level and they are within 5 dB of the scattering level by branch group 2. As a result the total scattering from HH polarization is higher than VV. To generate branch-ground interaction the incident wave must penetrate to the ground. This means that the amount of attenuation by the vegetation canopy cannot be too large at 1.25 GHz. In cross polarization the secondary contributors are dominated entirely by branch-ground interaction terms of branch groups 2, 3 and 4. The major cause of depolarization in single scattering is from the orientation of the branches. Theoretically the calculation shown in Figure 11.16 is incomplete because crown-crown interaction has not been included. However, at 1.25 GHz the albedo of the vegetated medium is small and hence the effect of crown-crown interaction is not expected to be significant. Unlike the coniferous vegetation the trunk-ground interaction term is not of importance here. The major reason for this difference appears to be due to the presence of large branches with a much broader orientation distribution than the trunk in the deciduous vegetation. In addition, the depth of the crown layer is taken to be 3 m, while that of the trunk layer is 1.7 m here. Thus, scattering from the crown layer is generally stronger than that from the trunk layer even if the orientation distributions are similar in the two layers.

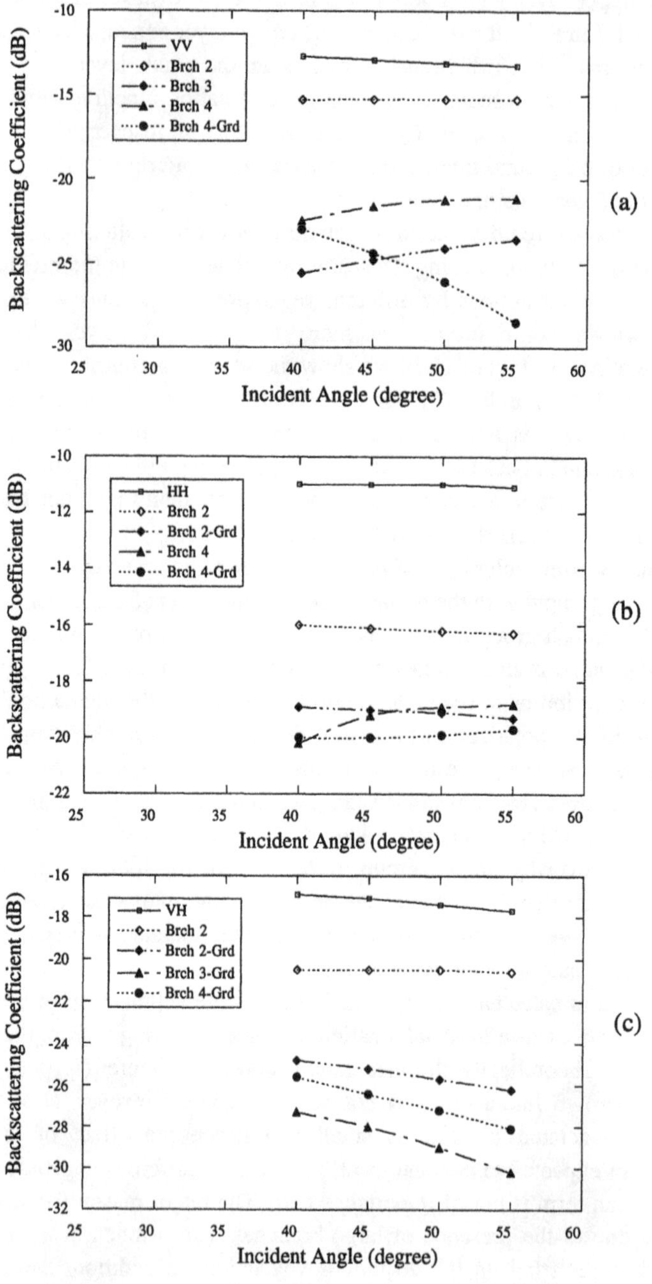

Figure 11.16 Scattering contributions by tree components froma walnut canopy: (a) VV, (b) HH and (c) VH at 1.25 GHz.

To examine the effect of crown-crown interaction on cross polarization we plot in 11.17 the calculations of cross polarized return with and without the crown-crown interaction term. At 1.25 GHz the difference is small as expected. However, the difference becomes quite large at 9.6 GHz.

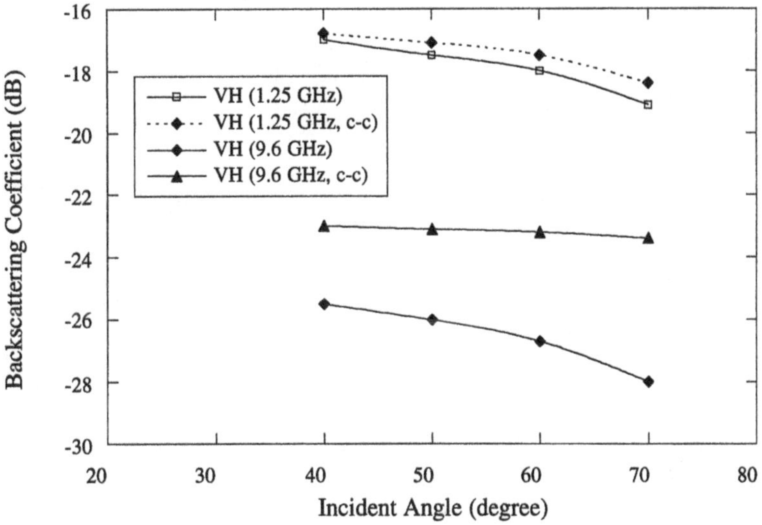

Figure 11.17 Comparisons between cross polarized calculations with and without the crown-crown interaction term at 1.25 GHz and 9.6 GHz.

Next, we consider the scattering contributions by tree components at 9.6 GHz. The dominant scatterers are expected to be the smaller scatterers as opposed to the large branches. Backscattering characteristics are shown in Figure 11.18 for VV, HH and VH polarizations without including the crown-crown interaction term, because this term mixes all branch groups together and will not allow us to identify the individual branch groups. Furthermore, it is significant mainly for cross polarization.

Figure 11.18 shows that the tree components contributing to both VV and HH polarizations are the same. The first three scatterers in order of importance are the leaves, branch group 1 (stems) and branch group 2. The reason for leaves and stems to become important is due to their increased value of albedo at 9.6 GHz. The branch-ground interaction terms are no longer important because the attenuation due to the crown layer is very high. Unlike at 1.25 GHz, VV is now higher than HH. The cross polarized return is dominated only by the stems in single scattering. One possible reason why the leaves are not contributing like the needles to cross polarization is because its leaf area is larger than a wavelength in diameter and hence can be important only in double scattering.

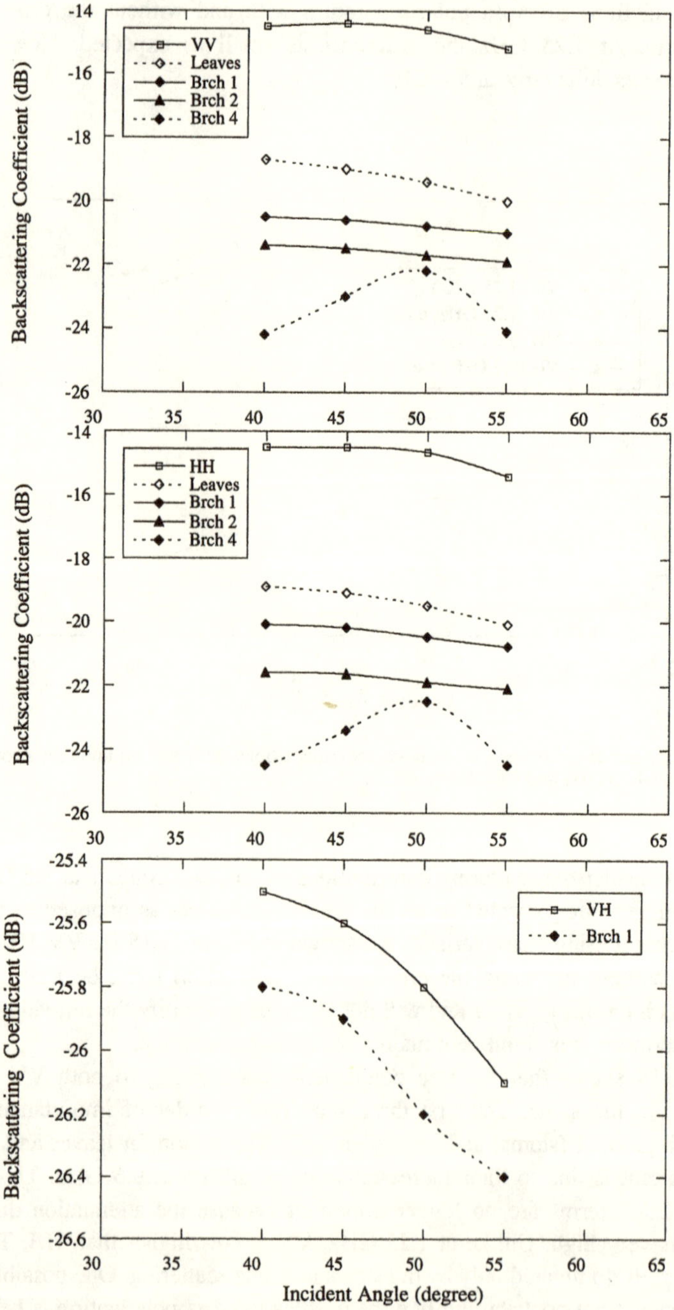

Figure 11.18 Scattering contributions by tree components from a walnut canopy: (a) VV, (b) HH and (c) VH at 9.6 GHz.

11.6 COMPARISONS WITH MEASUREMENTS

In the previous section we have examined the backscattering behavior of coniferous and deciduous vegetation along with the ground truths for three different sites in the frequency range from P- to X-band. We have found that the importance of a scatterer at a given frequency is determined by the product of its albedo and number density. Whether it can contribute along a specific direction depends on its orientation distribution. Now we are ready to see how the simple model given in Section 11.4 compares with measurements taken from the sites illustrated in the previous section and in the literature. We shall consider backscattering behaviors versus angle, biomass and age.

11.6.1 Angular Characteristics

In Figure 11.19 we show comparisons between the backscatter model of Section 11.4 and backscattering measurements from the walnut canopy described in Section 11.5 for VV, HH and VH polarizations at 1.25 GHz. As expected HH is higher than VV due to branch-ground interaction. There is a general agreement in level for all polarizations and angular trends except for VH polarization where the angular trend is different.

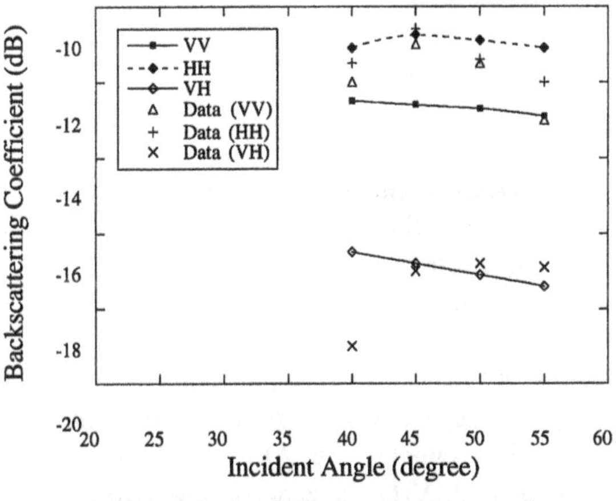

Figure 11.19 Comparisons between model and measurements from a walnut canopy at 1.25 GHz.

At 9.6 GHz we know the branch-ground interaction is not important. Thus, VV and HH should be fairly close to each other. As shown in Figure 11.20 the agreement in level for all three polarizations is very good. Due to an obvious drop in the data level at 40 degrees it is not clear whether the angular trend should be flat or upward. A possible reason for the drop in data at 40 degrees is that the measured value at an angle is averaged over an azimuthal sweep and different angles are viewed at different ranges. Hence, there is the possibility that the statistics of the canopy geometry at 40 degrees are not the same as at other angles reported in the graph. In general, when only one type of scatterers is illuminated by a scatterometer, the scattering should increase with frequency. The scattering level in Figure 11.20 is actually lower than those in Figure 11.19, because we are looking at different types of scatterers as explained in the previous section.

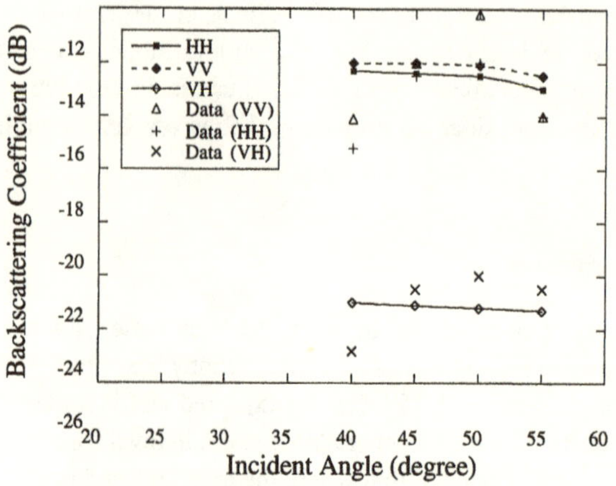

Figure 11.20 Comparisons between model and measurements from a walnut canopy at 9.6 GHz.

11.6.2 Effects of Age

In this section we consider backscattering measurements taken at the Landes site versus age. As reported in the previous section the geometry of the canopy changes with age. Hence, scattering also changes with age. Computations are performed at P-, L- and C-bands at 45 degrees incidence for VV, HH and VH polarizations and results are plotted in Figures 11.21 and 11.22.

Figure 11.21 shows that at P-band the model predictions are in very good agreement with measurements. In particular, the VH polarization shows more sensitivity to age than the like polarizations. For example, up to 20 years VV polarization only increases from about -19 dB to -14 dB, while VH polarization changes from -28 dB to -18 dB. The change in HH polarization is from -19 dB to -11 dB, which is in between VV and VH polarization. This seems to indicate that sensitivity to age requires (a) penetration of the incident wave and (b) dominance of scattering by major components of the canopy. VV polarization is weaker in penetration capability due to the presence of vertically oriented branches and trunks. HH polarization has the best penetration capability but is influenced somewhat by ground effects, because it has more canopy-ground interaction than at other polarizations. In VH polarization the interaction terms are weaker relative to the major contributing term (see Figure 11.16). This appears to be the reason why the cross polarization is the most sensitive at P-band to age.

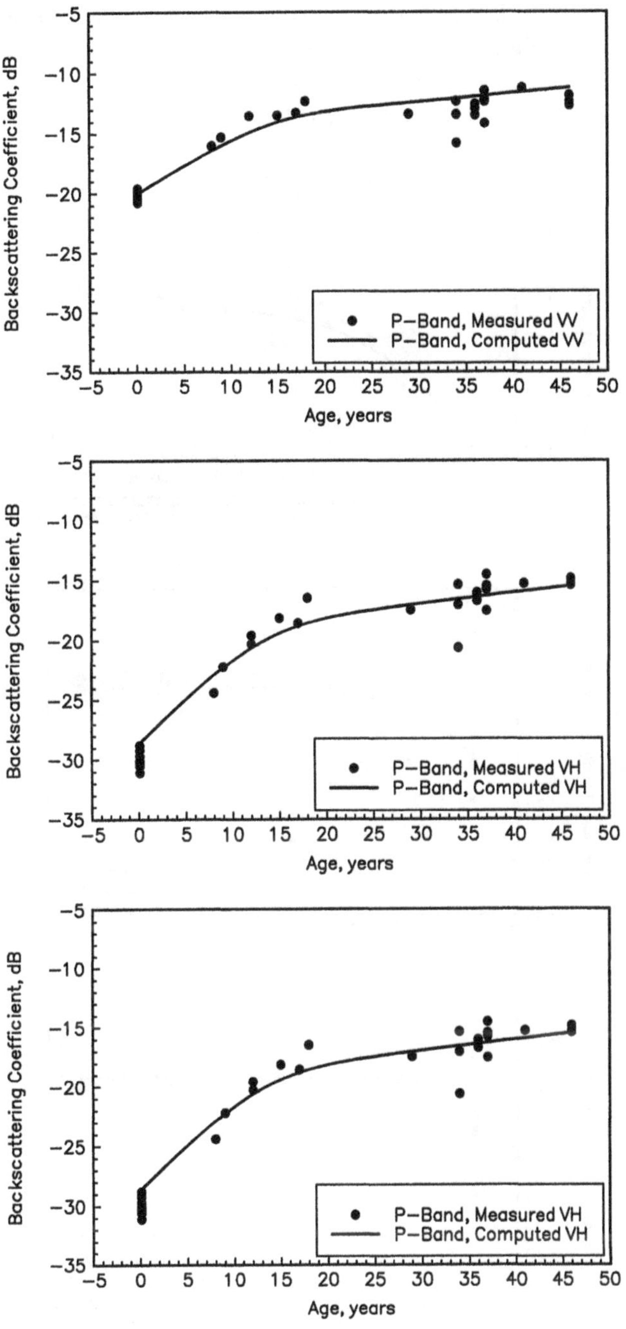

Figure 11.21 A comparison between model and measurements versus age at P-band for VV, HH, and VH polarizations.

It is clear that, if the incident signal did not penetrate through the canopy, the return signal should not correlate well with age. It stands to reason that, as frequency increases, the amount of penetration will decrease and so too the sensitivity to age. To demonstrate this point we plot in Figure 11.23 the cross polarized scattering at P-, L- and C-bands.

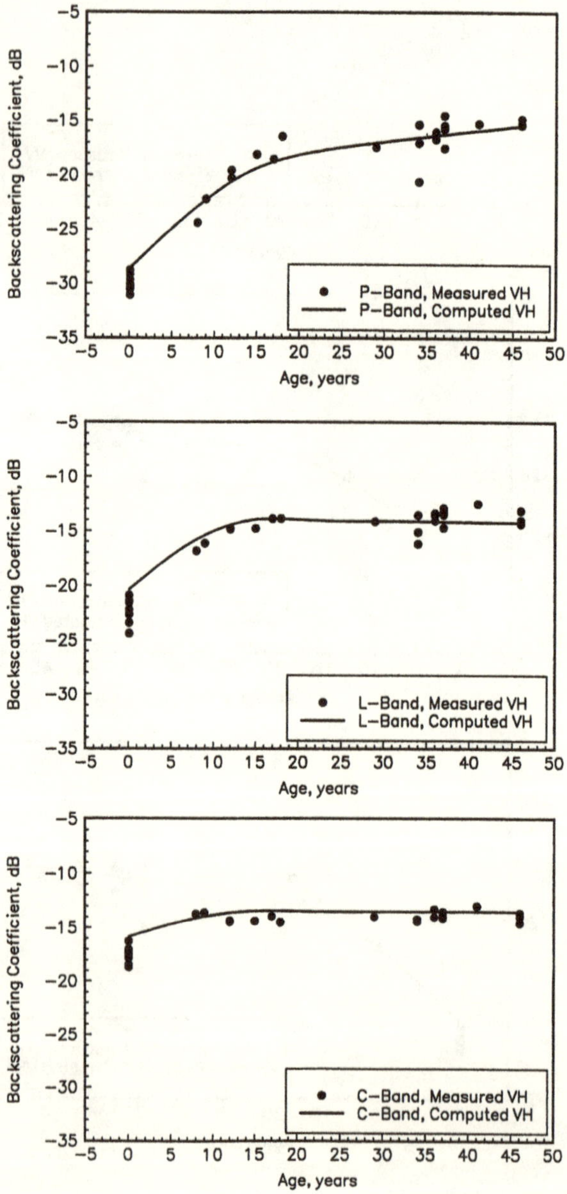

Figure 11.22A comparison between model and measurements at P, L and C bands illustrating loss of sensitivity to age at higher frequencies.

11.6.3 Effects of Biomass

In experimental studies a relation between the biomass and age has been found and a regression curve was used to describe the functional relation as shown in Figure 11.23.

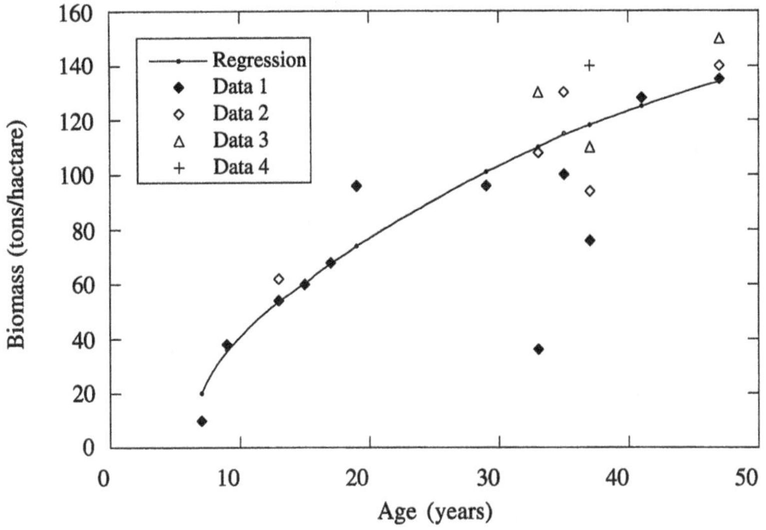

Figure 11.23 Measured biomass as a function of age.

By using the functional relation given in Figure 11.23 we can plot backscattering versus biomass instead of age. Model calculations are shown along with measurements at P-band for VV, HH and VH in Figure 11.24. Upon examining Figure 11.24 we see that there is a very good agreement between the model and measurements in VH polarization. In VV polarization the model seems to predict an increase with the biomass that is a little faster than the data. Similar remarks are applicable to HH. Again, the results indicate that it is the cross polarization that is more sensitive to biomass or age. In view of the positive correlation between biomass and age and the results, it is clear that there will be a loss of sensitivity to biomass if we perform the same calculations at L- or C-bands. Indeed, we should anticipate an answer similar to those illustrated in Figure 11.22. Hence, further calculations at L- and C-bands will be unnecessary.

Theoretically, we know that, due to many vertical structures within a coniferous forest canopy, VV polarization experiences more attenuation and hence should have less sensitivity than HH to the overall canopy structure. On the other hand, HH having the best penetration capability is significantly affected by ground-scattering and, therefore, its ability to represent the biomass. Thus, we expect VH to be the best polarization to use in sensing forest biomass and HH as the second best at P-band. There is the possibility that at some frequency between L-band and P-band the volume-ground interaction terms are not significant and HH becomes a better estimator than it did at P-band.

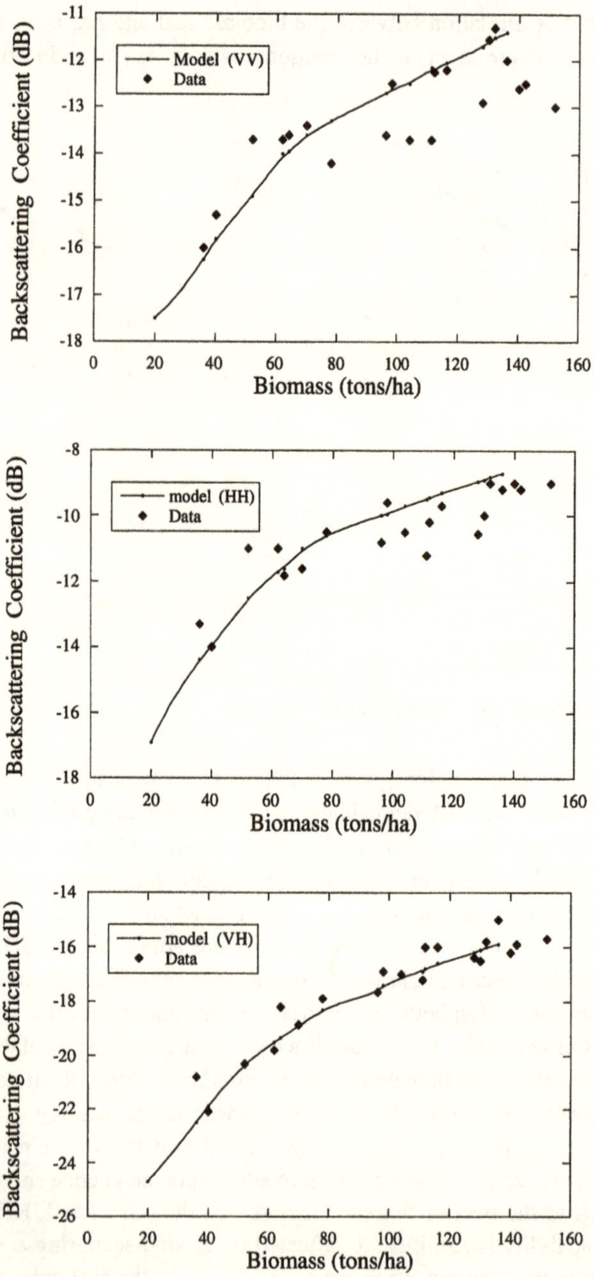

Figure 11.24 A comparison between model and measurements versus biomass at P-band for VV, HH and VH polarizations.

11.7 EFFECT OF SLANTED TERRAIN

In previous sections we have assumed that a vegetation layer is located above a flat ground plane. Many naturally occurring forested areas are situated on the slope of a mountain. In this section we would like to consider the effect of such a slope on radar backscattering measurements. The consideration here is restricted to a first-order effect using the simple model given in Section 11.4. The geometry of the problem is depicted in Figure 11.25.

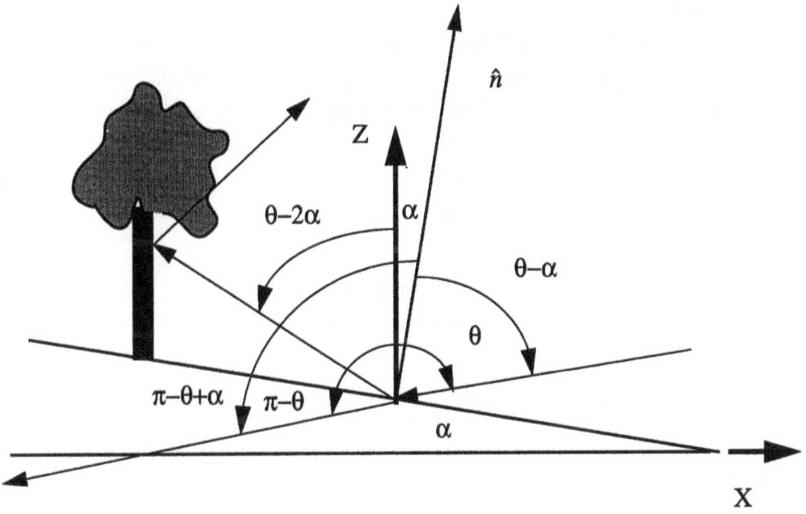

Figure 11.25 Geometry of backscattering from a slanted ground plane.

In view of the geometry of the problem we note that the tree continues to grow vertical with respect to the ground plane. Thus, any single scattering from a tree component in the canopy remains unchanged. The only effect is on the surface scattering and the canopy-ground interaction terms [Amar et al., 1993]. In what follows we consider these two types of terms.

For the surface backscattering term its original representation in the reference coordinates is re-referenced to the local normal $\hat{n}$ instead of the z-axis due to the local tilt-angle as follows:

$$S_{pq}(\theta, \pi; \pi-\theta, 0) \rightarrow S_{pq}(\theta-\alpha, \pi; \pi-\theta+\alpha, 0) \tag{11.110}$$

For the volume-ground interaction term both the surface scattering part and the subsequent volume scattering part are affected. Due to reciprocity in backscattering the reverse order of scattering should generate the same result, and hence we do not have to consider separately. From the geometry in the figure the changes for this type of term are as shown below:

$$S_{pq}(\theta, \phi_s; \pi - \theta, \phi) \rightarrow S_{pq}(\theta - \alpha, \phi_s; \pi - \theta + \alpha, \phi) \tag{11.111}$$

$$Q_{pq}(\theta, \phi_s; \theta, \phi) \rightarrow Q_{pq}(\theta, \phi_s; \theta - 2\alpha, \phi) \tag{11.112}$$

Other than the changes indicated above the same model given in Section 11.4 is applicable. There is only a re-interpretation of the applicable angles in the scattering phase functions.

From previous studies we know that volume-ground interaction is important mainly at P-band. Hence, we shall illustrate the impact of a slanted ground plane only at P-band using a set of parameters for a coniferous canopy similar to the Freiburg site surface given in Table 11.10. The orientation distribution function has been given by (11.109) in Section 11.5. The parameter values chosen for this function for different tree components are given in Table 11.11.

Table 11.10
Simulation Parameters

Crown-layer height = 11.5 m	Radius (md brch) = 2.1 mm
Trunk-layer height = 13.0 m	Length (md brch) = 1.3 m
Surface rms height = 2 cm	Gravimetric moisture (md brch) = 0.55
Exponential surface correlation, length = 15 cm	Number density (md brch) = 2.9 /m^3
Radius (needle) = 1 mm	Radius (lg brch) = 8.5 mm
Length (needle) = 1.5 cm	Length (lg brch) = 2.2 m
Gravimetric moisture (needle) = 0.65	Gravimetric moisture (lg brch) = 0.55
Number density (needle) = 3000 /m^3	Number density (sm brch) = 0.89 /m^3
Radius (sm brch) = 2 mm	Radius (trunk) = 15 cm
Length (sm brch) = 20 cm	Length (trunk) = 13 m
Gravimetric moisture (sm brch) = 0.55	Gravimetric moisture (trunk) = 0.5
Number density (sm brch) = 6.4 /m^3	Number density (trunk) = 0.005 /m^3

With the vegetation parameters known calculations are performed at P-band with the slope angle varying from -15 degrees to +15 degrees. Figures 11.26 and 11.27 show results of theoretical calculations at incident angles 20 degrees and 40 degrees, respectively, for VV, HH and VH polarizations. For each polarization the contributions from individual tree components are also shown. In figures the small, medium and large branches are denoted as branch 1, 2 and 3 respectively. The anticipated effect of the slope on scattering is that it is more at 20 degrees than at 40 degrees because there is more attenuation by the canopy at 40 degrees. Terms expected to play a significant role are from the direct ground scattering, large branch-ground interaction and the trunk-ground interaction.

Table 11.11
Parameters for Angular Distribution

Tree Components	β_0, *deg*	β_1, *deg*	β_2, *deg*	β_m, *deg*	*m*
Needle	0	0	90	45	0
Small branch	0	20	90	55	1
Medium branch	20	20	110	65	1
Large branch	30	30	110	70	1
Trunk	10	0	10	5	4

With the vegetation parameters known calculations are performed at P-band with the slope angle varying from -15 degrees to +15 degrees. Figures 11.26 and 11.27 show results of theoretical calculations at incident angles 20 degrees and 40 degrees, respectively, for VV, HH and VH polarizations. For each polarization the contributions from individual tree components are also shown. In figures the small, medium and large branches are denoted as branch 1, 2 and 3, respectively. The anticipated effect of the slope on scattering is that it is more at 20 degrees than at 40 degrees because there is more attenuation by the canopy at 40 degrees. Terms expected to play a significant role are from the direct ground scattering, large branch-ground interaction and the trunk-ground interaction.

For VV polarization the effects of ground scattering and trunk-ground interaction are equally important (see Figure 11.26). An increase in slope towards the radar raises the level of overall scattering to more than 5 dB, since ground scattering increases as the local incident angle decreases. On the other hand, the trunk-ground interaction term has an oscillatory behavior due to the combination of canopy attenuation and its need to see the corner formed by the trunk and the ground in order to generate interaction. Thus, the increase of VV with the slope is not necessarily monotone. Indeed, the trunk-ground term does not peak at the same location for different polarizations. For example, the VH polarization tends to peak towards positive slope while HH peaks very nearly around the 0 degree slope. Overall, scattering behavior is very non-symmetric around the 0 degree slope mainly because of the ground scattering term for like polarizations. For VH polarization, the trunk-ground interaction term itself is not symmetric around the 0 degree slope. The large branch-ground scattering term is generally smaller than the trunk-ground term in this study.

At 40 degrees incidence, the amount of canopy attenuation increases. As shown in Figure 11.27, the slope has a negligible effect on VH polarization. Its effect on VV and

HH are similar in that it generates a peak around the 0 degree slope. This peak is larger for HH than VV as expected because of the difference in the attenuation properties. Furthermore, the peak is almost symmetric around the 0 degree slope because of strong canopy attenuation.

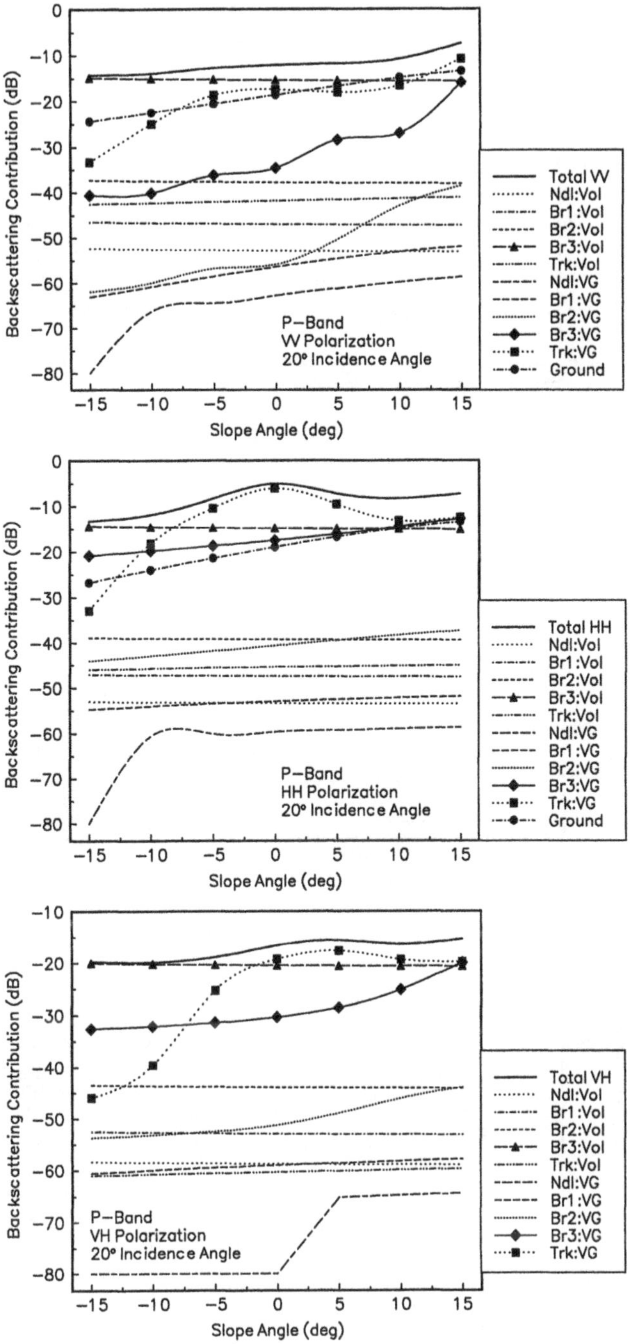

Figure 11.26 Model calculations of backscattering versus the slope angle from a slanted ground plane at 20 degrees incidence.

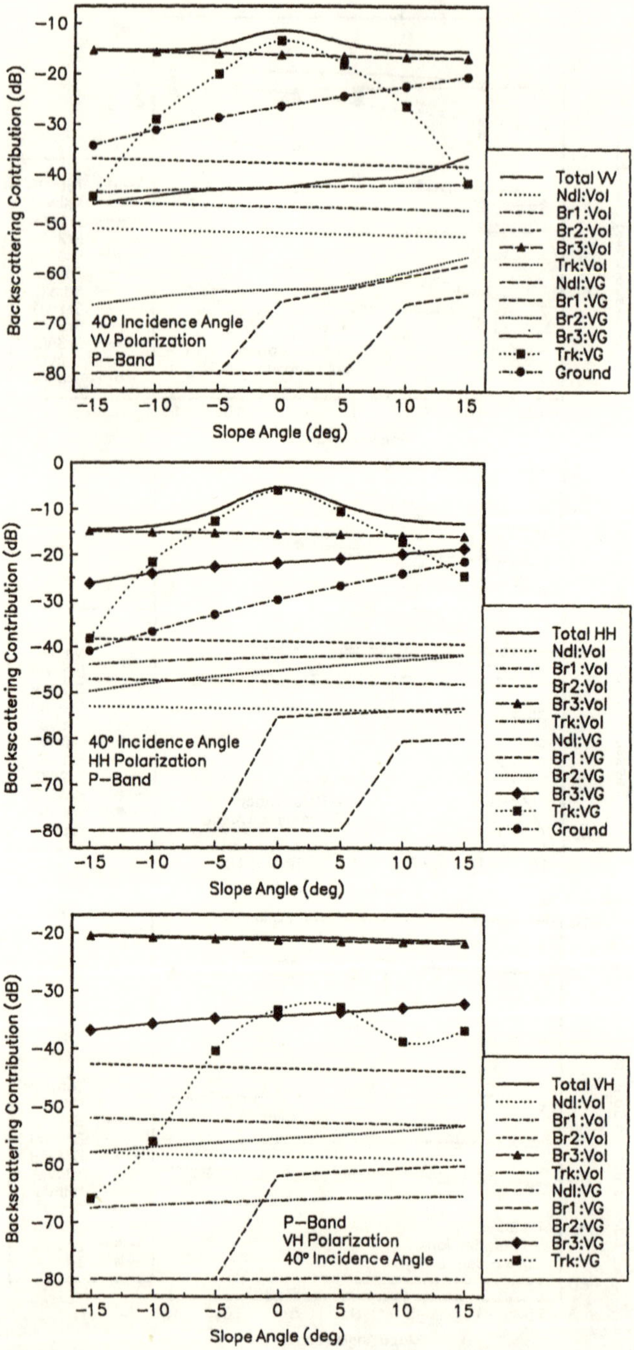

Figure 11.27 Model calculations of backscattering versus the slope angle from a slanted ground plane at 40 degrees incidence.

REFERENCES

Amar, F., A.K. Fung, G. DeGrandi, C. Lavalle and A. Sieber, "Backscattering from Forest Canopies over Slanted Terrain," IGARSS'93, Tokyo, Japan, 1993, pp. 576–579.

Amar, F., A.K. Fung, G. DeGrandi and A. Sieber, "Effect of Component Attenuation on the Total Backscattering from a Forest Canopy," 2nd Annual MAC Europe Meeting, Gras, Austria, 1993.

Balanis, A., *Advanced Engineering Electromagnetics*, John Wiley and Sons, 1989, Chapters 7 and 11.

Barton, I.J., "A Case Study Comparison of Microwave Radiometer Measurements over Bare and Vegetated Surfaces," *J. Geophys. Res.*, Vol. 83, 1978, pp. 3513–3517.

Cimino, J., M.C. Dobson, D. Gates, E. Kasischke, R. Lang, J. Norman, J. Paris, F.T. Ulaby, S. Ustin, V. Vanderbilt and J. Weber, " EOS Synergism Study 1987 Field Experiment Data Report," JPL Technical Report, May 1988

Durden, S. L., J.J. Van Zyl and H. A. Zebker, "Modeling and Observation of Radar Polarization Signature of Forested Areas," *IEEE Trans. Geosci. Remote Sensing*, Vol. 27, no. 3, 1989, pp. 290–301.

Eom, H.J., and A.K. Fung, "Scattering from a Random Layer Embedded with Dielectric Needles," *Remote Sensing of Environment*, Vol. 19, 1986, pp. 139–149.

Fung, A.K, M.F. Chen and K.K. Lee , "Fresnel Field Interaction Applied to Scattering from a Vegetation Layer," *Remote Sensing of Environment,* Vol. 23, 1987, pp. 35–50.

Goedecke, G.H., "Radiative Transfer in Closely Packed Media" GOSA, Vol. 67, no. 10, 1977, pp. 1339–1348.

Hallikaninen, M.T, F.T. Ulaby, M.C. Dobson, M.A. El-Rayes and L. Wu, "Microwave Dielectric Behavior of Wet Soil- part I : Empirical Models and Experimental Observations," *IEEE Trans. on Geoscience and Remote Sensing*, Vol. 23, no. 1, 1985, pp. 25–34.

Ishimaru, A., *Wave Propagation and Scattering in Random Media*, Academic Press, New York, 1978, p. 17.

Ishimaru, A., and R. Cheung, "Multiple Scattering Effects of Wave Propagation due to Rain," *Ann Telcommmun.*, Vol. 35, no. 11-12, 1980, pp. 373–379.

Karam, M.A., F. Amar, D.M. LeVine and A.K. Fung, "Understanding the Relation Between the Forest Biomass and the Radar Backscattered Signals," IGARSS'93, Tokyo, Japan, 1993, pp. 573–577.

Karam, M.A., and A.K. Fung, "Vector Forward Scattering Theorem," *Radio Science*, Vol. 17, no. 4, 1982, pp. 752–756.

Karam, M.A., and A.K. Fung, "Scattering from Randomly Oriented Circular Discs with Application to Vegetation," *Radio Science*, Vol. 18, 1983, pp. 557– 563.

Karam, M.A., and A.K. Fung, "Electromagnetic Scattering from a Layer of Finite Randomly Oriented, Dielectric, Circular Cylinders over a Rough Interface with Application to Vegetation," *Int. J. Remote Sensing*, Vol. 9, no. 6, 1988, pp. 1109–1134.

Karam, M.A., and A.K. Fung, "Leaf Shape Effects in Electromagnetic Wave Scattering from Vegetation," *IEEE Trans. Geosci. Remote Sensing*, Vol. 27, no. 4, 1989, pp. 687–697.

Karam, M.A., A.K. Fung and Y.M.M. Antar, "Electromagnetic Wave Scattering from Some Vegetation

Samples, " *IEEE Trans. Geosci. Remote Sensing*, Vol. 26, no. 6, 1988, pp. 799–808.

Karam, M.A., A.K. Fung, R. H. Lang and N.S. Chuahan, "A Microwave Scattering Model for Layered Vegetation, " *IEEE Trans. Geosci. Remote Sensing*, Vol. 30, no. 6, 1992, pp. 767–784.

Karam, M.A., A.K. Fung, A. Lopes and E. Mougin, "A Fully Polarimetric Scattering Model for a Coniferous Forest, " *Proceedings of the International Geoscience and Remote Sensing Symposium*, Vol. 1, 1991, pp. 19–22.

Lang, R.H., and J.Sidhu, "Electromagnetic from a Layer of Vegetation: A Discrete Approach, " *IEEE Trans. Geosci. Remote Sensing* , Vol. 21, 1983, pp. 62–71.

McDonald, K.C., M.C. Dobson and F.T. Ulaby, "Modeling Multi-Frequency Diurnal Backscatter from a Walnut Orchard, " *IEEE Trans. Geosci. Remote Sensing*, Vol. 29, 1991, pp. 62–71.

Seker, S., "Microwave Backscattering from a Layer of Randomly Oriented Discs with Application to Scattering from Vegetation, " *IEEE Proceedings*, Vol. 133, pt. H., 1986, pp. 497–502.

Stratton, J.A., *Electromagnetic Theory*, New York, McGraw-Hill, 1941.

Tsang, L., J.A. Kong and R.T. Shin, "Radiative Transfer Theory for Active Remote Sensing of a Layer of Small Ellipsoidal Scatterers, " *Radio Science*, Vol. 19, 1984, pp. 629–642.

Ulaby, F.T., and M.A. El-Rayes, "Microwave Dielectric Spectrum of Vegetation Part II: Dual Dispersion Model," *IEEE Trans. Geosci Remote Sensing*, Vol. 25, no. 5, 1987, pp. 550–557.

Ulaby, F.T., R.K. Moore and A .K. Fung, *Microwave Remote Sensing: Active and Passive*, Vol. III, Artech House, Dedham, MA, 1986.

Ulaby, F.T., K. Sarabandi, K. McDonald, M. Whitt and M.C. Dobson, "Michigan Microwave Canopy Scattering Model, " *Int. J. Remote Sensing*, Vol. 11, no. 7, 1990, pp. 1223–1253.

Wait, J.R., "Scattering of Plane Wave from a Circular Dielectreic Cylinder at Oblique Incidence, " *Canadian Journal of Physics*, Vol. 33, 1955, pp. 189–195.

Appendix 11A

An Integral Formulation for the Scattered Field

11A.1 INTRODUCTION

An integral formulation for the scattered field from an object can be derived based on the volume equivalent theorem in electromagnetic theory [Balanis, 1989]. Such a formulation is in terms of the field inside the object. For an object thin in one dimension it is possible to estimate the field inside of it by a static approximation [Stratton, 1949], because the field can be viewed as a constant over a very short distance.

11A.2 EQUATION GOVERNING THE SCATTERED FIELD

From Balanis [1989] the differential equations satisfied by the scattered fields, $\vec{E}^s$ and $\vec{H}^s$, from an object are

$$\begin{aligned} \nabla \times \vec{E}^s &= -\vec{M} - j\omega\mu_0\vec{H}^s \\ \nabla \times \vec{H}^s &= \vec{J} + j\omega\varepsilon_0\vec{E}^s \end{aligned} \tag{11A.1}$$

where $\vec{M} = j\omega(\mu - \mu_0)\vec{H}$ and $\vec{J} = j\omega(\varepsilon - \varepsilon_0)\vec{E}$ and $\vec{E}, \vec{H}$ are the fields inside the object. It is assumed that ε and μ are the permittivity and permeability of the object at points inside the object and become ε_0 and μ_0 at points outside the object.

To find the governing equation for $\vec{E}^s$ of a dielectric scatterer for which $\vec{M} = 0$, we take the curl of the first equation in (11A.1) and substitute into it the curl of the magnetic field from the second equation, yielding

$$\begin{aligned} \nabla \times \nabla \times \vec{E}^s &= -j\omega\mu_0\left(\vec{J} + j\omega\varepsilon_0\vec{E}^s\right) \\ &= \omega^2\mu_0\varepsilon_0\vec{E}^s - j\omega\mu_0\vec{J} \equiv k_0^2\vec{E}^s - j\omega\mu_0\vec{J} \end{aligned} \tag{11A.2}$$

In the charge-free region the above equation reduces to

$$\nabla^2\vec{E}^s + k_0^2\vec{E}^s = j\omega\mu_0\vec{J} \tag{11A.3}$$

which is the governing equation for the scattered electric field. Due to the dual property of the electromagnetic fields the scattered magnetic field must satisfy a mathematically similar equation. An integral representation for $\vec{E}^s$ can be found by integrating (11A.3) using the Green's second theorem as described in the next section.

11A.3 AN INTEGRAL REPRESENTATION FOR THE SCATTERED FIELD

The vector Green's second theorem [Stratton, 1941; Ulaby et al., 1982] can be written as,

$$\int_V [\vec{E}\cdot \nabla\times\nabla\times\mathbf{G}-\mathbf{G}\cdot\nabla\times\nabla\times\vec{E}]\,dV = \oint_S [\mathbf{G}\times\nabla\times\vec{E}-\vec{E}\times\nabla\times\mathbf{G}]\cdot d\vec{S} \quad (11A.4)$$

where $\mathbf{G}$ is the dyadic Green's function related to the scaler Green's function $G(\vec{r}-\vec{r}')$ by [Ulaby et al, 1982]

$$\mathbf{G} = [I+\nabla\nabla/k^2]\,G(\vec{r}-\vec{r}')$$

where $G = -\exp(-jk|\vec{r}-\vec{r}'|)/(4\pi|\vec{r}-\vec{r}'|)$ and $\mathbf{G}$ satisfies the equation below:

$$\nabla\times\nabla\times\mathbf{G}-k^2\mathbf{G} = -\mathbf{I}\delta(\vec{r}-\vec{r}') \quad (11A.5)$$

Upon substituting (11A.2) and (11A.5) into (11A.4) and let the enclosing surface of the volume V recede to infinity we have

$$\int_V \{\vec{E}^s\cdot[k^2\mathbf{G}-\mathbf{I}\delta(\vec{r}-\vec{r}')]-\mathbf{G}\cdot\left[k_0^2\vec{E}^s-j\omega\mu_0\vec{J}\right]\}\,dV = -\vec{E}^s(\vec{r})+j\omega\mu_0\int_V \mathbf{G}\cdot\vec{J}dV$$

$$= 0 \quad (11A.6)$$

Hence, the scattered field is given by

$$\begin{aligned}\vec{E}^s(\vec{r}) &= j\omega\mu_0\int_V \mathbf{G}\cdot\vec{J}dV \\ &= j\omega\mu_0\int_V \mathbf{G}\cdot j\omega(\varepsilon-\varepsilon_0)\vec{E}dV \\ &= -\omega^2\mu_0\varepsilon_0(\varepsilon_r-1)\int_V \mathbf{G}\cdot\vec{E}dV = -k^2(\varepsilon_r-1)\int_V \mathbf{G}\cdot\vec{E}dV \\ &= -k^2(\varepsilon_r-1)\int_V [I+\nabla\nabla/k^2]\,G(\vec{r}-\vec{r}')\cdot\vec{E}dV \\ &= \frac{k^2(\varepsilon_r-1)}{4\pi}\int_V \left[\left(I+\frac{\nabla\nabla}{k^2}\right)\left(\frac{exp(-jk|\vec{r}-\vec{r}'|)}{|\vec{r}-\vec{r}'|}\right)\right]\cdot\vec{E}dV\end{aligned} \quad (11A.7)$$

where $\vec{E}$ is the field inside the scatterer. Since our final objective is to consider either the vertically or horizontally polarized scattered field component that is orthogonal to the direction of observation, we can drop the longitudinal term in (11A.7) and the expression for the $\hat{p}$ polarized scattered field component simplifies to

$$\hat{p}\cdot\vec{E}^{s}(\vec{r}) = \frac{k_0^2(\varepsilon_r - 1)}{4\pi}\int_V \frac{\exp\left(-jk\left|\vec{r}-\vec{r}'\right|\right)}{\left|\vec{r}-\vec{r}'\right|}\left(\vec{E}\cdot\hat{p}\right)dV \qquad (11A.8)$$

where $\hat{p}$ is the unit polarization vector transverse to the direction of observation.

Appendix 11B
Permittivity of Vegetation

11B.1 INTRODUCTION

A formula for the permittivity of vegetation has been developed by Ulaby and El-Rayes [1987] in terms of the gravimetric moisture content, M_g. They also provide another formula in terms of the volumetric moisture content, M_v. Since these two quantities are related through the density of the dry matter, ρ, different values may be obtained from these formulae depending upon ρ. For both formulae the conductivity σ is taken to be unity.

11B.2 PERMITTIVITY OF VEGETATION GIVEN GMC

When the gravimetric moisture constant (GMC) is given, the nondispersive residual component of the dielectric constant is

$$\varepsilon_n = 1.7 - 0.74M_g + 6.16M_g^2 \tag{11B.1}$$

The free-water volume fraction is

$$vf_f = M_g(0.55M_g - 0.076) \tag{11B.2}$$

while the volume fraction for the bound water is

$$vf_b = (4.64M_g^2)/(1 + 7.36M_g^2) \tag{11B.3}$$

With the above quantities known, the permittivity of vegetation is given by

$$\begin{aligned}\varepsilon(M_g) = {} & \varepsilon_n + vf_f\{4.9 + 75/[1 + j(f/18)] - j(18\sigma/f)\} \\ & + vf_b\{2.9 + 55/[1 + (jf/0.18)^{0.5}]\}\end{aligned} \tag{11B.4}$$

11B.3 PERMITTIVITY OF VEGETATION GIVEN VMC

The volumetric moisture constant is related to the gravimetric moisture content through the dry matter density ρ as follows:

$$M_v = \rho M_g/[1 - M_g(1 - \rho)] \tag{11B.5}$$

The nondispersive residual component of the dielectric constant is

$$\varepsilon_n = 1.7 + 3.2M_v + 6.5M_v^2 \tag{11B.6}$$

The free-water volume fraction is

$$vf_f = M_v(0.82M_v + 0.166) \tag{11B.7}$$

while the volume fraction for bound water is

$$vf_b = (31.4M_v^2)/(1 + 59.5M_v^2) \tag{11B.8}$$

With the above quantities known, the permittivity of vegetation is given by

$$\begin{aligned}\varepsilon(M_v) &= \varepsilon_n + vf_f\{4.9 + 75/[1 + j(f/18)] - j(18\sigma/f)\} \\ &\quad + vf_b\{2.9 + 55/[1 + (jf/0.18)^{0.5}]\}\end{aligned} \tag{11B.9}$$

Note that except for a re-interpretation of the volume fractions and the residual dielectric component the expressions in (11B.4) and (11B.9) are the same.

Chapter 12

Applications of Electromagnetic Scattering Models to Parameter Retrieval and Classification

Michael S. Dawson

12.1 INTRODUCTION

Up to this point we have considered a class of problems in which we know the geometric and physical characteristics of the medium and its boundary and wish to determine the far-zone scattered fields and thus the scattering coefficient or emitted brightness temperature. We now consider a different class of problems in which the measured brightness temperatures or scattering coefficients are known and we wish to determine the characteristics (permittivity, surface roughness, volume scattering attributes etc.) of the medium that could produce such measurements. This problem is referred to as parameter *inversion*. Another closely related problem considered is that given a series of measurements, we would like to determine which of several types or classes of object could be responsible for the measurements (a classification problem). This section examines how electromagnetic emission and scattering models such as those developed in Chapters 2 and 3 can be used to aide the retrieval of parameters and the classification of measured targets. Examples of how the scattering models (SM) can be used in conjunction with neural networks (NN) and unsupervised classification methods to perform retrieval of parameters and classification from remote sensing data are given.

12.2 BACKGROUND

In the last two decades a new emphasis has been placed on looking at different components of the Earth not as individual constituents but rather as highly interconnected processes. The increasing need to monitor and evaluate geophysical information on a continuous global basis has made it necessary to derive methods to extract specific data products from satellite-based platforms such as EOS. Inversion and classification of remote sensing data are commonly employed to retrieve this information. With the ever-increasing volume of data available to the remote sensing and scientific community, new processing algorithms are being explored in an attempt to synergize data sets and/or simplify the data handling requirements for classification and inversion of geophysical parameters. This work examines the combined use of electromagnetic scattering and emission models with neural networks and/or unsupervised classifiers to perform parameter retrieval and classification from remotely sensed data.

The retrieval of scattering parameters can be viewed as a mapping problem from the domain of measured signals to the range of surface/medium characteristics that quantify the observed medium. One problem that may not be obvious is that the mapping of scattering parameters to the scattering coefficient or emissivity is not always unique. That is, in some circumstances, equivalent scattering coefficients or emissivities can be generated from distinctly different media that have very dissimilar scattering mechanisms and scattering parameters. This can become problematic for closed form inversion methods. Explicit inverse electromagnetic models have been examined for some years [Tsang et al., 1987]. However, the complexity of the problem has limited its usefulness. To date, parameter inversion has been based largely on empirical models [Mo et al., 1987; Westwater and Cohen, 1973]. Empirical models have usually avoided the nonuniqueness problem by limiting the validity of the model to a single parameter and a narrow range. This limit on the range of validity requires that multiple empirical models be created— one model for each parameter.

The inversion method shown in this work addresses the problem of nonunique parameter inversion in two ways. First, the method selects useful inputs from all available data whereas the empirical methods generally use only a few channels of the available data. This omission of data can clearly dispose of useful information that may remove some ambiguities within the data. Second, the method described here uses the electromagnetic scattering models of Chapters 2 and 3 to (a) determine an effective range for inversion via a *sensitivity analysis* and (b) train a neural network (NN) that is used to perform the actual inversion. The incorporation of the NN to perform inversion is an important advancement in current retrieval techniques. The combination of a scattering model (SM) and NN makes it possible to perform inversion with higher accuracy and in real time. A more detailed introduction of the neural network will be given in Section 12.3.

In addition to the use of SM-NN techniques to perform inversion, we also consider the use of electromagnetic scattering models and neural networks to perform classification. Classification of remote sensing data is commonly performed using conventional statistical or empirical discriminants [Grody, 1991; Nystuen and Garcia, 1992]. Here, we show how the neural network can be trained with the scattering models to perform classification. Another use of the combined SM-NN method for classification comes in the identification and labeling of classes from unsupervised classification methods. For situations in which ground truth is not available, researchers have used the output of unsupervised clustering algorithms as training data for neural networks [Heermann and Khazenie, 1992]. The resultant classified images are based solely upon signal statistics that, generally, have not been correlated with ground truth information.

There is as yet no standard way to identify classes. The use of the SM is a possible way to identify classes derived from an unsupervised classifier or clustering algorithm.

The neural network (NN) is much different from conventional programmed computing in that programmed computing requires that the inversion algorithm be known exactly. If the algorithm is not known, the costly and time-consuming process of developing the rules of the algorithm must be undertaken. For the cases of parameter retrieval and classification from remote sensing data, this may be quite difficult due to the many nonlinear and poorly understood factors involved. The NN, in contrast to conventional methods, does not require that the relationship between the input vectors and the output vectors be known. The NN determines the relationship between the inputs to the network and the outputs from the network directly from the training data.

Classification of remote sensing data has traditionally been performed by Bayesian and other statistically based classifiers. The drawback of these classification methods is that the underlying distribution needs to be assumed and that the classifier is theoretically optimal if and only if this assumption concerning the probability density functions (PDF's) is correct [Kay, 1993]. The neural network, in contrast, does not require that a particular form of a PDF be assumed nor does the NN approach require that the relation between the estimator inputs and outputs be known. Instead the NN draws its own input-output discriminant relations directly from the data. For the case of classification, the NN uses data it has seen to draw decision boundaries. In addition, the need for the user to determine the weighting on feature importance is also not required in the application of the NN as it is in conventional multi-spectral classification applications. Another key factor in the use and implementation of the NN is that it is inherently a parallel processing structure and thus potentially has a high throughput. In time, hardware implementations of neural networks should allow training of NN and its applications to classification and inversion to be performed in real time.

The neural network has recently received considerable attention in the last few years by researchers looking for alternative methods to perform classification in many areas including character recognition, speech recognition, and feature extraction to name a few. The neural network has been used by many experimenters in the remote sensing community. Bischof et al. [1992], Heermann and Khazenie [1992], and Lure et al. [1992] have used neural networks to perform classification of Landsat and SAR data. Fitch et al. [1991] used an optimized neural network to detect ship wakes from SAR images. Similarly, numerous researchers including Chen et al. [1992] have used the neural networks to perform inversion for snow parameters. One problem, however, has been that the training time of the networks was quite large. Heermann and Khazenie reported training times as large as three weeks for large data sets and as small as three days for smaller data sets. In addition, Hammerstrom [1993] reported training times for some networks as long as a *year* on a Sparc workstation. In a comparison of neural network and statistical classification of remote sensing data, Benediktsson et al. [1990] noted that the neural network was superior to the statistical method in terms of classification accuracy; however, the training time was the largest problem in its use. Recent work in the area reported by Dawson et al. [1992, 1993], and Manry et al. [1993]

shows that optimized training methods can drastically reduce the training times by one to two orders of magnitude. The results given in this work employ this optimized training method.

After giving a brief overview of the neural network, several different applications of the SM to perform inversion and classification will be examined: (1) inversion of surface scattering parameters from simulated and measured active C-band and X-band data, and (2) classification of simulated active sea ice and measured SSM/I radiometric data. We begin by first discussing the background of the NN.

12.3 OVERVIEW OF ARTIFICIAL NEURAL NETWORKS

Artificial neural networks, or simply neural networks, are mathematical models originally designed in the mid 1950's to mimic the first-order behavior of the human nervous system. During subsequent studies, researchers found that these mathematical models had great potential to approximate arbitrary functions. In the late 1960's, Minsky and Papert [1969] showed that the single layer perceptron of Rosenblatt [1959] was able to perform some linear discriminant functions for simple problems. However, it was unable to perform slightly more complex problems such as the XOR logic function. The solution to this once formidable problem was to add additional layers of the processing units or *neurons*. Another important milestone was the development of learning algorithms that could effectively train these multilayer networks. The best known method for training the multilayer perceptron (MLP) is the back-propagation (BP) algorithm which was originally introduced by Werbos [1974] and later rediscovered by Rumelhart and McClelland [1986]. The reintroduction of the BP learning algorithm by Rumelhart and McClelland marks the beginning of the resurgence of neural networks. Since that time, many variations of the original BP algorithm have emerged.

Another important area of study has been in the theory of the inherent capabilities of the neural network to perform function approximation. Hornik et al [1989], Hartman et al. [1990], and Gallant and White [1988] have shown that the MLP with a single hidden layer having nonlinear activation is capable of approximating any real-valued continuous function provided a sufficient number of units within this hidden layer exist. In this sense, multilayer feedforward networks form a class of *universal approximators*.

When discussing the capabilities of neural networks to perform classification, Ruck et al. [1990] have shown that the MLP trained with the back-propagation algorithm approximates the Bayes optimal discriminant function. Yau and Manry [1991] have also shown the equivalence between Gaussian classifiers and sigma-pi neural networks. In studies by Kohonen et al. [1988], Draper et al. [1987], and Benediktsson et al. [1990], it was shown that the neural network performed as well as or outperformed conventional classification methods while making data handling simpler and faster. The primary difference, as noted by Benediktsson et al., was that the neural network does not require any prior knowledge about the statistical distributions in the data sources in order to apply the neural network for classification.

Although there are many types of neural networks including adaptive resonance theory (ART), bidirectional associative memories (BAM), and self-organizing maps (SOM), the emphasis of this work will be on the feedforward MLP, which has been found to be the best suited network for classification and inversion [Hsu et al., 1992]. The differences in these many types of networks are dependent primarily on their interconnecting *topology* and the method by which they learn [Lippmann, 1987]. The term *topology* refers to the structure of the network as a whole: the number of inputs, the number of outputs, the number of hidden layers (i.e., the layers between input and output layer) and the number of neurons in each hidden layer. The networks used here will all have multiple layers and will be used to represent functional mappings $f{:}R^n \rightarrow R^m$ between the *n*-space inputs and the *m*-space outputs. Because the actual function of the neural network is mapping of one space to another, there is no difference in neural network applications between forward mapping and inversion. Thus a physical parameter can always be determined using neural networks, if the measurement used as input to the neural network is sensitive to it.

As shown in Figure 12.1, the MLP consists of multiple layers of basic processing units, which are commonly referred to as *neurons*. The individual neuron is the elemental building block of each layer within the network. This neuron can be conceptually viewed as an element which processes one or more input signals $\boldsymbol{x}$ by (1) taking its inner product with a weight vector $\boldsymbol{w}$ and then adding a *bias term* θ, and (2) putting the resultant number through a generally nonlinear *activation function* to produce a single output termed the *activation level* of that neuron.

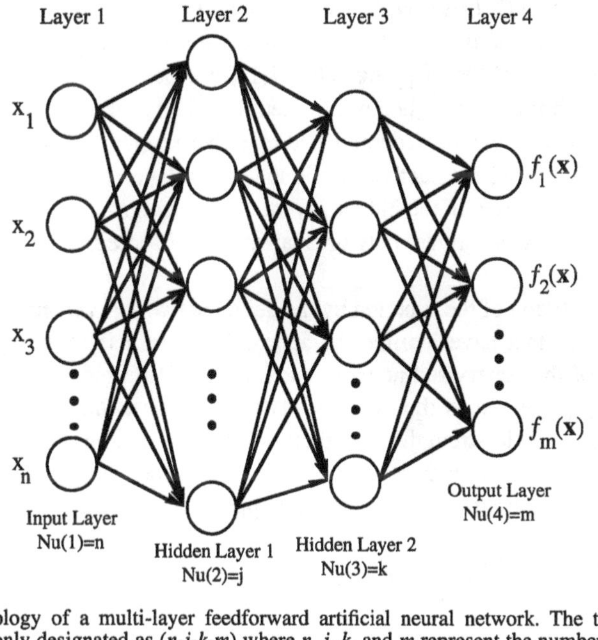

Figure 12.1 Topology of a multi-layer feedforward artificial neural network. The topology of the network is commonly designated as (*n-j-k-m*) where *n, j, k,* and *m* represent the number of elements in the first, second, third, and fourth layers in the networks respectively.

The sum $x \cdot w + \theta$ is commonly termed the *Net* of the neuron. Figure 12.2 indicates the arrangement of an individual neuron. The input to the neuron can either be the actual input to the system x or the output from other neurons in preceding layers as shown in Figure 12.1.

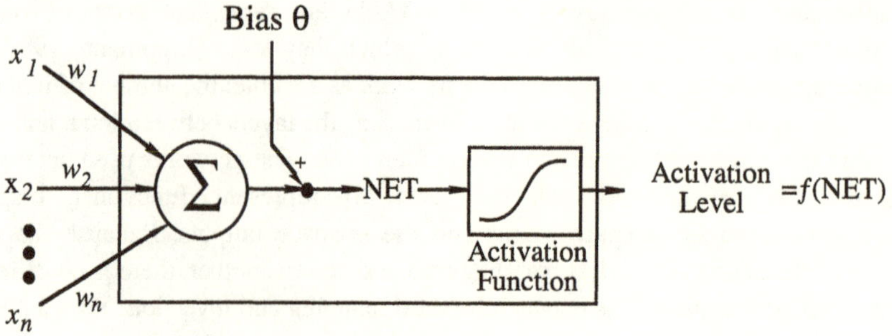

Figure 12.2 Conceptual view of an artificial neuron.

The activation function, commonly represented by *f*(Net), can have many forms. The most common activation function is the nonlinear sigmoid function given by

$$f\,(\text{Net}) = \frac{1}{1+e^{-\text{Net}}} = \frac{1}{1+e^{-(w \cdot x + \theta)}} \tag{12.1}$$

Other functions such as the simple linear activation function, the threshold activation, and the hyperbolic tangent activation are used. However, the sigmoid activation given by (12.1) is by far the most prevalent activation used. Activation functions used at the output of each neuron yield values typically in the range [-1, +1]. Since the output units of the mapping network must generally produce an estimate of a parameter with arbitrary range (i.e., not limited to [-1, +1]), the output range restriction must be removed. This can be accomplished by either scaling the inputs and outputs of the network or by using linear activations in the output layer, as opposed to the sigmoid activations, which are used in the hidden layers. Scale changes in the input and output of the network may, for feedforward networks, always be absorbed in the weight and biases and vice versa.

The representation of the function by the neural network is accomplished by a set of individual neurons that have *learned* the appropriate response to an input. During the training phase of the supervised network such as the MLP used here, training patterns are sequentially presented to the network. After all patterns have been presented, the interconnecting synaptic strengths (weights), w, of each neuron are adjusted such that the functional approximation created by the neural network minimizes the global error between the desired output and the summed output created by the network given by

$$\text{Error} = \sum_p E_p = \frac{1}{2}\sum_p \sum_i [T_{pi} - a_{pi}]^2 \tag{12.2}$$

where T_{pi} is the ith desired output of the pth training pattern, and a_{pi} is the corresponding activation level of the ith unit in the output layer. In (12.2), the values of i are summed only for output units. We can think of the trained neural network as a type of nonlinear, least mean squares interpolation formula for the discrete set of data points in the training set. Clearly, the accuracy of the approximation is dependent on the training data.

The modification of the interconnecting weights within the network during the training phase both positions the decision boundaries and determines the sharpness of transition. This essentially allows the neurons to modify their output in response to their input. Thus, the network is able to "learn" a response. Learning in the network can be viewed as minimizing the sum of the squared errors between the desired outputs and the computed network outputs as given in (12.2). One problem with the back-propagation learning method is that the resultant error is highly dependent on the initial weights and may not always be able to find the *global* minima of the error function. The fast-learning (FL) algorithm [Manry et al., 1993] used in this work has been shown to be less sensitive to this problem. The FL method differs from BP in that the output of the FL-trained MLP is expressed as a linear function of the output weight vector $\boldsymbol{W_O}$, and linear equations can be solved to minimize the training error. Readers are referred to Rumelhart and McClelland [1986] and Manry et al. [1993] or Dawson et al. [1993] for information regarding the specifics of the back-propagation and fast-learning algorithms respectively.

Before we go on to examples of SM-NN inversion and classification applications, some relevant differences between inversion and classification training data representation and NN/conventional training data selection should be pointed out.

When considering training data for neural networks that are to perform classification, it is necessary that the classification training data fully represent all of the domain of the input set rather than primarily the means as in a k-means classification. This is due to the fact that the NN needs to know where to position the discrimination planes in n-dimensional space. The performance of the neural network in classification problems is more dependent on having representative training samples, whereas the statistical classifiers need to have an appropriate distribution model for each class. This is an important point and highlights the difference between neural network and conventional statistical classification.

The form of the network outputs really depends on the application of the NN. For example, in mapping problems where we are interested in the relation of a specific input signal to an output such as the surface roughness or effective optical depth of a layer, the output of the NN is a continuous real value. For classification applications, however, the outputs will be binary representations for each class. When trained, the network should produce 0's (1's) for all classes except the true class and a 1 (0) for the correct class.

12.4 APPLICATION OF SM AND NN TO PARAMETER RETRIEVAL

This section presents the use of a combined SM-NN method to retrieve geophysical information from remote sensing data. Previous methods to extract information have, for the most part, been based largely on empirical models or signal statistics. Conventional inversion methods commonly require that after one collects the data, one must analyze the data by performing correlation and statistical analysis, make initial mapping formulae, check and double-check the models and then adjust and fine-tune the empirical models until the final form is derived. The development of the rules in these inversion models can take a considerable amount of time. Another problem is that the conditions under which the empirical models were developed depend directly on the data which may vary seasonally or regionally or with system parameters and for a specific set of target conditions. Thus, many empirical models may have to be developed each with a limitation of its own.

The neural network, in contrast, does not require the derivation of rules or signal statistics. Instead, the neural network determines the inverse mappings and functional relations *directly from a validated scattering or emission model.* This combined SM-NN technique eliminates the need for the development of an algorithmic procedure and/or a set of rules — an undertaking that, in general, has been found to be costly and time consuming. Thus, the SM-NN inversion technique avoids a large amount of the preprocessing required in the derivation of other inversion algorithms. The ease with which the SM-NN technique can be implemented and the quality of results makes it a powerful tool whose advantages have not been fully utilized by the remote sensing community.

The training of any neural network begins by defining the network inputs and outputs that are controlled by the type of application. Once the inputs and outputs of the neural network have been determined, the BP or FL training methods can be used to train the NN. The FL method of Manry et al [1993] is preferred due to its rapid convergence and fewer problems with local minima. The SM-NN inversion technique can be broken down into five basic steps.

1. *Selecting the bounds of the SM parameters via sensitivity analysis.* This phase of the analysis deals with the sensitivity of a validated scattering model to the parameters that are to be recovered and determines the bounds within which the parameters can be recovered unambiguously.
2. *Generation of NN training data using the scattering model.* A set of SM input-output pairs for training the neural network are generated from the SM. The inputs to the SM are randomly selected from within the bounds determined in step 1 above.
3. *Selection of optimal NN training data.* The analysis of the sensitivity of neural networks to different input-output training pairs (channels of SM data) is performed to determine the optimal set of inputs to the neural network. Although not always necessary, this step can remove ambiguous data or channels that are highly correlated with other data and provide no additional information. The training time of the neural network is dependent on the number of inputs to the network and the reduction of the number of network inputs can reduce the amount of time required for training the network. This step is desirable whenever there is a choice of the type and the number of inputs.
4. *Training the NN using FL or BP algorithms.* (See Manry et al. [1993] and Rumelhart and McClelland [1986] for learning algorithms.)

5. *Using the trained NN to retrieve the parameters of interest.* Once the NN is trained, it can be applied to estimate scene parameters form remotely sensed data.

12.4.1 Selection of Network Inputs via SM Sensitivity Analysis

Before the neural networks can be trained for parameter retrieval, a number of considerations regarding the information content of the training data need to be made. This is independent of the source of the training data whether the data come from measurements or from a model. By performing a *sensitivity study* on the training data, we recognize the range of parameter inputs to which the data is sensitive. Data that have been found to be sensitive to a parameter of interest can then be applied as inputs to the neural network to effectively retrieve that parameter. Selecting the inputs to the network in this manner eliminates unnecessary or misleading inputs that may confuse the network. If training data cannot be found that are sensitive to a parameter of interest, inversion for that parameter would be ineffective, if not futile.

In this study, a closed-form solution of the forward mapping relation (i.e., a scattering model) exists. Thus, a sensitivity analysis can be performed by examining the dependence of the output of the SM to the input parameters. As the output of the SM becomes saturated or insensitive to a parameter such as the layer permittivity or surface roughness, the parameter inversion range will need to be narrowed so that the forward mapping yields an unambiguous value. Once the sensitivity of the model to each parameter is known and the range of nonambiguous relation between the inputs and the model outputs has been determined, we are ready to use this range of parameters to create training data for the network. The SM is repeatedly run with the inputs to the model being selected randomly from within the bounds determined from the sensitivity analysis. Note that if this step were ignored and inputs not relevant to the parameter to be retrieved were used in training, the network would be confused and would not produce satisfactory results [Pierce, 1992].

Up to this point we have discussed the sensitivity of the model to parameters such as the layer permittivity and surface roughness but have not discussed how angular or frequency dependence information can be used during the training of the neural network. Some authors choose to use the incident angle or frequency as an input parameter to the network just as the scattering coefficients are inputs to the network. The problem, however, is that the input-output relation of the SM model to frequency or the incident angle is very complex and the unambiguous mapping problem is more of an issue. If the NN is used for this set of inputs, the number of hidden units and training time can become quite large. This problem can be avoided if a different input configuration is used. Rather than trying to perform the inversion using a single input measurement at, for example, 40 degrees and vertical polarization and having 40 as an input to the network, experimental results indicate that it is much more advantageous to apply several measured signals representing the trend of an angular or frequency response which has been digitized. In such a configuration, input 1, for example, is always vertical polarization at 15 degrees incidence and input 2 is always horizontal polarization at 30 degrees. This form of inputs to the network is also easier to implement when considering measured data.

In remote sensing, two forms of measured data are usually encountered; single-frequency multi-angle data and single-angle multi-frequency data. Figure 12.3 illustrates the input configuration for digitized multi-angle or multi-frequency data applied as inputs to a neural network. Data having an angular variation are more sensitive to changes in surface roughness. Thus, data that have angular variation are preferred for extracting parameters relating to surface roughness characteristics. Multi-frequency data, on the other hand, are more sensitive to changes in the dielectric value of a scene and are preferred for inversion of parameters that are frequency sensitive, such as water content.

12.4.2 Inversion of Surface Roughness Parameters Using a Neural Network Trained With Scattering Models

The combined SM-NN technique is applied to the problem of retrieval of parameters from a statistically rough surface. The geometry of the scattering problem for a single randomly rough dielectric half space is shown in Figure 12.4. Simulated data sets based on the IEM surface scattering model of Chapter 2 [Fung et al., 1992a] are used to train the neural network so that the training data may be viewed as taken from a completely known randomly rough surface. The IEM surface model is chosen over other surface models to aid in the surface inversion due to its wider range of validity than the Kirchhoff and SPM models and its computational advantage over more complex models such as the phase pertubation model (PPM) [Winebrenner and Ishimaru,1985] and the full wave model (FWM) [Bahar, 1980]. A study by Chen and Fung [1992] has demonstrated that in comparison to the PPM and the FWM, the IEM model is the least numerically intensive model while providing full polarimetric results.

The training of the neural network with the SM allows us to *(a) vary the surface parameters freely within the established bounds* and *(b) ensure the consistency of the training data.* If we use measured data for the training phase, it would be difficult to find the data corresponding to a variety of known surface conditions. In addition, we would have to account for the different time-varying and system-related effects that the data would contain. For the testing phase of inversion of parameters from a randomly rough surface, SM outputs will also be used to test the network so that the results may be validated more easily.

For backscattering from a randomly rough dielectric surface, the backscatter coefficient, $\sigma°$, is given by Fung et al., [1992a, or Chapter 2, (2.26)] and is repeated below for convenience:

$$\sigma°_{qp}(\theta, k\sigma, kl, \varepsilon_r) = \frac{k^2}{2} e^{-2k_z^2\sigma^2} \sum_{n=1}^{\infty} |I_{qp}^n|^2 \frac{W^n(-2k_x, 0)}{n!} \tag{12.3}$$

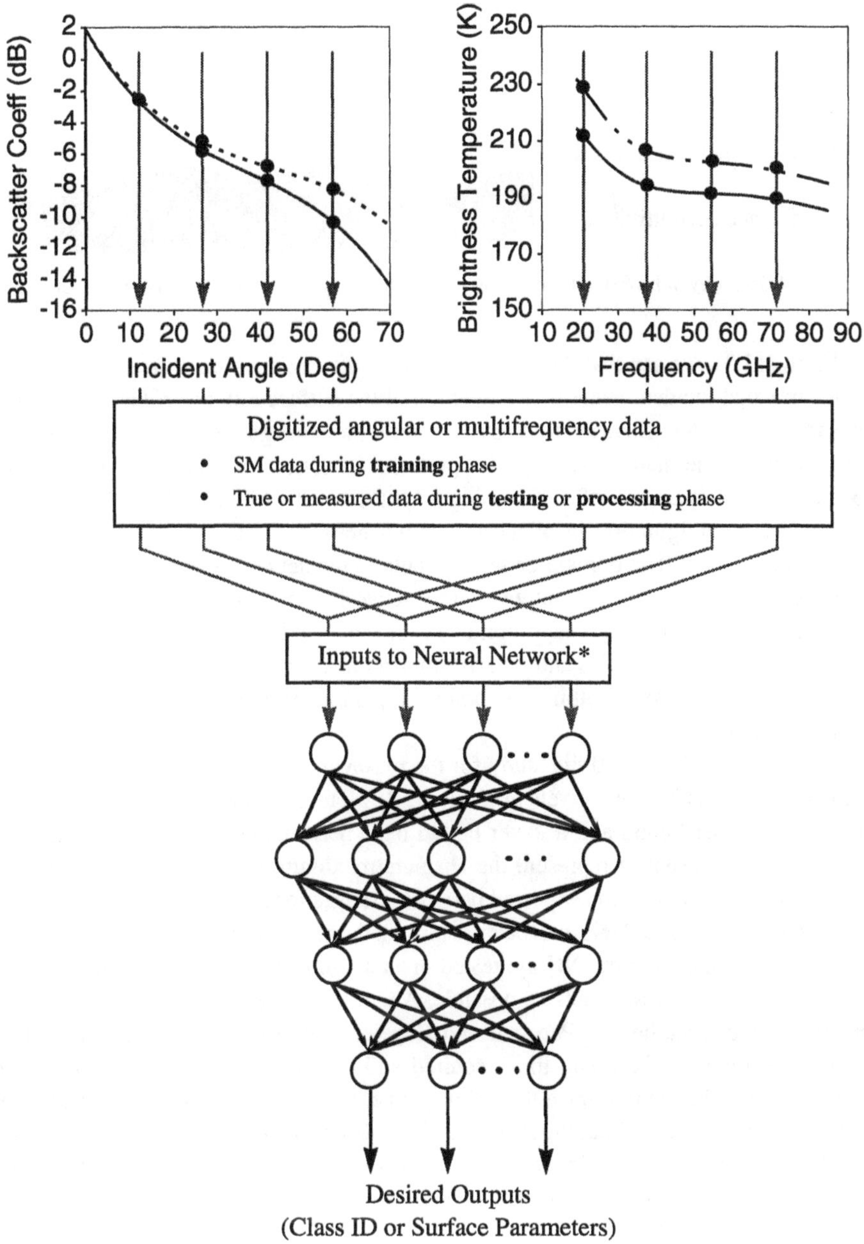

Figure 12.3 Multiangular and multifrequency configuration input to a neural network.

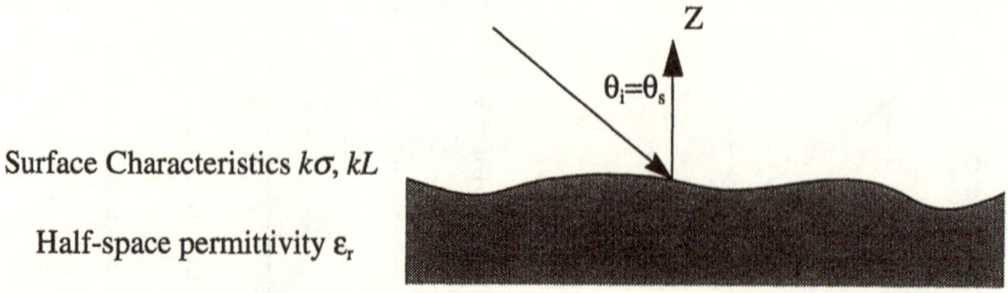

Figure 12.4 Geometry of inversion from a homogeneous half space with irregularly rough surface.

From (12.3), we see that the measured backscattering coefficient from a randomly rough surface at incident angle θ is a function of the surface roughness $k\sigma$, the surface correlation function characterized by a correlation length kl, and the relative permittivity of the half space ε_r. To perform inversion of the input parameters $(k\sigma, kl, \varepsilon_r)$ from measured $\sigma^\circ_{qp}(\theta)$ using a MLP trained with the IEM surface model, we consider a configuration in which the neural network will be used as a mapping function $f{:}R^n \rightarrow R^m$. The domain or set of inputs to the network is the set of measured values (σ°) and the range or output of the network is the set of surface scattering parameters. The selection of an appropriate set of inputs to train the network amounts to knowing the number of angles and polarizations required to unambiguously retrieve the surface parameters from within the bounded parameter range determined from a sensitivity study.

The results of the sensitivity study for the retrieval of parameters from a randomly rough dielectric surface are given in Table 12.1. Note that for each parameter to be retrieved, an upper bound and a lower bound have been defined. Although the ranges indicated are not meant to represent the absolute maximum range possible, the values are fairly representative of the expected parameter ranges for the experiment considered. Once the range of parameters has been determined, many different angle/polarization combinations of inputs to the NN are tested in an attempt to minimize the rms training error while having as few inputs as possible. Although the network can handle a large number of inputs, studies by Perugini and Engeler [1989] have shown a correlation between the amount of training time required and the number of inputs applied to a neural network. Thus to minimize the training time, we also would like to minimize the number of inputs. The optimal inputs required are determined via trial and error. Figure 12.5 illustrates the rms training error as a function of input polarization and angle combinations for a similar problem in which only the surface roughness parameters retrieval were considered.

Table 12.1
Range of Parameters for Training Single Surface Inversion Network

	Lower Bound	*Upper Bound*
ε	1.5	4.0
kσ	0.02	0.9
kL	1.2	4

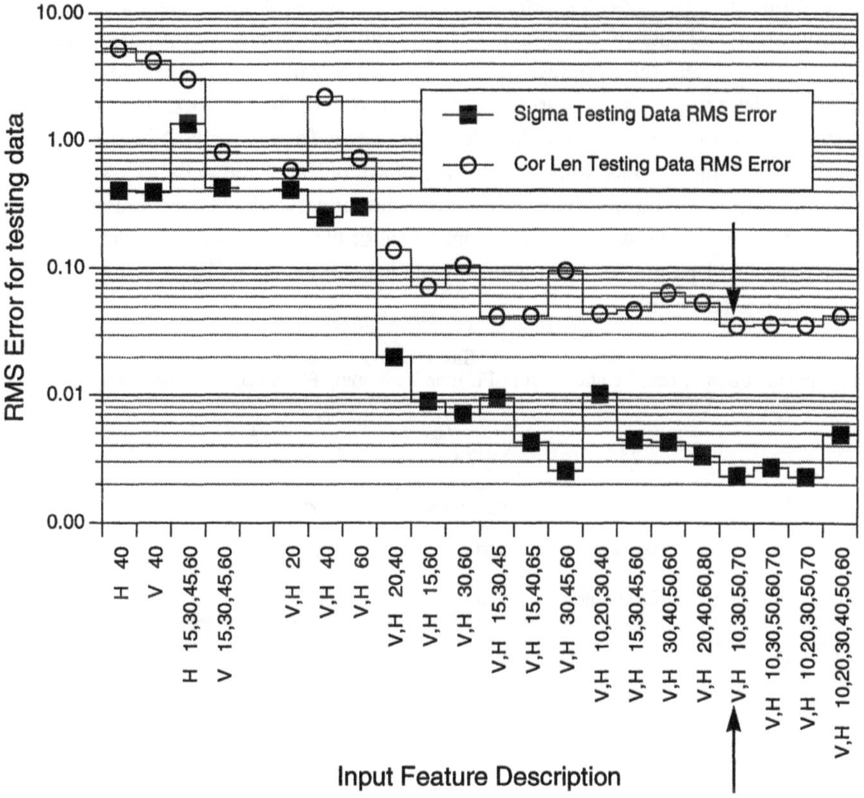

Figure 12.5 NN input analysis for inversion of parameters from randomly rough dielectric surface. V and H represent vertical and horizontal polarizations, respectively.

A satisfactory set of input-output training pairs for the neural network was found to be an input set that consisted of both vertical and horizontal polarizations at incidence angles of 10, 30, 50 and 70 degrees (a total of eight inputs—see Figure 12.5). Once the sets of inputs have been determined, the 8-input 3-output training sets are then applied to a four-layer NN having a 8-40-40-3 topology. All layers in the network were fully connected, which means that the eight inputs had synaptic connections to layer 2 as well as layer 3 and the output layer. Other authors including Rumelhart and McClelland use a simple connection framework in which the inputs to a layer come only from the previous layer and connections to all other layers in the network are not permitted. Networks that are fully connected are generally able to provide higher degree approximations using a smaller number of hidden units in comparison to the simply connected networks.

The NN is then trained with the fast learning (FL) and back-propagation (BP) learning algorithms. A comparison of FL and BP training error and the time required to achieve this training error is given in Table 12.2. Another comparison showing the rms training error and rate of convergence of FL and BP training algorithms is shown in Figure 12.6. Once the training of the 8-40-40-3 fully interconnected network is complete, the network is ready to be used. For testing of the network, the SM is used to create another set of input-output data that is used only for testing the NN. It is important that the inputs and outputs during the testing phase of the neural network have the same meaning as they did during the training of the neural network. The estimated parameter output of the neural network for each applied set of backscatter inputs is then compared to the actual SM parameter and the difference is summed as in (12.2). Table 12.3 gives the numerical test results.

Table 12.2
Comparison of RMS Error, Iterations, and CPU time for training FL and BP for Single Surface Inversion

	Fast Learning	*Back Propagation*
rms error	0.000286	0.00667
Iterations	100	500
CPU time	755 sec.	5337 sec.

Table 12.3
Single Surface Inversion Test Results

		ε	$k\sigma$	kL
Fast Learning	rms error	0.01878	0.00188	0.00345
	Relative std. dev.	0.02573	0.00736	0.00469
Back Propagation	rms error	0.07326	0.02369	0.02535
	Relative std. dev.	0.10038	0.09293	0.03442

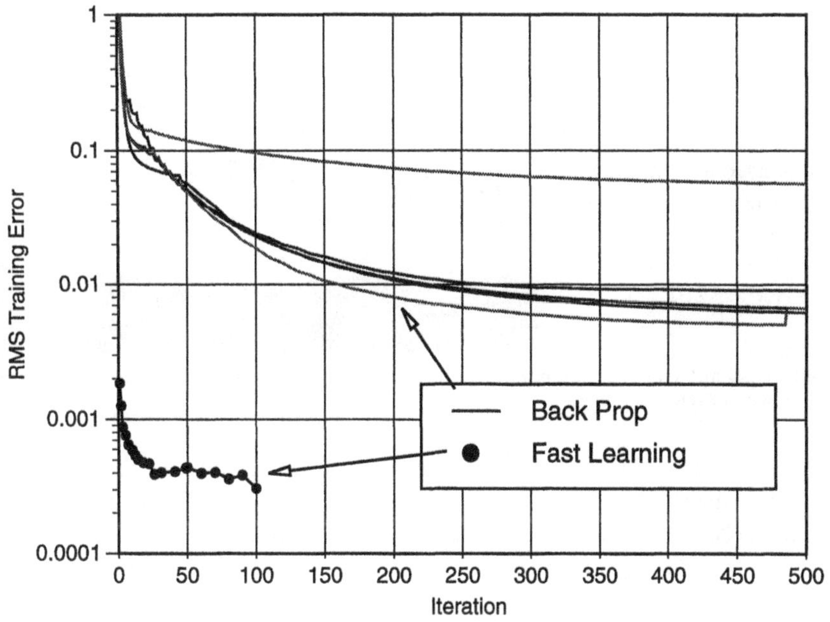

Figure 12.6 Comparison of BP and FL rms error during training of neural network for inversion of parameters from a single randomly rough dielectric surface.

12.4.3 Inversion of Parameters from a Multilayer Medium Using the Combined SM-NN Technique

The real test of the SM-NN inversion technique comes when we consider a more complex scattering medium in which scattering is not dominated by a single boundary but is the result of scattering from the upper and lower boundaries as well as the volume of an inhomogeneous layer. In such a configuration, the measured scattering coefficient is given by

$$\sigma^\circ_{qp}(\theta, \varepsilon_1, \varepsilon_2, k\sigma_1, kl_1, k\sigma_2, kl_2, a, \tau) = \sigma^\circ_{TopSurf} + \sigma^\circ_{BottomSurf} + \sigma^\circ_{Volume} \quad (12.4)$$

where the first term represents the contribution from the upper interface, the second is from the lower interface attributed by the layer above it and the third component on the right side of (12.4) accounts for volume scattering caused by inhomogeneities in the scattering layer. An example of this situation is available from the CRREL 1988 experiment (Bredow, 1989; Onstott, 1990; Fung et al., 1992b; Tjuatja et al., 1993). Figure 12.7 illustrates the geometry of the problem considered.

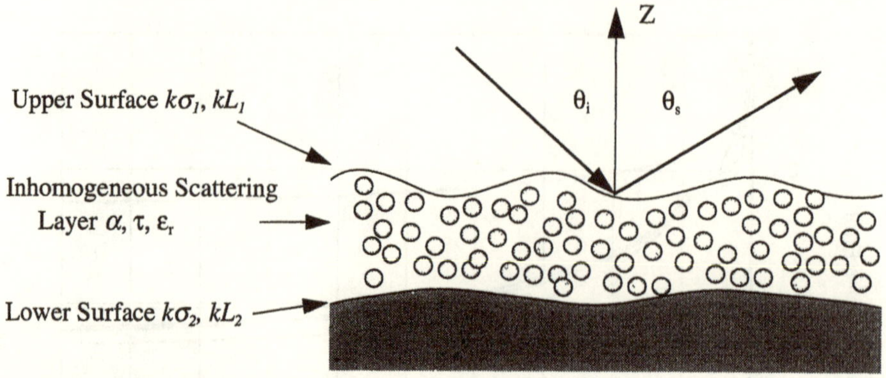

Figure 12.7 Geometry of inversion from an inhomogeneous layer above a homogeneous half space. Both interfaces are randomly rough.

As in the case of inversion of parameters from a randomly rough surface, the application of the combined SM-NN inversion method starts by defining the inputs and outputs of the NN. The input to the network for this single frequency, multi-angle, multi-polarization scattering experiment is the surface scattering coefficient $\sigma^\circ_{qp}(\theta)$ digitized at different angles. The outputs of the network will be the inverted surface parameters $\{\varepsilon_1, \varepsilon_2, k\sigma_1, kl_1, k\sigma_2, kl_2, a, \tau\}$. The single layer scattering model of Chapter 2 is used to generate training data. The SM input parameters were uniformly generated from a bounded set whose upper and lower limits were selected as to fully encompass the possible parameters given in the ground truth data. Subsequent analyses were performed to ensure that the SM was sensitive to all parameters over the entire range defined. Different network configurations were tested to determine the best network topology.

Note that the depth of the scattering layer and the size of the scatterers within the inhomogeneous layer are not explicit products of inversion. This is because these physical properties are not *direct inputs* to the scattering models. They are defined as *indirect inputs* to the scattering models available through the scattering layer albedo a and the optical thickness of the layer τ, which are inputs to the SM. If the intent of inversion is to give an estimate of an indirect variable such as the scatterer radius or the physical depth, it would be necessary to perform a second inversion of these variables from the albedo or optical depth.

During the training phase of the neural network, 10000 training patterns, which consist of vertical (V) and horizontal (H) theoretical backscattering values, σ°, at 10, 30, 50, and 70 degrees as inputs and the corresponding surface parameters $\{\varepsilon_1, \varepsilon_2, k\sigma_1, kl_1, \text{etc.}\}$ as outputs, are used as the supervised FL training algorithm adjusts the network weights. After the training of the network is complete, measured backscattering coefficient from the 1988 CRREL experiment are input into the network while the corresponding estimated surface parameters are output from the network. A comparison of the reported ground truth parameters and the retrieved parameters with the NN trained using the SM are given in Table 12.4. The optical depth and the albedo retrieved are also reported in Table 12.4 although no corresponding ground truth for these parameters are available. After reviewing the ground truth data, the inverted values easily fall into the bounds of the ground truth data.

Table 12.4
Comparison of Parameters Reported in Ground Truth of a CRREL Experiment and Parameters Retrieved by Neural Network.

	CRREL C Band		*CRREL X Band*	
	Ground Truth	*Neural Net Inverted*	*Ground Truth*	*Neural Net Inverted*
Layer permittivity	~3.405	3.415	~3.5	3.73
Upper interface $k\sigma$ (cm)	0.0503 ± 0.01	0.045	0.1005 ± 0.02	0.118
Upper interface kL (cm)	0.701 ± 0.3	1.026	1.4012 ± 0.6	0.769
Scattering albedo a	NA	0.000008	NA	0.00097
Optical depth τ	NA	0.957	NA	1.52
Lower interface $k\sigma$ (cm)	0.0503 ± 0.01	0.042	0.1005 ± 0.02	0.076
Lower interface kL (cm)	0.860 ± 0.3	1.041	1.72 ± 0.6	1.81

As a check of the estimated physical and electrical parameters (the outputs of the neural network), these inverted parameter values were then put into the SM, which predicted the levels as shown in Figure 12.8. The original CRREL measured data are also given for reference. As can be seen in the figure, the theoretical values of $\sigma°$ using the retrieved parameters closely match the original measured data.

It is important to note that because the network was trained with inputs at 10, 30, 50, and 70 degrees, the CRREL data input to the trained network must also be at these angles and polarizations. Once a network has been trained with a particular set of inputs and outputs, the same form of inputs must be used for retrieval. If the original CRREL data did not contain measurement values at 10 and 70 degrees, thus, values for 10 and 70 degrees should be extrapolated from the original measured data.

The previous two examples illustrate how the combined SM-NN can be used to retrieve parameters from remote sensing data. Results of retrieval of parameters using neural networks trained by both the BP and FL training algorithms were illustrated. The results indicate that the FL NN that was trained with SM data is very effective in retrieving the physical properties of a medium that includes both volume and surface roughness scattering contributions. The SM-NN inversion method can be applied to other data including sea surface data or tree canopy parameter inversion as reported by Amar and Dawson [1993].

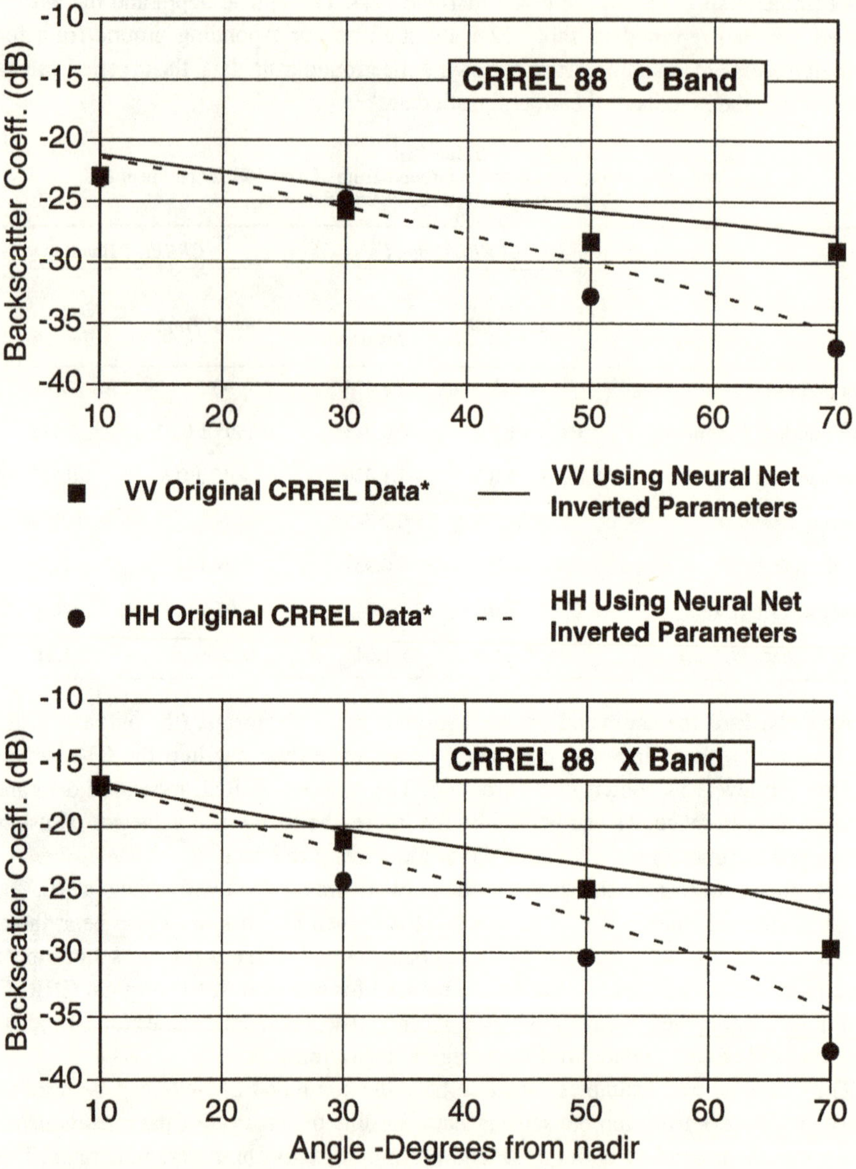

*Values for σ° at 10° and 70° were extrapolated from original data.

Figure 12.8 Comparison of theoretical backscatter response and actual response. Inputs to the theoretical model are parameters inverted from the combined SM-NN inversion technique, which used CRREL 1988 data to estimate parameters.

12.5 CLASSIFICATION OF REMOTE SENSING DATA AIDED BY THE SM

In this section we discuss how the scattering models can aid in the classification of remote sensing data. Two applications of the SM to aid in classification are illustrated in this section. The first case examined is what can be described as a *direct application* of the combined SM-NN method in which a neural network is trained with the scattering models. Once the NN is fully trained using SM input-output training pairs, the trained NN is then used to perform classification of measured data. This direct application of the combined SM-NN method is appropriate for applications in which there is a wide separation between the classes and each of the classes is well characterized by the scattering model. It is also assumed that the bounds for the parameters that characterize each of the classes are fully known.

The second use of the SM for classification examined in this section comes in the identification of classes from unsupervised classification methods. This application is more general in that it is more appropriate for situations in which no ground truth exists or where classes are not well separated. Under circumstances where no ground-truth exists such as in satellite data, an unsupervised classification algorithm is usually used to perform clustering of the raw data. The problem, however, is that the classes generated by this unsupervised classifier are based on statistical measures and the identification of what exactly are we looking at in any cluster in the absence of ground-truth information may be difficult. The SM can play an important role here by providing information concerning the identity of an unlabeled cluster. Note that the unsupervised clustering algorithm is used to form the class discriminants and the SM is used to identify the result of the clustering. Thus we are using the greatest strengths of both the unsupervised classifier and the SM. We begin first by considering the direct application of SM-NN to perform classification.

12.5.1 Classification of Sea Ice Using the Combined SM-NN Method

We consider the classification of multi-angle backscatter data at a single frequency using the SM-NN classification method. The use of the combined SM-NN method to perform classification is very similar to the process used to perform retrieval of parameters in that training data with *known* parameter bounds are generated using the SM model. The distinction between this supervised classification problem and the parameter retrieval problem is that the training bounds for *each* class included in the classification training set needs to be determined. These bounds are selected such that they characterize the expected physical attributes that delineate the dominant scattering mechanisms of a class. The scattering model used account for both surface and volume contributions as well as surface-volume interactions (see Chapters 8 and 9). Once the physical attributes that represent the dominant scattering mechanisms of a class of target have been identified and determined, this information can be used to generate training data for each of the classes of interest. After the generation of the training data is complete, FL or BP can be used to train the NN.

Once the training of the MLP NN is complete, the trained network is tested with simulated data sets based on the known dominant scattering characteristics of each class. Consider the problem where a set of experimental data is to be classified as one of four types: open water, thick first-year ice, thin saline ice, and multiyear ice. Each of the four classes is considered to have a unique angular signature. The case of open water is treated as a homogeneous half space with a randomly rough surface boundary and a known permittivity given by Ulaby et al. [1986]. Thick first-year ice (Class II) is modeled as an optically opaque homogeneous half space with a rough surface boundary and a permittivity around 3.4 - *j*0.22. Fung et al. [1992b] have shown that the return from a thin saline ice layer (Class III) is dominated by the upper and lower surface boundaries at low frequencies when the scattering albedo is small. The case of multiyear ice is treated as a highly inhomogeneous half space with a rough surface boundary. A summary of the four classes is given in Table 12.5.

Table 12.5
Four Types of Ice To Be Classified

Class I	Open water	Highly lossy half space with permittivity of the order 50 - *j*40
Class II	Thick FY ice	Very lossy ice with salinity about 12‰. Treated as a half space with permittivity ~3.4 - *j*0.22
Class III	Thin saline ice	Ice whose return is dominated by the upper and lower interfaces. Scattering albedo is small.
Class IV	Multiyear ice	Ice with a very high-volume scattering component. Permittivity relatively low.

A total of 500 backscatter responses to each of the four classes were generated using the SM requiring approximately fifty minutes on a NeXT workstation running *Mathematica*. The scattering layer parameters ($k\sigma$, kL, ε_r albedo and optical depth) were chosen randomly from a bounded set. The constraints on this bounded set were typical ranges of parameter for each of the four classes. The outputs of the scattering model are then used as training data for both the BP and the FL networks.

The same network topology (8-20-12-4) was used for both BP and FL tests. The training patterns consisted of theoretical VV and HH backscattering coefficients calculated at 15, 30, 45, and 60 degrees with an associated binary class code. Thus, we have eight inputs and four outputs (one binary output for each class). Two hidden layers were used with 20 units in the first hidden layer and 12 units in the second hidden layer. This particular topology was chosen through a trial-and-error method during which many topologies were tested. The initial interconnecting synaptic strengths (weights) were randomly generated using a Gaussian distribution. The classification error of the output during the training phase of the network is plotted as a function of epoch in Figure 12.9. As noted in Figure 12.9, the FL method converged to an acceptable error (error < 1%) in a single iteration while 50 iterations of the BP method yielded 20% to 35% classification error. The BP method was able to converge to the level achieved by the FL method only one time in six tests. It is interesting to note that the same training parameters (identical learning coefficient and momentum constant) were used on some of the BP training attempts, and the only difference was the initial weights ***W*** of the system. This illustrates that the convergence of the back-propagation trained network depending highly on the initial state of the network. During the training of the networks, the MLP trained by the back-propagation algorithm was unable to *consistently* converge to less than 25% training error while the FL method yielded an average error of approximately 1% on the first iteration of training.

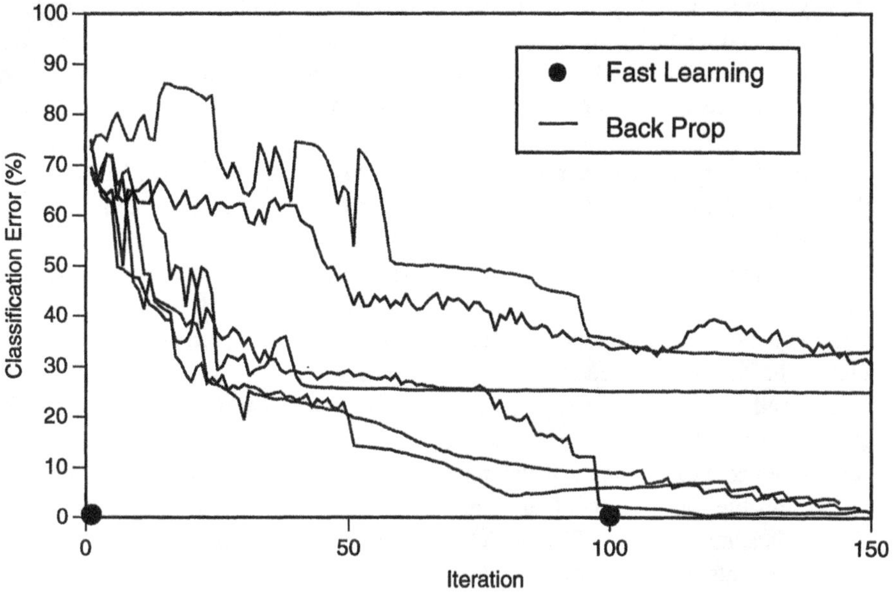

Figure 12.9 Classification training results for simulated backscattering measurements.

After the networks have been sufficiently trained, the trained network is tested by applying a different set of data consisting of theoretical VV and HH backscatter coefficients calculated at the same angles as the training data {15, 30, 45, and 60 degrees}. The resultant class code output from the network is then compared to the actual class ID of the pattern and the classification error is calculated. The testing error for the test data is almost equal to the training data error, which indicates that the neural network successfully formed an internal representation for the functional relations rather than simply memorizing the training data, as can occur in small data sets. The fast-learning method is able to significantly reduce the CPU time necessary to train a neural network as well as consistently yield higher classification accuracies than BP trained networks.

12.5.2 The Use of SM in the Unsupervised Classification Schemes

A large amount of data currently available to the remote sensing community originates from satellite-based platforms. This satellite data usually has little or no ground-truth information that can serve as training data as required in a supervised classification scheme. Because there is no true "answer" to compare the results of a classifier with, an unsupervised clustering algorithm or classifier is commonly employed. Although the unsupervised classifiers or clustering algorithms are well suited for grouping of data into logical sets that have similar features, these classifiers are unable to identify what specific type of scene each cluster represents. Thus the output of the classifier is usually labeled or binned manually either by an intuitive estimate or by an outright guess. This is where a validated SM can be very useful. The SM allows us to identify the unlabeled output from an unsupervised classifier in situations where ground-truth information is not available.

The previous example of Section 12.5.1 illustrates the use of the combined SM-NN method for the classification of sea ice from backscatter data. This direct application of the SM-NN method for classification requires that the dominant scattering characteristics of the medium be *fully* characterized and models that relate the physical characteristics of a medium to the electrical parameter inputs for the SM be available. One problem, however, is that the models that relate the electrical properties required as inputs to the scattering models such as the permittivity of wet snow medium to frequency and wetness, for example, are limited to special cases or a small range of frequencies. In addition, physical descriptions that characterize a class are usually limited to idealized cases and the development of physical models that fully characterize every imaginable physical configuration encountered for a class are usually not available. Thus, it would be difficult to define the bounds of parameters that establish the class discriminants for separating each class.

An alternative to the direct SM-NN classification method is to use the SM in a *post*-classification mode in which the SM is used to label or identify the results from an unsupervised algorithm. In this section we describe how the results from an unsupervised clustering or classification algorithm can be identified using the SM. This verification process ensures that the unsupervised means labeled or identified as a given class are truly representative of that class. In addition, this check may be useful to ensure the quality or consistency of the data.

The application of the SM for verification of classification is illustrated by an example. In this example, the results of an unsupervised ISODATA clustering algorithm [Pao, 1989] are to be used to train a neural network that can be used for mass processing of satellite data. Without the use of the SM to establish the identity of each class before training the NN, we cannot be certain that the data used to train the NN for first-year ice, for example, is representative of that class.

Consider the classification of radiometric data from the space-borne SSM/I platform. The SSM/I is a seven channel, four frequency, linearly polarized, passive microwave radiometer system [Hollinger et al., 1987; Peirce, 1980]. The instrument measures atmospheric and ocean surface brightness temperatures at 19.3, 22.2, 37.9, and 85.5 GHz. All data are at 53 degrees from nadir. In this experiment, data from the Arctic polar region are studied to determine the temporal and spatial characteristics of the polar ice cap. The calibrated seven-band images are first clustered with an unsupervised ISODATA algorithm. The results from the unsupervised clustering algorithm will later serve as pseudo-ground truth for the training of the NN. Before the data can be used in the NN, we use the SM to establish the identity of each of the classes defined by the unsupervised classifier.

Data from March 1, 1988 to March 15, 1988 are averaged and processed to remove any bad data points. The averaged data is then clustered into fifteen clusters using an ISODATA clustering algorithm. One of the first objectives will be to perform a grouping of clusters that are believed to be open water (OW). Out of fifteen clusters generated by the ISODATA algorithm, eight of the clusters have been identified that have very low brightness temperatures and are very likely to be OW. The use of the a validated SM will verify that our assumption as to the identity of these eight clusters is correct. We know from experimental data as well as modeled results that the emissivity of OW at lower frequencies (19 and 22 GHz) is fairly low and rises steadily as we increase frequency. This is due to the increase in dielectric loss factor as a function of frequency. We now use the SM to verify that these eight classes are truly OW by computing emissivity of saline water at different frequencies. The only input to the SM that is unknown at this point is the dielectric constant of saline water which is known to have dependence on frequency, salinity, and temperature. The surface roughness will have minimal affect on the measured brightness temperature. Values for the permittivity of saline water are computed from the empirical models given in Appendix E of Ulaby et al. [1986]. The levels and trends predicted by the SM agree with the trend and level of the OW class centroids given by the clustering algorithm. Since we are not interested in water which is classified differently due to differences in salinity, temperature or other physical traits, these eight classes are all grouped into a single cluster and labeled OW. The centroid of this grouped OW class along with the remaining seven class centroids from the ISODATA algorithm are given in Table 12.6 and plotted as a function of frequency in Figure 12.10. These eight classes of ice are to be identified using the SM. For clarity, we have already labeled the eight classes in Table 12.6 and Figure 12.10. Class 6 is an intermediate mix of second-year ice (2nd year-class 5) and multiyear (MY-class 7).

As in the case of OW, the remaining seven class centroids are to be fitted with the SM to determine each class identity. For the purpose of this work, we consider only clusters 1, 2, 4, and 7. The emission models of Chapter 3 are used to fit these four clusters to illustrate the method. The first step in the fitting of a data set is to estimate the type of ice and thus the associated permittivities and dominant scattering mechanisms. The determination of each ice type is based primarily on data from other experiments that have ground truth as well as past geophysical surveys of the Arctic region under similar conditions. The ice type can then be used to fix the physical structure and the permittivities of the ice layer. The process of using a SM to fit data sets with ground truth serves to calibrate as well as validate the model. The resulting fits for clusters 1, 2,

Table 12.6 Summary of radiometric averages for the eight classes considered. All values given are in K.

Channel	*Open Water*	*Frazil Ice*	*Thin Ice*	*FY*	*2nd Year*	*Inter. (mix)*	*MY*	*Old MY*
19V	197.6	241.5	252.6	246.7	239.7	234	231.6	223.7
22V	206.1	241.8	251	244.2	236.3	229.6	226.4	217.7
37V	214.5	241.4	247.4	235.8	223.7	216.2	206.7	195.9
85V	240.9	244	237.1	218.7	210.8	208.7	195.1	193.1
19H	136.5	208.6	233.6	228.7	220.7	214.5	214.4	206.4
37H	160.3	215.5	232.5	221.4	208.8	199.4	194.3	183.7
85H	203.6	226.7	225.9	207.9	199.8	197.2	185.6	183.5

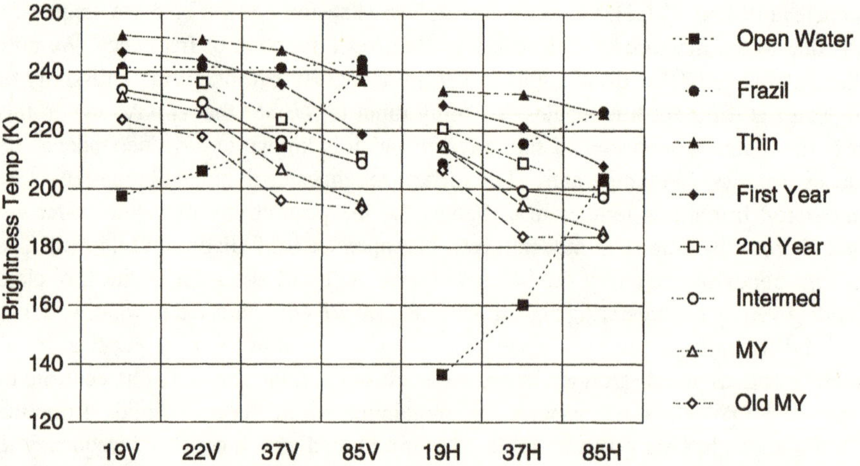

Figure 12.10 Class centroids as a function of frequency and polarization for each of the eight classes output from the unsupervised ISODATA clustering algorithm.

4, and 7 are given in Figure 12.11. The four clusters were found to be OW, frazil ice, first-year ice (FY), and multiyear ice (MY), respectively.

Now that the original images have been classified using the unsupervised method and the identity of each class determined by the SM, the final classified image can be used as pseudo ground-truth for a supervised neural network classification scheme. The NN classifier illustrated is to be used for batch processing of satellite data. Representative data from each of the eight classes is selected from the classified image for training the NN. The representative data and an associated class identification tag are then dumped to a file which is to be used to train the NN. For the example considered,

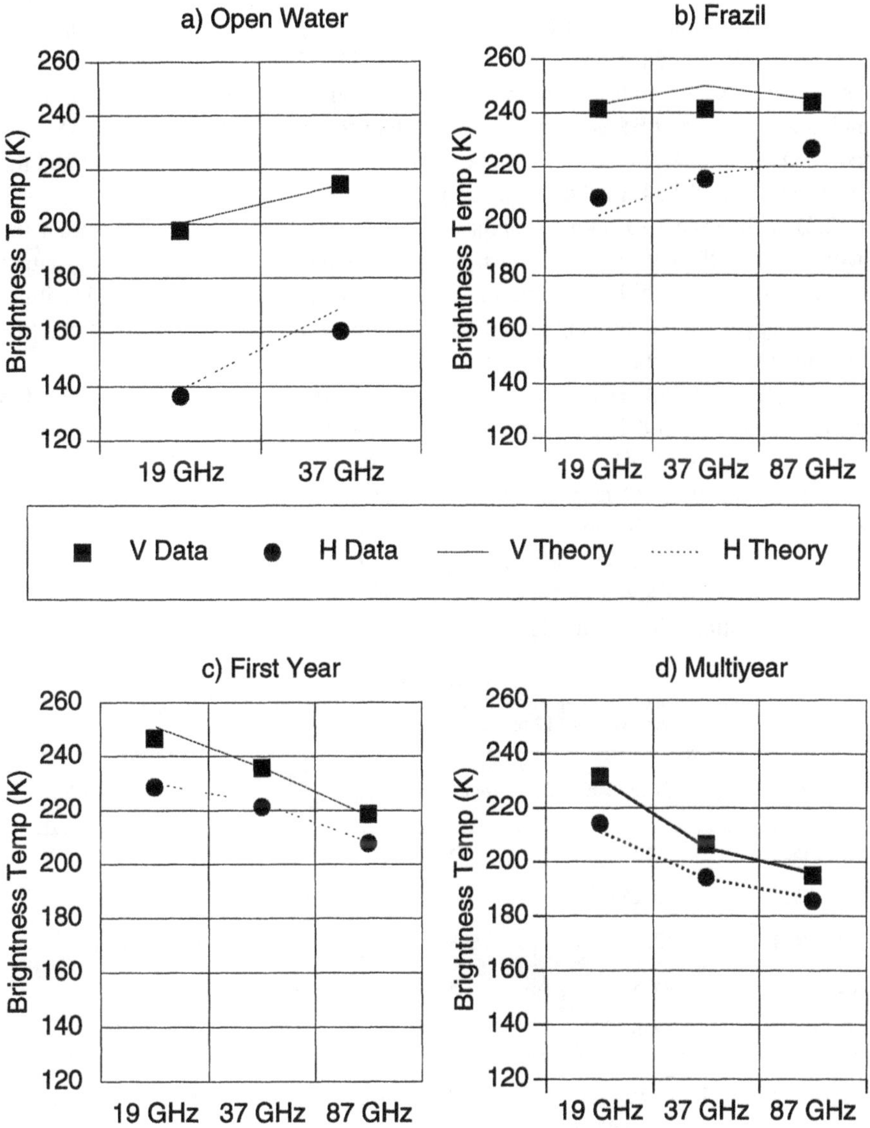

Figure 12.11 Identification of unsupervised classification results using emission models from Chapter 3.

the NN will need to have seven inputs (one for each radiometric channel) and eight outputs (one binary output for each class).

The training data are used to train a fully connected NN with 7-15-8-8 topology. Both the conventional back-propagation (BP) learning algorithm [Rumelhart and McClelland, 1986] and the fast learning algorithm (FL) [Manry et al., 1993] were used to train the NN. In addition to these two methods, a third method based on a combination of FL and BP is used. This third algorithm will be denoted as FL+BP. As noted in Dawson et al. [1993], the primary difference between the FL method and the BP method is that the FL method finds the global minimum of the error function given by (12.2) with respect to output weights, not with respect to all weights within the network as the BP method does. Thus, the FL method is optimizing the N_{out}-output weights rather than all hidden weights as required by the back-propagation algorithm. After solving for the output weights as performed by the FL method, the combined FL-BP method uses BP to correct the hidden weights.

All three algorithms (BP, FL, and FL+BP) were tested on the same training file using the same 7-15-8-8 fully interconnected topology. The training set consisted of 30512 training patterns. Three separate attempts to train the network using BP were made; each with different learning rates and momentum factors [see Rumelhart and McClelland for a description of the learning rate-z_o and momentum-a]. The time required for training the network, the final classification error and the number of iterations required are given in Table 12.7. The classification error for all methods is plotted as a function of epoch in Figure 12.12.

Table 12.7
Comparison of Training Time, Iterations, Error Using 7-15-8-8 Topology
(There were 30512 patterns in the classification training file)

Job Name	*Time Required*	*Iterations*	*Avg Time(sec) / Iteration*	*Min Error%*
BP-18522	19558.8	500	39.12	16.76
BP-18513	19526.9	500	39.05	14.86
BP-18494	19521.07	500	39.04	34.51
FL	4478.3	150	29.85	3.805
FL w/BP	6736.9	150	44.91	3.382

From Table 12.7 we see that the BP training algorithm did very poorly in comparison to the FL algorithms. Note that even in cases where BP was allowed to run three times as long as FL, the BP training error is a full magnitude greater. Although the combined FL-BP did slightly better than FL alone, the additional training time required may not be justified. Thus, the FL is the method of choice.

In summary, we have seen that the SM can be combined with other techniques to effectively perform classification. In this last example studied, we used a clustering algorithm that is better suited for data without ground truth and then applied the SM to

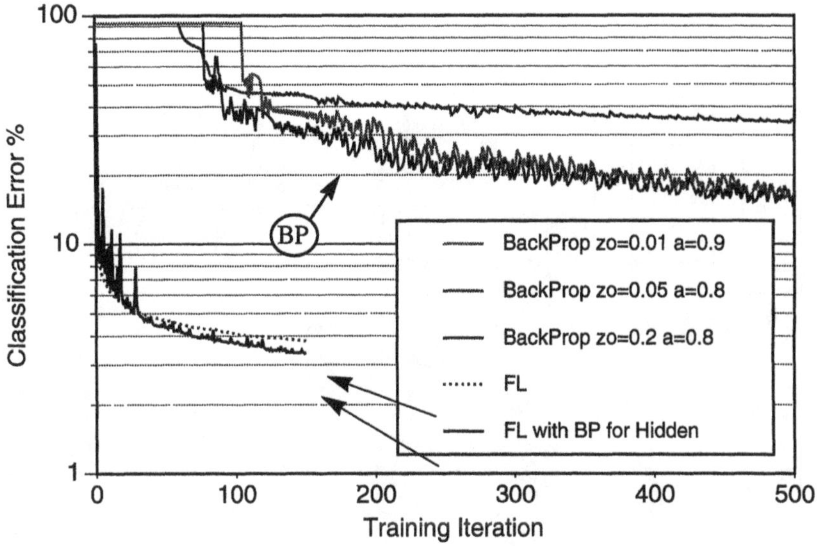

Figure 12.12 Comparison of training classification error as a function of iteration for the five cases listed. Results shown are for 880301.avg.trainb training data set for small topology (7-15-8-8). The training set contained 30512 patterns. All networks are fully connected.

determine class identity. These two methods combined allow us to use the results of the clustering algorithm to train neural networks which can then be used to process data in large volumes. The methodology used is summarized in Figure 12.13.

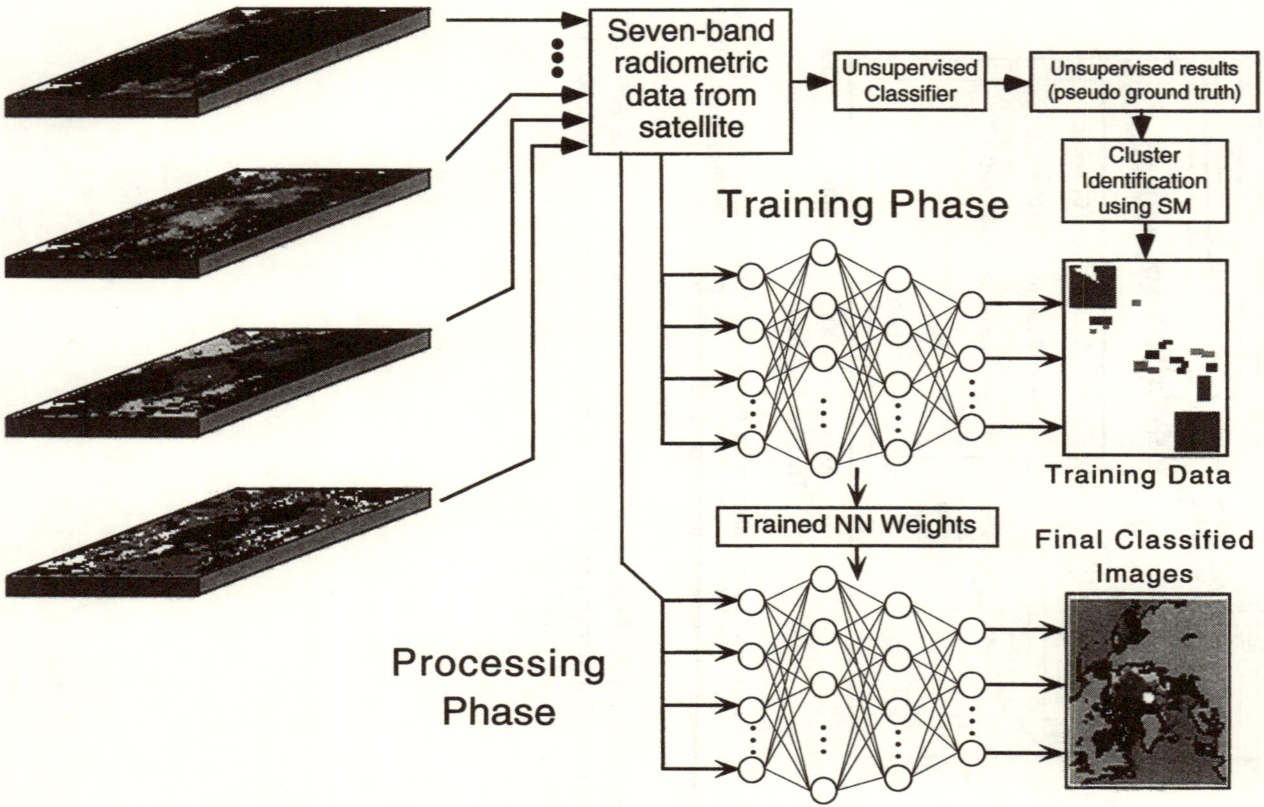

Figure 12.13 Methodology used to perform classification on multi-spectral satellite data using a neural network.

REFERENCES

Amar, F., M.S. Dawson and A.K. Fung, "Inversion of the Relevant Forest and Vegetation Parameters Using Neural Networks," in *Proc. of PIERS 1993*, Pasadena, CA.

Bahar, E., "Full Wave Solutions for the Scattered Radiation Fields from Rough Surfaces with Arbitrary Slope and Frequency," *IEEE Trans. on Ant. and Prop.*, Vol. AP-28, 1980, pp. 11–21.

Benediktsson, J.A., P.H. Swain and O.K. Ersoy, "Neural Network Approaches Versus Statistical Methods in Classification of Multisource Remote Sensing Data," *IEEE Trans. Geosci. and Remote Sensing,* Vol. GE-28, 1990, pp. 540–552.

Bischof, H., W. Schneider and A.J. Pinz, "Multispectral Classification of Landsat Images Using Neural Networks," *IEEE Trans. Geosci. and Remote Sensing*, Vol. GE-30, 1992, pp. 482–489.

Bredow, J.W., "A Laboratory Investigation into Microwave Backscattering From Sea Ice," RSL Tech. Report 8240-1, August, 1989.

Chen, K.S., and A.K. Fung, "A Comparison Between Backscattering Models for Rough Surfaces," in *Proc. of IGARSS 1992*, pp. 907–909.

Chen, Z., D. Davis, L. Tsang, V. Hwang and A.T.C. Chang, "Inversion of Snow Parameters by Neural Network with Iterative Inversion," in *Proc. of IGARSS 1992*, pp. 1061–1063.

Dawson, M.S., A.K. Fung and M.T. Manry, "Sea Ice Classification Using Fast Learning Neural Networks," in *Proc. of IGARSS 1992*, pp. 1070–1071.

Dawson, M.S., A.K. Fung and M.T. Manry, "Surface Parameter Retrieval Using Fast Learning Neural Networks," *Remote Sensing Reviews*, Vol. 7, 1993, pp. 1–18.

Draper, J., D. Frankel, H. Hancock and A. Mize, "A Microcomputer Neural Net Benchmarked Against Standard Classification Techniques," in *Proc. IJCNN 1987*, Vol. IV, pp. 651–658.

Fitch, J.P., S.K. Lehman, F.U. Dowla, S.Y. Lu, E.M. Johansson and D.M. Goodman, "Ship Wake Detection Procedure Using Conjugate Gradient Trained Artificial Neural Networks," *IEEE Trans. Geo. and Remote Sensing*, Vol. 29, no. 5, Sept. 1991, pp. 718–725.

Fung, A.K., L. Zongqian and K.S. Chen, "Backscattering from a Randomly Rough Dielectric Surface," *IEEE Trans. Geo. and Remote Sensing*, Vol. 30, no. 2, March 1992.

Fung, A.K., M. Dawson and S. Tjuatja, "An Analysis of Scattering From a Thin Saline Ice layer," in *Proc. of IGARSS 1992*, pp. 1262-1264.

Gallant, A., and H. White, "There Exists a Neural Network That Does Not Make Avoidable Mistakes," in *Proc. IJCNN 1988*, Vol. I, pp. 657–664.

Grody, N.C., "Classification of Snow Cover and Precipitation Using the SSMI," *Journal of Geo. Res.*, Vol. 96, no. D4, 1991, pp. 7423–7435.

Hammerstrom, D., "Neural Networks at Work," *IEEE Spectrum*, 1993, pp. 26–32.

Hartman, E. J., J.D. Keeler, and J.M. Kowalski, "Layered Neural Networks with Gaussian Hidden Units as Universal Approximations," *Neural Computation,* Vol. 2, no. 2, 1990, pp. 210–215.

Heermann, P.D., and N. Khazenie, "Classification of Multispectral Remote Sensing Data Using a Back-Propagation Neural Network," *IEEE Trans. Geosci. and Remote Sensing*, Vol. GE-30, 1992, pp. 81–88.

Hsu, Shin-Yi, T. Masters, M. Olson, M. Tenorio and T. Grogan, "Comparative Analysis of Five Neural Network Models," *Remote Sensing Reviews*, Vol. 6, 1992, pp. 319–329.

Hollinger, J., R. Lo, G. Poe, R. Savage and J. Peirce, "Special Sensor Microwave/Imager User's Guide," Naval Research Laboratories, 1987.

Hornik, K., M. Stincombe and H. White, "Multilayer Feedforward Networks are Universal Approximators," *Neural Networks,* Vol. 2, 1989, pp. 359–366.

Kay, S., *Fundamentals of Statistical Signal Processing*, Prentice-Hall, Englewood Cliffs, NJ, 1993.

Kohonen, T., G. Barna and R. Chrisley, "Statistical Pattern Recognition with Neural Networks: Benchmark Studies," *IEEE International Conference on Neural Networks*, San Diego, CA, 1988, Vol. I, pp. 61–68.

Lippmann, R.P., "An Introduction to Computing With Neural Nets," *IEEE ASSP Magazine*, 1987, pp. 4–22.

Lure, Y., Y. Chiou, N. Grody and H. Yeh, "Classification of Earth Surface from SSMI Using Artificial Neural Network Data Fusion," *in Proc. of IGARSS 1992*, pp. 833–835.

Manry, M.T., S. Apollo, L. Allen, W. Lyle, W. Gong, M.S. Dawson and A.K. Fung, "Fast Training of Neural Networks for Remote Sensing," *Remote Sensing Reviews,* accepted 1993.

Minsky, M., and S. Papert, *Perceptrons*, MIT Press, Cambridge MA, 1969.

Mo, T., J. Wang and T. Schmugge, "Estimation of Surface Roughness Parameters from Dual-Frequency Measurements of Radar Backscattering Coefficients," *IEEE Trans. on Geo. and Rem. Sens.*, Vol. 26, no. 5, 1988.

Nystuen, J.A. and F.W. Garcia, "Sea Ice Classification Using SAR Backscatter Statistics," *IEEE Trans. Geosci. and Remote Sensing*, Vol. GE-30, 1992, pp. 502–509.

Onstott, R.G. "Polarimetric Radar Measurements of Artificial Sea Ice During CRRELEX' 88," ΣRIM Tech. Report 196100-23-T, 1990.

Pao, Yoh-Han, *Adaptive Pattern Recognition and Neural Networks*, Addison Wesley, Reading, MA, 1989.

Peirce, J.L., ed., "Special Sensor Microwave/Imager Critical Design Review: Vol. 1. Space Segment," Hughes Aircraft Co (Contract No. F04701-79-C-0061), 1980.

Pierce, L.E., K. Sarabandi and F.T. Ulaby, "Application of an Artificial Neural Network in Canopy Scattering Inversion," *in Proc. of IGARSS 1992*, pp. 1067–1069.

Perugini, N., and W. Engeler, "Neural Network Learning Time: Effects of Network and Training Set Size," *Proc. IJCNN 89*, 1989, Washington D.C., Vol. II, pp. 395–401.

Rosenblatt, R., *Principles of Neurodynamics*, Spartan Books, New York, 1959.

Ruck, D., S.K. Rogers, R. Kabrisky, M. Oxley and B.W. Suter, "The Multilayer Perceptron as an Approximation to a Bayes Optimal Discriminant Function," *IEEE Trans. on Neural Networks*, Vol. 1, no. 4, 1990, pp. 296–298.

Rumelhart, D.E., and J.L. McClelland, eds. *Parallel Distributed Processing,* Vols. 1 and 2, MIT Press, Cambridge, MA, 1986.

Tjuatja, S., A.K. Fung and M.S. Dawson, "An Analysis of Scattering and Emission from Sea Ice," *Remote Sensing Reviews,* Vol. 7, 1993, pp. 83–106.

Tsang, L., A. Ishimaru, R. Porter and D. Rouseff, "Holography and the Inverse Source Problem," *J. Opt. Cos, Am.*, Vol. A4, 1987, pp. 1783–1787.

Ulaby, F.T., R.K. Moore and A.K. Fung, *Microwave Remote Sensing: Active and Passive,* Vol. III, Artech House, Norwood, MA, 1986.

Weinbrenner, D., and A. Ishimaru, "Application of the Phase-Pertubation Technique to Randomly Rough Surfaces," *Journal of the Optical Society of America, A*, Vol. 2, no. 12, 1985, pp. 2285–2293.

Werbos, P., "Beyond Regression: New Tools For Prediction And Analysis In The Behavioral Sciences," Ph.D. Dissertation, Appl. Math., Harvard University, November 1974.

Westwater, E.R., and A. Cohen, "Application of Backus-Gilbert Inversion Technique in Determination of Aerosol Size Distributions from Optical Scattering Measurements," *Appl. Opt.*, Vol. 12, 1973, pp. 1340–1348.

Yau, H.C., and M.T. Manry, "Shape Recognition with Nearest Neighbor Isomorphic Network," *Proceedings of the First IEEE-SP Workshop on Neural Networks for Signal Processing*, Princeton, NJ., Sept. 29 – Oct. 2, 1991, pp. 246–255.

Index